Volume 5

Mechanisms of Inorganic and Organometallic Reactions

Volume 5

Mechanisms of Inorganic and Organometallic Reactions

Edited by

M. V. Twigg

Imperial Chemical Industries P.L.C.
Billingham, Cleveland, United Kingdom

PLENUM PRESS • NEW YORK AND LONDON

no anal

QD
501
M426
v.5
Chem

SD
5-19-88
Cy

Library of Congress Catalog Card Number 87-648073

ISBN 0-306-42841-5

© 1988 Plenum Press, New York
A Division of Plenum Publishing Corporation
233 Spring Street, New York, N.Y. 10013

Printed in the United States of America

Contributors

A. Bakác   Ames Laboratory and Department of Chemistry, Iowa State University, Ames, Iowa 50011

J. Burgess   Chemistry Department, The University, Leicester LE1 7RH, U.K.

R. D. Cannon   Chemistry Department, University of East Anglia, University Plain, Norwich NR4 7TH, U.K.

R. J. Cross   Department of Chemistry, The University, Glasgow G12 8QQ, Scotland, U.K.

D. J. Darensbourg   Department of Chemistry, Texas A&M University, College Station, Texas 77843

R. van Eldik   Institute für Anorganische Chemie, Universität Witten/ Herdecke, 5810 Witten-Annen, Germany

J. H. Espenson   Ames Laboratory and Department of Chemistry, Iowa State University, Ames, Iowa 50011

R. W. Hay   Department of Chemistry, University of Stirling, Stirling FK9 4LA, Scotland, U.K.

D. A. House   Department of Chemistry, University of Canterbury, Christchurch 1, New Zealand

L. A. P. Kane-Maguire   Chemistry Department, Wollongong University, P.O. Box 1144, Wollongong, NSW 2500, Australia

A. G. Lappin   Chemistry Department, University of Notre Dame, Notre Dame, Indiana 46556

S. F. Lincoln   Department of Physical and Inorganic Chemistry, University of Adelaide, South Australia 5001, Australia

D. J. Mangold   Department of Chemistry, Texas A&M University, College Station, Texas 77843

B. E. Mann   Department of Chemistry, The University of Sheffield, Sheffield S3 7HF, U.K.

D. P. Riley   Monsanto Company, Central Research Laboratories, 800 N. Lindbergh Boulevard, St. Louis, Missouri 63167

S. J. Tremont   Monsanto Company, Central Research Laboratories, 800 N. Lindbergh Boulevard, St. Louis, Missouri 63167

N. J. Stone   Department of Chemistry, Brown University, Providence, Rhode Island 02912

D. A. Sweigart   Department of Chemistry, Brown University, Providence, Rhode Island 02912

N. Winterton   ICI PLC, Mond Division, The Heath, Runcorn, Cheshire, WA7 4QE, U.K.

Preface

This series provides a continuing critical review of published work concerned with mechanistic aspects of inorganic and organometallic reactions in solution, with each volume covering the whole area. Covered in this volume are papers published during the eighteen-month period from July 1985 through December 1986, together with some earlier work where it is appropriate to make comparisons. Results selected for discussion are chosen for their relevance to the elucidation of reaction mechanisms, and increasingly data of a nonkinetic nature are being included because of the mechanistic information that can be deduced. As always, numerical data are usually reported in the units used by the original authors, except where results from different papers are compared and conversion to common units is necessary.

The amount of relevant published work continues to increase, and it is becoming increasingly necessary to give less space to the less important results in order to keep the size of the volume within sensible limits. Nonetheless, efforts are being made to maintain the comprehensive nature of the series. Thus in this volume the first chapter (on general and theoretical aspects of electron transfer) has been deliberately shortened compared with its size in previous volumes, and in future volumes it may be necessary to modify slightly some other chapters. However, it is not intended to dramatically change the overall format since the present format makes it easy for readers to follow progress in a particular area over several years. The material is grouped into four parts: Electron Transfer Reactions, Substitution and Related Reactions, Reactions of Organometallic Compounds, and Compilations of Numerical Data. The last section, when used in conjunction with previous volumes, now provides a particularly useful data base of all of the relevant activation volume results published since 1980.

The continuing support of all of the people involved in the writing and production of the series is one of its strengths. This is very much appreciated, as are the many helpful and always welcome suggestions from readers.

Contents

Part 2. Substitution and Related Reactions

Chapter 4. Reactions of Compounds of the Nonmetallic Elements

N. Winterton

*Chapter 5. Substitution Reactions of Inert-Metal Complexes—
Coordination Numbers 4 and 5*

R. J. Cross

*Chapter 6. Substitution Reactions of Inert-Metal Complexes—
Coordination Numbers 6 and Above: Chromium*

D. A. House

Chapter 7. Substitution Reactions of Inert-Metal Complexes— Coordination Numbers 6 and Above: Cobalt

R. W. Hay

Part 3. Reactions of Organometallic Compounds

Chapter 10. Substitution and Insertion Reactions

D. J. Darensbourg and D. J. Mangold

Chapter 11. Metal-Alkyl and Metal-Hydride Bond Formation and Fission; Oxidative Addition and Reductive Elimination

D. A. Sweigart and N. J. Stone

Chapter 14. Homogeneous Catalysis of Organic Reactions by Complexes of Metal Ions

D. P. Riley and S. J. Tremont

Part 4. Compilations of Numerical Data

Part 1
Electron Transfer
Reactions

Chapter 1
Electron Transfer: General and Theoretical

1.1. Introduction

In the last few years, interest in the theory of electron transfer, which originally centered largely on problems of electron transfer between metal complex ions, has shifted steadily toward other areas, notably organic and bioinorganic systems on the one hand, and solid-state materials on the other. This reflects the trend of technological interest toward such areas as solar-energy conversion and superconductivity. It means, however, that for the purpose of reviewing developments more closely related to the inorganic systems discussed in other chapters of this book, the time-scale of eighteen months is rather short. The usual survey of electron-transfer theory is therefore held over until the next volume in the present series, and in this chapter only recent reviews, and theoretical contributions with particularly extensive bibliographies, are noted.

1.2. Reviews

A recent issue of the *Journal of Physical Chemistry*[1] is dedicated to Professor R. A. Marcus on the thirtieth anniversary of his first two papers on electron transfer.[2] It contains a bibliography of all his papers[3] and a personal reminiscence outlining the development of his ideas in electron and atom transfer, and generalized theories of the transition state, of collision processes, and of unimolecular reactions.[4] A total of sixty-one papers by other workers are gathered into two sections on Chemical Reaction Dynamics[5] and Electron-, Proton-, and Group-Transfer Reactions.[6] The latter includes a wide-ranging article on the Marcus theory in organic chemistry.[7] In two subsequent articles, Sumi and Marcus consider the effects of solvent molecule motion on electron-transfer rates.[8]

Electron-transfer reactions of cytochrome-*c* have been reviewed,[9] and biological aspects of electron transfer generally are considered in two major theoretical surveys. Marcus and Sutin briefly outline classical Marcus theory with its extensions to cover quantum-mechanical aspects (nonadiabaticity), effects of orientation of reactants, and nuclear and electron tunneling; and they review the relevant experimental data.[10] Guarr and McLendon concentrate on quantum effects, with particular emphasis on transfer over long distances in proteins, on the inverted region, and on isotope effects.[11] Distance effects in the inverted region are treated further by Closs *et al.*[12] Two recent conferences on photochemical and photophysical processes have dealt with biological electron transfer and solar-energy storage[13,14] with further contributions on distance effects,[15,16] and on the use of micelles, microemulsions, and vesicles,[17] all with a view to inhibiting the back-reaction to the primary photolysis. Photochemically-induced redox reactions in binuclear metal complexes and ion pairs have been reviewed experimentally and theoretically.[18,19] There continues to be interest in the problem of relating thermal, optical, and photochemical electron-transfer pathways for the same overall reaction, with articles on solvent reorganization,[20] on the relationship of different expressions—in terms of enthalpy and free energy—for the various energy barriers,[21] and a review of the experimental data for intramolecular reactions of ion pairs in solution.[22] In view of the experimental effort expended on this field over the past twenty years, it is salutary to be reminded that there is still only one system for which both the thermal and optical electron-transfer processes can confidently be said to have been measured.[22]

A comprehensive review deals with structural changes accompanying electrode reactions of metal complexes.[23] The author challenges the widely-held view that when a reaction is found to be reversible by a technique such as cyclic voltammetry, reactants and products must be isostructural or nearly so. The proper inference is that any structural changes must be rapid within the time-scale of the experiments. Typical mechanisms are classified and the conditions for reversibility or quasi-reversibility are examined with special reference to organometallic and metal cluster complexes. A short review of "electroprotic phenomena" deals with coupled electron and proton transfer reactions in the electrochemistry of metal hydroxamate complexes.[24]

Solid-state electron transfer is covered in three reviews of contrasting areas. A survey of recent neutron inelastic scattering studies of mixed valence materials covers mainly rare-earth alloys with the Kondo type of fluctuating valence.[25] A review of mixed-valence iron oxides deals with structural and magnetic interactions, as well as electron transfer.[26] The latter is described in terms of the familiar limiting models of "itinerant" electrons, fully delocalized and describable by band theory, and "small polarons" moving by tunneling or thermal hopping. These are the two limits known otherwise as Robin and Day classes III and II. The intermediate case is also emphasized, however, namely systems which pass from polaron to band behavior with increasing temperature.[26] Finally, and closest to the electron-transfer models applicable in solution, a survey by Hendrickson and co-workers of recent work on mixed valence iron(III,III,II) clusters of the type $[Fe_3O(OOCR)_6L_3]$ highlights two features not previously recognized: the profound effects of noncoordinated molecules of crystallization (effects somewhat analogous to the effects of solvent reorganization in solution reactions) and cooperative effects due to intermolecular interaction, leading to sharp transitions between phases composed of electron-localized and delocalized molecules.[27]

In the area of gas-phase electron transfer, we note here a theoretical discussion of potential-energy surfaces,[28] and a further summary of detailed experimental and theoretical work on one system, namely the reaction $N_2^+ + Ar \rightarrow N_2 + Ar^+$, and the reverse process, with the diatomic molecules in selected vibrational states.[29]

Chapter 2

Redox Reactions between Two Metal Complexes

2.1. Introduction

The study of electron-transfer reactions between metal-ion complexes appears to be undergoing a transition. While work continues on detailed mechanistic examination of reactions in aqueous media, there is more emphasis on the chemistry of novel complexes and on the study of reactions in nonaqueous media. The newer probes, principally activation volume studies, have also been applied more extensively. The format of the chapter differs slightly from previous issues in an attempt to improve the readability of what is essentially a compilation of rate data. Broader subheadings have been employed and articles which address similar problems are gathered together as much as possible. Rate data are presented in the text or in Tables 2.1 and 2.2, which deal with reactions between metal ion complexes and reactions involving metalloproteins, respectively. As in previous volumes, these tables are arranged in the order of increasing atomic number of the central metal ion in the reductant.

2.2. Reactions of Aqua-Metal Ions

Studies continue on the reductions of metal-ion complexes by aqua-metal ions. The reaction of $[(NH_3)_5CoNCacac]^{3+}$ with Ti^{3+} involves[1] the rapid, reversible formation of a precursor complex $[(NH_3)_5CoNCacacTi(OH)(OH_2)_3]^{4+}$ with the loss of two protons, and the rate of intramolecular electron transfer is estimated to be $\geq 6\ s^{-1}$, which is much slower than the rate in the corresponding[2] ruthenium(III)–titanium(III) system. In the former case there is mismatch between the symmetry of the donor and acceptor orbitals (t_{2g}-e_g) while, in the latter system, both donor and acceptor use t_{2g} orbitals. The effect of oxalate on the redox interaction between ruthenium(III) and titanium(III) has been investigated[3] with $[Ru(ox)_3]^{3-}$. Reduction by $TiOH^{2+}$ is almost six orders of magnitude faster than the calculated outer-sphere rate, implying an oxalate bridged inner-sphere mechanism. When free oxalate is added to the system, the increased thermodynamic driving force for the reductant, $[Ti(ox)_2]^-$ causes rate enhancement. Peroxotitanium(IV), TiO_+^{2+}, oxidizes Fe^{2+} and Ti^{3+} either directly[4] or by a pathway which involves prior dissociation of H_2O_2. For both metal-ion reductants, direct reaction is slower than the corresponding reactions with H_2O_2 itself. While the available data do not allow distinction between outer-sphere and inner-sphere mechanisms, the latter seems more plausible since it might be expected that TiO_2^{2+} would be a poor Lewis base. The corresponding zirconium(IV) complex[5] is more difficult to characterize but there is good evidence that it exists as a tetramer, $[Zr_4(O_2)_2(OH)_4]^{8+}$, in acidic media. Reaction with cerium(IV) follows the stoichiometry shown in equation (1). The rate is independent of cerium(IV)

$$[Zr_4(O_2)_2(OH)_4]^{8+} + 4\,Ce^{4+} \rightarrow 4\,Zr^{4+} + 4\,Ce^{3+} + 2\,O_2 \qquad (1)$$

concentration with a rate constant of $2.4 \times 10^{-3}\,s^{-1}$ at 25 °C but faster than the rate of dissociation of the complex. It is thought that the rate-determining step is a ring-opening process which exposes a bound peroxide to attack by cerium(IV). The fact that cerium(IV) reacts very rapidly with H_2O_2 and that no direct reaction with the complex is detected suggests that the coordinated peroxides are bridging and well protected from attack.

The oxidation of vanadium(IV) by outer-sphere reagents proceeds[6] by two pathways, differing in their acidity dependence. Direct reaction of VO^{2+} is acid independent but the hydrolysis product, $VOOH^+$, is more reactive by around five orders of magnitude. Two pathways are also detected in the outer-sphere reduction of vanadium(V), and the acid-dependent term relates to $VOOH^{2+}$, allowing the self-exchange rate of this species to be probed. The calculated rate constant, about $10^{-3}\,M^{-1}\,s^{-1}$, is suggestive of moderate bond-length changes and compares closely with the isoelectronic $TiOH^{3+/2+}$ system. A previously undetected acid-independent outer-sphere pathway is noted[7] in the chromium(II) reduction of vanadium(III) in perchlorate media. Care has been taken to eliminate the possibility that this term in the rate law is the result of a medium effect and the rate constant is in agreement with the predictions of Marcus theory. As with the cobalt(III)–titanium(III) reactions, this system represents a t_{2g}-e_g change though in the opposite sense and the inner-sphere advantage χ is estimated to be in excess of 12,000, somewhat greater (though not excessively so) than values for an e_g-e_g system. Product analysis is highlighted in two studies of the Cr^{2+} reduction of cobalt(III) complexes. With $[(NH_3)_5CoNCCHCHCN]^{3+}$ as oxidant,[8] the initial step is inner-sphere giving $[NCCHCHCNCr(H_2O)_5]^{3+}$ as product. This product is subject to aquation or to further attack by Cr^{2+} in a reaction which is second-order in [chromium(II)] to give the $Cr-C$ bonded $[(H_2O)_5CrCH-(CH_2CN)CN]^{2+}$. Reduction of $[(NH_3)_5CoNO_2]^{2+}$ has been reexamined.[9] Again the immediate product $[(H_2O)_5CrONO]^{2+}$ is subject to aquation or to reaction with Cr^{2+} to give the $Cr-NO$ product and chromium(IV). This latter species leads to the formation of the μ-hydroxy chromium(III) dimer. A number of outer-sphere reductants, tris-polypyridine complexes, have been used[10] to probe the electron-transfer behavior of Mn^{3+}. The calculated self-exchange rates are consistent with an adiabatic process but are four or five orders of magnitude smaller than the rate measured by other means. In this respect the behavior is similar to that shown by the $Co^{3+/2+}$ system. To give the appropriate trends in rate with reaction driving force, a model is considered in which there is interaction between the transition state for the electron transfer with low-lying excited states, leading to a reduction in the barrier to electron transfer. The explanation fits better for the cobalt(III)/(II) system where ligand field excited states are of sufficiently low energy.

There has been some activity in the preparation and characterization of aqua-ions of the second- and third-row transition series. Two dimeric iridium(III) complexes are reported[11] as a result of electrochemical preparation. The rhodium(III) aqua-ion $[Rh(H_2O)_6]^{3+}$ is reduced[12] by chromium(II) by an inner-sphere pathway involving $[Rh(H_2O)_5(OH)]^{2+}$ as reactive species. For the corresponding halo-ions, $[Rh(H_2O)_5X]^{2+}$ (X = Br, Cl), detection of halide transfer to chromium(III) confirms the mechanism. The other product of the reactions is the Rh_2^{4+} dimer. Attempts to scavenge monomeric Rh^{2+} were unsuccessful and this may provide an explanation for the observation that reduction of $[Rh(H_2O)_6]^{3+}$ by vanadium(II), in which monomer results from outer-sphere electron transfer, leads to the formation of rhodium metal. Further studies are reported[13] for reactions of the $[Rh_2(OAc)_4(OH_2)_2]^{+/0}$ dimer with outer-sphere reagents, allowing evaluation of the self-exchange rate, $1.5 \times 10^5 \ M^{-1} s^{-1}$, much higher than previously estimated.[14] However, it is pointed out that inner-sphere reorganization is small and the low charge leads to little solvent reorganization consistent with the higher value. The system is of interest since the electron exchanged is in a δ^* or π^* orbital.

The chromium(V) complex, bis(2-ethyl-2-hydroxybutyrato)oxochromate(V), $[L_2CrO]^-$, has a reduction potential around 0.43 V (vs N.H.E.)[15] and reductions by both iron(II)[16] and vanadium(IV)[17] involve formation of the same transient chromium(IV) complex which absorbs at 515 nm with an extinction coefficient of $1.5 \times 10^3 \ M^{-1} cm^{-1}$. In the presence of excess reductant, the final product is $[CrL_2(H_2O)_2]^+$, a chromium(III) complex in which two ligand molecules remain coordinated. Addition of free LH^- leads to the establishment of a slow equilibrium between $[L_2CrO]^-$ and $[L_3CrO]^{2-}$, both of which react by a proposed inner-sphere mechanism to give the initial chromium(IV) product, complicating the kinetic behavior. Reduction of $[L_2CrO]^-$ by uranium(IV) on the other hand[18] gives formation of the chromium(III) product and UO_2^{2+} in a reaction catalyzed by the addition of free LH^- to this system. The explanation involves formation of $[ULOH]^{2+}$ which reacts to give the chromium(IV) intermediate and uranium(V) which in turn rapidly disproportionates. From the relative rates of reaction of $[ULOH]^{2+}$ with chromium(V) and chromium(IV), these reactions are deduced to be outer-sphere. Further complex formation has an inhibitory effect on the rate. Outer-sphere reactions without formation of measurable amounts of the chromium (IV) intermediate are proposed[19] for the corresponding reductions by titanium(III) species. Inner-sphere mechanisms are proposed[20,21] for the reduction of a series of mixed-ligand cobalt(III) malonate complexes by Fe^{2+} although precursor complexes are not detected. The cobalt(II) macrocyclic complex $[(H_2O)_2Co(dmgBF_2)_2]$ is oxidized[22] by Fe^{3+} in aqueous perchlorate media and the rate shows two pathways, an acid independent path due to Fe^{3+} and a dominant path with an inverse acid dependence due to $FeOH^{2+}$. The best estimate of the mechanism is inner-sphere with intermolecular electron transfer in the rate-limiting step. Reduction of Fe^{3+} by $[Co(sep)]^{2+}$ must be outer-sphere in nature[23] and the rate data provide information on the collision frequency for highly charged cationic species in aqueous media.

^{31}P NMR has been used[24] to measure the rate of electronic self-exchange between the heteropoly-tungstate ions $\alpha\text{-}[PW_{12}O_{40}]^{3-}$ and $\alpha\text{-}[P_2W_{18}O_{62}]^{6-}$ and their heteropoly blue one-electron reduced forms. The rates are $1.2 \times 10^7 \ M^{-1} s^{-1}$ and $1.4 \times 10^4 \ M^{-1} s^{-1}$ respectively and calculations show that they represent diffusion control for the substantially different hydrodynamic radii for these highly charged ions. The rate of intramolecular tungsten–tungsten hopping is estimated to be $1.7 \times 10^{11} \ s^{-1}$. For the $\alpha\text{-}[P_2MoW_{17}O_{62}]^{6-/7-}$ system, the electronic change in the heteropoly-blue form is localized on the molybdenum and the self-exchange rate is an order of magnitude smaller than for $\alpha\text{-}[P_2W_{18}O_{62}]^{6-/7-}$, ascribed to the steric factor imposed by requiring close contact of the molybdenum caps. Redox reactions of a number of molybdenum complexes have been examined.[25,26] Oxidation of $[Mo_2O_4]^{2+}$ by Fe^{3+} is dominated by the substitution-controlled inner-sphere reaction of $FeOH^{2+}$ to yield, as initial product, the mixed valence [Mo(V)Mo(VI)] dimer although the overall stoichiometry is 1:2. The same is true of outer-sphere oxidation by $[Fe(phen)_3]^{3+}$ and of the corresponding reaction of $[Mo_2O_2S_2]^{2+}$ but in this latter case the kinetic behavior is complex. Outer-sphere oxidation of $[Mo_3O_4]^{4+}$ by $[Fe(phen)_3]^{3+}$ is also reported and again the initial product is the one-electron oxidized mixed-valence species. The rate, measured in chloride media, is two orders of magnitude faster than the previous measurement[27] for perchlorate conditions. Cerium(IV) has been used as an oxidant for substrates bound to cobalt(III)[28] and for a number of derivatives of $[Fe(bipy)_3]^{2+}$.[29] In nitrate media, $[Fe(4,4'\text{-}(CO_2Et)_2bipy)_3]^{2+}$ reacts more rapidly with Ce^{4+} than with its hydrolysis product, $CeOH^{3+}$, by almost two orders of magnitude, presumably by an outer-sphere mechanism. The effects of micelles on this reaction have also been investigated.[30]

Table 2.1. Rate Constants for Electron-Transfer Reactions between Metal-Ion Complexes at 25 °C

Oxidant	Reductant	Medium M	k $(M^{-1}\,s^{-1})$	$\Delta H^{\ddagger}$ (kcal mol^{-1})	$\Delta S^{\ddagger}$ (cal K mol^{-1})	Ref.
Titanium(III)						
$[Ru(ox)_3]^{3-}$	$TiOH^{2+}$	1.0	300			3
$[Ru(ox)_3]^{3-}$	$[Ti(ox)]^+$	1.0	<1			3
$[Ru(ox)_3]^{3-}$	$[Ti(ox)_2]^-$	1.0	3.2×10^3			3
$[TiO_2]^{2+}$	Ti^{3+}	1.0	$11.1\ s^{-1}\ (1/[H^+])$			4
$[L_2CrO]^-$	$TiOH^{2+}$	0.5^e	5×10^5			19
$[L_2CrO]^-$	$[Ti(edta)]^-$	0.5^e	$>10^6$			19
Vanadium(II)						
$[Co(NH_3)_5py\text{-}3\text{-}CN]^{3+}$	V^{2+}	1.0	2.83			53
$[Co(NH_3)_5py\text{-}3\text{-}Cl]^{3+}$	V^{2+}	1.0	0.76			53
$[Co(NH_3)_5py]^{3+}$	V^{2+}	1.0	0.27			53
$[Co(NH_3)_5py\text{-}4\text{-}C(CH_3)]^{3+}$	V^{2+}	1.0	0.18			53
$[Co(NH_3)_5py\text{-}4\text{-}N(CH_3)_2]^{3+}$	V^{2+}	1.0	0.032			53
$[Co(pic)_3]$	$[V(pic)_3]^-$	0.5	7.4×10^6			36
$[Co(sep)]^{3+}$	$[V(pic)_3]^-$	0.1	3.5×10^5			37
$[Co([9]aneN_3)_2]^{3+}$	$[V(pic)_3]^-$	0.1	1.1×10^5			37
Vanadium(IV)						
$[Ru(Me_2bipy)_3]^{3+}$	VO^{2+}	1.0	0.22			6
$[Ru(Me_2bipy)_3]^{3+}$	$VOOH^+$	1.0	5.2×10^5			6
$[Ru(bipy)_3]^{3+}$	VO^{2+}	1.0	1.0			6
$[Ru(bipy)_3]^{3+}$	$VOOH^+$	1.0	6.4×10^5			6
$[Ni(Me_2bipy)_3]^{3+}$	VO^{2+}	1.0	2.0			6
$[Ni(Me_2bipy)_3]^{3+}$	$VOOH^+$	1.0	6.4×10^5			6
$[Ni(bipy)_3]^{3+}$	VO^{2+}	1.0	61			6
$[Ni(bipy)_3]^{3+}$	$VOOH^+$	1.0	3.6×10^6			6
$[L_2CrO]^-$	VO^{2+}	1.0^f	2.2×10^3			17
$[L_3CrO]^{2-}$	VO^{2+}	1.0^f	2.8×10^3			17
$Cr(IV)$	VO^{2+}	1.0^f	7.7×10^2			17
$[IrCl_6]^{2-}$	$[Vo(pida)OH_2]^{2-}$	0.1	1.13×10^4	6.9	-15.5	71
$[IrCl_6]^{2-}$	$[VO(pida)OH]^{3-}$	0.1	2.9×10^6	10.0	5.0	71
$[IrCl_6]^{2-}$	$[VO(pidaH)OH_2]^-$	0.1	$\sim 7 \times 10^2$			71
Chromium(II)						
V^{3+}	Cr^{2+}	3.0	0.20			7
VOH^{2+}	Cr^{2+}	3.0	370			7
$[(NH_3)_5CoNCCHCHCN]^{3+}$	Cr^{2+}	0.5	1.9×10^5			8
$[Rh(H_2O)_5OH]^{2+}$	Cr^{2+}	2.0	2.3×10^5			12
$[Rh(H_2O)_5Cl]^{2+}$	Cr^{2+}	2.0	2.2×10^3			12
$[Rh(H_2O)_5Br]^{2+}$	Cr^{2+}	2.0	$>8 \times 10^3$			12
$[Co(sep)]^{3+}$	$[Cr[15]aneN_4]^{2+}$	0.1	70			37
$[Co([9]aneN_3)_2]^{3+}$	$[Cr[15]aneN_4]^{2+}$	0.1	9			37
Fe^{3+}	$[Cr(pic)_3]^-$	2.3	14.1			38
$[Co(NH_3)_6]^{3+}$	$[Cr(phen)_3]^{2+}$	0.1	440			50
$[Co(NH_3)_5cha]^{3+}$	$[Cr(phen)_3]^{2+}$	0.1	3.5×10^3			50
$[cis\text{-}Co(en)_2(cha)Cl]^{2+}$	$[Cr(phen)_3]^{2+}$	0.1	5.7×10^3			50
$[Ru(bipy)_3]^{3+}$	$[Cr(H_2O)_5CH_2C_6H_4\text{-}4CH_3]^{2+}$	0.1	2.9×10^8			55
$[Ru(bipy)_3]^{3+}$	$[Cr(H_2O)_5CH_2C_6H_5]^{2+}$	0.1	5.3×10^8			55
$[Ru(bipy)_3]^{3+}$	$[Cr(H_2O)_5CH_2C_6H_4\text{-}4CF_3]^{2+}$	0.1	1.71×10^7			55
$[Ru(bipy)_3]^{3+}$	$[Cr(H_2O)_5CH_2C_6H_4\text{-}4\text{-}CN]^{2+}$	0.1	9.5×10^6			55
$[Ru(bipy)_3]^{3+}$	$[Cr(H_2O)_5CH_2OCH_3]^{2+}$	0.1	1.04×10^7			55
$[Ru(bipy)_3]^{3+}$	$[Cr(H_2O)_5CH_2CH(CH_3)_2]^{2+}$	0.1	4.25×10^7			55
$[Ru(bipy)_3]^{3+}$	$[Cr(H_2O)_5CH_2CH_3]^{2+}$	0.1	2.00×10^5			55
$[Ru(bipy)_3]^{3+}$	$[Cr(H_2O)_5CH_3]^{2+}$	0.1	$<10^3$			55

(*continued*)

Table 2.1. (continued)

Oxidant	Reductant	Medium M	k $(M^{-1}\,s^{-1})$	$\Delta H^{\ddagger}$ (kcal mol^{-1})	$\Delta S^{\ddagger}$ (cal K mol^{-1})	Ref.
Iron(II)						
$[TiO_2]^{2+}$	Fe^{2+}	1.0	1.1			
			$0.16\,s^{-1}\ (1/[H^+])$			4
Mn^{3+}	Fe^{2+}	2.0	5.2×10^3			10
$MnOH^{2+}$	Fe^{2+}	2.0	2.0×10^4			10
$[L_2CrO]^-$	Fe^{2+}	0.5^f	$5.3\,s^{-1}\ (1/[H^+])$			16
Cr(IV)	Fe^{2+}	0.5^f	2.7×10^4			16
$[Co(gly)(mal)_2]^{2-}$	Fe^{2+}	1.0^a	0.11			
			$1.1\ M^{-2}\,s^{-1}\ ([H^+])$			20
$[Co(mal)_2py_2]^-$	Fe^{2+}	1.0^a	0.10			
			$0.9\ M^{-2}\,s^{-1}\ ([H^+])$			20
$[Co(nta)mal)]^{2-}$	Fe^{2+}	1.0^a	0.34	9.8	-28	20
$[Co(nta)(ox)]^{2-}$	Fe^{2+}	1.0^a	0.44	9.8	-27	20
$[Co(ida)_2]^-$	Fe^{2+}	1.0^a	1.47×10^{-3}	14.1	-29	21
$[MoO_2Cl_2]$	Fe^{2+}	8.0^b	3.6×10^3			26
Mn^{3+}	$[Fe(5\text{-Mephen})_3]^{2+}$	2.0	3.3×10^4			10
Mn^{3+}	$[Fe(bipy)_3]^{2+}$	2.0	9.1×10^3			10
Mn^{3+}	$[Fe(phen)_3]^{2+}$	2.0	2.0×10^4			10
Mn^{3+}	$[Fe(5\text{-Clphen})_3]^{2+}$	2.0	3.8×10^3			10
Mn^{3+}	$[Fe(5\text{-NO}_2phen)_3]^{2+}$	2.0	5.5×10^2			10
Ce^{4+}	$[Fe(4,4'\text{-}(EtOCO)_2bipy)_3]^{2+}$	2.0	7.3×10^6			29
$CeOH^{3+}$	$[Fe(4,4'\text{-}(EtOCO)_2bipy)_3]^{2+}$	2.0	1×10^5			29
$[Co(pic)]$	$[Fe(edta)]^{2-}$	0.1	1.8×10^3			36
$[CoTPPS]^{3-}$	$[Fe(edta)]^{2-}$	0.5	2.0×10^3			
			$<10\ M^{-1}$			67
$[CoTAP]^{5+}$	$[Fe(edta)]^{2-}$	0.5	7.3×10^3			
			$460\ M^{-1}$			67
$[CoTMPy(4)P]^{5+}$	$[Fe(edta)]^{2-}$	0.5	7.8×10^4			
			$530\ M^{-1}$			67
$[CoTMPy(3)P]^{5+}$	$[Fe(edta)]^{2-}$	0.5	8.7×10^4			
			$470\ M^{-1}$			67
$[CoTMPy(2)P]^{5+}$	$[Fe(edta)]^{2-}$	0.5	1.4×10^4			
			$770\ M^{-1}$			67
$[AgTPPS]^{3-}$	$[Fe(edta)]^{2-}$	0.1	3×10^7			68
$[Co(pic)_3]$	$[Fe(pic)_2]$	0.1	2.0×10^2			36
$[Co(pic)_3]$	$[Fe(pic)_3]^-$	0.1	3.5			36
$[Co(ox)_3]^{3-}$	$[Fe(pic)_2]$	0.1	1.6×10^2			36
$[Co(ox)_3]^{3-}$	$[Fe(pic)_3]^-$	0.1	3.4			36
$[Co(NH_3)_5H_2O]^{3+}$	$[Fe(CN)_6]^{4-}$	0.5	$483\ M^{-1}$			
			$0.1\,s^{-1}$	21.4	9	45
$[Co(NH_3)_5py]^{3+}$	$[Fe(CN)_6]^{4-}$	1.0	$168\ M^{-1}$			
			$8.9 \times 10^{-3}\,s^{-1}$	28.3	27	45
$[Co(NH_3)_5Me_2SO]^{3+}$	$[Fe(CN)_6]^{4-}$	1.0	$34\ M^{-1}$			
			$0.2\,s^{-1}$	20.2	6	45
$[Co(NH_3)_5H_2O]^{3+}$	$[Fe(CN)_6]^{4-}$	0.5	$194\ M^{-1}$			
			$9.31 \times 10^{-2}\,s^{-1}$		29	46
$[AgTPPS]^{3-}$	$[Fe(CN)_6]^{4-}$	0.1	5.6×10^5	14.5	17	68
$[Ag(OH)_4]^-$	$[Fe(CN)_6]^{4-}$	1.2	1.1×10^5			69
$[Ni(H_{-2}G_3)_2]^{3-}$	$[Fe(CN)_6]^{4-}$	0.1	380			72
$[Ni(H_{-2}G_3)(H_{-1}G_3)]^{2-}$	$[Fe(CN)_6]^{4-}$	0.1	1.6×10^4			72
$[Co(phen)_3]^{3+}$	$[Fe(cp)_2]$	0.02^d	4.4×10^4			63
$[Co(5\text{-Mephen})_3]^{3+}$	$[Fe(cp)_2]$	0.02^d	2.0×10^4			63
$[Co(5,6\text{-Me}_2phen)_3]^{3+}$	$[Fe(cp)_2]$	0.02^d	6.5×10^3			63
$[Co(bpy)_3]^{3+}$	$[Fe(cp)_2]$	0.02^d	4.9×10^3			63
$[Co(dmg)_3(BF_2)_2]^+$	$[Fe(cp)_2]$	CH_3CN	1.22×10^4	8.5	-11.5	64
$[Co(dmg)_3(BF_2)_2]^+$	$[Fe(Mecp)_2]$	CH_3CN	9.8×10^4	6.4	-14.5	64
$[Co(dmg)_3(BPh_2)_2]^+$	$[Fe(cp)_2]$	CH_3CN	3.92×10^3	9.0	-12.3	64

(continued)

Table 2.1. (continued)

Oxidant	Reductant	Medium M	k ($M^{-1}\,s^{-1}$)	$\Delta H^{\ddagger}$ (kcal mol^{-1})	$\Delta S^{\ddagger}$ (cal K^{-1} mol^{-1})	Ref.
[Co(dmg)$_3$(BC$_4$H$_9$)$_2$]$^+$	[Fe(cp)$_2$]	CH$_3$CN	1.58×10^6	5.9	-10.4	64
[Co(nox)$_3$(BF)$_2$]$^+$	[Fe(cp)$_2$]	CH$_3$CN	2.52×10^4	7.1	-14.9	64
[Co(nox)$_3$(BF)$_2$]$^+$	[Fe(Mecp)$_2$]	CH$_3$CN	2.04×10^5	6.4	-13.1	64
[Co(nox)$_3$(BPh)$_2$]$^+$	[Fe(Mecp)$_2$]	CH$_3$CN	7.67×10^3	7.6	-15.6	64
[Co(nox)$_3$(BC$_4$H$_9$)$_2$]$^+$	[Fe(Me$_5$cp)$_2$]	CH$_3$CN	3.86×10^6	4.0	-15.0	64
[Co(dpg)$_3$(BPh)$_2$]$^+$	[Fe(cp)$_2$]	CH$_3$CN	8.07×10^3	7.9	-14.4	64
[Co(dpg)$_3$(BPh)$_2$]$^+$	[Fe(Mecp)$_2$]	CH$_3$CN	6.78×10^4	6.2	-15.1	64
[FeTPP(1-MeIm)$_2$]$^+$	[FeTPP(1-MeIm)$_2$]	CH$_2$Cl$_2$	8.1×10^7			64
[Fe3-MeTPP(1-MeIm)$_2$]$^+$	[Fe3-MeTPP(1-MeIm)$_2$]	CH$_2$Cl$_2$	5.3×10^7			64
[Fe4-MeTPP(1-MeIm)$_2$]$^+$	[Fe4-MeTPP(1-MeIm)$_2$]	CH$_2$Cl$_2$	9.7×10^7			64
[Fe4-MeOTPP(1-MeIm)$_2$]$^+$	[Fe4-MeOTPP(1-MeIm)$_2$]	CH$_2$Cl$_2$	6.8×10^7			64
[Fe2,4,6-Me$_3$TPP(1-MeIm)$_2$]$^+$	[Fe2,4,6-Me$_3$TPP(1-MeIm)$_2$]	CH$_2$Cl$_2$	1.6×10^8			64
[FeTPP(1-Bu^{n}Im)$_2$]$^+$	[FeTPP(1-Bu^{n}Im)$_2$]	CH$_2$Cl$_2$	7.0×10^7			64
[FeTPP(1-Bu^{t}Im)$_2$]$^+$	[FeTPP(1-Bu^{t}Im)$_2$]	CH$_2$Cl$_2$	6.8×10^7			64
[Fe3MeTPP(5-MeIm)$_2$]$^+$	[Fe3MeTPP(5-MeIm)$_2$]	CH$_2$Cl$_2$	2.3×10^7			64
[Fe3MeTPP(5-Bu^{t}Im)$_2$]$^+$	[Fe3MeTPP(5-Bu^{t}Im)$_2$]	CH$_2$Cl$_2$	1.6×10^7			64
Cobalt(II) and Cobalt(III)						
Fe^{3+}	[Co(dmgBF$_2$)$_2$]	0.5	2.9×10^2			22
FeOH^{2+}	[Co(dmgBF$_2$)$_2$]	0.5	6.2×10^4			22
Fe^{3+}	[Co(sep)]$^{2+}$	0.23	1.75×10^5			23
[Co(edta)]$^-$	[Co(sep)]$^{2+}$	0.1	8.26×10^4			37
[Co(cydta)]$^-$	[Co(sep)]$^{2+}$	0.1	8.30×10^4			37
[Co(terp)$_2$]$^{3+}$	[Co(sep)]$^{2+}$	0.1	1.48×10^5			37
[Co(dmgH)$_2$(PhNH$_2$)$_2$]$^+$	[Co(sep)]$^{2+}$	0.1	8.26×10^4			37
[Co(NH$_3$)$_6$]$^{3+}$	[Co(sep)]$^{2+}$	0.1	0.15			50
[Co(NH$_3$)$_5$cha]$^{3+}$	[Co(sep)]$^{2+}$	0.1	0.14			50
[Co(NH$_3$)$_5$NH$_2$Ph]$^{3+}$	[Co(sep)]$^{2+}$	0.1	0.60			50
[Co(NH$_3$)$_5$CN]$^{2+}$	[Co(sep)]$^{2+}$	0.1	0.58			50
[Co(NH$_3$)$_5$Cl]$^{2+}$	[Co(sep)]$^{2+}$	0.1	58			50
[Co(NH$_3$)$_5$Br]$^{2+}$	[Co(sep)]$^{2+}$	0.1	1.6×10^3			50
[Co(NH$_3$)$_5$I]$^{2+}$	[Co(sep)]$^{2+}$	0.1	1.2×10^4			50
[Co(NH$_3$)$_5$N$_3$]$^{2+}$	[Co(sep)]$^{2+}$	0.1	7.4			50
[Co(NH$_3$)$_5$NO$_2$]$^{2+}$	[Co(sep)]$^{2+}$	0.1	2.1×10^3			50
[cis-Co(en)$_2$(cha)Cl]$^{2+}$	[Co(sep)]$^{2+}$	0.1	0.82			50
[cis-Co(en)$_2$(NH$_2$Ph)Cl]$^{2+}$	[Co(sep)]$^{2+}$	0.1	3.3			50
[cis-Co(en)$_2$(NH$_2$CH$_2$Ph)Cl]$^{2+}$	[Co(sep)]$^{2+}$	0.1	85			50
[cis-Co(en)$_2$(CN)Cl]$^+$	[Co(sep)]$^{2+}$	0.1	1.2			50
[cis-Co(phen)$_2$(CN)$_2$]$^+$	[Co(sep)]$^{2+}$	0.1	24			50
[Co(phen)$_3$]$^{3+}$	[Co(sep)]$^{2+}$	0.1	4.8×10^3			50
[Co[14]aneN$_4$(NH$_3$)$_2$]$^{3+}$	[Co(sep)]$^{2+}$	0.1	0.10			50
[CoMe$_4$[14]tetraeneN$_4$(NH$_3$)$_2$]$^{3+}$	[Co(sep)]$^{2+}$	0.1	36			50
[Co[14]aneN$_4$(OH$_2$)$_2$]$^{3+}$	[Co(sep)]$^{2+}$	0.1	1.6×10^3			50
[CoMe$_4$[14]tetraeneN$_4$(OH$_2$)$_2$]$^{3+}$	[Co(sep)]$^{2+}$	0.1	1.7×10^4			50
[$meso$-Co(bzo$_3$[12]hexaeneN$_3$)$_2$]$^{3+}$	[Co(sep)]$^{2+}$	0.1	100			50
[rac-Co(bzo$_3$[12]hexaeneN$_3$)$_2$]$^{3+}$	[Co(sep)]$^{2+}$	0.1	780			50
[Co(edta)]$^-$	[Co([9]aneN$_3$)$_2$]$^{2+}$	0.1	2.76×10^4			37
[Co(cydta)]$^-$	[Co([9]aneN$_3$)$_2$]$^{2+}$	0.1	3.00×10^4			37
[Co(terpy)$_2$]$^{3+}$	[Co([9]aneN$_3$)$_2$]$^{2+}$	0.1	3.28×10^4			37
[Co(dmgH)$_2$(PhNH$_2$)$_2$]$^+$	[Co([9]aneN$_3$)$_2$]$^{2+}$	0.1	2.76×10^4			37
[Co(sep)]$^{3+}$	[Co([9]aneN$_3$)$_2$]$^{2+}$	0.1	4.62			37
[Tl([9]aneN$_3$)$_2$]$^{3+}$	[Co([9]aneN$_3$)$_2$]$^{2+}$	0.5	3.4×10^{-2}	16	-11	39
[Ru$_2$(NH$_3$)$_{10}$pz]$^{6+}$	[Co(bipy)$_3$]$^{2+}$	0.1	1.1×10^5			34
[Ru$_2$(NH$_3$)$_{10}$pz]$^{5+}$	[Co(bipy)$_3$]$^{2+}$	0.1	3.2×10^3			34
[RuRh(NH$_3$)$_{10}$pz]$^{6+}$	[Co(bipy)$_3$]$^{2+}$	0.1	4.7×10^4			34
[Ru(NH$_3$)$_5$pz]$^{3+}$	[Co(bipy)$_3$]$^{2+}$	0.1	7.3×10^3			34
[Co(NH$_3$)$_5$pzMe]$^{4+}$	[Co(bipy)$_3$]$^{2+}$	0.1	1.7×10^5			34

(continued)

Table 2.1. (continued)

Oxidant	Reductant	Medium M	k ($M^{-1}s^{-1}$)	$\Delta H^{\ddagger}$ (kcal mol^{-1})	$\Delta S^{\ddagger}$ (cal K mol^{-1})	Ref.
$[Co(phen)_3]^{3+}$	$[Co(terpy)_2]^{2+}$	0.057	1.14×10^2	6.5	-27.5	59
$[Co(phenSO_3)_3]$	$[Co(terpy)_2]^{2+}$	0.013	1.20×10^4	4.8	-23.9	59
$[Co(bipy)_3]^{3+}$	$[Co(terpy)_2]^{2+}$	0.057	20.7	7.6	-26.0	59
$[Co(bipySO_3)_3]$	$[Co(terpy)_2]^{2+}$	0.0053	3.03×10^4	2.6	-28.7	59
$[NiMe_2L]^{2+}$	$[Co(pdta)]^{2-}$	0.1	48			61
$[NiMe_2LH]^{2+}$	$[Co(pdta)]^{2-}$	0.1	310			61
			$7 \times 10^{-2}\,s^{-1}$			61
$[NiMe_2L]^{2+}$	$[Co(cdta)]^{2-}$	0.1	45			61
$[NiMe_2LH]^{2+}$	$[Co(cdta)]^{2-}$	0.1	$280\,M^{-1}$			61
			$7 \times 10^{-2}\,s^{-1}$			61
$[PtCl_6]^{2-}$	MeB_{12}(base on)	1.0	$580\,M^{-1}$			
			$4 \times 10^{-3}\,s^{-1}$			59
$[PtCl_6]^{2-}$	MeB_{12}(base off)	1.0	$15\,M^{-1}$			
			$4 \times 10^{-3}\,s^{-1}$			59
Nickel(II) and Nickel(III)						
Mn^{3+}	$[Ni[14]aneN_4]^{2+}$	2.0	8.2×10^3			10
$MnOH^{2+}$	$[Ni[14]aneN_4]^{2+}$	2.0	1.8×10^4			10
Mn^{3+}	$[NiMe_6[14]4.11\text{-}dieneN_4]^{2+}$	2.0	28			10
$MnOH^{2+}$	$[NiMe_6[14]4,11\text{-}dieneN_4]^{2+}$	2.0	48			10
Mn^{3+}	$[NiMe_2LH_2]^{2+}$	2.0	780			
			$86\,s^{-1}\,(1/[H^+])$			10
$[Rh_2(OAc)_4(OH_2)_2]^+$	$[Ni[14]aneN_4]^{2+}$	1.0	1.59×10^6	0.9	-27	13
$[Rh_2(OAc)_4(OH_2)_2]^+$	$[Ni([9]aneN_3)_2]^{2+}$	1.0	5.0×10^6			13
$[Rh_2(OAc)_4(OH_2)_2]^+$	$[NiMe_2[14]dieneN_4]^{2+}$	1.0	1.54×10^5	1.4	-30	13
$[Rh_2(OAc)_4(OH_2)_2]^+$	$[NiMe_2L]^{2+}$	1.0	1.60×10^4	4.8	-23	13
$[Rh_2(OAc)_4(OH_2)_2]^+$	$[Nimeso\text{-}Me_6[14]aneN_4]^{2+}$	1.0	2.3×10^3			13
$[Rh_2(OAc)_4(OH_2)_2]^+$	$[Nirac\text{-}Me_6[14]aneN_4]^{2+}$	1.0	1.0×10^3			13
Co^{3+}	$[Ni([10]aneN_3)_2]^{2+}$	1.0	550			
			$579\,s^{-1}\,([H^+])$			43
$[Ni([10]aneN_3)_2]^{3+}$	$[Ni([9]aneN_3)_2]^{2+}$	1.0	2.2×10^4			43
$[Ni[14]aneN_4]^{3+}$	$[Ni([10]aneN_3)_2]^{2+}$	1.0	1.3×10^3			43
$[Ni([10]aneN_3)_2]^{3+}$	$[Ni[14]aneN_4]^{2+}$	1.0	6.1×10^3			43
$[Ni([10]aneN_3)_2]^{3+}$	$[Nimeso\text{-}Me_2[14]aneN_4]^{2+}$	1.0	1.08×10^4			43
$[Nimeso\text{-}Me_2[14]aneN_4]^{3+}$	$[Ni([10]aneN_3)_2]^{2+}$	1.0	2.2×10^4			43
$[NiMe_2[14]dieneN_4]^{2+}$	$[Ni([10]aneN_3)_2]^{2+}$	1.0	3.4×10^3			43
$[Ni([10]aneN_3)_2]^{3+}$	$[NiMe_2[14]dieneN_4]^{2+}$	1.0	1.13×10^3			43
$[Ni(tet\ c)]^{3+}$	$[Ni([10]aneN_3)_2]^{2+}$	1.0	1.16×10^5			43
$[Ni(tet\ d)]^{3+}$	$[Ni([10]aneN_3)_2]^{2+}$	1.0	1.38×10^5			43
$[NiMe_2LH_2]^{2+}$	$[Ni([10]aneN_3)_2]^{2+}$	1.0	23.1			43
$[Nirac\text{-}Me[14]aneN_4]^{3+}$	$[Ni([10]aneN_3)_2]^{2+}$	1.0	1.4×10^4			43
$\alpha\text{-}[Nimeso\text{-}5,12\text{-}Me_2[14]aneN_4]^{3+}$	$[Ni([9]aneN_3)_2]^{2+}$	1.0	1.74×10^4			44
$\beta\text{-}[Nimeso\text{-}5,12\text{-}Me_2[14]aneN_4]^{3+}$	$[Ni([9]aneN_3)_2]^{2+}$	1.0	8.9×10^3			44
$\delta\text{-}[Nimeso\text{-}5,12\text{-}Me_2[14]aneN_4]^{3+}$	$[Ni([9]aneN_3)_2]^{2+}$	1.0	1.55×10^4			44
$[Nirac\text{-}5,12\text{-}Me_2[14]aneN_4]^{3+}$	$[Ni([9]aneN_3)_2]^{2+}$	1.0	2.2×10^4			44
$[Nimeso5,12\text{-}Et_2[14]aneN_4]^{3+}$	$[Ni([9]aneN_3)_2]^{2+}$	1.0	2.5×10^4			44
$[NiMe_2L]^{2+}$	$[NiMe_2LH_2]^{2+}$	0.1	3.5×10^2			60
$[NiMe_2L]^{2+}$	$[NiMe_2LH]^+$	0.1	3.8×10^3			60
$[NiMe_2L]^{2+}$	$[NiMe_2L]$	0.1	1.24×10^6			60
$[NiMe_2LH]^{2+}$	$[NiMe_2LH]^{2+}$	0.1	4			60
$[NiMe_2L]^+$	$[NiMe_2LH_2]^{2+}$	0.1	3.4×10^3			60
Copper(II)						
$[Ni(H_{-2}G_3)_2]^{3-}$	$[CuH_{-3}Aib_3a]^-$	0.1	8.19			72
$[Ni(H_{-2}G_3)(H_{-1}G_3)]^{2-}$	$[CuH_{-3}Aib_3a]^-$	0.1	3.7×10^3			72
$[Ni(H_{-2}G_3)_2]^{3-}$	$[CuH_{-3}Aib_3]^-$	0.1	7.67×10^{-2}			72
$[Ni(H_{-2}G_3)(H_{-1}G_3)]^{2-}$	$[CuH_{-3}Aib_3]^-$	0.1	4.5×10^3			72

(continued)

Table 2.1. (continued)

Oxidant	Reductant	Medium M	k $(M^{-1}s^{-1})$	$\Delta H^{\ddagger}$ (kcal mol^{-1})	$\Delta S^{\ddagger}$ (cal K mol^{-1})	Ref.
Molybdenum(V)						
FeOH^{2+}	[Mo$_2$O$_4$]$^{2+}$	2.0	1.3×10^3			25
Fe^{3+}	[Mo$_2$O$_4$]$^{2+}$	2.0	≤ 0.2			25
[Fe(phen)$_3$]$^{3+}$	[Mo$_2$O$_4$]$^{2+}$	2.0	13			25
[Fe(phen)$_3$]$^{3+}$	[Mo$_2$O$_2$S$_2$]$^{2+}$	2.0	0.8			25
Molybdenum(III)						
Fe^{3+}	Mo^{3+}	1.0	1.3×10^3			26
Molybdenum(IV)						
[Fe(phen)$_3$]$^{3+}$	[Mo$_3$O$_4$]$^{4+}$	2.0	51			25
[AgTPPS]$^{3-}$	[Mo(CN)$_8$]$^{4-}$	0.1	2.5×10^3			68
[Ag(OH)$_4$]$^-$	[Mo(CN)$_8$]$^{4-}$	1.2	80			70
Tungsten(IV)						
[Ag(OH)$_4$]$^-$	[W(CN)$_8$]$^{4-}$	1.2	2.9×10^4			70
[Ni(H$_{-2}$G$_3$)(H$_{-1}$G$_3$)]$^{2-}$	[W(CN)$_8$]$^{4-}$	0.1	3.7×10^3			72
[Ni(H$_{-2}$G$_3$)$_2$]$^{3-}$	[W(CN)$_8$]$^{4-}$	0.1	147			72
Tungsten(V)						
[α-PW$_{12}$O$_{40}$]$^{3-}$	[α-PW$_{12}$O$_{40}$]$^{4-}$	—	$\sim 10^7$			24
[α-P$_2$W$_{18}$O$_{62}$]$^{6-}$	[α-P$_2$W$_{18}$O$_{62}$]$^{7-}$	—	1.4×10^4			24
[α-P$_2$MoW$_{17}$O$_{62}$]$^{6-}$	[α-P$_2$MoW$_{17}$O$_{62}$]$^{7-}$	—	2×10^3			24
Ruthenium(II)						
[Rh$_2$(OAc)$_4$(OH$_2$)$_2$]$^+$	[Ru(bipy)$_3$]$^{2+}$	1.0	1.1×10^7			13
Mn^{3+}	[Ru(4,4'-Me$_2$bipy)$_3$]$^{2+}$	2.0	6.1×10^3			10
Mn^{3+}	[Ru(bipy)$_3$]$^{2+}$	2.0	7.2×10^2			10
Mn^{3+}	[Ru(5-NO$_2$phen)$_3$]$^{2+}$	2.0	22			10
[Ru(NH$_3$)$_6$]$^{3+}$	[Ru(NH$_3$)$_6$]$^{2+}$	0.5^c	3.3×10^3	5		31
[Ru(en)$_3$]$^{3+}$	[Ru(en)$_3$]$^{2+}$	0.75	3.1×10^4	6.0	−17	32
[Ru(bipy)$_2$acac]$^{2+}$	[Ru(bipy)$_2$acac]$^+$	CH$_3$CN, 0	1.5×10^7	2.2		33
[Ru(bipy)$_2$hfac]$^{2+}$	[Ru(bipy)$_2$hfac]$^+$	CH$_3$CN, 0	4.3×10^6	4.1		33
[Ru$_2$(NH$_3$)$_{10}$pz]$^{6+}$	[Ru(NH$_3$)$_5$pz]$^{2+}$	0.1	1.2×10^7			34
[Ru$_2$(NH$_3$)$_{10}$pz]$^{6+}$	[Ru(NH$_3$)$_4$bipy]$^{2+}$	0.1	1.9×10^6			34
[Ru$_2$(NH$_3$)$_{10}$pz]$^{6+}$	[Ru$_2$(NH$_3$)$_{10}$pz]$^{4+}$	0.1	4.8×10^6			34
[Ru(NH$_3$)$_{10}$pz]$^{5+}$	[Ru(NH$_3$)$_5$pz]$^{2+}$	0.1	2.6×10^5			34
[RuRh(NH$_3$)$_{10}$pz]$^{6+}$	[Ru(NH$_3$)$_5$pz]$^{2+}$	0.1	4.2×10^6			34
[RuRh(NH$_3$)$_{10}$pz]$^{6+}$	[Ru(NH$_3$)$_4$bipy]$^{2+}$	0.1	7.4×10^5			34
[Ru(NH$_3$)$_4$bipy]$^{3+}$	[Ru$_2$(NH$_3$)$_{10}$pz]$^{4+}$	0.1	8.9×10^5			34
[Ru(NH$_3$)$_5$pz]$^{3+}$	[Ru$_2$(NH$_3$)$_{10}$pz]$^{4+}$	0.1	$3.7 \times 10^{5\cdot}$			34
[Ru(NH$_3$)$_5$pz]$^{3+}$	[Ru(NH$_3$)$_5$py]$^{2+}$	0.1	1.2×10^6			34
[Ru(NH$_3$)$_5$pz]$^{3+}$	[Ru(NH$_3$)$_4$bipy]$^{4+}$	0.1	1.5×10^5			34
[Ru(NH$_3$)$_5$pzMe]$^{4+}$	[Ru(NH$_3$)$_4$bipy]$^{4+}$	0.1	6.1×10^6			34
[Co(ox)$_3$]$^{3-}$	[Ru(NH$_3$)$_6$]$^{2+}$	0.2	0.178	10.9	−25.7	35
[Co(ox)$_3$]$^{3-}$	[Ru(en)$_3$]$^{2+}$	0.2	0.055			35
[Co(ox)$_3$]$^{3-}$	[Ru(NH$_3$)$_5$OH$_2$]$^{2+}$	0.2	0.033	23.5	13.5	35
[Co(NH$_3$)$_5$OH$_2$]$^{3+}$	[Ru(NH$_3$)$_5$OH$_2$]$^{2+}$	0.2	0.43			49
[Co(NH$_3$)$_5$OH]$^{2+}$	[Ru(NH$_3$)$_5$OH$_2$]$^{2+}$	0.2	0.87			49
[Co(CN)$_5$OH$_2$]$^{2-}$	[Ru(NH$_3$)$_5$OH$_2$]$^{2+}$	0.2	0.96			49
[Co(CN)$_5$OH]$^{3-}$	[Ru(NH$_3$)$_5$OH$_2$]$^{2+}$	0.2	0.60			49
[Co(NH$_3$)$_5$cha]$^{3+}$	[Ru(NH$_3$)$_6$]$^{2+}$	0.1	0.07			50
[Co(NH$_3$)$_5$NH$_2$Ph]$^{3+}$	[Ru(NH$_3$)$_6$]$^{2+}$	0.1	0.11			50
[Co(NH$_3$)$_5$CN]$^{2+}$	[Ru(NH$_3$)$_6$]$^{2+}$	0.1	4.6			50
[Co(NH$_3$)$_5$NO$_2$]$^{2+}$	[Ru(NH$_3$)$_6$]$^{2+}$	0.1	80			50
[cis-Co(en)$_2$(CN)Cl]$^+$	[Ru(NH$_3$)$_6$]$^{2+}$	0.1	0.10			50
[cis-Co(phen)$_2$(CN)$_2$]$^+$	[Ru(NH$_3$)$_6$]$^{2+}$	0.1	11.4			50
[Co(phen)$_3$]$^{3+}$	[Ru(NH$_3$)$_6$]$^{2+}$	0.1	9.3×10^3			50

(*continued*)

Table 2.1. *(continued)*

Oxidant	Reductant	Medium M	k $(M^{-1}\,s^{-1})$	$\Delta H^{\ddagger}$ (kcal mol^{-1})	$\Delta S^{\ddagger}$ (cal K mol^{-1})	Ref.
$[Co[14]aneN_4(NH_3)_2]^{3+}$	$[Ru(NH_3)_6]^{2+}$	0.1	4.5×10^{-2}			50
$[CoMe_4[14]tetraeneN_4(NH_3)_2]^{3+}$	$[Ru(NH_3)_6]^{2+}$	0.1	1.1			50
$[Co[14]aneN_4(OH_2)_2]^{3+}$	$[Ru(NH_3)_6]^{2+}$	0.1	1.1×10^3			50
$[CoMe_4[14]tetraeneN_4(OH_2)_2]^{3+}$	$[Ru(NH_3)_6]^{2+}$	0.1	1.1×10^5			50
$[meso\text{-}Co(bzo_3[12]hexaeneN_3)_2]^{3+}$	$[Ru(NH_3)_6]^{2+}$	0.1	55			50
$[rac\text{-}Co(bzo_3[12]hexaeneN_3)_2]^{3+}$	$[Ru(NH_3)_6]^{2+}$	0.1	1.4×10^3			50
$[Co(NH_3)_5py\text{-}3\text{-}CN]^{3+}$	$[Ru(NH_3)_6]^{2+}$	1.0	20.5			53
$[Co(NH_3)_5py\text{-}3\text{-}Cl]^{3+}$	$[Ru(NH_3)_6]^{2+}$	1.0	7.0			53
$[Co(NH_3)_5py]^{3+}$	$[Ru(NH_3)_6]^{2+}$	1.0	1.15			53
$[Co(NH_3)_5py\text{-}4\text{-}C(CH_3)]^{3+}$	$[Ru(NH_3)_6]^{2+}$	1.0	0.98			53
$[Co(NH_3)_5py\text{-}4\text{-}N(CH_3)_2]^{3+}$	$[Ru(NH_3)_6]^{2+}$	1.0	0.16			53
Ruthenium(I)						
$[Co(NH_3)_6]^{3+}$	$[Ru(bipy)_3]^+$	1.0	2.70×10^9			54
$[Co(en)_3]^{3+}$	$[Ru(bipy)_3]^+$	1.0	2.29×10^9			54
$[Co(Me)(dmgBF_2)_2OH_2]$	$[Ru(bipy)_3]^+$	1.0	3.2×10^9			54
$[Co(sep)]^{3+}$	$[Ru(bipy)_3]^+$	1.0	$>2 \times 10^9$			54
$[Cr(H_2O)_5CF_3]^{2+}$	$[Ru(bipy)_2]^+$	1.0	$<5 \times 10^5$			54
$[Cr(H_2O)_5CH_2OCH_3]^{2+}$	$[Ru(bipy)_3]^+$	1.0	2.2×10^6			54
$[Cr(H_2O)_5CHCl_2]^{2+}$	$[Ru(bipy)_3]^+$	1.0	2.12×10^7			54
$[Cr(H_2O)_5CH_2C_6H_5]^{2+}$	$[Ru(bipy)_3]^+$	1.0	3.2×10^7			54
$[Cr(H_2O)_5CH_2\text{-}4\text{-}C_5H_4NH]^{3+}$	$[Ru(bipy)_3]^+$	1.0	1.39×10^9			54
$[Cr(H_2O)_5NC_5H_4\text{-}4\text{-}CH_3]^{3+}$	$[Ru(bipy)_3]^+$	1.0	4.2×10^8			54
$[Cr(H_2O)_5NC_5H_5]^{3+}$	$[Ru(bipy)_3]^+$	1.0	5.5×10^8			54
$[Cr(H_2O)_5NC_5H_4\text{-}3\text{-}Cl]^{3+}$	$[Ru(bipy)_3]^+$	1.0	1.29×10^9			54
$[Cr(H_2O)_5NC_5H_4\text{-}3\text{-}CN]^{3+}$	$[Ru(bipy)_3]^+$	1.0	2.64×10^9			54
$[cis\text{-}Cr(H_2O)_4py_2]^{3+}$	$[Ru(bipy)_3]^+$	1.0	1.54×10^9			54
$[Cr(H_2O)_5bipy]^{3+}$	$[Ru(bipy)_3]^+$	1.0	1.8×10^9			54
Eu^{3+}	$[Ru(bipy)_3]^+$	1.0	2.7×10^7			54
Yb^{3+}	$[Ru(bipy)_3]^+$	1.0	1.5×10^5			54
Sm^{3+}	$[Ru(bipy)_3]^+$	1.0	$<2 \times 10^4$			54
Cr^{3+}	$[Ru(bipy)_3]^+$	1.0	4.59×10^6			54
$[Nitmc]^{2+}$	$[Ru(bipy)_3]^+$	1.0	5.1×10^8			54
$[Eu2.2.1]^{3+}$	$[Ru(bipy)_3]^+$	1.0^e	7.0×10^8			81
$[*Ru(bipy)_3]^{2+}$						
$[Co(NH_3)_6]^{3+}$	$[*Ru(bipy)_3]^{2+}$	0.1	6.8×10^6			78
$[Co(en)_3]^{3+}$	$[*Ru(bipy)_3]^{2+}$	0.1	3.7×10^6			78
$[Co(NH_3)_5Cl]^{2+}$	$[*Ru(bipy)_3]^{2+}$	0.1	3.8×10^8			78
$[Eu2.2.1]^{3+}$	$[*Ru(bipy)_3]^{2+}$	1.0^e	4.8×10^7			81
Rhodium(II)						
$[Nimeso\text{-}Me_6[14]aneN_4]^{3+}$	$[Rh_2(OAc)_4(OH_2)_2]$	1.0	1.5×10^4			13
$[Nirac\text{-}Me_6[14]aneN_4]^{3+}$	$[Rh_2(OAc)_4(OH_2)_2]$	1.0	8.7×10^3			13
$[NiMe_6[14]dieneN_4]^{3+}$	$[Rh_2(OAc)_4(OH_2)_2]$	1.0	1.64×10^3	13.2	0.6	13
$[Ru(bipy)_3]^{3+}$	$[Rh_2(OAc)_4(OH_2)_2]$	1.0	2.5×10^7			13
$[Ni(Me_2bipy)_3]^{3+}$	$[Rh_2(OAc)_4(OH_2)_2]$	1.0	1.2×10^7			13
Europium(II)						
$[*Ru(bipy)_3]^{2+}$	$[Eu(2.2.1)]^{2+}$	1.0	1.3×10^9			81
$[Ru(bipy)_3]^{3+}$	$[Eu(2.2.1)]^{2+}$	1.0	1.3×10^9			81
Osmium(II)						
Mn^{3+}	$[Os(4,4'\text{-}Me_2bipy)_3]^{2+}$	2.0	2.5×10^6			10
Mn^{3+}	$[Os(phen)_3]^{2+}$	2.0	7.0×10^5			10
Mn^{3+}	$[Os(bipy)_3]^{2+}$	2.0	5.2×10^5			10
Mn^{3+}	$[Os(5\text{-}Clphen)_3]^{2+}$	2.0	1.2×10^5			10

(continued)

Table 2.1. (continued)

Oxidant	Reductant	Medium M	k $(M^{-1}s^{-1})$	$\Delta H^{\ddagger}$ (kcal mol^{-1})	$\Delta S^{\ddagger}$ (cal K mol^{-1})	Ref.
VO_2^+	$[Os(4,4'-Me_2bipy)_3]^{2+}$	1.0	21			
			185 M^{-2} s^{-1} ([H$^+$])			6
VO_2^+	$[Os(4,4'-Ph_2bipy)_3]^{2+}$	1.0	15			
			75 M^{-2} s^{-1} ([H$^+$])			6
VO_2^+	$[Os(5,6-Me_2phen)_3]^{2+}$	1.0	7.5			
			41 M^{-2} s^{-1} ([H$^+$])			6
VO_2^+	$[Os(4,7-Ph_2phen)_3]^{2+}$	1.0	9.2			
			39 M^{-2} s^{-1} ([H$^+$])			6
VO_2^+	$[Os(5-Mephen)_3]^{2+}$	1.0	3.2			
			8.6 M^{-2} s^{-1} ([H$^+$])			6
VO_2^+	$[Os(phen)_3]^{2+}$	1.0	1.7			
			6.4 M^{-2} s^{-1} ([H$^+$])			6
VO_2^+	$[Os(bipy)_3]^{2+}$	1.0	2.7			
			7.5 M^{-2} s^{-1} ([H$^+$])			6
VO_2^+	$[Os(5-Clphen)_3]^{2+}$	1.0	0.14			
			0.91 M^{-2} s^{-1} ([H$^+$])			6
$[Fe(bipy)_3]^{3+}$	$[Os(bipy)_3]^{2+}$	CH$_3$CN	1.3×10^5			60
$[Fe(bipy)_3]^{3+}$, PF$_6^-$	$[Os(bipy)_3]^{2+}$	CH$_3$CN	4×10^6			60
$[Fe(bipy)_3]^{3+}$, ClO$_4^-$	$[Os(bipy)_3]^{2+}$	CH$_3$CN	9×10^6			60
$[Fe(Me_2bipy)_3]^{3+}$	$[Os(Me_2bipy)_3]^{2+}$	CH$_3$CN	5.2×10^5			60
$[Fe(Me_2bipy)_3]^{3+}$, PF$_6^-$	$[Os(Me_2bipy)_3]^{2+}$	CH$_3$CH	1.0×10^7			60
$[Fe(Me_2bipy)_3]^{3+}$, ClO$_4^-$	$[Os(Me_2bipy)_3]^{2+}$	CH$_3$CN	1.2×10^7			60
$[Fe(Ph_2bipy)_3]^{3+}$	$[Os(Me_2bipy)_3]^{2+}$	CH$_3$CN	2.2×10^6			60
$[Fe(Ph_2bipy)_3]^{3+}$, ClO$_4^-$	$[Os(Me_2bipy)_3]^{2+}$	CH$_3$CN	1.7×10^8			60
$[Os(Me_2bipy)_3]^{3+}$	$[Os(Me_2bipy)_3]^{2+}$	CH$_3$CN	1.9×10^5			61
$[Os(Me_2bipy)_3]^{3+}$, PF$_6^-$	$[Os(Me_2bipy)_3]^{2+}$	CH$_3$CN	9×10^5			61
Uranium(IV)						
$[L_2CrO]^-$	$[ULOH]^{2+}$	0.5^f	1.9×10^2 M^{-1} s^{-1}			18

^{a} 30 °C. ^{b} 20 °C. ^{c} 4 °C. ^{d} 40 °C. ^{e} 23 °C. ^{f} 21 °C.

2.3. *Reactions of Metal-Ion Complexes*

The self-exchange rates of $[Ru(NH_3)_6]^{3+/2+}$ [31] and $[Ru(en)_3]^{3+/2+}$ [32] have been determined by NMR in triflate media. While the former rate is in good agreement with previous estimates, the latter is somewhat higher than is predicted by Marcus theory. Both reactions show a strong chloride ion dependence on the rate. In acetonitrile the self-exchange rates of $[Ru(bpy)_2(acac)]^{2+/+}$ and $[Ru(bpy)_2(hfac)]^{2+/+}$ have been determined[33] as 1.5×10^7 M^{-1} s^{-1} and 4.3×10^6 M^{-1} s^{-1} respectively at 25 °C and ionic strength approaching zero. Comparisons with current theory reveal that small reaction separation distances are required to model the processes. Marcus theory has been used[34] to calculate self-exchange rates for a number of pyrazine bridged binuclear Ru$_2$ and RhRu complexes. The rates show a strong dependence on the charges on the complexes. The reductions of $[Co(ox)_3]^{3-}$ by $[Ru(NH_3)_6]^{2+}$ and $[Ru(en)_3]^{2+}$ are outer-sphere in nature allowing evaluation[35] of a self-exchange rate for $[Co(ox)_3]^{3-/4-}$ as 1.4×10^{-12} M^{-1} s^{-1} at 25 °C and 0.2 M ionic strength, five orders of magnitude lower than the currently accepted value. Reduction by $[Ru(NH_3)_5(OH_2)]^{2+}$ on the other hand is inner-sphere, revealed by the detection of $[Ru(NH_3)_5ox]^+$ as a reaction product and by markedly different activation parameters from the other two reductants. The oxidant $[Co(ox)_3]^{3-}$ has also been used[36] to study redox behavior with labile iron(II) picolinate complexes. The equilibrium (2) between the bis- and tris-complexes is rapidly established, and reaction with the bis-species is two orders

$$[Fe(pic)_2] + pic^- \rightleftharpoons [Fe(pic)_3]^- \qquad (2)$$

of magnitude faster than reaction with $[Fe(pic)_3]^-$. Although an outer-sphere mechanism is proposed for these reactions, the similarity of the rate data for oxidations by both $[Co(ox)_3]^{3-}$ and $[Co(pic)_3]$ suggests substitution control and an inner-sphere process. A variety of reactions of metal picolinate reductants have been reported. The potent reductant $[V(pic)_3]^-$ undergoes outer-sphere electron transfer with $[Co(pic)_3]$,[36] $[Co(sep)]^{3+}$,[37] and $[Co([9]aneN_3)_2]^{3+}$,[37] contrasting markedly with the behavior[38] of the corresponding chromium(II) species $[Cr(pic)_3]^-$. In the reaction with Fe^{3+}, this latter reagent shows evidence for a long-lived binuclear intermediate, absorbing maximally at 540 nm, which decays with a rate constant of $5.6 \times 10^{-4} s^{-1}$.

The electron-transfer reactions[37] of the $[Co([9]aneN_3)_2]^{2+}$ (k_{12}) and $[Co(sep)]^{2+}$ (k_{13}) with a number of reaction partners, encompassing a range of driving potentials from -0.6 V to $+0.4$ V, have been compared. The two cobalt complexes have similar size and charge such that electrostatic interactions should cancel. It is found that the ratio of rate constants, k_{12}/k_{13}, varies considerably with the thermodynamic driving force rather than yielding a constant value as might be predicted by a simple Marcus analysis. Redox chemistry of a number of other metal complex derivatives containing [9]aneN₃-type ligands have been reported. Two equivalents of $[Co([9]aneN_3)_2]^{2+}$ are oxidized[39] by $[Tl([9]aneN_3)_2]^{3+}$ in an outer-sphere reaction which is first order in both oxidant and reductant. A thalium(II) intermediate is proposed but not detected as cobalt(IV) is considered improbable, and while the immediate thalium(I) product is uncertain, the structure of $[Tl[9]aneN_3]^+$ has been determined. X-ray data for iron(III), (II)[40] and nickel(III)[41] complexes are now available. Outer-sphere reactions[42,43] of the related $[Ni([10]aneN_3)_2]^{3+/2+}$ species have allowed estimation of the self-exchange rate from Marcus calculations as $2 \times 10^4 M^{-1} s^{-1}$. The reagent $[Ni([9]aneN_4)_2]^{2+}$ has been used[44] to estimate self-exchange rates for a number of isomers of $[Ni(5,12-Me_2[14]aneN_4)]^{3+}$. In the α form, the methyl substituents are both equatorial and the self-exchange rate is high, $8 \times 10^3 M^{-1} s^{-1}$, while for the β form, they are axial and the self-exchange rate constant is $6 M^{-1} s^{-1}$. The difference lies in the inner-sphere reorganizational energy which is larger for the more sterically congested axial isomers.

Reduction of $[Co(NH_3)_5X]^{3+}$ (X = H_2O, py, DMSO) by $[Fe(CN)_6]^{4-}$ involves kinetically detectable precursor complex formation prior to electron transfer and the system with this separation of these two processes has attracted two studies[45,46] of the volume of activation, $\Delta V^{\ddagger}$. The activation volume for the electron-transfer step is large and positive and little dependent on the nature of X, suggesting that there are no mechanistic differences between the three systems.[45] This casts doubt on a previous interpretation[47] of the data in which the large positive value was ascribed to intrinsic changes at the cobalt center. The current suggestion involves a strong solvational contribution. The second study[46] is a careful reexamination of both precursor formation and electron-transfer data. The precursor formation constant decreases with increasing pressure in contrast to earlier work.[48] This makes the activation volume for the electron-transfer step even more positive. A comparison[49] of the reductions of aquo- and hydroxo-complexes, $[Co(NH_3)_5OH_2]^{3+}$ and $[Co(NH_3)_5OH]^{2+}$, and $[Co(CN)_5OH_2]^{2-}$ and $[Co(CN)_5OH]^{3-}$, by $[Ru(NH_3)_5OH_2]^{2+}$ reveals very little difference on the extent of hydrolysis, implying outer-sphere mechanisms. Adiabaticity in the reduction of a number of cobalt(III) complexes by $[Co(sep)]^{2+}$ is probed[50] by comparing the reaction rates with those for the corresponding reductions by $[Ru(NH_3)_6]^{2+}$. The ratio of the rate constants varies over three orders of magnitude and can be rationalized in terms of the extent of mixing of ligand-field excited states between reactants and products. For oxidants of the type $[Co(NH_3)_5X]^{2+}$, there is also a charge-transfer component. The theoretical treatment used in estimating the extent of nonadiabaticity in electron-tranfer reactions has also been the subject of a dispute between two groups.[51,52] Central to the arguments is the question of whether self-exchange reactions should be assigned adiabatic character or not.

Effects of pyridine ring substitution have been used[53] to probe the mechanism of reduction of complexes of the type $[Co(NH_3)_5py]^{3+}$ by V^{2+} and $[Ru(NH_3)_6]^{2+}$. The substitutions cause little change in rate and parallel the effects on the corresponding reductions by aliphatic radicals which, in contrast, show marked substituent effects in the reduction of free pyridinium ions. Hence it is argued that electron transfer is direct to the cobalt(III) center and does not involve a chemical mechanism with a reduced coordinated pyridine as an intermediate. Similar conclusions are reached[54] in the reduction of $[Co(NH_3)_5py]^{3+}$ and its derivatives by the ruthenium(I) complex $[Ru(bipy)_3]^+$ which is generated by reductive quenching of $[*Ru(bipy)_3]^{2+}$ using Eu^{2+}, although the rates are close to the diffusion limit. With this reagent, outer-sphere reductions of a variety of cobalt(III) and other complexes show reasonably good agreement with the predictions of Marcus theory. Reactions of chromium(III)–alkyl and –aryl complexes, $[Cr(H_2O)_5R]^{2+}$, have also been examined. Oxidative quenching of $[*Ru(bipy)_3]^{2+}$ by $[Co(NH_3)Br]^{2+}$ yields $[Ru(bipy)_3]^{3+}$ which has been used[55] as a one-electron oxidant for $[Cr(H_2O)_5R]^{2+}$ complexes. The immediate products are $[Cr(H_2O)_5R]^{3+}$ species, which decay by homolysis or by intramolecular redox processes. Rate data parallel the rates of formation of the organic cation radicals. Platinum(IV) has also been used[56] for oxidative dealkylation of metal complexes. Reaction of $[PtCl_6]^{2-}$ with methyl-cobalamin, Me-B_{12}, involves limiting first-order kinetics consistent with the formation of a strong precursor complex. Initial electron transfer yields a [platinum(III),Me-B_{12}^+] radical pair which decomposes in a choride-ion-dependent step to give, eventually, methyl chloride, platinum(II) and aquo-cobalamin. While the reaction, overall, is pH-dependent due to the equilibrium between base-on and base-off forms of Me-B_{12}, the electron-transfer rate is essentially independent of this equilibrium.

Effects of structure on the activation volume for the self-exchange $[Mn(CNR)_6]^{2+/+}$ have been investigated[57] by NMR in acetonitrile at 5 °C. The values range from -2.4 cm^3 mol^{-1} for R = Me to -19.7 cm^3 mol^{-1} for R = Bu^n and comparisons with theory are unhelpful. The most plausible explanation is that the larger complexes must distort to allow closer approach of the manganese centers and there is a correlation with ligand flexibility. Other thermodynamic parameters support this argument; the $\Delta S^{\ddagger}$ values do not change significantly with increasing steric bulk, as might be expected if the electron-transfer rate dropped off with distance. Two different aspects of the reduction of $[Co(bipy)_3]^{3+}$ by $[Co(terpy)_2]^{2+}$ have been investigated. The volume of activation, $\Delta V^{\ddagger}$, is -9 cm^3 mol^{-1} at 25 °C and 0.01 M ionic strength, close to the value predicted for the Marcus–Hush model.[58] Calculations show that this arises primarily from the coulombic interaction and solvent reorganization terms. Similar negative volumes of activation are reported for acetonitrile and DMF as solvents. The effects of solvent on the electrostatic work terms have been investigated[59] by comparing the reduction of $[Co(bipy)_3]^{3+}$ with corresponding reduction of the neutral $[Co(5-SO_3-bipy)_3]$ derivative. The lack of electrostatic repulsion results in a rate acceleration between two and three orders of magnitude and it is shown that the dielectric continuum model is inadequate in explaining the effects of changing solvent. A model with localized features is required to replace the structureless continuum model. The effects of ion pairing at low electrolyte concentrations in acetonitrile have been studied for the reduction of $[Fe(bipy)_3]^{3+}$ [60] by $[Os(bipy)_3]^{2+}$ [61] and for self-exchange in the osmium system. Comparisons with theory suggest that electrostatic interactions are best modeled by a minimum contact distance between the reactants, suggesting a compact precursor complex. Solvent[62] and surfactant[63] effects have also been investigated for the oxidation of ferrocene by $[Co(bipy)_3]^{3+}$ and related species. Ion pairing generally reduces reactivity in the oxidation of ferrocene by cobalt(III) clathrochelates in acetonitrile solution.[64] Different added anions have been used to probe the reaction and although some specific interactions have been detected, there is no evidence for a superexchange pathway. There is an interesting study[65] of the gas-phase self-exchange of various metallocenes by ion cyclotron resonance. The rates are best considered by comparing the efficiencies of the collisions in allowing electron transfer which are $[MnCp_2]^{+/0}$, 0.006, $[FeCp_2]^{+/0}$, 0.12, $[CoCp_2]^{+/0}$, 0.36, and $[RuCp_2]^{+/0}$, 0.13. The low efficiency in the manganese reaction represents a higher reorganizational barrier.

Self-exchange rate constants for a series of bis(imidazole) iron(III)/(II) prophyrins have been determined[66] by NMR in CD_2Cl_2 at $-21\,°C$. There is little dependence of the rate on substitution at the porphyrin periphery or on increasing steric constraints on the 1-position of the imidazole. The use of 5-methylimidazole in which the imidazole nitrogen is protonated causes some rate reduction and is cited as evidence that the $N-H$ might be important in controlling electron transfer. Hydrogen bonding is reputed to be stronger in the iron(III) form than in the iron(II) form, requiring some additional reorganization, though the exact nature of this interaction is obscure. The kinetics and mechanism of reduction of a series of cationic cobalt(III) porphyrins by $[Fe(edta)]^{2-}$ have been reported.[67] Choice of the reductant is determined by the desire to avoid reduction of the porphyrin ring. The reactions show limiting first-order kinetic behavior, evidence for strong precursor complex formation but the authors prefer a "dead-end" explanation as it gives better agreement with Marcus theory. Marcus theory has also been employed[68] to calculate a self-exchange rate for the silver(III)/(II) porphyrin, $[AgTPPS]^{-/2-}$, from reactions of the oxidized form with cationic reductants. The self-exchange rate is $4 \times 10^5\ M^{-1}\,s^{-1}$ at $25\,°C$ in $0.1\ M$ NaCl, but the reactions show a strong anion dependence. Square-planar $[Ag(OH)_4]^-$ is a potent oxidant and reactions with $[Mo(CN)_8]^{4-}$ and $[W(CN)_8]^{4-}$ have been reported.[69,70] In the former reaction an equilibrium is established, giving a value of $0.87\ V$ for the silver(III)/siver(II)-dimer reduction potential in $1.2\ M$ $[OH^-]$. Although the rates of these and other reactions show a correlation with thermodynamic driving force, it is pointed out that an outer-sphere mechanism cannot be unambiguously assigned and, indeed, an inner-sphere mechanism with axial attack of coordinated cyanide on the square-planar complex is preferred. Reduction of $[Mo(CN)_8]^{3-}$ by Fe^{2+} involves[26] initial rapid formation of a cyanide bridged intermediate with a formation constant of $1.1 \times 10^6\ M^{-1}$ at $3\,°C$ and $1.0\ M$ ionic strength. Electron transfer is rapid, $>5 \times 10^6\ s^{-1}$, to give a stable successor which decays with a rate constant of $160\ s^{-1}$.

Oxidation of a vanadium(IV) complex with [(phosphonomethyl)imino]-diacetate, a phosphato-derivative of nta as ligand, $[VO(pida)H_2O]^{2-}$, by $[IrCl_6]^{2-}$ occurs[71] at a rate in excess of the rate of substitution of the coordinated water molecule by SCN^- and is most likely outer-sphere in nature. Protonation of the phosphate residue causes rate retardation, and the detection of a D_2O isotope effect of 1.6 suggests the presence of a hydrogen bond between this residue and the *cis*-coordinated water molecule. Reactions of bis-tripeptide complexes of nickel(III), $[Ni(H_{-2}G_3)_2]^{3-}$, with $[Fe(CN)_6]^{4-}$, $[W(CN)_8]^{4+}$, and copper(II) peptide complexes must be outer-sphere in nature[72] and are much slower than the corresponding reactions of the mono-complexes. Protonation, with a pK_a around 11, leads to rate enhancement. It is calculated that the reduction potentials of the six-coordinate bis-complexes are significantly lower than those of the mono-complexes. Disproportionation of the nickel(III) complex, $[NiMe_2LH]^{2+}$, where Me_2LH_2 is a sexidentate bis(oxime-imine) ligand, to give nickel(IV), $[NiMe_2L]^{2+}$, and nickel(II), $[NiMe_2LH_2]^{2+}$, has been examined[73] in acidic media. The bulk of the reaction is carried by the outer-sphere electron transfer between $[NiMe_2L]^+$ and $[NiMe_2LH]^{2+}$ which is less thermodynamically favorable than disproportionation of the protonated species but does not require H-atom transfer to form thermodynamically stable products. Comproportionation of the corresponding nickel(IV) and nickel(II) complexes has also been examined and again the reactions are outer-sphere, showing good agreement with Marcus theory. The nickel(IV)/(III) self-exchange rate has been estimated by measuring the pseudo-self-exchange between a chiral analogue $[\Lambda\text{-}Me_2NiL']^{2+}$ and $[NiMe_2L]^+$ and is found to be $6 \times 10^4 \ M^{-1} s^{-1}$, in excellent agreement with the predictions from cross-reaction data which give a value of $4 \times 10^{-4} \ M^{-1} s^{-1}$ at $25 \, ^\circ C$ and $0.1 \ M$ ionic strength. Reductions of $[NiMe_2L]^{2+}$ by $[Co(edta)]^{2-}$ and its substituted derivatives, $[Co(pdta)]^{2-}$ and $[Co(cdta)]^{2-}$, are biphasic[74] with an initial rapid, pH-independent, outer-sphere single electron transfer to form nickel(III) and a slower step, reduction of $[NiMe_2LH]^{2+}$, to give nickel(II), which shows kinetic evidence for precursor complex formation. When the chiral nickel(IV) complex, $[\Lambda\text{-}Me_2NiL']^{2+}$, is used, the $[Co(edta)]^-$ formed is optically active with a 10% enantiomeric excess of the Δ isomer. Comparison with ion-pairing studies allows the source of the stereoselectivity to be pinpointed to the precursor ion pair. Changes in the structure of the reductant have little effect on either the kinetic behavior or the stereoselectivity and this has been used to produce a detailed model for the precursor ion pair in which the carboxylate face of the reductant interacts with the C_2 oxime axis of the nickel(IV) complex. Stereoselectivity has been investigated in several other reactions. In the oxidation of $[Co((\pm)chxn)_3]^{2+}$ by $[\Delta\text{-}Co(edta)]^-$, the enantiomeric excess of the cobalt(III) product is dependent on the chelate ring conformation.[75] Addition of optically active, cationic cobalt(III) complexes to the oxidation of the molybdenum(V) dimer $[Mo_2O_4(R,S\text{-}pdta)]^{2-}$ by $[IrCl_6]^{2-}$ induces[76] stereoselectivity due to the formation of ion pairs. The effect is small, but much greater stereoselectivity is possible in the reduction[77] of $[Co(phen)_3]^{3+}$ by $[^*Ru(phen)_3]^{2+}$ in the presence of DNA, the result of stereospecific binding. DNA, a medium of reduced dimension, catalyzes the reaction which is estimated to have a rate constant of $2 \times 10^{11} \ M^{-1} s^{-1}$ along a DNA strand.

Quenching of $[^*Ru(bipy)_3]^{2+}$ by cobalt(III) complexes has been reinvestigated[78] and an exclusive electron-transfer mechanism[79] for quenching by $[Co(NH_3)_6]^{3+}$ and $[Co(en)_3]^{3+}$ has been disproven by measurements of the quantum yields for cobalt(II) production which are 0.45 and 0.11, respectively. The dominant quenching mechanism is energy transfer and the quantum yields allow determination of the electron-transfer rates as $3.7 \times 10^6 \ M^{-1} s^{-1}$ for $[Co(en)_3]^{3+}$ and $6.8 \times 10^6 \ M^{-1} s^{-1}$ for $[Co(NH_3)_6]^{3+}$, in the expected order for outer-sphere reactions. An electron-transfer mechanism is confirmed for

$[Co(NH_3)_5Cl]^{2+}$. A similar investigation[80] for the quenching mechanism has been carried out for $[Co(sep)]^{3+}$ and its derivatives as oxidants. Europium complexes of the cryptate [2.2.1] have been used as quenchers for $[*Ru(bipy)_3]^{2+}$ and its derivatives. It is argued[81] that reduction of $[*Ru(bipy)_3]^{2+}$ by $[Eu(2.2.1)]^{2+}$ and oxidation of $[*Ru(bipy)_3]^{2+}$ by $[Eu(2.2.1)]^{3+}$ are outer-sphere electron-transfer processes but while the latter shows a free-energy dependence, the former does not. This is difficult to explain since the coordination environments are similar for the europium species. Other reactions of $[*Ru(bipy)_3]^{2+}$ have been reported.[82] The highly exorgenic reaction between $[Ru(bipy)_3]^{3+}$ and $[Co(bipy)_3]^+$ in acetonitrile solution gives chemiluminescence from $[*Ru(bipy)_3]^{2+}$ formed as an immediate product.[83] Formation of ground-state products lies in the inverted region and is slow. Excited state quenching of $[*V(phen)_3]^{2+}$ by Eu^{3+} proceeds[84] by electron transfer, leading to Eu^{2+} and a vanadium(II) hydroxy-dimer. In ethanol solution, reduction of a number of cobalt(III) complexes by $[*Cu(dmp)$-

Table 2.2. *Electron-Transfer Reactions Involving Metalloproteins at 25 °C*

Oxidant	Reductant	Medium (M)	k ($M^{-1}s^{-1}$)	Ref.
$[Co(sep)]^{3+}$	Parsley ferredoxin	0.1 (pH 7.5)	2.6×10^4	37
Parsley ferredoxin	$[Co([9]aneN_4)_2]^{2+}$	0.1 (pH 7.5)	1.06×10^3	37
Cytochrome c(III)	$[Co([9]aneN_3)_2]^{2+}$	0.1 (pH 7.5)	4.50×10^4	37
Cytochrome c(III)	$[Co(sep)]^{2+}$	0.1 (pH 7.5)	3.04×10^5	37
T. zostericola (Hr(III))$_8$	$[Co(sep)]^{2+}$	0.15 (pK_a 7.6)	255 (acid) 129 (base)	86 86
T. zostericola (Hr(III))$_8$	$[Co(sarCl_2)]^{2+}$	0.15 (pK_a 7.6)	114 (acid) 56 (base)	86 86
T. zostericola (Hr(III))$_8$	$[Co([9]aneN_3)_2]^{2+}$	0.15 (pK_a 7.6)	12.8 (acid) 7.3 (base)	86 86
T. zostericola (Hr(III))$_8$	$[Cr(bipy)_3]^{2+}$	0.15	2.5×10^5 (acid)	86
T. zostericola (Hr(III))	$[Co(sep)]^{2+}$	0.15 (pH 8.2)	1.4×10^3	87
T. zostericola (Hr(III))	$[Co([9]aneN_3)_3]^{2+}$	0.15 (pH 8.2)	33	87
T. zostericola (Hr(III))	$[V(pic)_3]^-$	0.15 (pH 8.2)	4.4×10^3	87
Horse heart cytochrome c(III)	$[Fe(edta)]^{2-}$	0.1 (pH 7.5)	2.80×10^4	89
Horse heart cytochrome c(III)	$[Co(sep)]^{2+}$	0.1 (pH 7.5)	2.60×10^5	89
Horse heat cytochrome c(III)	Stellacyanin(I)	0.1 (pH 7.0)	2.4×10^3	90
a. variabilis plastocyanin(II)	Horse heart cytochrome c(II)	0.1 (pH 7.5)	3.3×10^4	90
Parsley plastocyanin(II)	$[Co(phen)_3]^{2+}$	0.1 (pK_a 5.05)	1.11×10^3 (acid) 2.49×10^3 (base)	93 93
Parsely plastocyanin(II)	$[Co(terpy)_2]^{2+}$	0.1 (pK_a 5.05)	3.56×10^4 (acid) 7.32×10^4 (base)	93 93

(continued)

Table 2.2. (continued)

Oxidant	Reductant	Medium (M)	k $(M^{-1} s^{-1})$	Ref.
Parsley plastocyanin(II)	$[Ru(NH_3)_5py]^{2+}$	0.1	2.19×10^5 (acid)	93
		(pK_a 5.05)	4.44×10^5 (base)	93
Spinach plastocyanin(II)	$[Ru(NH_3)_5py]^{2+}$	0.1	3.1×10^5 (acid)	95
		(pK_a 5.33)	5.8×10^5 (base)	95
Spinach plastocyanin(II)	$[Co(terpy)_2]^{2+}$	0.1	4.4×10^4 (acid)	95
		(pK_a 5.33)	8.2×10^4 (base)	95
$[Fe(CN)_6]^{3-}$	Spinach plastocyanin(I)	0.1	8.0×10^4	95
$[Co(phen)_3]^{3+}$	Spinach plastocyanin(I)	0.1	2.5×10^3	95
$[Co(dipic)_6]^-$	Spinach plastocyanin(I)	0.1	4.0×10^2	95
$[Ru(NH_3)_5MeCN]^{3+}$	Spinach plastocyanin(I)	0.1	1.82×10^5	95
p. vulgaris plastocyanin(II)	$[*Cr(phen)_3]^{3+}$	0.1 (pH 7.0)	6.0×10^8	96
p. vulgaris plastocyanin(II)	$[Cr(phen)_3]^{2+}$	0.1 (pH 7.0)	2.3×10^9	96
p. vulgaris plastocyanin(II)	$[*Ru(bipy)_3]^{2+}$	0.1 (pH 7.0)	4.2×10^9	96
p. aeruginosa azurin(II)	$[*Cr(bipy)_3]^{3+}$	0.1 (pH 7.0)	1.0×10^8	96
p. aeruginosa azurin(II)	$[*Ru(bipy)_3]^{2+}$	0.1 (pH 7.0)	1.2×10^9	96
Stellacyanin(II)	$[*Cr(bipy)_3]^{3+}$	0.1 (pH 7.0)	4.9×10^7	96
Stellacyanin(II)	$[*Ru(bipy)_3]^{2+}$	0.1 (pH 7.0)	1.3×10^9	96
$[Mn(cydta)H_2O]^-$	Rusticyanin(I)	0.1	554 (acid)	97
		(pK_a 4.17)	104 (base)	97
$[Co(dipic)_2]^-$	Rusticyanin(I)	0.1	11.4 (acid)	97
		(pK_a 4.10)	2.5 (base)	97
Rusticyanin(II)	$[Ru(NH_3)_4phen]^{2+}$	0.1 (15 °C)	3×10^5 (acid)	97
		(pK_a 4.06)	1.6×10^5 (base)	97
$[Co(dipic)_2]^-$	Umecyanin(I)	0.1	8×10^4	98
$[Co(ox)_3]^{3-}$	Umecyanin(I)	0.1	580 (acid)	98
		(pK_a 9.68)	223 (base)	98
Unmecyanin(II)	$[Ru(NH_3)_5py]^{2+}$	0.1	6.3×10^3 (acid)	98
		(pK_a 9.50)	3.2×10^3 (base)	98
Umecyanin(II)	$[Co(phen)_3]^{2+}$	0.1 (pH 7.55, 5.8 °C)	301	98
Umecyanin(II)	$[Fe(CN)_6]^{3-}$	0.1 (pH 7.55, 5.8 °C)	2.8×10^6	98

$(PPh_3)_2]^+$ and its derivatives has been studied.[85] The order of the rates is $[Co(acac)_3] > [Co(edta)]^- > [Co(NH_3)_5Cl]^{2+}$ and does not follow the trend in reduction potentials.

2.4. Reactions of Metalloproteins

Electron-transfer reactions between metal-ion complexes and metalloproteins continue to provide a fruitful area for study. Reductions of the two-iron ferredoxin from spinach[37] and both octameric[86] and monomeric[87] forms of met-hemerythrin from *themiste zostericola* by a variety of outer-sphere metal-ion complexes have been examined. Both forms of hemerythrin show the same triphasic behavior. The first phase is first-order in reductant concentration and involves reduction of one of the iron(III) atoms in the dimeric site, while the rate of the second phase is independent of reductant and is ascribed to slow intramolecular electron transfer within the dimeric unit followed by rapid reduction. A previous explanation of this step for the octamer involved long-range intramolecular electron transfer between the semireduced subunits[88] and although this fits with the reaction stoichiometry [half of the remaining iron(III) is reduced in this step] the similarity of the rate constant for the monomeric protein, $4 \times 10^{-3}\,\mathrm{s}^{-1}$, to that for the octamer, $1.2 \times 10^{-3}\,\mathrm{s}^{-1}$, casts doubt on this possibility. The third phase involves a slow isomerization in the reduced protein. The rates of reduction of native and eight lysine modified derivatives of horse heart cytochrome c by $[\mathrm{Fe(edta)}]^{2-}$ and $[\mathrm{Co(sep)}]^{2+}$ have been measured.[89] Compared with the native protein, the largest rate differences, around a factor of four, occur with lysine 72 for the anionic reductant and with lysine 27 for the cationic reductant. These regions of the protein straddle the heme edge region, consistent with electron transfer at that site. Similar studies have been carried out[90] with *anabaena variabilis* plastocyanin as oxidant and stellacyanin as reductant for the cytochrome c. Again the effects reflect a maximal interaction at lysine 72 close to the heme edge, but they are relatively small, covering an order of magnitude and indicating much less specificity than with the physiological partners. Sykes has reviewed[91] the interactions of the blue copper proteins with metal-ion complexes, making substantial sense of the large body of information now available in this area. Inhibition studies using cationic complexes to block the negative patch near tyrosine-83 in parsley plastocyanin reveal[92] that the extent of retardation of the oxidation by $[\mathrm{Co(phen)_3}]^{3+}$ is 53%, independent of the nature of the blocking reagent. On reduction by $[\mathrm{Co(phen)}]^{2+}$ and its derivatives, the same tyrosine-83 site is also used[93] for around 50% of the reaction. Recent NMR studies, however, reveal[94] that the cation binding sites on spinach platocyanin are more extensive and less specific than previously determined. A comparison[95] of the spinach and parsley proteins reveals very close similarities.

Electron-transfer quenching of $[\mathrm{^*Cr(phen)_3}]^{3+}$ and $[\mathrm{^*Ru(phen)_3}]^{2+}$ by the oxidized forms of the blue copper proteins, plastocyanin, azurin, and stellacyanin, has been reported.[96] The studies are of considerable interest since the protein is the reagent in excess, contrasting with previous investigations of the reactions of metalloproteins with metal-ion complexes. They also cast more light on the multiplicity of reaction pathways noted in the previous paragraph. Two pathways for electron transfer are detected, an adjacent pathway resulting from diffusion of free protein and metal complex together and electron transfer at the shortest reductant–copper(II) distance and a remote pathway derived from intramolecular electron transfer within a 1:1 or 1:2 reductant:protein complex. The sites of association of the metal-ion complexes with the protein plastocyanin are relatively well defined from NMR studies to be remote from the copper center and close to the so-called "negative patch" near tyrosine-83. The distance between the reacting centers, determined from molecular modeling, is 10.3 Å, close to the value predicted on the basis of the distance dependence of the rate constant. The adjacent pathway is diffusion controlled and dominates the reaction. It is pointed out that, to be competitive, binding at the remote site must overcome the difference in distance which amounts to about five orders of magnitude. This clearly has implications for the assignment of "dead-end" complex mechanisms where limiting kinetic data are observed. Electron-transfer reactions of the blue copper proteins, rusticyanin[97] and umecyanin,[98] with outer-sphere reagents have been reported, and show evidence for protein protonation sites.

Long-range electron transfer between metalloprotein redox centers and covalently bound metal-ion complexes has been reviewed.[99] Electron transfer[100] over 11.8 Å from ruthenium(II) to iron(III) in [$(NH_3)_5Ru(II)$his-33-cytochrome c(III)] is independent of pH above pH 5 with a rate constant of 52 s^{-1} but this increases to 500 s^{-1} below this pH. Addition of imidazole which displaces methionine-80 as an axial ligand for iron causes a reduction in rate to 1.2 s^{-1}. It is important to point out that interpretation of these first-order rates remains rather ambiguous since there is no readily available means of distinguishing between electron transfer and a rate-limiting conformational change within the protein. In the corresponding [$(NH_3)_5Ru(II)$his-45-myoglobin(III)] system[101] where the metal centers are 13.3 Å apart and the protein's function is not electron transfer, the reorganizational barrier is much higher than in the cytochrome system, and likely involves dissociation of a coordinated water molecule from the met-protein prior to reduction. Temperature-dependence data for intramolecular electron-transfer quenching over 25 Å in ^{3}ZnFe(III) hemoglobin derivatives have been published.[102] At room temperature the rates for [αFe(III)H_2O,βZnHb] and [αZn,βFe(III)H_2OHb] are similar and are coupled to vibrations in the $\alpha\beta$ complex and with the solvent. Below 160 K, again both rates are comparable and temperature independent with a rate constant around 9 s^{-1} ascribed to a nonadiabatic tunneling process. However, between these two extremes, the behavior of the two complexes differs due to replacement of coordinated water in [αFe(III)H_2O,βZnHb] by a protein imidazole group. Triplet quenching of zinc porphyrins by cytochrome c(III) is also reported[103] to follow an electron-transfer mechanism. There have been a number of reports of electron transfer within the 1:1 complex formed between yeast cytochrome c peroxidase (ccp) and cytochrome c. The native process[104] involves transfer of an electron from yeast ccp(II) to yeast cytochrome c(III) with a rate constant of 3.4 s^{-1}, which is slower for other sources of the cytochrome showing species specificity. Removal of the iron from the cytochrome allows examination of the rate from the reduced porphyrin radical generated pulse radiolytically to ccp(III) with a rate constant of $150\,s^{-1}$. Other systems include ^{3}Zncyt c:ccp(III)[105] and ^{3}Znccp:cyt c(III)[106] with rate constants $30\,s^{-1}$ and $25\,s^{-1}$, respectively, which are ascribed mainly to electron transfer rather than energy transfer. The latter interaction is also strongly species-dependent revealing differences in binding. In contrast, quenching of the iron-free pophyrin triplet of cytochrome c is primarily energy transfer. The dependence of the electron-transfer rate on the nature of the donor and acceptor, and hence on the thermodynamic driving force for the reactions, is an important observation and suggests a significant barrier for intracomplex redox.

Chapter 3

Metal–Ligand
Redox Reactions

3.1. Introduction

This chapter reviews the literature during the period July 1985–December 1986. We have considered the subject to include the reactions of a nonmetallic with a metallic substrate, whether or not the two are coordinated during the reactions. The chapter is organized such that like substrates are grouped together by their central element. Many of these are "noncomplementary" reactions, and thus proceed by way of radical intermediates. Reports of the direct investigations of the reactivities of inorganic and organic radicals are given in Sections 3.7 and 3.12, respectively. Topics in the area of metal–ligand redox reactions that have been reviewed during this period include an account of the reactions of coordinated azide,[1] reductions of an oxochromium(V) carboxylato complex,[2] two reviews concerning intramolecular electron transfer in mixed-valence complexes,[3,4] reactions of hemoglobin and oxyhemoglobin with oxidants and reductants,[5] reactive oxygen intermediates in biochemistry, especially as regards cytochrome oxidase,[6] and activation of molecular oxygen by metal complexes.[7] A critical summary of numerical data concerning the reactivity of hydroperoxyl and superoxide radicals ($HO_2^{\cdot}$ and $O_2^{\cdot-}$) in aqueous solution has appeared.[8] A review of "second-sphere" photochemistry includes many examples in which electron transfer occurs between cation–anion pairs.[9] Accounts of the quenching of photoexcited states by electron transfer[10] and of transient organometallic radicals[11] provide many examples of reactions in this area. The reduction of carbon dioxide by cobalt(I) complexes has been reviewed.[12]

As well as the usual abbreviations the following are used in this chapter:

amp	2-(aminoethyl)pyridine
bpz	2,2'-bipyrazine
denc	*N,N*-diethylnicotinamide
dien	diethylenetriamine
dmc	1,4,5,7,7,8,11,12,14,14-decamethyl-1,4,8,11-tetraazacyclo-tetradecane
dpnH	3,3'-(trimethylenediimino)bis(butan-2-one)oximato
dppm	bis(diphenylphosphino)methane
H$_2$A	ascorbic acid
HisH	histidine
Me$_6$-14-ane	meso-5,7,7,12,14,14-hexamethyl-1,4,8,11-tetraazacyclotetra-decane
MV^{2+}	1,1'-dimethyl-4,4'-bipyridinium ion
nta	nitrilotriacetate
OEP	octaethylporphyrin
pbim	2-(2-pyridyl)benzimidazole
pd	2,4-pentadionate
pdma	*o*-phenylenebis(dimethylarsine)
pn	1,2-propanediamine
PPIX	protoporphyrin IX
taab	tetrabenzo(1,5,9,13)tetraazacyclohexadecane
TCPP	tetrakis(4-carboxyphenyl)porphyrin
TMAP	meso-tetrakis-(N,N,N-timethylanilinium-4-yl)porphine
tmc	1,4,8,11-tetramethyl-1,4,8,11-tetraazacyclotetradecane
TMP	tetramesitylporphyrin
TMPyP	tetrakis(N-methylpyridinium-4-yl)porphine
TPPS	tetrakis(4-sulfonatophenyl)porphine anion
tptp	5,10,15,20-tetra-*p*-tolylporphyrin
trpH	tryptophan

3.2. Nitrogen Compounds and Oxoanions

3.2.1. Hydroxylamine, Hydrazine, and Their Derivatives

3.2.1.1. Hydroxylamine

Complexes with coordinated NO$^+$ commonly yield N$_2$O when treated with hydroxylamine, while both *trans*-[RuX(NO)(py)$_4$]$^{2+}$ and *trans*-[FeCl(NO)(pdma)$_2$]$^{2+}$ (X = Cl or Br) yield [MX(NO)L$_4$]$^+$. An electrochemically reversible wave is seen in acetonitrile starting with either of the separately prepared complexes [RuCl(NO)(py)$_4$]$^+$PF$_6^-$ or [RuCl(NO)(py)$_4$]$^{2+}$(ClO$_4^-$)$_2$.[13] The well-characterized reaction between Fe$^{3+}_{aq}$

and hydroxylamine has been reexamined[14] under conditions of low $[NH_2OH]$ and over a wide pH range. At very low $[NH_2OH]$, 80–380 nM, products other than N_2O are obtained and even ambient light levels in the laboratory affect the rate and yield. At $[NH_2OH] > 380$ nM the pH profile suggests that the initial step of the principal pathway for N_2O formation consists of a reaction between $Fe(OH)_2^+$ and NH_2OH. The intermediacy of *trans* hyponitrite can be ruled out, but the *cis* isomer or a complex such as $[HOFeNHOH]^+$ may intervene. Rate constants and activation energies for the oxidation of $[Cr(H_2O)_6]^{2+}$ by NH_2OH in H_2O and by NH_2OD (ND_2OD?) in D_2O have been reported.[15]

3.2.1.2. Hydrazine

The intriguing observation[16] that alkylation of the metal arises from the ternary mixture of $[Co(salen)]$ (and similar complexes), alkyl hydrazines $(RNHNH_2)$ and dioxygen[17] may involve a superoxocobalt(III) intermediate. The analogous *tert*-butylsuperoxo species is indicated when *t*-BuOOH is used in place of O_2. Ultimately the reaction is completed by formation of a diazene, that in turn leads to N_2 and an alkyl radical that alkylates the cobalt(II) center. The oxidation of hydrazinium ions by O_2 occurs slowly, but the species CrO_2^{2+} formed in the presence of Cr^{2+} reacts efficiently.[18] The principal path is suggested to occur by the electron transfer step of equation (1). The protonated hydrazyl radical cation is subject to subsequent disproportionation as in equation (2).

$$N_2H_5^+ + [CrO_2H]^{3+} \rightarrow {}^{\cdot}N_2H_5^{2+} + [CrO_2H]^{2+} \tag{1}$$

$$2\,{}^{\cdot}N_2H_5^+ \rightarrow N_2 + 2\,NH_4^+ \tag{2}$$

The oxidation of hydrazine by peroxodiphosphate is catalyzed by Ag^+. The kinetic data are consistent with reactive Ag^+ complexes of $H_2P_2O_8^{2-}$ and/or N_2H_4, but permit no definitive decisions as to the reacting partners.[19]

3.2.2. Nitrous Acid, Nitrite, and Nitrosonium (NO$^+$) Ions

3.2.2.1. Nitrite Ions

A complex series of transformations accompanies the reduction of coordinated NO_2^- in the complex $[M(terpy)(bipy)(NO_2)]^+$ (M = Ru, Os) to $[M(terpy)(bipy)(NH_3)]^{2+}$.[20] Acid-base chemistry leads to $[L_5M(NO)]^{3+}$, a species that proceeds to products by stepwise, one-electron reductions via $[L_5M^{IV}(NH)]^{2+}$, $[L_5M^V(N)]^{2+}$ or $[L_5M^{II}(NH_2O)]^{2+}$, $[L_5M^{II}(NHO)]^{2+}$, and $[L_5M^{II}(NO^{\cdot})]^{2+}$. The reduction of NO_2^- by formic acid yields N_2O, and is catalyzed by $[MoO(S_2CNEt_2)_2]$ or $[MoO_2(S_2CNEt_2)_2]$. The critical interconversions are "oxygen atom transfers," as in equations (3) and (4).[21]

$$MoO + NO_2^- \rightarrow MoO_2 + NO^- (\rightarrow \tfrac{1}{2}N_2O_2^{2-}) \tag{3}$$

$$MoO_2 + HCO_2H \rightarrow MoO + H_2O + CO_2 \tag{4}$$

3.2.2.2. Nitrous Acid

As often happens, the nitrosonium ion is the important reactant in nitrous acid reactions. The reactions of a series of $[(H_2O)_5CrR]^{2+}$ complexes illustrate the dual oxidizing capacity and electrophilicity of NO^+.[22] An exceptionally detailed study of the reactions of $[IrCl_6]^{2-/3-}$ in nitrous acid solutions allowed identification of the electron-transfer step $NO^+ + [IrCl_6]^{3-} \rightarrow NO + [IrCl_6]^{2-}$.[23] This reaction is diffusion controlled despite the calculated NO^+/NO self-exchange rate of $\sim 10^{-2}\ M^{-1}\ s^{-1}$.[24] These authors[23] also present data giving $E^0_{NO_2/NO_2^-} = 1.04$ V, in support of an older value. The rate of oxidation of In^+ by HONO is proportional to $[In^+][HONO][H^+]$, also implicating NO^+ as the electron-transfer reagent.[25] In aprotic solvents the $NO^+/HONO$ interconversion is avoided; thus $NOPF_6$ inserts directly into a chromium–methyl bond.[26] The reaction between the nitrite ions and the dinuclear platinum(II) pyrophosphito complex $[Pt_2(P_2O_5H_2)_4]^{4-}$ yields NO_2 and $[Pt_2(P_2O_5H_2)_4(NO_2)_2]^{4-}$. The latter species reductively eliminates NO_2 and forms $[Pt_2(P_2O_5H_2)_4X(NO_2)]^{4-}$ upon reaction with halide ions.[27] Evidence points to one-electron and two-electron reactions of the dinuclear complex. The nitro group of the cobalt(III) cage complex $[Co(diNOsar)]^{3+}$ is reduced to a hydroxyamino group by $SO_2^{\cdot -}$ (from dithionite ions).[28] The new data for $S_2O_4^{2-}$ agree with those obtained before.[29]

3.2.3. Trioxodinitrate and Other Oxoanions

3.2.3.1. Trioxodinitrate

The generally accepted[30-33] view that the decomposition of the monoanion $HN_2O_3^-$ (pK_a 9.2) leads initially to HNO and NO_2^- was challenged,[34] the suggestion being made that NO and $HONO^-$ are produced, since hemoglobin (Hb) yields HbNO and Hb^+. This alternative was subjected to direct tests and disproven.[35,36] Indeed, these authors find a reaction between $HN_2O_3^-$ and Hb that is zero-order with respect to [Hb], and thus infer that initial and rate-limiting tautomerization of $HN_2O_3^-$ precedes any involvement of Hb. The same tautomerized intermediate (T) occurs during decomposition, as in equation (5).

$$HN_2O_3^- \xrightarrow[\text{(pH 7, 25 °C)}]{6.6\times10^{-4}\,s^{-1}} T \quad \begin{cases} \longrightarrow NO_2^- + N_2O[\rightarrow HNO[\rightarrow N_2O + H_2O] \\[1em] \xrightarrow{2Hb} Hb^+, HbNO \end{cases} \qquad (5)$$

The reduction of $[Ru(NH_3)_6]^{3+}$ by $N_2O_3^{2-}$ ($k = 45.8\ M^{-1}\ s^{-1}$ at 2 °C) leads principally to NO_2^- and NO, with traces of N_2O and N_2. Labeling experiments (^{15}N) suggest the most likely formulation of the major path is outer-sphere electron transfer yielding *NO and NO_2^- (from $O*NNO_2^-$).[37]

3.2.3.2. Nitrate

The reaction between HNO_3 and Fe^{2+} forms traveling brown waves of $FeNO^{2+}$ in an initially homogeneous mixture when a nitrous acid scavenger such as hydrazine or azide is present in a flow stirred reactor.[38] When a batch reactor is used, however, a normal "clock" reaction occurs. The reduction of NO_3^- by the molybdenum(III) dimer $[(H_2OMoL)_2(\mu\text{-}OH)_2]^{2+}$ (with L = [9]aneN$_3$) yields NO_2^- and the molybdenum(V) dimer, *anti*-$[(LMo(=O))_2$-$(\mu\text{-}O)_2]^{2+}$, ($k = 0.10\ M^{-1}\,s^{-1}$ at 25 °C).[39] With $N(^{18}O)_3^-$, the product is exclusively $[(LMo(=^{18}O))_2(\mu\text{-}O)_2]^{2+}$.

3.2.4. Azide Ions and Hydrazoic Acid

The reactions of coordinated azide have been reviewed.[1] Reactions in which free azide ions are oxidized to N_2 by $[IrCl_6]^{2-}$, $[IrBr_6]^{2-}$, and $[Fe(bipy)_3]^{3+}$ all involve the production of the azidyl radical, $N_3^\bullet$; the respective second-order rate constants at 25 °C are 1.8×10^2, 6.6×10^1, and $8.4 \times 10^4\ M^{-1}\,s^{-1}$.[40] The reaction of $N_3^\bullet$ with $[IrCl_6]^{3-}$ has (by pulse radiolysis) $k = 5.8 \times 10^8\ M^{-1}\,s^{-1}$.[40] These data lead to $E^0(N_3^\bullet/N_3^-)$ 1.33 V and to a self-exchange rate constant $\sim 4 \times 10^4\ M^{-1}\,s^{-1}$. Other information pertaining to $N_3^\bullet$ and N_6^- has been obtained by pulse radiolysis.[41]

Complexes containing coordinated azide ions are often photosensitive. The main group anions $[M^{IV}(N_3)_6]^{2-}$ (M = Sn, Pb) are efficiently photolyzed

$$[M(N_3)_6]^{2-} \xrightarrow{h\nu} [M(N_3)_4]^{2-} + 3N_2 \qquad (6)$$

in acetonitrile, as in equation (6).[42] Irradiation (at $\lambda > 335$ nm) of solutions of cytochrome c containing azide ions yields Fe(II)-cyt c and N_2, the latter formed from $N_3^\bullet$, released from an $[N_3^\bullet$-Fe(II)] intermediate.[43]

3.2.5. Organic Amines

Iron(III) porphyrins oxidize amines to imines by two successive reversible ligation steps followed by two one-electron reductions in which the resulting $[PFe^{III}(RNH_2)_2]$ intermediate reacts with additional RNH_2. The rate-limiting step in benzene is the second ligation reaction and in dmf the first electron-transfer reaction.[44] The data suggest that the first reduction is a reversible outer-sphere electron transfer, reaction (7), to the porphyrin.

$$PFe^3 + RNH_2 \rightleftharpoons PFe^{II}\cdots NH_2R^{+\bullet} \rightleftharpoons PFe^{II} + RNH_2^{+\bullet} \qquad (7)$$

N-substituted-*p*-phenylene diamines, *p*-$(RR'N)_2C_6H_4$, are oxidized rapidly by $[Fe(CN)_6]^{3-}$ and slowly by $[Fe(H_2O)_6]^{3+}$. Both reactions involve sequential electron-transfer steps, and are autocatalytic. Even at 10^{-8} M, $[Fe(CN)_6]^{3-}$ is an effective catalyst for the $[Fe(H_2O)_6]^{3+}$ reactions.[45] The oxidation of L-cysteine by TcO_4^- in strong acid yields yellow and violet species, respectively, Tc(V)–cystine and Tc(V)–cysteine complexes.[46] Aryl amines quench the triplet excited state of Rh(III) compounds by electron transfer.[47] This includes $[*RhL_3]^{3+}$ (L = Ph_2 phen and phen) and $[*RhCl_2L_2]^+$. Quenching of $[*Ru(bipy)_3]^{2+}$ (and its analogues) by $ArNH_2$ also occurs by electron transfer:[48]

$$[*RuL_3]^{2+} + ArNH_2 \rightarrow [RuL_3]^+ + ArNH_2^{\ddot{+}} \qquad (8)$$

3.2.6. Pyridinium and Bipyridinium Ions

3.2.6.1. Pyridinium Ions

A series of substituted pyridinium ions are effective electron-transfer quenchers for $[*Ru(bipy)_3]^{2+}$.[49] The substance 10-methyl-9-phenylacridan is oxidized to the acridinium ion by $[Co(bipy)_3]^{3+}$ in acetonitrile by a one-electron mechanism and, in contrast to hydride-abstracting reagents, shows no kinetic deuterium isotope effect.[50]

3.2.6.2. Bipyridinium Ions

The quenching of excited-state ruthenium(II)–polypyridine complexes (e.g., $[*Ru(bipy)_3]^{2+}$ and analogues) by bipyridinium ions such as MV^{2+} has been well-established. Recent studies have extended this to include (1) data taken in mixed solvents and in the presence of different anions,[51] (2) a dipolar, zwitterionic viologen,[52] (3) polyelectrolyte-anchored $[Ru(bipy)_3]^{2+}$ derivatives,[53] (4) the bipyrazine analogue, $[Ru(bpz)_3]^{2+}$,[54] and (5) variations of the donor–acceptor separations by use of $[Ru\{(C_nH_{2n+1})_2bipy\}_3]^{2+}$.[55] The authors of the latter publications consider the transposition of the Marcus equation for electron transfer into the Rehm–Weller equation, and the implications of this for the "inverted region." Solvent effects on the photocatalytic reduction of MV^{2+} have also been explored for $[Cu(NN)(PPh_3)_2]^+$ complexes.[56] The photogenerated $[(MeNC)_3Pd]^+$ radical reduces viologens,[57] as does $[Ir(C^3, N'-Hbpy)(bipy)_2]^{3+}$.[58] The respective rate constants are $k = 3 \times 10^7$ M^{-1} s^{-1} in acetonitrile and 1.5×10^6 M^{-1} s^{-1} in water for the reduction of MV^{2+} by the iridium complex.

3.2.7. Miscellaneous N-Containing Compounds

3.2.7.1. Dinitrogen

Treatment of *trans*-$[Mo(N_2)_2(triphos)PR_3]$ (where PR_3 = PPh_3 and $PMePh_2$) with anhydrous HCl and HBr releases ammonia in THF and ammonia + hydrazine in benzene. Intermediate hydrazido complexes, $[MoX(NNH_2)(triphos)(PR_3)]$, were identified.[59]

3.2.7.2. Nitrous Oxide

Nickel(I) complexes of saturated tetraazamacrocycles react very rapidly with N_2O. The reaction is presented to be a two-electron process: $N_2O + [NiL]^+ + 2H^+ \rightarrow [NiL]^{3+} + N_2 + H_2O$, although $[NiL]^{3+}$ was not detected in every case. The rate constants were evaluated by pulse radiolysis.[60,61]

3.2.7.3. Aryldiazonium Ions

Both ferrocyanide ions and hemoglobin react with ArN_2^+, but with notable differences. The $[Fe(CN)_6]^{4-}$ oxidations are electron-transfer processes $\{Fe(II) + ArN_2^+ + H_2O \rightarrow Fe(III) + ArH + N_2 + OH^-\}$, and well-correlated by Hammett σ-constants ($\rho = +4.7$). The reactions of hemoglobin show, in competition with that, formation of σ-aryls $\{2Hb(II) + ArN_2^+ \rightarrow Ar\text{-}Hb(III) + Hb(III)^+ + N_2\}$. The authors suggest that this may occur by way of an $ArN_2^\cdot$ intermediate that, inside the hydrophobic pocket, is more likely to release N_2 and permit capture of $Ar^\cdot$ by $Hb(II)$.[62]

3.2.7.4. Nitriles

Azide ions attack cobalt(III) nitriles to yield N-1 bonded tetrazoles as in equation (9).[63] The reaction rates are first-order with respect to each reagent; k_{298} (R = Ph) is 4.4×10^{-3} $M^{-1}s^{-1}$. The values for a series of substituted benzonitriles follow the Hammett equation, $k = 2.42\sigma - 2.44$ (25 °C, ionic strength 1.0 M).[63]

$$[L_5Co^{III}-N\equiv CR] + N_3^- \longrightarrow R-C \underset{\underset{CoL_5}{\overset{|}{N}}}{\overset{N-N}{\diagdown N}} \qquad (9)$$

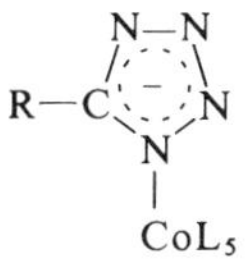

3.2.7.5. Pyridine N-Oxides

The oxomolybdenum complex $[MoO(L\text{-}NS_2)(dmf)]$, where L-NS$_2$ = 2,6-bis(2,2-diphenyl-2-thioethyl(pyridinate)(2-), reacts with a series of N-oxides to form $[MoO_2(L\text{-}NS_2)]$; these reactions model those of molybdenum hydroxylases.[64]

3.3. Oxygen, Peroxides, Water, and Other Oxygen Compounds

3.3.1. Oxygen

The kinetic data and $[V^{2+}]$-dependent stoichiometry of the reaction of V^{2+} with O_2 are explained[65] by the mechanism in equation (10).

$$V^{2+} + O_2 \xrightarrow{2\times10^3\,M^{-1}s^{-1}} VO_2^{2+} \underset{20\,s^{-1}}{\overset{+V^{2+},\,3.7\times10^3\,M^{-1}s^{-1}}{\rightleftharpoons}} VO_2V^{4+}$$

$$\Big\downarrow{\sim}100\,s^{-1} \qquad\qquad\qquad\qquad \Big\downarrow 35\,s^{-1} \tag{10}$$

$$VO^{2+} + H_2O_2 \qquad\qquad\qquad\qquad 2VO^{2+}$$

Flash photolysis of Co(III)–amine complexes $[Co(am)_2X_2]^+$ (am = en, pn; X = Cl, NO_2) produces the corresponding Co(II) amines, $[Co(am)_2X]^+$. The reduced lability of the latter in nonaqueous solvents (CH_3CN, CH_3OH) was utilized in the study of oxygenation reactions.[66] The rate constants in CH_3CN for the transient formation of monomeric superoxocobalt(III) complexes, $[Co(am)_2(X)O_2]^+$, fall in the range 10^4–$10^5\ M^{-1}\,s^{-1}$ at 25 °C. Subsequent reactions yield the stable μ-peroxo dimers. $[Mo(H_2O)_6]^{3+}$ reacts with O_2 in acidic aqueous solutions to form di-μ-oxomolybdenum(V) ion, $[Mo_2O_4(H_2O)_6]^{2+}$. In the presence of excess $[Mo(H_2O)_6]^{3+}$ the mechanism in equations (11), (12) and (13) holds[67] (units M, s; 25 °C; $\mu = 2.0\ M$).

$$Mo^{3+} + O_2 \overset{180}{\rightleftharpoons} MoO_2^{3+}, \qquad K_1 = 360 \tag{11}$$

$$MoO_2^{3+} + Mo^{3+} \overset{42}{\rightleftharpoons} MoO_2Mo^{6+}, \qquad K_2 = 535 \tag{12}$$

$$MoO_2Mo^{6+} + 2H_2O \overset{0.23}{\longrightarrow} Mo_2O_4^{2+} + 4H^+ \tag{13}$$

The rate constant for the oxidation of $[Fe(CN)_6]^{4-}$ by O_2 in aqueous-organic solvent mixtures increases with the electron-donating ability of the solvent and decreases in the presence of free CN^-.[68] This is attributed to the rate-limiting dissociation of CN^-, followed by the rapid oxidation of $[Fe(CN)_5H_2O]^{3-}$ by O_2. Similarly, the complex $[Fe(dmgH)_2(im)_2]$ is oxidized by O_2 only after the substitution of one of the coordinated imidazoles by H_2O,[69] as shown in equations (14) and (15). The parameters at 25 °C are $K = 2.1 \times 10^{-3}\ M$ and $k = 20\ M^{-1}s^{-1}$.

$$[Fe(dmgH)_2(Im)_2] + H_2O \overset{K}{\rightleftharpoons} [Fe(dmgH)_2(Im)(H_2O)] + Im \tag{14}$$

$$[Fe(dmgH)_2(Im)(H_2O)] + O_2 \overset{k}{\rightarrow} products \tag{15}$$

The oxidations of $[Fe^{II}(TMPyP)]$ by O_2 and H_2O_2 at pH 8 follow the rate laws of equations (16) and (17), respectively,[70] and the proposed mechanism is given in equations (18)–(22).

$$-d[Fe(II)]/dt = 3.5 \times 10^2[Fe(II)]^2[O_2] \tag{16}$$

$$-d[Fe(II)]/dt = 6.0 \times 10^6 [Fe(II)][H_2O_2] \tag{17}$$

$$[(P)Fe] + O_2 \overset{K}{\rightleftharpoons} [PFeO_2] \tag{18}$$

$$[(P)Fe] + [(P)FeO_2] \overset{k'}{\rightarrow} [(P)FeO_2Fe(P)] \tag{19}$$

$$[(P)FeO_2Fe(P)] + 2H_2O \rightarrow [2(P)FeOH] + H_2O_2 \tag{20}$$

$$[(P)Fe] + H_2O_2 \rightarrow [(P)FeOH] + OH^{\cdot} \tag{21}$$

$$[(P)Fe] + OH^{\cdot} \rightarrow [(P)FeOH] \tag{22}$$

The tetranuclear Pt(III) complex $[Pt_4(NH_3)_8(C_4H_6NO)_4]^{6+}$ (C_4H_6NO = deprotonated α-pyrrolidone)[71] is oxidized by O_2 in strongly acidic aqueous solutions[72] to $[Pt_4(NH_3)_8(C_4H_6NO)_4]^{8+}$. The latter is reduced by H_2O, even in the solid state, to regenerate the 6+ species according to equation (23). The

$$2(Pt_4)^{6+} + O_2 + 4H^+ \rightleftharpoons 2(Pt_4)^{8+} + 2H_2O \tag{23}$$

mixed-valence complexes $[Mo^{IV}Mo^VO_4(edta)]^{3-}$ and $[Mo^{IV}Mo^VO_4(C_2O_4)-(H_2O)_2]^{3-}$ are oxidized by O_2 to the Mo(V) dimers.[73] The respective rate constants are approximately 2×10^8 and $1 \times 10^8\ M^{-1}\ s^{-1}$. Outer-sphere oxidation of $[Ru(bpz)_3]^+$ by O_2[74] takes place with a rate constant $k = 5.8 \times 10^8\ M^{-1}\ s^{-1}$. Oxidation of U(IV) by O_2 is a chain reaction involving $OH^{\cdot}$, $HO_2^{\cdot}$, and U(V) as chain carrying species.[75] Cyclopentadienyl complexes of U(IV) are oxidized by O_2 to inorganic uranyl derivatives and products of oxidation of the cyclopentadienyl ligands.[76]

The oxidation of $[(P)Co(II)]$ (P = PPIX) by O_2 proceeds readily in alcohols in the presence of amines (L),[77] as in equations (24)–(26). The

$$[(P)Co(II)(ROH)] + L \rightleftharpoons [(P)Co(II)L] + ROH \tag{24}$$

$$[(P)Co(II)L] + O_2 \rightleftharpoons [(P)Co(II)(L)(O_2)] \tag{25}$$

$$[(P)Co(II)(L)(O_2)] + HL^+ \rightarrow [(P)Co(III)(L)_2]^+ + HO_2^{\cdot} \tag{26}$$

formation of $[(P)Co(L)_2]^+$ requires O_2 and protons for elimination of superoxide. Accordingly, no $[(P)Co(L)_2]^+$ is observed in aprotic coordinating solvents. The complex $[Cu^I(pbim)_2]ClO_4$ is oxidized by O_2 in CH_3CN.[78] A second-order dependence on $[Cu(pbim)_2^+]$ was assumed, but not verified. The quoted rate constant is $30\ M^{-2}\ s^{-1}$, independent of temperature (16–28 °C) and the presence of the free ligand. The reaction of $[(H_2O)_2Co(pd)_2]$ with O_2 in the presence of PPh_3 yields $[Co(pd)_3]$ and $[Co(pd)(O_2)(PPh_3)(OH_2)]$.[79] The key features of the mechanism include coordination of PPh_3 and O_2 to $[Co(pd)_2]$, one-electron oxidation and loss of one pd ligand, and rearrangement of the end-on oxygen adduct to the side-on peroxide.

Oxidation of macrocyclic nickel(I) complexes with O_2 ($k = 10^7$–$10^9\ M^{-1}\ s^{-1}$) yields O_2^-.[80] The rates of oxidation of *cis*-$[R_2Co(bpy)_2]ClO_4$ (R = C_2H_5, $C_6H_5CH_2$) by O_2 in acetonitrile are strongly dependent on the concentration of added $HClO_4$.[81] This was attributed to protonation of O_2

to give the highly reactive (and most unusual) species, HO_2^+. The benzylnickel complexes $[Ni(X)(CH_2C_6H_5)(PCy_3)]$ (X = Cl, CN; Cy = cyclohexyl) react with O_2 to give benzaldehyde and benzyl alcohol as the main oxidation products.[82] Isotopic labeling shows that the extra hydrogen in benzyl alcohol comes from the cyclohexyl groups. The thiolato complexes $[M(II)(tdt)_2]^{2-}$ (M = Zn, Cu, Ni, Co, Fe, Mn) are rapidly oxidized by air to the monoanionic species.[83] The site of electron transfer has been assigned to a bound sulfide group, which spin pairs with an unpaired d electron of the metal and yields a stable covalent bond. Co(II) complexes of pentadentate polyamines react with O_2 to form μ-peroxo bridged complexes.[84] These decompose by either oxidative dehydrogenation of the coordinated ligand or metal-centered oxidation to form Co(III) complexes of the unchanged ligand and H_2O_2.

The reaction of the hydridoiridium(III) sulfoxide complex *trans, mer*-$[IrHCl_2(Me_2SO)_3]$ with O_2 occurs with a 1:1 stoichiometry[85] according to equation (27). Tetranuclear complexes $[(denc)_3Cu_3M(NS)X_4]$ (M = Ni, Co,

$$3[IrHCl_2(Me_2SO)_3] + 3\,O_2 \rightarrow$$

$$[Ir(OH)Cl_2(Me_2SO)H_2O] + 2[IrCl_2(O_2)(Me_2SO)_2] \cdot Me_2SO + Me_2SO \qquad (27)$$

Zn; X = Cl, Br; NS = monoanionic Schiff base) are oxidized by O_2 in aprotic solvents[86] to $[(denc)_3Cu_3M(H_2O)X_4O_2]$. Kinetic data ($k$ = 0.7 to 16 $M^{-1}\,s^{-1}$ at 25 °C, $\Delta H^{\ddagger}$ = 2 to 8 kcal mol^{-1}, $\Delta S^{\ddagger}$ = −26 to −48 cal mol^{-1} K^{-1}) are indicative of the rate-determining insertion of O_2 into the tetranuclear halo core. The oxidation of bis(μ-halo)-bridged complexes $[L_2Cu_2X_2]$ (L = tetraalkyldiamine; X = Cl, Br) by O_2 yields dimeric oxocopper(II) products.[87] The kinetic data in nitrobenzene and CH_2Cl_2 obey the mixed third-order rate law (28). The value of k is sensitive to the steric and electronic properties of L and X.

$$d[L_2Cu_2X_2O]/dt = k[L_2Cu_2X_2]^2[O_2] \qquad (28)$$

Dinuclear Cu(I) complexes containing two bridged pyridyl-amine-pyridyl tridentate units react with O_2 to give peroxo–dicopper(II) complexes.[88] Copper(I) complexes of a series of acyclic Schiff bases react[89] with O_2 in a partially reversible reaction (29). A parallel irreversible oxidation produces $[CuL_2]^{2+}$. Mixed ligand complexes of Co(II) containing 2,2'-bipyridine and glycylglycine react with O_2 to give binuclear peroxo complexes.[90] Several Co(II) Schiff base complexes add O_2 reversibly in dmf at <0 °C, but irreversible decomposition takes place at room temperature.[91]

$$4[CuL]^+ + O_2 \rightleftharpoons [Cu_4L_4O_2]^{4+} \qquad (29)$$

Solutions of $[Cu_2(NN)_2I_2]$ (NN = bipy or phen) add O_2 rapidly and irreversibly in the presence of imidazole.[92] The peroxo complex $[Cu_2(NN)_2Im_2O_2]I_2$ thus formed produces H_2O_2 on acidification with $HClO_4$. Reaction with CO forms the highly insoluble carbonate $[Cu_2(NN)_2Im_2CO_3]I_2$.

The activation parameters for the oxygenation of Mn(II)–Schiff base complexes are strongly dependent on the solvent.[93]

3.3.2. Hydrogen Peroxide

H_2O_2 oxidizes V^{2+} to V^{3+} with a rate constant $k = 17.2 \ M^{-1} s^{-1}$ (25 °C, 0.12 M HClO$_4$, excess V^{2+}).[65] No VO^{2+} is produced, consistent with the one-electron mechanism shown in equations (30) and (31).

$$V^{2+} + H_2O_2 \rightarrow V^{3+} + \text{·OH} \tag{30}$$

$$V^{2+} + \text{·OH} \rightarrow V^{3+} \tag{31}$$

Radiolysis of aqueous solutions containing CH_3OH, Cu^{2+}, and H_2O_2 produces CH_2O in a chain reaction.[94] The low yields of CH_2O and the half-order kinetic dependence on CH_3OH and dose rate were taken as evidence against ·OH radical formation in the reaction of Cu^+ with H_2O_2. The proposed chain-carrying species in the oxidation of CH_3OH are Cu(III) and ·CH_2OH as in equations (32)–(34). The results of similar experiments with U(IV)[75] are considered indicative of the ·OH radical formation at pH 0.7, but not at pH 2.

$$Cu(I) + H_2O_2 \rightarrow Cu(III) \tag{32}$$

$$Cu(III) + CH_3OH \rightarrow \text{·}CH_2OH + Cu(II) + H^+ \tag{33}$$

$$Cu(II) + \text{·}CH_2OH \rightarrow Cu(I) + CH_2O + H^+ \tag{34}$$

Mixtures of H_2O_2 and edta complexes of Fe(II) and Fe(III) oxidize ferrocytochrome *c* without any loss of cytochrome *c*.[95] This is presented as evidence against the formation of free ·OH radicals, since the latter are known to degrade cytochrome *c*. In a related study[96] of the oxidation of Fe^{II} (edta) by excess H_2O_2 two distinctly different intermediates were observed; see equations (35) and (36). I_1 was identified as an edta complex of FeO^{2+} based

$$Fe(II) + H_2O_2 \rightarrow I_1 \tag{35}$$

$$I_1 + H_2O_2 \rightarrow I_2 \tag{36}$$

on its reactivity toward a number of substrates. Intermediate I_2 is most likely the hydroxyl radical formed in a reaction of I_1 with H_2O_2, $k = 3.2 \times 10^3 \ M^{-1} s^{-1}$. H_2O_2 oxidizes the mixed valence and the fully reduced cytochrome *c* oxidase in two-electron processes.[97] The rate constants for both reactions are $\sim 3 \times 10^4 \ M^{-1} s^{-1}$ (temperature not specified). The oxidation of tetraazamacrocyclic complexes of Fe(II) by H_2O_2 is inhibited in the presence of nitrogen bases that coordinate to Fe(II).[98]

Oxidation of deoxyhemerythrin to hydroxomethemerythrin is a two-electron process,[99] independent of pH in the range 6.3–9.5, as in equation (37).

$$\text{dexoyHr}(\mu\text{-OH}) + H_2O_2 \rightarrow \text{metHr}(OH)(\mu\text{-O}) + H_2O \tag{37}$$

The values of k, $\Delta H^{\ddagger}$, and $\Delta S^{\ddagger}$ are 5.5 $M^{-1}\,s^{-1}$ (25 °C, 0.15 M ionic strength), 7.6 kcal mol^{-1}, and -2.9 cal K^{-1} mol^{-1}, respectively.

The rate constants for the oxidation of $[CuL_2(DNA)]^+$ by H_2O_2[100] are similar to those for the analogous reactions of $[CuL_2]^+$. The values for the ternary complexes are 1450 (L = phen), $\leq$1240 (L = bipy), and 270 (L = 5-nitro-1,10-phenanthroline). The complex $[Mo^VO(TMPyP)(OH_2)]^{5+}$ is oxidized by H_2O_2 in acidic aqueous solutions,[101] as in equation (38), according to rate law (39). The kinetic parameters have the following values: $k = 3.27 \times 10^{-4}\ M^{-2}\,s^{-1}$ (25 °C), $\Delta H^{\ddagger} = 74\ kJ\ mol^{-1}$, $\Delta S^{\ddagger} = -63\ J\ mol^{-1}\ K^{-1}$, $\Delta V^{\ddagger} = 1.3\ cm^3\ mol^{-1}$. The substitution of the oxo group in $[Mo^VO(TMPyP)OH_2]^{5+}$ by H_2O_2 is thought to precede the rate-determining oxidation by the second H_2O_2.

$$[Mo^VO(TMPyP)OH_2]^{5+} + 2H_2O_2 + H^+ \rightarrow [Mo^{VI}(O_2)(TMPyP)(OH_2)]^{6+} + 2H_2O + OH^{\cdot} \quad (38)$$

$$-d[Mo(V)]/dt = k[Mo(V)][H_2O_2]^2 \quad (39)$$

A standard reduction potential of >0.44 V and the formation of a complex with H_2O_2 seem to be important requirements for mineral compounds that react with H_2O_2 to yield $^1O_2$ in alkaline aqueous solution.[102] The chromium(V) complex bis(ethyl-2-hydroxybutyrato)oxochromate(V) oxidizes H_2O_2 to O_2 with a rate constant of 0.055 $M^{-1}\,s^{-1}$ at 25 °C and 0.40 M ionic strength.[103] The reaction is independent of pH (in the range 2.5–4.5) and [ligand anion], consistent with the proposed coordination of H_2O_2 to the Cr(V) center and subsequent intramolecular transfer of two electrons to form Cr(III) and O_2. The initial phase of the reaction of H_2O_2 with MnO_4^- takes place according to the stoichiometry of equation (40) and rate law (41),[104] where k_1 and k_2

$$2MnO_4^- + 5H_2O_2 + 6H^+ \rightarrow 2Mn^{2+} + 5O_2 + 8H_2O \quad (40)$$

$$-d[MnO_4^-]/2dt = (k_1 + k_2K[H^+])[MnO_4^-][H_2O_2] \quad (41)$$

represent the rate constants for the reactions of MnO_4^- and $HMnO_4$, respectively, and K is the protonation constant for MnO_4^-. The values are $k_1 = 23.2\ M^{-1}\,s^{-1}$ and $k_2K = 310\ M^{-2}\,s^{-1}$ at 25 °C and 1 M ionic strength. The kinetics of oxidation of H_2O_2 to O_2 by 5,10,15,20-tetrakis(2,6-dimethyl-3-sulfonatophenyl)porphyriniron(III) are pH dependent.[105] At 30 °C the values of the rate constants ($M^{-1}\,s^{-1}$) for the three identified pathways are: 44 for $[(P)Fe^{III}(H_2O)_2] + H_2O_2$, 1.67×10^3 for $[(P)Fe^{III}(H_2O)(OH)] + H_2O_2$, and 1.05×10^5 for $[(P)Fe^{III}(OH)(H_2O)] + HO_2^-$. The p$K_a$ of $[(P)Fe^{III}(OH_2)_2]$ is 6.8.

Low concentrations of H_2O_2 react with superoxide dismutase from *E. coli* predominantly by reducing Fe(III) to Fe(II).[106] At higher concentrations tryptophan residues are modified. H_2O_2 reduces $[Cu(Me_2phen)_2]^{2+}$ to $[Cu(Me_2phen)_2]^+$, which is destroyed in further reactions with H_2O_2.[107] Solutions of $[Fe^{II}(CH_3CN)_4(ClO_4)_2]$ in dry acetonitrile catalyze the rapid disproportionation of H_2O_2, as in equation (42).[108] In the proposed mechan-

$$2H_2O_2 \xrightarrow{Fe^{II}} 2H_2O + O_2 \quad (42)$$

ism the formation of a side-on adduct, $Fe(H_2O_2)^{2+}$, is followed by a concerted transfer of the two hydrogen atoms from the second molecule of H_2O_2. The adduct $Fe(H_2O_2)^{2+}$ promotes dehydrogenation and monoxygenation of a large number of organic substrates. Fe^{3+} catalyzes the disproportionation of H_2O_2 in acidic aqueous solutions. The radical and nonradical mechanisms for this reaction were recently compared.[109] Cu^{2+} [110] and complexes of Cu(II) containing polypyridines and amino acids[111] catalyze photochemical[110] and thermal[111] disproportionation of H_2O_2. Ce(IV)–edta complexes catalyze the thermal reaction in alkaline solutions.[112,113] The decomposition of H_2O_2 in CH_3COOH and CF_3COOH in the presence of $VO(acac)_2$ yields small amounts (1–10%) of O_3.[114] The Cu^{2+}-catalyzed reaction between H_2O_2 and SCN^- exhibits bistability.[115]

3.3.3. Alkyl Hydroperoxides, Peroxy Acids, and Metal Peroxides

The nickel(I) complexes R,R,S,S-$Ni(tmc)^+$ and R,S,R,S-$Ni(tmc)^+$ react in alkaline aqueous solution with t-butylhydroperoxide to form $[CH_3Ni(tmc)]^+$ [116,117] according to the mechanism in equations (43)–(45). The rate constant k_{43} is the same for the two nickel complexes, $3.8 \times 10^5 \, M^{-1} \, s^{-1}$ at 25 °C and 0.10 M ionic strength.

$$[Ni(tmc)]^+ + (CH_3)_3COOH \rightarrow [HONi(tmc)]^{2+} + (CH_3)_3CO^{\cdot} \qquad (43)$$

$$(CH_3)_3CO^{\cdot} \rightarrow CH_3^{\cdot} + (CH_3)_2CO \qquad (44)$$

$$[Ni(tmc)]^+ + CH_3^{\cdot} \rightarrow [CH_3Ni(tmc)]^+ \qquad (45)$$

The oxidation of $[(edta)Fe(III)]$ by m-chloroperbenzoic acid,[118] $ArCO_3H$, in CH_3OH at 30 °C yields $[(edta)Fe^VO]$ as in equations (46) and (47). The oxo complex, a potent oxidant, can be trapped by reducing species

$$[(edta)Fe(III)] + ArCO_3H \rightleftharpoons [(edta)Fe^{III}(ArCO_3H)], \qquad K = 12 \qquad (46)$$

$$[(edta)Fe^{III}(ArCO_3H)] \rightarrow [(edta)Fe^VO]^{5+} + ArCO_2H, \qquad k = 6.2 \, s^{-1} \qquad (47)$$

such as 2,4,6-tri-*tert*-butylphenol ($k = 5.5 \times 10^3 \, M^{-1} \, s^{-1}$) and CH_3OH ($k = 48 \, M^{-1} \, s^{-1}$). The rate constants for oxygen transfer from percarboxylic acids and alkyl hydroperoxides to $[(TPP)Mn^{III}Cl]$ are sensitive to the presence of nitrogen bases.[119] The two-electron oxidation of $[(TPP)Co^{III}Cl]$ by percarboxylic acids and hydroperoxides yields a Co(III) porphyrin dication.[120] A change in mechanism from heterolytic (percarboxylic acids) to homolytic (hydroperoxides) is signalled by a break in the plot of log k vs. pK_a of the leaving group (carboxylic acid or alcohol).

Cumene hydroperoxide reacts with $[Co(acac)_2]$ in $CDCl_3$ containing 0.2 M pyridine according to equation (48). Both $[Co(acac)_2(O_2R)py]$ and ROH were identified by NMR.[121] The reaction takes place by a series of one-electron steps via $RO^{\cdot}$ and $RO_2^{\cdot}$ intermediates.

$$[Co(acac)_2py] + 2ROOH \rightarrow [Co(acac)_2(O_2R)py] + ROH + \tfrac{1}{2}H_2O + \tfrac{1}{4}O_2 \qquad (48)$$

Styrene ozonide reacts with [(TPP)Cr^{III}Cl] in CH$_2$Cl$_2$ to yield the chromium tetraphenylisoporphyrin.[122] The reaction, which requires two moles of ozonide per mole of [(TPP)CrCl], is completely suppressed by addition of 1 M CH$_3$OH. This suggests that ozonide coordinates to Cr(III) and transfers an oxygen to a *meso* position of the porphyrin, followed by abstraction of a proton from the second ozonide and formation of hydroxyisoporphyrin. Oxygen transfer from bis(2-ethyl-2-hydroxybutyrato)oxochromate(V) to PPh$_3$ is catalyzed by copper(II) salts.[123] Diperoxo amine complexes of Cr(IV) decompose to give Cr(VI) and Cr(III) products.[124] The kinetics and products fit a scheme involving a monoperoxochromium(VI), 6-coordinate chromium(IV), and HO$_2^{\cdot}$. The latter was detected in scavenging experiments with tetranitromethane.

The mixed-ligand peroxo complex [{Co^{III}(dien)amp)}$_2$(O$_2$)]$^{4+}$ undergoes oxidative dehydrogenation of coordinated amp in alkaline aqueous solutions.[125] The first-order dependence on [OH$^-$] and the large kinetic deuterium isotope effect (k_H/k_D = 13–16) are consistent with deprotonation of the coordinated amino group followed by a concerted homolytic fisson of dioxygen and transfer of the α-hydrogen from the aminomethyl group to the coordinated oxygen. The resulting imino complex hydrolyzes to aldehyde and ammonia. The H$^+$-assisted decompostion of a μ-peroxo cobalt(III) complex, [(en)$_2$Co(μ-OH)(μ-O$_2$)Co(en)$_2$]$^{3+}$, yields Co^{2+} and O$_2$.[126] The data are consistent with the rapid protonation of the complex followed by the rate-determining cleavage of the Co—O bond. The same complex liberates O$_2$ photochemically in alkaline solutions.[127] The decomposition of (acyl-peroxo)manganese(III) porphyrins in alkaline CH$_2$Cl$_2$ occurs by O—O bond homolysis, while acidic conditions favor the heterolytic process.[128]

3.3.4. Water and Other Oxygen Compounds

The reduction of H$_2$O to H$_2$ by [Co^I(TMAP)][129] takes place with the rate constant $k > 10^4$ M^{-1} s^{-1}. The Ru(IV) complexes [(bipy)$_2$(py)RuO]$^{2+}$ (**1**) and [(terpy)(phen)RuO]$^{2+}$ (**2**) undergo self-reduction in basic solutions yielding Ru(II) products with both unmodified and partially oxidized ligands.[130] The rate law for **2** is given by $-d[\text{Ru(IV)}]/dt = (k_a[\text{OH}^-] + k_b[\text{OH}^-]^2)$-[Ru(IV)], $k_a = 0.21$ M^{-1} s^{-1} ($\Delta H^{\ddagger} = 12$ kcal mol^{-1}, $\Delta S^{\ddagger} = -20$ cal mol^{-1} K^{-1}), $k_b = 0.08$ M^{-2} s^{-1} at 25 °C and 1.0 M ionic strength (Na$_2$SO$_4$). The reduction of **1** yields Ru(III), which subsequently undergoes further reduction. Two studies of the self-reduction of [Ru(bipy)$_3$]$^{3+}$ have appeared. Optical rotation data[131] on the Ru(II) products are quoted in support of the pseudobase formation.[132] In the alternative mechanism[133] the reaction is initiated by nucleophilic attack at the metal center M (M = Fe, Ru, Os). Further reactions of the seven-coordinate intermediate yield O$_2$ and oxidized ligands via M^{IV} = O and μ-oxo intermediates.

Pyridine solutions of [Fe(III)(TPP)] are reduced by OH^- to give Fe(II) porphyrin and polymeric hydroxypyridine.[134] The reaction is initiated by nucleophilic attack by OH^- at the C-4 carbon of a bound pyridine, followed by electron transfer from the pyridine nitrogen to Fe(III) as in equation (49).

$$[pyFe^{III}(TPP)py] + OH^- \rightarrow [(4\text{-}OHpy)Fe^{III}(TPP)py]$$

$$\xrightarrow{py} \frac{1}{n}(HO \cdot py)_n + [pyFe^{II}(TPP)py] \tag{49}$$

The reduction of $[Pt(CN)_4Br_2]^{2-}$ by OH^- yields $[Pt(CN)_4]^{2-}$,[135] according to equation (50) with rate law (51). The mechanism involves formation of

$$[Pt(CN)_4Br_2]^{2-} + 2OH^- \rightarrow [Pt(CN)_4]^{2-} + BrO^- + Br^- + H_2O \tag{50}$$

$$d[Pt(CN)_4^{2-}]/dt = 26[OH^-][Pt(CN_4Br_2^{2-}] \tag{51}$$

the bridged intermediate $[Br\cdots Pt(CN)_4\cdots Br\cdots OH]^{3-}$ and rapid transfer of Br^+ from the platinum to OH^-. Subsequent oxidation of $[Pt(CN)_4]^{2-}$ by HOBr slowly generates $[Pt(CN)_4Br(OH)]^{2-}$. Silver(II) ions, complexed by 2,2'-bipyridine, oxidize H_2O[136] according to rate law (52). At 1 M ionic strength

$$-d[Ag(II)]/dt = k[Ag(II)]^2[H^+]^0[Ag(I)]^{-x} \quad (0 \le x \le 1) \tag{52}$$

$(HNO_3 + NaNO_3)$, 4.0 mM excess $Hbipy^+$, 0.05 M H^+ and 0.10 M Ag(I), the values of $\Delta H^{\ddagger}$, $\Delta S^{\ddagger}$, and k (25 °C) are 83 ± 15 kJ mol^{-1}, -55 ± 50 J K^{-1} mol^{-1}, and 2.5×10^{-4} M^{-1} s^{-1}, respectively. The suggested mechanism is shown in equations (53) and (54).

$$2[Ag(bipy)OH]^+ \xrightarrow{H^+} 2Ag(I) + H_2O_2 + 2Hbipy^+ \tag{53}$$

$$2\,Ag(II) + H_2O_2 \rightarrow 2\,Ag(I) + 2H^+ + O_2 \tag{54}$$

Aqueous solutions of vanadium(II) complexes of cysteine, cysteamine, and cysteine methyl ester evolve hydrogen[137] according to rate law (55). The

$$d[H_2]/dt = k[V(II)][H^+]^0, \quad pH\ 7.5\text{-}8.5 \tag{55}$$

value of k is 2.3×10^{-3} s^{-1} (21 °C), $E_a = 54$ kJ mol^{-1}, $A = 5 \times 10^6$ s^{-1}. Aqueous solutions of $[\{MnL(H_2O)\}_2]^{2+}$ (L = dianion of O_2N_2 tetradentate Schiff base) liberate O_2 and reduce p-benzoquinone when irradiated with visible light.[138] Acidic aqueous solutions of tris(2,2-bipyridine-4,4'-dicarboxylic acid)ruthenium(II) containing $S_2O_8^{2-}$ as electron acceptor generate O_2 under illumination by visible light.[139]

The catalytic decompostion of p-cyano-N,N-dimethylaniline N-oxide in the presence of Fe(III) porphyrins[140-142] is initiated by the complex formation between the substrate and Fe(III). The rate-determining oxygen transfer to iron ensues to give p-cyano-N,N-dimethylaniline and the Fe(IV)-oxo porphyrin π-cation radical as in equation (56). Stepwise oxidation of p-CN-C_6H_4-$N(CH_3)_2$ by $[(P)Fe^{IV}O]$ completes the sequence. Oxygen transfer

$$p\text{-}CN\text{-}C_6H_4\text{-}N(CH_3)_2O + [(P)Fe(III)] \rightarrow p\text{-}CN\text{-}C_6H_4\text{-}N(CH_3)_2 + [(P)Fe^{IV}O] \tag{56}$$

from p-cyano-N,N-dimethylaniline-N-oxide to [TPPCr^{III}Cl] occurs only by photocatalysis.[143] The oxochromium(V) thus formed readily oxidizes 1-phenyl-1,2-ethanediol to benzaldehyde and formaldehyde. Solutions of heteropolyanions [SiMoO$_{39}$Cr(OH$_2$)]$^{5-}$ and [SiW$_{11}$O$_{39}$Cr(OH)]$^{6-}$ in toluene, benzene or acetonitrile react with C$_6$H$_5$IO to form the corresponding oxochromium(V) species.[144]

3.4. Halogens, Halides, and Halogen Oxoacids

3.4.1. Halogens

The reactions of benzyl cobaloximes, [XC$_6$H$_4$CH$_2$Co(dmgH)$_2$(py)], with Cl$_2$ or Br$_2$ in chloroform or acetic acids form benzyl halides and benzyl ethers of dimethylglyoxime.[145] The distribution of products and the faster reactions of Cl$_2$ are taken to support the previously-suggested oxidative pathway (i.e., the initial step yields [RCo^{IV}(dmgH)$_2$py]$^+$ and X$_2^-$), rather than an electrophilic attack. The dealkylations of methyl (and ethyl) cobalamin with iodine in aqueous solutions also proceed by electron transfer.[146] The intermediacy of the one-electron oxidized species [CH$_3$-B$_{12}$]$^+$ is suggested by the production of CH$_3$Cl and not CH$_3$I when the iodine reaction is carried out in the presence of chloride ions.[146] Oxidative dealkylation reactions are also indicated for reactions in acetonitrile between iodine and organocobalt(III) compounds, such as *cis*-[R$_2$Co(bipy)$_2$]$^+$, *trans*-[Me$_2$Co(dpnH)], and [RCo(dmgH)$_2$py], on the basis of products and kinetic data. The rate constants for these complexes parallel those for iodine oxidations of ferrocenes (Fc + I$_2$ → Fc$^+$ + I$_2^-$), and both correlate with the electrode potentials.[147]

The coordinated ligands of the complexes [(NH$_3$)$_5$Co(ImH)]$^{3+}$ and [(NH$_3$)$_5$CoOC(NH$_2$)$_2$]$^{3+}$ are oxidized by halogens. The bromination of imidazole in acidic, aqueous bromide solutions proceeds by way of a bimolecular reaction between [(NH$_3$)$_5$Co^{III}Im]$^{2+}$ and Br$_2$ ($k = 6 \times 10^9$ M^{-1}s^{-1} at 25 °C, $\mu = 0.13$ M). Successive but slower reactions of [(NH$_3$)$_5$Co(4,5-Br$_2$ImH)]$^{3+}$ and [(NH$_3$)$_5$Co(2,4,5-Br$_3$ImH)]$^{3+}$ were observed.[148] The O-urea cation reacts with Cl$_2$ in two steps, both with a first-order dependence on [Cl$_2$] (the values of k are 14 and 4.1 M^{-1}s^{-1} in 1 M H$_2$SO$_4$ at 25 °C). The intermediate, although not fully characterized, appears on the basis of spectroscopic data to contain N-urea, and may be [(NH$_3$)$_5$Co-NH$_2$C(O)NHCl]$^{3+}$.[149] Hypochlorous acid reacts similarly but more slowly with the O-urea complex. Indeed [Pt(CN)$_4$]$^{2-}$ is also oxidized by both Cl$_2$ and HOCl in the same pattern, with respective second-order rate constants 1.1×10^7 and 1.0×10^2 M^{-1}s^{-1} (25 °C, 1 M perchlorate medium).[150]

Oxidation of [Ru^{II}(OEP)] dianion to the paramagnetic *trans*-[X$_2$Ru^{IV}(OEP)], X = F, Cl, Br, occurs upon treatment with anhydrous HX in dichloromethane. Oxidation is attributed to traces of X$_2$ in the gaseous HX reagents. Halogens themselves can be used, but side products are obtained.[151] The reaction between dimeric chromium(II) acetate and iodine was studied in acetic acid, which minimizes overall dissociation of the dimer, as a function of concentration variables including water and sodium iodide. There is a balance struck between monomerization {$Cr(II)_2 \rightleftharpoons 2Cr(II)$} and oxidation {$Cr(II) + X_2 \rightarrow Cr(III)$-products}, the effects of sodium iodide attributed to axial coordination of iodide ions to the dimer.[152] The reaction between iron(III) dithiocarbamates and iodine in dichloromethane yields [$IFe(S_2CNR_2)_2$] and ($S_2CNR_2)_2$, following an initial electron-transfer reaction that yields [$Fe(S_2CNR_2)_3$]$^+$ (for R = CH_2Ph or Et, NR_2 = morpholino).[153] An initial electron-transfer reaction also occurs when η^6-arene complexes of zero-valent molybdenum are treated with iodine in toluene or methylene chloride. The cation [$IMo(CO)_3(arene)$]$^+$ is always formed initially, which ultimately leads to other compounds depending on the reaction conditions.[154]

The oxidation of [$Mn^{III}(TMPyP)$] by Br_2 was studied *in alkaline solution* (pH 7-11) by the stopped-flow technique.[155] The reaction was carried out by mixing solutions of the reagent buffered at the same pH and ionic strength. A strong acceleration with alkalinity was noted with $k = 1 \times 10^2, 4 \times 10^5$, and $1.5 \times 10^7 \, M^{-1} \, s^{-1}$ (at pH 7, 9, and 11, 0.050 M ionic strength, and unspecified but presumably ambient temperature). This effect was attributed to the hydrolysis of Br_2 at alkaline pH: $3Br_2 + 3OH^- \rightleftharpoons BrO_3^- + 5Br^- + 3H^+$).[155]

3.4.2. Halide Ions

The peroxotitanium(2+) ion oxidizes iodide ions by a direct reaction ($k = 3.0 \times 10^{-3} \, M^{-1} \, s^{-1}$ in 1.00 M perchloric acid at 25 °C) as well as by an indirect pathway in which the reactive species is hydrogen peroxide from the solvolysis of TiO_2^{2+}.[156] The rate constants for the two are in the ratio 1:9 for I^- oxidation; this is emerging as a rather general pattern, H_2O_2 being more reactive than TiO_2^{2+} toward a variety of substrates. The industrially important oxidations of aromatic hydrocarbons by cobalt salts in carboxylic acids are accelerated by bromide ions. The results of a kinetic study of the oxidation of bromide ions by cobalt(III) acetate in acetic acid–pyridine solutions at 60-93 °C suggest that Br_2^- is formed as an intermediate, as in equations (57) and (58). Pyridine or other amines function to control the concentrations of bromide ions and the dissociation of the active Co(III) dimer.[157]

$$[Co(III)_2Br] + Br^- \rightarrow [Co(III)Br_2]^- + Co(II) \qquad (57)$$

$$[Co(III)Br_2]^- \rightarrow Co(II) + Br_2 \qquad (58)$$

Pertechnetate ions in concentrated HBr solutions are reduced first to the Te(V) complex [$TcOBr_3$]$^{2-}$ and then over many hours to [$TcBr_6$]$^{2-}$. At

[Tc(V)] > 10^{-3} M, the pseudo-first-order rate constant at 0° and 8.7 M HBr is 5.3×10^{-3} h^{-1} $(E_a = 81$ kJ $mol^{-1})$.[158] Deviations at lower [Tc(V)] are accounted for by a second term with a zeroth-order dependence on [Tc(V)], arising perhaps from "some material absorption or from a surface reaction." The μ-oxo dimeric iron complex $[\{Fe(TPP)\}_2O]^+$ rapidly oxidizes iodide ions in dichloromethane; the second-order rate constant at 273 K is 8×10^4 $M^{-1}s^{-1}$ $(E_a = 13$ kJ $mol^{-1})$.[159] The reaction partners appear to be an iodo complex and iodide ions, with the mechanism being electron transfer to I^- via bound I^-, such that I_2^- results. The fate of I_2^- in this solvent is considered but not resolved. Various gold(III) complexes such as $[Au(NH_3)_4]^{3+}$, $[Au(NH_3)_3I]^{2+}$, and *trans*-$[Au(NH_3)_2X_2]^+$ (X = Cl, Br, I) react with iodide ions.[160] In every case the kinetic step is ligand substitution, with intramolecular electron transfer (i.e., reductive elimination of I_2 or IX) being fast. The single exception may be the reaction of *trans*-$[Au(NH_3)_2Br_2]^+$, where the data indicate a direct reduction step in which noncoordinated I^- reduces the complex via one of the bromide ligands as in equation (59).

$$[Au^{III}(NH_3)_2Br_2]^+ + I^- \rightarrow [Au^I(NH_3)_2]^+ + Br^- + IBr, \qquad k > 10^7 \ M^{-1}s^{-1} \qquad (59)$$

The formation of ion pairs may in general[8] be important in quenching photoexcited states, as has been demonstrated in the specific case[161] $[Co(sep)]^{3+} - X^-$ (X = Br, I). Quenching involves X_2^- formation. This phenomenon gives rise (from quenching by peroxodisulfate ions) to the species $[Ir((C^3, N'\text{-bipy})(bipy)_2]^{4+}$, a potent acceptor with considerable stability (>2 h at pH < 1.5). This complex oxidizes Cl^- to Cl_2^- $(k = 8 \times 10^8 \ M^{-1}s^{-1}$ at 20 °C and $\mu = 0.2 \ M)$.[58]

3.4.3. Perhalate Ions

The Mo(III) complexes $[Mo(H_2O)_6]^{3+}$ and $[\{H_2O)_4Mo\}_2(\mu\text{-OH})_2]^{4+}$ are oxidized to the Mo(V) dimer $[Mo_2O_4(H_2O)_6]^{2+}$ by perchlorate ions. The reactions occur in several stages, and suddenly accelerate after a long induction or initiation period. The chemical mechanism has not been unraveled, although a study of the phenomena[162] suggests a pattern rather different from that taken by other perchlorate oxidations. Molybdenum(V) species, both dimeric and monomeric, are also oxidized by perchlorate ions.[163] The oxidation of the monomer in 14 M H_2SO_4 follows second-order kinetics $(k = 0.16 \ M^{-1}s^{-1}$ at 25 °C); the rate constant is approximately 10^3 lower in 9 M H_2SO_4. Solvent effects on the oxidation of several cobalt complexes by periodate ions have been reported. The studies include the oxidations of $[Co^{II}(edta)]^{2-}$ in H_2O/EtOH mixture,[164] of $[Co^{III}(en)_2(SCH_2CO_2)]^+$ in H_2O/t-BuOH and H_2O/MeOH mixtures,[165] and of $[Co^{III}(en)_2(cysteinsulphenate)]^+$ in H_2O/t-BuOH and H_2O/i-PrOH mixtures.[166]

3.4.4. Halate Ions and Other Halogen Oxoanions; Oscillating Reactions

3.4.4.1. Halates

See also Section 3.4.4.3. The Co(I) state of Vitamin B_{12} reduces XO_3^- to X^- ($X = Cl, Br, I$) at rates that increase with acidity.[167] The $[H_3O^+]$-independent rate constants (22°, $\mu = 0.5\ M$) are 4.4×10^5, 1.1×10^3, and $8.1 \times 10^1\ M^{-1}\ s^{-1}$, respectively, for IO_3^-, BrO_3^-, and ClO_3^-. The well-known ferrocyanide–iodide reaction has been studied in H_2O–MeOH solvent mixtures.[168] Both Mo(VI) and V(V) catalyze the BrO_3^-–I^-–ascorbic acid "clock" reaction, although the metals intervene at different steps in the sequence.[169] The Mo(VI) catalysis affects the $BrO_3^- + I^-$ component, while V(V) catalyzes the direct reaction between BrO_3^- and ascorbic acid. Americium(III) is oxidized by BrO_3^- in the presence of phosphotungstates yielding Am(IV).[170]

3.4.4.2. Other Halogen Oxoanions

The oxidation of Vitamin B_{12s} by ClO_2^- is similar to the halate oxidations cited above, although it occurs much more rapidly than the analogous ClO_3^- oxidation ($k = 5.1 \times 10^4$ vs $8.1 \times 10^1\ M^{-1}\ s^{-1}$).[167] Ferricyanide reacts with alkaline hypobromite solution resulting in cyanate ions and $Fe(OH)_3$ at low $[OH^-]$, 0.05–0.5 M. On the other hand, the reaction with BrO^- is said to cause *reduction* to $[Fe(CN)_6]^{4-}$ at $[OH^-] > 1\ M$.[171]

3.4.4.3. Oscillating Reactions

The oxidation of Mn(II) in aqueous sulfuric acid, one of the important steps in the BZ oscillating reaction, was shown to be a complex process with an induction period.[172] Both the length of the induction period and the maximum rate depend on concentrations of Mn(II), BrO_3^-, Br_2, Br^-, H_3O^+, and Mn(III). The system BrO_3^-–$[Fe(phen)_3]^{2+}$–Br^- exhibits bistability,[173] and is described by the Field–Körös–Noyes (FKN) or Oregonator model.[174] In this model, there is still a question as to which step(s) produce Br^-, the control intermediate, but it appears the principal path, but perhaps not the only one, is the reaction between bromate ions and an oxidation product of malonic acid.[175] Other investigations of the BZ-type systems include (1) an investigation of the perturbing effect of Hg(II), a bromide-ion remover;[176] (2) an examination of a peculiar result obtained when acetylacetone is used as the organic substrate, such that Ag/AgBr electrodes show large-amplitude oscillations, while shiny platinum electrodes show no response[177]; (3) calorimetric studies of the malic acid BZ reaction as compared to that for malonic acid[178]; and (4) the mixed substrate system of acetone–malonic acid, wherein with $H_2O_2/IO_3^-/Mn(II)$ there are two oscillatory phases corresponding to I_2

production and consumption.[179] A nickel complex of a tetraazamacrocycle catalyzes a BZ-type oscillating reaction by virtue of NiL^{3+}/NiL^{2+} oxidations, although ligand oxidation reduces the efficiency of the process.[180]

3.5. *Sulfur Compounds and Oxoanions*

3.5.1. *Peroxosulfates*

3.5.1.1. *Peroxomonosulfate*

The oxidation of Mn^{2+}_{aq} by HSO_5^- yields MnO_2; it is autocatalytic, with an apparent second-order rate constant (in acetate buffers, pH 4.4–5.1) given by $Rate/[Mn^{2+}][HSO_5^-] = k_{app} = k_0/[H^+] + k_1[MnO_x]/[H^+]$.[181]

3.5.1.2. *Peroxodisulfate*

The well-known reactions in which $[Mo(CN)_8]^{4-}$ and $[Fe(CN)_6]^{4-}$ are oxidized by $S_2O_8^{2-}$ have been reexamined. The effect of alkali–metal cations on the first of these is quite pronounced, and is described by the rate law $k[Mo(CN)_8^{4-}][S_2O_8^{2-}][M^+]$. Values of k increase regularly from Li^+ $(3.8 \times 10^{-3} \ M^{-2} \ s^{-1}$ at 40 °C) to Cs^+ (64×10^{-3}), and vary linearly with the polarizability of the cation, which is suggested to bridge the anionic centers in the transition state.[182] Ion-pair formation with K^+ is also a prevalent feature of the reaction between $[KFe(CN)_6]^{3-}$ and $KS_2O_8^-$, as assessed in a variety of water–cosolvent mixtures.[183] The oxidation of U^{4+}_{aq} by $S_2O_8^{2-}$ proceeds by three pathways: direct oxidation, reaction with the products of thermal decomposition of $S_2O_8^{2-}$ (which we presume means SO_4^-), and by intramolecular electron transfer within a U^{4+}, $S_2O_8^{2-}$ association complex.[184] The excited state $[*Ir\{(C^3, N'-Hbipy)(bipy)_2\}]^{3+}$ is oxidatively quenched by $S_2O_8^{2-}$, yielding SO_4^- and a potent electron acceptor, M^{4+}.[58]

3.5.2. *Sulfur Dioxide and Sulfite Ions*

3.5.2.1. *Sulfur Dioxide*

The peroxotitanium(IV) ion does not react directly with SO_2; rather, rate-determining dissociation of TiO_2^{2+} yields H_2O_2, which reacts rapidly.[185] The reduction of the tetrameric peroxozirconium(IV) complex by SO_2 is also independent of $[SO_2]$, suggesting that the ring-opening on the peroxozirconium(IV) complex is rate-limiting.[186]

3.5.2.2. Coordinated Sulfite

Certain O-bonded sulfito complexes of Co(III) undergo intramolecular electron transfer. The complex $[(en)_2Co(H_2O)(OSO_2H)]^+$ undergoes self-reduction to Co(II) at pH 2.1–5.5 with a complicated $[H_3O^+]$-dependence (e.g., $k = 2.1 \times 10^{-3} s^{-1}$ at pH 5.5, 25 °C), while at pH < 6 it rearranges to the S-sulfito analogue.[187] Unlike all other aquo-amine Co(III) complexes, *cis*-$[CoL_2(OH_2)_2]^{3+}$ (L = phen, bipy) reacts with SO_2 to yield S-sulfito complexes without a detectable O-sulfito intermediate. The internal conversion to Co(II) and SO_4^{2-} occurs only slowly, requring temperatures above 45 °C and high $[H_3O^+]$.[188]

3.5.2.3. Bisulfite and Sulfite Ions

The Cr(V) complex bis(2-ethyl-2-hydroxybutyrato)oxochromate(V) oxidizes HSO_3^- in a "clock reaction," which proceeds through a Cr(IV) complex that is detectable by its absorption at 600 nm.[189] The proposed mechanism involves oxidations of HSO_3^- to SO_3^- by both Cr(V) and Cr(IV), and then competition for SO_3^- between Cr(V) and Cr(IV). Permanganate ions oxidize sulfite ions via a manganate(VI) intermediate, which is formed by electron transfer and disproportionates to permanganate and a soluble Mn(IV) product. Direct oxidation of SO_3^{2-} by MnO_4^{2-} occurs slowly enough to compete with disproportionation of MnO_4^{2-}.[190] The well-studied oxidation of SO_3^{2-} by $[Fe(CN)_6]^{3-}$ has been reexamined in several concentrated salt solutions and in several isodielectric water–cosolvent mixtures.[191]

3.5.3. Dithionite Ions

This powerful reducing reagent often reacts after homolytic dissociation ($S_2O_4^{2-} \rightleftharpoons 2SO_2^-$, $K_D = 4 \times 10^{-9} M$). Such is the case for manganese(III) myoglobin, where kinetic data suggest that SO_2^- reacts in parallel, bimolecular processes with Mn(III)-Mb ($k - 1.3 \times 10^4 M^{-1} s^{-1}$) and with a protonated species Mn(III)-Mb-H$^+$ ($k = 5.2 \times 10^6 M^{-1} s^{-1}$ at 25 °C).[192] The same pattern applies to vesicle-bound cytochrome b_5, where the SO_2^- reagent is suggested to transfer an electron via the partially exposed heme edge.[193]

3.5.4. Thiosulfate Ions

The oxidation of $S_2O_3^{2-}$ by $[W(CN)_8]^{3-}$ follows the rate law Rate $= k[H^+][S_2O_3^{2-}]^2$ ($k = 0.26 M^{-2} s^{-1}$ in water at 25 °C and pH 3.9–5.0).[194] The zero-order dependence on $[W(CN)_8]^{3-}$, which contrasts with that found for $[Mo(CN)_8]^{3-}$ but agrees in form with that for $[Fe(CN)_6]^{3-}$, suggests that the rate-limiting step is the disproportionation of thiosulfate ions, $HS_2O_3^- + S_2O_3^{2-} \rightarrow HSO_3^- + SO_3^{2-} + 2S$. This view is strengthened by the numerical

Table 3.1. *Rate Constants ($M^{-1} s^{-1}$) for the Reactions of Pt(IV) Complexes with Anions*[195]

Complex	SCN⁻	CN⁻	SO_3^{2-}
$[Pt(CN)_4Br_2]^{2-}$	1.1×10^2	3.6×10^5	1.8×10^8
$[Pt(CN)_4Cl_2]^{2-}$	3.1×10^{-1}	$\sim 10^3$	4.3×10^5

agreement among the rate constants for the oxidations of $[W(CN)_8]^{3-}$ and $[Fe(CN)_6]^{3-}$, and for the acid decomposition of thiosulfate ions. The complex *trans*-$[Pt(CN)_4Cl_2]^{2-}$ is reduced by $S_2O_3^{2-}$ to $[Pt(CN)_4]^{2-}$ via equations (60) and (61) with $k_{60} = 8.5 \times 10^3 \, M^{-1} s^{-1}$ (25 °C, $\mu = 0.10 \, M$).[195] Analogous

$$[Pt(CN)_4Cl_2]^{2-} + S_2O_3^{2-} \xrightarrow{k} [Pt(CN)_4]^{2-} + ClS_2O_3^- + Cl^- \qquad (60)$$

$$ClS_2O_3^- + S_2O_3^{2-} \rightarrow S_4O_6^{2-} + Cl^- \qquad (61)$$

schemes apply when other anions are used in place of $S_2O_3^{2-}$, the second-order rate constants under the same conditions being as in Table 3.1.

Complexes between thiosulfate ions and cobalt(III) undergo internal redox decomposition, e.g., at pH 5.6 $[(NH_3)_5CoS_2O_3]^+$ yields Co^{2+} and (ultimately) $S_4O_6^{2-}$. The reaction is retarded at pH < 4 presumably because the anionic ligand is protonated.[196] The closely related complexes *cis*-$[(en)_2Co(NO_2)(S_2O_3)]$ and *cis*-$[(en)_2Co(S_2O_3)_2]^-$ are oxidized by aqueous iodine to produce respectively *cis*-$[(en)_2Co(NO_2)(S_3O_3)]$ and *trans*-$[(en)_2Co(H_2O)(S_3O_3)]^+$, containing the ligand disulfanemonosulfonate.[197]

3.5.5. Sulfides, Thiols, and Disulfides

3.5.5.1. Hydrogen Sulfide and its Anions

Manganese(IV) porphyrins produced by oxidation of $[Mn(III)P]$ (Section 3.4.1) are reasonably stable ($t_{1/2} > 2 \, h$ at pH 13). They are rereduced rapidly with hydrogen sulfide $\{2[Mn(IV)P] + H_2S \rightarrow 2[Mn(III)P] + 2H^+ + S\}$.[155] The second-order rate constant for "H_2S at pH 13" is cited as $2.8 \times 10^4 \, M^{-1} s^{-1}$, but it is not stated whether $[H_2S]$, $[HS^-]$ or $[S^{2-}]$ is the species used to define the numerical value given. In neutral and mildly alkaline solutions Np(VI) and Pu(VI) are reduced in an intramolecular reaction ($k_{Pu} = 27.4 \pm 4.1 \, s^{-1}$, $k_{Np} = 139 \pm 30 \, s^{-1}$).[198] The dinuclear palladium(I) complexes of a bridging diphosphine ligand, $[Pd_2X_2(\mu\text{-dppm})_2]$ with X = Cl, Br, and I, quantitatively liberate H_2 from H_2S[199] as in equation (62).[199]

$$[Pd_2X_2L_2] + H_2S \rightarrow [Pd_2X_2(\mu\text{-dppm})_2(\mu\text{-S})] + H_2 \qquad (62)$$

3.5.5.2. Dialkyl Sulfides

Ferrate ions in alkaline phosphate buffers oxidize R_2S compounds such as diethyl sulfide to the sulfoxide.[200] The quenching of $[*UO_2]^{2+}$ by dialkyl sulfides has been reexamined,[201] since the literature descriptions were strongly at odds. The system is characterized by a rate constant for the luminescence quenching ($k = 3.1 \times 10^8 \, M^{-1} \, s^{-1}$ for di-*tert*-butyl sulfide with 0.1 M $HClO_4$, 0.1 M $UO_2(NO_3)_2$ in 3:1 CH_3CN-H_2O), which is not matched by a large quantum yield for the photoredox reaction. This suggests that only a small proportion of the exciplexes $[*UO_2 \cdot SR_2]^{2+}$ decay by electron transfer ($\rightarrow UO_2^+ + R_2S^+$), the dominant pathway being dissociation.

3.5.5.3. Alkyl Thiols

The photoexcited state $[*Ru(bipy)_3]^{2+}$ is quenched reductively at pH values greater than the pK_a of the RSH group by certain water-soluble alkyl thiols such as 2-mercaptobenzoate ions and 2-mercaptopyridine, which have[202] respective rate constants 1.7×10^9 and $5 \times 10^7 \, M^{-1} \, s^{-1}$ for reaction (63). Other thiols, such as cysteine and 2-mercaptoethanol, did not quench

$$[*Ru(bipy)_3]^{2+} + ArS^- \rightarrow [Ru(bipy)_3]^+ + ArS^\cdot \qquad (63)$$

the excited-state emission. Thiomalic acid is oxidized to the disulfide by $[Fe(CN)_6]^{3-}$ in acidic aqueous solution. The second-order rate constant is $0.122 \, M^{-1} \, s^{-1}$ (at 25 °C in 0.10 M HCl), and the reaction shows an inverse dependence on $[H_3O^+]$. The reaction occurs by a one-electron oxidation to the doubly protonated thiyl radical anion, which was detected by ESR; this radical rapidly dimerizes.[203]

3.5.5.4. L-Cysteine and Glutathione

The reaction of $[Fe(CN)_5(NO)]^{2-}$ with cysteine (cysSH) is strongly accelerated by CN^-, and occurs by way of the intermediate $[Fe(CN)_5(NOScys)]^{3-}$ and a tetra-cyano analogue.[204] The first-order dependence on $[CN^-]$ is accounted for by the reaction of the cystine product with CN^-, which leads to partial regeneration of the intermediate, and additional consumption of the iron reagent as shown in equation (64).

$$[Fe(CN)_5(NO)^{2-}] + cysS^- \rightarrow [Fe(CN)_5(NOScys)]^{3-} \xrightarrow{cysS^-} [Fe(CN)_5(NO)]^{3-} + cysSScys \qquad (64)$$

$$[Fe(CN)_5(NO)]^{2-}$$

$$cysS^- + cysSCN \xleftarrow{\quad CN^- \quad}$$

L-Cysteine reacts with TcO_4^- via yellow and violet species that are complexes between Tc(V) and, respectively, cystine and cysteine.[205] With excess cysSH, the more stable violet $[cysSTcO_3]^-$ is formed through ligand exchange.

This is an esterification process, as postulated in the reaction between cysSH and chromate ions. Chromium(V) species of considerable stability ($t_{1/2} \sim 300$ s at 21 °C) can be generated in the reduction of Cr(VI) by glutathione (GSH) at pH 7. The concurrent disappearance of the 650 nm absorption and the EPR signal suggest they arise from the same species, and the hyperfine structure and satelites due to ^{53}Cr (9.55% abundance, $I = 3/2$) suggest that a d^1 Cr(V) species is responsible.[206] The lifetime decreases sharply at lower pH.

3.5.5.5. Disulfides

Vitamin B_{12s} [cobalt(I) cobalamin] reduces disulfides to thiols. Rate constants were obtained for eight compounds and ranged from 0.27–12 M^{-1} s^{-1} (23 °C, 0.20 M ionic strength), including the value 0.40 M^{-1} s^{-1} for cystine.[207] The reactions are accelerated by monoprotonation of a basic site on the disulfide.

3.5.6. Miscellaneous Sulfur-Containing Compounds

Thiourea and N-substituted thioureas are oxidized by [Fe(CN)$_6$]$^{3-}$ according to a rate law with first-order and second-order dependences on [TU] in 50% aq acetic acid.[208] The oxidation of dmso to the sulfone by lead dioxide is characterized at 50–70 °C by an activation energy of 58.92 kJ mol^{-1}.[209] The chemiluminescence in the ternary system S/$Cr_2O_7^{2-}$/UO_2^{2+} is characterized by "quasiharmonic oscillations." Deactivation of [*UO_2^{2+}] involves participation of Cr(V), and the chemiluminescence is attributed to the peroxosulfur radical $HS_2O_6^{\bullet}$.[210]

3.6. Compounds of Phosphorus and Arsenic, and Their Oxoanions

3.6.1. Hypophosphorous Acid

The Cr(V) oxoanion [L_2CrO]$^-$, L = 2-ethyl-2-hydroxybutyrato, oxidizes hypophosphate to phosphite. The simplicity of the stoichiometric equation, [L_2CrO]$^-$ + H_3PO_2 → [$L_2Cr(III)$]$^-$ + H_3PO_3, belies the complexity of the process, for the reaction occurs in at least two stages via a mixture of intermediates that are easily detected by their strong visible absorption.[211] The process is an autocatalytic one in which it appears that a Cr(IV) complex competes with [L_2CrO]$^-$ for oxidation of $H_2PO_2^-$ to $H_2PO_2^{\bullet}$, for which Cr(V) and Cr(IV) then compete. [The situation is analogous to the oxidation of HSO_3^- by Cr(V); see Section 3.5.2.] The Bi(V) species present in a medium of 1.0 M HClO$_4$ and 1.5 M HF oxidizes H_3PO_2. The kinetic effects of Bi(III) are attributed to association between Bi(V) and Bi(III).[212]

3.6.2. Peroxodiphosphate Ions

Ag(I) catalyzes oxidations by peroxodiphosphate, including those of hydroxylamine,[213] phosphorous acid[214], and hydrazine.[215]

3.7. Inorganic Radicals

Kinetic data for the reactions of inorganic radicals with transition-metal complexes[216-228] are summarized in Table 3.2. The reactions of O_2^- and H_2O_2 with colloidal MnO_2 in alkaline solutions are autocatalytic owing to the buildup of Mn^{3+} on the surface of the colloid.[229] Cationic ruthenium(II) complexes containing a cyclic dienyl ligand react with O_2^- at the terminal position of the dienyl to form a coordinated dienone.[230] The complex $[Mo^VO(TPP)X]$ (X = Cl, Br, NCS) reacts with O_2^- in aprotic solvents to yield $[Mo^{IV}O(TPP)]$ via $[Mo^VO(TPP)(O_2^{2-})]^-$.[231] The latter was isolated as a solid and characterized by elemental analysis and spectroscopically at low temperatures. The reaction of O_2^- with $[Co(II)(tptp)]$ yields the superoxo-complex $[Co(tptp)(O_2)]^-$, which undergoes intramolecular electron transfer to give $[Co(tptp)]^+$ and HO_2^-.[232] The reductions of Mn(III) and Fe(III) porphyrins by O_2^- and reducing C-centered radicals was studied by pulse radiolysis.[233,234]

3.8. Hydrogen

Solutions of Cr(II) chloride in aqueous HCl evolve H_2 according to equation (65) in the presence of trace concentrations of the cobalt(II) macrocyle $[(H_2O)_2Co(dmgBF_2)_2]$.[235] The kinetics are described by the Michaelis–Menten scheme. The pre-steady-state phase produces a dark blue intermediate, $[(H_2O)_5CrClCo(dmgBF_2)_2]^+$, which dissociates to the hydrogen forming $[HCo(dmgBF_2)_2]$.

$$Cr^{2+} + Cl^- + H^+ \rightarrow CrCl^{2+} + \tfrac{1}{2}H_2 \qquad (65)$$

The reduction of $[Co(dmgH)_2]$ and $[Co(dpnH)]^+$ by H_2 in $H_2O/2$-PrOH (1:1, v/v) produces the respective hydridocobalt complexes[236] according to rate-law equation (66). The kinetics were determined in the presence of benzil

$$d[CoH]/dt = k_1[H_2][Co]^2 \qquad (66)$$

which rapidly removes the hydridocobalt and eliminates complicated side reactions. The activation parameters in unbuffered solutions have the following values: $[Co(dpnH)]^+$, $\Delta H^{\ddagger} = 46.7 \text{ kJ mol}^{-1}$; $\Delta S^{\ddagger} = -78.6 \text{ J mol}^{-1} \text{ K}^{-1}$; and $[Co(dmgH)_2]$, $\Delta H^{\ddagger} = 48.4 \text{ kJ mol}^{-1}$, $\Delta S^{\ddagger} = -57.4 \text{ J mol}^{-1} \text{ K}^{-1}$. The irradiation of $[Co(bpy)(PEt_2Ph)_2H_2]^+$ with 436 nm light produces H_2 with a quantum yield of 0.1.[237] The acylcobalt tetracarbonyl complexes $[n$-$C_3H_7C(O)Co(CO)_4]$ and $[i$-$C_3H_7C(O)Co(CO)_4]$ react with H_2 to yield the

Table 3.2. *Kinetics of the Reactions of Inorganic Radicals with Metal Complexes in Aqueous Solution at 25 °C*

Radical	Metal complex	pH	Reaction type/product[a]	$\log k / M^{-1} s^{-1}$	Ref.
O_2^-	$[Ru(bipy)_3]^{3+}$	≥7	$[Ru(bipy)_3]^{2+} + {}^1O_2$	10.2	216
	Fe^{2+}	1–7	O	7.0	217
	$FeSO_4^+$	0–1.7	R	8.18	217
	$FeOH^{2+}$	6.8	R	8.18	217
	Ceruloplasmin	7.8	$O_2 + H_2O_2$	7.0	218
	Superoxide Dismutase	7.8	$O_2 + H_2O_2$	9.32	218
$HO_2^{\cdot}$	Fe^{2+}	1–7	O	6.08	217
	$FeSO_4^+$	0–1.7	R	≤3	217
$^{\cdot}OH$	$[Pt(NH_3)_4]^{2+}$	5.6	A	9.82	219
	$[Pt(NH_3)_4(OH)_2]^{2+}$	5.6	—	≤8.48	219
	$[Ni(13\text{-}aneN_4)]^{2+}$	3.3	O	9.54	220
	$[Ni(15\text{-}aneN_4)]^{2+}$	3.3	O	9.60	220
	$[CoL(OH_2)_2]^{3+\,b}$	4	L	9.08	221
	$[CuL]^{2+\,b}$	4	L	9.54	221
	$[NiL]^{2+\,b}$	4	L	9.65	221
	$[CoL]^{2+\,b}$	4	L	10.0	221
	$[Fe(phen)_3]^{3+}$	1	L	9.84	222
	$[Ru(bipy)_3]^{2+}$	7	L	9.83	228
$^{\cdot}OH/O^-$	$[Fe(OH)_4]^-$	13.3	O	7.93	223
$H^{\cdot}$	$[Fe(phen)_3]^{3+}$	1	R or L	9.60	222
	$[Pt(NH_3)_4]^{2+}$	2	A	10.4	219
Cl_2^-	$[Pt(NH_3)_4]^{2+}$	0.3	O or A	9.95	219
	Bi(III)	0	O	9.95	224
Br_2^-	$[Mn(IV)TSPP]$	14	O	9.04	225
		12.9	L	8.78	225
		11.1	L	8.89	225
	$[Mn(IV)(TCPP)]$	14	O	9.26	225
		12.9	L	9.20	225
	$[Mn(IV)(TMPyP)]$	14	L	8.65	225
	$[Fe(II)(TrpH)]$	>8	L	8.84	226
	$[Fe(II)(HisH)]$	7–10	O	8.26	226
	$[Fe(II)(HisH)]_2$	7–10	O	8.93	226
CO_3^-	$[PuO_2(CO_3)(OH)_2]^{2-}$	12.5	A	7.16	227

[a] Reactions take place by oxidation (O) or reduction (R) of the metal complex, formation of the ligand centered radical (L), or addition to the metal center (A).
[b] $L = Me_2\text{-}3,4,5\text{-pyo-}[14]trieneN_4$.

respective aldehydes.[238] The proposed mechanism, given in equations (67)–(70), is consistent with the inhibiting effect of CO and direct observation of $[HCo(CO)_4]$ at high $[H_2]$.

$$[RC(O)Co(CO)_4] \rightleftharpoons [RC(O)Co(CO)_3] + CO \tag{67}$$

$$[RC(O)Co(CO)_3] + H_2 \rightleftharpoons [RC(O)CoH_2(CO)_3] \tag{68}$$

$$[RC(O)CoH_2(CO)_3] \xrightarrow{CO} RCHO + [HCo(CO)_4] \tag{69}$$

$$[RC(O)Co(CO)_3] + [HCo(CO)_4] \xrightarrow{CO} RCHO + [Co_2(CO)_8] \tag{70}$$

$NaBH_4$ is reported to reduce $Cu(II)$ to $Cu(I)^{(239)}$ according to equation (71). The kinetics are first order in $[BH_4^-]$ and $[Cu^{2+}]$. The rate constant

$$BH_4^- + 8\,Cu^{2+} + 3\,H_2O \rightarrow B(OH)_3 + 8\,Cu^+ + 7\,H^+ \tag{71}$$

increases by a factor of 5.6 as the pH changes from 12.5 to 8. This was interpreted as a first-order dependence on $[H^+]$, and $H^+[BH_4^-]$ is the proposed reactive species. The kinetics of reduction of $Ni(II)^{(240)}$ and $Cr(VI)^{(241)}$ by $NaBH_4$ were also reported.

3.9. Alkyl and Aryl Halides

$[Cr(H_2O)_6]^{2+}$ reacts with iodoacetamide$^{(242)}$ to give $[(H_2O)_5CrI]^{2+}$ (50%), $[(H_2O)_5CrO(NH_2)CCH_3]^{3+}$ (18%), and $[(H_2O)_5CrCH_2CO(NH_2)]^{2+}$ (30%). The reaction is first order in each reactant and independent of $[H^+]$ with $k = 0.56\ M^{-1}\,s^{-1}$ (25 °C, 1.0 M ionic strength), $\Delta H^\ddagger = 6.7$ kcal mol^{-1}, $\Delta S^\ddagger = -37.2$ cal mol^{-1} K^{-1}. The mechanism shown in equations (72) and (73) is proposed. Similarly both $[CrO_2CCH_3]^{2+}$ and $[CrCH_2COOH]^{2+}$ are formed in the reaction of Cr^{2+} with ICH_2COOH.$^{(243)}$ The organochromium complexes, $[CrCH_2COOH]^{2+}$ and $[CrCH_2C(O)NH_2]^{2+}$, polymerize on Dowex ion exchange resins.

$$Cr^{2+} + ICH_2CO(NH_2) \rightarrow [CrI]^{2+} + {}^\bullet CH_2CO(NH_2) \tag{72}$$

$${}^\bullet CH_2CO(NH_2) + Cr^{2+} + H^+ \rightarrow [CrOC(NH_2)CH_3]^{3+}\ (36\%)$$
$$+ [Cr(CH_2C(O)NH_2]^{2+}\ (60\%) \tag{73}$$

The cationic nickel(I) macrocycles, $[R,R,S,S\text{-}Ni(tmc)]^+$ and $[R,S,R,S\text{-}Ni(tmc)]^+$, react in aqueous alkaline solutions on the stopped-flow time scale with alkyl halides to form organonickel complexes.$^{(116,117)}$ The trend in the rate constants (benzyl > allyl ≫ secondary > primary > methyl > cyclopropyl, and RI > RBr ≫ RCl) and other evidence presented are consistent with the mechanism given in equations (74) and (75). The alkyl nickel complexes $R,R,S,S\text{-}[RNi(tmc)]^+$, formed in reaction (75), are further oxidized by

$$[Ni(tmc)]^+ + RX \rightarrow [XNi(tmc)]^{2+} + R^\bullet \tag{74}$$

$$[Ni(tmc)]^+ + R^\bullet \rightarrow [RNi(tmc)]^+ \tag{75}$$

alkyl halides R'X to yield combination and disproportionation products of R and R'.$^{(244)}$ The rate constants ($M^{-1}\,s^{-1}$) for the reaction of $[C_2H_5Ni(tmc)]^+$ are 38.1 (CH_3I), 63.1 (C_2H_5I), 4680 (2-C_3H_7I), 18 (C_2H_5Br), and 498 (2-C_3H_7Br). The proposed one-electron reduction of R'X in the rate-determining step produces an organonickel(III) complex and R'X$^-$. Both products rapidly eliminate radicals, whose self-reactions yield the final products. (π-Allyl) nickel complexes react with organic halides RX[245] to give the coupled products $RCH_2CH=CH_2$. In the proposed chain reaction Ni(I) intermediates are the chain-carrying species and [(allyl)Ni(III)(R)(Br)] the immediate precursor of the coupled product. The involvement of free carbon-centered radicals is ruled out and the formation of cyclic/linear products from linear/cyclic alkyl halides is explained by rearrangements of the groups coordinated to Ni(III).

Oxidative addition of alkyl halides RX to a four-coordinate Rh(I) macrocycle, $[RhC_2(DO)(DOBF_2)pn]$, to yield six-coordinate Rh(III) alkyl products occurs by a nucleophilic attack by Rh(I) for all but the most hindered alkyl iodides.[246] The latter react by electron transfer. Dihalides[247] yield mononuclear and binulcear alkylrhodium(III) products. *cis*- and *trans*-Dibromocyclohexanes yield cyclohexene, with the *trans* isomer reacting about 10^5 times more rapidly than the *cis* isomer. The electrochemical reduction of $[(TPP)RhL_2]^+$ (L = dimethylamine) in CH_2Cl_2 is followed by a chemical reaction with the solvent to give $[(TPP)Rh(CH_2Cl)]$,[248] which was characterized by IR, UV-visible and ^1H NMR spectroscopy and cyclic voltammetry.

The radicals $[Re(CO_4)L]^·$ (L = CO, P(OR)$_3$, PR$_3$, AsEt$_3$) radicals, generated by laser flash photolysis of $[L(CO)_4Re]_2$, react with CH_2Br_2, $CHCl_3$, and CCl_4 by halogen atom transfer[249] to produce *cis*-$[Re(CO)_4LX]$ (X = Cl, Br). The rate constants fit a two-parameter free-energy relationship which can account for both electronic and steric effects. Laser flash photolysis of a series of bridged dirhenium octacarbonyls yields organometallic biradicals $^·(OC)_4ReL-LRe(CO)^·_4$ [L-L = $Ph_2P(CH_2)_2PPh_2$, $Ph_2P(CH_2)_3PPh_2$, etc.]. Halogen atom abstraction from halomethanes by the biradicals ($k_{RX} = 10^7-10^9\ M^{-1}s^{-1}$) competes with intramolecular radical recombination ($k = 10^6-10^8\ s^{-1}$).[250]

The triplet $[^*Pt_2(P_2O_5H_2)_5]^{4-}$ is oxidatively quenched[251] by $CCl_4(k = 2 \times 10^9\ M^{-1}s^{-1})$ and $CHCl_3(k = 6 \times 10^7\ M^{-1}s^{-1})$ in CH_3OH. The initial step in the reaction of $[^*Pt_2(\mu-P_2O_5H_2)_4]^{4-}$ with C_6F_5Br is bromine atom abstraction.[252] The data for C_6H_5Br[253] and $BrCH_2CH_2Br$[252] are consistent with both an $S_{RN}1$ mechanism and bromine atom abstraction. The reaction of *trans*-$[PdAr_2L_2]$ (Ar = *m*-tolyl, L = PEt_2Ph) with CH_3I in benzene yields *m*-xylene (70%) and 3,3'-bitolyl (30%),[254] as in equations (76) and (77). The

$$[PdAr_2L_2] + MeI \rightarrow ArMe + [PdArL_2I] \tag{76}$$

$$[PdAr_2L_2] + MeI \rightarrow ArAr + [PdMeL_2I] \tag{77}$$

results of kinetic and deuterium labeling experiments indicate autocatalytic formation of *m*-xylene in a bimolecular reaction between *trans*-$[PdAr_2L_2]$ and the product $[PdMeL_2I]$. The reaction of Fe(II) deuteroporphyrin IX with CH_3I yields the Fe(II) complex $[(P)Fe(CH_3I)]$, and not the organoiron(III), $[(P)FeCH_3]$.[255]

The ruthenium complex $[Ru(NO)_2(PPh_3)_2]$ reacts with benzyl bromide to give the transient nitrosobenzyl intermediate.[256] Rapid tautomerization yields the oxime species $[RuBr(NO)\{N(OH) = CHPh\}(PPh_3)_2]$, which quickly dehydrates to the nitrile $[RuBr(NO)(NCPh)(PPh_3)_2]$. Subsequent reactions yield several products. Ruthenium and osmium porphyrin complexes $K_2[Ru(TTP)]$, $K_2[Ru(OEP)]$, and $K_2[Os(TTP)]$ react with CH_3I in THF to form the respective *trans*-dimethyl complexes.[257] Similarly, the reaction of $K_2[Ru(TTP)]$ with CH_3CH_2I yields $[(CH_3CH_2)_2Ru(TTP)]$, which slowly decomposes to the ethylidene complex $[(CH_3CH)Ru(TTP)]$ and finally to $[(CH_2 = CH_2)Ru(TTP)]$. The latter is also produced in the reactions of $K_2[Ru(TTP)]$ with $BrCH_2CH_2Br$, CH_2Cl_2, CH_2I_2, and CH_2N_2.

Heptamethyl cob(I)yrinate reacts with methyl *p*-toluenesulfonate and methyl iodide to yield β-and α-methylcobalt(III) products, respectively.[258] The formation of different isomers in the two reactions was explained by the change in mechanism from S_N2 substitution (sulfonate) to a radical mechanism (CH_3I). Photohomolysis of *cis*-$[R_2Co(bipy)_2]^+$ (R = Me, Et, $PhCH_2$) yields alkyl radicals and $[RCo(bipy)_2]^+$. The latter yields coupled products in the reactions with benzyl and allyl bromides.[259] The metal–metal bonded dimer $[CpCr(CO)_3]_2$ reacts with CH_3I to give $[CpCr(CO)_3I]$ and $[CpCr(CO)_3CH_3]$. The reaction appears to involve the monomeric $[CpCr(CO)_3]^·$.[260] The sterically hindered analogue, $[CpCr(CO)_2PPh_3]^·$, was prepared and its monomeric nature confirmed by X-ray crystallography. Primary alkyl halides are believed to react with $NaSn(CH_3)_3$ by electron transfer and to involve free radicals.[261] The reactions of alkyl halides with two highly reduced iron porphyrins, formally Fe(I) and Fe(O) complexes, yield the same Fe(II)R product[262]; see equations (78) and (79). No ring alkylation takes place.

$$Fe(I) + RX \rightarrow Fe(III)R + X^- \tag{78a}$$

$$Fe(III)R + Fe(I) \rightarrow Fe(II)R + Fe(II) \tag{78b}$$

$$Fe(O) + RX \rightarrow Fe(II)R + X^- \tag{79}$$

3.10. Quinols, Catechols, Quinones, Alcohols, and Diols

3.10.1. Aliphatic

Oxoruthenium(IV) complexes bearing a phosphine ligand, e.g., *cis*-$[(bipy)_2Ru^{IV}O(PR_3)]^{2+}$, oxidize a variety of substrates: 2-propyl alcohol to acetone, propionaldehyde to propionic acid, and benzyl alcohol to benzaldehyde. The rate constants in aqueous solution for the oxidation of $PhCH_2OH$ and concomitant formation of $[(bipy)_2Ru(OH_2)(PR_3)]^{2+}$ are sensitive to the phosphine ligand; values (*at unspecified temperature*) are $k = 1.9 \times 10^{-2}$ and 1.1×10^{-3}, respectively, for PPh_3 and PEt_3.[263] Oxometallates such as $[PW_{12}O_{40}]^{3-}$ photooxidize organic substrates, including primary and secondary alcohols.[264] Of several mechanisms considered for the oxidation of alcohols by cobalt(III) Schiff base complexes, β-elimination as in equation (80) is

$$[LCo(III)OH] + RCH_2OH \rightarrow [LCo(III)(OCH_2R)] \rightarrow [LCo(III)H] + RCHO \tag{80}$$

preferred.[265] The oxidation of alcohols (e.g., ethanol) by O_2 in the presence of KOH is catalyzed by one of a series of copper(II) Schiff base complexes. Many others are ineffective catalysts, however, presumably because they cannot adopt distorted coordination geometries about the metal.[266] The rates of ethanol oxidation by vanadium(V) in sulfuric acid were determined,[267] as were the rates of vanadium(V) oxidation of 20 heterocyclic alcohols.[268] Secondary alcohols are oxidized to ketones by coordinated peroxides such as the anionic molybdenum(VI) peroxo complex $[OMo(O_2)pic]^-$.[269] The irradiation of iron(III) protoporphyrin IX in deoxygenated ethanol/water solutions containing pyrazine leads to the formation of $[(pyz)(PPIX)Fe^{II}(pyz)]$. The homolytic cleavage of the $Fe^{III}—OC_2H_5$ bond in the proposed reactive species, $[(C_2H_5O)PFe^{III}(OH)]$, is followed by displacement of the coordinated OC_2H_5 radical by pyrazine.[270]

3.10.2. Phenols

A reinvestigation showed that the rate of electron transfer from PhO^- to $[Fe(CN)_6]^{3-}$ has a second-order dependence on $[Fe(CN)_6^{3-}]$ as well as complex inhibitory dependences on $[Fe(CN)_6]^{3-}$ and $[Fe(CN)_6^{4-}]$. The data are interpreted in terms of rate-limiting electron transfer to $[Fe(CN)_6]^{3-}$ from a phenoxy radical produced in a preequilibrium step[271] as in equations (81) and (82).

$$PhO^- + [Fe(CN)_6]^{3-} \rightleftharpoons PhO^• + [Fe(CN)_6]^{4-} \tag{81}$$

$$PhO^• + [Fe(CN)_6]^{3-} \rightleftharpoons [PhO\cdots Fe(CN)_6]^{3-} \rightarrow PhO^+ + [Fe(CN)_6]^{4-} \tag{82}$$

A reexamination[272] of the equilibrium between $Na_2[Fe(CN)_5NH_3]$ and *para*-aminophenol confirmed the 1:1 formula originally attributed[273] to the formation of the species $[Fe(CN)_5(H_2N\cdot C_6H_4O)]^{3-}$. The EPR spectrum shows that this blue complex contains the *para*-aminophenoxyl radical, indicating that the initial step is a one-electron oxidation of the phenol. The iron Mössbauer spectrum also supports an $[Fe^{II}—NH_2PhO\cdot]$ electronic structure.[272] Nitrophenols are oxidized by the bis(tellurato)cuprate(III) anion according to a rate law that is consistent with phenol complexation by copper as one of the kinetic steps.[274] In 96% sulfuric acid Mo(VI) is reduced by phenol to Mo(V), while a 1:1 complex between Mo(VI) and phenol forms at lower acidity.[275] Phenols are oxidized in benzene solvent by $[Co^{III}(TPP)Cl]$ and even more readily by the π-cation radical complex $[Co^{III}(TPP^+)Cl_2]$[276]; see equation (83). To

$$[Co^{III}(TPP^+)Cl_2] + ArOH \rightarrow [Co^{III}(TPP)Cl] + ArO^• + H^+ + Cl^- \tag{83}$$

obtain a consistent kinetic treatment the author assumed that the phenolate anion rather than the phenol reacts, *and that* $[ArO^-]$ *is constant*. This treatment seems questionable, although it affords "rate constants" that vary linearly with the reduction potentials of the $ArO^•/ArO^-$ couples.

3.10.3. Diols

The preferred mechanism for the oxidative cleavage of glycols by cobalt(III) acetate in acetic acid is suggested to be diol chelation to the metal, because in the oxidation of five-membered cyclic diols the *cis*-diols were generally oxidized faster than the *trans*-isomers.[277] Experimental data (stoichiometry, kinetics, products) for the oxidation of 1,4-C_4-diols by cerium(IV) are consistent with theoretical calculations of electron density in the butane, butene, and butyne diols.[278]

3.10.4. Hydroquinols, Catechols, and Quinones

Quinones oxidize dialkylcobalt(III) complexes, *cis*-[$R_2Co(bipy)_2$](ClO_4) (R = CH_3, C_2H_5), to give R_2 exclusively. The reaction with *p*-benzoquinone is catalyzed by $Mg(ClO_4)_2$ and $HClO_4$.[279]

3.11. Ascorbic Acid

3.11.1. Iron(III) Complexes

Solutions resulting from the oxidation of ascorbic acid (H_2A) by $FeCl_3$ in water, methanol, and in mixtures of the two have been examined at different pH values. The formation of Fe(II) is evident as are intensely-absorbing intermediates that have been examined by Mössbauer spectroscopy and stopped-flow spectrophotometry.[280] The kinetics of reduction of iron(III)–picolinate complexes by H_2A show that [$Fe(pic)_2OH$] and HA^- react readily ($k = 6.4 \times 10^3$ M^{-1} s^{-1} at 25 °C and $\mu = 0.10$ M), but that [$Fe(pic)_3$] is unreactive.[281] The trinuclear acetate complex [$Fe_3(OH)_2(OAc)_6$]$^+$ reacts with HA^- via reaction intermediates.[282] The mononuclear porphyrin [$Fe^{III}(TMpyP)$-(OH)]$^{4+}$ reacts with HA^- in a biphasic process. The first step is binding of $2HA^-$; the second, proceeding at a rate independent of [HA^-] ($k = 4 \times 10^{-3}$ s^{-1} at pH 8.0; 1.3×10^{-2} s^{-1} at pH 7.0), may correspond to slow intramolecular electron transfer, although this is not unequivocal.[70] Several modified metmyoglobin complexes have been examined; the values of log k_{HA^-} correlate with the pK_a for the acid dissociation of the porphyrin monocation, suggesting the predominance of electronic factors.[283]

3.11.2. Other Oxidants

The μ-oxo complex [{(H_2O)$_5$Cr}$_2$(μ-O)]$^{4+}$ is reduced by H_2A and HA^- ($k = 6$ and 7.2×10^5 M^{-1} s^{-1}, respectively, at 25 °C and $\mu = 1.0$ M). A Marcus-theory correlation gives a self-exchange rate constant for [{(H_2O)$_5$Cr}$_2$O]$^{4+/3+}$ of about 10 M^{-1} s^{-1}, some 10^6 times larger than for [$Cr(H_2O)_6$]$^{3+/2+}$, attributed to the π^* character of the μ-oxo redox orbital.[284] The reaction of

[IrBr$_5$(OH$_2$)]$^-$ with H$_2$A and HA$^-$ also greatly favors the latter ($k = 1.1 \times 10^4$ and 1.4×10^8 M^{-1}s^{-1}, respectively, at 20 °C and $\mu = 1.0\ M$). The results are correlated with those for other Ir(IV) reactions, and with the potentials and self-exchange rates of the species involved.[285] The copper(II) macrocyclic complex [Cu(taab)]$^{2+}$ is reduced by HA$^-$ ($k = 7.1\ M^{-1}$s^{-1} at 25.0 °C, $\mu = 0.1\ M$).[286] The oxidation of H$_2$A by TcO$_4^-$ in strongly acidic solution leads to an intense red coloration, presumably a complex of a reduced technetium and H$_2$A or A. Addition of Sn(II) prevents the formation of this species.[287]

3.12. Organic Radicals

The kinetic data for the reactions of organic radicals with metal complexes[61,80,222,224,244,288-307] are summarized in Table 3.3. Trichloromethyl-peroxyl radicals, CCl$_3$OO$^•$, react with [ZnTPP] to form the porphyrin radical

Table 3.3. *Kinetic Data for the Reactions of Organic Radicals with Transition-Metal Complexes at 25 °C*

Radical	Metal complex^a	Log$_k$ (M^{-1}s^{-1})	Reaction type	Ref.
$^•$CH$_3$	Cu$^+$	9.54	M–R	289
	[Fe(phen)$_3$]$^{3+}$	8.48	R	222
	[Cu(nta)]$^-$	7.54	M–R	291
$^•$C$_2$H$_5$	V^{2+}	5.8	O	244
	[Fe(phen)$_3$]$^{3+}$	9.0	R	222
$^•$CH$_2$OH	[Cr(15-aneN$_4$)]$^{2+}$	8.08	M–R	290
	[Cr(nta)]$^-$	8.34	M–R	290
	[Cu(nta)]$^-$	8.28	M–R	291
	[Fe(phen)$_3$]$^{3+}$	9.70	R	222
	[Ni(IV)L]$^{2+\ b}$	9.55	R	297
$^•$CH(CH$_3$)OH	[Cr(nta)]$^-$	8.0	M–R	290
	[Cr(edta)]$^{2-}$	7.59	M–R	290
	[Cu(nta)]$^-$	7.79	M–R	291
	[Fe(phen)$_3$]$^{3+}$	9.70	R	222
	[Ni(IV)L]$^{2+\ b}$	9.55	R	297
$^•$CH$_2$CH$_2$OH	[Fe(phen)$_3$]$^{3+}$	8.18	R	222
$^•$C(CH$_3$)$_2$OH	[Cr(15-aneN$_4$)]$^{2+}$	7.69	M–R	290
	[Cr(nta)$^-$]	7.92	M–R	290
	[Cr(edta)]$^{2-}$	7.41	M–R	290
	[Cu(nta)]$^-$	7.34	M–R	291
	[Fe(phen)$_3$]$^{3+}$	9.61	R	222
	[UO$_2$]$^{2+}$	7.61	R	75
	MnO$_2$(colloidal)	6.90	R	292
	[PhCH$_2$Co(dmgH)$_2$]	6.70	R–R	293
	[CH$_3$Co(dmgH)$_2$]	5.84	R–R	293

continued

Table 3.3. (continued)

Radical	Metal complex[a]	Log_k $(M^{-1}\mathrm{s}^{-1})$	Reaction type	Ref.
	$[CH_3CH_2Co(dmgH)_2]$	5.78	R–R	293
	$[(CH_3)_2CHCo(dmgH)_2]$	5.60	R–R	293
	$[BrCo(dmgH)_2]$	8.18	R	288
	$[ClCo(dmgH)_2]$	7.95	R	288
	$[Cr(C_5H_5N)]^{3+}$	<4.70	L	294
	$[Cr(4\text{-}CN\text{-}C_5H_4N)]^{3+}$	8.72	L	294
	$[Cr(3\text{-}CN\text{-}C_5H_4N)]^{3+}$	8.26	L	294
	$[Cr(3\text{-}Cl\text{-}C_5H_4N)]^{3+}$	6.67	L	294
	$[Cr(4\text{-}Cl\text{-}C_5H_4N)]^{3+}$	5.71	L	294
	$[cis\text{-}Cr(C_5H_5N)_2]^{3+}$	~5	L	294
	$[(NH_3)_5Co(C_5H_5N)]^{3+}$	7.08	R	295
	$[(NH_3)_5Co(3\text{-}CN\text{-}C_5H_4N)]^{3+}$	8.15	R	295
	$[(NH_3)_5Co(3\text{-}Cl\text{-}C_5H_4N)]^{3+}$	7.49	R	295
	$[(NH_3)_5Co(4\text{-}CH_3\text{-}C_5H_4N)]^{3+}$	6.97	R	295
	$[(NH_3)_5Co(4\text{-}C(CH_3)_3\text{-}C_5H_4N)]^{3+}$	6.94	R	295
	$[(NH_3)_5Co(4\text{-}N(CH_3)_2\text{-}C_5H_4N)]^{3+}$	6.53	R	295
	MP^c	9.3	L	296
	Bi(III)	5.23	R	224
	$[Ni(IV)L]^{2+\ b}$	9.42	R	297
	$[(NH_3)_5CoYXPhNO_2]^d$	9.23–9.60	L	298
	$[Ru(bpz)_3]^{2+}$	9.54	R	299
$^\bullet CH(OH)CH_2CH_3$	$[Fe(phen)_3]^{3+}$	9.50	R	222
$^\bullet CH_2C(CH_3)_2OH$	$[Fe(phen)_3]^{3+}$	7.18	R	222
$^\bullet CH(OH)_2$	Cr^{2+}	8.11	M–R	300
$^\bullet CH(CH_3)OC_2H_5$	$[PhCH_2Co(dmgH)_2]$	7.0	R–R	293
	$[CH_3Co(dmgH)_2]$	6.15	R–R	293
	$[CH_3CH_2Co(dmgH)_2]$	5.90	R–R	293
	$[(CH_3)_2CHCo(dmgH)_2]$	5.84	R–R	293
$^\bullet CH_2CHO$	$[Fe(phen)_3]^{3+}$	<6	—	222
$^\bullet CH_2COOH$	$[Fe(phen)_3]^{3+}$	6.08	R	222
$^\bullet CH_2C(CH_3)_2COOH$	$[Fe(phen)_3]^{3+}$	7.86	R	222
$\{^\bullet C(CH_3)_2COOH +$ $^\bullet CH_2CH(CH_3)COOH\}$	$[Fe(phen)_3]^{3+}$	7.80	R	222
$\{^\bullet CH(CH_3)Br +$ $^\bullet CH_2CH_2Br\}$	$[Fe(phen)_3]^{3+}$	~9	R	222
$C_6H_6{}^\bullet H$ or $C_6H_6{}^\bullet OH^e$	$[Fe(phen_3]^{3+}$	9.38	R	222
MV^+	MnO_2 (colloid)	7.40	R	292
	$[Ru(bpz)_3]^{2+}$	7.23^f	R	299
$^\bullet CH_2O^-$	$M(P)^c$	8.9–9.9	L	296
$C(CH_3)_2O^-$	$[Ni(IV)(dmg)_3]^{2-}$	9.24	R	297
CO_2^-	$[Ni(IV)(dmg)_3]^{2-}$	9.24	R	297
	$[Ni(13\text{-}aneN_4)]^{2+}$	9.23	R	61
	$[Ni(14\text{-}aneN_4)]^{2+}$	9.72	R	80
	$R,R,S,S\text{-}[Ni(tmc)]^{2+}$	9.18	R	80
	$[Ni(Me_6\text{-}14\text{-}ane)]^{2+}$	9.76	R	80

continued

Table 3.3. (continued)

Radical	Metal complex[a]	Log_k $(M^{-1}\,s^{-1})$	Reaction type	Ref.
	$[\mathrm{Ni(dmc)}]^{2+}$	6.60	R	80
	$[\mathrm{Fe(III)TPPS}]^{3-}$	9.26	R	301
	$[(\mathrm{Fe(III)TPPS})_2\mathrm{O}]^{8-}$	9.3	R	301
	$[\mathrm{Fe(III)TMPyP(OH)}]^{4+\,g}$	10.1	R, L	301
	$[\mathrm{Ru(bpz)_3}]^{2+}$	10.1	R	299
	$[(\mathrm{NH_3})_5\mathrm{CoYXPhNO_2}]^{d}$	9.1–9.6	L	298
$\mathrm{CH_3CN}^-$	$[\mathrm{VCl_3}(\gamma\text{-picolin})]_3$	11.0	R	302
	$\mathrm{Ta_2Cl_6(4\text{-}CH_3\text{-}C_5H_4N)_4}$	11.1	R	303
$\mathrm{C_6H_5NO_2^-}$	$\mathrm{Pt(colloid)}$	3.43	$(\mathrm{Pt})(\mathrm{C_6H_5NO_2^-})$	304
$1,4\text{-}(\mathrm{NO_2})_2\mathrm{C_6H_4^-}$	$[\mathrm{Mo(NCS)_6}]^{2-}$	8.90	R	305
$^\cdot\mathrm{CH_2(OH)OO}$	$\mathrm{PFe(III)}^{h}$	7.0	L	306
$(\mathrm{CH_3})_2\mathrm{C(OH)OO}$	$\mathrm{PFe(III)}^{h}$	7.78^{i}	A	306
$\mathrm{CH_3CH_2S^\cdot}$	$\mathrm{Cu(I)(CN)_{2\,or\,3}(CH_3CH_2S)}$	9.52(pH 9) 9.32(pH 7)	O	307
$\mathrm{RSSR}^{-\,j}$	$\mathrm{Cu(II)(RS)_2}$	≥ 8.81	R	307

[a] Coordinated waters are not shown. Reactions take place by oxidation (O) or reduction (R) of the metal, formation of the metal–carbon (M–R) or carbon–carbon (R–R) bond, formation of a ligand-centered radical (L) or an adduct (A).
[b] L = 3,14-dimethyl-4,7,10,13-tetraazahexadeca-3,13-diene-2,15-dione dioximate.
[c] Metal porphyrin complexes: M = Ga(III), Ge(IV), Bi(III), Pb(II), and Sb(V); P = tetra(3-pyridyl)porphyrin, and TMPyP.
[d] X = $\mathrm{CH_2}$, CH=CH, $(\mathrm{CH_2})_3$, $\mathrm{CONHCH_2}$, $(\mathrm{CONHCH_2})_2$; Y = $\mathrm{O_2C}$, CN, $\mathrm{NHSO_2}$, $\mathrm{OSO_2}$, O.
[e] Cyclohexadienyl radicals formed by addition of H$^\cdot$ and OH$^\cdot$ to benzene.
[f] Reverse reaction has log k = 8.11.
[g] At pH 8; the rate constant shows 0.5 order dependence on [H$^+$].
[h] P = Deuterioporphyrin dimethyl ester.
[i] Reverse reaction has $k \simeq 2 \times 10^8\,s^{-1}$.
[j] RSH = penicillamine.

cation,[309,309] as in equation (84). The rate constants $(10^{-7}\,k/M^{-1}\,s^{-1})$ were determined in several solvents: $\mathrm{CCl_4}$ (170), $c\text{-}\mathrm{C_6H_{12}}$ (160), $c\text{-}\mathrm{C_6H_{10}}$ (300),

$$\mathrm{CCl_3O_2^\cdot} + [\mathrm{ZnTPP}] \rightarrow \mathrm{CCl_3O_2^-} + [\mathrm{ZnTPP}]^{\ddagger} \qquad (84)$$

$\mathrm{CH_3CN}$ (120), $(\mathrm{CH_3})_2\mathrm{CO}$ (40), $(\mathrm{CH_3})_2\mathrm{CHOH}$ (26), $(\mathrm{CH_3})_2\mathrm{SO}$ (7.7), and $\mathrm{C_5H_5N}$ (3.5). Carbon-centered radicals $(\mathrm{C_6H_5^\cdot},\ (\mathrm{CH_3})_2\mathrm{C{=}CH^\cdot},\ c\text{-}\mathrm{C_3H_5^\cdot},\ (\mathrm{CH_3})_3\mathrm{C^\cdot},$ etc.) react with tri-n-butylstannane and tri-n-butylgermane in aprotic solvents[310] by hydrogen atom abstraction; see equation (85). The rate constants decrease with a decrease in the R—H bond strength and, for any given

$$\mathrm{R^\cdot} + \mathrm{Bu_3MH} \rightarrow \mathrm{RH} + \mathrm{Bu_3M^\cdot} \qquad (85)$$

radical, $k_{\mathrm{Sn}} > k_{\mathrm{Ge}}$. A cobalt cage complex containing two nitro groups is reduced by viologen radicals to a complex with one nitro and one hydroxyamino group.[28] Viologen radicals, derived from diquaternary salts of 4,4'-bipy and 2,2'-bipy, reduced a number of Co(III) complexes.[311] The rate constants fall in the range 1.5×10^2 to $6 \times 10^8\ M^{-1}\,s^{-1}$. One-electron reduction of bis(viologen) species, X^{4+}, yields the radical cations X^{3+} and X^{2+}. Oxidation of these radicals by Co(III) complexes is reported to conform to Marcus theory.[312] Photosensitized reduction of $[Cu(acac)]_2$[313] and $[Ni(acac)_2]$[314] occurs only in the presence of scavengers for the acetylacetonyl radical formed in the initial electron-transfer step. The adducts of e_{aq}^- and H$^{\cdot}$ with uracil, U$^-$ and UH, reduce the cobalt(III) complexes of edta and nta.[315] The OH adduct, UOH, does not. Triptophan radicals oxidize aquo and tryptophan complexes of Fe(II)[226] with rate constants of $(8-450) \times 10^5 \ M^{-1} \ s^{-1}$. Complexes of Co(II) and Mn(II) are oxidized by alkylperoxy radicals.[316] Co(II) complexes are reduced to Co(I) by radicals formed during polymerization of methyl-methacrylate.[317] The anion radical SPV$^{\mp}$ (sulfonatopropyl viologen) decomposes in the presence of colloidal platinum.[318]

3.13. Carboxylic Acids and Carboxylates, Aldehydes, Ketones, and Carbon Dioxide

3.13.1. Carboxylic Acids and Carboxylates

Cob(I)alamin reduces maleic and fumaric acids[318] to succinic acid, according to equation (86), and acetylenedicarboxylic acid to fumaric acid as in equation (87). The acid dependence of the rate constants and the ratio

$$\text{cis- or trans-HOOC}-\text{CH}{=}\text{CH}-\text{COOH} + 2\text{Co(I)} + 2\text{H}^+$$

$$\rightarrow \text{HOOC}-\text{CH}_2\text{CH}_2-\text{COOH} + 2\text{Co(II)} \qquad (86)$$

$$\text{HOOC}-\text{C}{\equiv}\text{C}-\text{COOH} + 2\text{Co(I)} + 2\text{H}^+ \rightarrow \text{trans-HOOC}-\text{CH}{=}\text{CH}-\text{COOH} \qquad (87)$$

observed for the alkyne- and alkene-derived acids are consistent with rate-determining nuclophilic attack by the protonated Co(I) on the multiple bond, followed by cleavage of the Co(III)$-$C bond and a rapid Co(III)$-$Co(I) comproportionation reaction. The doubly protonated cob(I)alamin is the most reactive form in the reduction of unsaturated esters.[319]

Rate law (89) applies to reaction (88) in the presence of excess Cr^{2+}. With the olefinic acids in excess, rate law (90) is observed. The change in the rate

$$\text{XRC}{=}\text{CRY} + 2\text{Cr}^{2+} + 2\text{H}^+ \rightarrow \text{XRHC}-\text{CHRY}$$
$$\text{X} = \text{COOH}; \text{Y} = \text{H, Cl, CH}_3 \qquad (88)$$

$$\text{Rate} = k[\text{Cr}^{2+}]^2[\text{L}] \qquad (89)$$

$$\text{Rate} = k'[\text{Cr}^{2+}][\text{L}] + k''[\text{Cr}^{2+}][\text{L}]^2 \qquad (90)$$

law is explained by the different fate of the intermediate Cr(III)$-$L$^{\mp}$ under the two sets of conditions.[320] The oxidation of malonic acid, HA, and its substituted analogues by $[IrCl_6]^{2-}$ takes place with a 4:1 stoichiometry,[321] as in equation (91). The kinetic data fit rate law (92), where K_a denotes the acidity constants of malonic acids and k the rate constant for the reaction of the deprotonated form. The values of $10^2 \, k/M^{-1} \, s^{-1}$ at 25 °C and 0.5 M ionic strength are 0.093 (R = H), 1.39 (R = n-butyl), 5.4 (R = benzyl), and 8.9

$$4[IrCl_6]^{2-} + RCH(COOH)_2 \rightarrow 4[IrCl_6]^{3-} + RC(O)COOH \tag{91}$$

$$-d[IrCl_6^{2-}]/dt = 4kK_a[MA]_T[IrCl_6^{2-}]/(K_a + [H^+]) \tag{92}$$

(R = phenyl). The oxidation of lactic acid (LA) to pyruvic acid by Cr(VI) produces Cr(V),[322] which subsequently reacts with additional lactic acid as in equations (93) and (94). The buildup and decay of Cr(V) was monitored

$$2\,Cr(VI) + 2\,CH_3CHOHCOOH \xrightarrow{2k_6} 2\,CH_3COCOOH + Cr(V) + Cr(III) \tag{93}$$

$$Cr(V) + CH_3CHOHCOOH \xrightarrow{k_5} Cr(III) + CH_3COCOOH \tag{94}$$

directly at 750 nm. At 25 °C the two reactions follow the rate laws of equations (95) and (96), respectively.

$$2k_6/s^{-1} = [LA] \, (0.0007 + 0.013 \, [H^+] + 0.0021 \, [LA]) \tag{95}$$

$$k_5/s^{-1} = [LA] \, (0.0023 + 0.027 \, [H^+]^2) \tag{96}$$

The Ru(IV) complex $[(bipy)(py)RuO]^{2+}$ oxidizes HCOOH and HCO_2^- with rate constants $k/M^{-1} \, s^{-1}$ (25 °C, $\mu = 0.1 \, M$) of 0.01 and 4.2, respectively.[323] The path involving HCO_2^- exhibits a large kinetic isotope effect, $k_{HCO_2}/k_{DCO_2} = 19$. It is suggested that the reaction occurs by a two-electron hydride transfer. The Ru(III) product, $[(bipy)_2pyRu(OH)]^{2+}$, reacts further with HCO_2^-, $k = 0.01 \, M^{-1} \, s^{-1}$. $[Ru(bpz)_3]^{2+}$ oxidizes edta[74] in alkaline solutions with a rate constant of $6.9 \times 10^8 \, M^{-1} \, s^{-1}$ (pH 8.7) and 7.7×10^8 (pH 11). Vanadium(V) oxidizes 4-oxopentanoic acid to CH_3COOH and CO_2 with the intermediate formation of acetoin[324] as in equations (97) and (98). The initial step in the proposed mechanism is the slow outer-sphere electron transfer to give the carbon-centered radical. Final products are formed in a series of subsequent rapid steps.

$$CH_3CO(CH_2)COOH + 2\,V(V) + H_2O \rightarrow CH_3COCH(OH)CH_3 + CO_2 + 2\,V(IV) + 2\,H^+ \tag{97}$$

$$CH_3COCH(OH)CH_3 + 4\,V(V) + 2\,H_2O \rightarrow 2\,CH_3COOH + 4\,V(IV) + 4\,H^+ \tag{98}$$

In aqueous sulfuric acid $Ce(SO_4)_2$ oxidizes glyoxylic acid to formic acid and CO_2,[325] according to equation (99). The value of the rate constant at

$$CHOCOOH + 2\,Ce(IV) + H_2O \rightarrow 2\,Ce(III) + HCOOH + 2\,H^+ + CO_2 \tag{99}$$

40 °C is 457 $M^{-1} \, s^{-1}$, $\Delta H^{\ddagger} = 15.5$ kcal mol^{-1}, $\Delta S^{\ddagger} = 3$ cal mol^{-1} K^{-1}. In another report[327] $CeSO_4^{2+}$ is suggested as the most reactive Ce(IV) species in the oxidation of α-keto acids. The formation of a complex between Ce(IV) and the substrate is considered crucial in these and other[327–329] oxidations by Ce(IV). Hexacyanoferrate(III) oxidizes diethylenetriaminepentaacetic acid.[330] The oxidation of α,β-unsaturated carboxylic acids by Mn(III)[332] in

the presence of Cl^- involves the association of Mn(III) and Cl^-, formation of $Cl^\cdot$, decarboxylation, and oxidation of chloroethenes formed in the reaction. Glycolic and lactic acids are oxidized by Mn(III) in aqueous H_2SO_4[332] as in equations (100) and (101). Both reactions involve formation of a Mn(III)–substrate complex and intramolecular electron transfer to give a carbon-centered radical. Cyclic intermediates[333] are formed in the process of oxidation of unsaturated dicarboxylates by Mn(VI) in alkaline solutions.

$$CH_2OHCOOH + 6\,Mn(III) + H_2O \rightarrow 2CO_2 + 6\,Mn(II) + 6\,H^+ \quad (100)$$

$$CH_3CHOHCOOH + 4\,Mn(III) + H_2O \rightarrow CO_2 + CH_3COOH + 4\,Mn(II) + 4H^+ \quad (101)$$

The initial stage of the oxidation of lactic acid by $KMnO_4$ in aqueous H_2SO_4 follows a mixed second-order rate law.[334] The reaction is accelerated by Mn^{2+} owing to the formation of the more reactive Mn(IV). $HMnO_4$ is the active oxidant in the reaction with phenoxyacetic acid.[335] L-Proline and *l*-hydroxyproline are oxidized by V(V) in aqueous sulfuric acid.[336] Oxalic acid reduces Am(VI) to Am(IV) in aqueous solutions containing 0.03–$1.0\ M$ $HClO_4$.[337] The Am(IV) is stabilized in the presence of heteropolyanions, but readily disproportionates in their absence to Am(III) and Am(V). The kinetics of reduction of dodecatungstocobaltate(III) by oxalate are affected by the nature and concentration of the cations in solution.[338] A cation-bridged outer-sphere electron transfer is proposed.

3.13.2. Carbonyl Compounds

Benzaldehyde is oxidized by vanadium(V) to benzoic acid in a reaction which shows a large kinetic isotope effect upon substitution of benzylic hydrogen by deuterium, $k_H/k_D = 4.8$.[340] The reaction is proposed to take place by formation of the V(V)–benzaldehyde complex, followed by hydrogen transfer from benzylic carbon to bridging oxygen, and dissociation of V(IV). SeO_2 oxidizes ethyl acetoacetate to α,β-diketo ethyl butyrate,[340] as in equation (102). The reported activation parameters in aqueous acetic acid are $E_a = 75.0\ kJ\ mol^{-1}$, $\Delta S^{\ddagger} = -100\ J\ mol^{-1}\ K^{-1}$. 2-Methylbenzophenone reacts with CH_3MgBr and $PhMgBr$ by single electron transfer.[341] Photogenerated manganese radicals $[Mn(CO)_{6-n}(S)_n]^{\cdot}$ (S = solvent) react with *para*-quinones by electron transfer, while *ortho*-quinones oxidatively add to $[Mn(CO)_5]^{\cdot}$.[342]

$$CH_3COCH_2COOC_2H_5 + SeO_2 \rightarrow CH_3COCOCOOC_2H_5 + Se \quad (102)$$

3.13.3. Carbon Dioxide

The stoichiometric reduction of CO_2 to coordinated methoxide by $[cp_2Zr(H)Cl]_n$ in THF is a two-step process, shown in equations (103) and

$$2[cp_2Zr(H)Cl] + CO_2 \rightarrow [cp_2Zr(Cl)]_2O + CH_2O \quad (103)$$

(104). The oxophilicity of the metal and the high stability of the μ-oxo complex provide the driving force for the reaction.[343]

$$[cp_2Zr(ClH] + CH_2O \rightarrow [cp_2Zr(Cl)OCH_3] \tag{104}$$

3.14. Alkenes, Alkynes, and Arenes

3.14.1. Alkenes

Aqueous Cr^{2+} reduces fumaronitrile to succinonitrile (20%) and $[(H_2O)_5CrCH(CN)CH_2CNCr(OH_2)_5]^{5+}$.[344] At 25 °C and 0.5 M ionic

$$-d[NCCH=CHCN]/dt = 2.7[NCCH=CHCN][Cr^{2+}]^2 \tag{105}$$

strength the kinetics follow rate law (105). The reaction takes place by prior coordination of Cr^{2+} to nitrogen at one end of succinonitrile, followed by the attack of the second molecule of Cr^{2+} at either the second nitrogen or the =CH– group; see equations (106) and (107).

$$[CrNCCH=CHCN]^{2+} + Cr^{2+} \rightarrow [CrNCCH_2CH_2CNCr]^{6+} + [CrNCCH_2CH(Cr)CN]^{5+} \tag{106}$$

20%   70%

$$[CrNCCH_2CH_2CNCr]^{6+} \rightarrow 2\,Cr^{3+} + NCCH_2CH_2CN \tag{107}$$

One-electron oxidation of [Cr(TPP)O] yields the Cr(V) oxo-porphyrin, which epoxidizes norbornene.[345] Iodosobenzene oxidizes [Cr(TPP)Cl] to an oxo-bridged Cr(IV) dimer. The disproportionation of the latter yields the same Cr(V) species. Oxidation of terminal olefins by $[Pd(Cl)(NO_2)(CH_3CN)_2]$ in acetic acid affords equal amounts of 2-acetoxy-1-alkanol and 1-acetoxy-2-alkanol.[346,347] An intermediate, described as either a π-olefin complex or an oxametallacycle, was detected in the reaction of $[OFe^{IV}(TMP)]^+$ with olefins.[348] The intermediate subsequently decomposed to the epoxide. The manganese-containing product of the reaction of $[MeBu_3N]^+MnO_4^-$ with alkenes in CH_2Cl_2 is colloidal MnO_2 and not the Mn(V) cyclic diester.[349]

3.14.2. Alkynes

The σ-bonded ethynyl complexes of Rh(III) and Co(III), $RC_2M(salen)py$, are formed from alkynes and metal complexes in either 1+ or 3+ oxidation state,[350] as in equations (108) and (109).

$$M(I) + HC\equiv CR \rightarrow M(H)(C\equiv CR) \rightarrow M-C\equiv CR + H^-$$

M = Rh, R = C_6H_5

M = Co, R = CF_3 (108)

$$XM(III) + HC\equiv CR \rightarrow MC\equiv CR + HX$$

M = Co, R = C_6H_5 (109)

3.14.3. Arenes

[Fe(phen)$_3$]$^{3+}$ reacts with methylarenes ArCH$_3$ to afford the unstable radical cations ArCH$_3^{+}$ [351] as in equation (110). The cations react with bases such as pyridine and yield ArCH$_2$py$^+$ according to equations (111) to (113).

$$ArCH_3 + [FeL]_3^{3+} \rightleftharpoons ArCH_3^{+} + [FeL_3]^{2+} \qquad k_1, k_{-1} \qquad (110)$$

$$ArCH_3^{+} + py \rightarrow ArCH_2^{.} + Hpy^+ \qquad k_2 \qquad (111)$$

$$ArCH_2 + [FeL_3]^{3+} \rightarrow [FeL_3]^{2+} + ArCH_2^{+} \qquad (112)$$

$$ArCH_2^{+} + py \rightarrow ArCH_2py^+ \qquad (113)$$

The rate constants k_1, k_{-1} and k_2 were evaluated for a series of methylarenes and pyridines. Facile insertion of styrene into the Rh—Rh and Rh—H bonds of [Rh$_2$(OEP)$_2$] and [(OEP)RhH], respectively, is a radical chain reaction [352] propagated by reactions (114) and (115). Analogous chain mechanisms can

$$[(OEP)Rh] + PhCH=CH_2 \rightarrow [(OEP)RhCH_2\dot{C}HPh] \qquad (114)$$

$$[(OEP)RhCH_2\dot{C}HPh] + [Rh_2(OEP)_2] \rightarrow [(OEP)RhCH_2CH(Ph)Rh(OEP)] + [(OEP)Rh] \qquad (115)$$

accommodate insertion of CO into [(OEP)RhH] and oxidative addition of PhCH$_2$Br to [Rh$_2$(OEP)$_2$]. (Octaethylporphyrin)rhodium(II) dimer, [(RhOEP)$_2$], reacts with toluene [353] to form a benzylrhodium(III) complex as in equation (116). In analogous reactions ethylbenzene and 2-propylbenzene

$$[(RhOEP)_2] + CH_3C_6H_5 \rightarrow [(OEP)RhCH_2C_6H_5] + [(OEP)RhH] \qquad (116)$$

yield [(OEP)RhCH(C$_6$H$_5$)CH$_3$] and [(OEP)RhCH$_2$CH(C$_6$H$_5$)CH$_3$], respectively. [(OEP)RhH], formed in reaction (116), slowly eliminates hydrogen. Excited states of Rh(III) polypyridine complexes undergo rapid reductive quenching with di- and trimethoxybenzenes, and aromatic amines. [354] The kinetics and products of the oxidation of β-methylstyrenes by Co(III) acetate are sensitive to the nature of ring substituents. [355] The results are explained by the substituent effect on the stability of the radical cation produced in the first step of the reaction. Similar observations were made with other aromatic olefins, [356] while steric effects dominate in the reactions of alkyl cyclohexenes. [357] The molybdenum(VI) complex [Mo(tdt)$_3$] oxidizes polynuclear hydrocarbons, amines, purines, aldehydes, and riboflavin to form the one-electron reduced species [Mo(tdt)$_3$]$^-$ and organic radicals. [358] The excited state [*Ru(bipy)$_2$(CN)$_2$] reduces nitrobenzene, $k = 7.5 \times 10^9\ M^{-1}\ s^{-1}$. [359] Paladium(II) oxidizes methylbenzenes in CF$_3$COOH by single electron transfer. [360]

Part 2

Substitution and Related Reactions

Chapter 4

Reactions of Compounds of the Nonmetallic Elements

4.1. Boron

^{11}B NMR has shown[1-4] that the equilibrium constants, K_1 and K_2, for the 1:1 and 1:2 complexes between $B(OH)_4^-$ and various polyols have values different from those determined by other methods. Increasing the number of OH groups in the polyol leads to an increase in both K_1 and K_2. Complexation was observed with both 1,2- and 1,3-diols, with $K_1 = 2.2$ (temperature unspecified)[2] or 1.4 (25 °C)[3] M^{-1} for propan-1,2-diol and $1.1^{[2]}$ or $0.9^{[3]}$ for propan-1,3-diol. The values of K for groups containing negatively charged substituents were generally less than for those without. For D-mannitol, D-glucitol, D-fructose, and D-glucose $K_1 \cdot K_2$ was, respectively, 16, 55, 37, and $3 \times 10^3 \ M^{-2}$.

The effects of 3- and 4-ring substituents on the rates of hydrolysis of quinoline-borane complexes are reported[5] to be similar to those observed in earlier studies of substituted aniline-boranes. Hydrolysis proceeds by two pathways, one acid independent (k_1, rate-determining loss of BH_3 via a D or I_d process) and the other first order in $[H^+]$ (k_2, bimolecular electrophilic displacement of BH_3 by a solvated proton). No evidence was reported for solvated BH_3 as a kinetically important intermediate. A 2-Me substituent increases k_1 by about 35-fold and k_2 by only 2.5-fold. Activation parameters for the hydrolytic decomposition of N-alkyl-2-phenyl-1,3,6,2-perhydro-dioxaborocines (1) have been studied using ^{11}B and ^{15}N NMR.[6] Bulky alkyl substituents, R, have been shown to affect the reactivity of the boranes, R_2BX (R = i-Pr, t-Bu; X = F, Cl, Br, I, OR', SMe, NR_2'').[7] Cragg, Miller, and Smith have used variable-temperature ^{13}C NMR to obtain $\Delta G^\ddagger$ for restricted rotation about B—N bonds in aminoboranes.[8-10] In a series of 2- and 3-methyl-piperidinophenylboranes, $\Delta G^{\ddagger}$ for the B—OMe derivative was 3 kcal mol^{-1} greater than for the B—SMe analogue, ascribed to greater p_{π}-p_{π} bonding between B and O. $\Delta G^{\ddagger}$ values for the 2-methylpiperidino compounds were lower than for those with the 3-methyl analogues, consistent with steric inhibition of the attainment of planarity and maximum p_{π}-p_{π} B—N bonding. Similar values have been obtained for $\Delta G^{\ddagger}$ for rotation about the B—N bonds in the cations $(RR'N)_2B^+$.[11] Steric factors govern the relative rates of addition of HX to the BN triple bond of the aminoiminoborane (2), the reported order[12] being $NH_3 > NH_2R > NHR_2$; $HOR > HSR > HNR_2$.

1   2

By monitoring all volatile components present during the initial stages of the thermal decomposition of B_4H_{10}, a dependence of the rate on $[B_4H_{10}]$ to the first order was observed,[13] rather than $\frac{3}{2}$ order previously reported. B_5H_{11} and H_2 were formed initially with B_2H_6, B_6H_{12}, and $B_{10}H_{14}$ formed subsequently. Two groups[14,15] have examined the pyrolysis of $(Et_4N)(BH_4)$, under vacuum or in suspension in refluxing hydrocarbon. Both conclude that $Et_3N \cdot BH_3$ was formed as a volatile intermediate on route to $(Et_4N)(B_{10}H_{10})$ and other polyhydroborate anions. Using pulse radiolytic generation of $N_3^{\cdot}$, estimates of the rate constants for the oxidation of BR_4^-, R = H, Ph, to $BR_4^{\cdot}$ [16] have been made. At 250 rad/pulse, first-order kinetics are observed, with $k_1 = 5.6 \times 10^3$ s^{-1}. Roberts and his group continue studies of transient amine- and phosphine ligated boryl radicals, $R_3E \rightarrow \dot{B}H_2$, (3) R = H, Me, Et and other alkyl, Ph; E = N, P.[17-24] Compound (3) (R = Et, E = B) is thought to have a pyramidal geometry at B and to undergo a second-order self-reaction at 190 K with $2k = 8 \pm 2 \times 10^8$ M^{-1} s^{-1}.[17] The radical reacts rapidly to abstract halogen from alkyl bromides, but more slowly with alkyl chlorides. Depending on the nature of the N-alkyl substituent, (3) may be transformed[21,22] into the isomeric aminyl-borane radical, $R_2\dot{N} \cdot BH_3$, a process which deuterium labeling suggests occurs by other than 1,2-hydrogen migration.[19] For (3) containing secondary and tertiary alkyl groups, β-scission as shown for (4) in equation (1) occurs with a surprisingly rapid rate even at 195 K.[18] In cyclopropane-t-pentyl alcohol-di-t-butyl peroxide at 224 K, (4) displays alkyl radical formation

$$i\text{-}Pr_2EtN \cdot BH_3 \rightarrow i\text{-}Pr_2EtN \cdot BH_2^{\cdot} \rightarrow i\text{-}Pr^{\cdot} + i\text{-}PrEtN=BH_2 \qquad (1)$$

4

by a first-order process with $k = 7.45 \times 10^4$ s^{-1}. Kinetics of the addition of (3) to the spin traps, e.g., t-BuNO$^{\cdot}$, have been reported.[20] Secondary phosphine boranes give the phosphine centered radical, $R_2\dot{P} \cdot BH_3$, as the major product upon hydrogen atom abstraction. No boron centered radical was detected.[23] Stepwise S_H2 alkoxydealkylation processes have been observed in the reaction of a series of 1,3,2-dioxaboroles [e.g., (**5**)] with alkoxy radicals. RO˙ addition at the boron center is followed by an α-scission of a B-alkyl substituent.[24]

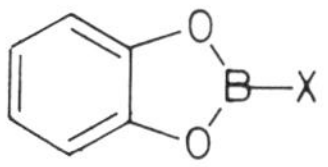

5

H. C. Brown and co-workers have investigated hydroboration reactivity differences between carbocyclic monoolefins and their oxygen- and sulfur-containing analogues.[25] For instance, in tetrahydrofuran, 2,3-dihydrofuran reacts with 9-borabicyclo[3.3.1]nonane (**6**) 106 times faster than cyclopentene, while 2,5-dihydrofuran reacts 4.2 times more slowly. On the other hand, the seven-membered 2,3,4,5-tetrahydrooxepin reacts at rates only 0.034 times that of cycloheptene. In contrast, the sulfur heterocycles, 2,3-dihydrothiophene and Δ^2-dihydrothiopyran, react with (**6**) much more slowly than their carbocyclic analogues. The precise origins of these differences are not clear, though it is believed that mesomeric effects in the ground state are not responsible. Regio- and stereoselectivity of alkene hydroboration has also been investigated.[26-29] Qualitative rate studies have been reported for the reaction of 2,2,3,3-tetramethylethylchloroborane-methyl sulfide complex with 56 organic compounds in CH_2Cl_2 at 0 °C.[30] Rates of reductions of carbonyl compounds by organoboranes R_3B may be increased by increasing the Lewis acidity of boron by replacement of R by Cl or Br. For instance, diisopinocampheylchloroborane (**7**, X = Cl) reacts according to equation (2) almost instantly at room temperature while the diisopinocampheyl-*n*-hexylborane (**7**, X = *n*-hexyl) requires *ca* 6 days.[31] By using specifically vinylically and allylically

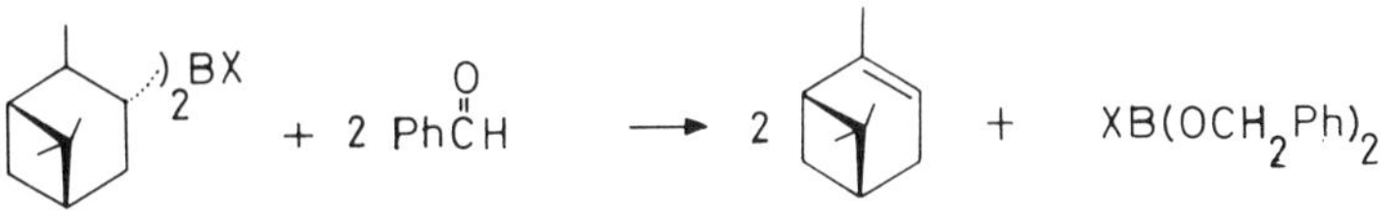

$$2 \, \text{PhCH} \longrightarrow 2 \quad + \quad XB(OCH_2Ph)_2 \qquad (2)$$

7

deuterated (Z)-alkenes to investigate enantioselective hydroborations with (+)-diisopiniocampheylborane (**7**, X = H) it has been concluded[32] that there is no significant steric compression at the vinylic hydrogens but very strong compression at the allylic positions. In theoretical studies, Nelson and Cooper[33] report a correlation between the MNDO HOMO energies corresponding to the π-MOs of a series of vinyl and allyl compounds with their hydroboration rates with (**6**). McKee[34] has calculated the relative energies of transition states for a series of pathways between the known C_{2v} structure for B_4H_{10} and the unknown bis(diboranyl) C_2 structure.

Acyclic trialkylboranes react with trimethylamine *N*-oxide to give, sequentially, borinates, boronates, and borates, with the order of alkyl group reactivity being tertiary > secondary > primary. The leaving Me_3N adopts an antiperiplanar conformation with respect to the $B-C$ bond undergoing cleavage.[35] A similar reactivity order is described for oxidations involving molecular oxygen.[36] The reactivity of other oxidants has also been studied.[37-39] Protonolysis by carboxylic acids of R_3B is influenced by steric factors, particularly those resulting from β-Me substitution.[40] Isomerizations and substitutions in B-X-closo-2,4-$C_2B_5H_6$ (8) and B,B'-X_2-closo-2,4-$C_2B_5H_5$ (9) have been reported.[41-44] For the B-halo substituted (8), halide exchange appears possible only when the leaving X^- is larger than the entering halide. Rates of substitution follow the qualitative order $F^- > Cl^- > Br^-$.[41] All but one of the isomers for (8) and (9), X = Cl,[42] Br, I,[43] have been obtained and the thermal rearrangements of selected species studied at 295 °C. The following statistically corrected stability orders were obtained: 3-Br > 5-Br > 1-Br and 5-I > 3-I > 1-I for (8) and 3,5- ≥ 5,6- > 1,3- > 1,5- > 1,7- for (9) with X = Br and 5,6- > 3,5- > 1,5- ≥ 1,3- > 1,7- for (9) with X = I. The results are interpreted in terms of Lipscomb's Diamond-Square-Diamond (DSD) mechanism and rate constants for individual rearrangement processes derived by a curve-fitting approach. In similar studies on (9) with X = Me, Cl,[44] the rates of rearrangement, compared with the related mono- and disubstituted compounds, 5-X- (8); X = Cl, Me and 5,6-X_2- (9); X = Cl, Me, suggest that the presence of the second substituent accelerates the apparent rearrangement of the first substituent and that Cl has a bigger effect than Me. The relative stabilities of the eight B-Me-B'-Cl- (9) isomers have been determined. Isomerizations of trigonal bipyramidal $C_2B_3H_5$ by the DSD mechanism are forbidden according to orbital symmetry conservation considerations.[45] A similar conclusion is drawn for $B_9H_9^{2-}$ and $C_2B_7H_9$ isomerizations though a double DSD process is symmetry allowed.[46]

4.2. Carbon

The ^{13}C kinetic isotope effects on the dehydration of HCO_3^- in D_2O and H_2O have been compared[47] with values computed theoretically. A stepwise mechanism for the dehydration is preferred to a concerted process. The effect of temperature on the isotope effect is also consistent with $C-O$ bond breaking in the rate-determining step. HCO_3^- is first protonated to form the zwitterionic intermediate (10) which gives CO_2 according to equation (3). Theoretical

$$^-O_2COH + H^+ \rightleftharpoons {}^-O_2C\text{-}OH_2^+ \rightleftharpoons [H_2O\cdots CO_2]^\ddagger \rightleftharpoons CO_2 + H_2O \qquad (3)$$
$$\textbf{10}$$

aspects of the hydration of CO_2 have also been reported.[48,49] Raman spectra of solutions of CO_3^{2-} and HCO_3^- in 4 M hydrogen peroxide provide direct evidence for the presence of a peroxocarbonate species, suggested to be HCO_4^-.[50] It is estimated that, for the equilibrium, $O_2H^- + CO_2 \rightleftharpoons HOOCO_2^-$,

k_f at 25 °C = $2.2 \times 10^5 \ M^{-1} \ s^{-1}$ and k_r = $2.8 \times 10^{-1} \ M^{-1} \ s^{-1}$ and K_{eq} = 7.9×10^3 M^{-1}.[51] The rate constant for the reaction of CO_3^{2-} with (p-methoxy-phenyl)tropylium cation at 23 °C ($2.5 \times 10^4 \ M^{-1} \ s^{-1}$) is only slightly less than the value for OH^- under the same conditions (4.0×10^4).[52] k_{am} at 25 °C for the reaction, $RR'NH + CO_2 \rightarrow RR'N{-}CO_2^- + H^+$, in the formation of carba-mates from CO_2 and aqueous alkanolamine are $4700 \pm 630 \ M^{-1} \ s^{-1}$ for monoethanolamine, 110 ± 15 for diethanolamine, and 54 ± 7 for diisopropanolamine.[53] These values and associated activation enthalpies highlight the importance of steric effects in the base-catalyzed removal of H^+ from the zwitterionic intermediate, $RR'N^+HCO_2^-$. By exploiting a method for the rapid synthesis of HNCO, a kinetic analysis has been made of the proton catalyzed reaction, $MeOH + HNCO \rightarrow MeOC(O)NH_2$.[54] Carbonate radicals, produced from CO_3^{2-} and $SO_4^{-\cdot}$, react with amines, either by hydrogen atom abstraction (primary amines) or by electron transfer (tertiary amines).[55] The primary radicals formed from oxalate and $OH^\cdot$ may be intercepted by methyl-viologen in competition with their rapid ($k = 2 \times 10^6 \ s^{-1}$) conversion to CO_2^-. This species, which may also be obtained from HCO_2^- and $H^\cdot$ or $OH^\cdot$, reacts with methylviologen with $k \sim 10^{10} \ M^{-1} \ s^{-1}$.[56] The kinetics of the reaction of CO_2^- with 2-thioriboflavin have also been described.[57] At 25 °C in water, the equilibrium constant between glyoxal and its dimer is reported[58] to be $0.56 \ M^{-1}$, the value being little affected by pH. Dissociation of the dimer was first order in [dimer], with the pseudo-first-order rate constants taking the form $a[H_3O^+] + b + c[OH^-]/(1 + d[OH^-]) + e[OH^-]$. Disproportionation studies[59] gave a rate form in contradiction to earlier work. A mono- and dihydrated form of glyoxal is thought to be in equilibrium, with intramolecular hydride transfers to the unhydrated carbonyl carbon of glyoxalmonohydrate mono- and dianions being rate limiting. The hydrolysis rate for ClCN is independent of pH (1–5) with activation parameters consistent with attack by water.[51] For pH 5–10.5, the rate is first order in [ClCN] and [OH^-]. Oxoanion buffers accelerate Cl^- release. Nitrogen nucleophiles were found to be more reactive than oxygen nucleophiles. 2,3-dihydroxypropen-2-al gives a radical adduct on reaction with $OH^\cdot$ with a second-order rate constant reported to be $9.9 \times 10^9 \ M^{-1} \ s^{-1}$.[60] This intermediate, and those derived by reaction with X_2^- (X = Cl, Br, I, SCN)[61] decay by second-order kinetics.

Lifetimes and reaction rates with nucleophiles have been estimated for carbocations generated by olefin hydration,[62] orthoester hydrolysis,[63] photo-lysis of benzyl[64] or trityl[65] compounds, and oxidation of carbon-centered radicals.[66] Rate studies on the addition of water to cis- and trans-cyclooctene (**11**) and 2,3-dimethyl-2-butene (**12**) to the corresponding alcohols show that the tertiary carbocation derived from (**12**) has a lifetime $\tau \simeq 10^{-10}$ s in dilute aqueous solution, indicating that the ion may be a solvationally equilibrated intermediate in the addition of water to (**12**). Hydration of (**11**) in dilute acids leads to an estimate of τ of approximately 5×10^{-12} s suggesting that the cyclooctyl cation intermediate may not be equilibrated with the solvent medium.[62] Benzyl carbocation is reported[64] to have a lifetime of 7×10^{-10} s in 30% methanol–water at 0 °C and to react with halide ions at rates approaching those for diffusion control. The rate constant for its reaction with water is $2.4 \times 10^7 \, M^{-1} s^{-1}$. Trityl carbocation, in 1:2 MeCN–H$_2$O, reacts at 20 °C with the solvent with $k = 1.5 \times 10^5 \, s^{-1}$. Rate constants for a series of other nucleophiles reach a limiting rate with the more reactive species (e.g., N_3^-) of significantly less than the limit of diffusional control.[65] Trialkoxymethyl carbocations can be generated in aqueous solution by hydrogen atom abstraction from $(RO)_3CH$ followed by electron transfer.[66] Conductance measurements gave a hydrolysis rate for $(MeO)_3CH$ of $1.4 \pm 0.2 \times 10^3 \, s^{-1}$ with $E_A = 7.6 \, \text{kcal mol}^{-1}$ and $\Delta S^{\ddagger} = -20.2$ eu. Reaction with OH^- has $k = 7 \pm 1 \times 10^7 \, M^{-1} s^{-1}$. In examination of the kinetics of addition of diarylcarbocations to alkenes, constant, proportional, and inversely proportional reactivity–selectivity behavior have been seen[67–69] in a closely related series of reactions. Ta-Shma and Jencks have reported[70] structure–reactivity correlations for general base catalysis of the addition of alcohols to 1-[4-(dimethylamino)phenyl]ethyl and 1-(4-methoxyphenyl)ethyl carbocations in 5:4:1 $H_2O/CF_3CH_2OH/ROH$ in attempts to detect changes from a stepwise to a concerted mechanism. For ROH additions in the presence of water, there is an increased dependency on ROH basicity as the stability of the 1-phenylethyl carbocations is increased. A small involvement of proton transfer in the transition state (associated with ROH···B) is proposed. Rates of the racemization, k_{rac}, and radiobromide exchange, k_{exch}, for 1-arylbromoethanes[71] support an ion-pair process for S_N at saturated carbon, with the carbocation from 1-(3,4-dimethylphenyl)bromoethane being sufficiently stable that its $k_{rac} \ll 2k_{exch}$. Rate constants for the electron-transfer reactions between n-$C_{16}H_{34}^+$ and cycloalkanes, alkenes, and aromatics increase from $3.0 \pm 1.0 \times 10^9 \, M^{-1} s^{-1}$ for pentylcyclohexane to a limiting value of $9.0 \pm 1.0 \times 10^9$ for the more exothermic processes involving, for example, benzene.[72] The rate constant for dioxygen addition to the biadamantylidene radical cation (13) at room

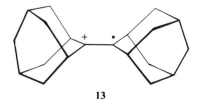

13

temperature is found[73] to be $5600 \, M^{-1} s^{-1}$. Substitution of Cl in the adamantane skeleton results in a reduced reactivity. The biadamantylidene dioxetane radical cation dissociates to (13) and dioxygen with $k > 800 \, s^{-1}$.

Hydrolysis of N-[$^2H_3$]-methyl-N-nitrosoethyl carbamate, in undeuteriated phosphate buffer at pH 7.4, produced a distribution of all isotopic methanols and demonstrated a significant primary kinetic isotope effect

$(k_H/k_D$ ca 2.7) for the exchange of H and D in the methyldiazonium ion intermediate. No exchange was observed in the undecomposed starting material. Almost identical isotopic distributions were seen from three other precursors of MeN_2^+.[74] Methyl H–D exchange has also been described[75] for $CH_3(CH_2O)_3NH_2D^+$ in $DFSO_3$-SbF_5. Using the INADEQUATE pulse sequence in natural abundance ^{13}C NMR to study the degenerate rearrangement of the 2,3-dimethyl-2-butyl cation, a secondary ^{13}C isotopic perturbation of the degenerate equilibrium has been observed.[76] These effects were shown to be significantly greater than the intrinsic ^{13}C isotope effect on the ^{13}C chemical shift in, for example, the *tert*-butyl cation. 1-Alkyl-[77] and 1-phenyl-cyclopropanediazonium ions[78] decompose via allyl rearrangement and nucleophilic addition processes. Ring cleavage and addition range from the stereospecific to the completely unselective, indicating a gradual change from a concerted to stepwise mechanism.[77] π-Donor effects in aryl substituents[78] favor retention of the carbocyclic ring while diminishing reaction stereoselectivity.

Studies relevant to the pathways involved in the zeolite catalyzed conversion of methanol to hydrocarbons have been reported. A proposal that C—C bond formation may occur via methylene insertion into methyloxonium species has not been substantiated in model experiments.[79] Support for a radical process is claimed[80] from the observation of radical initiated formation of 1,2-dimethoxyethane from dimethyl ether in fluorosulphonic acid solutions of peroxodisulfuryl difluoride, ascribed to the stabilization of the methoxymethyl radical by O-protonation. Methyl loss from a gem-dimethyl group has been observed[81] during a cyclic ether formation from (14) in FSO_3H-SO_2 at $-60\,°C$, favored by steric crowding in the transition state.

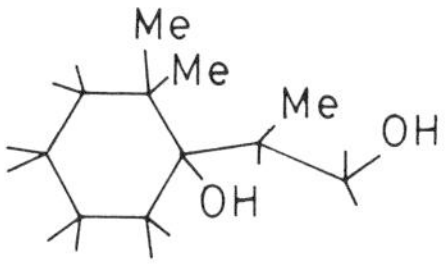

14

Tritium exchange in $[^3H_1]$-$CHBrCl_2$ [82] in carbonate buffers is subject to special salt effects. It was concluded that cation concentrations should be maintained constant, in addition to ionic strength, when determining rates in a series of carbonate buffers. A study of $[^2H_1]$-$CHCl_3$,[83] carried out in a two-phase liquid system, gave an activation energy for base-catalyzed deuteron abstraction of 87 kJ mol^{-1}. *Ab initio* SCF-MO calculations on the gas-phase interaction of F^- with ethyl fluoride suggest[84] that the E2 reaction proceeds via four elementary steps (cluster formation, net E2 reaction to give second cluster, decomposition of successor cluster, and product formation with synchronous C—H and C—F bond fission in the net E2 step). The S_N2 reaction proceeds by three steps (precursor cluster, net reaction to give second cluster, decomposition of successor cluster). Calculations suggest that E2 is preferred in the gas phase. The effect of addition of up to three molecules of D_2O, MeOH, and EtOH to F^- and of MeOH, EtOH, Me_2CO, $HC(O)OH$, and $MeC(O)OH$ to Cl^- in the gas-phase nucleophilic displacement reactions with MeCl and MeBr have been studied using the flowing afterglow technique.[85] *Ab initio* calculations suggest[86] that alkyl lithiums add to the archetypal carbonyl compound, HCHO, in processes in which the electrophile plays a decisive role.

Using modulated photochemical generation of radicals with phase-sensitive detection, Wayner and Griller[87] have measured redox potentials of radicals with average concentrations of 10^{-7} to 10^{-8} M and lifetimes of $ca\ 10^{-3}$ s. For instance, the benzyl radical has $E_{1/2}^{ox}$ of 0.40 ± 0.03 V (vs $Ag/Ag^+/MeCN$) in MeCN and $E_{1/2}^{red}$ of -1.78 ± 0.02 V. Ingold and co-workers, in studies of the photochlorination of alkanes in halocarbon solution (e.g., CCl_4) containing benzene, have been able, for the first time, to obtain absolute rate constants for hydrogen atom abstraction by a "free" chlorine atom and its π-complex with benzene.[88] Solution rate constants for H· abstraction from alkanes by trichloromethyl radicals were similar to those measured in the gas phase while the rate constants for the Cl· abstraction by alkyl radicals from CCl_4 was $ca\ 10^2$ greater in solution than in the gas phase.[89] The latter differences are suggested to arise either from polar transition-state solvation or errors in the values determined for the gas phase. Closs and Redwine have reported[90] that biradicals, observed following flash photolysis of 2,2-dimethyl-cyclooctanone and differing only in their nuclear spin states, may react at different rates, demonstrated from time-dependent CIDNP of the biradical, $O=\dot{C}(CH_2)_6\dot{C}Me_2$. Rate constants for the reaction of free radicals and a nitroxide spin trap have been obtained from competition reactions involving hex-5-enyl radical (for which accurate kinetic data for the competing ring closure are known).[91] Absolute rate constants for H· abstraction reactions of benzyl radical,[92] for the reaction of phenyl and alkyl radicals with n-Bu_3SnH,[93] for the chlorine atom abstraction from α-toluenesulphonyl chloride by alkyl and phenyl radicals,[94] and for the reaction of methylcyclopropyl radical with alkenes and halocarbons[95] have been reported. The latter study also demonstrated that the radical is nonplanar and inverts (at 71 °C) with $k = 2.1 \pm 0.8 \times 10^{11}$ s^{-1}. Laser flash photolysis has generated an excited state of 1-naphthylmethyl radical[96] which reacts with dioxygen ($k = 4.7 \pm 8 \times 10^9$ M^{-1} s^{-1}) and CCl_4 ($k = 4 \times 10^6$ M^{-1} s^{-1}). The excited radical has poor hydrogen-atom abstraction properties. Using ethyl radicals generated by five independent methods, a mean value for the rate ratio (k_d/k_c) for the disproportionation vs combination reactions in aqueous solution at $ca\ 25$ °C of 0.35 ± 0.04 has been determined.[97]

Benson[98] has proposed a "contact transition state" model for the disproportionation of alkyl free radicals which leads to calculated rate constants which are within $\pm 30\%$ of the experimentally determined values. Aldehydes may be comproportionately reduced by alcohols (other than tertiary) in a thermally induced redox radical chain reaction as in equation (4) ($R = 4\text{-}i\text{-}PrC_6H_4$). This reaction is first order in oxidant and reductant, and is characterized by a very slow initiation and large chain length (*ca* 10^7 at 200 °C),[99,100]

$$RCHO + Me_2CHOH \rightarrow R\dot{C}HOH + Me_2\dot{C}OH \rightarrow RCH_2OH + Me_2C=O \qquad (4)$$

with kinetic isotope effects for both O- and C-deuteriated reductant favoring the radical process. Metzger[101] claims to have observed experimentally for the first time the β-hydrogen atom abstraction from an alkyl radical by an alkene. ESR has been used to obtain activation parameters for the unimolecular homolysis of $t\text{-BuOC(O)}^{\bullet}$ and the hydrogen atom abstraction reactions of $t\text{-BuO}^{\bullet}$ from cyclopentane and $t\text{-BuOMe}$.[102] 1-Naphthylcarbene reacts readily with MeCN to give a nitrile ylide with $k = 4.6 \times 10^5 \, M^{-1} \text{s}^{-1}$ at 300 K and with oxygen to yield the carbonyl oxide with $k = 3.5 \times 10^9 \, M^{-1} \text{s}^{-1}$.[103] Other reactions of C-centered species with oxygen and the mechanistic studies of reactions of peroxo compounds are discussed in Section 4.8. Solvent effects on carbene reactivity have been reported.[104,105] Anion radicals derived, for example, from azobenzil $PhC(O)C(N_2)Ph$ (**15**) lose N_2 at 273.2 K with $k = 125 \, \text{s}^{-1}$, with the hydrogen bonded complex (**15**)$^{\bar{}} \cdot$ HOH decomposing with $k = 1700 \, \text{s}^{-1}$ in aqueous MeCN at the same temperature.[106] Reactions of the radical anion $Ph_2CN_2^{\bar{}}$ and the dianion $Ph_2CN_2^{2-}$ which involve N_2 loss have also been described.[107] Solvent effects on the reaction of CS_2 with Et_3P in nitrile solvents have been shown[108] to be quite small.

4.3. *Silicon*

Results from ^{29}Si NMR studies[109-113] in aqueous silicate solutions continue to be controversial. An ICI group has shown[109] that the spin–spin relaxation times (T_2) decrease with increasing temperature, rendering invalid exchange rates between Si-containing species obtained from line-width determinations. The rates of silicate monomer exchange in 1.0 M sodium silicate, k_{exch} ($SiO_2\text{:}Na_2O = 1.0$) $\leq 2.8 \, \text{s}^{-1}$ at 320 K and k_{exch} ($SiO_2\text{:}Na_2O = 0.33$) $\leq 24 \, \text{s}^{-1}$ at 277 K and 25–44 s^{-1} at 320 K, were higher than those determined by spin inversion recovery techniques. Kinrade and Swaddle[110] suggest that protonation equilibria influence both the distribution of polymeric silicate species in solution and the rates of silicon exchange between them, with the temperature dependence of the line broadening being the result of a $H_3SiO_4^{-}$ mediated process, as opposed to one mediated by $H_2SiO_4^{2-}$. The same group has reported[111] that ^{29}Si longitudinal relaxation times (T_1), measured by inversion recovery, are dominated by dipole–dipole interactions between the ^{29}Si and M^+ nuclei rather than between ^{29}Si and ^{1}H nuclei. The role of paramagnetic impurities seems to be less important than previously proposed.

In an investigation of aqueous alkaline tetramethylammonium silicate solutions, equilibria were found to be reestablished slowly after a perturbation, unlike the behavior of analogous solutions containing alkali metals.[112] Trimethylsilylation has been used to study the effect of tetraalkylammonium ion addition on silicate ion equilibria in solutions obtained from $Si(OEt)_4$ hydrolysis.[114] The tendency for Me_3NR^+ to promote the formation of the double four-membered cage octamer $Si_8O_{20}^{8-}$ is independent of R. Natural abundance ^{29}Si NMR and Raman spectroscopy have been used[115] to study gelation processes in partially hydrolyzed $Si(OMe)_4$ under conditions where hydrolysis and condensation occur on comparable time scales. It was tentatively suggested that formamide inhibition of the hydrolysis process arises from the greater viscosity of the formamide sols and the stronger hydrogen bonding between formamide and the silicate species. Regression analysis of the rates of hydrolysis in 45% aqueous acetone buffer solutions, for ten R_3SiOR' compounds, reveals good correlations with the electronic (Taft's σ^*) and steric (Taft's E_s) characteristics of R and R'.[116] In the acid hydrolysis, electron withdrawing R' and R both slow the reaction, with R' doing so to a greater extent than R. Rates are reduced by bulky groups, particularly those on Si. Electron-releasing and bulky R and R' all reduce the rate of base hydrolysis. The hydrolysis of a series of methoxysilanes in a two-phase system (pentane:aqueous buffer) gave the unexpected order of reactivity $Me_3SiOMe >$ $Me_2Si(OMe)_2 > MeSi(OMe)_3$.[117] The ease of cleavage of the $Si-R_f$, $R_f =$ $CF_2CF_2Oi\text{-}C_3F_7$ in a series of perfluoroalkyl ether substituted silanes in aqueous media, follow the order $HSi(R_f)_3 > MeSi(R_f)_3 > Me_2Si(R_f)_2 >$ Me_3SiR_f.[118] Reaction of Me_3SiOH in dioxane:water and toluene:water[119] and of $Me_2Si(OH)_2$ and four methylsiloxanediols in toluene:water[120,121] have been discussed in terms of competing silanol condensation processes and siloxane fission. At 30 °C in dioxane containing an equimolar amount of water, second-order rate constants for the hydrolysis of the chlorodisiloxanes, $PhMe_2SiOSiRArCl$, are in the range 0.3×10^{-2} 1 mol^{-1} s^{-1}, for $RAr = Ph_2$, to 2.5×10^{-2}, for $RAr = Me(p\text{-}ClC_6H_4)$.[122] Chloropentasiloxanes were also studied. Solvolytic ring cleavage in a series of 1,3,5-trisila-2,4-diaza-6-oxa-cyclohexanes in buffered aqueous methanol was shown to be faster than in analogues containing only one Si–N–Si unit, though both series reacted by similar mechanisms.[123] Silyperoxides, such as $Me_3SiOOSiMe_3$, are much more reactive than $t\text{-}BuOOt\text{-}Bu$ in oxidations proceeding by nucleophilic attack on the peroxy group.[124] H_3SiOH–H_2O interactions have been studied theoretically[125] and by high-pressure mass spectrometry.[126] Lukevics and Dzintara have reviewed the alcoholysis of hydrosilanes.[127] Steric effects in the acetolysis of $RSiMe_2Cl$ (R = Me, Et, n-Pr, n-Bu, c-Hex) in acetic acid/acetic anhydride are better correlated[128] with Taft's E_s than Cartledge's $E_s(Si)$ values. Acetolysis of the configurational isomers of 1-chloro-1,3,5-trimethyl-3,5-(bis-trimethylsiloxy)cyclotrisiloxane have also been studied.[129]

Bassindale and Stout[139-132] determined equilibrium constants, K, for reactions involving salt formation between Me_3SiX (16), X = Cl, Br, I, OAc, ClO_4, CF_3SO_3, and sixteen nucleophiles. In competition experiments in CH_2Cl_2, K for equation (5) (relative to Nu = N-methylpyridone) spanned a

$$Nu + Nu'SiMe_3^+ \rightarrow Nu' + NuSiMe_3^+ \qquad (5)$$

range from 0.1 (pyridine) to 32800 for N-methylimidazole.[130] For Nu = N-trimethylsilylimadazole ion, associated bis(trimethylsilyl)imidazolium salts are formed,[131] with K for their formation decreasing in the order $ClO_4 > CF_3SO_3 > I > Br > Cl$. The exchange reaction of these salts with the parent species can be followed by NMR, with reactivity decreasing along the series $Cl > Br > CF_3SO_3 > I$, ClO_4.[132] NMR data on the product of reaction of trityl perchlorate with Ph_3SiH suggest that Ph_3Si^+,[133] formed in acetonitrile or 1,2-dichloroethane, is uncomplexed by solvent. (16) with X = CF_3SO_3 may be synthesized by the thermal rearrangement of $Me_2Si(H)CH_2OS(O)_2CF_3$, with first-order kinetics.[134] F^- exchange in the pentacoordinate anions $Ph_2SiF_3^-$ (17), $PhMeSiF_3^-$ (18), $PhSiF_4^-$, and $Ph_3SiF_2^-$ has been studied by ^{19}F NMR. Replacement of Ph in (17) by Me to give (18) lowers $\Delta G^{\ddagger}$ for F site exchange from 11.7 to 9.9 kcal mol^{-1}, with both of these values being much less than that for the isoelectronic Ph_2PF_3.[135]

Si-H reactivity in pentacoordinate dihydrosilicon compounds [e.g., (19), R = H] is greater than that in tetracoordinate analogues, such as α-Np(Ph)SiH_2 (20). (19) reacts with methanol to give the methoxide α-Np($Me_2NCHRC_6H_4$)Si(OMe)$_2$ in 94% yield after 12 h at 25 °C. (20) fails to produce any methoxide after 48 h at 70 °C, even in the presence of Et_3N.[136] (21) with X, Y, Z = H_2Ph, in which the Si—H bonds occupies an equatorial site (established by X-ray crystallography[137]), reacts more quickly with RR'CO than derivatives in which H is apical. It is concluded that nucleophilic displacement of H^- from Si with retention of configuration takes place by apical attack of the incoming nucleophile without electrophilic assistance, with Si—H in the equatorial site. [Electrophilic attack by Hg(II) on apical Si—C in a series of silatranes appears particularly facile according to kinetic studies.[138]] Pseudorotation in (19), R = Me, does not take place in $CDCl_3$ over the temperature range -100 °C to $+20$ °C, while the corresponding trihydrosilane (21) with X, Y, Z = H_3 undergoes such a process which is rapid

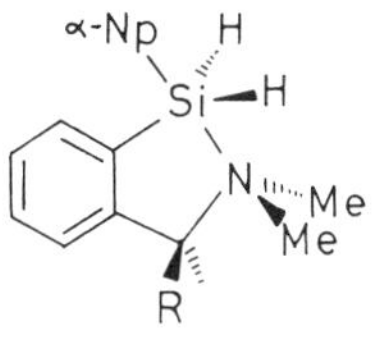

19

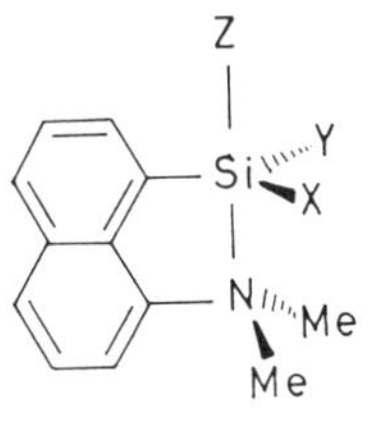

21

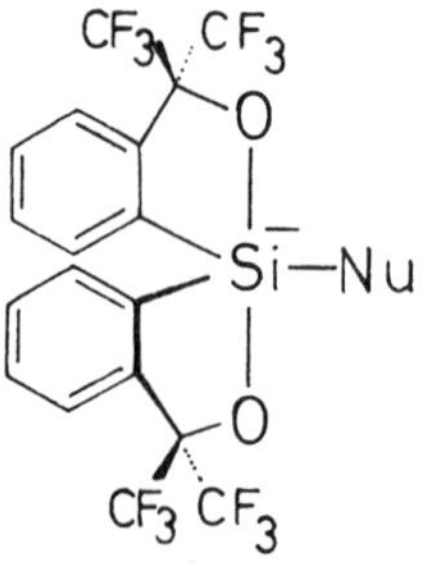

even at $-100\,°C$.[139] NMR studies of the pentacoordinate (8-dimethyl-aminonaphthyl)silanes (21)[140] show that for X, Y, Z = $(OR)_3$, pseudorotation has a low activation free energy. For X = Y = Cl, Br, OR pseudorotation is less favored. When X, Y and Z are all different [e.g., X, Y, Z = PhMeH, MeHCl, PhMe(OMe)] pseudorotation occurs with a barrier of >20 kcal mol^{-1} and may occur with the breaking of the Si$-$NMe$_2$ bond. Corriu has also reported important synthetic studies which have mechanistic implications.[141,142] Both penta- and hexacoordinate anionic silicon complexes [e.g., potassium tris(catecholato)silicate(IV)] (22) may react with nucleophiles (such as RMgX) to form Si$-$C bonds. For example, (22) reacts with EtMgBr to give Et$_3$SiOC$_6$H$_4$OH at $0\,°C$. R$_4$Si and R$_3$SiH may also be prepared. Extensive studies of pseudorotation and other reactions of 10-Si-5 siliconates, both as stable species and reactive intermediates derived from four-coordinate (8-Si-4) in the presence of catalytic quantities of weak nucleophiles, have been reported by Martin, Stevenson, Wilson, and Farnham.[143,144] For a series of anions (23), undergoing a Berry pseudorotation,[143] a good linear correlation has been observed between the free energy of activation $\Delta G^{\ddagger}$ with σ^* for ten monodentate groups, Nu in (23), including Nu = n-Bu, Ph, OC$_6$F$_5$, CN, F. Electron-withdrawing groups favor pseudorotation. (23) with Nu = Ph (Et$_4$N$^+$ countercation) inverts with $\Delta G^{\ddagger}$ = 26.0 kcal mol^{-1}, which is 2.3 kcal mol^{-1} less than the isoelectronic isostructural phosphorane. The rate of inversion of (24) in CD$_2$Cl$_2$ in the presence of catalysts, RC$_6$H$_4$CHO (25), R = H, Me, MeO, Cl, NO$_2$, NMe$_2$, are found to be first order in catalyst with electron-releasing R being the most effective and to have a large negative entropy of activation.[144] The equilibrium between (24) and (25) (R = NMe$_2$) was studied directly by NMR at low temperatures and found to have a $\Delta H^{\ddagger}$ = -2.1 kcal mol^{-1} and $\Delta S^{\ddagger}$ of -47.7 eu. The intermediate 10-Si-5 species formed from (23) and (25) (R = NMe$_2$) was shown to invert by a nondissociative pseudorotation with $\Delta G^{\ddagger}$ of 10.2 kcal mol^{-1}. Other structural[145] and theoretical[146] aspects relevant to pentacoordination on silicon and nucleophilic substitution have been reported.

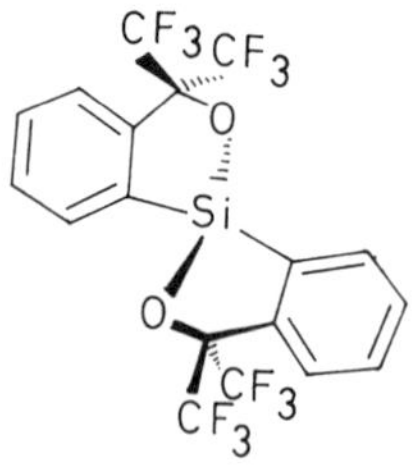

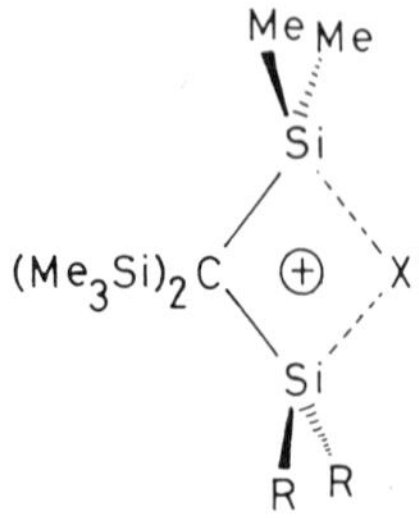

Pentacoordinate intermediates have been implicated in various fluoride catalyzed transformations.[147-152] The dependence on R and X of the relative proportions of Si$-$F products from the rearrangement of Me$_2$RSiX have been further studied[147] and the importance of the capacity of the migrating group R to support a negative charge confirmed. F^--catalyzed reduction of ketones by $PhMe_2SiH$ has been shown to involve hydride transfer with reactive substrates such as $CF_3C(O)Ph$.[148] Cleavage of the vinyl group from $R_3Si-CH=CH_2$ by F^-,[150] reaction of $MeC(O)SiMe_2Ph$ to give $PhC(Me)HOH$ and $[Me_2SiO]_x$ using $F^-/THF/H_2O$,[151] as well as the addition of silyl enol ethers to nitroaromatics[152] have been reported. Substitution and abstraction reactions of $F^{\cdot}$ and F^- on CH_4 and $SiH_4^{\cdot}$ have been subjected to theoretical study.[153] Rozell and Jones have described a LiCl-catalyzed cleavage of Me_3Si from $C_5H_4SiMe_3^-$.[149]

Eaborn and his group have continued their synthetic studies on organosilicon compounds containing highly hindered silicon centers such as $(Me_3Si)_3$-$CSiR_3$, $(Me_3Si)_2C(SiR_3)_2$, or $(Me_3Si)_2[(CH_2=CH)Me_2Si]CSiR_3$[154-164] some of which contain data of mechanistic interest. In contrast with earlier work in which compounds of the type $TsiSiRR'X$ [$Tsi = (Me_3Si)_3C$] were found to undergo 1,3 Si to Si methyl migrations on reaction with electrophilic reagents such as silver salts, AgY, Eaborn and Reed[154] now report that the compounds $TsiSiMe(OMe)I$ and $TsiSi(OMe)_2I$ give only the unrearranged products on reaction with AgY, $Y = NO_3$, ClO_4, CF_3SO_3. A reinterpretation of the role of anchimeric assistance [such as that seen in (26) with $R = X = Me$] for more electronegative groups in the 1-Si position is evidently required. $(Me_3Si)_2C(SiMe_2X)(SiMe_2Y)$ (27), required for solvolysis and substitution reaction kinetic studies, have been obtained either by rearrangement of $TsiSiMeICl$ using AgY, or by reaction of (27) ($X = Y = H$) with halogen, X_2, to give (27) ($X = Y = Cl$, Br, I).[155] $C(SiMe_2X)_4$, $X = H$, F, Cl, Br, I, OR, $OC(O)R$ have also been prepared.[156] To cast further light on the solvolysis of (27) ($X = Me$, $Y = I$, ClO_4, CF_3SO_3) in methanol, which was thought not to involve the intermediacy of cations analogous to (26), attempts have been made to study systems involving unimolecular $Si-X$ cleavage without nucleophilic involvement of the solvent.[157] It was shown that $(Me_3Si)_2C(SiMe_2OMe)(SiMe_2Cl)$ solvolyzes in 9:1 MeOH–dioxane at a rate independent of [NaOMe]. Solvolysis occurred more rapidly in 8:1 CF_3CH_2OH–dioxan to give $(Me_3Si)_2C(SiMe_2OMe)(SiMe_2OCH_2CF_3)$. $(Me_3Si)_2C(SiMe_2OMe)(SiPh_2Br)$ gave the rearranged product $(Me_3Si)_2$-$C(SiMe_2OEt)(SiPh_2OMe)$ on reaction in EtOH, indicative of the involvement of cations such as (26) ($R = Ph$, $X = OMe$) in which subsequent attack of the solvent at the less hindered Si is proposed.[157] This anchimeric assistance is so great that hydride may be displaced from $(Me_3Si)_2C(SiMe_2OMe)$-$(SiMe_2H)$ by AgO_2CCF_3 to give $(Me_3Si)_2C(SiMe_2OMe)(SiMe_2O_2CF_3)$. Other studies suggest that γ-O_2CCF_3 provides less anchimeric assistance than γ-OMe.[158] Reaction of $(Me_3Si)_2C(SiMe_2X)(SiEt_2Y)$ (28) with $X = Cl$, $Y = I$, and $X = I$, $Y = Cl$, both give mixtures of (28), $X = Cl$, $Y = F$ and (28), $X = F$, $Y = Cl$, on treatment with $AgBF_4$[159] consistent with anchimeric assistance arguments. Vinyl substituents in $TsiSi(CH=CH_2)_2Cl$ (29) appear to affect only slightly the reactivity with nucleophiles in MeOH or MeCN compared with those observed for the saturated analogue TsiSiMe$_2$Cl.[160] The formation of the simple trifluoroacetate substitution product and the rearranged product (Me$_3$Si)$_2$C[Si(CH=CH$_2$)$_2$Me](SiMe$_2$O$_2$CCF$_3$) from reaction of (29) with AgO$_2$CCF$_3$ in CF$_3$CO$_2$H is taken to support the involvement of cationic intermediates such as (26) (R = Ph, X = OMe). (29) reacts with other nucleophiles in methanol without any evidence of solvolysis. Nucleophilic attack on the O bonded Si in VsiSiMe$_2$OC(O)R, Vsi = (Me$_3$Si)$_2$[(CH$_2$=CH)Me$_2$Si]C, R = CH$_3$, CF$_3$ is sufficiently sterically hindered to permit preferential attack at the carboxylate carbonyl carbon to give VsiSiMe$_2$OH to be seen.[161,162] Ease of attack on Si followed the order F$^-$ > SCN$^-$ > N$_3^-$ and on C, OCN$^-$ > N$_3^-$ > F$^-$ > SCN$^-$. Reaction of TsiSiMe$_2$R, R = CH=CH$_2$, CH$_2$CH=CH$_2$, C≡CPh with electrophiles such as HX and X$_2$ (X = Cl, Br, I) were not diverted toward attack on R as a consequence of steric hindrance toward attack on Si. Si—R cleavage to give TsiSiMe$_2$X was the sole reaction. No kinetic data were reported.[163] A remarkably thermally stable silane triol, TsiSi(OH)$_3$, has been characterized.[164,165] The dependence on n of the rates of cleavage R (m-ClC$_6$H$_4$CH$_2$, PhC≡C and Cl$_2$CH) from a series of RSiMe$_n$(OMe)$_{3-n}$, n = 0, 1, 2, 3, in methanolic NaOMe [in the presence of Si(OMe)$_4$ to scavenge water] results from the lowering of the energy of a pentacoordinate intermediate and the raising of the energy barrier on the way to it or from it by opposing polar and steric effects.[166] Specific second-order rate constants for the cleavage of RCH$_2$SiMe$_3$ in 2.00 M NaOMe in MeOH at 50 °C are 7.0 × 10^{-7} M^{-1} s^{-1} for R = furan-2-yl and 2.6 × 10^{-6} for R = 2-thienyl. Isotope effects in MeOD point to rate-determining loss of RCH$_2^-$ and allow estimates of pK_a values of RCH$_3$ to be made, viz. 40.6 and 39.7, respectively, for 2-methylfuran and 2-methylthiophen.[167] Rates of anionic catalyzed cleavage of Si—C in substituted aryl-SiMe$_3$ in the presence of PhCHO as carbanion scavenger[168] correlate with Hammett's σ constants of the aryl substituent rather than with carbanion stability. The effect of silyl substituents on neighboring carbon radical centers[169,170] or of carbon cations β-[171,172] or γ-[173,174] to silicon have also been investigated.

Photolysis of benzyltrimethylsilane leads to an inefficient isomerization to o-tolyltrimethylsilane. A Si—C homolysis[175] leading to a Me$_3$Si˙ PhCH$_2^˙$ radical pair is proposed. Formation of PhCH$_2$C(R)OCH$_2$Si(Me)$_2$O on the photolysis of [(acyloxy)methyl]benzyldimethylsilanes arises from a 1,2 C—Si acyloxy migration of the radical formed on Si-benzyl Si—C homolysis followed by benzyl radical capture within the solvent cage. CIDEP was observed for two products from the addition of R$_3$Si˙ to the carbonyl oxygen and a ring C=C of substituted p-benzoquinones.[176] Evidence for the isomerization of Me$_2$Si(H)CH$_2^˙$ to Me$_3$Si˙ has come from trapping studies on the pyrolysis of 4-(dimethylsilyl)-1-butene.[177] The gas-phase isomerization of ClCH$_2$SiMe$_2$H to Me$_3$SiCl has been studied by two groups.[178,179] Pentamethyldisilane reacts with t-BuO˙ with a second-order rate constant at 294 ± 1 K of 1.7 × 10^7 M^{-1}

s^{-1}, which is *ca* 3 times that for Et_3SiH.[180] The reactivity of the $Me_3SiSiMe_2$ product (**30**) with alkyl halides, olefins, and carbonyl compounds were studied and the k_2 obtained compared with values seen for reaction of these same substrates with $Et_3Si^•$, n-$Bu_3Ge^•$, and n-$Bu_3Sn^•$. (**30**) abstracts Br from t-BuBr at 294 ± 2 K with $k_2 = 2.6 \pm 0.2 \times 10^6$ M^{-1} s^{-1} and adds to tetrachloroethylene with $k_2 = 4.0 \pm 0.15 \times 10^6$ M^{-1} s^{-1}. Di and trisilanes, such as Si_2H_6, $Me_3SiSiMe_2H$, Si_2Cl_6, and Me_8Si_3, react with hydrogen atoms by a S_H process while monosilanes are unreactive.[181] Electron-transfer processes have been proposed[182-184] for the oxidative ring-cleavage reactions of cyclopolysilanes and the photochemical chlorination of dodecamethylcyclohexasilane with CCl_4/CH_2Cl_2. Blinka and West propose cyclosilane–MCl_3 (M = Fe, Al) intermediates from studies of the Lewis acid catalyzed rearrangements of the permethylcyclosilanes, $(Me_2Si)_n$, $n = 5$-12.[185] Russian workers report studies of the oxidation of polysilanes by peroxyacids.[186,187] The long-lived radical, $[t$-Bu(Cy-Hex)Si=Si(Cy-Hex)t-Bu]$^{\cdot}$, has been characterized[188] and intramolecular aryl group exchange in tetraaryldisilenes studied by ^{29}Si NMR.[189]

Gas kinetic studies on the pyrolysis of diallydimethylsilane,[190] various alkyl-substituted 1,3-disilacyclobutanes,[191] (trimethylsilyl)cyclopropane,[192] ethynylsilane,[193] 1-silapropene,[194] dimethyl-*cis*-1-propenylvinylsilane,[195] 1,1-dimethyl-1-silacyclopent-3-enes,[196] 7-silabicyclo[2.2.1]heptadienes,[197] and Si_2F_6 [198,199] have been reported. The reactivity and further transformation of intermediates, such as silylenes,[200-202] silenes[203-206] (and silene–silylene isomerization)[207] and silylsilenes and disilenes[208] formed in these thermolytic processes, have been reported, as well as the gas-phase chemistry of dimethylsilanone[209-212] and related theoretical studies on silanone and silanethione.[213,214] Silylene–silanone interconversions have also been investigated.[215,216] Raabe *et al.*[217] and Brook[218,219] have reviewed this area of chemistry. The stabilities of silenes, silylenes, and silicon containing cations have been discussed using a thermochemical kinetic approach.[220]

4.4. Germanium

The use of ^{29}Si 2D-COSY NMR on aqueous tetramethylammonium germanosilicate solutions obtained by the addition of GeO_2 to 1.1 M Me_4N^+ silicate solutions (Si : N = 0.5) permits the unequivocal structural assignment of a double four-membered ring (a cubic octameric cage) and provides support for the existence of a double three-membered cage structure (prismatic hexameric cage).[221]

Alkyl germanium compounds [such as $(CF_3)_3GeX$, X = F] yield both five- and six-coordinate species on reaction with F^- in $MeCN/H_2O$ solutions.[222] α-Halosubstituted alkyl groups (R = CHX_2, CX_3; X = Cl, Br) may be cleaved from Me_3GeR in aqueous alcoholic ammonia/ammonium chloride at 25 °C, though the nature of the germanium-containing product was not described. Rates are dependent on [buffer], though the order to reactivity (which spans a range of *ca* 10^6) is $CBr_3 > CCl_3 > CHBr_2 > CHCl_2$. Rate-determining decomposition of five-coordinate intermediates, e.g., $ROGeMe_3CHX_2^-$, was proposed.[223] The mechanisms of the photochemical reaction of polyhaloalkanes with allyltriorganogermanes (and -silanes) have been investigated using ^1H CIDNP.[224] The effect of changes in X in (31) and Y in (32), as well as solvent polarity, point to a mechanism for the reaction of substituted arylthiotrimethylgermanes with aroyl chlorides shown in equation (6), in which rate-determining ArS addition to carbonyl carbon occurs

$$Me_3GeSC_6H_4X + YC_6H_4C(O)Cl \rightarrow Me_3GeCl + YC_6H_4C(O)SC_6H_4X \qquad (6)$$
$$\mathbf{31} \qquad\qquad \mathbf{32}$$

following Ge–S heterolysis of an initially formed five-coordinate intermediate.[225] Studies of the formation[226] and reactions[227,228] of germylene intermediates, $R_2Ge:$, and of germanone and germathiones[229,230] have been described.

4.5. Nitrogen

Two processes have been proposed for the formation of N_2O and nitrite from the trioxodinitrate dianion, $N_2O_3^{2-}$, namely, via nitrosyl hydride and nitrite and via nitric oxide and $HONO^-$. The N isotope distribution in the N_2O formed (pH 7.0, 25 °C, anaerobic) in the presence of excess ^{15}NO supports[231] the involvement of HNO. Photochemical decomposition of $N_2O_3^{2-}$ is also found[232] to proceed by a similar mechanism and to give singlet and triplet NO^- as reactive intermediates. The formation of peroxonitrite in the presence of O_2 is taken as evidence of triplet NO^- formation. Dimerization of HNO is thought to be diffusion controlled with elimination of water from $H_2N_2O_2$, presumed to occur via $N_2O_2^-$. The kinetics of reaction between $N_2O_3^{2-}$ and $[Ru(NH_3)_6]^{3+}$ have also been studied.[233] NO has been shown to react with superoxide in alkaline solution to form peroxonitrite.[234] In aqueous base, NO reacts with hydroxylamine with a rate law $-dP_{NO}/dt = k_2[NH_2O^-]P_{NO}$ confirming that the rate-determining step is the abstraction of H from the substrate by NO.[235] Trace amounts of oxygen catalyze the reaction and affect the mole ratio of N_2 to N_2O in the products. The intermediacy of N_2O_3 is implicated. The reactivity of the HNO intermediate appears to be different from that derived from the trioxodinitrate ion, possibly arising from $HON=NOH$ equilibria and electronic-state differences. NO has been shown[236] to abstract H from the N—H in MeNHOH, NH_2OMe, and MeNHOMe though from the α-C—H with Me_2NHOH. O-Methylation inhibits the abstraction reaction by suppressing anion formation. A rate ratio for NH_2O^- vs NH_2OH of 3.38×10^5 was determined. In the case of MeNHOH, reaction with NO leads to the stable nitrosohydroxylamine, MeN(NO)OH. By monitoring the formation of N-nitrosohydroxylamine-N-sulphonate (NHAS) the kinetics of the reaction of predissolved nitric oxide with aqueous sulfite in the pH range 4–10 have been shown[237] to follow the rate expression, $d[NHAS]/dt = k_a[NO][HSO_3^-] + k_b[NO][SO_3^{2-}]$, where $k_a = 32 \pm 10 \ M^{-1} s^{-1}$ and $k_b = 620 \pm 100 \ M^{-1} s^{-1}$ at 25 °C. The question of the state of the NO in solution ("dissolved" or "hydrolyzed") was not clarified. The pH and ionic-strength dependence on the acid hydrolysis of nitridotrisulfate suggests that $N(SO_3)_3^{3-}$ and a proton react together in the rate-determining step.[238]

Nottingham and Sutter[239] have compared reactivity differences between gas phase (k_g) and solution (k_s), by studying the kinetics of the reaction of NO and chlorine in CCl_4 and acetic acid and those of NO and dioxygen in CCl_4. The observed third-order rate ratio k_s/k_g obtained for NO/Cl_2 was 140 in CCl_4 and 766 in acetic acid, compared with 450 derived from transition-state theory for termolecular gas-phase processes. IR and Raman spectra of anhydrous HNO_3 reveal[240] the presence of oligomeric chains $[HNO_3]_x$, NO_2^+, $[NO_3 \cdot 3HNO_3]^-$, two forms of $[HNO_3 \cdot H_2O]$, the ion pair $H_2OH^+ \cdots ONO_2^-$, and the association $H_2O \cdots HONO_2$. Values for K_{eq}(298 K) for the ionic dissociation $N_2O_4 \rightleftharpoons NO^+ + NO_3^-$, corrected for homolysis to give NO_2, are 6.3×10^{-10} (nitromethane) and $5.0 \times 10^{-8} \ M$ (propylene carbonate), with the values correlating with the basic properties of the solvents.[241] The effect of added metal nitrates on these equilibria in sulfolane has also been studied.[242] The forward and reverse rate constants and dissociation constants for the reactions of N_2O_3 in sulfolane were determined.[241,243] N_2O_3 dissociates with k_b(303 K) = 8.9 s^{-1}. NO and N_2O_4 react together with $k_f = 1350 \ M^{-1/2} \ s^{-1}$. For NO with NO_2, $k_f = 3.6 \times 10^5 \ M^{-1} \ s^{-1}$. The value of K_{eq}(303 K) for the reaction $N_2O_3 \rightleftharpoons NO + NO_2$ is $2.5 \times 10^{-5} \ M$. Similar rate studies for the ionic and molecular dissociation of N_2O_4 have also been described.[244] K_{eq}(298 K) for the ionic dissociation of NO_2Cl has been found to be $4.0 \times 10^{-19} \ M$ (nitromethane) and $2.0 \times 10^{-18} \ M$ (propylene carbonate) and K_{eq}(303 K) = $6.3 \times 10^{-18} \ M$ (sulfolane).[245] By using rotating disk voltammetry, it proved possible to obtain estimates of the equilibrium constants for the dissociation of NO_2Cl to give Cl_2 and NO_2, though the reactions were complicated by the presence of trace amounts of water. Additional mechanistic studies of ipso-nitration have been reported.[246–252] ^{15}N NMR spectra of the nitration of durene displays[246] nuclear polarization for nitrodurene signals both under conditions of nitrous-acid catalysis as well as under conditions leading to reaction via nitronium ions. These and other observations point to the partition-ing, between dissociation and recombination, of radical pairs $[ArH^+, NO_2^+]$ thought to arise, in the durene/NO_2^+ reaction, via C—N bond homolysis in the intermediate formed by ipso attack. The nitrous-acid catalysed nitration of p-nitrophenol in aqueous nitric acid remains first order in $[p$-nitrophenol$]$ as $[HNO_2]$ is increased to a point where the reaction becomes zeroth order in $[HNO_2]$, consistent with p-nitrophenoxide oxidation to give p-nitrophenoxyl radical which then reacts with NO_2 to give 2,4-dinitrophenol.[250] Rearrangement studies of nitroaromatics in CF_3SO_3H avoid complicating degradative and sulfonation processes seen with sulfuric acid.[248] For example, 2,3-dinitrophenol reacts cleanly with first-order kinetics to give 98–99% 2,5-dinitrophenol, confirming the importance of partial thermodynamic control in certain high-temperature nitrations. A reversible nitro-nitrito rearrangement has been observed for a Wheland intermediate derived from the ipso-nitration of the substituted phenol as in equation (7). Dissolution of (33) in $CHCl_3$ leads to (34) and (35).[249] (33) can be formed via (34) on the addition of nitrous acid to authentic (35). Other workers have recently inferred such a

$$
\begin{array}{ccc}
\text{Cl} \overset{O}{\underset{Me}{\bigcirc}} \text{Cl} & \text{Cl} \overset{O}{\underset{Me}{\bigcirc}} \text{Cl} & \text{Cl} \overset{O}{\underset{Me}{\bigcirc}} \text{Cl} \\
\text{Me } NO_2 & \text{Me } ONO & \text{Me } OH
\end{array} \qquad (7)
$$

33   **34**   **35**

rearrangement without providing direct evidence.[252] An association–dissociation mechanism has been favored over outer-sphere electron transfer for aromatic radical cation and NO_2 formation from the reaction of NO_2^+ and ArH. Aspects of the important equilibrium between Ar^+ and NO_2 and $Ar(H)NO_2^+$ have been studied in CH_2Cl_2[253] for methylnaphthalene radical cations. However, the nitro-product isomer distributions are different from those obtained in electrophilic nitrations and nitrous-acid catalyzed nitrations and indicate that reaction between free radical cation and free NO_2 is not an elementary step in these processes. Solvent effects on the nitration of fluorene, ArH, with N_2O_5[254] suggest that, in CCl_4, initial formation of a radical pair, $[Ar(H)ONO_2^{\cdot}, NO_2^{\cdot}]$, occurs, though such processes are suppressed in more polar media where an ionic electrophilic mechanism is favored. The mechanisms of nitric-acid oxidation of acetophenone[255] and of the photohydroxonitration of biphenyl by nitrate ion[256] have been investigated. The oxidation of benzyl alcohols by nitrous or nitric acid in sulfuric acid occurs via the common oxidant NO^+ with hydride abstraction from the alcohol being the rate-determining step.[257] Two groups[258–260] have studied redox reactions of NO_2 with organic and inorganic free radicals. Huie and Neta[258,259] report data for oxidations by $OH^{\cdot}$, $Br_2^{\cdot}$ ($k = 2 \times 10^7$ M^{-1} s^{-1}), BrO_2, $CO_3^{\cdot}$, $PhNH_2^+$ ($k = <10^5$ M^{-1} s^{-1}) and reductions by phenols, anilines and $O_2^{\cdot}$, HO_2, $O_3^{\cdot}$, and SO_3^{2-}. These occur via simple electron transfer, except for SO_3^{2-}, which reacts via oxygen atom transfer. For NO_2-phenoxides,[259] marked substituent effects are seen, with rates in the range $>10^9$ M^{-1} s^{-1} for hydroquinone to $<10^4$ for phenols containing electron-withdrawing substituents. Kinetics of the reaction of NO_2 with the hindered phenol, 2,6-di-*t*-butyl-4-methylphenol, to form 2,6-di-*t*-butyl-4-methyl-4-nitrocyclohexa-2,5-dienone have been reported.[260]

$NO_3^{\cdot}$, formed by pulse radiolysis, may react either by hydrogen atom abstraction [e.g., with methanol ($k = 2.1 \times 10^5 \, M^{-1} s^{-1}$)], by alkene addition, or by electron transfer [e.g., with Cl^- ($k = 7.1 \times 10^7 \, M^{-1} s^{-1}$)].[261] NO_3^{2-}, formed by pulse radiolytic reduction of nitrate, reacts with dioxygen, giving nitrate and O_2^-, with a rate constant in reasonable agreement with published data.[262] Between 70-100 °C, N_2H_4 is oxidized by HNO_3 to form N_2, N_2O, and NH_4^+ in a complex process via HN_3, HNO_2, and N_2H_2 according to the rate law $-d\ln[N_2H_4]/dt = k[NO_3^-][H^+]^2$.[263] Monolabeled diazonium ion, $^{15}N \equiv ^{14}NH^+$, could not be detected in the reaction of $(Me_3Si)_2NH$ with $^{15}NO^+$ in CH_2Cl_2 at -40 °C.[264] Although $MeON_2^+$ could be prepared from $N_2O +$ $MeF/SbF_5/SO_2F_2$, no HON_2^+ could be detected in the reaction of N_2O with superacids.[265] HNO_2 reacts with m-chlorophenylhydrazine according to the rate equation,[266] $v_0 = k_1[HNO_2]^2[ArNHNH_3^+]/[H^+]$, consistent with rate-determining electrophilic attack by N_2O_3 on the free base form of the hydrazine, with the second-order rate constant being close to that expected for encounter control. It was not possible to observe nitrosation of hydrazine itself by N_2O_3. Similar studies of the N_2O_3–azide reaction yield a rate constant for the process $N_2O_3 + N_3^- \rightarrow N_3NO + NO_2^-$ that is similarly close to the encounter rate. The activation energy for the overall process, $2HNO_2 \rightarrow N_2O_3 + H_2O$, was obtained under conditions for which the rate expression for the azide/nitrous acid reaction reduced to rate $= k_4[HNO_2]^2$, with the value obtained ($67.3 \pm 2.0 \, kJ \, mol^{-1}$) being in reasonable agreement with that obtained from the reaction of thiosulfate with nitrous acid under conditions where N_2O_3 formation is also rate determining. Nitrosation of N-methylaniline in the presence of thiosulfate involves nitrosothiosulfate (36) as the nitrosating species.[267] This is significantly less reactive by factors of 10^4-10^5 than NOX (X = Cl, Br, SCN) probably as a consequence of the formal negative charge on the electrophile. (36) decomposes to give dithionite and NO according to the rate law $-d[(36)]/dt = k_1[(36)] + k_2[(36)]^2$. Two parallel reaction pathways are proposed: first, homolysis of (36) to give $S_2O_3^-$ followed by rapid dimerization, and second, bimolecular reaction of (36) with concurrent S–N fission and S–S formation. Mechanisms of other organic nitrosation processes and the reactions of nitrosated substrates have been described.[268-277] Studies of bromide catalysis of the nitrosation of a series of aliphatic secondary amines show[270] that the rate of the NOBr–amine reaction is insensitive to the nature of the substrate. Two studies focus on nitrosation reactions of amino acids.[268,272] 1-Naphthylamine[269] and benzenesulfinic acid[271] have been used as substrates. Denitrosation of N-nitroso-2-imidazolidone,[273] N-acetyl-N^1-nitrosotryptophan,[272] and a series of 3-alkyl-1-methyl-1-nitrosothioureas[274] has been studied.

Sonolysis of water under an atmosphere of argon and 1:1 $^{14}N \equiv ^{14}N$ and $^{15}N \equiv ^{15}N$ produce the nitrogen-containing products $^{14}N \equiv ^{15}N$, NO_2^-, NO_3^-, NH_3, and N_2O.[275] N_2O is itself readily decomposed to N_2, O_2, NO_2^-, and NO_3^- under similar conditions.[278] The mechanisms of the acid-catalyzed hydrolysis of 3-hydroxy-1-phenyl-3-alkyltriazenes,[279] 1,3-diphenyl-3-methyl-triazenes,[280] and 1,3-dialkyltriazenes (37)[281] have been reported. The buffer and pH dependence of the reaction rate and the solvent isotope effects for (37) (R = Me, Et, i-Pr) suggest that the reaction pathways are more complex than previously reported, being different for the three compounds. N_3^- is oxidized reversibly to $N_3^{\cdot}$ by $[IrCl_6]^{2-}$ at 25 °C with an apparent second-order rate constant of $k = 1.8 \times 10^2 \ M^{-1} \ s^{-1}$, with the azidyl radical dimerizing to N_6. Estimates of 1.33 V for the reduction potential of azidyl radical[282,283] and a self-exchange rate for the $N_3^{\cdot}/N_3^-$ couple of about $4 \times 10^4 \ M^{-1} \ s^{-1}$ have been made. $N_3^{\cdot}$, generated from azide ion and $OH^{\cdot}$, reacts with aniline and phenoxide by electron transfer with second- order rate constants of $(3-5) \times 10^9$ $M^{-1} \ s^{-1}$.[284] PhX, X = H, Me, MeO, are unreactive and phenol is much less reactive than phenoxide. The rate constant for reaction of azide with hydrogen atoms has been shown by ESR to be $1.85 \times 10^9 \ M^{-1} \ s^{-1}$.[285] In acid, $HN_3^{\cdot}$, produced by attachment of e_{eq}^- to HN_3, decomposes to give $N_3^{\cdot}$.[286] Photolysis of azide ion in ammonia leads to the diazene radical anion,[287] probably via the addition of an initially formed $N_3^{\cdot}$ to NH_3 with subsequent N_2 loss, hydrogen shift, and deprotonation.

The pressure dependence of the rates of cis to $trans$ isomerization of azanorbornane[288] seems to be more consistent with an inversion rather than a radical-pair mechanism, though, in the absence of studies on the effect of solvation, this conclusion was drawn only tentatively. Inversion was also favored from the linear relationship established[289] between $\Delta V^{\ddagger}$ for the thermal cis–$trans$ isomerization of 4-dimethylamino-4'-nitroazobenzene in a range of solvents of varying polarity and the partial molar volumes of the $trans$ isomer in the same solvents. First-order rate constant for the decomposition of secondary C-nitroso dimers, such as for $trans$-α-alkyl substituted 2-nitroso-1-phenylethane, decrease with increasing bulk of the alkyl substituent.[290] Decompositions of C-nitroso dimers with one α-H were more rapid in EtOH than in nonpolar solvents, while those dimers without α-H were relatively indifferent to the solvent medium. Dissociation of arylnitroso compounds have also been studied,[291] with $K(293 \ K)$ in MeCN for 2,4,6-trimethylnitrosobenzene equal to $8.26 \pm 0.80 \times 10^{-3}$ with $k_f = 1.07 \pm 0.05 \times 10^{-3} \ s^{-1}$. Activation parameters have been reported[292,293] for the homolytic decomposition of a series of 14 $trans$-hyponitrites (RONNOR) in t-butylbenzene. The data for a small series of benzyl derivatives could not be correlated satisfactorily with either Hammett's σ or σ^+.

Hydrazine oxidation by triiodide has been reexamined kinetically.[294] Reaction is thought to proceed via a 1:1 charge transfer complex of I_2 and NH_2NH_2 prior to electron transfer and via HOI and NH_2NH_2, rather than via the predominant components of the system, I_3^- and $NH_3NH_2^+$. Formation of protonated hydrazyl radical cation $N_2H_5^+$ is thought to be rate limiting in the reaction of hydrazinium ions with CrO_2^{2+}.[295] Electron-transfer processes involving hexaalkylhydrazine dications,[296] tetraarylhydrazine dications,[297] and tetraalkylhydrazine radical cations[298,299] have been reported. Solvent effects on the self-exchange rates for the hydrazine/hydrazine radical cation pairs derived from 8,8′-bi-8-azabicyclo[3.2.1]octane and related compounds have been studied.

Reaction of chlorine (as $HOCl/OCl^-$) with *tert*-alcoholamines[300] displays a complicated dependence on [amine], though at pH 11.5 and 12.2 the formation of the N-chloro-quaternary ammonium ion, e.g. $ClN(Me)(CH_2CH_2OH)_2^+$, is rate determining. A good correlation between $\log k_{exp}$ and $\Sigma\sigma^*$ is found, with ρ^* equal to -2.7 ± 0.1 being consistent with rate-controlling electrophilic attack on amine N by HOCl. The related N-bromoalcoholamine, e.g. $BrN(CH_2CH_2OH)_2$, decomposes at pH 8.0, 25 °C, with $t_{1/2} = 1110$ min to give imines which hydrolyze to aldehydic products.[301]

$$RNHCl + 3I^- + H^+ \rightarrow I_3^- + RNH_2 + Cl^- \tag{8}$$

$$RNCl_2 + 6I^- + 2H^+ \rightarrow 2I_3^- + RNH_2 + 2Cl^- \tag{9}$$

The rate expression for the reaction of I^- with monochloroamines given in equation (8) and dichloroamines as in equation (9) is $-d[\text{substrate}]/dt = k_{HA}[HA][\text{substrate}][I^-]$. Cl^+ transfer to I^- is proposed,[302] followed by the rapid reaction of the ICl so produced with I^- to give I_3^-. From the role of added acids it was concluded that, for chloramine, proton transfer from HA in a preequilibrium step was unlikely since this would have led to specific rather than general acid assistance. For HA = H_3O^+, the third-order rate constant at 25 °C for NH_2Cl is 2.4×10^{10} M^{-2} s^{-1} and for MeNHCl is 4.04×10^{10}. For the dichloramines, $NHCl_2$ and $MeNCl_2$, the third-order rate constants were much lower at 9.3×10^5 and 1.1×10^6 M^{-2} s^{-1}, respectively. The behavior of NCl_3 is more complex. Kinetics of the reaction of NH_2OH with NH_2Cl[303] and bromamine-T[304] have also been reported. Photochlorination studies of *n*-alkylammonium salts by Cl_2 or Me_2NCl reveal a strong directing effect on the site of substitution by the positive center.[305] The kinetics of HCl elimination from $XC_6H_4CH_2N(Me)Cl$ by reaction with secondary amines in MeCN suggest[306] that the transition state has similar extents of C_β—H and N—Cl bond cleavage and significant π bond formation. Similar studies of the reaction of N-halobenzylalkylamines with NaOMe/MeOH show that for the N-chloro compounds only benzylidenemethylamines are formed, while for N-bromo analogues competing bimolecular elimination and nucleophilic substitution at N are seen,[307] the former being favored with electron-withdrawing aryl groups.

Halogen interchange reactions between N-halosuccinimides and quaternary ammonium halides during the formation of 1:1 and 2:1 complexes (implicated in olefin bromination reactions)[308,309] are thought[310] to proceed as follows:

$$[SX,Y] \rightleftharpoons [>S^{\delta-}\cdots X^{\delta+}\cdots Y^{\delta-}]^{\ddagger} \rightleftharpoons S^- + XY \rightleftharpoons [SY,X^-]$$

Other aspects of the chemistry of N-haloimides and derived imidyl radicals,[311-321] of amidyl radicals,[322-325] and of hydrazyl radicals[326] have been reported. ESR has been employed[327,328] to examine the spin-trapped nitrogen center radicals obtained from the acid decomposition of chloramine-T and from its photolytic decomposition in alkali. Under photolytic conditions sulfur- and carbon-centered radicals are also formed. The kinetics have been studied[329,330] of two processes, one pH independent the other acid catalyzed, for the elimination of OH^- from the primary adduct formed from the reaction of $OH^•$ with Ph_2NH. The amino radical cation, Ph_2NH^+, was found to have a pK_a of 4.2. Me_3N reacts with $OH^•$ in aqueous solution to give the aminoalkyl radical $Me_2NCH_2^•$, its conjugate acid ($pK_a \simeq 3.6$), and the radical cation $Me_3N^+(pK_a \simeq 8.0)$.[331] Kinetic studies on the reaction of Et_2NOH with dioxygen in aqueous media [to give $EtN(O)=CHMe + H_2O_2$] point to a radical chain mechanism in which HO_2 and O_2^- are chain carriers.[332] $N—N$ three-electron σ-bonded species have been further studied.[333-335] Rate constants have been measured[333] for oxidation and reduction reactions of the $N \therefore N$ compound 1,6-diazabicyclo[4.4.4]dodecane radical cation with a range of free radicals and found to be below that expected for diffusion control.

Many studies have been reported[336-362] on the reactions of N-halogeno compounds with a variety of organic substrates. The kinetics of proton-transfer reactions involving nitrogen substrates have been studied.[363-368] Bednar and Jencks have made an extensive study into reactions of HCN, taking advantage of its pK_a to study proton transfers with bases having $\Delta pK_a = 0$ for which the intrinsic barriers against transfer are expected to be most significant compared with barriers for diffusional processes. The rate constant for exchange of a proton between HCN and water in aqueous solution ([H^+] independent) is $40\ s^{-1}$ at 20 °C, $\mu = 1.0\ M$, corresponding to $4 \times 10^{10}\ M^{-1}\ s^{-1}$ for protonation of CN^- by H_3O^+.[364] Involvement of water in a H-bonded $H_2O\cdots CN^-$ intermediate was ruled out. Similar direct transfer of proton from HCN to Me_3N and $MeOCH_2CH_2NH_2$ without the mediation of water was concluded from NMR selective saturation[365] and is the process expected when $\Delta pK_a \leq 5$. Bednar and Jencks propose that the weaker hydrogen bonding to carbon favors direct transfer of protons with respect to processes involving more electronegative acids and bases since, for the latter, desolvation rates are lower and proton transfer rates via water higher. Rate constants and equilibrium constants have been determined[366] using T-jump for the ring opening and closing for the intramolecular hydrogen-bond formation in phenylazoresorcinol anions. Steric effects on the kinetic isotope effects of proton transfers between amines and nitriles have been investigated in MeCN.[367] From the weak dependence on the nature of the leaving group of $^{14}N/^{15}N$ kinetic isotope effect data for transfer of Me to pyridines in aqueous solution, it has been concluded[369] that an early transition state is involved, with a $N—Me$ bond order of 0.2-0.3. In aprotic solvents (MeCN, $ClCH_2CH_2Cl$) values are smaller, but by values which suggest a change in transition-state structure or solvation.

4.6. Phosphorus

Very small quantities of undissociated H_3PO_4 may effect general acid-catalyzed processes in aqueous phosphate buffers, such as the hydrolysis of a series of 2-alkoxy-2-phenyl-1,3-dioxolanes,[370] particularly when the Brønsted exponent of the reaction, α, is close to unity. The effect becomes insignificant as $\alpha \to 0.6$–0.7. Mechanistic proposals from product studies have been made for the reactions of cyclotriphosphate with ethanolamines[371] and amino acids.[372] The role of metal ions and their complexes in the hydrolyses of polyphosphates,[373] phosphate triesters,[374-376] monoesters,[377,378] phosphonate esters,[376] phosphinite esters,[379] adenosine 5′-triphosphate,[380-382] and acetyl phosphate[383,384] has been studied. Pyrophosphate is formed from acetylphosphate, $MeC(O)OPO_3^{2-}$, and orthophosphate in concentrated aqueous $NaClO_4$ at rates dependent to the first order on each reagent and with a pH dependence consistent with a concerted mechanism rather than one involving trapping of a free PO_3^-.[385] The hydrolysis of acetylphosphate is mediated by the hexaazadioxo macrocycle [24]-N_6O_2 (38) to give 30% pyrophosphate[386] via phosphoryl transfer to a phosphate substrate from phosphorylated (38), formed from an acetylphosphate–macrocycle precursor complex. Using ATP as a substrate, pyrophosphate formation is observed[382] using the same macrocycle complexed with Ca^{2+}, Mg^{2+} or Zn^{2+}, with a metal-ion-stabilized phosphoramidate cited as a key intermediate. Pyrophosphate obtained by the oxidation of hypophosphate $O_2P(O)P(O)O_2^{4-}$ by bromine in ^{18}O labeled water was found[387] to be exclusively labeled in the nonbridging position, supporting a mechanism involving attack by hypophosphate O on $H^{18}OBr$ followed by $^{18}OH^-$ attack on the intermediate so produced with subsequent 1,2-nucleophilic shift of P and expulsion of Br^-. The effect of metal ions on the one bond coupling constant $^1J_{PH}$ in the phosphinate ion $H_2PO_2^-$ has revealed that ions with small radii have large effects ascribed to bidentate coordination.[388] The kinetics of the oxidation of H_3PO_2 in solutions of Bi(V) and F^- have been reported.[389]

The ^{17}O- and ^{18}O-labeled thiophosphonates, $PhCMe[*OC(O)CF_3]P(S)$-$(OEt)_2$ (39), rearrange in acetic acid to give $PhCMe[SC(O)CF_3]P(O)(OEt)_2$ (40) in which 80% of the label was incorporated into the P(O) group and 20% into the C(O) group.[390,391] A mechanism shown in equation (10) is favored in which an ion pair (41) leads to an intermediate (42). Studies of the apparent distribution of acetic anhydride (43) and tetraethyl pyrophosphate (44) between water and chloroform as a function of time have led[392] to values of K_{eq} at 293 K for the hydrolysis of (43). K_{eq} is less in wet chloroform than water by a factor of 17 and, for the hydrolysis of (44), less by a factor of 3070. Solvent and substituent effects on the unimolecular fragmentation of anhydrides derived from carboxylic and amidophosphoric acids, $(RO)(R_2'N)P(O)OC(O)R''$, to give $R''C(O)NR_2'$ and metaphosphate esters $ROPO_2$, suggest[393] that there is little charge development in the transition state. Rates of disproportionation of phosphinic carboxylic mixed anhydrides derived from protected amino acids have also been studied.[394] Alkylation reactions of phosphate esters are subject to significant medium effects. For example, dimethyl 2-pyridylethyl phosphate is unreactive toward methylation at pyridyl nitrogen in MeCN while in aqueous solution, methyl 2-(N-methyl)pyridylmethyl phosphate is readily formed.[395]

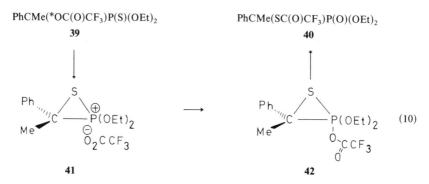

PhCMe(*OC(O)CF$_3$)P(S)(OEt)$_2$ PhCMe(SC(O)CF$_3$)P(O)(OEt)$_2$
              39   **40**

              41   **42**   (10)

Equilibrium ^{18}O isotope effects, $^{18}K_{eq}$, on the deprotonation of phosphate and phosphate esters enable the isotope effect on P—O bond cleavage to be ascertained.[396] A ^{31}P NMR technique has been employed to derive values of 1.019 ± 0.001 for $^{18}K_{eq}$ for HPO$_4^{2-}$ and H$_2$PO$_4^-$ at 27 °C with Na$^+$ as counterion. An extrapolated value of $^{18}K_{eq}$ for glycerol 3-phosphate of 1.0125 at 100 °C was used to distinguish between dissociative and associative processes for phosphate transfer reactions.[397] A value of 1.0004 was estimated for the ^{18}O isotope effect on P—O cleavage in glucose 6-[^{18}O$_3$]-phosphate in aqueous phthalate buffer, pH 4.5, 100 °C. The Brønsted correlations of rate with pK_a of the entering or leaving pyridine for the bimolecular reactions of pyridines with pyridinium-N-phosphonates (and sulfonates)[398] have been used to calculate the rate constants for the identity exchange processes. In addition to providing experimental support for the validity of the Marcus theory of group transfer to these processes, subtle "microscopic medium effects" were detected. Related kinetic studies of the reaction of substituted quinuclidines with phosphate monoester dianions[399] lead to plots of log k vs pK_a with negative slopes, attributed to the requirement for nucleophile desolvation prior to attack. Such desolvation is likely to be more difficult the greater the nucleophile basicity.

     The first cases have been established[400-403] of racemization at phosphorus during phosphoryl transfer. Transfer of the phosphoryl group from the dianion of phenyl (R)-[^{16}O, ^{17}O, ^{18}O]phosphate to *tert*-butyl alcohol in MeCN subject to conditions under which the *tert*-butyl phosphate did not racemize resulted in an almost complete loss of stereochemical labeling at P.[400] The ester dianion thus releases free PO$_3^-$ which, in the absence of an unhindered nucleophile, has a lifetime long enough to suffer loss of chirality before capture by the alcohol. The possibility could not be excluded, however, that loss of stereochemistry resulted from multiple transfers among MeCN molecules before capture by the alcohol. Similar conclusions were reported for phosphoryl transfer from adenosine 5'-[β-(S)-^{16}O, ^{17}O, ^{18}O]diphosphate to 2-O-benzyl-(S)-propan-1,2-diol in MeCN.[402] Positional isotope exchange for adenosine 5'-[β-^{18}O$_4$]diphosphate in acetonitrile has also been described[401] and taken to be consistent with stepwise preassociative processes. Phosphoryl transfer from the P^1,P^1-disubstituted pyrophosphate (45) to 2-O-benzyl-(S)-propan-1,2-diol in dichloromethane gave extensive racemization as determined from a stereochemical analysis of the resulting 1-[^{16}O, ^{17}O, ^{18}O]phospho-(S)-propan-1,2-diol.[403] The authors of the latter study suggest that the observations could also be in accord with a preassociative mechanism if the incoming nucleophile and the leaving group are not required to move along the same axis. The aqueous hydrolysis of 2,4-dinitrophenyl phosphate dianion (46) is accelerated by the application of pressure. The value of $\Delta V^{\ddagger}$ (-4.8 cm^3 mol^{-1}) cannot be reconciled[404] with the formation of free PO$_3^-$. Solvolysis in EtOH of p-nitrophenyl (R)-[^{16}O, ^{18}O]thiophosphate gave 80% racemic and 20% inverted ethyl [^{16}O, ^{18}O]thiophosphate under conditions where neither starting material nor product suffer racemization. This was taken[405] to support dissociative formation of PSO$_2^-$. Additional approaches have been described in the synthesis and analysis of configurationally assigned phosphorus centers for use in investigating the mechanisms of phosphoryl transfer in 2'-deoxyadenosine 5'-[^{16}O, ^{18}O]phosphorothioate.[406] In addition, the enantiomers of [^{16}O, ^{17}O, ^{18}O]thiophosphate have been synthesized.[407] Abell and Kirby ascribe[408] the 10^6–10^7-fold acceleration for the hydrolysis of (46) in >95% DMSO or HMPA (a factor similar to that seen in MeCN containing 0.02 M water[409]) to the decrease in H-bonding of water to the phosphate dianion. Analysis of differences in P–O separations and OPO bond angles between phosphate monoester monoanions and phosphate triesters has led Modro[410] to conclude that the anions display structural characteristics compatible with an early stage of fragmentation leading to alcohol and the metaphosphate anion. PO$_3^-$ is also thought to be generated by photochemical C—P bond cleavage in o-, m-, or p-nitrobenzylphosphonic acid in alcohol in the presence of base.[411,412] The observed rate law for the formation of Ph$_2$CS from Ph$_2$CO and Lawesson's reagent, anisyldithiophosphinic anhydride (47), points[413] to the symmetrical cleavage of (47) to give the aryl metathiophosphate ArPS$_2$[414] as a reactive intermediate. The monomeric pyramidal trithiometaphosphate anion has been isolated as the Ph$_4$As$^+$ salt from the reaction of P$_4$S$_{10}$ with KCN in MeCN.[415] The diastereoisomers of the phosphonamidic chloride, t-BuP(O)(Cl)NHR [R = (S)-PhMeCH—], both give a 55:45 mixture of the diastereoisomers of t-BuP(O)(NHt-Bu)(NHR) with t-BuNH$_2$ in MeCN, consistent with the formation of the monomeric metaphosphonimidate, t-BuP(O)(NR).[416] Isotope effects on ^{31}P chemical shifts have been employed in the estimation of P—S bond orders in a series of phosphorothioate anions.[417] The effect of changing P=O to P=S and P—OR to P—SR on the alkaline hydrolysis of phosphinate esters has been reported.[418]

Rate ratios for the fission of one ester linkage in the hydrolysis of the bicyclic phosphate $MeC(CH_2O)_3PO$ compared with the acyclic analogue $(EtO)_3PO$ and for the bicyclic phosphorothionate $MeC(CH_2O)_3PS$ with its acyclic analogue $(EtO)_3PS$[419] were found to be 5.2×10^3 and 8.1×10^2, respectively. These were ascribed, at least in part, to the stereoelectronic effect, associated with the facilitating of P—O cleavage by antiperiplanar interaction with oxygen lone pairs with P—O antibonding orbitals. Two studies by Kluger and Thatcher[420,421] challenge earlier conclusions (see Vol. 4, p. 135) that such stereoelectronic effects are important in endo- vs exocyclic cleavage processes in the alkaline hydrolysis of methyl ethylenephosphate (48). In contrast to earlier studies, (48) was found[420] to undergo alkaline hydrolysis to give both exocyclic as well as endocyclic cleavage products. Both ethylene phosphate and 2-hydroxyethyl phosphate were detected by ^{31}P NMR. The extent of exocyclic cleavage was found to increase with added base. The rate enhancement of approximately 10^6 for the hydrolysis of (48) compared with the acyclic analogue $(MeO)_3PO$ was ascribed solely to the relief of ring strain in the pentavalent transition state. ^{18}O label incorporation into the products of exocyclic cleavage of (48) in D_2O enriched with ^{18}O has also been investigated.[421] Single label incorporation into ethylene phosphate and hydroxyethyl phosphate is thought to rule out hexavalent intermediates. A reinvestigation of the reaction of 2-methoxy-5,5-dimethyl-1,3,2-dioxaphosphorinan-2-one (49, X = OMe) with the 2-chloro analogue (49, X = Cl),[422] using reagents labeled in the P=O group with ^{18}O, give the pyrophosphate (50) with the label scrambled completely between all three exocyclic oxygens. The involvement of traces of (51) is proposed to explain the scrambling.

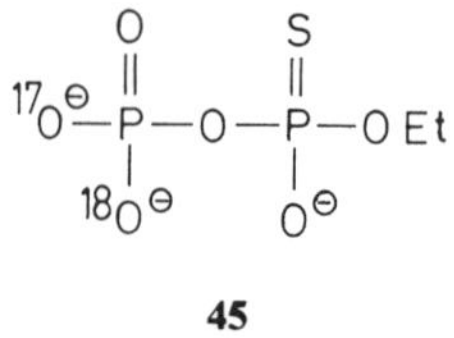

45

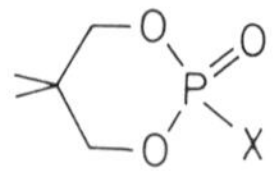

49

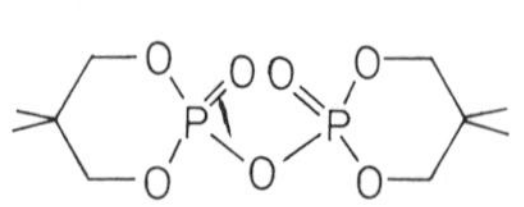

50

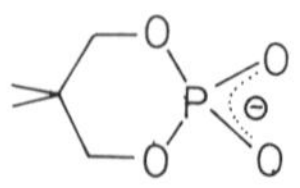

51

Kinetic and equilibrium parameters for the reactions of the hydroxyphosphorane (52) in aqueous solution shown in equation (11) suggest[423] that (52) is in rapid equilibrium with its conjugate base (53) and the isomeric phosphinate ester (54). Slower equilibria connect (53) and (54) with an acyclic hydrolysis product (55) and its conjugate base (56). The interconversion of (54) and (55) is presumed to be the key reaction between these species, with K (25 °C, $\mu = 0.1$) estimated to be about $3\text{-}4 \times 10^7$. Furthermore, the equilibrium between the conjugate bases (**53**) and (**56**) favors the ring-opened product, with $K = 1.3 \times 10^3$. The observed rate constants for the hydrolysis of a series of acyclic alkoxyphosphonium ions including $4\text{-}XC_6H_4P(Me)(OMe)_2^+$ (X = H, Cl, Me), $PhP(OR)_3^+$ (R = Me, Et)[424] take the form $k_{obs} = k_0 + k_{OH}[OH^-] + k_A[A^-]$ revealing reactions with water, hydroxide, and the buffer anion (associated with general base catalysis of the hydrolysis involving H_2O). Reaction with OH^- and A^- is believed to occur solely with P—O cleavage (with rate-determining formation of a pentacoordinated hydroxyphosphorane), while that with water also involves reaction at C. Rate data from competition

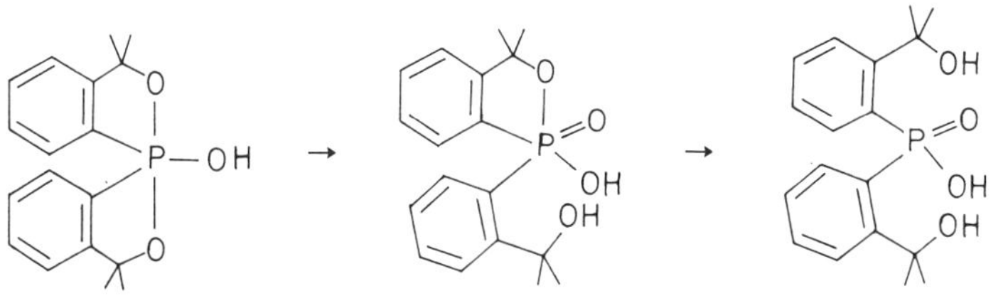

$$\text{(11)}$$

52 **54** **55**

studies on the neutral hydrolyses of $Ar_nP(OCF_2CF_3)_{5-n}$[425] point to a change in mechanism, from a $S_N2(P)$ for $n = 0$ to a $S_N1(P)$ process for $n = 1, 2, 3$. The kinetics of the hydrolysis of spirooxyphosphoranes (**57**) with two six-membered rings have been compared with earlier data for compounds with five-membered rings.[426] At pH < 9 a ring-opened phosphonium ion intermediate (**58**) is formed as in equation (12), with $K_{eq} = 3 \times 10^{-9}$. The acid-catalyzed ring opening occurs with a second-order rate constant of $4 \times 10^9\ M^{-1}\ s^{-1}$. The hydrolysis proceeds via rate-determining capture of this cation by water at low pH and hydroxide at higher pH. An A2 process involving a pentacoordinate intermediate has been proposed[427] for the hydrolysis of methyl dimethylphosphinothionate, $Me_2P(S)(OMe)$, in 25–85% sulfuric acid. A pK_a of -5.53 was determined for this ester, being 2.5 units less than those of the P=O analogues.

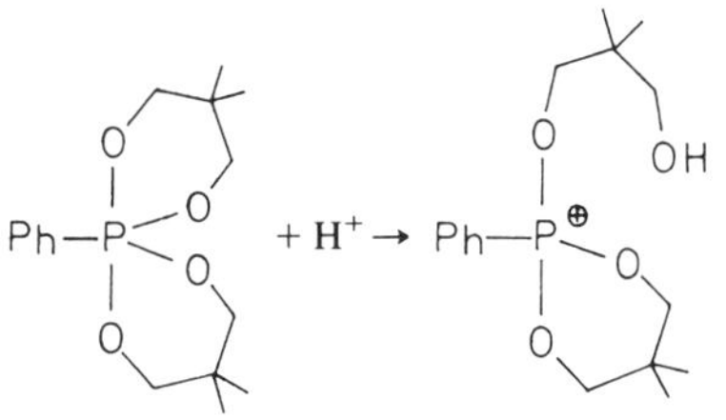

$$\text{(12)}$$

57 **58**

Kinetic studies of nucleophilic displacements at P in 2-chloro-2-oxo-1,3,2-dioxophospholanes, leading to either retention or inversion at P, have been interpreted in terms of HOMO–LUMO interactions between nucleophile and substrate.[428] Mechanisms suggested from phosphate hydrolysis cannot be generalized to include these processes. The order of reactivity of aminolysis in MeCN of aryl diphenylphosphinates $Ph_2P(E)(E'Ar)$ (E = O,S; E' = O,S) is diamines > butylamine > sec-amines.[429] Reaction with butylamine, to give $Ph_2P(O)(NHBu)$ and ArOH, follows a two-term rate law, $k_{obs} = k_1[amine] + k_2[amine]^2$, with the second-order term predominating. This results from general base catalyzed collapse of a zwitterionic pentacoordinated intermediate. For diamines, the dominance of the first-order term suggests that this process occurs with intramolecular general base catalysis.

The chemistry of phosphorus azides, and reactive intermediates formed during their transformations, has been reviewed.[430] Geometric isomer product distribution from the reactions of pentachloro- and pentafluoro-(triphenylphosphazenyl)cyclo-triphosphazene with NaOMe[431] and reactivity differences between Cl and F displacement from $P(NPPh_3)X$ centers have been rationalized in terms of a change of mechanism from $S_N2(P)$ to $S_N1(P)$ as Cls are replaced with OMe and the operation of an $S_N2(P)$ process throughout all stages of F replacement by OMe. Reactivity patterns of hexachloro-cyclotriphosphazenes and cyclothiaphosphazenes toward alkyl lithium reagents[432,433] and 2-substituted ethylamines[434] have also been interpreted mechanistically. The influence of alkyl substitution on the tautomeric equilibrium, $H_2P≡N \rightleftharpoons HP=NH$, has been exploited synthetically.[435] The iminophosphane, RN=PR, R = t-Bu, is reported to dimerize to $RP(=NR)NRPR$ by a [2 + 1] cycloaddition, with the product itself susceptible to a series of acid-catalyzed transformations.[436] The fluxional process connecting inverted forms of the 8-P-3 compound (59) via the transition state (60) has been studied by NMR magnetization transfer.[437] $P(OPh)(OC_6H_4-p-Cl)-(OC_6H_4-p-Me)$ has been resolved and found to be remarkably stable to substituent exchange.[438] The labeled phenylphosphonic acid $PhP^{18}O_2H_2$ reacts with ethyl chloroformate to give the ethyl ester containing the label only on the phosphoryl position. A four-membered cyclic transition state is proposed

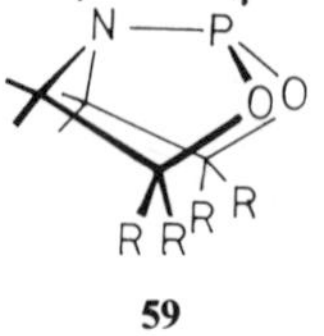

59 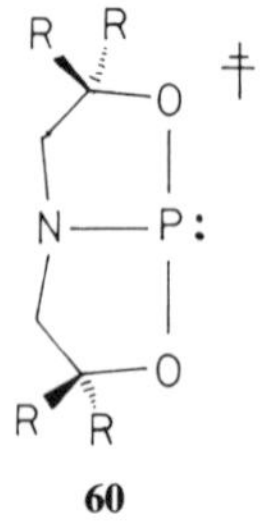   60

with CO_2 elimination from an intermediate P(III) carbonic anhydride as in equation (13).[439]

The role of lithium salts, the effect of varying reagent concentrations, and other aspects of Wittig reactions have been investigated.[440-444] Particular

$$PhP(H)(=O)O^- + ClCO_2Et \rightarrow PhP(H)(=O)OCO_2Et$$

(13)

$$PhP(H)(=O)OEt \leftarrow Ph-P-O$$
with OH above and Et-O⁺C=O below

attention has been given to the extent to which the stereochemical outcome of the reaction is influenced by the reversibility of the formation of diastereomeric 1,2-oxaphosphetane intermediates. In THF, and in the presence of LiBr, increased proportions of *trans*-oxaphosphetane and (*E*)-alkene were observed as concentrations of butylidenetriphenylphosphorane and benzaldehyde were increased. The effect was significantly lessened in the more solvating DMSO. By using diastereomerically pure 1,2-oxaphosphetanes, Maryanoff and Reitz[442] provide additional evidence that the *cis* isomer is significantly more prone to reversible formation of the Wittig reagent and aldehyde than the *trans* isomer and that these processes are affected by the presence of lithium salts. On the other hand, Cairns and McEwen conclude[443] that oxaphosphetanes derived from benzaldehyde and benzylidene(4-N,N-dimethlyaminophenyl)diphenylphosphorane in THF react to form stilbenes and the correspondig phosphine oxide at rates much faster than their dissociation to starting materials. Electron-withdrawing substituents on both benzaldehyde and the benzylidenetriphenylphosphine lead to rate enhancements on the Wittig reactions.[445] The chemistry of other phosphaalkene and phosphacarbene species has been studied.[446-450]

The Hammett ρ value (+3.33) derived from studies of the kinetic acidity of *p*-substituted benzyltributylphosphonium ions (X = H, Cl, Me) in OD^-/D_2O at 25 °C indicates significant negative-charge development and C—H bond lengthening in the transition state.[451] Volumes of activation of the alkaline decomposition of $p\text{-}XC_6H_4CH_2PPh_3^+$ (X = H, NO₂) support a scheme in which proton loss from an initially formed hydroxyphosphorane precedes expulsion of the benzyl carbanion.[452] The reaction of $Ph_4P^+Br^-$ with OH^- and OMe^- has also been studied kinetically.[453] A stable radical monocation is formed on one-electron oxidation of cyclotetraphosphanes, $(RP)_4$ (e.g. R = $i\text{-}Pr_2N$), while the dications rearrange rapidly to white phosphorus and $(i\text{-}Pr_2N)_2P^+$ even at -100 °C.[454] A facile 1,2-shift in the cyclic dication is proposed from theoretical calculations. The intermediates so generated then decompose by ring fission. The reactivity of the phosphenium ions, e.g. $(R_2N)_2P^+$, toward 1,3- and 1,4-dienes has been examined qualitatively.[455] The stereochemistry at P in the rearrangement of *S*-phosphorylisothioureas to *N*-phosphorylthioureas has been investigated.[456] A reexamination of the reactivity of P(III) species $Ar_nP(OR)_{3-n}$ with cyclooctasulfur in toluene reveals[457] that for n = 0, R = Me, Et, *n*-Pr, *i*-Pr, *n*-Bu, $\Delta H^\ddagger$ decreases along the *n*-alkyl series, while $\Delta S^\ddagger$ shows compensating increasingly negative values, associated with a decrease in S—S cleavage relative to P—S bond formation along this series. The phosphonites and phosphinites are markedly more reactive than the phosphites. Rates of reaction for $ArP(Oi\text{-}Pr)_2$ and $Ar_2P(Oi\text{-}Pr)$ correlate with Hammett σ constants, giving ρ values consistent with P developing a high degree of positive charge in the transition state. Substitutions remote from the phosphorus center affect the decomposition of a series of ozonides derived from 4-hydroxymethyl-1-phospha-2,6,7-trioxabicyclo[2.2.2]octane.[458] Triplet benzophenone reacts with tetraethyl-pyrophosphite in benzene in a process thought to resemble S_H2, with a second-order rate constant at 300 K of $7.96 \pm 0.04 \times 10^8\ M^{-1}\,s^{-1}$ [459] to yield the radical $Ph_2\dot{C}OP(O)(OEt)_2$ arising from the addition of the phosphorus-centered radical $(EtO)_2(O)P^{\cdot}$ to benzophenone. Ph_4P_2 also reacted rapidly. The generation, structure, and reactivity of the diarylphosphonyl radical, $Ar_2\dot{P}=O$, $Ar = 2,4,6\text{-}t\text{-}BuC_6H_2$, have been described.[460]

4.7. Arsenic and Antimony

^{121}Sb, ^{13}C, and ^1H NMR have been used to monitor the (very limited) extent of the autoionization, $2SbCl_5 \cdot L \rightleftharpoons SbCl_4L_2^+ + SbCl_6^-$, in dichloromethane, to determine the equilibrium constants $K_{a\text{-}b}$ for the exchange processes, $SbCl_5 \cdot L_a + L_b \rightleftharpoons SbCl_5 \cdot L_b + L_a$,[461] and to monitor the kinetics of 17 such exchange processes using variable temperature and pressure NMR.[462] For example, the rate of displacement of tetramethylurea from $SbCl_5 \cdot L$ by THF in ethylene dichloride is first order with respect to [THF], with $\Delta H^{\ddagger} = 28.9 \pm 1.6$ kcal mol^{-1} and $\Delta S^{\ddagger} = 21.7 \pm 4.2$ cal K^{-1} mol^{-1}. The volumes of activation and compressibility coefficients for activation were determined for the ligand exchange process involving $L = Me_2O$, Me_2CO, $MeCN$, and other derivatives. The first-order rate law and large positive values of the activation parameters are consistent with a dissociative process as is the slope of -1 for the linear correlation of the free-energy changes associated with adduct formation and the free energy of activation (relative to $L = MeCN$).

The equilibria, Me_4E_2 (**61**) + $Me_4E_2' \rightleftharpoons Me_2EE'Me_2$, E', E = P, As, Sb, Bi, have been studied in benzene solution by ^1H NMR.[463] K_{eq} at 25 °C for the exchange involving E = P and E' = As was 0.26, for Sb and Bi 1.2, for Sb and As 0.9, and for Bi and As 0.009. The exchange reactions of (**61**) with dimethyldichalcogenides, Me_2A_2, A = S, Se, Te, to give Me_2EAMe have been studied qualitatively.[464] Reactions leading to Me_4As_2 from Me_2AsH with Me_2AsNMe_2 or from $Me_2AsNMe_2 \cdot BH_3$ have been followed by NMR.[465] The first 10-Sb-3 compound, 5-aza-2,8-dioxa-3,7-di-*tert*-butyl-1-stibabicyclo[3.3.0]-octa-3,6-diene (**62**), has been characterized and found to have the planar geometry of the analogous P and As compounds.[466] In contrast to the 10-P-3 compound, (**62**) does not undergo rehybridization on coordination to metals such as Pt.[467] Further studies have been described of the debromination of phenacyl and benzylic bromides with tertiary stibines.[468]

4.8. Oxygen

Deuterium spin-lattice relaxation time measurements on H_2O/D_2O mixtures have been used to probe rotational motions of water molecules.[469] Proton relaxation studies of H_2O in mixtures with pyridine or DMF, with or without electrolytes of different charge types, have also been described.[470] Isotope fractionation factors have been obtained by NMR for substrates undergoing rapid proton–deuteron exchange with solvent water[471] and for water–methanol[472] and water–3,3-dimethylbutan-2-one[473] mixtures. H–D isotopic fractionation between "free" water molecules and those involved in the hydration sphere of urea in aqueous urea suggest[474] enrichment of deuterium in the weaker H-bonding urea–water cluster. Proton-transfer and ion-association rates in aqueous sec-butylamine have been studied by ultrasonic methods.[475] The effect of pressure on proton transfers in ethanol–water mixtures suggest[476] that these occur in water-rich regions rather than in ethanol-rich regions. The structure of hydration shells for nonpolar solutes in water, such as noble gases and hydrocarbons, has been analyzed experimentally and values for the hydration numbers of these solutes determined.[477] Monte Carlo simulation methods have been used[478] to calculate hydration numbers for a series of alkanes in aqueous solution. H/D exchange in D_2/H_2O[479] and in HD/H_2O[480] under 300 kHz sonolysis involves $H^{\cdot}$ and $OH^{\cdot}$, formed during cavitation. Deprotonation rates of excited states of 1- or 2-naphthol by water in water/alcohol mixtures decrease nonlinearly with alcohol concentration.[481] These variations were used to probe the local solvation environment at the reaction site. Rates of protonation[482] of water by the excited states of 6-methoxyquinoline and acridine have also been studied. The kinetic product from the acid dissociation of H_2O in pure water is $H_9O_4^+$ which arises from the quasi-tetrahedral environment around oxygen in pure liquid water.[483] Proton transfer in $H_5O_2^+$ and $H_5O_2^+ \cdot (H_2O)_4$[484] and the hydration of H_3O^+[485] have been studied theoretically. Rate constants of both proton transfers[486] and nucleophilic displacements[487] in the gas phase have been studied as a function of the number of water molecules ($n = 0, 1, 2, 3$) associated with OH^- acting as either base or nucleophile. MO studies of the S_N2 process involving the cluster $OH^- \cdot (H_2O)_n$ ($n = 0, 1, 2$) with $MeCl \cdot (H_2O)_m$ ($m = 0, 1, 2$) have also been reported.[488] Theoretical studies have been reported of nucleophilic addition of OH^- to H_2CO both in aqueous solution and the gas phase,[489] of the proton transfers between carbonyl and hydroxyl oxygens[490] and those involving carboxyl groups,[491] and of the structure of liquid water itself.[492-494] The observed slow H–D exchange between $D_2H^{17}O^+$ and $DH_2^{17}O^+$ in $HF:SbF_5/SO_2$ at $-15\,°C$ is thought[495] to occur via the protonated hydronium dication, whose structure and stability were estimated by ab initio methods.

The intermediate from the decay of hydrated electrons forms at a rate which increases with pH and decays by a second-order process with $k \leq 5 \times 10^9\ M^{-1}\ s^{-1}$.[496] Christensen and Sehested record[497] a similar value for this rate at 20 °C which they find is independent of pH. The equilibrium $2e^-_{aq} \rightleftharpoons (e^-_{aq})_2 \rightarrow H_2$ is proposed as the simplest scheme to account for the kinetic data. Atomic hydrogen and OH^- react in aqueous solution to give e^-_{aq} and water. Using pulse radiolytic techniques, the rate of formation of e^-_{aq} by this route, at pH 11.7, 40 °C, is reported[498] to be $4.5 \pm 0.9 \times 10^7 \, M^{-1} s^{-1}$ with an activation energy of $26.4 \pm 2.5 \, kJ \, mol^{-1}$. The entropy of solvation for e^-_{aq} was calculated to be $49 \pm 22.6 \, J \, mol^{-1} \, K^{-1}$. 300 kHz sonolysis of water,[499] in the presence of mixtures of argon and $^{18}O_2$, leads to the production of H_2O_2, with $H_2^{16}O_2$ being produced in yields much higher than $H_2^{18}O_2$. $^{16}O_2$ and $^{18}O^{16}O$ were also detected as products. Reactions similar to those seen under combustion conditions are proposed. The products of sonolysis of aqueous solutions of potassium iodide and sodium formate under argon-oxygen[500] derive from the formation of $H^{\bullet}$ and $OH^{\bullet}$, with HO_2 and oxygen atoms being formed in the presence of oxygen. Oxygen atoms are scavenged by formate and iodide. HO_2 is not scavenged by the solutes but yields H_2O_2. Further reductions are reported in the rate of the purported uncatalyzed decomposition of hydrogen peroxide in aqueous base. Evans and Upton[501] obtain a rate constant of $4.7 \times 10^{-7} \, M^{-1} s^{-1}$ at pH 11.6 and 35 °C in the presence of diethylenetriamine-N,N,N',N'',N''-penta(methylphosphonic acid), with the spontaneous process leading to singlet oxygen and a catalyzed reaction leading to triplet oxygen. By correcting for the latter, a value of $3.7 \times 10^{-8} \, M^{-1} s^{-1}$ for the "true" spontaneous rate was proposed. The dependence of singlet oxygen formation in H_2O_2 decomposition under basic conditions on the presence of a range of inorganic derivatives has been described, using tetrapotassium rubrene-2,3,8,9-tetracarboxylate as a $^1O_2$ trap.[502] The effect of chelating agents was not investigated. The role of other inorganic species in the reactions of H_2O_2 and other peroxy derivatives has been described.[503-510] Tracer studies using $^{18}O_2$ reveal that the hydroxylation of benzene using Fenton's reagent proceeds via the hydroxycyclohexadienyl radical with isotope incorporation into phenol *ca* 20% and that into benzoquinone being 100%.[511] The rate law derived for the oxidation of 2-mercaptoethanol by H_2O_2, pH 9–13, to give 2-hydroxyethyl disulfide[512] supports a mechanism in which mercaptoethanol displaces OH^- from either H_2O_2 or HO_2^- to give $HOCH_2CH_2SOH$ as intermediate. The rates of reaction between O_2^{-} [513] and H_2O_2 in acetonitrile have been reported. In anhydrous acetonitrile, HO_2, generated by oxidation of H_2O_2, disproportionates to H_2O_2 and O_2 with a second-order rate constant of $1.0 \times 10^7 \, M^{-1} s^{-1}$.[514,515] In further studies of tetramethylammonium superoxide, a previously reported dimer of $[Me_4N]O_2$ has been found in fact to be a peroxide adduct of acetonitrile, $MeC(OO^-)NH^+$.[516] Hydroxide has been reported to reduce anthraquinone in acetonitrile in a one-electron process leading to H_2O_2.[517] Using pulse radiolysis to generate HO_2 and O_2^- in aqueous solutions[518] of bromine/bromide and iodine/iodide, Schwarz and Bielski obtain[519] rate constants for reaction of $HO_2^{\bullet}$ with I_2 ($1.8 \times 10^7 \, M^{-1} s^{-1}$) and Br_2 ($1.3 \times 10^8 \, M^{-1} s^{-1}$) and of O_2^- with I_2 ($6.0 \times 10^9 \, M^{-1} s^{-1}$), I_3^- ($0.25 \times$

$10^9\ M^{-1}\,s^{-1}$), Br_2 ($5 \times 10^9\ M^{-1}\,s^{-1}$), Br_3^- ($1.5 \times 10^9\ M^{-1}\,s^{-1}$), and HOBr ($3.5 \times 10^9\ M^{-1}\,s^{-1}$). Similar techniques were used on aqueous solutions of Fe^{2+} to obtain rate constants for a variety of redox processes involving HO_2 and O_2^- with iron species.[520] Using chemiluminescence at 1268 nm to detect its formation, a 12.4% yield of $^1O_2$ was estimated[521] from the reaction of ceric ions with 13-hydroperoxylinoleic acid. Similar approaches were used[522] to monitor $^1O_2$ formation from the reaction of potassium superoxide with CBr_4, $CHCl_3$, and CCl_4. Ozonides derived from a series of phosphites gave quantitative formation of adducts of $^1O_2$ in the presence of scavengers.[523] Singlet oxygen is formed from ozone and Et_3SiH[524] and from ozone and amines, sulfides and phosphites in the gas phase.[525] The reaction of $^1O_2$ with isomeric 1,4-di-*tert*-butoxy-1,3-butadienes and other conjugated dienes,[526–529] 3,4-dihydro-6-methyl-$2H$-pyran-5-carboxylic acid ethyl ester,[530] 4-[2-(*N,N*-dimethylhydrazono)ethylidene]-2,6-diphenyl-$4H$-pyran,[531] aliphatic amines,[532] azines,[533] 1,3,5-triaryl-2-pyrazolines,[534] tetramethylallene,[535] $5H,7H$-dibenzo[b.g][1.5]dithiocin,[536] biadamantylidene thiirane,[537] 2-[1-phenylselenyl]ethyl-4,5-diphenyl oxazole,[538] 1,1,3,3-tetramethyl-2-indanone triphenylphosphazine,[539] and enol ester[540] have been described. The chemistry of peroxonium ions and dioxygen ylides has been reviewed.[541] Triplet oxygen oxygenation of alkylated 1,3-dienes occurs to give endoperoxides by cation radical chain processes[542] that are less susceptible to steric effects compared with analogous reactions involving singlet oxygen. A ground-state oxygen–benzophenone complex, stable on the 25 ps time scale, has been detected in addition to triplet benzophenone on 266 nm irradiation in nondeoxygenated solutions of Ph_2CO in cyclohexane.[543] Reactions of diphenylmethanes with O_2^- in benzene containing 18-crown-6 are first order in the two reagents,[544] with the primary isotope effect supporting rate-determining proton transfer from substrate to O_2^-. The resulting carbanion then undergoes oxygenation to the product ketone. Gas-phase reactions of O_2^- with MeCl and MeBr[545] and of HO_2^- and $HC(O)O_2^-$ with a range of substrates[546] have been described. The first direct observation has been reported[547] of benzoyloxy radicals formed in the photodecomposition of dibenzoylperoxides in CCl_4. Temperature dependence of the lifetime measurements of the benzoyloxy radical led to an estimate of the rate of its decarboxylation, *ca* $4 \times 10^7\ s^{-1}$ at 130 °C, which is quite different from the often quoted value. Further evidence for acyloxy intermediates comes from the photolysis of *tert*-butyl 9-methyl-fluorene-9-peroxycarboxylate (**63**)[548] in oxygen-free acetonitrile. Pulsed laser irradiation of (**63**) at 266 nm generates the fluorene radical, formed from a precursor with a 55 ± 15 ps lifetime. Rate constants for the one-electron oxidation of $MeOO^{\cdot}$, $HOCH_2OO^{\cdot}$, and $CCl_3OO^{\cdot}$ in aqueous solution fall in the range 3×10^5 to $6 \times 10^8\ M^{-1}\,s^{-1}$.[549] The yield of docosane from the thermal decomposition of dilauryl peroxide in octane is increased by 3–6% by the application of a magnetic field (0.15–1 T).[550] Mechanistic studies are reported on the decomposition of di-*tert*-butyl peroxydicarbonate in cumene,[551] on the decomposition of 2,2-bis(tert-butyldioxy)-3-methylbutane and tert-butyl peracetate in Ph_2CH_2,[552] of solvation effects on the concerted decomposition of tert-butyl peroxypivalate[553] and of tert-butyl heptafluoroperoxybutyrate,[554] and on the induced thermolyses of $CH_2{=}CH(CH_2)_nOOt$-Bu (n = 1, 2, 3, 4).[555]

In 0.01 M OH^-, ozone decomposes according to the rate law $-d[O_3]/dt = k_1[O_3] + k_2[O_3]^2$, with the second-order term being suppressed by a radical scavenger.[556] The rate law changes at higher $[OH^-]$. The results favor O_3 + $OH^- \rightarrow HO_2^- + O_2$ as the initiation step with values for the rate constants for a range of propagation and termination steps involving O_3, HO_2^-, H_2O_2, HO_2, O_2, O_2^-, OH, and OH^- being obtained. In the oxidation of ferrous ion by ozone, the latter is reported to be completely consumed within 0.1 s while the consumption of Fe^{2+} continues for up to 15 min. This was rationalized[557] in terms of reactions involving secondary oxidizing species, though their precise role could not be elucidated. Ozone may also be rapidly decomposed in 0.1 M $HClO_4$ on 300 kHz sonolysis under an atmosphere of $O_2/O_3/Ar$.[558] Formation of H_2O_2, at a maximum at 80% argon, is much slower than ozone decomposition, which is confined to pulsating gas bubbles. Ozone is known not to be formed during sonolysis of water under oxygen atmospheres.[500] Reaction pathways have been studied in the ozonolysis of 2,3-dimethyl-1,3-butadiene in methanol,[559] of enol ethers,[560] of 2,3,4,5-tetramethyl-2,4-hexadiene,[561] of styrene in the presence of a series of substituted benzaldehydes,[562] of 3-methyl-1-phenyl-4-phenylazo-5-pyrazolone,[563] and of 2,5-dimethylthiophene.[564] The first example has been described of stereospecificity (implying[560] concertedness) for a reaction of a carbonyl oxide with a substrate (in this case, ethyl vinyl-2-d₁ ether) to give ethoxy-1,2-dioxolane with retention of the configuration of the alkene. Further studies on the formation[565,566] and reactions[567–569] of carbonyl oxide intermediates in the Criegee process for olefin ozonolysis have been described. It has been established, for instance, that the carbonyl oxide, generated from PhCMe: and $^{18}O_2$ tagged oxygen, does not isomerize to give the doubly labeled ester PhC(O)OMe.[569] Complete scrambling of the label rules out an intramolecular process and reaction via a cyclic tetroxide is proposed. Rates of reaction of ozone with methyl oleate and methyl linoleate in CCl_4 solution have been compared with those for α-tocopherol (vitamin E).[570] The rate constant for the α-tocopherol reaction of 5500 M^{-1} s^{-1} at 25 °C is ca 10^2 less than values for the fatty-acid esters. The detection of the α-tocopheroxy radical supports initial electron transfer from the substrate to ozone. Rates and activation parameters for the reaction in CCl_4 of ozone with the C—H bonds in 14 organic materials are reported[571] to be inconsistent with hydrogen-atom abstraction or prior complexation of O_3 with an α-oxygen atom. Hydride abstraction and the concerted insertion of O_3 into C—H is preferred, though the 2–3-fold acceleration of these processes in MeCN points to limited charge separation.

4.9. Sulfur

The rate constant, k_f, for the reaction $H_3O^+ + SO_4^{2-} \rightarrow HSO_4^- + H_2O$ has been found by IR line broadening[572] at 23 °C to be $1.8 \pm 0.5 \times 10^{12} \, M^{-1} \, s^{-1}$ and to be unaffected by the changes in water structure resulting from the addition of electrolyte. Proton transfer between SO_4^{2-} and HSO_4^- was ruled out owing to the absence of line broadening for the latter. MO calculations on the reaction between SO_3 and H_2O to form H_2SO_4 reveal[573] that a large barrier exists to the formation of the adduct $H_2O \cdot SO_3$. Values of K_{eq} for the formation of 1:1 complexes of sulfate with tri- and tetra-protonated 1,4,7,10,13,16-hexaazacyclooctadecane and a neutral 2:1 complex of sulfate with the tetraprotonated species suggest[574] that desolvation of the reagent species provides the thermodynamic driving force.

The absence of curvature in plots of pK_a versus rate constant for sulfuryl group transfer between isoquinoline-N-sulfonate and a series of pyridines with pK_a values covering a span of 8 pK_a units points to a single transition state in a concerted, symmetrical mechanism.[575] Estimates of the effective charge on the entering and leaving nitrogen atoms in an analogous study suggest that the transition state possesses considerable SO_3 character. The markedly different ^{17}O and ^{18}O isotopic shifts of the symmetric and antisymmetric SO_2 stretching modes of cyclic sulfate esters have been exploited in the analysis of the absolute configuration of chiral $[^{16}O, ^{17}O, ^{18}O]$sulfate monoesters.[576] The hydrolytic release of ArO^- from aryl salicyl sulfates is catalyzed by the neighboring carboxylate group, with the intermediate cyclic acyl sulfate being trapped by hydroxylamine.[577] The effect of $C_{10}-C_{14}$ 1-alkanols on the acid hydrolysis of sodium dodecyl sulfate below the critical micelle concentration suggests the formation of 1:1 or 2:1 complexes between SDS and the alkanol.[578]

In moderately acidic aqueous solutions, peroxodisulfate decomposition via the acid-catalyzed route releases HSO_5^-, which suffers a chain decomposition induced by $OH^{\cdot}$ generated from the uncatalyzed decomposition of $S_2O_8^{2-}$.[579] Mechanisms have been investigated of the oxidation by peroxodisulfate of Ph_3E ($E = P$, As, Sb),[580] of o-substituted N,N-dimethylanilines,[581] of formate ion[582] and CO,[583] and of octacyanomolybdate[584] and the oxidation by peroxomonosulfate of aldehydes,[585-587] ketones,[588] and carboxylic acids.[589] Direct ^{17}O NMR evidence is reported[590] for the existence of two isomers of bisulfite in aqueous solution. Peak area measurements lead to an equilibrium quotient for the isomerization between the $O-H$ bonded and $S-H$ bonded species, $HSO_3^- \rightleftharpoons SO_3H^-$, of 4.9 ± 0.1 at 298 K, with the more abundant $O-H$ isomer being tentatively identified from its greater rate of oxygen exchange with water. Studies of the polarographic reduction of aqueous SO_2[591] show four reduction waves, arising from $SO_2 \cdot H_2O$, bisulfite ion, HSO_2, and SO_2^-. Hydroxymethanesulfonate, formed in the equilibrium between bisulfite and formaldehyde, dissociates at 25 °C and pH 5.6 with a rate constant of $1.1 \times 10^{-5} \, s^{-1}$. This is a composite constant arising from reactions with solvent water and OH^-.[592] A three-term rate law has been derived for the reaction of bisulfite with benzaldehyde[593] with the terms being associated with rate-determining nucleophilic attack of SO_3^{2-} and HSO_3^- on benzaldehyde carbonyl carbon and the attack of HSO_3^- on the nonring C of the protonated species $PhC^+H(OH)$, with the latter process important only at pH < 1. Addition of sulfite to *p*-benzoquinone to give hydroquinone monosulfonate (**64**) follows kinetics which are first order in [quinone],[594] though the reversible formation of sulfite carbonyl adducts which precede the irreversible C—S bond formation give a complex dependence both on pH and [sulfite]$_{tot}$. Direct formation of (**64**) from free *p*-benzoquinone (298 K, $\mu = 0.1$) and sulfite has $k = 7.7 \times 10^4 \, M^{-1} \, s^{-1}$. Rate and equilibrium constants for the reaction of sulfite with a series of substituted dinitro- and chlorodinitrobenzenes have been reported.[595] Rate constants for the reactions of sulfite radicals, $SO_3^{\bar{}}$, and peroxysulfate radicals, $SO_5^{\bar{}}$, with hydroxyphenols have been reported by Huie and Neta.[596,597] Additions of $SO_3^{\bar{}}$ to olefins in aqueous media are found to be more rapid in acidic than in alkaline solution.[598] $SO_3^{\bar{}}$ is also found[599] to add to the C=N bond of the aci-anion form of nitroalkanes. Kinetic studies have been reported of the reaction of sulfite and permanganate,[600] of peroxotitanium(IV) and aqueous S(IV) species,[601] of sulfite and hexacyanoferrate(III),[602] of bisulfite and chromium(V)–carboxylato complexes,[603] and of the reactions of sulfito-complexes of cobalt(III).[604,605] The reaction of SO_2 with H_2S in diglyme, kinetically first order in each reagent,[606] is accelerated by the addition of *N,N*-dimethylaniline.

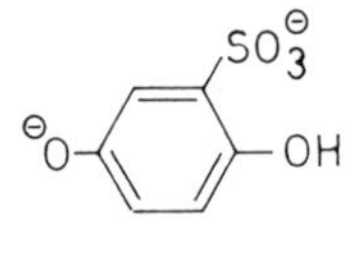

64 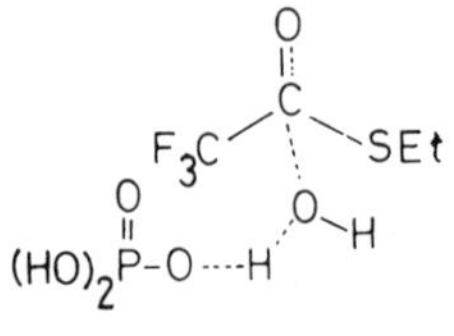   65

Using an improved method for ^{18}O analysis, Paradisi and Bunnett[607] have shown, assuming that oxygen equilibration in the solvolyses of *sec*-alkyl benzenesulfonates, $PhS(O)_2OR$, occurs via internal return from intimate ion pairs, that internal return is extensive for R = 2-adamantyl, negligible for R = *t*-BuCHMe, and significant for R = *i*-Pr. Studies of the solvolysis of cumyl sulfones and sulfinate esters reveal[608] that the $CF_3SO_2^-$ leaving group lies between Cl^- and *p*-nitrobenzoate in its carbocation-forming reactivity. $CH_3SO_2^-$ and $PhSO_2^-$ are *ca* 10^7 times less reactive. Hammett ρ constants, labeling and activation parameter data for the alkaline hydrolyses or aryl 2-(acylamino)benzenesulfonates are consistent with the formation of an intermediate benzoxathiazane S,S-dioxide formed by intramolecular attack by the amido group.[609] Hydrolysis or aryl (methylsulfonyl)methanesulfonates, studied by similar methods, is reported[610] to proceed via an $E_{1c}B$ mechanism. The k_a term in the rate law $k_{obs} = (k_a + k_b[OH^-])/(1 + [H^+]/K_a)$, where K_a is the ionization constant of the ester, involves expulsion of ArO^- from the ionized ester to give a sulfene ($CH_3SO_2CH=SO_2$). The k_b term is associated with the formation of the anionic sulfene ($CH_3SO_2C^-=SO_2$) from a dianionic form of the substrate. The anion $^-CH_2SO_2CH=SO_2$ was shown not to be kinetically important. In related studies on aryl sulfamates,[611] the inter-mediacy of sulfonylamine ($HN=SO_2$) and the sulfonylamide anion ($^-N=SO_2$) is also proposed. The $H_2PO_4^-$-catalyzed hydrolysis of $EtSC(O)CF_3$ is thought[612] to occur via (65).

By direct measurement of $[H_2S]$, $[HDS]$, and $[D_2S]$ using ^1H and ^2H NMR, K_{eq} for the liquid-phase disproportionation, $H_2S + D_2S \rightleftharpoons 2HDS$, at 25.7 °C was found to be 3.99 ± 0.11, compared with a theoretical value of 3.92 calcu-lated from ideal-gas data.[613] A numerical approach to the calculation of equilibrium constants involving the species OH^-, H^+, H_2S, HS^-, S_2^-, S_2^{2-}, S_3^{2-}, S_4^{2-}, and S_5^{2-}, water and Na^+ give values which are an order of magnitude different from those reported by other workers.[614] Rates and activation parameters of the bimolecular exchange of deuterium between propane-2-thiol and 2-methylpropane-2-thiol in a series of solvents of differing polarity[615] point to reaction via a cyclic hydrogen bonded dimer with the proton transfers facilitated by increasing the dielectric constant of the solvent.

Autoxidation of aqueous sodium polysulfide ($Na_2S_{2.0} - Na_2S_{4.6}$) has been studied.[616] Oxidation of prochiral sulfides by optically active hydroperoxides gives chiral sulfoxides in moderate optical yields.[617] Solvent effects on the kinetics of the base-catalyzed oxidation of thiols by dioxygen have also been reported.[618] Pillai and Gould have reported kinetics (first order in each reagent) for the reduction of organic disulfides by vitamin B_{12_s}.[619] Rate studies have been reported on the formation and dissociation of the three-electron bonded disulfide radicals.[620–622] Oxidation of thiolate anions with both $R_2S^{\ddagger}$ and $R_2 \therefore SR_2^+$ (66) (to give a disulfide and a thiyl radical) occurs at essentially diffusion-controlled rates,[620] while (66) are much less reactive toward undis-sociated thiols than the monosulfide radical cations. For example, Me_2S^+ reacts with EtSH in aqueous acid with a rate constant of $1.8 \times 10^9 \, M^{-1} \, s^{-1}$ while the corresponding reaction involving $Me_2S \therefore SMe_2^+$ is slower by several orders of magnitude. The actual rate constant is governed by the stability of the thiyl radical produced. Studies of the equilibrium between the three-electron bonded (66) and its constituent sulfide and sulfide radical cation provides[621] equilibrium and dissociation rate constants (R = Me) of $2.0 \pm 0.3 \times 10^5 \, mol^{-1} \, dm^3$ and $1.5 \pm 0.2 \times 10^4 \, s^{-1}$ at 20 °C. Other electron-transfer reactions of organosulfur radical cations[623] and solvolyses[624] and other[625,626] reactions have been described. ESR and NMR spectroscopy have been used to determine the equilibrium constants and associated thermody-namic parameters for dimers of a 9-S-3 sulfuranyl radical (67).[627,628] Compar-able quantities of a S(IV)–S(IV) bonded bisulfuranyl and a S(IV)–O bonded isomer are formed. Formation of perthiyl radicals have been unequivocally established from triplet sensitized laser photolysis of t-BuSSH.[629] Triplet diphenylcarbene gives radical addition products, $Ph_2CSR^•$, on reaction with sulfides and disulfides (rate constants ca 10^6 $M^{-1}s^{-1}$) in contrast to the reaction of triplet fluorenylidene which proceeds via an ylide with rate constants 10^2–10^3 greater.[630] At 298 K, carbon-centered radicals such as n-Bu$^•$ react with α-toluenesulfonyl chloride to give $ArSO_2^•$ and n-BuCl[631] with a second-order rate constant of 1.3×10^6 $M^{-1}s^{-1}$.

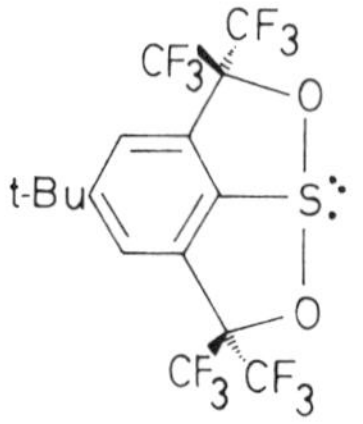

67 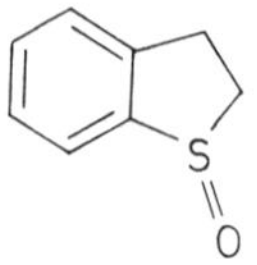   68

Intramolecular homolytic substitution at S in the sulfoxide group of (**68**) has been shown[632] to occur with complete inversion of configuration. Optically active p-toluenesulfinimides, $ArS(O)NR_2$, react with thiols RSH in CF_3CO_2H to give $ArS(O)SR$[633] and with alcohols ROH to give $ArS(O)OR$[634] with predominant inversion of configuration. The stereospecific hydrogenolysis of vinylic sulfones with sodium dithionite proceeds via a β-addition–elimination mechanism, in which the intermediate formed by the addition of HSO_2^- (from dithionite) could be trapped by alkylation to give a 1,2-bis-sulfone.[635]

Reductions of sulfilimines, R_2SNHAr^+, by iodide[636,637] are first order in $[H^+]$ and are accelerated by electron-releasing groups in Ar. Rate-limiting partitioning of a common tetracoordinate sulfurane intermediate $R_2S(NHAr)I$ via one of two channels is proposed. One involves protonation followed by S–N cleavage to release $ArNH_2$ and the other involving proton transfer to amine concerted with iodide attack and S–N cleavage to release the amine, with the relative importance of these two processes being dependent on the pK_a of the leaving amino group. The very small secondary isotope effects in the proton-catalyzed reaction with N-aryl-S,S-di(methyl-d_3)sulfilimines[637] were ascribed to the canceling of effects arising from the formation of the sulfurane intermediate and its breakdown. In the analogous reductions involving thiocarboxylate anions,[638] a concerted substitution process is preferred which does not involve the sulfurane species, $R_2S(NH_2Ar)SR'$, as an intermediate. The mechanism of formation of $(SN)_x$ from the thermal decomposition of S_4N_4 has been studied by mass spectrometry[639] and by photoelectron spectroscopic gas analysis.[640] Characterization and reactions of other S–N species have been reported[641,642] and structure–reactivity relationships reviewed.[643] Stoichiometric studies[641] show that the reaction of Me_2NCN with $(NSCl)_3$ in CCl_4 proceeds via $(NSCl)_2$ rather than the NSCl monomer.

4.10. Selenium and Tellurium

Equilibrium and rate constants of reactions of Se(V) species, thought to be SeO_3^- and $HSeO_4^{2-}$, formed in the pulse radiolysis of aqueous selenite and selenate solutions, are shown in equations (14)-(18).[644] ^{76}Se, ^{80}Se, and ^{82}Se

$$SeO_3^- + OH^\cdot \rightarrow HSeO_4^{2-} \qquad k_f = (3.5 \pm 0.2) \times 10^9 \, M^{-1} s^{-1} \quad (14)$$
$$k_b = (7.3 \pm 0.5) \times 10^5 \, s^{-1}$$

$$HSeO_3^- + OH^\cdot \rightarrow (H_2SeO_4^-) \rightarrow SeO_3^- + H_2O \qquad k = (1.6 \pm 0.1) \times 10^8 \, M^{-1} s^{-1} \quad (15)$$

$$H_2SeO_3 + OH^\cdot \rightarrow (H_3SeO_4) \rightarrow SeO_3^- + H_3O^+ \qquad k = (1.0 \pm 0.1) \times 10^9 \, M^{-1} s^{-1} \quad (16)$$

$$2\,SeO_3^- \rightarrow Se(IV) + Se(VI) \qquad k = (5.2 \pm 0.5) \times 10^8 \, M^{-1} s^{-1} \quad (17)$$

$$CO_3^{2-} + SeO_3^- \rightarrow CO_3^- + SeO_3^{2-} \qquad k = (6 \pm 1) \times 10^6 \, M^{-1} s^{-1} \quad (18)$$

isotope fractionation, observed in the reduction of SeO_3^{2-} by hydroxylamine to $Se(0)$,[645] proceeds via a two-step process, with the second step $ca\ 10^2$ faster than the first. The isotopic rate ratios k_{76}/k_{82} and k_{76}/k_{80} are 1.007 and 1.005, respectively, for the first step and 1.020 and 1.013 for the second step.

PhSeBr adds to substituted styrenes to give $ArCH(Br)CH_2SePh$ and $ArCH(SePh)CH_2Br$ with kinetics which are first order in each reagent.[646] Hammett reaction constants suggest that little positive charge is developed at the styrene α-C and that C—Se bond making lags behind Se—Br bond breaking in the rate-determining transition state. Rates of reaction of ArSeCl with 2-chloroalkyl phenyl selenides involving nucleophilic displacement at selenenyl Se are little affected by changes at either the selenide or selenenyl chloride.[647] Rates and equilibria of methyl transfer for a series of reactions of ArSeMe with p-$ClC_6H_4SeMe_2^+$ have been used[648] to probe the application of the Marcus equation to soft nucleophiles.

Kinetics of oxidation of o-nitrobenzeneselenenyl compounds o-$O_2NC_6H_4SeX$ (69), X = H, OH, SeAr, OSeAr, OEt,[649,650] and o-benzoylbenzeneselenenic acid o-$PhC(O)C_6H_4SeX$ (70), X = OH, with peracids, hydroperoxides, and H_2O_2 yield second-order rate constants which, for the reaction of (69) with m-chloroperbenzoic acid at 25 °C in ethanol, are $1.1 \times 10^4\ M^{-1} s^{-1}$ (X = H), 80 (X = OH), 12 (X = OSeAr), and 0.15 (X = SeAr). This reaction and that of (69) and (70) with other perbenzoic acids are thought to involve nucleophilic attack by Se on peroxide oxygen. With hydroperoxides and H_2O_2, (70) with X = OH reacts in an acid-catalyzed process $ca\ 10^4$ times faster than (69) with X = OH. This is thought to result from a reversible formation of a peroxyhemiketal (71) from (70) with X = OH, which suffers intramolecular attack of Se on a peroxidic oxygen of protonated (71). Substituted o-nitrobenzeneselenenic acids, 4-Y-2-$O_2NC_6H_3SeOH$, Y = H, Cl, Me, MeO, have been shown to be surprisingly weak acids, with pK_a in the range 10.2-10.8.[651] The value of K_{eq} for the equilibrium, $ArSeOSeAr + H_2O \rightleftharpoons 2ArSeOH$, in aqueous dioxan, is 1.8 ± 0.8 for Ar = o-benzoylphenyl and is ca 10 times larger than for *o*-nitrophenyl $(0.16, 25 \,°C)$.[652] The selenoxides, $ArSe(O)Et$, eliminate ethylene to give the selenenic acids (**69**), X = OH and (**70**), X = OH with rates at 25 °C of $1.23 \pm 0.01 \times 10^{-3} \, s^{-1}$ and $0.96 \pm 0.01 \times 10^{3} \, s^{-1}$, respectively. The selenenic acid (**69**) with X = OH undergoes rapid acid-catalyzed esterification by alcohols to give the corresponding esters (**69**) with X = OR.[653] The hydrolyses of these esters are general, rather than specific, acid-catalyzed. It is now suggested that, in acid-catalyzed nucleophilic substitution reactions, the association of the ester with the general acid HA is followed by loss of the coordinated nitro group, with the intermediate so formed reacting reversibly with the incoming nucleophile, R'OH. It is also proposed that transfer of proton from the general acid HA to the departing RO group is accompanied by transfer of a proton from $SeOR'H^{+}$ to the incipient A^{-}. Stereoselective intermolecular fluorine exchange between $Ph_3TeF_2^{+}$ and *mer*-$Ph_3TeF^aF_2^b$ has been demonstrated[654] by ^{19}F and ^{125}Te NMR. Only F^a undergoes exchange, pointing to a fluorine bridge intermediate, $Ph_3F_2^bTe-F^a-TeF_2^bPh_3$. The bond-breaking process appears not to lead to intramolecular F^a-F^b scrambling. The fluxional behavior of the 10-Te-4 telluranes resulting from oxidative addition of halogens to a series of tellurapyrylium compounds[655] has been ascribed to halogen exchange. The kinetics are first order in substrate concentration in MeCN, second order in $CHCl_2CHCl_2$, and mixed in $CHCl_3$. The chemistry of 12-Te-5 partelluranes and 10-Te-3 telluranes has also been investigated.[656] Preliminary qualitative rate data for ligand exchange in the Te(IV) compounds $[Te(S_2CNR_2)_4]$[657] suggest a reversible disulfide elimination reaction leading to Te(II). Novel chemistry of selenium and tellurium anions and cations has also been reported.[658-662]

4.11. Halogens

Mechanisms of reactions of the halogens and of oxohalogen species in aqueous acid have been reviewed by Thompson.[663]

4.11.1. Fluorine

The simple displacement reaction, $K_2MnF_6 + 2SbF_5 \rightarrow 2KSbF_6 + MnF_3 + \frac{1}{2}F_2$, has been used[664] to effect the chemical synthesis of elemental fluorine in >40% yield. Fluorination of uracil with $MeC(O)OF$ or F_2 in acetic acid gives *trans*-(**72**) and *cis*-5-fluoro-6-acetoxy-5,6-dihydrouracil in the absence

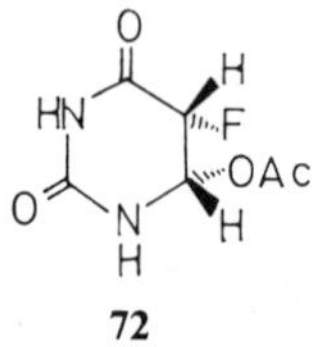

72

of acetate and (72) and 5-fluorouracil in the presence of acetate.[665] The novel fluoride bridged species, $[C_6F_5I\text{-}F\text{-}IC_6F_5]^-$, has been claimed[666] from the reaction of perfluoroalkyl carbanions and fluoroorganoiodides.

4.11.2. Chlorine

The kinetics of oxidation of $[Pt(CN)_4]^{2-}$ in aqueous chlorine at pH < 7 give a rate law consistent with two parallel reactions, one involving Cl_2 as oxidant $[k_{Cl_2}$ (25 °C, $\mu = 1.00\ M$ NaClO$_4$) $= 1.08 \pm 0.10 \times 10^7\ M^{-1}\,s^{-1}$; dominant in weakly acidic solutions buffered with HCl], the other involving HOCl ($k_{HOCl} = 97.9 \pm 1.1\ M^{-1}\,s^{-1}$; dominant in unbuffered solutions of NaClO$_4$).[667] Under certain conditions, the displacement of the hydrolytic equilibrium, $Cl_2 + H_2O \rightleftharpoons HOCl + Cl^- + H^+$, is rate determining. The value of K_{eq} for this hydrolysis, determined from the forward ($k_f = 8.7 \pm 0.2\ s^{-1}$) and reverse rate constant ($k_r = 2.66 \pm 0.03 \times 10^4\ M^{-1}\,s^{-1}$), was found to be $3.27 \pm 0.11 \times 10^{-4}\ M^2$.

Substituent and solvent effects suggest that the reaction of p-substituted styrenes with aqueous ClO_2 at pH 7.0 to give ClO_2^- and the olefin radical cation[668] proceeds via an electron-transfer mechanism. The kinetics of the oxidation of thiosulfate by aqueous ClO_2^- at 90 °C, $S_2O_3^{2-} + 2ClO_2^- + 2OH^- \rightarrow 2SO_4^{2-} + 2Cl^- + H_2O$, are complex.[669] The rate of disappearance of ClO_2^- up to a conversion of 70% follows the autocatalytic rate expression $-d[ClO_2^-]dt = k[S_2O_3^{2-}][ClO_2^-][H^+]$ with $k = 1.3 \pm 0.2 \times 10^8\ M^{-2}\,s^{-1}$. An intermediate, $S_2O_3ClO^-$, is formed by nucleophilic displacement of OH^- from $HClO_2$ by $S_2O_3^{2-}$. ClO_2^- reacts with bromine to give ClO_2 and Br^-. The intermediate $BrClO_2$ is formed in a rapid process, with its further reaction with chlorite to give ClO_2 and Br^- being rate determining with $k(25\,°C, \mu = 0.66\ M) = 2.94 \pm 0.25 \times 10^3\ M^{-1}\,s^{-1}$.[670] The p$K_a$ for $BrClO_2H^+$ was found to be 1.9, which provided an estimate of the rate constant for the reaction of BrOH and ClO_2^- of $20.6\ M^{-1}\,s^{-1}$, suggesting that BrOH is less reactive toward Cl(III) than Br_2. In a related study, ClO_2 was shown to react with Br^- to give Br_3^- and Cl^- (with excess Br^-) and ClO_2 and Br_2 [in excess Cl(III)].[671] The rate law for the reaction, $HClO_2 + 6Br^- + 3H^+ \rightarrow 2Br_3^- + Cl^- + 2H_2O$, is $\frac{1}{2}d[Br_3^-]/dt = k[H^+][Br^-][Cl(III)]$, where $k = 9.51 \pm 0.14 \times 10^{-2}\ M^{-2}\,s^{-1}$. In excess Cl(III), the stoichiometry and kinetics are complex and are interpreted in terms of a 16-step mechanism. Simoyi,[672] in studying the same system, reports that the reaction proceeds by two routes, one catalyzed by product Br_2 and suppressed by reactant Br^-, with the rate expression $\frac{1}{2}d[Br_2]/dt = k_1[ClO_2^-][Br^-][H^+] + k_2[ClO_2^-][Br_2]/[Br^-]$ ($k_1 = 1.39 \pm 0.05 \times 10^{-1}\ M^{-2}\,s^{-1}$), with the first term predominating at high $[Br^-]$. k_{35}/k_{37} rate ratios for the solvolysis of a series of 1-phenylethyl chlorides[673] are consistent with a S_N1 heterolysis mechanism with internal return. MO calculations show[674] that the transformation of HOCl into its valence isomer HClO is endothermic with an activation energy of 74 ± 5 kcal mol^{-1}. The structure and fluxionality of the unknown anion ClF_6^- have been modeled.[675] Monte Carlo methods have been employed[676,677] to simulate the energetics of the hydration and dehydration of Cl^- and Br^-. Activation-energy difference estimates for the reactions of Cl_2^- and Br_2^- with aliphatic dipeptides and other substrates have been determined[678] by flash photolysis.

4.11.3. Bromine

Schmitz and Rooze[679] have extended their use of o-toluidine (73) to simplify the kinetics of redox processes to those of BrO_3^-, particularly the reaction with Br^-, $BrO_3^- + 5Br^- + 6H^+ \rightarrow 3Br_2 + 3H_2O$. A fourth-order rate law is derived in the presence of (73) with rate of formation of (73)-$2H = 3k_4[BrO_3^-][Br^-][H^+]^2$ with $k_4 = 1.54\ M^{-3}\ s^{-1}$ at 25 °C. The pK_a of bromic acid was estimated to be 2.9. The rate constant for the reaction of BrO_3^- with ClO_2^- in the presence of (73) from the rate law $-d[Br_3^-]/dt = k[BrO_3^-][HClO_2][H^+]$ was itself found to be acid dependent, with $k = 0.83 + 0.76[H^+]$. The reaction, $Br^- + HBrO_2 + H^+ \rightarrow 2HOBr$, has a second-order rate constant, k_2' [680] (24 °C, 0.5 M H_2SO_4), from the rate equation $d[Br^-]/dt = -3k_2'[Br^-][HBrO_2]$ ($k_2' = k_2[H^+] = 4.2 \times 10^5\ M^{-1}\ s^{-1}$), which is about 10^3 greater than that calculated in a computer simulation of the Belousov–Zhabotinskii (BZ) reaction. Likewise, the rate constant for the disproportionation reaction, $HBrO_2 \rightarrow HBrO_3 + HOBr$, is found[681] from kinetic studies to be $2.2 \times 10^3\ M^{-1}\ s^{-1}$, some 10^4 smaller than the value obtained by the same BZ simulation. Field and Fösterling[682] have measured a value of $1 \times 10^{-6}\ M^{-1}$ for the equilibrium, important for the BZ reaction, $HBrO_2 + BrO_3^- + H^+ \rightarrow 2BrO_2 + H_2O$. A reinvestigation of the kinetics of the bromate–iodide reaction was conducted in both excess I^- ($BrO_3^- + 9I^- + 6H^+ \rightarrow Br^- + 3I_3^- + 3H_2O$) and excess BrO_3^- ($BrO_3^- + 6I^- + 6H^+ \rightarrow Br^- + 3I_2 + 3H_2O$). A rate law, $-d[BrO_3^-]/dt = k_0[BrO_3^-][H^+]^2[I^-]$, is proposed at high $[I^-]$ and $[Br^-]$, with k_0 [25 °C, $\mu = 0.2\ M\ (NaClO_4)] = 44.3 \pm 1.1\ M^{-3}\ s^{-1}$. A mechanism is proposed in which nucleophilic attack of I^- on $HOBrO_2$ occurs, to give $HOBrO_2I^-$, from which HOI is then eliminated.[683]

The forward rate constants for the equilibrium $X^- + X_2 \rightleftharpoons X_3^-$, X = Br, I, in water and methanol, have been found by relaxation methods ($\tau = 20$–75 ns)[684,685] to be ca $10^9\ M^{-1}\ s^{-1}$ with the reverse rate constants varying between 6×10^5 and $5 \times 10^7\ s^{-1}$ depending on solvent and halogen. Ion solvation dynamics are thought to be responsible for the nondiffusion controlled reaction for X = Br. The Br_3^-/Br_5^- equilibrium has also been studied.[686] Bromination of cyclohexene in chlorinated hydrocarbon solvents is third order (second order in $[Br_2]$) when Br_2 is reactant and second order (first order in $[Br_3^-]$) when bromine/$[Bu_4N]Br$ is the reactant.[687] The solvent effects point to a highly polarized transition state. The reaction of Br_3^- displays a kinetic solvent isotope effect $k_H/k_D = 1.175$ in $CHCl_3/CDCl_3$ while that with Br_2 displays no such effect. An ionic mechanism is proposed involving the rate-determining formation of a $Br^+ \cdot Br^{3-}$ ion-pair intermediate for the free bromine addition, with the formation of an olefin–bromine $1:1$ charge-transfer complex followed by rate-determining nucleophilic attack by Br^- through a charge-delocalized transition state. The product of bromination of adamantylideneadamantane in CH_2Cl_2 contains a three-membered cyclic bromonium ion with close contact of the tribromide counteranion in an intimate ion pair.[688]

4.11.4. Iodine

Pseudo-first-order rate constants in the range $3800–70,000 \ s^{-1}$ were obtained from a study of the reaction, $ICl_2^- + 2I^- \rightarrow I_3^- + 2Cl^-$, using the pulsed accelerated-flow technique, allowing a value for the second-order rate constant for the reaction $ICl + I^- \rightarrow I_2 + Cl^-$ of $1.1 \pm 0.1 \times 10^9 \ M^{-1} s^{-1}$ at $25 \ °C$, $\mu = 1.0 \ M$ to be obtained.[689] Palmer and van Eldik[690] have studied the hydrolysis of iodine by T-jump methods. Studies of the disproportionation of the key intermediate, $I_2OH^- \rightarrow HOI + I^-$, formed either by reaction of I_2 with H_2O or OH^-, yield the forward k ($994 \pm 36 \ s^{-1}$) with the reverse k ($3.2 \times 10^5 \ M^{-1} s^{-1}$) being computed from the published value for the equilibrium constant. Disproportionation of I_2, $3I_2 + 3H_2O \rightarrow 5I^- + IO_3^- + 6H^+$, follows the rate law (at $25 \ °C$, $1.0 \ M$ NaOH) $-d\Sigma[I]/dt = (0.05 + 2.60[I^-])[\Sigma[I]]^2$, where $\Sigma[I] = ([I_2] + [I_3^-] + [IO^-] + [I_2OH^-])$ and involves[691,692] the elementary steps shown in equations (19)–(23). The fact that, in the reaction of I^-

$$2IO^- \rightarrow IO_2^- + I^- \qquad\qquad k \le 7.2 \times 10 - 2 \ M^{-1} s^{-1} \qquad (19)$$

$$IO^- + HOI \rightarrow IO_2^- + I^- + H^+ \qquad k = 40 \pm 7 \ M^{-1} s^{-1} \qquad (20)$$

$$IO^- + I_2OH^- \rightarrow IO_2^- + 2I^- + H^+ \qquad k = 6.0 \pm 0.2 \ M^{-1} s^{-1} \qquad (21)$$

$$IO^- + IO_2^- \rightarrow IO_3^- + 2I^- + H^+ \qquad k = 0.5 \pm 0.1 \ M^{-1} s^{-1} \qquad (22)$$

$$I_2OH^- + IO_2^- \rightarrow IO_3^- + 2I^- + H^+ \qquad k = 26 \pm 2 \ M^{-1} s^{-1} \qquad (23)$$

with ClO_2^-, $[I_2]$ increases to a maximum, then decreases to a minimum, and thereafter increases to its final value is accounted for[693] by the formation and hydrolytic disproportionation of ICl. The rate expression, $\frac{5}{2}d[I_2]/dt = k[ICl]^2$, for the reaction $5ICl + 3H_2O \rightarrow IO_3^- + 5Cl^- + 2I_2 + 6H^+$ leads to a value of k (at pH 3.40, $[Cl^-] = 0$, by extrapolation) equal to $40 \ M^{-1} s^{-1}$.

In aqueous acid, the kinetics of the oxidation of amino acids with periodate[694] are first order in periodate and amino acid, with the acid dependence suggesting that the reaction occurs between the zwitterionic form of the acid and $H_4IO_6^-$ or $H_3IO_6^{2-}$. The thermochemistry of the hydrolysis of I_2X_6, $X = Cl, MeCO_2$, has been described.[695] The effect on the volatilization rate of iodine from aqueous solutions of the presence of α-, β-, and γ-cyclodextrins has been used[696] to estimate formation constants of $1:1$ complexes between I_2 and the cyclodextrin. The first examples have been reported[666,697] of the hypervalent 10-I-2 iodine species R_2I^+, $R = C_6F_5$, $(CF_3)_3C$. Further evidence is reported[698,699] on the rates and solvent effects for the reaction of iodine with R_3SnI under photochemical conditions to give RI and R_2SnI_2 via weak complexes formed between iodine atoms and R_3SnI.

4.11.5. Oscillating Reactions

Some reports describing oscillating reactions involving halogen-containing species have been discussed in the previous section.[669–672,680–683,693] Papers have appeared dealing with the Belousov–Zhabotinskii (BZ) and BZ-type reactions,[700–720] other bromate,[721–723] iodate,[724–728] and chlorite[729–731] oscillators, models of oscillatory behavior,[701,708,713–715,718,720,733] and gas evolution oscillators.[734,735]

Br^- control in the presence of Ag^+,[700,701] the source of Br^- (using [82]Br-labeled bromomalonic acid and Ag^+),[702] Br_2 hydrolysis control,[703] and rate of Br_2 transfer from liquid to vapor phase[704] in the BZ reaction have been investigated. BZ-type reactions have been reported based on Mn(II) catalysis using saccharides as oxidizable substrates.[705] Spiral waves are obtained from a two-dimensional simulation of a ferroin-catalyzed BZ-type reaction.[706] Bistability and Br^--controlled oscillation has been described[707] for a ferroin-catalyzed BZ reaction in a continuous-flow stirred tank reactor (CSTR). Further refinements have been proposed for the assignment of all the rate constants for the oxidation of Ce^{3+} by BrO_3^-.[708] Induced oscillations can be generated by the introduction of solutions of Br^- or I^- into a classical BZ system during its induction period in the oxidizable steady state.[709] Crowley and Field have studied electrically coupled BZ oscillators in attempts to avoid problems resulting from mass-transfer coupling.[710] Spatially distributed concentration fluctuations have been reported[711] in a BZ reaction in a stirred batch reactor, though Ruoff and Noyes[712] suggest that the observed noise spectrum was an experimental artifact. McKinnon and Field conclude[713] that the Explodator model is not applicable to any oscillating reaction that exhibits bistability, though this view has been challenged.[714,715] A new bromate oscillator based on thiourea has been studied[721] both in a closed system and a CSTR. No bistability was seen. Mo(VI) has been shown[722] to catalyze the BrO_3^-/I^- process and V(V) the BrO_3^-/ascorbic acid reaction in a BrO_3^-/I^-/ascorbic acid clock reaction. Citri and Epstein[723] have proposed a 13 elementary step mechanism for the reaction between BrO_3^- and iodide in aqueous acid. IBr is involved but no radical species. Oscillations and oligo-oscillations have been described for the first time in the systems iodate–sulfite-malonic acid, iodate/hydroxylamine, and iodate/thiourea in weakly acidic solution.[724,725] Temporal pH oscillation was also noted in unbuffered chlorite/thiosulfate in a CSTR. Large changes of $[H^+]$ have been observed[726] in unbuffered solutions of iodate and arsenous acid. Studies of added Cl^- or I^- on the bistability of the IO_3^-/H_3AsO_3 reaction has permitted rate constants for key steps to be obtained.[727] By using ferrocyanide to consume I_2 and operating in a CSTR at $>30\,°C$, the bistable sulfite–iodate ('Landolt') reaction is converted into a new iodate oscillator.[728] Alamgir and Epstein report the new chlorite oscillator, ClO_2^-/Br^-, which possesses bistability, and ClO_2^-/SCN^-, which displays no bistability, in a CSTR.[729] Chaotic behavior is seen when an external flux of ClO_2^- is introduced into a chlorite/iodide oscillator.[730] Convective effects on traveling wave velocities in the chlorite/thiosulfate and the Fe(II)/nitric acid reactions have been studied experimentally and modeled qualitatively.[731,732] A new model for oscillatory behavior in closed systems, the 'Autocatalator', has been proposed.[733] Oscillations in gas evolution from the formic acid/concentrated sulfuric acid have been studied experimentally[734] and a quantitative model proposed.[735]

Chapter 5
Substitution Reactions of Inert-Metal Complexes— Coordination Numbers 4 and 5

5.1. Introduction

The period covered has again been a fruitful one, with useful advances made in most of the usual active research areas. Studies on square-planar complexes of platinum(II) and palladium(II) continue to dominate the field.

Several review articles are of direct relevance to the subject. An examination of ligand exchange (and isomerization) reactions of square-planar molecules initiated by nucleophilic attack is extended to other recognized exchange mechanisms, including electrophilic attack and oxidative addition followed by reductive elimination. The point is brought out that these, and other routes such as dissociatively-controlled ligand exchange, might best be regarded not as independent individual pathways, but as formal descriptions representing points on a continuum of reaction pathways. The route followed in any given case will largely depend on the initial interaction at the square-planar molecule, and this can range between the extremes of nucleophilic and electrophilic attack.[1]

A review by Skibsted[2] on ligand substitution and redox reactions of gold(III) complexes summarizes and discusses the mechanisms involved, and provides a short but useful comparison with related reactions of square-planar compounds of other elements. An assessment of the application of high-pressure techniques to the elucidation of mechanisms in coordination chemistry by van Eldik[3] includes examples of ligand substitution at square-planar complexes.

Of great interest also is a review by Stoddart and co-workers[4] on second-sphere coordination. The structures of many adducts of metal complexes, both square-planar and tetrahedral, with macrocyclic molecules of the crown-ether variety are discussed. The interactions are predominantly hydrogen-bonding between O or N on the macrocycles and protic ligands of the first coordination sphere, and direct coordinations of the macrocycles are not involved. Relevant to this is a report of several complexes *cis*- and *trans*-[MX$_2$(BiL)], where M is Pd or Pt, X is halide, and BiL is Ph$_2$PC$_2$H$_4$OC$_2$H$_4$OC$_2$H$_4$PPh$_2$. None of the oxygen atoms coordinate.[5] Conversely, however, platinum(II) or palladium(II) complexes of the macrocyclic thioether (1) reveal two, long axial M—S bonds as well as the usual square-planar links,[6] emphasizing the difficulty of predicting coordination modes. (See also Section 5.8.)

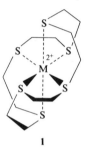

1

A review of bonding modes of β-diketonates in metal complexes[7] is dominated by Pd and Pt. Ligand–metal bond energies in palladium and platinum compounds have been reviewed,[8] as have reactions of platinum complexes with alkylcobalamins.[9]

Lastly, the elucidation of the molecular structure of tetrakis(1-norbornyl)-cobalt as tetrahedral is especially significant in that this represents the first low-spin tetrahedral molecule of a first-row transition element.[10] First prepared in 1972, its magnetic moment of 1.89 BM corresponds to the d^5 of cobalt(IV).

5.2.   Ligand Exchange at Square-Planar Platinum(II)

More details are available on the gas-phase study of reaction (1) by negative-ion mass spectrometry and Fourier-transform mass spectrometry.[11] The evidence consistently points to formation of 5-coordinate intermediates [Pt(X, Y)$_3$(PEt$_3$)$_2$], derived from the primary reactants or products of equation (1), with lifetimes greater than 10^{-6} s, but less than 10^{-3} s. Despite the fact

$$[PtX_2(PEt_3)_2] + Y^- \rightarrow [PtXY(PEt_3)_2] + X^- \tag{1}$$

that these studies are necessarily performed at low pressure, it seems probable that the intermediates closely resemble those encountered in solution reactions.

Base hydrolysis of $[Pt(dien)(dmso)]^{2+}$ reveals the operation of a conjugate base (CB) pathway, still a rarity for square-planar complexes.[12] The substrate is a weak acid ($pK_a = 11.94 \pm 0.02$ at 25 °C and $I = 0.10$) and addition of NaOH causes complete deprotonation of the secondary nitrogen. This contrasts with the known behavior of $[Pt(phen)_2]^{2+}$, which adds OH^- to form $[Pt(phen)_2(OH)]^+$, but resembles the behavior of $[Au(dien)Cl]^{2+}$. The conjugate base subsequently reacts with nucleophiles according to Scheme 1. The product of the reaction is more basic than the CB substrate, and rapidly takes a proton from the solvent. The $[Pt(dien)X]^+$ thus formed then reacts with OH^- at a rate strongly dependent on the nature of X^-, and independent of $[OH^-]$. A comparison of the k_2 values for step 1 of Scheme 1 with those obtained for

Scheme 1

$$[Pt(dien - H)(dmso)]^+ + X^- \xrightarrow{1} [Pt(dien - H)X] + dmso$$

$$\text{fast} \downarrow H_2O$$

$$[Pt(dien)OH]^+ + X^- \overset{2}{\rightleftharpoons} [Pt(dien)X]^+ + OH^-$$

reactions of $[Pt(dien)(dmso)]^{2+}$ itself with X^- reveals that removal of the proton has a remarkably small effect on the substrate reactivity, the CB being about half as reactive as the dicationic substrate (though it has to be kept in mind that the values relate to a high ionic strength of 1.0). The complexes $[Pt(en)(dmso)_2]^{2+}$ and $[Pt(NH_3)_3(dmso)]^{2+}$ are also acidic, and the combination of the 2+ charge and a *trans* dimethylsulfoxide ligand appears to be necessary for this condition and the operation of a CB mechanism.

Nucleophilic discrimination values, n^0_{Pt}, have been determined for N,N'-dimethylthiourea (7.02), N,N'-diphenylthiourea (3.05), N,N,N',N'-tetramethylthiourea (6.05), and the pseudohalide $N(CN)_2^-$ (3.05) using the standard substrate *trans*-$[Pt(py)_2Cl_2]$.[13] The k_Y data obtained from the one-term second-order substitution of $[Pt(OND)(H_2O)]^+$ {formed by complete hydrolysis of $[Pt(OND)(NO_3)]$ in MeOH:OND are phenolate anions of tridentate Schiff bases, for example N-(2-diethylaminoethyl)-salicyladimine} by various neutral or anionic nucleophiles Y, could not adequately be correlated with their n^0_{Pt} values using the usual relationship [equation (2)]. The deviations were greatest with bulky nucleophiles and substrates, and had rate-reducing *cis* effects of up to 2640 for Y = PPh_3. It was concluded that the steric crowding hinders formation of the 5-coordinate transition state in these cases.[13]

$$\log k_Y = sn^0_{Pt} + \log k_S \qquad (2)$$

Displacements of chloroacetate from $[Pt(dien)(ClCH_2CO_2)]^+$ in aqueous solution by solvent, Cl^-, Br^-, and SCN^- have revealed normal nucleophilic discrimination powers, comparable to those derived from the corresponding chloro complex.[14] The nucleophile-independent pathway makes only a small contribution here, but when this substrate is replaced by the neutral biscarbox-ylato complex cis-$[Pt(Pr^iNH_2)_2(ClCH_2CO_2)_2]$, the nucleophile-independent route dominates for substitution by chloride, and the strong nucleophile SCN^- is needed to enhance the value of the direct substitution pathway. This unexpec-ted change of behavior led the authors to suggest formation of a transiently-chelated acetate ligand to account for the change (Scheme 2). In the presence of acid, the reactions are catalyzed by a rapid prequilibrium protonation of the carboxylate ligands.[14]

Scheme 2

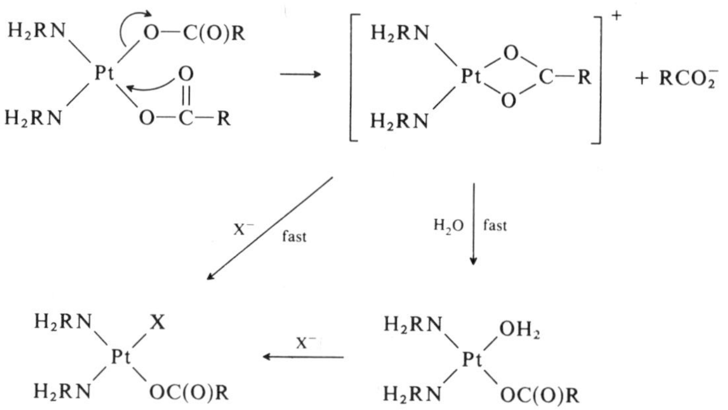

The reaction of Zeise's salt with β-alanine follows second-order kinetics and proceeds in two steps (Scheme 3)[15] A crystal structure of the final product discloses a close $N\cdots Cl$ approach, which may be due to $NH_2\cdots Cl$ hydrogen bonding. The reaction of $[PtCl_4]^{2-}$ with N-acetyl-L-methionine, on the other hand, results in displacement of one Cl^- only, by sulfur.[16] Followed by ^{195}Pt NMR spectroscopy, S,N-chelation was found not to occur.

The second-order rate constants for the replacement of Br^- from $[PtBr_4]^{2-}$ by several sulfoxides, RR'SO, are all about one order of magnitude greater than for the analogous replacements of Cl^- from $[PtCl_4]^{2-}$, reflecting the higher *trans* effect order $Br^- > Cl^-$. Although the rate constants are sensitive to the steric bulk of R and R', no enhancement of this effect arises from the change from $[PtCl_4]^{2-}$ to $[PtBr_4]^{2-}$.[17] This contrasts with related substitution reactions involving $[AuCl_4]^-$, $[AuBr_4]^-$, and amines, where only the reactions of the tetrabromoaurate anion are sensitive to steric hindrance (*vide infra*).

Scheme 3

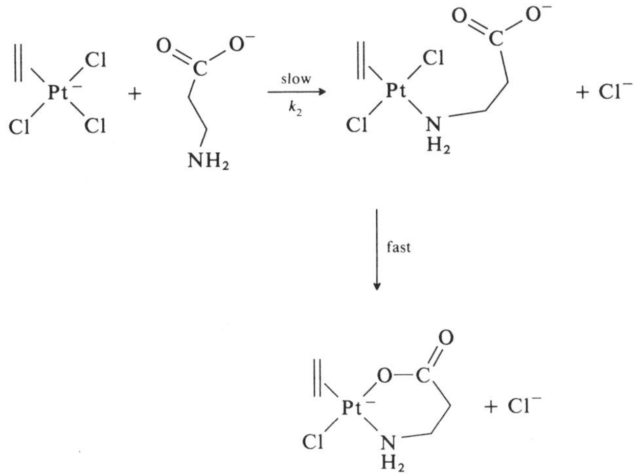

Variable temperature NMR spectroscopy has been used to determine $\Delta G^{\ddagger}$ for sulfur inversion in a number of O,S-chelate complexes, and assessments of the "dynamic *trans* effect" of the ligands opposite sulfur, including tertiary phosphines, has been made.[18] ^{195}Pt NMR has also been used for the first time for this purpose, as well as for following reaction rates.[16]

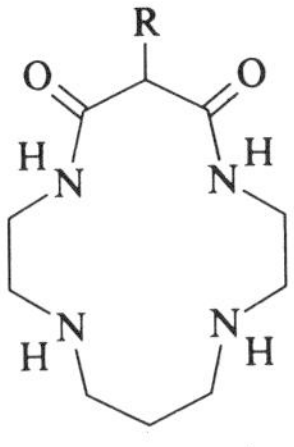

R = H or $C_{16}H_{33}$

2

The tetradentate macrocycles (**2**) (R = H or $C_{16}H_{33}$) are reported to complex selectively to platinum(II) at low pH and in the presence of $Na_2S_2O_3$.[19] Suggested as a possible method of removing Pt from biological systems, for example after *cisplatin* drug treatment, the method might also be of value for estimation and recovery of platinum.

5.3. Ligand Exchange at Square-Planar Palladium(II)

Solubility measurements on aqueous $PdI_2(s)$ have enabled comparisons to be made between the formation of iodide complexes of palladium(II) and platinum(II).[20] Formation constants for aqueous $[MI_4]^{2-}$ (M = Pd or Pt) are similar, the large values of both resulting mainly from the β_2 values for formation of cis-$[MI_2(OH_2)_2]$. The kinetic and activation parameters for each step are compatible with associative activation. Contributions from bond breaking to the formation of the transition states appear less important for palladium than for platinum.

More studies by van Eldik and co-workers add usefully to our knowledge of reactions of substituted dien (dien = diethylenetriamine) complexes of Pd(II). Solvolysis kinetics of the highly sterically hindered $[Pd(R_5dien)X]^{n-}$ (R_5 = 1,1,4,7,7-Et_2MeEt_2 or 1,1,4,7,7-Et_5; X = Cl^-, Br^-, I^-, py, NH_3, or oxalate^{2-}) at various temperatures and pressures reveal that the leaving groups have a marked effect on $\Delta V^{\ddagger}$, which vary between -3 and -12 $cm^3\,mol^{-1}$. The parameters are discussed in terms of an earlier suggestion of a gradual change from A to I_a mechanisms, but the authors point out that it is still possible to account for all the observed trends by A mechanisms alone.[21] The formation, aquation, and base hydrolysis of the carbonato complexes $[Pd(R_4dien)CO_3]$ (R_4 = 1,1,7,7-Me_4 or 1,1,7,7-Et_4) have also been investigated.[22] The evidence indicates that in the pH range 8–9, formation of the carbonato complexes does not proceed via uptake of CO_2 by the aquo species, as is commonly the case for octahedral complexes of other metals, but rather it proceeds by hydrogen-carbonate (bicarbonate) substitution [equation (3)]. Values of k_{obs} decrease

$$[Pd(R_4dien)OH_2]^{2+} + HCO_3^- \rightarrow [Pd(R_4dien)OCO_2] + H_2O + H^+ \tag{3}$$

with increasing pH, presumably due to formation of hydroxo complexes, which are known to be substitutionally inert. Attempted studies of anation by CO_3^{2-} at higher pH were complicated by this factor, and by possible participation of conjugate base formation by deprotonation at the central N atom (reported in Volume 4). Formation of the parent species, $[Pd(dien)CO_3]$ itself, proceeded too fast to be followed by the stopped-flow methods employed, in keeping with many reactions of Pd-dien compounds.

For the aquation reaction under acid conditions, only that of the Et_4dien compound proceeded slowly enough to be followed. Equation (4) is indicated

$$[Pd(Et_4dien)CO_3] + H^+ \longrightarrow [Pd(Et_4dien)OCO_2H]^+$$

$$\xrightarrow[H_2O]{k_2} [Pd(Et_4dien)OH_2]^{2+} + CO_2 + OH^- \tag{4}$$

as the overriding mechanism. Although the values of k_2 and the activation parameters are in agreement with those reported for decarboxylations of octahedral complexes, the microscopic reversibility of the formation reaction requires k_2 to represent a solvolysis reaction, not a CO_2 loss. Associative solvolysis by H_2O is also implicated in the base hydrolysis reaction shown in equation (5). Contributions from direct substitution by OH^- are negligible.[22] Values of k_3 [equation (5)] are significantly smaller than k_2 [equation (4)], suggesting that protonation of the carbonate weakens the Pd—O bond, an effect which may be related to protonation of the platinum acetate complexes previously mentioned.[14] The $\Delta V^{\ddagger}$ values for the solvolysis[22] are the first reported for a leaving group with a double negative charge.

$$[Pd(R_4dien)OCO_2] + H_2O \xrightarrow{k_3} [Pd(R_4dien)OH_2]^{2+} + CO_2$$
$$\longrightarrow [Pd(R_4dien)OH]^+ + H^+ \qquad (5)$$

Dissolving $Na_2[PdCl_4]$ in acetone quantitatively produces $Na[PdCl_3(acetone)]$, and the reaction of this species with a variety of dienes has been followed by UV and ^1H NMR spectroscopy.[23] The reactions proceed in two steps [equations (6) and (7)]. The rate of the "ring closure" reaction (7) is

$$[PdCl_3(acetone)] + diene \rightleftharpoons [PdCl_3(\eta^2\text{-}diene)]^- + acetone \qquad (6)$$

$$[PdCl_3(\eta^2\text{-}diene)]^- \rightleftharpoons [PdCl_2(\eta^4\text{-}diene)] + Cl^- \qquad (7)$$

very dependent on the diene, and with cyclooctadiene or norbornadiene was too fast even for the intermediate η^2-complex to be detected. The operation of step (7) was slower with diene as 1,5-hexadiene or 1,5-cyclononadiene, however.

The trimeric palladium acetate complex $[Pd_3(OAc)_6]$ reacts with excess N,N'-dimethylbenzylamine to form the complex *trans*-$[Pd(OAc)_2(NMe_2CH_2Ph)_2]$.[24] This reversibly loses $PhCH_2NMe_2$ to produce $[Pd(OAc)_2(NMe_2CH_2Ph)]$, which was described by the authors as a pseudo-3-coordinate 14-electron species, though the possibility of bidentate coordination of an acetate was not mentioned. This species then undergoes rate-determining *ortho*-palladation (Scheme 4), for which an intramolecular electrophilic character is indicated from a Hammett plot for the ring substituents when substituted benzylamines were employed. A nondissociative *ortho*-metalation path from the bis-dimethylbenzylamine complex also operates, but about 100 times more slowly.

Scheme 4

trans-$[Pd(OAc)_2(NMe_2CH_2Ph)_2] \rightleftharpoons [Pd(OAc)_2(NMe_2CH_2Ph)] + NMe_2CH_2Ph$

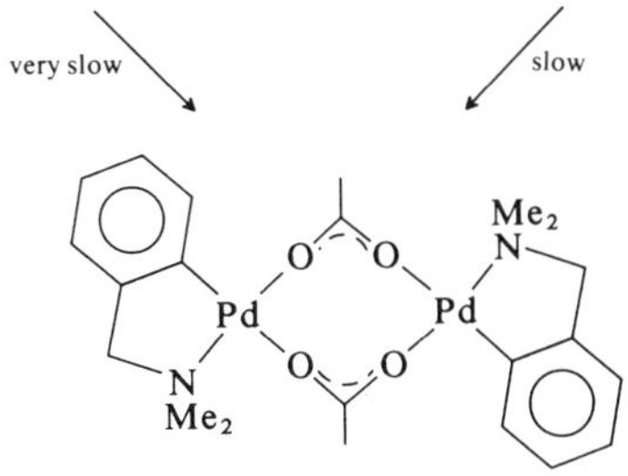

very slow   slow

Scheme 5.   L = PPh$_3$ or PMePh$_2$

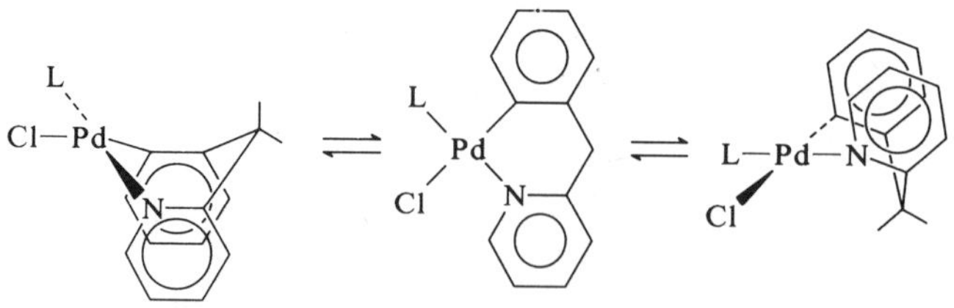

Inversion dynamics of the 6-membered palladocycles shown in Scheme 5 have been followed by ^{1}H NMR spectroscopy in CDCl$_3$ between 213 and 323 K.[25] The first-order rate constants are independent of the complex concentration, and of the concentration of added pyridine-type ligand. The implication is that a planar transition state is achieved without any bond breaking.

5.4.   Ligand Exchange at Square-Planar Gold(III) and Iridium(I)

The replacement of ammonia from [Au(NH$_3$)$_4$]$^{3+}$ by chloride ions proceeds sequentially via *trans*-[AuCl$_2$(NH$_3$)$_2$]$^+$ to produce, finally, [AuCl$_4$]$^-$. There is no significant water solvento pathway, and the rates fall by a dramatic 5 orders of magnitude after the first two substitutions,[26] a consequence of charge neutralisation, *trans* effect, and statistical factors. The generally negative entropies of activation are in accord with a greater importance of bond making, rather than bond breaking, at the transition state. Charge neutralization during these anation reactions makes the entropies of activation more positive than might otherwise be expected, however.

Kinetic studies on the forward and reverse reactions of equation (8) (am = pyridine or substituted pyridine) in 95% methanol/water[27] reveal significant changes in comparison with the analogous reactions of [AuCl$_4$]$^-$. Unlike the reactions of the chloro complexes, reactions (8) are sensitive to

$$[AuBr_4]^- + am \rightleftharpoons [AuBr_3(am)] + Br^- \qquad (8)$$

steric hindrance by 2- or 6-substituents on the pyridine rings, as well as to amine basicity. Further studies on the nature of the steric hindrance are promised.[27]

Ligand replacement reactions of iridium(I) complexes are even more rare than those of gold(III), and are thus the more welcome. A recent study on the reaction of *trans*-[Ir(OR)(CO)(PPh$_3$)$_2$] (R = Me, Pr, Bu^t, or Ph; made from the reaction of Vaska's complex with NaOR) with CO, and which finally produces $[Ir(COOR)(CO)_2(PPh_3)_2]$ by "CO insertion" into the Ir—O bond, shows that the alkoxide ligand is replaced by CO as the initial step, and the free RO$^-$ then attacks a coordinated CO.[28] The intermediate $[Ir(CO)_3(PPh_3)_2]^+OR^-$ was detected by IR and conductivity.

5.5. Nickel(II) Complexes

The availability of square-planar, tetrahedral, and octahedral geometries at similar energies complicates mechanistic studies as usual, though interesting relationships between these forms continue to emerge. The delicate energy balance between tetrahedral and square-planar forms of $[NiX_2(PR_3)_2]$ is emphasized by a crystal structure of $[NiCl_2(PPh_3)_2]\cdot2C_2H_4Cl_2$.[29] These triphenylphosphine complexes are normally tetrahedral, but this solvate is square-planar with the 1,2-dichloroethane molecules in clathrate-like cavities. The 4-coordinate complexes $[Ni\{Et_2P(S)NR\}_2]$, with 4-member chelate rings, are shown by ^{31}P NMR spectroscopy and magnetic-moment measurements to exist in solution as equilibria between tetrahedral and square-planar isomers.[30] High temperatures and bulky organic substituents, R, favor the tetrahedral form, and when R = Bu^t only the tetrahedral molecules are present in the solid or solutions.

$$Ph_2P \qquad O^-$$

3

The perfluoroalcoholate, (**3**), forms bis-chelate complexes $[M(BiL)_2]$ $[M = Pt(II), Pd(II), Ni(II),$ or $Co(II)]$.[31] ^{31}P and ^{19}F NMR spectroscopy show the platinum and palladium complexes to be *cis*-square-planar, but the nickel complex is *trans* in the solid state (confirmed by X-ray crystallography), and behaves as a rapidly interconverting *cis–trans* mixture in solution. Lower solvent polarity favors the *trans* form, the equilibrium being dominated by $\Delta S°$. The isomerization reaction is first order in complex and is unaffected by free ligand, indicating an intramolecular process. A geometry change via a tetrahedral intermediate is proposed[31] and the authors remind us that the activation barriers need not represent the difference in ground-state energies of those two forms, referring to related studies on $[NiX_2(PR_3)_2]$ molecules with almost equienergetic planar and tetrahedral forms, but for which $\Delta G^‡$ for isomerization is 35–45 kJ mol^{-1}.

Molecules of the type (**4**) on the other hand, have tetrahedral ground-state structures, but these tetrahedra undergo rapid inversion, this time probably via square-planar intermediates.[32] Use of prochiral groups, such as iso-propyl or benzyl, for R^1 or R^2 enabled the inversions to be followed by ^1H NMR spectroscopy (though the paramagnetism produced broad lines with $\Delta v_{1/2}^0 \simeq$

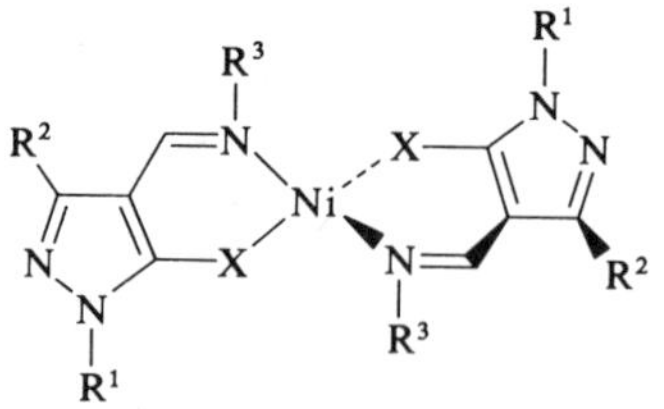

X = S or Se.

4

10 Hz). Rapid inversion led to averaging of the diasteriotopic groups. The $\Delta \rightleftharpoons \Lambda$ interconversion rate was found to be independent of concentration and solvent polarity, and the process presumably proceeds by an intramolecular digonal twist motion through a planar intermediate. Bulky groups R^3 make attainment of the planar configuration difficult to achieve, and slow the rate.

The nickel(II) complexes NiA_2 (A are the N,O-chelating N-alkylsalicyl-adiminate, N,N'-dialkyl-2-aminotropone iminate, and N-alkylbenzoylacetone iminate) are also subject to fast configurational planar $\rightleftharpoons$ tetrahedral equilibria.[33] The equilibrium position, determined by IR and NMR spectroscopy, depends on the nature of the solvent and the alkyl substituents. The kinetics of ligand substitution at these complexes by acetylacetone (HB) or N,N'-disalicylideneethylenediamine (H_2B) were followed by stopped-flow spectrophotometry at different temperatures. Substitution by HB in methanol takes place in two observable steps, each of which follows rate law (9). Substitution by H_2B in acetone allows observation of only one step [equation (10)].

$$\text{rate} = (k_S + k_{HB}[HB])[NiA_2] \tag{9}$$

$$\text{rate} = k_{H_2B}[H_2B][NiA_2] \tag{10}$$

Correlation of k_{H_2B}, the equilibrium constants for the planar $\rightleftharpoons$ tetrahedral equilibria, and the activation parameters $\Delta H^{\ddagger}$ and $\Delta S^{\ddagger}$, indicate that H_2B attacks only the planar configurational isomer, and that the tetrahedral isomer is substitution inert in these systems.[33] The alkylsalicylaldiminato complexes readily equilibrate with one and two pyridine molecules to form 5- and 6-coordinate species, respectively, but rate constants for H_2B substitution in the presence of pyridine proved that the reaction takes place exclusively via the planar, pyridine-free complex (Scheme 6). The substitutions probably proceed by the "double acid/base reaction" to give (**5**), followed by rate-determining rupture of a Ni—O bond.

The square-planar nickel(II) α-amino*iso*butyric acid complex, (**6**), reacts with CN^- to form $[Ni(TetraL)CN]^{2-}$, which then reacts with more cyanide with a $[CN^-]^2$ dependence to give, finally, $[Ni(CN)_4]^2$. Increased Ni—N (peptide) bond strength due to the α-carbon methyl groups probably accounts for the enhanced stability of $[Ni(TetraL)]^-$, which undergoes reactions orders of magnitude more slowly than comparable species.[34]

Scheme 6

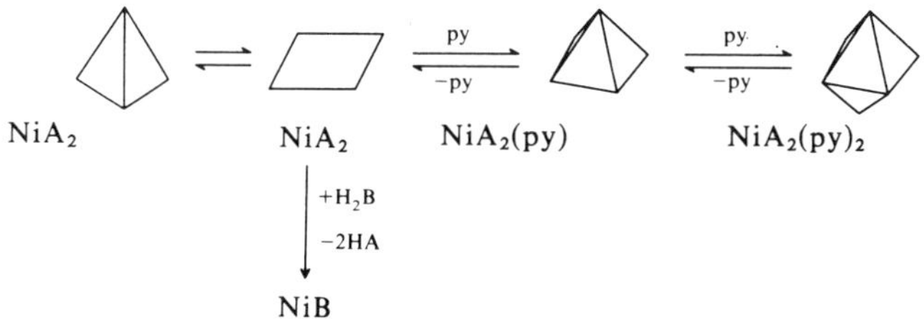

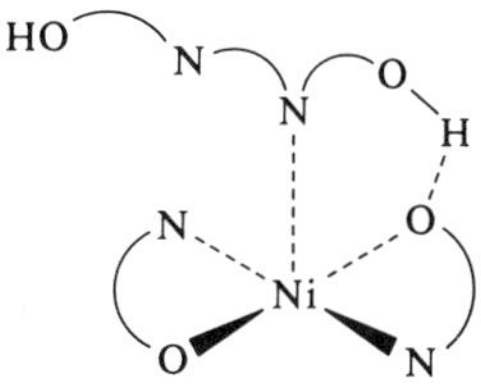

$$NiA_2 \rightleftharpoons NiA_2 \xrightarrow[-py]{py} NiA_2(py) \xrightarrow[-py]{py} NiA_2(py)_2$$

$$\downarrow \begin{array}{c} +H_2B \\ -2HA \end{array}$$

$$NiB$$

5 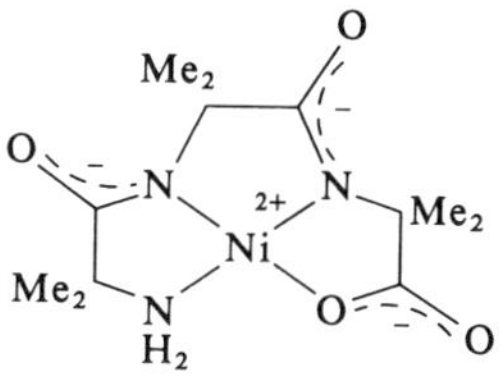   6

The kinetics of substitution reactions of *trans*-$[NiX(C_6Cl_5)(PMePh_2)_2]$ by pyridine and substituted pyridines have been determined.[35] Not surprisingly for these bulky molecules, the rates are determined primarily by steric effects, and not by the basicity of the entering group.

Interest in cyclam and tetramethylcyclam complexes of nickel(II) continues, and crystal structures on tetramethylcyclam complexes have confirmed the conformations of the *R,S,R,S* form[36] and the *R,R,R,R* form.[37] The various conformers are known to behave quite differently (the *R,S,R,S* isomer forms a monosolvento paramagnetic 5-coordinate adduct, whereas the *R,S,S,R* form adds two *trans* waters and the *R,R,R,R* two *cis* waters[37]), as well as to interconvert under certain conditions. The enthalpy change[37] for conversion of the planar *R,R,R,R*(aq) form to the more stable *R,R,S,S* form was a surprisingly small -1.6 kJ mol^{-1}. The conversion of the 1,4,7,10-tetraazacyclotridecane ([13]aneN$_4$) complex *cis*-$[Ni([13]aneN_4)(OH_2)_2]^{2+}$ to the planar $[Ni([13]aneN_4)]^{2+}$ has been studied kinetically.[38] $[Ni([13]aneN_4)]^{2+}$ appears to be unique out of the 12- to 16-member tetradentate macrocyclic nickel(II) complexes in not equilibrating with water to form an octahedral molecule. The *cis*-bis-aquo molecule was prepared by treating the ethylenediamine adduct $[Ni([13]aneN_4)en]^{2+}$ with HBr to make *cis*-$[Ni([13]aneN_4)Br_2]$, which then produces the blue bis-aquo ion in solution. Loss of the coordinated water readily occurs in basic solution, and follows the rate law (11). Cyclam tetraacetate, (7), normally complexes to Ni(II) by two O and two N atoms, and a

$$\text{rate} = k_{OH^-}[\text{complex}][OH^-] \tag{11}$$

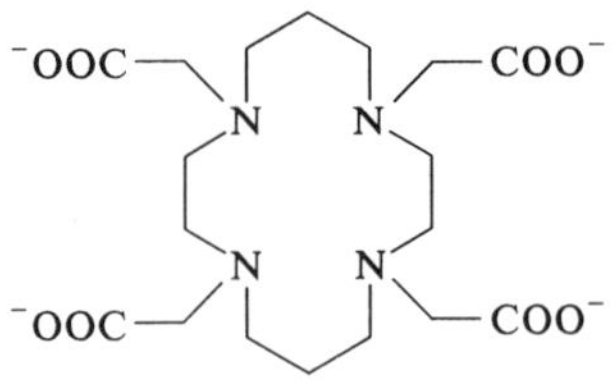

7

high-energy barrier has prevented movement of the nickel into the macrocycle to form the N_4-coordinated "Ni in" form. This barrier has been overcome by using reversible electrochemical oxidation, however, the Ni(III) ion having the ability to change coordination readily (Scheme 7).[39]

Scheme 7

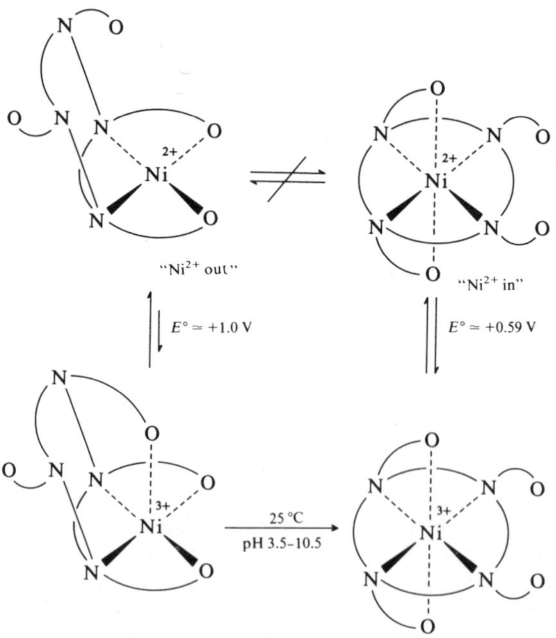

Crystal structures of thirteen nickel(II) complexes of hexahydro- and tetrahydro-porphinoids have been reported and compared.[40] Contraction of the "coordination hole," as determined by decreasing Ni—N bond lengths in the NiN_4 plane, leads to saddle-shaped deformation of the rest of the ligand system. The effects of the distortions on the residual metal electrophilicity are discussed.

5.6. Trans Effect

More consequences of this phenomenon have been reported, some predictable but others surprising. Molecules of (8) have been shown by ^{31}P and ^{195}Pt NMR spectroscopy to be fluxional,[41] but only one of the Pt—S bonds

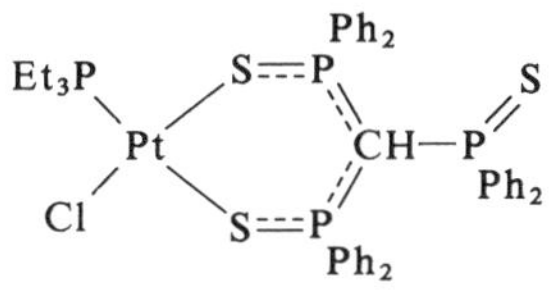

8

is labile, presumably reflecting the greater *trans* effect of Et$_3$P than Cl$^-$. The iodo-bridged complexes *trans*-[Pt$_2$(μ-I)$_2$I$_2$(am)$_2$] (am are various amines), made by treating *cis*-[PtI$_2$(am)$_2$] with HClO$_4$, undergo cleavage reactions by other amines to produce *cis*-[PtI$_2$(am)(am′)].[42] Cleavage reactions leading directly to *cis* products are quite uncommon, and this one results from the *trans* effect order I$^-$ > RNH$_2$. The more usual *trans* geometry was initially produced from the cleavage of [Pt$_2$(μ-Cl)$_2$Cl$_2$(PCy$_3$)$_2$] (PCy$_3$ is tricyclohexylphosphine) with dmso or dibenzylsulfoxide,[43] though interestingly while the dmso was in the usual S-bonded form, (PhCH$_2$)$_2$SO was oxygen bonded, probably due to steric hindrance.

The high *trans* effect of olefins at platinum is generally accepted as resulting from π back-donation at the 5-coordinate transition states. ^{1}H NMR studies this year on complexes *trans*-[PtCl$_2$(py)L] (L are CH$_2$=CHMe, *E*- and *Z*-CHMe=CHMe, and *E*- and *Z*-CHMe=CHPh) indicate from solution stabilities that olefin to Pt σ donation is more important than π back-donation,[44] in agreement with recent similar findings from ^{13}C NMR spectroscopy. These results run against the conventional picture of metal olefin bonding, and it would be of interest to discover if the same trend holds in 5-coordinate reaction intermediates. If so, some reinterpretation of the mechanism of the *trans* effect of such molecules may be necessary.

Measurements of relative *trans* influences, the ground-state σ-bonding effect, continue to emerge from NMR and X-ray crystallographic studies. The values of $^1J_{\text{Pt}-\text{P}}$ and $^1J_{\text{Pt}-\text{C}}$ in the complexes *trans*-[PtHL(PEt$_3$)$_2$]$^+$ and *trans*-[PtDL(PEt$_3$)$_2$]$^+$ (L is ^{13}CO or PEt$_3$) indicate that D has a significantly greater *trans* influence than H.[45] Measurements of ^{195}Pt, ^{119}Sn, ^{31}P, and ^{15}N NMR spectra, as well as a crystal structure on compound (9), also provide useful data for comparing *trans* influences.[46] A comparison of the molecular struc-

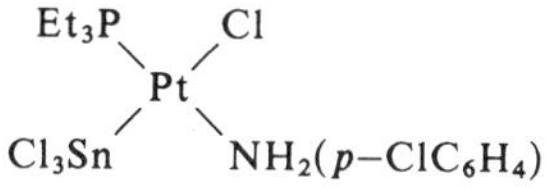

9

ture of $K_2[Pd(NO_3)_4]^{(47)}$ with that of *cis*-$[Pd(NO_3)_2(dmso)_2]$ indicates a strong ground-state *trans* influence of dmso in palladium(II) complexes.

^{195}Pt and ^{15}N NMR spectroscopic results on several ^{15}N labeled compounds $[Pt(NH_3)_3Z]^{n+}$, *cis*-$[Pt(NH_3)_2Z_2]^{m+}$, and *cis*-$[Pt(NH_3)_2XZ]^{m+}$ have been used to estimate *cis* influences, based on J_{Pt-N} and $\delta(N)$.[48] These are, as usual, smaller than the *trans* influences of the same groups, but are nevertheless significant.

NMR and X-ray crystallographic measurements on the formally platinum(I) metal–metal bonded complexes $[Pt_2X_2(\mu\text{-dmpm})_2]$ and $[Pt_2XY-(\mu\text{-dmpm})_2]^+$ [dmpm is bis(dimethylphospino)methane: X is Cl, Br, I, or Me; Y is $PPh_3]^{(49)}$ reveal an interesting distant *trans* influence. Although the influence of X in the linear X—Pt—Pt—Y system is felt mainly by the Pt—Pt bond, lengthening and reduction of $^1J_{Pt-P}$ for Y can also be detected. A long-range addition of *trans* influences has been claimed to operate across the Pt—Pt bonds of some binuclear platinum(III) complexes, also.[50]

5.7. Isomerization Reactions

Treatment of the pyridine-2-carboxylate complex (**10**) (Scheme 8) with dimethylsulfoxide produces isomers (**11**) and (**12**) by chloride displacement.[51] The kinetically-controlled initial ratio of 11:12 is 3.8, but slow isomerization catalyzed by dmso changes this ratio to 0.38, presumably the thermodynamic

Scheme 8

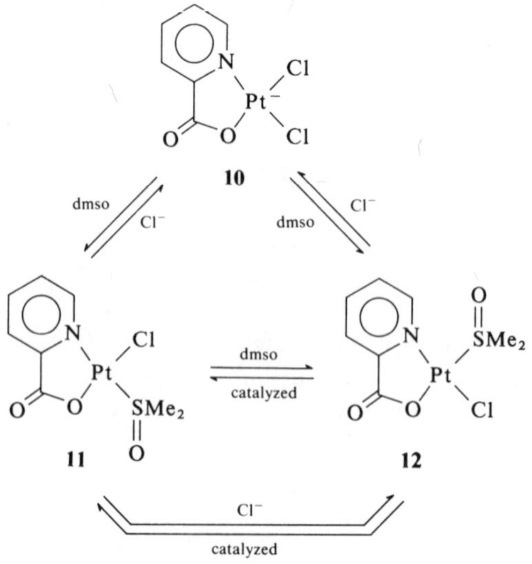

ratio. Cl^- also catalyzes the isomerization, but eventually displaces the dmso to reform (**10**). Reaction rates for all the steps (Scheme 8) were presented, and it is apparent that the (**11**) $\rightleftharpoons$ (**12**) isomerization cannot proceed by consecutive displacement via (**10**). An unspecified rearrangement of a 5-coordinate intermediate is presumably responsible for the geometry change.[51]

The reaction of *trans*-$[PtHCl(PPh_3)_2]$ with $SnCl_2$ and then C_2H_4 leads to *cis*-$[PtEt(SnCl_3)(PPh_3)_2]$. In solution, this complex slowly isomerizes to the *trans* form, and the process is markedly catalyzed by free carbon monoxide, a reaction rapid even at $-80\,°C$ in CD_2Cl_2. Here, too, consecutive displacement at the isomerization mechanism seems unlikely {a slower reaction leads to *trans*-$[PtEt(CO)(PPh_3)_2](SnCl_3)$}, and the possibility of pseudorotation, which has been proposed before for other CO-catalyzed isomerizations, must be kept in mind.

The complexes *trans*-$[PtCl_2(PCy_3)(dmso)]$ and *trans*-$[PtCl_2(PCy_3)$-$(\{PhCH_2\}_2SO)]$ both undergo nucleophile-catalyzed isomerizations, but extremely slowly compared to related compounds with less bulky phosphines.[43] Curiously, the O-bonded $(PhCH_2)_2SO$ ligand emerges S-bonded in the *cis* position, indicating that both electronic and steric influences effect its bonding.

More solid-state isomerizations have been reported, though in common with most such examples, the mechanisms remain unknown. DTA, XPS, and IR spectroscopy reveal that crystalline *Cisplatin*, *cis*-$[PtCl_2(NH_3)_2]$, changes to the *trans* isomer at $306\,°C$, before decomposing at $319\,°C$.[53] The complex *trans*-$[PtI_2L_2]$ (L = n-$C_3H_7NH_2$) appears to be more thermodynamically stable than *cis*-$[PtI_2L_2]$, $[PtIL_3]I$, or $[PtL_4]I_2$, all of which convert to it on heating.[54]

Scheme 9. X = Cl or I; R = Me or Et

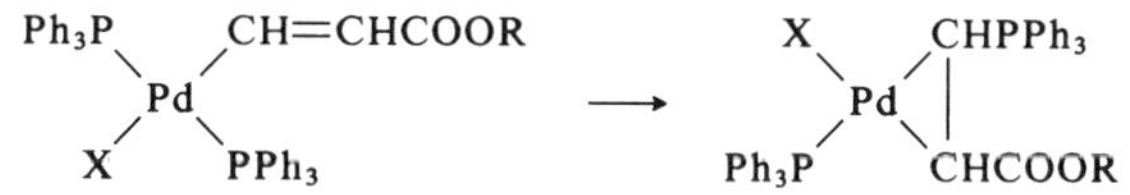

The reaction shown in Scheme 9 proceeds via PPh_3 loss. Since the groups X change site from *trans* to the C(H) bond to *cis* to it, it is conceivable that this process involves isomerization of a 3-coordinate intermediate.[55] The presence of water has a critical effect on the geometry of $[PdBr_2L_2]$, made from $PdBr_2/HBr/LiBr/L/BuOH$ (L is $P\{OPh\}_3$).[56] The more stable *cis* form is produced if $[H_2O] \geq 0.9\ M$, but *trans* if $[H_2O] \leq 0.1\ M$. This is shown to be the result of kinetic changes on the formation, however, and not an isomerization.

A totally new type of isomerization is implied from studies by H. C. Clark and co-workers. Variable-temperature ^1H NMR measurements on compounds

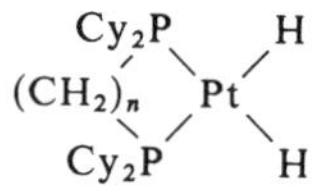

$$Cy_2P \diagdown \diagup H$$
$$(CH_2)_n \diagdown Pt$$
$$Cy_2P \diagup \diagdown H$$

13

(**13**) reveal a rapid (at 40 °C) site exchange of either the two phosphorus atoms or the two hydrogen atoms.[57] The process is intramolecular, and the recent recognition of several η^2-H_2 complexes of transition elements[57,58] led the authors to propose Scheme 10 to account for the geometry change. Similar NMR line-shape analyses for $[PtH(SiPh_3)(Cy_2P(CH_2)_7PCy_2)]$ and *cis*-$[PtH(SiPh_3)(PCy_3)_2]$[57] indicate rapid site exchange between H and $SiPh_3$, and suggest the possibility of η^2-bonded R_3Si—H intermediates in place of the η^2-H_2 of Scheme 10. This explanation could equally well apply to related phenomena observed in the similar complexes *cis*-$[PtH(SiPh_3)(PPh_3)_2]$ a few years ago.[59]

Scheme 10

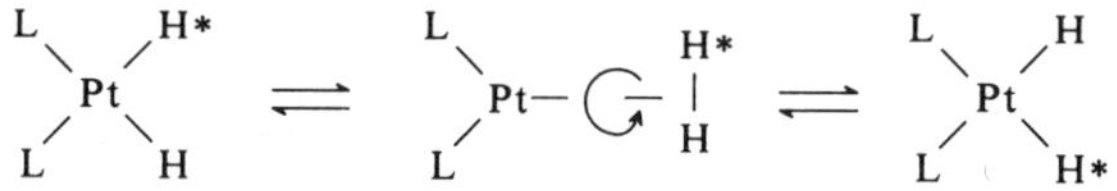

The site exchange phenomena observed previously[60] in the square-planar complex $[NiH\{P(O\text{-}p\text{-tolyl})_3\}_3]^+$ could proceed either through a square-planar to tetrahedral digonal twist like those described in Section 5.5, or via an η^2-phosphonium intermediate like those proposed for the silyl complexes. (R_3PH^+ are, of course, isoelectronic with R_3SiH.)

These site exchanges are not isomerizations as such, but if rotation of η^2-phosphonium intermediates is indeed a viable mechanism, then a similar application could account for an otherwise unexplained isomerization. *Cis*- and *trans*-$[PtH_2L_2]$ (L = PMe$_3$ or PEt$_3$) rapidly interconvert in solution.[61] Both isomers reversibly lose H_2, but this process is not fast enough to account

Scheme 11

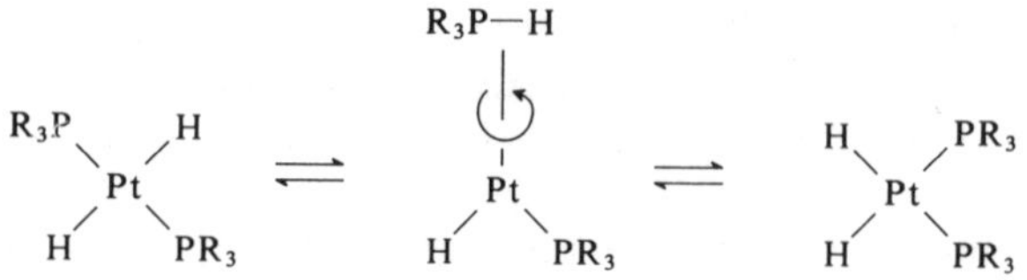

for the geometry change. Addition of H_2 to PtL_2 to form a *trans* isomer is, in any case, a symmetry-forbidden process.[62] Rotation of an η^2-phosphonium intermediate as in Scheme 11 remains a possible, though speculative, route for the isomerization.

5.8. Five-Coordinate Complexes

The cobalt(II) complex with two of the ligands (**3**) is *trans* square-planar, but unlike the Pt, Pd, or Ni analogues it readily forms an adduct with pyridine.[31] The tetradentate tripod ligand *tris-o-*(dimethylarsino)phenylphospine (ptas) forms a series of trigonal bipyramidal nickel(II) complexes $[Ni(ptas)X]^{n+}$, where arsenic atoms occupy the equatorial sites and X is axial, *trans* to O.[63] The degree of deshielding of the apex phosphorus atom, determined from ^{31}P NMR measurements, established a *trans* influence series at these 5-coordinate species as $X = I^- > P(OMe)_3 \simeq CN^- > Br^- > PEt_3 \simeq SCN^- \simeq PPh_3 > NCS^- > NO_3^-$.

A series of 5-coordinate square pyramidal organonickel(II) complexes has been synthesized by the free radical reaction between organic halides and planar (*R,S,R,S*-tetramethylcyclam)nickel(I).[64] The rate of hydrolysis of these $[Ni(tmc)R]^+$ complexes to $[Ni(tmc)OH]^+$ is independent of pH between 9 and 13, and insensitive to most organic groups, R.[65] Activation parameters also span a narrow range, with $\Delta H^{\ddagger} \simeq 50$ kJ mol^{-1} and $\Delta S^{\ddagger} \simeq -44$ J mol^{-1} K^{-1}. An appreciable solvent deuterium kinetic isotope effect indicates that a significant proton transfer from water to the incipient hydrocarbon product takes place at the transition state. No nucleophilic assistance from OH^-, OAc^-, or NH_3 was detected, but when R was $CH_2CH_2CH_2OH$, an enhanced hydrolysis rate was assigned to a cyclic transition state resulting from assistance by the pendant OH. In acid solution, direct protonolysis enhances the rate.

The 5-coordinate complex (**14**) involves an uncommon template synthesis of the tetradentate macrocyclic ligand.[66] The free ligand can be obtained by CN^- treatment.

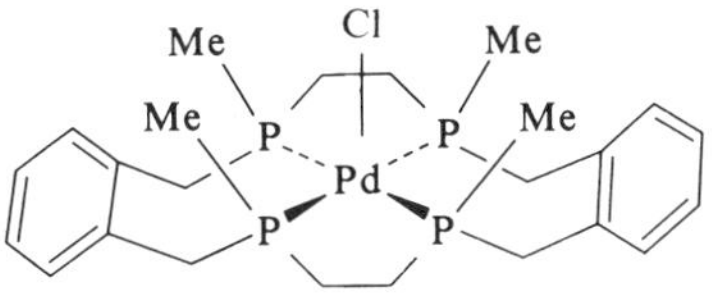

14

The delicate balance between some 4- and 5-coordinate palladium(II) complexes is illusrated this year by crystal structures of the imine complex (**15**)[67] and the cationic o-phenanthroline complex (**16**).[68] Both have long axis Pd···N contacts, although the closely related azobenzenyl derivative *trans*-[PdCl(*o*-C$_6$H$_4$N=NPh)(PEt$_3$)$_2$], reported several years ago, has none. The NMR spectra of (**16**) and its related 2,9-disubstituted phenanthroline

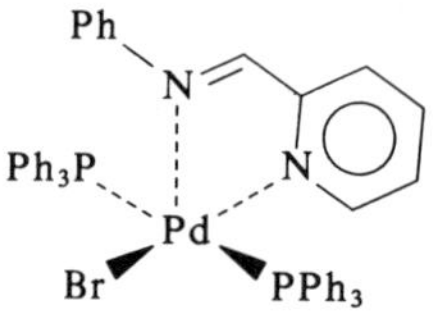

15 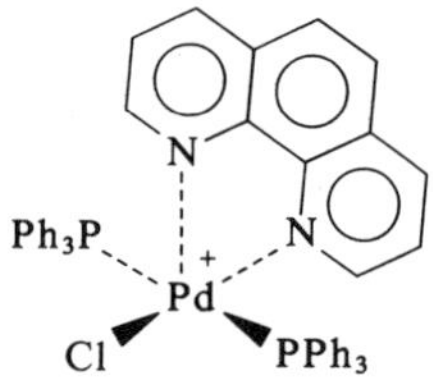   **16**

derivatives, as well as of $[PdCl(2,9-Me_2phen)_2]^+$ and $[Pd(PPh_3)(phen)_2]^{2+}$, show all the complexes to be fluxional in solution at ambient temperatures.[69] It seems probable that a rapid geometry change via a trigonal bipyramidal intermediate would be responsible for the apparent equivalence of the two nitrogen atoms. The ground state of molecule (**17**) is trigonal bipyramidal (Scheme 12).[70] The monodentate form of the ligand in complex (**18**), which equilibrates with (**17**), can bridge to a second platinum atom.

Scheme 12

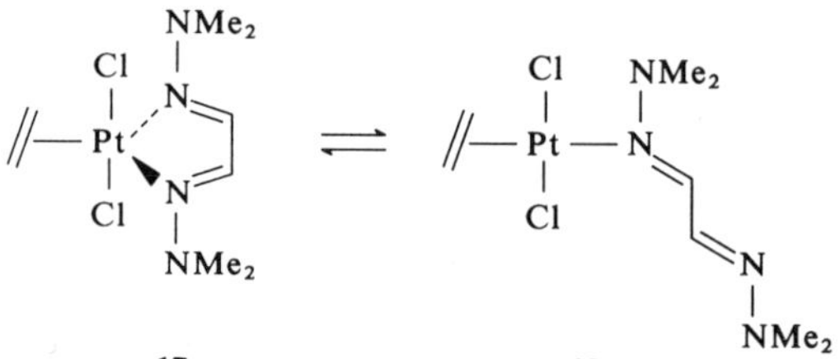

17   **18**

Stability constants for the reversible formation of 5-coordinate complexes from $[Pt(phen)_2]^{2+}$ or $[Pt(bipy)_2]^{2+}$ and one of the nucleophiles I^-, SO_3^{2-}, $S_2O_3^{2-}$, thiourea, OH^-, NH_3, NH_2Me, $NHMe_2$, NMe_3, piperidine and en have been determined.[71] Values were generally similar, except for NMe_3 which was lower than expected from basicity considerations.

The solution chemistry of the complexes $[PtX_2L_2]$, where X is Cl, Br, or I and L are various (2-cyanoethyl)phosphines, has been thoroughly examined.[72] Excess L can either promote *trans* to *cis* isomerization or add to form a 5-coordinate species. In a similar series with L = divinylphenylphosphine, the compounds are mainly *cis* in geometry, though when X = I a slow isomerization to *trans* takes place in $CDCl_3$ solution.[73] In this system also, excess L establishes the equilibrium shown in equation (12), which was examined by variable-temperature ^{31}P NMR spectroscopy.

$$[PtX_2L_2] + L \rightleftharpoons [PtX_2L_3] \tag{12}$$

More examples of the formation of 5-coordinate complexes by nucleophile attack preceding electrophile attack, the two resulting in oxidative addition to the square-planar compound, have been reported. Prior coordination of I^- or OAc^- (L) to $[RhI_2(CO)_2]^-$ catalyzes MeI oxidative addition[74] resulting, finally, in $[Rh(COMe)I_2(CO)L]^-$. (Curiously, the similar oxidative addition of 1-bromopropane to *trans*-$[RhBr(CO)\{P(\textit{p}\text{-}EtC_6H_4)_3\}_2]$ is not catalyzed by Br^-.[75]) Cl^- and *trans*-$[PtCl(neopentyl)(PEt_3)_2]$ equilibrate prior to HCl cleavage of the organic group in $9:1$ $MeOH/H_2O$, but in this example also does not assist the reaction (Scheme 13). No coordination of Br^- to the bromo analogue was observed.

Scheme 13

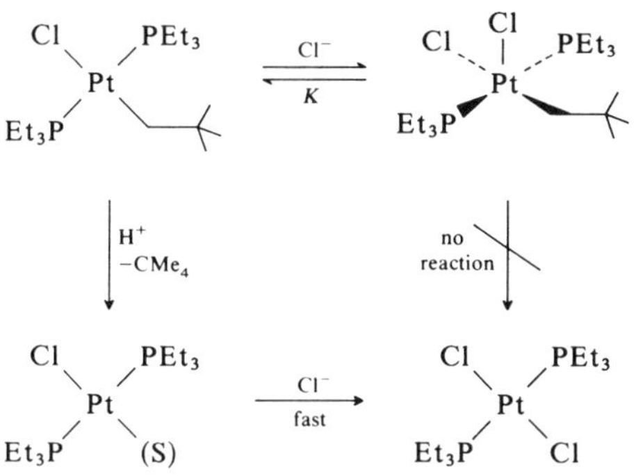

Into this section finally go a number of 5-coordinate complexes formed by addition of *electrophiles* to square-planar compounds. When the electrophile is a positively charged species such as H^+, then the step is formally an oxidation, and this is the case in Scheme 14, where the conversion of the carbene to the carbyne by H^+ has been shown to proceed by protonation at the metal.[77] Other cases are less distinct, however, including protonolysis of complex (**19**) by HCl.[78] Here it was not possible to discern which end of the Pd—C bond

Scheme 14.   L = PPr$_3^i$; R = H, Me, or Et

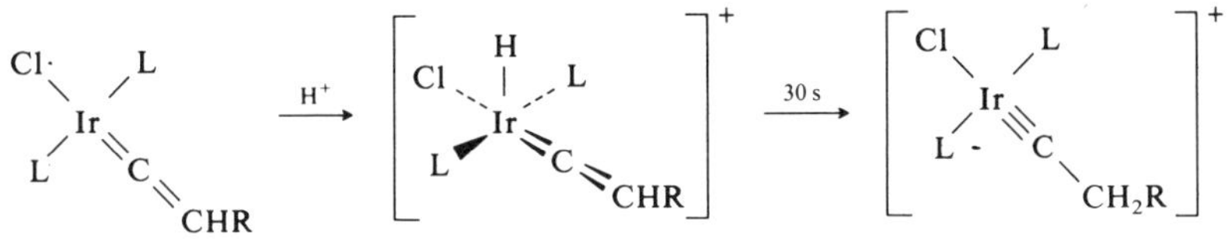

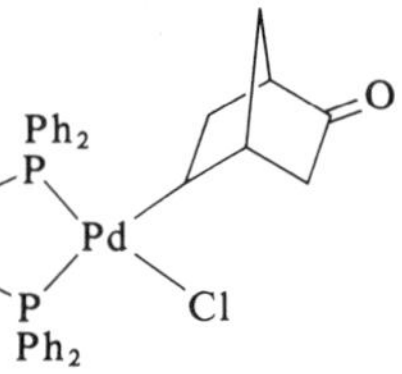

19 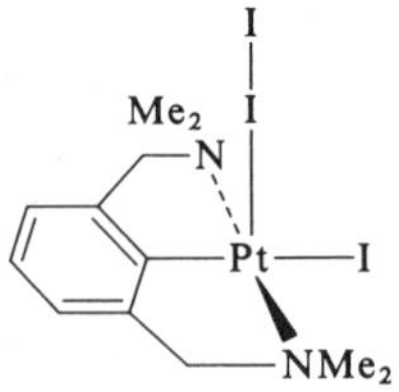   **20** 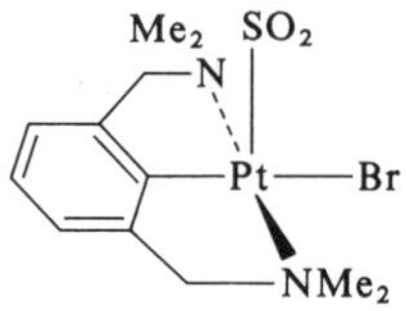   **21**

is approached by H^+. Compounds (**20**) and (**21**), both of which have been characterized crystallographically, are also described as resulting from electrophilic attack at platinum(II).[79,80] Evidence for regarding the bonding as σ donation from Pt to I_2 or SO_2, respectively, includes the NMR parameters [closer to those for Pt(IV) than Pt(II)] and the bond lengths.

5.9. Binuclear Complexes

Scheme 15 depicts bridge cleavage reactions where singly-bridged species are formed. The dimethylsulfide-bridged compound can be isolated,[81] and all three isomers have been obtained at low temperatures. These reactions, like those involving pyrazine molecules as bridge-splitting ligands[82] (see Volume 4), run counter to the general trend in that the cleavage of the first halide bridge is faster than that of the second.

^{1}H NMR measurements at high pressure have enabled $\Delta V^{\ddagger}$ to be determined for the reaction of $[Pd_2(\mu\text{-}Cl)_2(2\text{-methylallyl})_2]$ with $[PdCl(PPh_3)(2\text{-methylallyl})]$.[83] At 22 °C, at which temperature *syn–anti* exchange in the triphenylphosphine adduct is catalyzed by the dimer, $\Delta V^{\ddagger} = 0 \pm 2$ cm^3 mol^{-1}.

Scheme 15. L = 2,6-Dimethylpyridine or 2,4,6-Trimethylpyridine

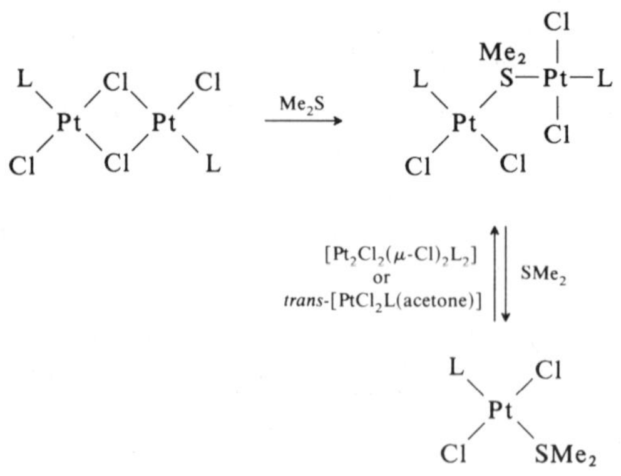

To account for this low value, the authors propose an overall canceling of opposite contributions from partial bridge opening of the dimer (ΔV positive) and its subsequent association with the monomer (ΔV negative). At 56 °C, when the methylallyl groups exchange readily, $\Delta V^{\ddagger} = 11 \pm 2$ cm^3 mol^{-1}. Total dimer cleavage to [PdCl(methylallyl)], followed by PPh$_3$ transfer, is proposed to account for the new value.

The crystal structure of [Pt$_2$Cl(PPh$_3$)(μ-dppm)$_2$]PF$_6$ (dppm is bis-diphenylphosphinomethane) reveals a significant distortion toward tetrahedral at the PPh$_3$-bonded platinum.[84] Probably steric in origin, this retards the rate of axial/equatorial interchange of substituents on the Pt$_2$(μ-dppm)$_2$ rings, as observed by ^{1}H NMR spectroscopy. The reactions of [Pt$_2$Cl$_2$(μ-dppm)$_2$] with S$_8$ or SO$_2$ in CH$_2$Cl$_2$, to form the Pt—S—Pt or Pt—S(O)$_2$—Pt A-frame complexes, are second order.[85] The SO$_2$ reaction involves preassociation of the reactants, probably as a charge-transfer complex. Kinetic data for both reactions are consistent with substantial involvement of Pt—Pt breaking in the activation process.

Chapter 6

Substitution Reactions of Inert-Metal Complexes— Coordination Numbers 6 and Above: Chromium

6.1. Introduction

This review covers the period between the end of the last report[1] (July 1985) and the literature available up to December 1986. General reviews[2,3] on the coordination chemistry of chromium have appeared and more specific reviews on Cr—Cr multiple bonded systems,[4] the unusual reactivity of Cr(III)–edta type complexes,[5] Cr(III) magnetochemistry,[6] and arene–chromium–tricarbonyl complexes[7,8] have been prepared. There has also been considerable interest in the use of the water-stable Cr(V) complexes, $[Cr(EABA)_2O]^-$ [EABA = 2-ethyl-2-hydroxy butanato (-2) and related ligands], as oxidizing agents.[9-16] Much of this work has been reviewed[17] and will not be further considered here. Other aspects of Cr(IV) and Cr(V) chemistry have also been reviewed.[18,19] Perhaps the most mechanistically significant information produced in this period comes from the discovery that the so-called *cis*-(hydroxo)- (aqua) ions are, in fact, dimeric in the solid state[20] and contain di-μ- (HO—H—OH) bridges as in structure (1). Similar structural units are found in "$[(H_2O)(NH_3)Cr(OH)_2Cr(NH_3)_3(OH)]^{3+}$" which consists of chains of the diaqua^{4+} and dihydroxo^{2+} cations linked by HO—H—OH bridges.[21] Thus a pathway for the facile thermal dehydration of *cis*-(hydroxo)(aqua) salts to give μ-diol species becomes clear, as does the reason for the low solubility of

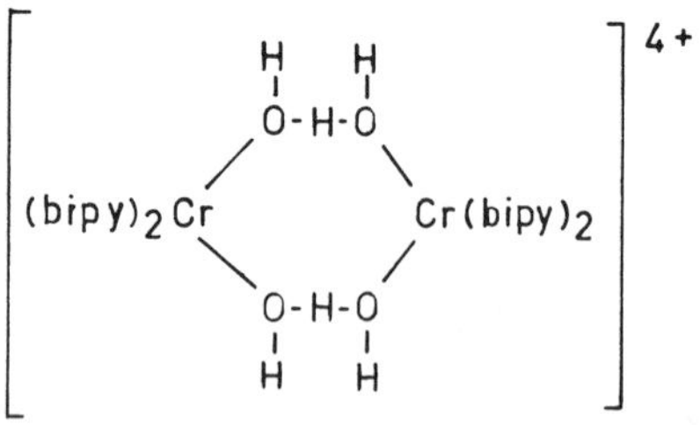

$$\left[\begin{array}{c} \overset{\displaystyle H\quad H}{\underset{\displaystyle |\quad |}{}} \\ \overset{\displaystyle O\text{-}H\text{-}O}{} \\ (bipy)_2Cr \qquad Cr(bipy)_2 \\ O\text{-}H\text{-}O \\ \underset{\displaystyle H\quad H}{\overset{\displaystyle |\quad |}{}} \end{array}\right]^{4+}$$

1

such salts when compared to the *cis*-dihydroxo or *cis*-diaqua analogs. Such dimeric hydroxo aqua bridged species could also account for the facile dimerization[22] and oligimerization of inert aqua ions, such as $[Cr(OH_2)_6]^{3+}$. They also bring the analogy between coordinated OH^- and coordinated $R-OH$ (alcohol) systems closer together, as both inter- and intramolecular $R-O-H-O-R$ (alcohol) bridges are known.[23]

The possibility of seven-coordinate Cr(III) in $[Cr(edta)(OH_2)]^-$, suggested[1] on the basis of ^2H NMR studies, has not received support from CD investigations.[24,25] However, ^2H NMR investigations[26-30] continue to provide important structural evidence for Cr(III) and the subject has been reviewed.[31] One further feature of chromium(III) chemistry that is quite striking during the period under consideration is the interest in complexes containing $Cr-F$ bonds. While only a start has been made on the rate of thermal[32,33] or photochemical[34] $Cr-F$ bond rupture in these systems, there has been considerable emphasis on synthesis[35-40] and characterization by single crystal X-ray data (Table 6.1).

Table 6.1. X-Ray Data for Cr(III)—F Complexes[a]

Complex	Cr–F (Å)	Ref.
$(enH_2)[CrF_4(en)]Cl$	1.90 (2)	42
$t\text{-}Li[CrF_2(1,3\text{-}pdda)]\cdot 2H_2O$	1.878 (2)	43
$[CrF_3(OH_2)_3]$	disordered	
$[CrF_3(OH_2)_3]\cdot 2H_2O$	disordered	
$t\text{-}[CrF_2(NH_3)_4]I\cdot H_2O$	1.894 (3)	
$c\text{-}[CrF_2(NH_3)_4]ClO_4$	1.887 (6)	
$t\text{-}[CrF_2(en)_2]ClO_4$	1.887 (6)	
$t\text{-}[CrF_2(en)_2]Cl$	1.904, 1.886	44
$t\text{-}[CrF_2(tn)_2]ClO_4$	1.876 (3)	45
$t\text{-}[CrF(OH)(en)_2]^+$	1.854 (4)	
$t\text{-}[CrF(OH)(tn)_2]^+$	1.829 (8)	
$t\text{-}[CrF(NH_3)(en)_2](ClO_4)_2$	1.862 (3)	46
$t\text{-}[CrF(NH_3)(tn)_2](ClO_4)_2$	1.872 (3)	
$t\text{-}[CrF(OH_2)(tn)_2]^{2+}$	1.900 (8)	
mean = 1.88 (2)		

[a] Data from Reference 41 unless otherwise stated.

In the following sections, $\Delta H^{\ddagger}$ and $\Delta S^{\ddagger}$ are quoted in units of kJ mol^{-1} and J K^{-1} mol^{-1} respectively.

6.2. Aquation and Solvolysis of Chromium(III) Complexes

6.2.1. Unidentate Leaving Groups

6.2.1.1. $[Cr(III)(O_5)X]^{n+}$ Systems

The rates of aquation of complexes of Cr(III) containing relatively basic ligands may be governed by rate laws with positive dependences upon the hydrogen ion concentration as well as terms that are acid-independent or have inverse acid dependency as in equation (1a).[47] The aquation of $[CrF(OH_2)_5]^{2+}$

$$k_{obs} = k_1 + k_2 K_a[H^+]^{-1} \tag{1a}$$

has been investigated[33] in 1–10 M HClO$_4$ at 30 °C (Table 6.2). Corrections for the acid independent pathway ($k_0 = 1.4 \times 10^{-9}$ s^{-1}, 30°) were not necessary and constant ionic strength was not maintained. Consequently a plot of $[H^+]$ vs. k_{obs} is parabolic. For $[(H_2O)_5Cr(\text{fumaronitrile})]^{3+}$ $k_1 = 1.19 \times 10^{-2}$ s^{-1} and $k_2 K_a = 4.14 \times 10^{-4}$ M^{-1} s^{-1} [equation (1a)] at 25 °C, $I = 0.5$ M.[47]

6.2.1.2. Cr(III)—C Bond Rupture

L$_5$Cr(III)—C bonded species (L usually H$_2$O) can be formed according to equation (1) (Table 6.3). The generation of R· can be achieved by photolysis,[48] radiolysis,[49] halogen atom abstraction,[50,51] or by Fentons

$$[L_5Cr(II)]^{2+} + R\cdot \xrightarrow{k} [L_5Cr(III)-R]^{2+} \tag{1}$$

Reagent,[48,49] as in equations (2) and (3). With $[L_5Cr(III)-R]^{2+}$ and H$^+$, the

$$Cr(II) + H_2O_2 \rightarrow Cr^{3+} + OH^- + \cdot OH \tag{2}$$

$$\cdot OH + RH \rightarrow R\cdot + H_2O \tag{3}$$

Table 6.2. *First-Order Rate Constants for the Aquation of $[CrF(OH_2)_5]^{2+}$ in HClO$_4$ at 30 °C (Reference 33)*

[HClO$_4$] (M)	$10^7 k_{obs}$ (s^{-1})	[HClO$_4$] (M)	$10^7 k_{obs}$ (s^{-1})
1.00	0.264	6.00	7.3
2.00	0.820	7.00	10.9
3.00	1.75	8.00	15.4
4.00	3.16	9.00	17.8
5.00	5.2	10.00	22.3

Table 6.3. *Rates of Formation of* $[L_5Cr(III)-R]^{n+}$ *Complexes According to Equation (1) at 25 °C (Reference 49)*

L	$-R$	$10^{-7}k\ (M^{-1}\,s^{-1})$
H_2O	$-CH_2OH$	16
trans-cyclam (H_2O)		12
cis-nta (H_2O)$_2$		22
H_2O	$-CH(CH_3)OH$	7.9
cis-nta (H_2O)$_2$		10
edta		3.9
H_2O	$-C(CH_3)_2OH$	5.1
trans-cyclam (H_2O)		4.9
cis-nta (H_2O)$_2$		8.4
edta		2.6
H_2O	$-CH(OH)_2$	1.3^a
H_2O	O_2	16^b

[a] Reference 53.
[b] Reference 60.

product is generally the parent RH and $[L_5Cr(OH_2)]^{3+}$ as in equation (4) and

$$[L_5Cr-R]^{2+} + H^+ \rightarrow [L_5Cr(OH_2)]^{3+} + HR \qquad (4)$$

with rate law (5), (6) but, with $[(H_2O)_5Cr-CH_2CH_2OEt]^{2+}$, EtOH and C_2H_4 are produced[52] (Table 6.4). Another reaction where "nonstandard" organic

Table 6.4. *Rate Data Associated with Equation (4) and Rate Law (5), (6)^a at 25.0 °C*

L_5	$-R$	$k\ (s^{-1})$	$k_H\ (M^{-1}\,s^{-1})$	Ref.
H_2O	THF		3.6×10^{-4}	48
H_2O	Dimethoxyethane		inert	48
H_2O	Dioxane		3.8×10^{-1}	48
H_2O	$-CH_2CH_2OEt^b$		$(0.8-4.6) \times 10^3$	52
H2O	$-CH_2OH$	6.6×10^{-4}	4.65×10^{-4}	49
H_2O	$-CH(CH_3)OH$	1.9×10^{-3}	1.22×10^{-3}	49
H_2O	$-C(CH_3)_2OH$	3.3×10^{-3}	4.9×10^{-3}	49
cyclam (H_2O)^c	$-C(CH_3)_2OH$	2.0×10^{-3}		49
nta (H_2O)$_2$^{d}	$-CH_2OH$	2.5×10^{-3}		49
nta (H_2O)$_2$^{d}	$-CH(CH_3)OH$	3.4×10^{-3}		49
nta (H_2O)$_2$^{d}	$-C(CH_3)_2OH$	3.8×10^{-3}		49
nta (H_2O)$_2$^{d}	$-CH_3$	$2.5 \times 10^{-2\ e}$		49
edta	$-CH_2OH$	2.25×10^{-4}	2.56×10^2	49
edta	$-CH(CH_3)OH$	1.35×10^{-3}	7.35×10	49
edta	$-C(CH_3)_2OH$	1.38×10^{-2}	1.76×10^2	49

[a] Acetate ion catalysis is frequently observed.[49]
[b] Organic products are C_2H_4 and EtOH.
[c] *trans* isomer.
[d] *cis* isomer.

Table 6.5. Kinetic Parameters for the Rate of Decomposition of Some
$[n\text{-}py\text{-}CH_2\text{-}Cr(polyamine)_x(H_2O)_y]^{n+}$ Complexes in 1.0 M $HClO_4$ (Reference 55)

			Cr—N rupture			Cr—C rupture		
(polyamine)$_x$	n	y	10^4k (s^{-1}) (°C)^b	$\Delta H^{\ddagger}$	$\Delta S^{\ddagger}$	10^4k (s^{-1}) (°C)^b	$\Delta H^{\ddagger}$	$\Delta S^{\ddagger}$
—	3	5	—	—	—	4.22 (55.0)	139	109
—	2	5	—	—	—	0.144 (55.0)	156	126
(en)$_2$ ^{a}	3	1	—	—	—	2.22 (55.0)	128	71
	2	1	~6 (27.0)	—	—	0.16 (55.0)	154	126
(tn)$_2$ ^{a}	3	1	9.15 (55.0)	99	11	4.00 (55.0)	133	107
	2	1	10.3 (26.0)	—	—	0.115 (54.0)	—	—
(dien)^a	3	2	10.6 (55.2)	183	267	1.61 (55.2)	108	24
	2	2	13 (27.0)	—	—	0.023 (27.0)	—	—
(trien)^a	3	1	16.3 (59.2)	96	2	6.81 (59.2)	103	15
	2	1	5.3 (26.0)	—	—	1.20 (68.0)	—	—
cyclam^a	3	1	—	—	—	0.838 (55.0)	97	−17
	2	1	—	—	—	0.252 (60.0)	134	81

^{a} Geometric configuration not established.
^{b} Rate constants at other temperature are presented in Reference 55.

$$-d[L_5CrR]/dt = k_{obs}[L_5CrR^{2+}] \tag{5}$$

$$k_{obs} = k + k_H[H^+] \tag{6}$$

compounds are found is reaction (7) with k_2 (25 °C) ~ 1 M^{-1} s^{-1}.[58] The Cr(III)—C bond also labilizes donor groups in the *trans* position.[51] Thus

$$(H_2O)_5CrCHO^{2+} + HCOH + H^+ \rightarrow CO + MeOH + Cr(OH_2)_6^{3+} \tag{7}$$

rate data for the reaction (8) give $k_1 = 85$ M^{-1} s^{-1} and $k_{-1} = 15$ s^{-1} at 25 °C.

$$trans\text{-}Cr(CHCl_2(acac)_2(CH_3OH)] + py \underset{k_{-1}}{\overset{k_1}{\rightleftharpoons}} trans\text{-}Cr(CHCl_3)(acac)_2(py)] + MeOH \tag{8}$$

These values are much greater than the rate constants for the ligand substitution of ordinary Cr(III) complexes. Also, studies on the rate of Cr(III)—C bond rupture in a series of $[py\text{-}CH_2\text{-}Cr(polyamine)_x(OH_2)_y]^{2+}$ complexes[54,55] were complicated by concurrent Cr—N bond rupture (Table 6.5).

Systems with Cr(III)—C bonds also have reducing properties.[48,56,57] For example, the reaction (9), (10) gives a rate law (11)[57] (25 °C, $I = 1.0$ M) with

$$[(hedtra)CrCH_2OH]^- + [CoCl(NH_3)_5]^{2+} + 5H^+ \rightarrow [intermediate] \tag{9}$$

$$[intermediate] \rightarrow [(hedtra)Cr(OH_2)] + Co^{2+} + 5HN_4^+ + Cl^- + EtOH(?) \tag{10}$$

$$-d[intermediate]/dt = \{k[H^+]\}\{K + [H^+]\}^{-1}[intermediate] \tag{11}$$

$k = 5.1 \times 10^{-2}$ s^{-1} and $K = 3.8 \times 10^{-5}$ M. I_2[48] and HNO$_2$ (effectively NO$^+$)[56] have also been used to react with Cr(III)—C systems (Table 6.6).

Table 6.6. Rate Constants for the
Reactions of HNO$_2$ with
[(H$_2$O)$_5$Cr-R]$^{2+}$ ^{a,b}

R	k_3 (M^{-2} s^{-1})
CH$_2$Br	0.12
CH$_2$I	0.247
CH$_2$Cl	0.46
4-CH$_2$C$_5$H$_4$NH$^+$	0.561
CH$_2$CH$_2$CH$_3$	15.1
CH(CH$_3$)$_2$	17.9
CH$_3$	38.8
CH(CH$_3$)OCH$_2$CH$_3$	45
CH$_2$CH$_3$	87
p-CH$_2$C$_6$H$_4$CN	92
p-CH$_2$C$_6$H$_4$CF$_3$	241
CH$_2$OCH$_3$	330
CH$_2$OH	745
CH$_2$C$_6$H$_5$	956
p-CH$_2$C$_6$H$_4$CH$_3$	1010

^{a} Data from Reference 56.
^{b} T = 23.4 °C, I = 0.1 M, with
$-d[\text{CrR}^{2+}]/dt = k_3[\text{CrR}^{2+}][\text{HONO}][\text{H}^+]$.

Finally, in this section, there are the reactions of [L$_5$Cr(III)—C]$^{2+}$ systems containing organic species with reactive functional groups, such as —CH$_2$CO$_2$H,[58] —CH$_2$CONH$_2$,[50] or —CH$_2$CN.[59] In these species, the Cr(III)—C bond rupture is assisted by the addition of Hg^{2+} (Table 6.7) but the resulting chemistry is often complicated.

6.2.2. Bridged Dichromium(III) Complexes

The hydrolysis of [(H$_2$O)$_5$CrOCr(OH$_2$)$_5$]$^{4+}$ is catalyzed[61] by reducing agents (Cr^{2+}, [Ru(NH$_3$)$_6$]$^{2+}$ and ascorbic acid)[62] and there are also specific

Table 6.7. Kinetic Parameters for the Reaction of Cr^{2+} with ICH$_2$Y (k_{Cr}) and Hg^{2+} with [(H$_2$O)$_5$CrCH$_2$Y]$^{n+}$ (k_{Hg}), both at 25.0 °C (References 50 and 58)

Y	k_{Cr} (M^{-1} s^{-1})	$\Delta H^\ddagger$	$\Delta S^\ddagger$	k_{Hg} (M^{-1} s^{-1})
—CO(NH$_2$)	0.56	28	−155	2.5 × 10^3
—CO$_2$H	1.1	22.6	−169	—
—CN	9.2	22.6	−150	8.7
—CNCo (NH$_3$)$_5$	45.2	15.9	−1160	—
—OCH$_3$	—			91
—CH$_2$CN	—			81

Table 6.8. Kinetic Parameters for the Aquation of Some $[CrX(L_5)]^{n+}$ Complexes

L_5	X^a	k_H (s^{-1}) (°C)	$\Delta H^{\ddagger}$	$\Delta S^{\ddagger}$	Ref.
$(NH_3)_5$	Cl	9.8×10^{-6} (25)	90.4		63
		2.93×10^{-5} (40)^b			64
	Br	Ionic strength effects			
		studied			64
	I	1.23×10^{-4} (7)			66
	DMSO	Solvent effects			
		studied			67
	OSO_2CF_3	1.24×10^{-2} (25)			68
		1.09×10^{-2} (25)			69
$(NH_2CH_3)_5$	OSO_2CF_3	6.43×10^{-4} (25)			69
mer-(en)(dpt)	Cl	5.34×10^{-7} (25)	99.2	-40	70
mer-(tn)(dpt)	Cl	5.04×10^{-7} (25)	82.8	-96	70
mer-(en)(2,3-tri)	Cl	2.87×10^{-7} (25)	90.9	-73	70
ufac-(N-Me-tn)(dien)	Cl	3.7×10^{-6} (25)	93	-45	71
cis-(en)$_2$(OH$_2$)	OSO_2CF_3	3.2×10^{-3} (25)			68
trans-(en)$_2$(OH$_2$)	F	2.63×10^{-5} (30)		^{c}	32
cis-(cyclen)(OH$_2$)	Cl	1.01×10^{-3} (25)	59	-103	72
cis-(tet *b*)(OH$_2$)	Cl	3.0×10^{-6} (25)	122	$+50$	63
(apdol-H$_2$)(OH$_2$)	Cl	1.31×10^{-4} (25)	81.7	-54	23
cis-(en)$_2$(DMSO)	Cl	2.44×10^{-5} (25)	82.6	-64	70
cis-(tn)$_2$(DMSO)	Cl	5.50×10^{-6} (25)	96.2	-31	70
sfac-(dien)(DMSO)$_2$	Cl	1.09×10^{-3} (25)	68.0	-100	70
cis-(NH$_3$)$_4$(oxH)	Cl	7×10^{-5} (50)			73
	Br	12×10^{-5} (50)			
	NH$_3$	0.8×10^{-5} (50)			73
$(NH_3)_5Co$-μ-edta^d					54

^{a} Leaving group.
^{b} Rate acceleration by dicarboxylate ions investigated.
^{c} $\Delta V^{\ddagger}$ positive.
^{d} See Table 6.12.

anion (ClO_4^-, Cl^-, Br^-, SO_4^{2-}) and cation effects (Fe^{3+}, Ce^{3+}), including an autocatalytic term from the $[Cr(OH_2)_6]^{3+}$ produced.[62] Dimeric $[(nta)Cr(OH)_2Cr(nta)]^{2-}$ has been detected in solution.[28]

6.2.3. Ammine and Amine Complexes

Tables 6.8 and 6.9 summarize the available data for thermal aquation (k_H) and Table 6.10 the data for Hg^{2+}-assisted chloride release (k_{Hg}). In most cases, the I_a mechanism can account for the variation in k_H with nonreplaced ligand. While there is an excellent correlation between k_H and k_{Hg} for Co(III) systems[80] (both D or I_d), until now there has been insufficient data available for Cr(III) comparisons. The information presented in Tables 6.8 and 6.10 is a start in this direction. The conclusion at present is that if the mechanism

Table 6.9. Kinetic Parameters for the Acid Hydrolysis of Some $[CrX_2(L_4)]^{2+}$
Complexes

L_4	X^a	k_H (s^{-1}) (°C)	$\Delta H^{\ddagger}$	$\Delta S^{\ddagger}$	Ref.
cis-(en)$_2$	OSO$_2$CF$_3$	5.7×10^{-3} (25)			68
	RCO$_2$				75
trans-(Me$_2$tn)$_2$	Cl	2.20×10^{-5} (25)	103.6	+5	76
trans-(dpt)(H$_2$O)	Cl	4.9×10^{-4} (40)			70
trans-(2,3-tri)(H$_2$O)	Cl	3.3×10^{-4} (40)			77
		7.25×10^{-5} (25)	78.5	−69	77
cis-cyclen	Cl	7.6×10^{-2} (25)	65	−59	72
		2.80×10^{-3} (20)			78
cis-isocyclam	Cl	1.01×10^{-3} (25)	59	−99	72
trans-isocyclam	Cl	5 (25)			72
trans-teta	CN	inert?			79

^{a} Leaving group.

Table 6.10. Kinetic Parameters for the Hg^{2+}-Assisted Chloride Release from Some
$[CrCl(N_5)]^{2+}$ Complexes ($I = 1.0$ M, HClO$_4$) at 25 °C

N_5	k_{Hg} (M^{-1} s^{-1})	$\Delta H^{\ddagger}$	$\Delta S^{\ddagger}$	Ref.
ufac-(N-Me-tn)(dien)	2.01×10^{-3}	64.2	−89	71
mer-(en)(dpt)	5.95×10^{-4}	78.1	−53	70
mer-(pn)(dpt)	5.24×10^{-4}	81.2	−44	70
mer-(chxn)(dpt)-A^a	5.85×10^{-4}	81.0	−41	70
mer-(chxn)(dpt)-B^a	1.41×10^{-3}	56.8	−115	70
mer-(tn)(dpt)	2.67×10^{-3}	85.7	−15	70
mer-(Me$_2$tn)(dpt)	2.18×10^{-3}	78.6	−41	70
mer-(en)(2,3-tri)	1.79×10^{-4}	80.1	−56	70

^{a} A = cis(meso)-chxn, B = trans(rac)-chxn.

for k_H in Cr(III) systems is I_a, then the mechanism for k_{Hg} is also I_a, in that slow k_H corresponds to slow k_{Hg} and vice versa.[70]

6.2.4. Multidentate Leaving Groups

6.2.4.1. Dechelation

The kinetics of the aquation of $[Cr(ox)_3]^{3-}$ to give cis-$[Cr(ox)_2(OH_2)_2]^-$ in dilute acid is catalyzed by metal ions.[61] The precise nature of the factors influencing the efficiency of any particular metal ion remain to be established but include the charge on the metal-ion catalyst, the association constant of the catalyst metal ion with coordinated or partially dechelated oxalate, and the hydration enthalpy of the catalytic ion. The catalytic effect of Pb^{2+} on this reaction has been reinvestigated[81] to give $10^3 k_{Pb}$ values of 3.05, 6.5, and 13.2 $M^{-1} s^{-1}$ at 50, 60, and 70 °C respectively in 0.02 M $HClO_4$ ($I = 1.0$ M). Associated kinetic parameters are $k_{Pb}(25) = 37.6 \times 10^{-5}$ $M^{-1} s^{-1}$ [cf. $k_H(25) = 1.62 \times 10^{-5}$ $M^{-1} s^{-1}$], $\Delta H^{\ddagger} = 65.2$, and $\Delta S^{\ddagger} = -93.2$. Two independent groups have reported preliminary studies on the rate of the first Cr—N bond rupture process in "*cis*"-$[Cr(dien)_2]^{3+}$[82] (more specifically, *s-fac-*[83]) in acid solution. Values of $t_{1/2} < 30$ min (1 M H^+, temperature unspecified)[82] and $t_{1/2}(25) = 8$ min (0.1 M or 3 M HCl)[83] are observed. The *mer*-isomer is much more inert.[83] Interestingly, partially dechelated polyamine Cr(III) complexes have now been characterized in the solid state, viz., chloro mercury(II) salts of *cis*-$CrCl(en_2)(Hen,N)]^{3+}$ and *s-fac*-$[CrCl(dien)(Hdien,N,N)]^{3+}$ [structures (**2**) and (**3**)].[70]

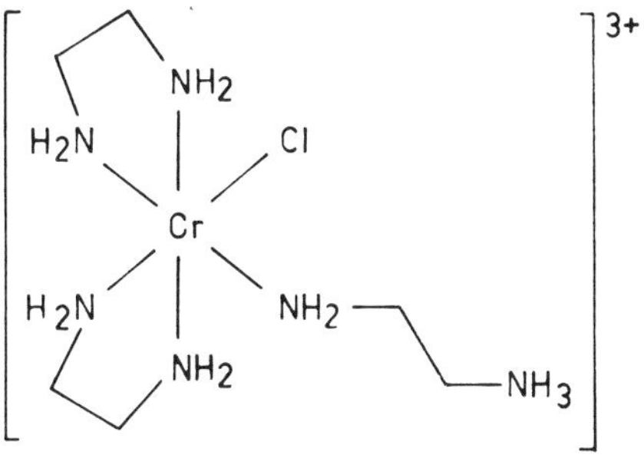

2

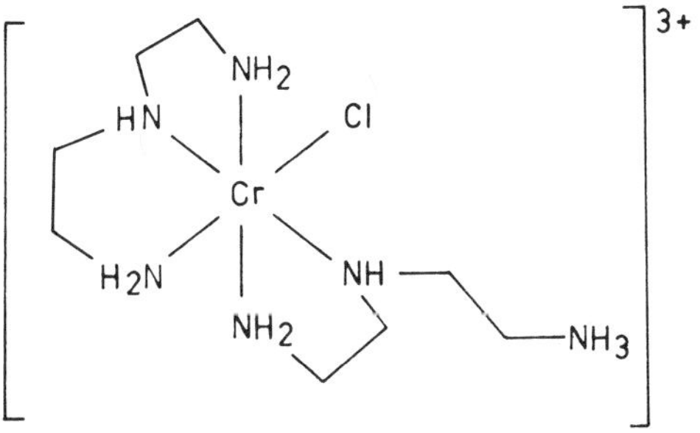

3

The kinetics of Cr—S bond rupture and subsequent reactions in the aquation of $[(en)_2Cr(S-X-CO_2)]^+$ [$X = -CH_2-$, $-(CH_2)_2-$, and $-C(CH_3)_2-$] complexes has been thoroughly investigated.[84] Following the initial Cr—S bond rupture, to give $[(en)_2Cr(OH_2)O_2C-X-SH)]^{2+}$, three subsequent reactions are possible; rechelation, Cr—O bond rupture to give $[Cr(en)_2(OH_2)_2]^{3+}$, or Cr—N bond rupture to give $[(en)(H_2O)_2-Cr(Hen,N)(O_2C-X-SH)]^{4+}$ which can then react further. Comparable rate parameters governing acid-catalyzed Cr—S bond rupture at 25° ($I = 1.0$ M)

are 19×10^{-4}, 11.1×10^{-4}, and $8.0 \times 10^{-4} M^{-1} s^{-1}$ for $X = -(CH_2)_2-$, $-CH_2-$, and $-C(CH_3)_2-$, respectively. Two isomers, red (*trans*-N) and violet (*cis*-N), of $[Cr(L\text{-asparto})_2]$ have been separated by chromatography on Sephadex cation exchange resin.[85] Spectal changes in acid solution are attributed to ring-opening processes and the rate of reaction is given by the rate law (12) with $k_H = 0.72$, 1.92, and 2.6 at 25, 35, and 45 °C, respectively.[85]

$$\text{rate} = k_H[\text{complex}][H^+] \tag{12}$$

6.3. Formation of Chromium(III) Complexes

6.3.1. Reactions of $[Cr(OH_2)_6]^{3+}$

Addition of OH^- to $[Cr(OH_2)_6]^{3+}$ leads to dark-green solutions containing $[Cr(OH_2)_5OH]^{2+}$ and $[Cr(OH_2)_4(OH)_2]^+$ in the pH range 4.0–5.0. These species further polymerize[22] and the rate of dimerization of the deprotonated aqua ion has been measured using pH-stat techniques (pH = 3.5–5.0, 25 °C, $I = 1.0$). Initial rate analysis gives the rate constants in equations (13)–(16) in which

$$Cr(OH)^{2+} + Cr^{3+} \rightarrow \text{dimer}, \qquad k_{10} \sim 6 \times 10^{-6} M^{-1} s^{-1} \tag{13}$$

$$Cr(OH)^{2+} + Cr(OH)^{2+} \rightarrow \text{dimer}, \qquad k_{11} = 2.0 \times 10^{-4} M^{-1} s^{-1} \tag{14}$$

$$Cr(OH)^{2+} + Cr(OH)_2^+ \rightarrow \text{dimer}, \qquad k_{12} = 3.8 \times 10^{-2} M^{-1} s^{-1} \tag{15}$$

$$Cr(OH)_2^+ + Cr(OH)_2^+ \rightarrow \text{dimer}, \qquad k_{22} = 1.8 \ M^{-1} s^{-1} \tag{16}$$

H_2O molecules are omitted from the Cr(III) species. At higher pH (≥ 5.0), pale green "active hydrated chromium hydroxide" precipitates immediately. The time-dependent aging of this precipitate has also been investigated[86] by dissolution in acid and determining the resulting monomer, low oligomer (dimer + trimer + tetremer), and high oligomer concentrations. Points of note are that aging is much faster in stirred than unstirred suspensions at pH = 5, and the rate of aging is at a minimum at pH 6–8. In earlier publications, "rather unorthodox"[87] structures were proposed for the tetramer.[1] The necessity for these has been questioned[87] and more classical structures, based on analogy with those determined for ammine and amine hydroxo-bridged polynuclear chromium(III) systems, have been suggested. Unfortunately, the untimely death of Professor Werner Marty may considerably delay the resolution of this debate.

Scheme 1 represents the general rate law observed for anation. However, if X has ionizable protons, e.g. H_nX, we can also have the equilibrium (17)

$$H_nX \rightleftharpoons H_{n-1}X^- + H^+ \rightleftharpoons H_{n-2}X^{2-} + H^+ \quad \text{etc.} \tag{17}$$

and $H_{n-y}X^{y-}$ can react with (4) or (5). There are also variations in the product (6) if X^{n-} or $H_{n-y}X^{y-}$ is chelating. For such a scheme we have (18) for both

$$\{k_{obs}\}^{-1} = \{k_{an}\}^{-1} + \{k_{an}K_{os}[X^{n-}]\}^{-1} \tag{18}$$

Scheme 1

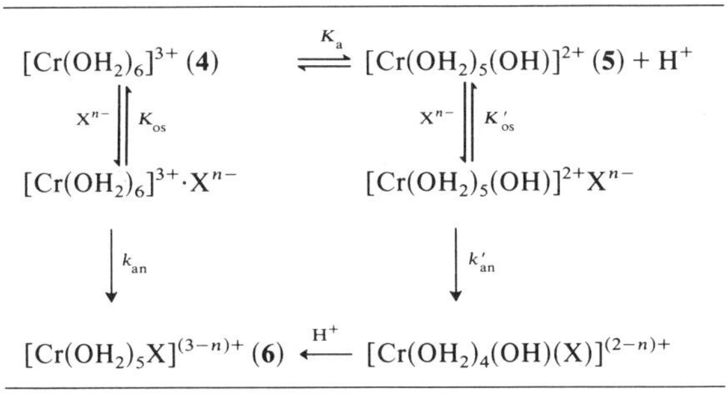

$$[Cr(OH_2)_6]^{3+} \text{ (4)} \quad \overset{K_a}{\rightleftharpoons} \quad [Cr(OH_2)_5(OH)]^{2+} \text{ (5)} + H^+$$

$$X^{n-} \Big\| K_{os} \qquad\qquad X^{n-} \Big\| K_{os}'$$

$$[Cr(OH_2)_6]^{3+} \cdot X^{n-} \qquad\qquad [Cr(OH_2)_5(OH)]^{2+} X^{n-}$$

$$\Big\downarrow k_{an} \qquad\qquad\qquad \Big\downarrow k_{an}'$$

$$[Cr(OH_2)_5 X]^{(3-n)+} \text{ (6)} \quad \overset{H^+}{\longleftarrow} \quad [Cr(OH_2)_4(OH)(X)]^{(2-n)+}$$

(4) and (5) paths, and combining we get equation (19). It follows that a plot

$$\{k_{obs}\}^{-1} = \{K_{os}[H^+] + K_{os}'K_a\}\{k_{an}K_{os}[H^+] + k_{an}'K_{os}'K_a\}^{-1}$$
$$+ \{[H^+] + K_a\}\{k_{an}K_{os}[H^+] + k_{an}'K_{os}'K_a[X^{n-}]\}^{-1} \quad (19)$$

of $1/k_{obs}$ vs $1/[X^{n-}]$ should be linear with intercept (20) and slope (21). Thus both slope and intercept will be H^+ dependent. The analysis is considerably

$$K_{os}[H^+] + \{K_{os}'K_a\}\{k_{an}K_{os}[H^+]\}^{-1} \quad (20)$$

$$\{[H^+] + K_a\}\{k_{an}K_{os}[H^+] + k_{an}'K_{os}'K_a\}^{-1} \quad (21)$$

simplified if the chosen pH is such that either (4) or (5) dominates and, similarly, only one of the $H_{n-y}X^{y-}$ species participates. Sometimes data are reported as in equation (22) or equation (23) at constant pH. Table 6.11

$$k_{obs} = k_0 + k_{-1}[H^+]^{-1} \quad (22)$$

$$\text{rate} = k_2[Cr(OH_2)_6][\text{ligand}] \quad (23)$$

summarizes the available data. The slight change in nomenclature from Scheme 1 should be noted in that k_{an} values are used throughout, rather than k_{an} for $[Cr(OH_2)_6]^{3+}$ and k_{an}' for $[Cr(OH)(OH_2)_5]^{2+}$.

A comparison of the water exchange rate k_{ex} with that of the anation rate constant k_{an} has been utilized as a means for distinguishing an associative from a dissociative interchange mechanism.[93] For an I_d process, k_{an} should be less than k_{ex} and have approximately the same value for all incoming ligans, while for an I_a process, k_{an} values will depend on the nucleophilic power of the incoming ligand toward Cr(III), and where this is greater than that of water, k_{an} should exceed k_{ex}. Inspection of the mechanistic assignments made in Table 6.11 shows that most, but not all, agree with these generalizations. There are also several studies reported[1,61] with rate laws differing from those discussed above. Thus, anation of $[(NH_3)_5Co\text{-}\mu\text{-}(edta)Cr(OH_2)]^{2+}$ (7) by OAc^-, N_3^-, and NCS^- follows pseudo-first-order kinetics according to equation

Table 6.11. Anation Reaction[a]

Substrate	Ligand	pH range	K_{os}	$10^5 k_{an}$ (s^{-1})[b]	$\Delta H^{\ddagger}$	$\Delta S^{\ddagger}$	Mechanism	Ref.
$[Cr(OH_2)_6]^{3+}$	H₂O			1.05 (k_{exch})	108.6	+11.6	I_a	88
	Glycine			8.8	52.9	−42.2		
				3.34	58	−129		
	nta		18.8	16.6	85.5	−36.4		90
	edda							
	Quinolinic acid			41.6	102	+18		
	(R)-Cysteine							91
	(RS)-Menthionine							91
	pic⁻		3.0	22.2	93.1	−14		91
	dipic⁻		50	79.3	80.6	−46.4		
	picNoxide⁻		13.6	12.2	88.3	−33.9		
	$[Mo(CN)_8]^{4-}$ c	3.4			129	−315	I_a	92[d]
	(S)-Asparagine[e]	3.25–3.9		40 (40)	46	−144	I_a	93
	Serine						I_a	61
	acac							1, 94
$[Cr(OH_2)_5(OH)]^{2+}$	H₂O			76.5 (k_{exch})	110	+55.6	I_d	88
	Glycine		10.9	51.7	41.2	−173		
	(S)-Aspartic acid		80.0	73.7	45.4	−156	I_a	

Complex	Ligand			$\Delta H^\ddagger$	$\Delta S^\ddagger$	Mechanism	Ref.
	(RS)-Alanine	4.35	39.9	33.7	−200	I_d	93
	(RS)-Phenylalanine	31.4	32.3	53.4	−135	I_d	92[d]
	β-Alanine	12.8	83.0	48.8	−144	I_a/I_d [g]	95
	(S)-Aspargine	3.25–3.9	49 (40)	170	+213	I_a/I_d [g]	65
	[Mo(CN)$_6$]$^{4-}$	3.4					65
[Cr(NH$_3$)$_5$(OH$_2$)]$^{3+}$	(RS)-Alanine	3–4	2.0 (40)	69	−96	I_a/I_d [g]	96
	Br	0.02	(25)				97
	Cl$^-$	0.4	(24)				
c-[Cr(ox)$_2$(OH$_2$)$_2$]$^-$	H$_2$O			72.2	−90.3	I_a	63
	dipicH^{-f}		0.778 (25) (k_{exch})	71.3	−73.6	I_a	63
c-[CrCl(tet$b$)(OH$_2$)]$^{2+}$	Cl$^-$						63
c[CrCl(OH)(tet$b$)]$^+$	Cl$^-$						63
[Cr(mal)(OH$_2$)$_4$]$^+$	dtpa						98

[a] Data from Reference 89 and at 35 °C unless otherwise stated.

[b] Obtained from plots of $1/k_{obs}$ vs. $1/$[ligand], equation (18) at constant [H$^+$] and T.

[c] Rate $\propto$ [Cr(OH$_2$)$_6$]$^{3+}$][Mo(CN)$_8$]$^{4-}$][H$^+$]$^{-1}$ with a positive salt effect. Plots of $1/k_{obs}$ vs. $1/$[Mo(CN)$_8$]$^{4-}$] are linear.[92]

[d] See also Reference 61 for an earlier investigation of this reaction.

[e] In ethanol/water mixtures, with a slight positive salt effect. Plots of $1/k_{obs}$ vs. $1/$[aspargine] are linear.[93]

[f] These activation parameters are associated with k_2 in equation (23) and $k_2(25) = 3.27 \times 10^{-4}$ M^{-1} s^{-1}.

[g] Less associative than for substitution in [Cr(OH$_2$)$_6$]$^{3+}$.

Table 6.12. Rate Constants (25 °C, I = 1.0 M) Associated with
Equation (24) (Reference 74)

Substrate/Reactant	k_{an} ($M^{-1}s^{-1}$)		
	OAc$^-$	N$_3^-$	NCS$^-$
(7)	44.5	676	133
[Cr(edta)(OH$_2$)]$^-$	3.3	98	13.7
[Cr(hedtra)(OH$_2$)]	7.6	19.8	3.32
	k_{aq} (s^{-1})		
	OAc$^-$ ^{a}	N$_3^-$	NCS$^-$
(7)	2.1	7.25	4.2
[Cr(edta)(OH$_2$)]$^-$	5.4	13.4	26.8
[Cr(hedtra)(OH$_2$)]	0.450	0.189	0.244

^{a} There is some [H$^+$] dependence for OAc$^-$.

(24) and k_{an} and k_{aq} are rate constants associated with equation (25).[74] For

$$k_{obs} = k_{an}[X^-] + k_{aq} \tag{24}$$

$$[L_5Cr(OH_2)]^{n+} + X^- \underset{k_{aq}}{\overset{k_{an}}{\rightleftarrows}} [L_5CrX^{(n-1)}]^+ \tag{25}$$

$$\underset{7}{}$$

X^- = OAc$^-$, there is some acid dependency but the rates of substitution (Table 6.12) proceed at 10^6–10^8 times faster than the corresponding reactions with [Cr(OH$_2$)$_6$]$^{3+}$. An associative mechanism is postulated.[74]

In a further investigation of the anation of Cr(III) by acac, excess Cr(III) was used and the reaction monitored by the disappearance of the acac (I = 0.5 M, T = 55 °C).[94] The enol tautomer is assumed to be the reactive ligand species and both [Cr(OH$_2$)$_6$]$^{3+}$ and [Cr(OH$_2$)$_5$OH]$^{2+}$ are involved in equations similar to equation (25) with k_{an} (55 °C) values of 1.05×10^{-2} $M^{-1}s^{-1}$ and 2.78×10^{-1} $M^{-1}s^{-1}$, respectively. An I_a mechanism is proposed.

6.4. Base Hydrolysis

Base hydrolysis of Cr(III) amine complexes (Table 6.13) is generally much slower than that for analogous Co(III) complexes. Assuming a common SN$_1$CB mechanism (and there is no reason to doubt this), one explanation is that, for Cr(III), the reduced strain in the ground state is reflected in a shorter lifetime for the five-coordinate conjugate base.[69] Alternatively, both the rate of formation of the conjugate base and its rate of dissociation could be smaller

Table 6.13.   Rate Constants for the Base Hydrolysis of Some Cr(III)
Complexes

Complex	k_{OH} $(M^{-1} s^{-1})$	Ref.
$[Cr(NH_3)_5(OSO_2CF_3)]^{2+}$	2.5 $(I = 1.0\ M)$	69
$[Cr(NH_2Me)_5(OSO_2CF_3)]^{2+}$	2.0×10^3 $(I = 1.0\ M)$	69
t-$[CrCl_2(cyclam)]^{+\ c}$	$1.11\ (20.40)^a$	99
c-$[CrCl_2(cyclam)]^{+\ d}$	$2.8 \times 10^{-2}\ (20.0)^b$	99

^{a} $\Delta H^{\ddagger} = 105$, $\Delta S^{\ddagger} = 113$; cf $k_{OH}(25) = 40\ M^{-1}\ s^{-1}$, $E_a = 100$, $\Delta S^{\ddagger} = 113.^{(100)}$
^{b} $\Delta H^{\ddagger} = 83$, $\Delta S^{\ddagger} = 46$.
^{c} Proton exchange rate (*all sec NH*) $= 9.8\ M^{-1}\ s^{-1}$ (0 °C).
^{d} Proton exchange rate (*trans-* to Cl) $= 8.2 \times 10^3\ M^{-1}\ s^{-1}$ (0 °C).

for Cr(III) relative to Co(III); this is apparently the case for *trans*-$[CrCl_2(cyclam)]^{+}$.[99]

The reaction (26)[101] has a rate law shown in equation (27), and kinetic parameters $k_b(25°) = 1.02\ s^{-1}$, $\Delta H^{\ddagger} = 71$, $\Delta S^{\ddagger} = -3$ $(I = 0.4\ M)$.[101] The rate

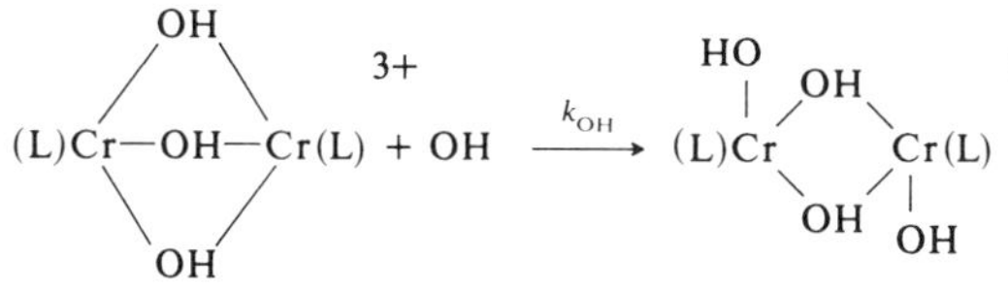

$$(L = 3,3,3\text{-tri})$$

$$k_{obs} = k_b(1 + K_0[OH^-]^{-1})^{-1} \tag{27}$$

of acid hydrolysis is very slow. Other base hydrolysis reactions investigated have used c-$[CrCl(OH)(tetb)]^{+}$,[63] *mer-* and *fac-*$[Cr(NCS)_3(DMSO)_3]^{-}$,[1,102] and t-$[Cr(NH_3)_2(teta)]^{3+}$.[79]

6.5. *Isomerization and Racemization*

The isomerism (*cis* → *trans*) of $[Cr(tfa)_3]$ has been observed using capillary gas chromatography. At temperatures of 458, 443, and 428 K, $10^4 k_{isom} = 240$, 76, and 20 s^{-1}, respectively, giving $E_a = 130$. A twist mechanism (rather than Cr—O bond rupture) is favored for this gas-phase interconversion.[103] On the other hand, a bond-rupture mechanism (one-ended dissociation) is favored for reaction (28) in chlorofrom solution[104] [bdtp = structure (8)].

$$\Lambda(-)(R,R)[Cr\{(-)\text{-bdtp}\}_3] \underset{k_2}{\overset{k_1}{\rightleftharpoons}} \Delta(+)(RR)- \tag{28}$$

Kinetic parameters associated with the rate constants are $k_1(25°) = 1.03 \times 10^{-3}\ s^{-1}$, $\Delta H^{\ddagger} = 101$, $\Delta S^{\ddagger} = 38$; $k_2(25°) = 5.82 \times 10^{-4}\ s^{-1}$, $\Delta H^{\ddagger} = 106$, $\Delta S^{\ddagger} = 49$.

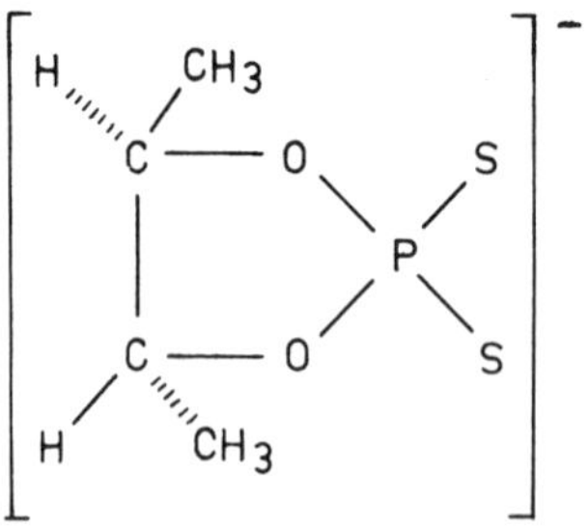

8

6.6. Photochemistry and Photophysics of Chromium(III) Complexes

The general sensitivity of Cr(III) complexes to electromagnetic radiation has long been appreciated by inorganic photochemists and a wide range of photochemical phenomena can be investigated. These include studies on picosecond transient formation,[105-107] photodissociation processes,[105,108-110] pressure dependence on emission lifetimes,[111,112] and quenching processes.[82,113-116] Among the Cr(III) complexes studied in more or less detail are those containing caged ligands,[117] macrocyclic ligands,[40,79,118] $[Cr(en)_3]^{3+}$,[119] $[Cr(tn)_3]^{3+}$,[110] $[CrI(NH_3)_5]^{2+}$,[66] *trans*-$[CrF_2(tn)_2]^+$,[34] $[Cr(CN)_x(OH_2)_{6-x}]^{(3-x)+}$,[120-122] and $[CrXY(N_4)]^{n+}$.[123,124] Where emission lifetimes have been measured under comparable conditions by different groups,[40,118,124] the agreement is generally satisfactory. The use of chiral $[Cr(en)_3]^{3+}$ as a chemical actinometer is anticipated.[40]

The photoaquation of $[Cr(AA)_3]^{3+}$ (AA = en, phen) has been studied at both high and low pH using conductivity to monitor the extent of reaction.[109] A six-coordinate intermediate with one of the heterocyclic ligands being monodentate is proposed. This accounts for the observation that "photoracemization" occurs in both acid and basic media, while photoaquation is very slow in acidic media.

Much of the data accumulated on emission lifetimes has been used to test theories on the modes of deactivation.[118,124] Early work showed a correlation between lifetime and the number of N—H bonds,[125] and lifetimes were considerably increased on deuteration.[126] More recent work has shown that there must be other factors operating among which are the transiton energy (the energy gap between ground state and the excited state), and geometric configurations.[118] In view of this latter factor, it is desirable that the precise isomeric form of any particular Cr(III) complex is known, before the effort of accumulating photochemical data is undertaken. The use of electronically excited complexes to study electron-transfer reactions is becoming increasingly popular. An oxidant or reductant that can be generated "instantaneously" by a laser pulse affords considerable advantage over conventional mixing techniques. $[Cr(phen)_3]^{3+}$ is a poor oxidant in the ground state, but is a strong oxidant ($E° = +1.42$ V) in its excited state.[115] The situation is complicated, however, by two possible modes of quenching—electron transfer or energy transfer, and corrections must be made for the energy-transfer processes.

6.7. The Solid State

6.7.1. Single-Crystal X-Ray Structures

The author has collected more than 50 "kinetically interesting" X-ray structures for chromium complexes reported during the period under review. Unfortunately space precludes even a brief mention of many of these but the Cr(III) amine systems include: α-cis-$[Cr(edda)(gly)]\cdot 2H_2O$,[30] mer-$[CrCl_3(py)_3]$,[12] trans-$[CrCl_2(NH_3)_4]Cl$,[128] cis-$[CrCl_2(cyclam)]Cl$,[129] trans-$[Cr(NCS)(2,3,3,3-tet)(OH_2)](NCS)_2$,[130] trans-$[CrCl(NH_3)_4(OH_2)]Cl_2$,[128] $[CrCl(NH_3)_5]Cl_2$,[131] $[Cr(diAMsar)]Cl_3\cdot H_2O$,[117,132] cis-$[Cr(bipy)_2(OH_2)_2]$-$(NO_3)_3$,[133] trans-$[Cr(H_2O)(NH_3)_3Cr(OH)_2Cr(NH_3)_3(OH)]Br_3\cdot 2H_2O$,[21] cis-$[Cr(bipy)_2Cr(OHOHO)_2Cr(bipy)_2]I_4\cdot 2H_2O$,[20] $[(3,3,3-tri)Cr(OH)_3(3,3,3-tri)]Br_3\cdot 2H_2O$,[101] $[(cyclen)Cr(OH)_2Cr(cyclen)](S_2O_6)_2\cdot 4H_2O$,[78] and $[(tri)Cr(OH)_2(SO_4)Cr(tri)](S_2O_6)\cdot 3H_2O$[134] [tri = structure (9)]. Data for complexes with Cr(III)—F bonds are given in Table 6.1.

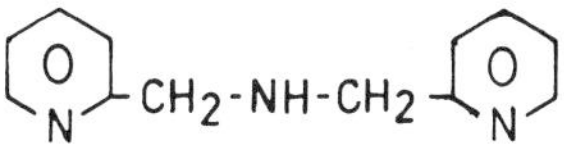

9

6.7.2. Thermal Decomposition

Thermally initiated solid-state substitution reactions, illustrated by equation (29), have been extensively investigated and the question of the

$$[Cr(NH_3)_5(H_2O)]X_3 \xrightarrow{\Delta} [CrX(NH_3)_5]X_2 + H_2O \qquad (29)$$

mechanism is under active consideration.[135] S_N1 processes begin with the coordinated water being displaced to an interstitial position in the lattice (formation of a nonionic Frenkel defect), followed by diffusion through the lattice. This latter step may well be anion dependent. Coordination of a lattice anion with the now five-coordinate complex completes the process. The S_N2 mechanism will require a lattice anion to move into the coordination sphere to form a seven-coordinate intermediate and to create an ionic Schottky defect. Such defect formation is regarded as a high-energy process and assignments based on S_N2 (associative) pathways have been questioned. More recently,[39,136] a dissociative pathway controlled by the "free space" in the lattice has been proposed, where the ease of dehydration can be related to the larger or smaller volume that the coordinated water has available to leave the lattice.

One major difficulty in this area, for the nonspecialist, is the variation of models used to calculate activation parameters. For example, four physical models, each with two or more integrated rate laws, can be used.[39] Among the systems studied in detail are $[CrF(AA)_2(OH_2)]X$[39,136] (AA = en, tn, chxn), $[Cr(en)_2(NH_2CH_3)_2]X$,[137] $[Cr(NH_2CH_3)_6]X$,[135] $[CrBr(NH_2CH_3)_5]X$,[135] $Na[Cr(edta)(N_2)]\cdot 2H_2O$,[138] $CuCrO_4$,[139] $[Cr(NH_3)_2(2,3,2\text{-tet})]Br_3$,[140] $[Cr(NH_3)_5(OH_2)]X$,[141,142] the racemization of *cis*-$[CrCl_2(en)_2]Cl$,[143] and the solid-state photochemistry of *trans*-$[CrX_2(en)_2]Y\cdot xH_2O$.[144]

6.8. Other Oxidation States

6.8.1. Chromium(II)

Acidic solutions of Cr(II) are thermodynamically unstable with respect to the formation of Cr(III) and H_2 as in equation (30). However, only rarely

$$2Cr^{2+} + 2H^+ \rightarrow 2Cr^{3+} + H_2 \tag{30}$$

does such a reaction occur at a finite rate. It has now been found that equation (30) is *catalyzed* by $[Co(II)(dmg\ BF_2)_2]$ (**10**).[145] In HCl (or HBr) medium

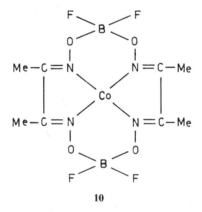

10

the stoichiometry corresponds to equation (31), but H_2 evolution *does not*

$$Cr^{2+}(aq) + Cl^- + H^+ \rightarrow [(H_2O)_5CrCl]^{2+} + \tfrac{1}{2}H_2 \tag{31}$$

occur in $HClO_4$. The proposed reaction sequence is shown in Scheme 2.[145]

Chromium(II) reacts by halogen abstraction with a wide range of halo-substituted organic compounds.[50,51,58] When the organic substrate has another substituent (Y) that can act as a donor atom toward the chromium, such as XCH_2Y (X = halogen, Y = Lewis base donor), then both $Cr\text{—}CH_2\text{—}Y$ and

Scheme 2

$$Cr^{2+} + Cl^- + Co(II)L \rightleftharpoons Cr-Cl-Co(II)L^+$$
$$Cr-Cl-Co(II)L^+ \rightarrow Co(I)L^- + Cr(III)Cl^{2+}$$
$$H^+ + Co(I)L^- \rightarrow HCo(I)L$$
$$2HCo(I)L \rightarrow H_2 + 2Co(II)L$$
$$HCo(I)L + H^+ \rightarrow H_2 + Co(III)L^+$$
$$Co(III)L^+ + Cr^{2+} + Cl^- \rightarrow Cr(III)Cl^{2+} + Co(II)L$$
$$L = (dmg\ BF_2)_2\ [structure\ (10)]$$

$Cr-Y-CH_3$ bonds can be formed. Thus kinetic data (Table 6.7) and product analysis for ICH_2CO_2H,[58] ICH_3CN,[59] and $ICH_2CO(NH_2)$[50] are consistent with equation (32) although there is the possibility of chelation in $[(H_2O)_5CrCH_2CO_2H]^{2+}$.[58]

$$Cr^{2+}(aq) + XCH_2Y \rightarrow [(H_2O)_5CrCH_2Y]^{2+} + [(H_2O)_5CrYCH_3]^{2+} + [(H_2O)_5CrX]^{2+} \quad (32)$$

For the reaction between Cr^{2+} and complex (11) (Table 6.14), rate law (33) is observed[146] where k_1 relates to the rate of reduction of (11) and K_a'

$$k_{obs} = \{k_1 K_a'\}\{K_a' + [H^+]\}^{-1} \quad (33)$$

is the acid dissociation constant for the protonated form of (11). Similar rate

Table 6.14. *Kinetic Data for the Reduction of Various Oxidants by $Cr^{2+}(aq)$*

Oxidant[a]	Comments	$k_2\ (M^{-1}\ s^{-1})$ (25 °C)	Mechanism[b]	Ref.
[M(fumaronitrile)]³⁺		1.9×10^5	I.S.	47
[M(4-acetylbenzonitrile)]³⁺		6×10^3	I.S.	146
[M(4-formylbenzonitrile)]³⁺		2.5×10^5	I.S.	146
[M(4-benzoylpyridine)]³⁺		4.7×10^4	I.S.	146
Structure (11)		2.9×10^4	I.S.	146
[M(4-acetoxybenzonitrile)]³⁺		0.021	O.S.	146
[M(benzonitrile)]³⁺		0.043	O.S.	146
[M(NO₂)]²⁺	Product analysis	—	—	147
[(H₂O)₅CrOCr(OH₂)₅]⁴⁺	$\Delta H^{\ddagger} = 25, \Delta S^{\ddagger} = -71$	2.5×10^4	O.S.	62
[(H₂O)₅Rh(OH)]²⁺	→ Rh₂⁴⁺ dimer	2.3×10^5	I.S.	148
[(H₂O)₅RhCl]²⁺		2.2×10^3	I.S.	148
[(H₂O)₅RhBr]²⁺		$>8 \times 10^3$	I.S.	148
[Co(bipy)₂(acac)]²⁺		$>3 \times 10^4$	—	149
Blue copper protein (Rusticyanin)		2.5×10^4	—	150
Hemerythrin (met octa)	[Cr(bipy)₃]²⁺ used	2.5×10^5	—	151
[(H₂O)₅Cr(fumaronitrile)]³⁺	Complex products			47

[a] $M = (NH_3)_5Co(III)$.
[b] I.S. = inner sphere, O.S. = outer sphere.

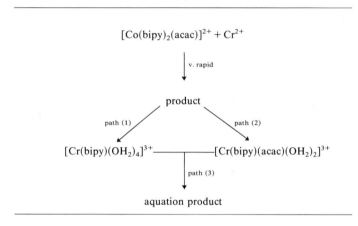

$$\left[(NH_3)_5CoNC-\langle O \rangle - \overset{O}{\overset{\|}{C}}-CH=\overset{OH}{\overset{|}{C}}-CH_3 \right]^{3+}$$

11

laws are observed for paths (1) and (2) in the Cr^{2+} reduction of $[Co(bipy)_2(acac)]^{2+}$ shown in Scheme 3.[149] These two paths are ascribed to ring closure of bipy on Cr(III). For path (1), $k_{obs} = 5.06 \times 10^{-1} s^{-1}$, $\Delta H^{\ddagger} = 41.3$, $\Delta S^{\ddagger} = -119$; for path (2), $k_{obs} = 5.83 \times 10^{-2} s^{-1}$, $\Delta H^{\ddagger} = 20.6$, $\Delta S^{\ddagger} = -199$; and for path (3), $k_{obs} = 4.6 \times 10^{-6} s^{-1}$ all at 25.0 °C, $[H^+] = 0.03 M$, and $I = 0.1 M$.[149]

Scheme 3

$$[Co(bipy)_2(acac)]^{2+} + Cr^{2+}$$

↓ v. rapid

product

path (1) ↙   ↘ path (2)

$[Cr(bipy)(OH_2)_4]^{3+}$ ——————— $[Cr(bipy)(acac)(OH_2)_2]^{3+}$

↓ path (3)

aquation product

The reaction between $Cr^{2+}(aq)$ and O_2 (Table 6.3) gives a species (CrO_2^{2+}) that is sufficiently long lived to be studied in the absence of Cr^{2+} and O_2.[60] $N_2H_5^+$ is oxidized by this species at a rate given by equation (34) with $k_0 = 7 \times 10^{-4} s^{-1}$ and $k_3 = 58.1 M^{-2} s^{-1}$ (25.0 °C, $I = 0.1 M$).[60]

$$-d[CrO_2^{2+}]/dt = k_0[CrO_2^{2+}] + k_3[N_2H_5^+][H^+][CrO_2^{2+}] \tag{34}$$

The kinetics of the Cr(II)/V(III) electron-transfer reaction has been reinvestigated[152] at high acidities. The previously accepted mechanism shown in equations (35)–(37) now needs an additional outer-sphere electron-transfer

$$V^{3+} + H_2O \rightleftharpoons VOH^{2+} + H^+ \qquad K_a \tag{35}$$

$$VOH^{2+} + Cr^{2+} \rightleftharpoons VOHCr^{4+} \qquad k_{IS}, k_{-IS} \tag{36}$$

$$VOHCr^{4+} + H^+ \rightarrow V^{2+} + Cr^{3+} + H_2O \qquad k_H \tag{37}$$

pathway as in equation (38), so k_{obs} is given by equation (39) with $k_{OS} = 0.2\ M^{-1}\,s^{-1}$ (25 °C, $I = 3.0\ M$).

$$V^{3+} + Cr^{2+} \rightarrow V^{2+} + Cr^{3+} \qquad k_{OS} \tag{38}$$

$$k_{obs} = k_{OS} + K_a k_{1S}[H^+]^{-1} \tag{39}$$

Turning now to other Cr(II) systems used as reducing agents, dimeric chromium(II) acetate $[Cr_2(Ac)_4 \cdot 2H_2O]$ reacts with a variety of oxidizing agents in one-electron steps via predissociation of the dimer into mononuclear Cr(II) species as in equations (40) and (41). Two limiting rate laws (42) and (43)

$$Cr(II)_2 \underset{k_{-1}}{\overset{k_1}{\rightleftharpoons}} 2Cr(II) \tag{40}$$

$$Cr(II) + X \xrightarrow{k_2} Cr(III) + products \tag{41}$$

$$-d[Cr(II)_2]/dt = k_1[Cr(II)_2] \tag{42}$$

$$-d[Cr(II)_2]/dt = k_2(k_1/k_{-1})^{1/2}[Cr(II)_2]^{1/2}[X] \tag{43}$$

are observed according to whether step (40) or (41) is rate determining. For $Cr_2(Ac)_4$ (excess) and I_3^- in acetic acid (0.4 M H_2O) containing NaI, the rate law changes from psuedo-zero order to pseudo-first order as the concentration of I_2 decreases with time and the rate constants in equations (42) and (43) are functions of $[NaI]$.[153]

$[Cr(II)(pic)_3]^-$ (**12**) and $[Cr(bipy)_3]^{2+}$ (**13**) have been used to investigate Fe(III) systems.[151,154] For (**12**) and Fe^{3+} the spectrophotometric changes are indicated in equation (44) with $k_1 = 14.1\ M^{-1}\,s^{-1}$ and $k_2 = 5.6 \times 10^{-4}\,s^{-1}$,

$$[Cr(pic)_3]^- + Fe^{3+} \xrightarrow{k_1} [Cr(pic)_3\text{-Fe}]^{2+} \xrightarrow{k_2} [Cr(pic)_3] + Fe^{2+} \tag{44}$$

$\Delta H^{\ddagger} = 87.4$, $\Delta S^{\ddagger} = -18.8$ at 25 °C and $I = 2.3\ M$.[154] It should be noted that $Cr(II)(pic)_3^-$ is relatively inert to air oxidation and is not readily dissociated.

Data for the rate of dissociation of $[Cr(II)(AA)_3]^{n+}$ complexes as in equation (45) at 25 °C are: AA = bipy,[155] $k_f = 0.343\ s^{-1}$; AA = Me_2-bipy,[155] $k_f = 0.353\ s^{-1}$; AA = pic,[154] $k_f = 0.5\ s^{-1}$.

$$[Cr(II)(AA)_3]^{n+} + 2H_2O \xrightarrow{k_f} [Cr(AA)_2(OH_2)_2]^{n+} + AA \tag{45}$$

Finally, a kinetic study on the formation of Cr(II) by the interaction of $[(H_2O)_5Cr(III)(py\text{-}X)]^{3+}$ with $\cdot C(CH_3)_2OH$ radicals has been reported [see equation (46)] and data are given in Table 6.15. There is evidence that the electron is transferred to the Cr(III) via the pyridine.[156]

$$[(H_2O)_5Cr(pyX)]^{3+} + \cdot C(CH_3)_2OH \xrightarrow{k_R} [(H_2O)_5Cr(pyX)]^{2+} + (CH_3)_2CO + H^+ \tag{46}$$

6.8.2. Chromium(IV)

Two studies on chromium(IV) species have been reported one involving the oxidation of organic sulfides using $[(HMPT)Cr(O)(O_2)_2]$ (**14**)

Table 6.15. Kinetic Data Associated
with Equation (46)^a

X	k_R (M^{-1} s^{-1})
4-CN	5.3×10^8
3-CN	1.8×10^8
3-Cl	4.7×10^6
4-Cl	5.1×10^5
py$_2$ ^{b}	$\sim 10^5$
H	$<5 \times 10^4$
4-t-Bu	$<1 \times 10^5$

^{a} Data from Reference 156 at 25 °C, $[H^+] = 0.1$ M,
$I = 1.0$ M in 1 M aqueous 2-propanol.
^{b} Complex is *cis*-$[(H_2O)_4Cr(py)_2]^{3+}$.

(characterized in CHCl$_3$ solution)[157] (HMPT = $[(CH_3)_2N]_3P{=}O$), and the other on the thermal decomposition of $[L_3Cr(O_2)_2]$ $[L_3 = (en)(H_2O)$ (15) or (dien) (16)] in aqueous acidic solution.[158] $[(HMPT)Cr(O)(O_2)_2]$ oxidized sulfides to sulfoxides according to equation (47) with rate law (48) and the

$$2R_2S + [(HMPT)Cr(O)(O_2)_2] \rightarrow 2R_2S{=}O + CrO_3 + HMPT \qquad (47)$$

$$\text{rate} = k_2[Cr(IV)][\text{sulfide}] \qquad (48)$$

reaction is inhibited by addition of free HMPT. (14) is more effective than t-butyl hydroperoxide as an oxidant, but less effective than m-chloroperoxy-benzoic acid.[157] The acid decomposition of the two chromium(IV) amine peroxo complexes (15) and (16) gives both Cr(VI) and Cr(III), the amount of Cr(VI) decreasing with increasing $[H^+]$. The Cr(III) is distributed between $[Cr(OH_2)_6]^{3+}$ and $[Cr(en)(OH_2)_4]^{3+}$ or $[Cr(Hdien)(OH_2)_4]^{4+}$, respectively. Two kinetic stages (k_d, k_s) are observed in the decomposition of (15) and both obey rate law (49) while the rate of decomposition of (16) k_d' obeys rate law (50).[158]

$$k_d \text{ or } k_s = A + B[H^+] + C[H^+]^2 \qquad (49)$$

$$k_d' = \{R + S[H^+] + T[H^+]^2\}\{1 + Q[H^+]\}^{-1} \qquad (50)$$

6.8.3. Chromium(V)

Although Cr(V) species have been proposed as intermediates in Cr(VI) oxidations,[59] this oxidation state has now been stabilized by the use of porphyrin[18,160,161] or salen[162,163] type ligands. The corresponding Cr(III) complexes can be used in catalytic oxidation of alkenes[161,162] or alcohols[18,160] involving an oxygen transfer agent such as iodosylbenzene[162] or KHSO$_5$[161] as in equation (51) in which RO = PhIO or SO$_5^-$, L = porphyrin or salen.

Table 6.16. Oxidations Involving Chromium(VI)

Reductant	Comments	Ref.
Tartaric acid[a]	$\Delta H^{\ddagger} = 93$, $\Delta S^{\ddagger} = -17$	164
Salicylic acid[a]	$\Delta H^{\ddagger} = 77$, $\Delta S^{\ddagger} = -57$	164
Phenylsalicilic acid[a]	$\Delta H^{\ddagger} = 74$, $\Delta S^{\ddagger} = -55$	164
Gallic acid[a]	$\Delta H^{\ddagger} = 65$, $\Delta S^{\ddagger} = -135$	164
Lactic acid[a]	Cr(V) intermediates investigated by ESR	159
Acrylic acid[a]		165
Sec. alcohols		166, 167[b]
Butanols[a]		168
D-Glucose[a]	$\Delta H^{\ddagger} = 56$, $\Delta S^{\ddagger} = -90$	16
D-Glucopyranose 6-phosphate[a]	$\Delta H^{\ddagger} = 50$, $\Delta S^{\ddagger} = -105$	169
D-Ribose[a]	$\Delta H^{\ddagger} = 49$, $\Delta S^{\ddagger} = -108$	169
D-Ribofuranose 5-phosphate[a]	$\Delta H^{\ddagger} = 35$, $\Delta S^{\ddagger} = -140$	169
Thiols[a]	Cr(VI) thio esters formed	170, 171
Cyclic sulfoxides	Rate varies with ring size	172
Naphthalene	$\Delta H^{\ddagger} = 12.6$, $\Delta S^{\ddagger} = 265$	173[b]
Phenanthrene	$\Delta H^{\ddagger} = 17.1$, $\Delta S^{\ddagger} = 248$	173[b]
Organic substrates	Heats of formation estimated	174[b]
BH_4^- [a]		175
N_2H_4	Heterogeneous reaction using $BaCrO_4$	176, 177
Diphenylcarbazide		178
I	Catalyzed by bipy	179

[a] [H$^+$] dependence observed.
[b] Pyridinium fluorochromate or pyridinium chlorochromate used.

Adduct formation as in equation (52) (L' = pyO) is common for many of these systems.[162,163]

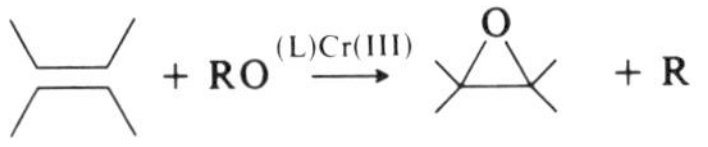

$$\diagdown\!\!\diagup\diagdown\!\!\diagup + RO \xrightarrow{(L)Cr(III)} \quad + R \qquad (51)$$

$$[(L)Cr(V) = O] + L' \rightleftharpoons [L'(L)Cr(V) = O] \qquad (52)$$

6.8.4. *Chromium(VI)*

Many oxidation studies involving Cr(VI) (Table 6.16) have been analyzed in terms of a preequilibrium, followed by a two-electron transfer process.[176] The oxidations are also pH dependent and $HCrO_4^-$ is a popular form of Cr(VI).[164,171,176] Unfortunately, Raman spectral studies[180] suggest that *only* CrO_4^{2-} and $Cr_2O_7^{2-}$ species are present in Cr(VI) solution between pH = 1 and 11 [equation (53)]. If this is the case, the acid dependence must involve the reductant species rather than Cr(VI).

$$2CrO_4^{2-} + 2H^+ \rightleftharpoons Cr_2O_7^{2-} + H_2O \qquad (53)$$

The reviewer's previous enthusiam[1] for compilations of the Cr(VI) species present in solution based on published equilibrium constants must now be tempered with caution. Partly because there are errors associated with the analysis,[181] the data are particularly sensitive to ionic-strength effects,[181] and the equilibrium constants used may not be correct.[180]

6.9. Catalysis by Chromium(III)

The decomposition of H_2O_2 is catalyzed by $[Cr(edta)]^{n-}$ with a complex pH dependence, as at high pH the Cr(III) species are oxidized to Cr(VI). In the pH range 5–8, both $[Cr(edta)(H_2O)]^-$ (k_2) and $[Cr(edta)(OH)]^{2-}$ (k_2') are believed to be effective with $k_2(30°) = 1.75 \times 10^{-2}\ M^{-1}\,s^{-1}$, $\Delta H^{\ddagger} = 58.9$, $\Delta S^{\ddagger} = -85$ and $k_2'(30°) = 1.754 \times 10^{-1}\ M^{-1}\,s^{-1}$, $\Delta H^{\ddagger} = 66.5$, $\Delta S^{\ddagger} = -40$. Substitution-controlled mechanisms are proposed.[182] In other studies, the photolytic decomposition of $S_2O_8^{2-}$ is catalyzed by $[Cr(OH_2)_6]^{3+}$ [183] and $[Zn_2Cr(OH)_6X\cdot nH_2O]$ (X = Cl, I) are shown to act as a heterogeneous catalyst for halide exchange in alkyl halides.[184]

6.10. Miscellaneous

Two studies have involved reactions of coordinated ligands. The rates of bromination and isomer distribution of $[Cr(acac)_3]$ have been determined[185] and the use of $[Cr(acac)_3]$ and $[Cr(dithiocarbamate)_3]$ as antiozonants is described.[186] Reaction (54) has been investigated[187] in the pH range 3.5–8.5

$$(NH_3)_5CrOCrO_3H^{2+} \underset{k_r}{\overset{k_f}{\rightleftharpoons}} (NH_3)_5Cr(OH_2)^{3+} + HCrO_4^- \qquad (54)$$

with values of $k_f(25°) = 2.2 \times 10^{-3}\ s^{-1}$ and $k_r(25°) = 2.3\ M^{-1}\,s^{-1}$ (see also Section 6.8.5). Measurement of the rate of extraction of Cr(III) from tanning waste-leather samples by chelating agents shows that citric, oxalic, and tartaric acids offer the best economic potential for ease of recycling.[188]

Chapter 7

Substitution Reactions of Inert-Metal Complexes— Coordination Numbers 6 and Above: Cobalt

7.1. Introduction

General reviews of interest include studies of the metal-ion-catalyzed aquation of transition-metal complexes[1] with an emphasis on Co(III) and Cr(III) complexes, coordinated trifluoromethanesulfonate and fluorosulfate complexes[2] which provide useful intermediates for the preparation of cobalt(III) complexes, while a recent issue of *Inorganic Synthesis*[3] describes the preparation of trifluoromethanesulfonato complexes. Other reviews of interest include a discussion on the mechanisms of formation and decomposition of μ-peroxo- and μ-superoxo-dicobalt(III) complexes[4] and an account of metal carbonato complexes.[5] A new text, *Inorganic and Organometallic Reaction Mechanisms*, has recently been published.[6]

7.2. Aquation

The syntheses of the cations $[Co(NH_2CH_3)_5L]^{3+}$ (L = urea, DMSO, DMF, trimethylphosphate, and MeCN) using the triflate precursor $[Co(NH_2CH_3)_5(OSO_2CF_3)](CF_3SO_3)_2$ proceed readily in high yield.[7] Acid aquation of these cations is at least 70-fold faster than the pentaammine analogues at 25 °C, a rate enhancement which is believed to be steric in origin. Values of $\Delta H^{\ddagger}$ are similar for the pentakis(methylamine) and pentaammine complexes, but values of $\Delta S^{\ddagger}$ and $\Delta V^{\ddagger}$ are more positive for the former complexes. A dissociative I_d mechanism is believed to apply with a diminished role for an incoming water molecule in the transition state of the pentakis(methylamine) complexes arising from steric crowding. Lay[8] has recently argued that a common I_d mechanism applies to the aquations of both pentaaminechlorocobalt(III) and -chromium(III) complexes, although it has been generally accepted that the Cr(III) reactions occur by an I_a pathway. The main evidence supporting an I_a mechanism for the aquation of Cr(III) complexes is that the rate of aquation of $[M(NH_2R)_5Cl]^{2+}$ is retarded on moving from R = H to a more bulky alkyl group,[9,10] although the reverse applies with Co(III).[7,11] Previously it was assumed that these kinetic differences reflected different transition states for the two metal ions. However, it now appears that the aquation rate of $[Cr(NH_2Me)_5Cl]^{2+}$ is slower than that for $[Cr(NH_3)_5Cl]^{2+}$ because of a shorter Cr—Cl bond in the ground state of $[Cr(NH_2Me)_5Cl]^{2+}$.

Kitamura and co-workers[12] have studied the acid-dependent aquation rates ($k_{obs} = k_0 + k_1/[H^+]$) for cis-$[CoX(NH_3)_4(H_2O)]^{2+}$ (X = Cl^-, Br^-, NO_3^-) ions as a function of pressure. Activation volumes for the k_0 and k_1 pathways ($\Delta V_0^{\ddagger}$ and $\Delta V_1^{\ddagger}$) and reaction volumes (ΔV) suggest that the k_0 pathway for aquation of cis-$[CoX(NH_3)_4(H_2O)]^{2+}$ proceeds by an interchange (I) mechanism, while the k_1 pathway for cis-$[CoX(NH_3)_4(OH)]^+$ is intermediate between a limiting dissociative (D) mechanism and the I mechanism. Activation volumes for aquation of cis-$[Co(en)_2(NO_2)Cl]^+$, $trans$-$[Co(en)_2(CN)Cl]^+$, and cis-α- and cis-β-$[Co(trien)Cl_2]^+$ have been determined[13] and Table 7.1 sum-

Table 7.1. Activation Parameters for Aquation of Chloroaminecobalt(III) Complexes[a]

Complex[b]	$k(25\,°C)$ (s^{-1})	$\Delta H^{\ddagger}$ $(kJ\,mol^{-1})$	$\Delta S^{\ddagger}$ $(J\,K^{-1}\,mol^{-1})$	$\Delta V^{\ddagger}$ $(cm^3\,mol^{-1})$
$[Co(NH_3)_5Cl]^{2+}$	1.7×10^{-6}	93 (±1)	-44 (±3)	-9.9 (±0.5)
$trans$-$[Co(NH_3)_4(NH_2CH_3)Cl]^{2+}$	1.1×10^{-5}	95 (±2)	-19 (±6)	-4.6 (±0.5)
$[Co(NH_2CH_3)_5Cl]^{2+}$	4.0×10^{-5}	95 (±1.5)	-10 (±4)	-2.3 (±0.4)
cis-α-$[Co(trien)Cl_2]^+$	$1.6\times10{-4}$	90 (±1.7)	-24 (±6)	-5.0 (±0.4)
cis-β-$[Co(trien)Cl_2]^+$	1.5×10^{-3}	88 (±1.7)	-11 (±6)	-2.0 (±0.6)
cis-$[Co(tren)Cl_2]^+$	3.0×10^{-3}	74 (±2)	-45 (±6)	$+7.3$ (±0.4)
cis-$[Co(en)_2NO_2)Cl]^+$	1.2×10^{-4}	90 (±0.5)	-17 (±2)	-2.9 (±0.3)
$trans$-$[Co(en)_2Cl_2]^+$	3.1×10^{-5}	115 (±5)	$+56$ (±15)	$+11.0$ (±0.6)
$trans$-$[Co(en)_2(CN)Cl]^+$	7.8×10^{-5}	94 (±0.6)	-9 (±2)	-2.0 (±0.4)
$trans$-$[Co(en)_2(N_3)Cl]^+$	2.2×10^{-4}	93 (±4)	$+2$ (±11)	$+0.7$ (±0.1)
$trans$-$[Co(dtcd)(N_3)Cl]^+$	4.6×10^{-4}	77 (±2)	-51 (±6)	-8.3 (±0.4)

[a] Data from G. A. Lawrance, *Polyhedron*, 5, 2113 (1986).
[b] en = ethane-1,2-diamine, trien = 1,8-diamino-3,6-diazaoctane, tren = N,N'-bis(2-aminoethyl)ethane-1,2-diamine, and dtcd = 5,12-dimethyl-1,4,8,11-tetraazacyclotetradeca-4,11-diene.

Table 7.2. *First-Order Rate Constants and Activation Parameters for Aquation of*
cis-[Co(NN)$_2$(CN)Cl]$^+$ Complexes (NN = en, bipy, and phen)^a

Complex	$10^7 k_{aq}$ (25 °C) (s^{-1})	$\Delta H^{\ddagger}$ (kJ mol^{-1})	$\Delta S^{\ddagger}$ (J K^{-1} mol^{-1})
cis-[Co(en)$_2$(CN)Cl]$^+$	6.2	103	−21
cis-[Co(phen)$_2$(CN)Cl]$^+$	2.7	82	−91
cis-[Co(bipy)$_2$(CN)Cl]$^+$	1.6	90.4	−73

^{a} Data from G. Schiavon and F. Marchetti, *Polyhedron*, **4**, 1143 (1985).

marizes the values reported for a range of chloroaminecobalt(III) complexes. Complexes such as *trans*-[Co(en)$_2$Cl$_2$]$^+$ with markedly positive values of $\Delta V^{\ddagger}$ may involve more substantial rearrangement in forming the transition state. Electrostriction of new charge centers are offset by displacement of electrostricted water molecules into the bulk solvent by motions of the nonleaving groups. The kinetics of aquation of *cis*-[Co(phen)(CN)Cl]$^+$ and *cis*-[Co(bipy)-(CN)Cl]$^+$ have recently been reported and activation parameters determined.[14] From the data it is possible to exclude a *trans*-labilizing effect of the aromatic α-diimines, when comparisons are made with the aquation of *cis*-[Co(en)$_2$(CN)Cl]$^+$; see Table 7.2.

The aquation of [Co(NH$_3$)$_5$OCrO$_3$H]$^{2+}$ and the anation of the aqua complex by chromate ions shown in equation (1) has been investigated at

$$[Co(NH_3)_5OCrO_3H)^{2+} + H_2O \underset{k_{an}}{\overset{k_{aq}}{\rightleftharpoons}} [Co(NH_3)_5OH_2]^{3+} + HCO_4^- \quad (1)$$

25 °C in the pH ranges 6.0 to 8.5 and 3.5 to 7.5, respectively.[15] The equilibrium constant $K = k_{aq}/k_{an}$ has a minimum value at pH *ca.* 6. The important reaction pathways are given in equations (2) and (3) for which $k_1 = 2.2 \times 10^{-3}$ s^{-1}, $k_{-1} = 2.3$ M^{-1} s^{-1}, $k_0 = 151$ M^{-1} s^{-1}, and $k_{-0} = 1.7 \times 10^5$ M^{-2} s^{-1} at 25 °C.

$$[Co(NH_3)_5OCrO_3H]^{2+} + H_2O \underset{k_{-1}}{\overset{k_1}{\rightleftharpoons}} [Co(NH_3)_5(OH_2)]^{3+} + HCrO_4^- \quad (2)$$

$$[Co(NH_3)_5OCrO_3H]^{2+} + H_2O + H^+ \underset{k_{-0}}{\overset{k_0}{\rightleftharpoons}} [Co(NH_3)_5(OH_2)]^{3+} + HCrO_4^- + H^+ \quad (3)$$

A number of Co(III) [and Cr(III)] complexes with the ligand 1,6-diamino-3-azahexane (**1** = L) have recently been characterized and a variety of

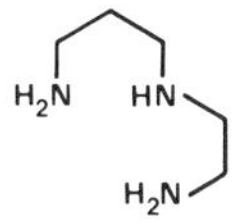

1

Table 7.3. Kinetic Parameters for the Aquation and mer → fac Isomerization of Some Co(III) Triamine Complexes [a]

Complex	Acid Hydrolysis		
	$10^5 k$ (25 °C) (s^{-1})	E_a (kJ mol^{-1})	$\Delta S^{\ddagger}_{298}$ (J K^{-1} mol^{-1})
trans-Dichloro-mer-[CoCl$_2$(NH$_3$)(dien)]$^+$	39.8	106	+36
trans-Dichloro-mer-[CoCl$_2$(NH$_3$)(L)]$^+$	30.9	103	+27
	Isomerization		
mer-[Co(dien)(OH$_2$)$_3$]$^{3+}$	24.7	109	+42
mer-[Co(L)(OH$_2$)$_3$]$^{3+}$	2.97	115	+46
trans-Diaqua-mer-[Co(NH$_3$)(dien)-(OH$_2$)$_2$]$^{3+}$	162	100	+29
trans-Diaqua-mer-[Co(NH$_3$)(L)(OH$_2$)$_2$]$^{3+}$	4.13	113	+52

[a] Data from D. A. House, *Inorg. Chim. Acta*, **121**, 167 (1986); L = 1,6-diamino-3-azahexane.

kinetic studies carried out.[16] The chiral forms of trans-dichloro-mer-[CoCl$_2$(NH$_3$)L]ClO$_4$ (2) (R or S at the *sec* NH center) were isolated. Aquation rates and isomerization rates for the mer → fac isomerization are summarized

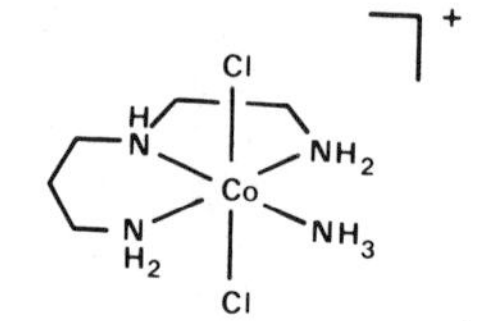

2

in Table 7.3. Two isomers of trans-[CoCl$_2$(ibn)$_2$]$^+$ (ibn = 2-methyl-1,2-diaminopropane) have been characterized by reaction of dioxygen with methanolic solutions of CoCl$_2$ · 6H$_2$O and the diamine.[17] The most abundant isomer is assigned the trans-gem-dimethyl configuration (3) and the least abundant has been characterized as the cis-gem-dimethyl isomer (4) by X-ray crystallography. At 25 °C, the aquation rates for loss of the first chloro-ligand from the trans-trans- and cis-trans-[CoCl$_2$(ibn)$_2$]$^+$ are 1×10^{-3} s^{-1} (E_a = 106 kJ mol^{-1}, $\Delta S^{\ddagger}_{298}$ = +44.5 J K^{-1} mol^{-1}) and 3.07×10^{-3} s^{-1} (E_a = 93.6 kJ mol^{-1}, $\Delta S^{\ddagger}_{298}$ = +12.8 J K^{-1} mol^{-1}), respectively.

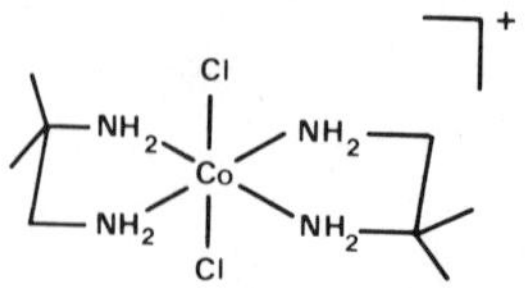

trans - trans

3

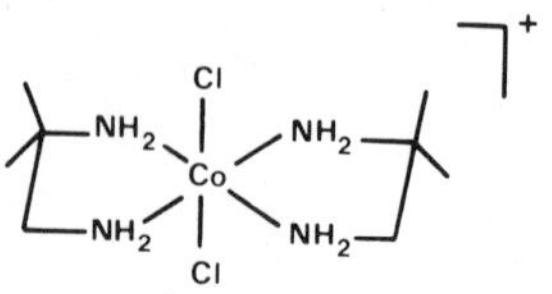

cis - trans

4

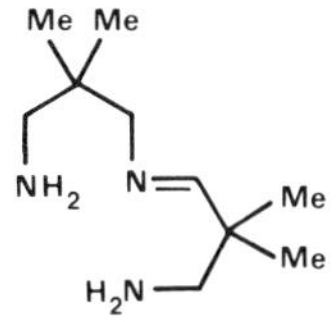

5

The *trans*-$[CoCl_2(Me_2tn)_2]^+$ and *mer*-$[CoCl(Me_2tn)(tri)]^+$ cations (Me_2tn = 2,2-dimethyl-1,3-diaminopropane; tri = **5**) have recently been characterized by House[18] as $ZnCl_4^{2-}$ salts. The latter product formed in small amount by dioxygen oxidation of $CoCl_2 \cdot 6H_2O$ and Me_2tn in methanol is the Me_2tn analogue of a series of recently isolated complexes of general formula *mer*-$[CoCl(diamine)(tridentate\ Schiff\ base)]^{2+}$, where the diamine can be en, ibn, or tn.[19-21] The tridentate Schiff base ligand is believed to be formed by oxidative deamination of one mole of diamine, followed by condensation of the resultant amino-aldehyde, either in situ or by using the metal ion as a template, with a second mole of diamine. Aquation rates for the loss of the first chloro-ligand in 0.1 M HNO_3 at 25 °C are *trans*-$[CoCl_2(Me_2tn)_2]^+$ (k_{aq} 5.57×10^{-3} s^{-1}, E_a = 96.8 kJ mol^{-1}, $\Delta S_{298}^{\ddagger}$ = +28 JK^{-1} mol^{-1}), *trans*-$[CoCl_2(Me_2tn)(en)]^+$ (k_{aq} 3.22×10^{-4} s^{-1}, E_a = 98 kJ mol^{-1}, $\Delta S_{298}^{\ddagger}$ = +8.5 JK^{-1} mol^{-1}), and *mer*-$[CoCl(Me_2tn)(tri)]^{2+}$ (k_{aq} 1.30×10^{-3} s^{-1}, E_a = 75 kJ mol^{-1}, $\Delta S_{298}^{\ddagger}$ = -57 JK^{-1} mol^{-1}). Rate constants were also obtained for the Hg(II)-assisted chloride release.

The spontaneous and acid-induced cleavage of $[(NH_3)_5CoOHCo(NH_3)_5]^{5+}$

$$[(NH_3)_5CoOHCo(NH_3)_5]^{5+} + H_2O \rightarrow [(NH_3)_5CoOH_2]^{3+} + [(NH_3)_5CoOH]^{2+} \quad (4)$$

as in equation (4) has been studied in detail,[22] using Cl^-/ClO_4^- and NO_3^-/ClO_4^- mixed-electrolyte media. The previous data[23,24] for acidic Cl^-/ClO_4^- media have been reinterpreted and the new data for NO_3^-/ClO_4^- media analyzed by the same procedure. This analysis removes an apparent discrepancy in the orders of magnitude of ion-pairing constants between the mono-ol and similar binuclear cations.

7.3. Catalyzed Aquation

The metal-ion-catalyzed aquation of transition-metal complexes with an emphasis on Co(III) and Cr(III) complexes has been reviewed.[1] Simple anions (X^-) such as $CH_3CO_2^-$, Cl^-, N_3^-, Br^-, SCN^-, and CN^- influence the Hg(II)-assisted aquation of $[CoCl(NH_3)_5]^{2+}$ and $[Co(edta)Cl]^{2-}$ by formation of HgX^+.[25] Aminopolycarboxylates such as $edta^{4-}$ and $cdta^{4-}$ (H_4cdta = cyclopentane-1,2-diamine-N,N,N',N'-tetraacetic acid) were especially effective. A feature of the latter reactions with $[CoCl(NH_3)_5]^{2+}$ was the formation of $[CoY(NH_3)_5]^-$ (Y = edta or cdta) in addition to $[Co(NH_3)_5(OH_2)]^{3+}$.

Kinetic evidence for the formation of $[Co(NH_3)_5NCSAg_3]^{5+}$ in the reaction of $[Co(NH_3)_5NCS]^{2+}$ with Ag(I) has been presented.[26] The pseudo-first-order rate constant (k) for reaction (5) is $0.158\ s^{-1}$ at 25 °C and $[Ag^+]$ in the

$$[Co(NH_3)_5NCS\,Ag_2]^{4+} + Ag^+ \rightarrow [Co(NH_3)_5NCS\,Ag_3]^{5+} \tag{5}$$

range $1.23–5.0 \times 10^{-2}$ M. The formation constant β_2 for $[Co(NH_3)_5NCS\,Ag_2]^{4+}$ has been determined as $\log \beta_2 = 4.717$. The kinetic effects of simple organic sulfonate ions on the Hg(II)-assisted aquation of *cis*-$[CoCl(OH_2)(bipy)_2]^{2+}$, *cis*-$[CoCl(OH_2)(phen)_2]^{2+}$, $[CoCl(NH_3)_5)]^{2+}$, *cis*-$[CoCl(NH_3)(en)_2]^{2+}$, and *cis*-$[CoCl(py)(en)_2]^{2+}$ have been studied.[27] Unusual deviations from the expected primary salt effects were observed in the reactions of the complexes containing bipy and phen ligands. The results are interpreted in terms of the Guggenheim equation for the activity coefficients of the reactants and the activated complexes. The specific effects are attributed to the hydrophobic interactions between the complex cations and the sulfonate anions.

Aquation of both *cis*- and *trans*-$[Co(1,2\text{-pn})_2(O_2CCH_3)_2]^+$ is catalyzed by H^+, in aqueous solution.[28] Values of k_{obs} show a linear dependence on $[H^+]$ and there is kinetic and, in the case of the *trans*-isomer, spectrophotometric evidence for a preequilibrium association of the complex cation with the proton. The *cis*-diacetato cation aquates with complete retention of configuration giving *cis*-$[Co(1,2\text{-pn})_2(O_2CCH_3)(OH_2)]^{2+}$. Aquation of the *trans*-cation gives an equilibrium mixture of *cis*- and *trans*-$[Co(1,2\text{-pn})-(O_2CCH_3)(OH_2)]^{2+}$ in the ratio of $3:1$, respectively, which isomerizes to an equilibrium mixture which is 75% *cis* and 25% *trans*. The isolation and characterization of $[CoCl(en)(L)]ZnCl_4$ (**6**) (L = $NH_2CH_2CH=NCH_2-CH_2NH_2$) as a biproduct (*ca.* 8%) in the synthesis of *trans*-$[CoCl_2(en)_2]^+$ by

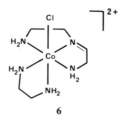

6

the H_2O_2 oxidation route has been described,[29] and the mechanism of this process is considered in Section 7.2. For Hg(II)-assisted chloride release from $[CoCl(en)L]^{2+}$, $k_{Hg} = 24.3 \times 10^{-3}\ M^{-1}s^{-1}$ at 25 °C and $I = 1.0$ M. The introduction of a C=N linkage in the tridentate ligand leads to a fivefold increase in the rate of Hg(II)-assisted chloride release compared with *mer*-$[CoCl(en)(dien)]^{2+}$, where $k_{Hg} = 5.21 \times 10^{-3}\ M^{-1}s^{-1}$ at 25 °C and $I = 1.0$ M. The complex $[CoCl(ibn)L]ZnCl_4$ [ibn = $NH_2CH_2C(CH_3)_2NH_2$, L = $NH_2CH_2C(CH_3)_2N=CHC(CH_3)_2NH_2$] has also been isolated from the reaction between dioxygen, $CoCl_2 \cdot 6H_2O$, and 2-methyl-1,2-diaminopropane in refluxing methanol.[30] The crystal structure of the dinitrate trihydrate salt

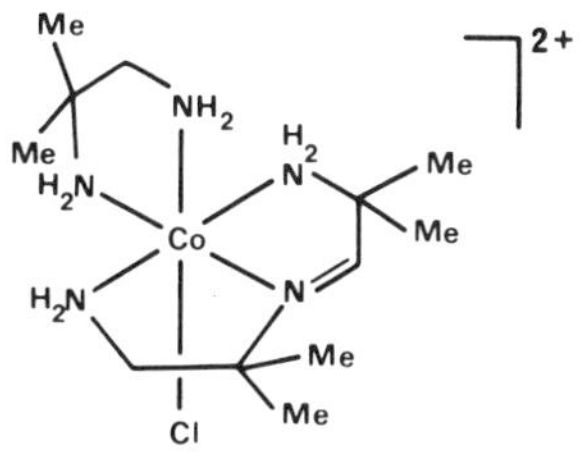

7

shows the complex to have the stereochemistry in (**7**). Kinetic parameters for the Hg(II)-assisted aquation of $[CoCl(ibn)(L)]^{2+}$ in $HClO_4$ solution at $I = 1.0\ M$ are $k_{Hg} = 1.33 \times 10^{-3}\ M^{-1} s^{-1}$ at 25 °C with $E_a = 73.8\ kJ\ mol^{-1}$ and $\Delta S_{298}^{\ddagger} = -60.7\ J\ K^{-1}\ mol^{-1}$.

7.4. Base Hydrolysis

The ratio of the rate constants for the solvolysis (k_2) and reprotonation (k_{-1}) of the amido-conjugate base of *trans*-$[M(RSSR$-cyclam)$Cl_2]^+$ [M = Cr(III), Ru(III), or Rh(III)] and *cis*-$\{M[RRRR(SSSS)$-cyclam]$Cl_2\}^+$ [M = Cr(III) or Rh(III)] (cyclam = 1,4,8,11-tetra-azacyclotetradecane) have been estimated from the proton exchange (k_1) and base hydrolysis (k_{OH}) rate constants.[31] The analogous data for complexes of cobalt(III) have previously been published.[32] All of the complexes are considerably less reactive toward base hydrolysis than the analogous cobalt(III) species. The lower reactivity of Cr(III) is due to a lowering of k_1 and k_2/k_{-1} by similar amounts, while that of Rh(III) arises mainly from a large reduction of k_2/k_{-1} in the *cis*-isomer.

Balt and co-workers have continued their studies of base-catalyzed ammoniation reactions in liquid ammonia. An investigation[33] of the kinetics and steric course of the base-catalyzed ammoniation of three asymmetric *trans*-$[Co(en)_2YX]ClO_4$ complexes [YX = $(N_3)Cl$, $(N_3)Br$, $(NCS)Cl$] provides evidence for rate-limiting deprotonation. Proton-exchange rates were obtained from ^1H NMR measurements and are compared with spectrophotometrically determined ammoniation rates. Base-catalyzed ammoniation has also been described[34] for the series *cis/trans*-$[Co(en)_2XY]^{n+}$ (X, Y = Cl^-, Br^-, Me_2SO, DMF) where both X and Y are readily substituted. Some members of the *trans* series exhibit rate-limiting deprotonation (X, Y = Cl^-, Cl^-; Br^-, Br^-; Cl^-, Me_2SO). The ^1H NMR spectra of the products established that in addition to the normal two-step ammoniation reaction, an additional pathway occurs involving simultaneous loss of both ligands X and Y. The percentage of double substitution varies from 23% (X, Y = Cl^-, DMF) to 65% (Br^-, Br^-). The $[Co(en)_2(NH_3)_2]^{3+}$ produced is a mixture of *cis* and *trans* isomers.

Base hydrolysis and aquation of $[M(NH_3)_5(OSO_2CF_3)]^{2+}$ [M(III) = Co, Rh, Ir, Cr] and $[M(NH_2Me)_5(OSO_2CF_3)]^{2+}$ (M = Co, Rh, Cr) complexes at 25 °C and $I = 1.0\ M$ have been studied.[35] *N*-Methylation of the ammine ligand causes a marked enhancement of the rate of base hydrolysis with k_{Me}/k_H of $>10^3$(Co), 150(Rh), and 800(Cr). Only minor rate enhancements occur in aquation with Co and Rh, while there is a slight rate decrease with Cr. The observation of positive values of $\Delta S^{\ddagger}$ for base hydrolysis of $[M(NH_3)_5(OSO_2CF_3)]^{2+}$ (M = Co, Ir), and competition experiments with N_3^- confirm that the D_{CB} mechanism occurs for all complexes. It is noteworthy that rate enhancements of *ca.* 10^3- to 10^6-fold occur for both aquation and base hydrolysis of triflates compared with those of halo analogues.

The synthesis of $[Co(pyDPT)Cl](ClO_4)_2$ (pyDPT = **8**) has been described and its base hydrolysis studied.[36] Aeration of $CoCl_2 \cdot 6H_2O$ and the ligand gives only one *cis*-isomer which, on the basis of ^1H NMR studies, is believed to have the configuration (**9**). Base hydrolysis is extremely rapid ($k_{OH} = 1.8 \times 10^6\ M^{-1}s^{-1}$ at 25 °C and $I = 0.1\ M$), with $\Delta H^{\ddagger} = 58\ kJ\ mol^{-1}$ and $\Delta S^{\ddagger} = +70\ J\ K^{-1}\ mol^{-1}$. It is noteworthy that the single site for conjugate base formation is *cis* to the leaving group.

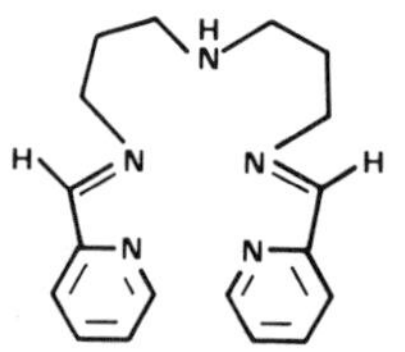

8 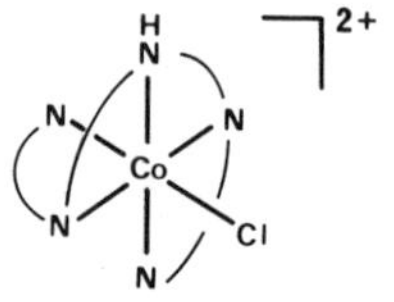　　　　　　　　　　**9**

The second-order rate constant for base hydrolysis of *cis*-[Co(en)$_2$] (1-MeIm)Cl]$^{2+}$ (1-MeIm = 1-methylimidazole = **10**) has been determined to be $25.1 \pm 2.5\ M^{-1}s^{-1}$ at 25 °C and $I = 0.1\ M$.[37] This value is similar to that ($k_{OH} = 28\ M^{-1}s^{-1}$) determined for the imidazole (ImH = **11**) complex *cis*-[Co(en)$_2$(ImH)Cl]$^{2+}$ under the same conditions. Both of these complexes are much less reactive than the benzimidazole (bzImH = **12**) derivative where

10 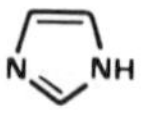　　　　　　**11** 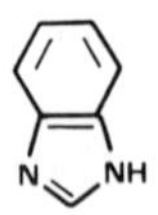　　　　　　**12**

$k_{OH} = 240 \pm 12\ M^{-1}s^{-1}$ at 25 °C. Unlike the 1-MeIm complex the imidazole and benzimidazole complexes undergo acid dissociation in the pH range of the kinetic studies with pK_a values of 10.25 and 8.67 respectively at 25 °C. Rate constants were also determined for base hydrolysis of the conjugate bases of these complexes and the kinetic data are summarized in Table 7.4.

Racemic *mer*-[CoCl(en)(NH$_2$CH$_2$CH=NCH$_2$CH$_2$)]ZnCl$_4$ (**13**) which contains no dissymmetric chelate rings, no asymmetric carbon centers, and no asymmetric nitrogen centers has been resolved using sodium arsenic(III)-(+)-tartrate.[38] The rate of base hydrolysis of the (+)-cation was studied over a

Table 4. Rate Constants k_{OH} for the Base
Hydrolysis of Imidazole, 1-Methyl Imidazole, and
Benzimidazole Complexes at 25 °C and
$I = 0.1\ M^a$

Complex	k_{OH} $(M^{-1} s^{-1})$
cis-[Co(en)$_2$(1-MeIm)Cl]$^{2+}$	25.1 ± 2.1
cis-[Co(en)$_2$(ImH)Cl]$^{2+}$	28.0 ± 1.5
cis-[Co(en)$_2$(Im)Cl]$^+$	4.0 ± 0.2
cis-[Co(en)$_2$(bzImH)Cl]$^{2+}$	257 ± 12
cis-[Co(en)$_2$(bzIm)Cl]$^+$	21.6 ± 1.5

^{a} Data from M. C. Gomez-Vaamonde and K. B. Nolan, *Inorg. Chim. Acta.*, **101**, 67 (1985). The pK_a values for the equilibria cis-[Co(en)$_2$(ImH)Cl]$^{2+} \rightleftharpoons cis$-[Co(en)$_2$(Im)Cl]$^+ +$ H$^+$ are 10.25 and 8.67 for the ImH and bzImH complexes, respectively, at 25 °C.

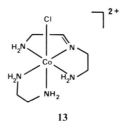

13

temperature range giving $k_{OH} = 1.28 \times 10^3\ M^{-1} s^{-1}$ at 25 °C, $E_a = 87.0 \pm 0.7$ kJ mol^{-1}, and $\Delta S^{\ddagger} = +98 \pm 1.4$ J K^{-1} mol^{-1}. Complete racemization accompanies the base hydrolysis reaction and the rate of loss of optical activity is 0.5 times that of base hydrolysis. These data are interpreted in terms of the formation of a symmetrical trigonal bipyramid intermediate generated from the conjugate base.

Base hydrolysis of O-bonded pentaammineglycinatocobalt(III) and the corresponding β-alaninato derivative have been studied over the temperature range 35–50 °C.[39] At 35 °C, $k_{OH} = 8 \times 10^{-4}\ M^{-1} s^{-1}$ ($\Delta H^{\ddagger} = 105$ kJ mol^{-1}, $\Delta S^{\ddagger} = 37.1$ J K^{-1} mol^{-1}) for the glycinato derivative and $k_{OH} = 4.67 \times 10^{-4}\ M^{-1} s^{-1}$ ($\Delta H^{\ddagger} = 132$ kJ mol^{-1}, $\Delta S^{\ddagger} = 119$ J K^{-1} mol^{-1}) for the β-alaninato complex. Anions retard the reaction in the order PO$_4^{3-} >$ SO$_4^{2-} >$ CO$_3^{3-}$. The cis-[Co(en)$_2$X(β-alaOR)]X$_2$ (X = Cl, Br; R = H, Me, i-Pr) complexes have been prepared and cis-[Co(en)$_2$Br(β-alaO-i-Pr)]$^{3+}$ resolved into its enantiomers.[40] For R = H, the pK_a values for the pendant carboxylic acid function are *ca.* 3.9. Mercury(II)-catalyzed removal of coordinated X(Cl, Br) occurs with retention of configuration on the metal giving 80% cis-[Co(en)$_2$(H$_2$O)(β-alaOR)]$^{3+}$ and 20% of the chelated ester complex

$[Co(en)_2(\beta\text{-alaOR})]^{3+}$ when R = H ($k_{Hg} = 2.9\ M^{-1}\ s^{-1}$ at 25 °C), and 90% and 10% respectively for R = Me and *i*-Pr ($k_{Hg} = 2.4\ M^{-1}\ s^{-1}$). Base hydrolysis occurs with ~60% racemization for R = *i*-Pr, X = Br and gives 90% *cis*-$[Co(en)_2(OH)(\beta\text{-alaO/R})]^{+/2}$ and 10% $[Co(en)_2(\beta\text{-alaO})]^{2+}$ for R = carboxylate ($k_{OH} = 46\ M^{-1}\ s^{-1}$ at 25 °C) and 32% and 68% respectively for R = *i*-Pr ($k_{OH} = 68\ M^{-1}\ s^{-1}$ at 25 °C). Competitive entry by N_3^- leads to a decrease in the percent of the chelated ester species produced for R = *i*-Pr. The work explores the consequences of increasing ring size on the capture of a competing intramolecular nucleophile (ester and carboxylate oxygen) by intermediates of reduced coordination number produced by base hydrolysis and by Hg(II) catalysis. Detailed work on the analogous glycine system has previously been published.[41]

In strong base (*ca.* 1 *M*) the six-membered β-alaninato chelate ring in $[Co(en)_2(\beta\text{-alaO})^{2+}$ (**14**) opens by Co—O bond fission to give a mixture of

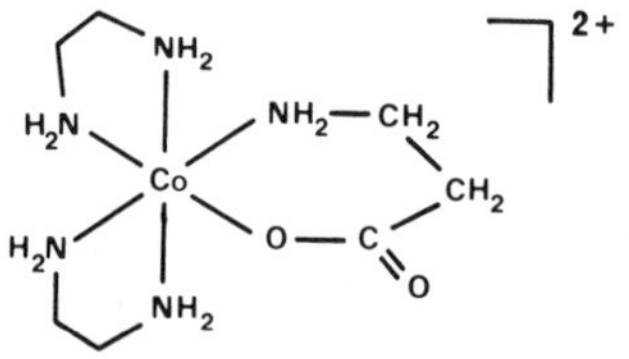

14

cis-$[Co(en)_2(OH)(\beta\text{-alaO})]^{3+}$ (70%) and 30% of the *trans*-isomer.[42] Kinetic studies indicate a first-order dependence on [OH$^-$]. Subsequently, both *cis*- and *trans*-$[Co(en)_2(OH)(\beta\text{-alaO})]^+$ ions hydrolyze to give predominantly *cis*-$[Co(en)_2(OH)_2]^+$, which then isomerizes to an equilibrium mixture of *cis*- and *trans*-$[Co(en)_2(OH)_2]^+$. Slow isomerization of *cis*- to *trans*-$[Co(en)_2(OH)(\beta\text{-alaO})]^+$ and chelation of $[Co(en)_2(\beta\text{-alaO})]^{2+}$ occur in dilute base.

A variety of *mer*-[Co(dien)(aminoacidate)X]$^+$ complexes with the *trans*-(O, X) configuration (**15** and **16**) have been prepared and characterized[43] (X = Cl, NO$_2$). Rapid base hydrolysis of Cl and NO$_2$ ligands occurs. (Table 7.5), a result which is expected for complexes with the *mer*-dien configuration.

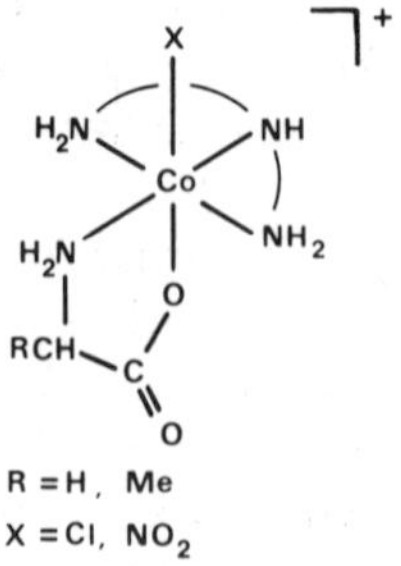

R = H, Me
X = Cl, NO$_2$

15

R = H, CH$_2$CO$_2$
CH$_2$CONHCH$_2$CO$_2^-$
X = Cl, NO$_2$

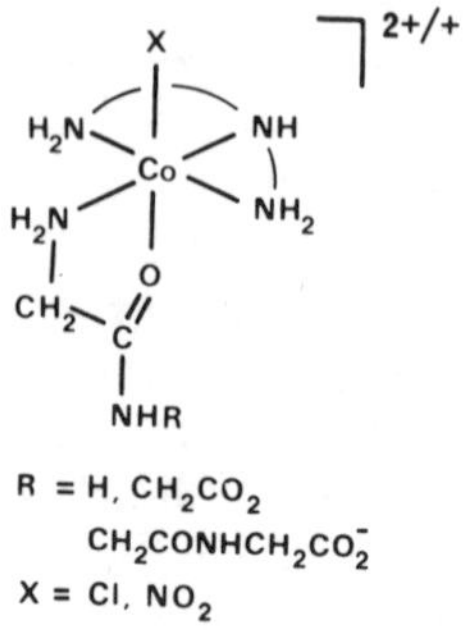

16

Table 7.5. Values of k_{OH} for Base Hydrolysis of
mer-[Co(dien)(Amino-Acidate)Cl]$^+$ Complexes
at 25 °C and I = 0.1 M^a

Complex	k_{OH} $(M^{-1} s^{-1})$
[CoCl(dien)(glyO)]$^+$	$1.3 \pm 0.1 \times 10^4$
[CoCl(dien)(α-alaO)]$^+$	$2.7 \pm 0.2 \times 10^4$
[CoCl(dpt)(glyO)$^+$	$3.4 \pm 0.2 \times 10^3$
[CoCl(dien)(glyglyOEt)]$^{2+}$	$1.1 \pm 0.1 \times 10^6$
[CoCl(dien)(glyNH$_2$)]$^{2+}$	$5.8 \pm 0.3 \times 10^5$

^{a} Data from R. W. Hay, V. M. C. Reid, and D. P. Piplani, *Transition Met. Chem.*, **11**, 302 (1986); dien = 1,5-diamino-3-azapentane, dpt = 1,7-diamino-4-azaheptane. The complexes have the *trans*-(O,Cl) configuration.

The *mer*-dien configuration provides a "flat" *sec*-NH to form the amido-group, and the amido-group is *cis* to the leaving group. The data are fully consistent with the criteria proposed by Henderson and Tobe[44] for the rapid base hydrolysis of chloropentamine cobalt(III) complexes.

Although the D_{CB} mechanism for the base hydrolysis of cobalt(III)–acido complexes is now well defined,[45] little information is available on the lifetime of the five-coordinate intermediate formed following the loss of X$^-$. Is any molecular event possible within this intermediate prior to the coordination of a water molecule or another entering group Y$^-$? Some of these problems have been discussed in a recent paper[46] dealing with the base hydrolysis of t-[Co(tren)NH$_3$)(SCN)]$^{2+}$ and other t- and p-[Co(tren)(NH$_3$)X]$^{2+}$ complexes. A significant feature of the base hydrolysis of t-[Co(tren)(NH$_3$)X]$^{2+}$ complexes is the large amount of stereochemical change to the p-configuration, as in Scheme 1 (Y = OH$^-$, N$_3^-$). Such stereochemical change can be regarded as mutarotation about the metal, and must occur in the five-coordinate intermediate or as part of the rate-determining loss of X$^-$, as it does not occur in the reactant, the conjugate base, or the t-[Co(tren)Y]$^{2+}$ products. Base hydrolysis of p-[Co(tren)(NH$_3$)SCN]$^{2+}$ ($k_{OH} = 1.1 \times 10^4$ $M^{-1} s^{-1}$ at 25 °C) gives only p-[Co(tren)(NH$_3$)OH]$^{2+}$, but hydrolysis of t-[Co(tren)(NH$_3$)SCN]$^{2+}$ gives t(SCN)-[Co(tren)(OH)SCN]$^+$ (35%), t-[Co(tren)(NH$_3$)NCS]$^{2+}$ (5%), in addition to p-[Co(tren)(NH$_3$)(OH)]$^{2+}$ (50%) and t-[Co(tren)(NH$_3$)(OH)]$^{2+}$ (10%). In basic solution t(SCN)-[Co(tren)(OH)SCN]$^+$ isomerizes to t(NCS)-[Co(tren)(OH)NCS]$^+$ ($k = 3 \times 10^{-5} s^{-1}$) while in aqueous acid isomerization of t(SCN)-[Co(tren)(OH$_2$)SCN]$^{2+}$ is slower with $k = 1.3 \times 10^{-6} s^{-1}$ at 25 °C. Base hydrolysis of t-[Co(tren)(NH$_3$)X]$^{2+}$ (X = NCS$^-$,N$_3$) gives only t-(X)-[Co(tren)(OH)X]$^+$ with $k_{OH} = 2.5 \times 10^{-3}$ $M^{-1} s^{-1}$ (NCS complex) and $k_{OH} = 1.2 \times 10^{-3}$ $M^{-1} s^{-1}$ (N$_3$ complex) at 25 °C and I = 1.0 M.

Scheme 1

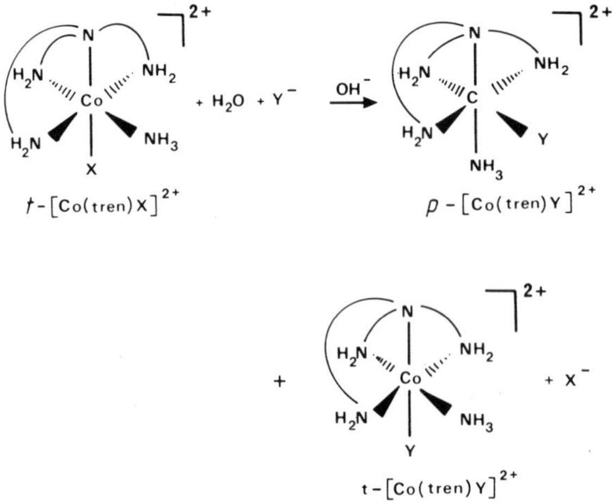

Base hydrolysis of $(\alpha\beta S)$-$[Co(tetren)(3NO_2\text{-salicylato})]^+$ has been studied in the presence and absence of cationic (CTAB) and anionic (SDS) surfactants at 30 °C.[47] The phenoxide form of the complex undergoes aquation in the micellar pseudophase of the surfactants much faster than in the aqueous pseudophase, the effect being more pronounced in the anionic surfactant. This result supports an internal conjugate base mechanism, the reactive amido conjugate base being produced by the intramolecular proton-transfer equilibrium (equation 6). The CTAB micelle, however, retards the second-order

$$[Co(tetren)O_2CC_6H_3(3NO_2)O]^+ \rightleftharpoons [Co(tetrenN\text{-}H)O_2CC_6H_3(3NO_2)OH]^+ \qquad (6)$$

base hydrolysis of the micelle-bound phenoxide species. The kinetics of base hydrolysis of $[Co(NH_3)_5Cl]^{2+}$ has also been studied at 25 °C in the presence and absence of the anionic surfactant sodium dodecyl sulfate over the surfactant concentration range 1×10^{-3} to 7.5×10^{-2} M. Base hydrolysis is strongly inhibited by the surfactant, fitting a model in which the cobalt(III) complex is distributed between water and the micellar pseudophase with a binding constant of 3.7×10^3 M.

Volumes of activation for the base hydrolysis of a series of complexes of the type $[Co(NH_3)_4(NH_2R)Cl]^{2+}$ (R = *cis*, CH_3; *trans*, CH_3; *trans*, C_2H_5; *cis*, $n\text{-}C_3H_7$; *trans*, $n\text{-}C_3H_7$; *trans*, $n\text{-}C_4H_9$; *trans*, $i\text{-}C_4H_9$) have been determined at 25 °C.[49] Values of $\Delta V^{\ddagger}$ fall within the range 26.4 to 29.9 cm^3 mol^{-1}. These data combined with the dilatometrically determined overall reaction volumes enable the construction of reaction volume profiles for the base hydrolysis. The partial molar of the five-coordinate species $[Co(NH_3)_3(NH_2)(NH_2R)]^{2+}$

increases linearly with the partial molar volume of RNH_2 and almost equals the partial molar volume of the $[Co(NH_3)_4(NH_2R)OH]^{2+}$ species. The results confirm the validity of the D_{CB} mechanism and demonstrate the additivity properties of the partial molar volumes of five- and six-coordinate species of similar charge.

A variety of cobalt(III) complexes of the pentadentate macrocycle [16]aneN$_5$ (17 = L) have been prepared and characterized, $[CoLX]^{n+}$ (X = Cl^-, H_2O, NO_2^-, DMF, HCO_2^-).[50] Both ^{13}C NMR[50] and X-ray data[51] confirm that $[CoLCl](ClO_4)_2$ has the *meso*-syn stereochemistry (18). The

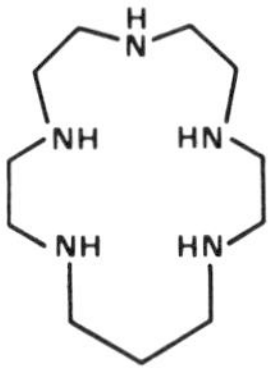

17 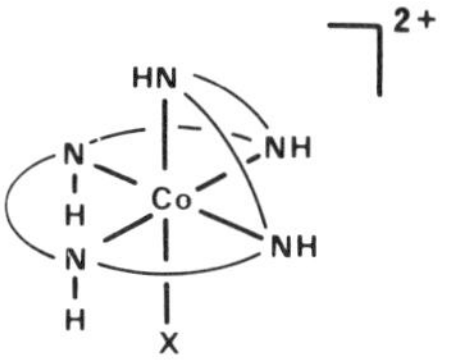   18

choloropentamine complex undergoes rapid base hydrolysis[50] with k_{OH} = $1.1 \times 10^5 \, M^{-1} s^{-1}$ at 25 °C and $I = 0.1 \, M$ ($\Delta H^{\ddagger} = 73 \, kJ \, mol^{-1}$ and $\Delta S^{\ddagger}_{298}$ = $98 \, J K^{-1} mol^{-1}$). Base hydrolysis of $[CoLCl]^{2+}$ is some 10^5-fold faster than $[Co(NH_3)_5Cl]^{2+}$ at 25 °C where $k_{OH} = 0.81 \, M^{-1} s^{-1}$ at 25 °C with $\Delta H^{\ddagger}$ = $120 \, kJ \, mol^{-1}$ and $\Delta S^{\ddagger}_{298} = 155 \, J K^{-1} mol^{-1}$. Very rapid base hydrolysis of $[CoL(NO_2)]^{2+}$ ($k_{OH} = 3.3 \, M^{-1} s^{-1}$) and $[CoL(O_2CH)]^{2+}$ ($k_{OH} = 1.7 \, M^{-1} s^{-1}$) at 25 °C are also observed. The very rapid base hydrolysis rates are consistent with Henderson and Tobe's criteria.[44]

The synthesis of the new amine ligand 3-(aminomethyl)-3-methylazetidine (ama) (19) and characterization of the complex *mer*-$[Co(ama)_3]^{3+}$ (20) by

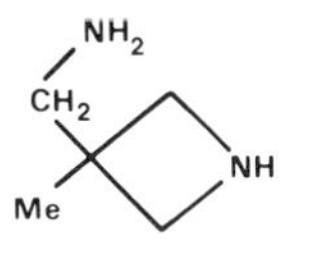

19    20

X-ray crystallography has been described.[52] The cobalt(III) complex is stable in 5 M DCl over several days but is susceptible to base hydrolysis (pH > 3). Coordination of the azetidine ring nitrogen to the inert Co(III) center might be expected to activate the adjacent strained-ring carbons to nucleophilic attack. However, the coordinated azetidine ring of the chelate does not rupture under the hydrolysis conditions and is stable to nucleophilic attack by OMe^- and CN^- prior to ligand substitution reactions. The complex *cis*-

$[Co(en)_2(O_2CCO_2)(OH)]PF_6$ has been isolated and characterized.[53] In basic solution the complex hydrolyses to cis-$[Co(en)_2(OH)_2]^+$ with Co—O bond cleavage with $k_{obs} = k_0 + k_{OH}[OH^-]$ ($k_0 = 5.2 \times 10^{-5}\,s^{-1}$, $k_{OH} = 6.3 \times 10^{-3}\,M^{-1}s^{-1}$ at $I = 1.0\,M$ and 25 °C). In the pH range 0 to 9, cis-$[Co(en)_2(O_2CCO_2)(OH_2)]^+$ and cis-$[Co(en)_2O_2CCO_2H)(OH_2)]^{2+}$ cyclize to the chelated oxalato complex $[Co(en)_2(O_2CCO_2)]^+$. Base hydrolysis of $[Co(en)_2(O_2CCO_2)]^+$ gives cis-$[Co(en)_2(OH)_2]^+$ with $k_{obs} = k_{OH}^1[OH^-] + k_{OH}^2[OH]^2$. Ring opening by the k_{OH}^2 pathway occurs by 100% O—C bond cleavage, while the k_{OH}^1 pathway involves 40% O—C cleavage and 60% Co—O bond cleavage.

7.5. Anation

Kinetic studies of the anation of cis-β-$[Co(trien)(OH_2)_2]^{3+}$ by NCS^- to give the dithiocyanato complex have been reported.[54] The reaction occurs in a stepwise manner, both steps having the common rate expression (7). The

$$k_{obs} = k' + \frac{kQ(NCS^-)}{1 + Q[NCS^-]} \tag{7}$$

variation of k_{obs} with pH shows that for the first step k increases with increase of pH (in the 1.8 to 2.8 range) while it remains fairly constant for the second step. The complex cis-$[Co(NH_3)_4(OH_2)_3]^{3+}$ reacts with acetate ion to give cis-$[Co(NH_3)_4(O_2CMe)_2]^+$; the reaction proceeds in a single step in the pH range 4.0 to 5.2 but becomes biphasic above pH 5.4.[55] Acetate anation follows the rate expression (8) where k_{obs}, k_0, k, and Q denote the observed rate

$$k_{obs} = k_0 + \frac{kQ[OAc^-]}{1 + Q[OAc^-]} \tag{8}$$

constant, the back aquation rate, the interchange rate, and the ion-pair formation constant, respectively. The complex also reacts with succinate ions ($HSuc^-$ and Suc^{2-}) to give $[Co(NH_3)_4(Suc)]^+$. No evidence was obtained for any significant buildup of the monodentate succinato intermediate and the chelation step is believed to be rapid.

The reaction of iminodiacetate ($Hida^-$) with cis-$[Co(NH_3)_4(OH_2)_2]^{2+}$ has also been studied spectrophotometrically in the pH range 3.0 to 4.15.[56] The rate law is as in equation (9) where k_{-1} and k_1 denote the reverse aquation

$$rate = \{k_{-1} + k_1[iminodiacetate]_T\}[Complex] \tag{9}$$

and anation rate constants, respectively. At 40 °C the k_{-1} rate constant increases with pH, reaching a limiting value at pH $ca.$ 4. The pH–rate profile for anation indicates that no reaction occurs below pH 1.5 and k_1 remains constant in the pH range 3.5 to 3.9. The pH dependence of k_1 can be expressed by the equation

$$k_1 = k_A K_{AH}/([H^+] + K_{AH})$$

where k_A is the anation rate arising from the reaction of the diaqua complex with Hida$^-$, and K_{AH} is the dissociation constant of Hida$^-$. The reaction is believed to occur by initial complexation via the carboxylate group, followed by rapid ring closure. *Trans*-[Co(en)$_2$(SO$_3$)(OH$_2$)]$^+$ reacts with imidazole (ImH) and imidazole-containing ligands (L) to give *trans*-[Co(en)$_2$(SO$_3$)L]$^+$ in the pH range 6.0 to 9.0.[57] The rate constant for the reaction of imidazole with the aqua ion is $6.0 \pm 0.7\ M^{-1}\,s^{-1}$ at 25 °C. The corresponding rate constant for reaction with the hydroxo complex is $4 \pm 1\ M^{-1}\,s^{-1}$. The apparent equilibrium constant for formation of the imidazole complex at pH 7 is *ca.* $3 \times 10^2\ M^{-1}$.

Double salts of the type [Co(NH$_3$)$_5$(OH$_2$)][Co(CN)$_5$X] (X = CN$^-$, Cl$^-$, Br$^-$, I$^-$, NO$_2^-$, N$_3^-$) have been prepared and characterized.[58] On heating dinuclear μ-CN complexes [(NH$_3$)$_5$CoNCCo(CN)$_4$X] can be synthesized. The solid-phase thermal deaquation–anation of the double complex salts has been studied by nonisothermal methods and several kinetic parameters calculated. The activation energies for these processes have been related to unit cell parameters. A kinetic study of the anation of *cis*-β-[Co(trien)(OH$_2$)$_2$]$^{3+}$ by thiourea has also been published.[59]

7.6. Solvolysis

The solvolysis of *trans*-[Co(4-Mepy)$_4$Cl$_2$]ClO$_4$ has been investigated using 0 to 70% v/v H$_2$O–MeOH mixtures at 40, 45, 50, and 55 °C.[60] High negative values of $\Delta S^{\ddagger}$ are observed compared with the analogous reactions of [Co(py)$_4$Cl$_2$]$^+$, a result attributed to the substituent methyl groups. The free energies of transfer of both the ground and transition states were calculated. The dominant effect of the solvent on the transition state is apparent. The solvolysis of *trans*-[Co(en)$_2$Cl$_2$]$^+$ has also been studied over a temperature range using H$_2$O–MeCN mixtures.[61] Acetonitrile has much less effect on the solvent structure than the alcohols used as cosolvents with water in previous investigations. However, the nonlinear relationship observed between log k and the reciprocal of the dielectric solvent shows that the effect of changing solvent structure is still important. The application of a free-energy cycle shows that changes in solvent structure affect the five-coordinate transition state more than the six-coordinate ground state.

7.7. Solvent Exchange, Racemization, Isomerization, and Ligand Exchange

A rather remarkable nitrate ion catalysis of various substitution and isomerization reactions involving Co(III) and Cr(III) complexes in aqueous solution has been reported by several groups in recent years. Thus nitrate catalysis of the anation of cis-$[Co(en)_2(H_2O)_2]^{3+}$, cis-α-$[Co(edda)(H_2O)_2]^+$, and $[Co(NH_3)_4(H_2O)]^{3+}$ by oxalic acid has been described.[62-64] A recent paper[65] discusses the background literature on the topic, and describes kinetic studies on the solvent exchange of cis-$[Co(en)_2(H_2O)_2]^{3+}$ in 2 M HClO$_4$ and 2 M HNO$_3$ studied by ^{17}O NMR measurements. The effect of nitrate on the solvent exchange is small, thus in 2 M HNO$_3$ $k_{ex} = 7.9 \times 10^{-6}\,s^{-1}$ compared with $k_{ex} = 5 \times 10^{-6}\,s^{-1}$ in 2M HClO$_4$ at 25 °C. Slight ^{17}O enrichment occurred for ClO$_4^-$, but not for NO$_3^{2-}$, during the exchange process indicating that ClO$_4^-$ may be coordinated to the cobalt(III) center at some stage during the reaction.

Many years ago, Chan and Tobe[66] reported that the reaction between $trans$-$[Co(en)_2Cl_2]Cl$ and KCN in aqueous solution gave $trans$-$[Co(en)_2(OH)Cl]^+$ or products resulting from the disproportionation or redox reactions. In no case was the substitution of one or both Cl ligands by cyanide observed. A recent paper[67] describes the preparation of cis-$[Co(bipy)_2(CN)Cl]^+$ and cis-$[Co(bipy)_2(CN)_2]^+$ by reaction of one and two equivalents of CN$^-$ respectively with the cis-dichloro complex in aqueous solution. The cis-$[Co(bipy)_2(CN)(OH_2)]^{2+}$ and the cis-$[Co(phen)_2(CN)(OH_2)]^{2+}$ cations have a v C$\equiv$N band in the IR at 2200 cm^{-1} shifted from the "normal value" of 2140 cm^{-1}. This effect is believed to be due to the cyano ligand bridging between the metal atom and one hydrogen atom of the water molecule in the cis-position.[68]

The reactions of the μ-amido-μ-hydroxo complex (**21**) with aqueous carbonate solutions have been studied kinetically.[69] Under the conditions

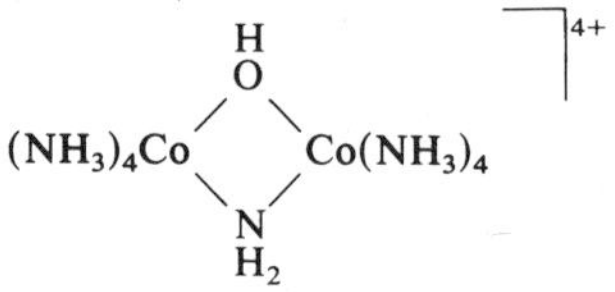

21

employed (pH 8-9.30, $[CO_3^{2-}]_T$ 0.05-0.20 M, temperature 30-45 °C) the reaction observed is the unexpected substitution of NH$_3$ by CO$_3^{2-}$ in one of the cobalt centers of the dinuclear complex. Loss of the two ammine ligands is biphasic with $k_1 = 5.25 \times 10^{-3}\,s^{-1}$ and $k_2 = 1.48 \times 10^{-3}\,s^{-1}$ at pH 8.62 and 40 °C. The final product of the reaction in carbonate solution is (**22**) which has previously been characterized by X-ray crystallography.

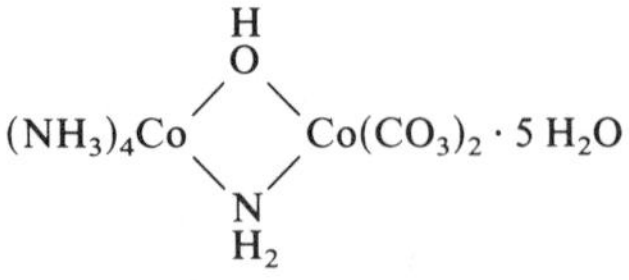

22

The substitution of coordinated glycinate in $[Co(en)_2(gly)]^{2+}$ by ethylenediamine has been studied kinetically at 60-70 °C and pH $ca.$ 12,[70] using both the racemic and optically active metal complexes. The reaction appears to proceed in three steps. The first step is independent of [en] and [OH⁻] but is accompanied by loss of optical activity. The second step is dependent on both [en] and [OH⁻]. Possible mechanisms are discussed. The X-ray crystal structures of two nonisomeric (α-amineoxime)dinitrocobalt(III) complexes, *trans*-[Co(ao-H-ao)(NO$_2$)$_2$] and *cis*-[Co(Hao)$_2$(NO$_2$)$_2$](NO$_3$) (Hao = 3-amino-3-methyl-2-butanone oxime), have been determined.[71] It is concluded that the kinetic lability of the nitro ligands as determined by ^{15}N isotopic exchange rates is not directly related to the metal–nitrogen bond length in the solid.

The acid-catalyzed substitution of [Co(CN)$_5$(N$_3$)]$^{3-}$ by NCS⁻ in aqueous solution gives [Co(CN)$_5$(OH$_2$)]$^{2-}$, [Co(CN)$_5$(SCN)]$^{3-}$, and [Co(CN)$_5$-(NCS)]$^{3-}$.[72] Spectroscopic and HPLC data are quantitatively consistent with an acid-catalyzed dissociative mechanism shown in equations (10) to (15) with

$$\text{[Co(CN)}_5\text{(N}_3\text{)]}^{3-} + \text{H}_3\text{O}^+ \overset{K_1}{\rightleftharpoons} \text{[Co(CN)}_5\text{(N}_3\text{H)]}^{2-} + \text{H}_2\text{O} \tag{10}$$

$$\text{[Co(CN)}_5\text{(N}_3\text{H)]}^{2-} \overset{k_2}{\longrightarrow} \text{[Co(CN)}_5\text{]}^{2-} + \text{HN}_3 \tag{11}$$

$$\text{[Co(CN)}_5\text{(OH}_2\text{)]}^{2-} \overset{k_3}{\longrightarrow} \text{[Co(CN)}_5\text{]}^{2-} + \text{H}_2\text{O} \tag{12}$$

$$\text{[Co(CN)}_5\text{]}^{2-} + \text{H}_2\text{O} \overset{k_4}{\longrightarrow} \text{[Co(CN)}_5\text{(OH}_2\text{)]}^{2-} \tag{13}$$

$$\text{[Co(CN)}_5\text{]}^{2-} + \text{NCS}^- \overset{k_5}{\longrightarrow} \text{[Co(CN)}_5\text{(SCN)]}^{3-} \tag{14}$$

$$\text{[Co(CN)}_5\text{]}^{2-} + \text{NCS}^- \overset{k_6}{\longrightarrow} \text{[Co(CN)}_5\text{(NCS)]}^{3-} \tag{15}$$

$K_1 = 4.47 \pm 0.22\ M^{-1}$ and $k_2 = 3.46 \times 10^{-3}\ \text{s}^{-1}$. The rate constant $k_3 = 6.1 \times 10^{-4}\ \text{s}^{-1}$ and $(k_5 + k_6)/k_4[\text{H}_2\text{O}] = 0.14\ M^{-1}$ with k_5/k_6 *ca.* 4 at 40 °C. Some 20% of the thiocyanate complex is N-bonded. Spectroscopically determined values of K_1 ($4.67 \pm 0.09\ M^{-1}$) agree well with the kinetically determined constant.

The crystal structure of *fac*-tris(S-methylisopropylidenehydrazinecarbodithioato)cobalt(III) (**23**) has been reported.[73] Kinetic studies establish that the *fac* to *mer* isomerization has a very large equilibrium constant ($K \geq 100$) in aprotic solvents and proceeds by a trigonal twist mechanism.

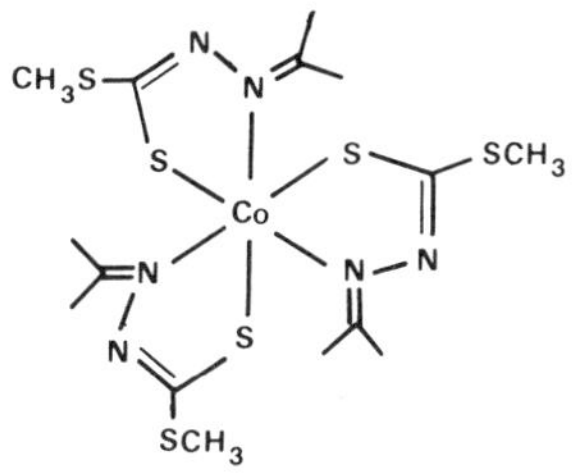

23

The activation volume ($\Delta V^{\ddagger} = 5.1 \pm 0.3$ cm^3 mol^{-1} in 0.01 M HClO$_4$) and reaction volume ($\Delta V = -2.9 \pm 0.2$ cm^3 mol^{-1} in 0.039 M HClO$_4$) have been determined at 3.15 °C for the *trans* $\rightleftharpoons$ *cis* isomerization of [Co(en)$_2$Cl(OH$_2$)]$^{2+}$.[74] The parameters are almost independent of ionic strength or temperature. The transition state is considered to be tetragonal-pyramidal with the water molecule outside the coordination sphere. The selenoether (selenide) complexes t-[Co(RSeCH$_2$CH$_2$NH$_2$)(tren)]$^{3+}$ (t indicates Se is *trans* to the *tert*-N of tren) (R = CH$_3$, C$_2$H$_5$, C$_6$H$_5$CH$_2$) have been prepared and resolved for the first time.[75] The complexes racemize in aqueous solution at rates which are independent of pH, suggesting an intramolecular inversion mechanism. The $\Delta G_{298}^{\ddagger}$ values for inversion of the complexes decrease in the order CH$_3$ (95.2 kJ mol^{-1}) > C$_2$H$_5$ (89.4) > C$_6$H$_4$CH$_2$ (87.7) while the $\Delta H^{\ddagger}$ values are almost the same (95–97 kJ mol^{-1}). A variable-temperature ^{13}C NMR study of p- and t-[Co(CH$_3$SeCH$_2$CH$_2$NH$_2$)(tren)]$^{3+}$, (24) and (25)

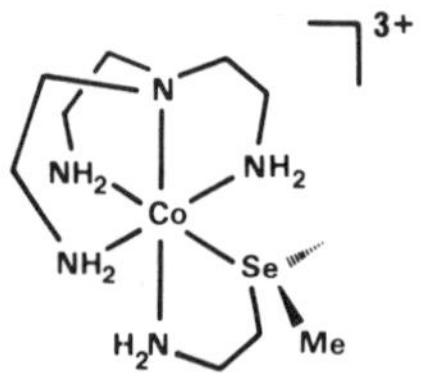

p-isomer

24

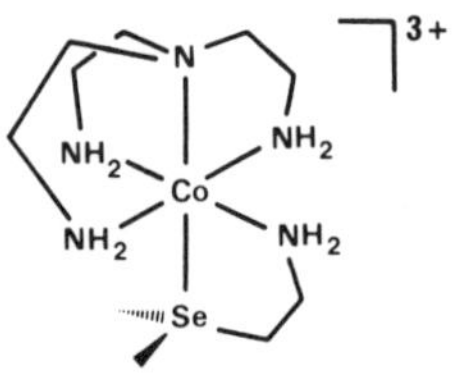

t-isomer

25

respectively, gave an inversion rate constant of 10 s^{-1} at the coalescence temperature (*ca.* 60 °C) which is much larger than that of the corresponding t-isomer (8.5×10^{-3} s^{-1} at 60 °C). In the solid state ($-$)-*cis*-[M(en)$_2$Cl$_2$]Cl [M = Co(III) or Cr(III)] undergoes smooth thermal racemization at 158 °C without any *cis* $\rightleftharpoons$ *trans* isomerization.[76] Values of k_{rac}, $\Delta H^{\ddagger}$, and $\Delta S^{\ddagger}$ are 6×10^{-6} s^{-1}, 218 kJ mol^{-1}, and 156.1 J K^{-1} mol^{-1} for the cobalt complex. The data is only consistent with a rhombic-twist mechanism of the type originally proposed by Ray and Dutt for [M(AA)$_3$] complexes. The rhombic twist involves movement of two of the metal-ligand chelate rings in their respective planes but in opposite directions. Less energy is required in the Ray and Dutt twist than in the Bailar twist.

The optically active complex ion [Co(HL)$_3$]$^{3+}$ (HL = 2,2'-dipyridylamine) is in equilibrium with its conjugate base [Co(HL)$_2$L]$^{2+}$ in aqueous solution due to ionization of a proton from the bridging amine group.[77] Both complexes decompose in aqueous solution by a first-order process ($k = 3.08 \times 10^{-5}$ s^{-1} for the protonated species and $k = 5.95 \times 10^{-5}$ s^{-1} for the conjugate base at 70 °C). Thermolysis appears to involve slow reduction of [Co(HL)$_3$]$^{3+}$ to a cobalt(II) species which then forms an inner-sphere complex stereospecifically with [Co(HL)$_2$L]$^{2+}$ via the bridging amido group. After electron transfer through the bridge a further optically active complex (as yet undefined) is formed with regeneration of the cobalt(II) complex. All four possible optical isomers [$\Delta(S)$, $\Delta(R)$, $\Lambda(S)$ and $\Lambda(R)$] of the bis(biguanide)sarcosinato-cobalt(III) ion [Co(sar)(Hbg)$_2$]$^{2+}$ (sarcosine = N-methyl glycine) have been isolated and characterized.[78] In basic solution the optically active [Co(sar)-(Hbg)$_2$]$^{2+}$ ion loses CD intensity due to racemization at both the chiral cobalt center ($\Delta \rightleftharpoons \Lambda$) and the asymmetric nitrogen center ($R \rightleftharpoons S$). The rate of the $R \rightleftharpoons S$ change is *ca.* 50 times slower than the $\Delta \rightleftharpoons \Lambda$ change. The rate of deuteration (H $\rightarrow$ D) at the chiral nitrogen center is slower than is observed in other sarcosinato complexes. The ratio k_{H-D}/k_{inv} is *ca.* 10^2–10^3. The Δ-α- and Λ-β-forms of [Co(edda)CO$_3$]$^-$ have been isolated as crystalline diastereoisomeric salts of the cation Δ-[Co(en)$_2$(ox)]$^+$.[79] When each salt is dissolved in 1.0 M aqueous Na$_2$CO$_3$ at 25 °C, an isomeric equilibration process occurs between the two species in which the Δ-α-form predominates before the eventual racemization of both complexes occurs. On proceeding to equilibrium the Λ-β-isomer inverts its absolute configuration at the metal center before isomerizing to the Δ-α-isomer. This is the first reported example of an inversion-isomerization involving a $\beta \rightarrow \alpha$ conversion in a chiral complex containing a linear tetradentate.

7.8. Carbonato Complexes

The decomposition of K[Co(NH$_3$)$_2$(CO$_3$)$_2$] in aqueous HClO$_4$ solutions has been investigated.[80] The complex is believed to have a dimeric structure with two carbonato bridges, each carbonato bridge using only one oxygen atom for bridging. The di-μ-carbonato complex is believed to decarboxylate with carbon–oxygen bond cleavage leaving the oxygen bridges intact to give [(NH$_3$)$_2$(H$_2$O)$_2$Co(OH)$_2$Co(H$_2$O)$_2$(NH$_3$)$_2$]$^{4+}$, which undergoes both water-induced and acid-catalyzed bridge cleavage in two successive steps to give *cis*-[Co(NH$_3$)$_2$(OH$_2$)$_4$]$^{3+}$ via a singly bridged intermediate. Under the experimental conditions employed ([H$^+$] = 0.1 to 2.0 M, 40 °C, I = 2.0 M), values of k_{obs} for the first and second steps respectively conform to equations (16) and (17). Three new binuclear cobalt(III) complexes containing bridging

$$k_{obs} = 2.2 \times 10^{-4} + 3.7 \times 10^{-4} [H^+] \tag{16}$$

$$k_{obs} = 2.9 \times 10^{-5} + 11.9 \times 10^{-5} [H^+] \tag{17}$$

carbonate have been prepared.[81] The complexes are of the type [(N)$_3$Co-(μ-CO$_3$)(μ-OH)(μ-X)Co(N)$_3$]$^{2+}$ where (N)$_3$ = dien or (NH$_3$)$_3$ and X = OH$^-$, NH$_2^-$, or NO$_2^-$. The complexes equilibrate rapidly with [(N)$_3$(H$_2$O)Co-(μ-CO$_3$)(μ-X)Co(N)$_3$(OH$_2$)]$^{3+}$ in acidic solution, and the thermodynamic parameters (K, $\Delta H°$ and $\Delta S°$) have been determined.

7.9. μ-Peroxo Complexes

The mechanisms of formation and decomposition of μ-peroxo- and μ-superoxo-dicobalt(III) complexes have been studied.[4] The decomposition of $[(en)_2Co^{III}(\mu\text{-}OH)(\mu\text{-}O_2^{2-})Co^{III}(en)_2]^{3+}$ (26) to cobalt(II) ions and O_2 in

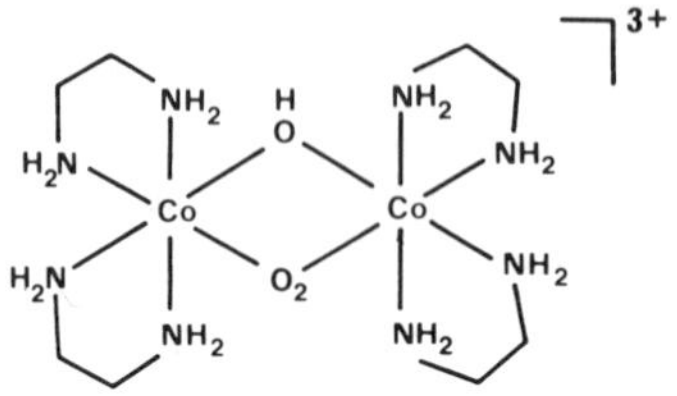

26

perchloric acid solutions ($[H^+] = 0.01$ to $1.5\ M$) has been observed to take place in two steps.[82] The initial fast step is completed within the stopped-flow mixing time ($<5\ ms$) and corresponds to the protonation equilibrium. The second slow step studied using 0.01 to 0.20 M $HClO_4$ solution gives rate constants of 0.02 to 0.3 s^{-1} at 35 °C and $I = 1.0\ M$ for decomposition of the binuclear complex. The decomposition occurs exclusively via the protonated complex by cleavage of the hydroxide bridge to give the singly bridged μ-peroxo complex which subsequently undergoes deoxygenation.

Only a few monomeric cobalt(III)–dioxygen complexes are known, in contrast to the large number of dimeric cobalt(III)–dioxygen complexes. Monomeric (yellow) and dimeric (red) cobalt(III)–dioxygen complexes containing 1,2-bis(dimethylphosphino)ethane [dmpe = $(CH_3)_2PCH_2CH_2P$-$(CH_3)_2$] have been prepared from $[Co(dmpe)_2]^+$ and $[Co(dmpe)_2]^{2+}$ respectively by air oxidation.[83] The corresponding 1,3-bis(dimethylphosphino)propane (dmpp) complexes gave only the monomeric dioxygen complexes. Interestingly the monomeric dioxygen complexes react immediately with SO_2 or concentrated HCl to give $[Co(SO_4)(dmpe\ or\ dmpp)_2]^+$ or cis-$[CoCl_2(dmpe\ or\ dmpp)_2]^+$.

7.10. Formation of Sulfito Complexes

A number of studies have appeared dealing with the uptake of SO_2 by cobalt(III) complexes. Dissolved SO_2 reacts rapidly with trans-$[Co(NH_3)_4(CN)OH]^+$ to give trans-$[Co(NH_3)_4(CN)(OSO_2)]$ which, on immediate acidification, loses SO_2 to give trans-$[Co(NH_3)_4(CN)(OH_2)]^{2+}$. The kinetics of the formation and acidification have been studied as a function of [total S], pH, and temperature.[84] It was found that $k(SO_2\ uptake) = 1.04 \times 10^8\ M^{-1}s^{-1}$ at 25 °C ($\Delta H^{\ddagger} = 41 \pm 6\ kJ\ mol^{-1}$ and $\Delta S^{\ddagger} = 50 \pm 20\ J\ K^{-1}\ mol^{-1}$) and $k(acidification) = 2.5 \times 10^6\ M^{-1}s^{-1}$ at 25 °C ($\Delta H^{\ddagger} = 61 \pm$

1 kJ mol^{-1} and $\Delta S^{\ddagger} = 83 \pm 4 \text{ J K}^{-1} \text{ mol}^{-1}$). The $trans$-$[\text{Co(NH}_3)_4(\text{CN})\text{OSO}_2]$ complex undergoes a subsequent slow linkage isomerization reaction to give $trans$-$[\text{Co(NH}_3)_4(\text{CN})\text{SO}_3]$ with $k = 1.6 \times 10^{-3} \text{ s}^{-1}$ at 40 °C independent of pH and [total S]. The fast reversible uptake of SO_2 by octahedral hydroxo-metal complexes is now well documented in the literature and occurs without metal–oxygen bond fission, a result which has been confirmed by ^{18}O NMR studies of the SO_2 uptake and elimination reactions.[85] The reaction can be represented as shown in Scheme 2. Rapid uptake of SO_2 by $trans$-$[\text{Co(en)}_2(\text{OH}_2)(\text{OH})]^{2+}$ and $trans$-$[\text{Co(en)}_2(\text{OH}_2)_2]^{+}$ occurs with $k = 7.5 \times 10^7 \text{ s}^{-1}$ ($\Delta H^{\ddagger} = 26.8 \pm 8 \text{ kJ mol}^{-1}$; $\Delta S^{\ddagger} = -46 \pm 6 \text{ J K}^{-1} \text{ mol}^{-1}$) and $k = 2.2 \times 10^9 \, M^{-1} \text{ s}^{-1}$ ($\Delta H^{\ddagger} = 31.8 \pm 12.5 \text{ kJ mol}^{-1}$; $\Delta S^{\ddagger} = -226 \pm 125 \text{ J K}^{-1} \text{ mol}^{-1}$)

Scheme 2

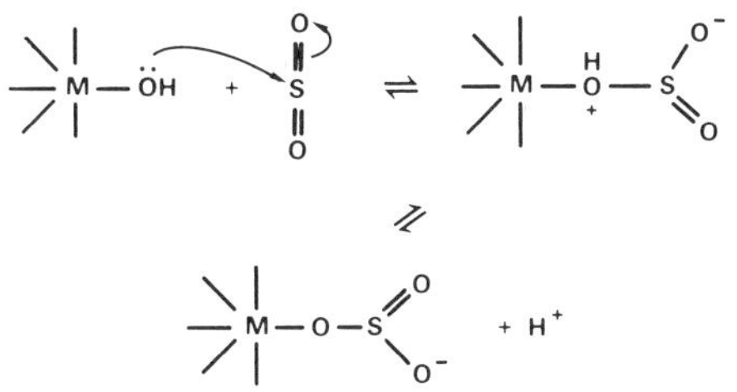

respectively.[86] The reaction of the resulting mono(sulfito) complexes with SO_2 to give the bis(sulfito) displays a hydrogen-ion dependence. The products obtained are shown by spectral studies to be the O-bonded bis(sulfito) complex which, in the unstable protonated form $[\text{Co(en)}_2(\text{OSO}_2\text{H}_2]^{+}$, eliminates SO_2 in a biphasic reaction.

Unlike other aquo amine cobalt(III) complexes so far studied, SO_2 uptake by cis-$[\text{Co(phen)}_2(\text{OH}_2)_2]^{3+}$ and cis-$[\text{Co(bipy)}_2(\text{OH}_2)_2]^{3+}$ in the pH range 2–7 does not lead to the formation of an observable O-bonded species, but gives the S-bonded aqua sulfito complex at rates too rapid to measure by stopped-flow techniques.[87] Formation of the cis-bis(sulfito) complexes occurs within the stopped-flow time scale. This latter reaction involves solely the ionic reactants HSO_3^{-} and SO_3^{2-} and appears to occur by a D mechanism. At higher pH values the only observable products are S-bonded cis-$[\text{Co(phen)}_2(\text{SO}_3)\text{(OH)}]$ and its (bipy)$_2$ analogue. Internal redox reactions to give Co(II) and SO_4^{2-} in a 2:1 ratio only occur at elevated temperatures and acidities.

7.11. Photochemistry

Cobalt(III) complexes are characterized by a substitutionally inert t_{2g}^6 ground state. However, temporarily changing the electronic structure to $t_{2g}^5 e_g^1$ by excitation, or to $t_{2g}^5 e_g^2$ by charge transfer, can lead to rapid ligand substitution. An interesting example of this effect has recently been reported.[88] A thermal ligand substitution reaction of ethylenediamine with $[Co(en)_2(gly)]^{2+}$ does not occur, however the reaction to give $[Co(en)_3]^{3+}$ occurs at pH 12 in the presence of photoexcited $[Ru(bipy)_3]^{2+}$ at 25 °C. The ligand substitution reaction between $[Co(NH_3)_5Cl]^{2+}$ and edta in the presence of $[Ru(bipy)_3]^{2+}$ has also been investigated in detail.[89] Irradiation with visible light of solutions of $[Co(NH_3)_5Cl]^{2+}$, edta, and $[Ru(bipy)_3]^{2+}$ in acetate buffer (pH 4.75) gives $[Co(edta)]^-$. The chain reaction is initiated by reaction of the photoexcited complex $[Ru(bipy)]^{2+}$ with $[Co(NH_3)_5Cl]^{2+}$.

The cobalt(II) complex cis-$[Co(1,2\text{-}pn)_2(NO_2)_2]^+$ after flash photolysis in pure acetonitrile has been shown to react[90] with dioxygen to give the μ-superoxo dinulcear cobalt(III) complex (27), which is subsequently converted to the stable μ-peroxo complex (28).

$$[(O_2N)(1,2\text{-}pn)_2CoO_2Co(1,2\text{-}pn)_2(NO_2)_2]^{2+}$$

27

$$[(O_2N)(1,2\text{-}pn)_2CoO_2Co(1,2\text{-}pn)_2(NO_2)]^{2+} + NO_2$$

28

The (μ-hydroxo)(μ-peroxo) ion $[(en)_2Co^{III}(\mu\text{-}OH,\mu\text{-}O_2^{2-})Co^{III}(en)_2]^{3+}$ in basic solution does not undergo deoxygenation in the dark, but does so on irradiation with ultraviolet light,[91] to set up the photoinduced equilibrium (18). Continuous irradiation leads to irreversible decomposition to give

$$[(en)_2Co^{III}(\mu\text{-}OH,\mu\text{-}O_2^{2-})Co^{III}(en)_2]^{3+} \underset{\Delta}{\overset{h\nu}{\rightleftharpoons}} 2[Co^{II}(en)_2(H_2O)_2]^{2+} + O_2 \qquad (18)$$

cobalt(III) species including $[Co^{III}(en)_2(H_2O)_2]^{3+}$. This irreverisble reaction is much slower in the dark (k ca. $5 \times 10^{-7} s^{-1}$ at pH 8.9 and 5 °C).

Adamson and co-workers[92,93] first reported a unique photochemical preparation of a sulfinato-O complex of Co(III). The S-bonded tris chelate complex $[Co(en)_2(SO_2CH_2CH_2NH_2)]^{2+}$ can be photolyzed to the thermodynamically unstable O-bonded isomer $[Co(en)_2(OS(O)CH_2NH_2)]^{2+}$. The S- and O-bonded isomers can be abbreviated to CoSOON and CoOSON respectively. A recent paper[94] discusses the effect of pressure on the photochemical formation and thermal back-isomerization of the CoOSON complex. Studies of the quantum yield as a function of pressure for the reaction CoSOON $\xrightarrow{h\nu}$ CoOSON give $\Delta V_{app}^{\ddagger} = +5.8 \pm 0.5$ cm^3 mol^{-1}, while for the thermal reaction CoOSON $\xrightarrow{\Delta}$ CoSOON, $\Delta V^{\ddagger} = -9.0 \pm 0.7$ cm^3 mol^{-1}.

7.12. B_{12} and B_{12} Models

The kinetics of the base-induced decomposition of five 2-alkoxyethyl-(aquo)cobaloximes, $[ROCH_2CH_2Co(dmgH)_2OH_2]$ $[R = C_6H_5, CF_3CH_2, CH_3,$ $CH_3CH_2, (CH_3)_2CH]$ have been studied manometrically in aqueous base at 25 °C and $I = 1.0\ M$ under argon.[95] For the complexes with good leaving groups ($R = C_6H_5$ and CF_3CH_2) the reactions are first order in cobaloxime and first order in hydroxide and produce stoichiometric amounts of ethylene and the leaving group alcohol (ROH). Endocyclic N-bound 2-aminopyridine (2-NH_2py) when coordinated to B_{12} coenzyme models has been found to simulate the very long Co—N bonds characteristic of axially bound 5,6-dimethylbenzimidazole in cobalamins.[96] Ambidentate isomerism has also been found for 2-NH_2py and related ligands coordinated to cobaloximes $[LCo(dmgH)_2R]$ (L = 2-NH_2py or substituted 2-NH_2py, R = alkyl). Rotation about the Co—N bond of endocyclic N-bound 2-NH_2py has $k_{rot} = 3.9 \times 10^2\ s^{-1}$ at −90 °C for $R = CH_2NO_2$. Apparent molar volumes ϕ_v have been measured at 25 °C, $I = 0.1\ M$ for aquocobalamin chloride, methylcobalamin, 5′-deoxyadenosylcobalamin, and aquanitrocobaloxime.[97] The base-off forms of the organocobalamins have the same volumes as the base-on forms. Vitamin B_{12s} [cob(I)alamin] reacts rapidly and completely with organic disulfides in aqueous solution to give the corresponding thiols and cob(II)alamin.[98] The reactions are first order in both B_{12s} and disulfide and are generally accelerated three- to sevenfold by monoprotonation of a basic site on the oxidant.

7.13. Reactions of Coordinated Ligands

7.13.1. Amino Acid Esters and Peptides

The potential of complexes of the type $[Co(en)_2(amino\ acid\ ester)]^{3+}$ as reagents for the mild and rapid synthesis of peptides has been recognized for several years.[99] Mensi and Isied[100] have recently studied the acylation of the chelated amino acid ester complex $[Co(en)_2(L-PheOMe)]^{3+}$ (29) with a variety of other amino acid esters as shown in Scheme 3. The reaction of the chelated ester complex 29 with L-PheOMe and L-PheOBu^t in Me_2SO gave 18% of the racemized peptide D-Phe-L-PheOR after removal of cobalt. The $[Co(en)_2(peptideOR)]^{3+}$ complexes (30) also undergo further racemization of the amino acid residue bound to Co(III) when they are dissolved in neutral aqueous solution. It thus appears that although the peptide bond formation reaction (Scheme 3) proceeds in 100% yield, the large degree of racemization observed makes the use of chelated amino acid ester complexes (29) unsuitable for the general synthesis of biologically active peptides. A different synthetic strategy involving complexes of the type $[Co(NH_3)_5O_2CCH(R)NH_2]^{2+}$ with

Scheme 3

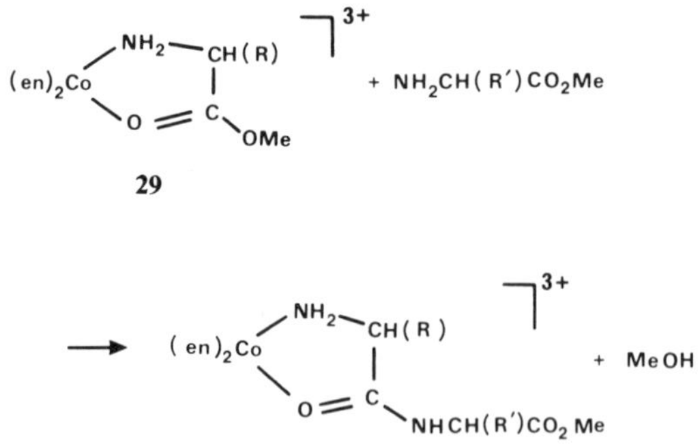

$$\text{(en)}_2\text{Co} \underset{O=C}{\overset{NH_2-CH(R)}{\diagup\!\!\!\!\diagdown}} \Bigg]^{3+} \quad + \; NH_2CH(R')CO_2Me$$

29

$$\longrightarrow \quad \text{(en)}_2\text{Co} \underset{O=C}{\overset{NH_2-CH(R)}{\diagup\!\!\!\!\diagdown}} \Bigg]^{3+} \quad + \; MeOH$$

30

Co(III) acting purely as a C-terminal protecting group leads to peptide bond formation without racemization.[101,102]

The aminolysis of the chelated, acyl-activated β-alanine isopropyl ester in $[Co(en)_2(\beta\text{-alaOPr}^i)]^{3+}$ by glycine ethyl ester in aqueous solution has been studied in detail.[103] The observed rate expression is $k_{obs} = k'[B] + k''[GlyOEt][B]$ with only the k'' terms resulting in the aminolysis product $[Co(en)_2(\beta\text{-ala-glyOEt})]^{3+}$. The k' terms result in hydrolysis of the ester to give $[Co(en)_2(\beta\text{-alaO})]^{2+}$. Values of k'' are 3.9×10^3 (B = OH$^-$), 3.5×10^{-2} (B = GlyOEt), and 2.9×10^{-2} (B = ImH) $M^{-2} s^{-1}$. The k' rate constants are 5.0×10^{-3} (B = OH$^-$), 3×10^{-3} (B = GlyOEt), and 4.5×10^{-3} (B = ImH) $M^{-1} s^{-1}$ at 25 °C and $I = 1.0 \; M$ (NaCl). The requirement for another catalyzing base to be present in the aminolysis reaction is interpreted as rate-determining deprotonation of an amine-alcohol intermediate.

Several years ago the hydrolysis of the cobalt(III) activated glycine esters (**31**) by H_2O and OH$^-$ shown in equation (19) was studied in detail.[104,105]

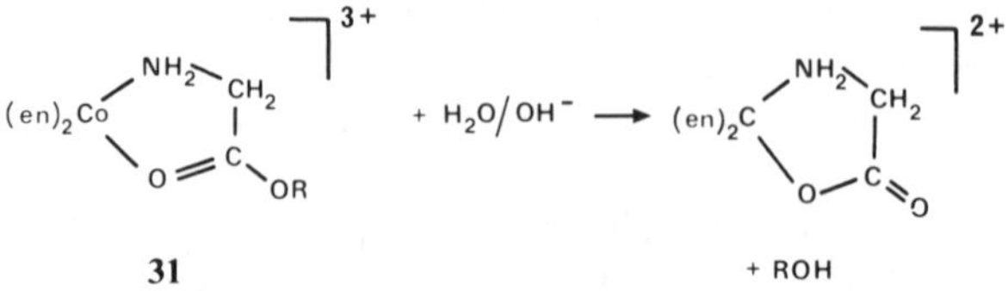

$$\text{(en)}_2\text{Co} \underset{O=C}{\overset{NH_2-CH_2}{\diagup\!\!\!\!\diagdown}} \Bigg]^{3+} \; + \; H_2O/OH^- \; \longrightarrow \; \text{(en)}_2\text{Co} \underset{O-C}{\overset{NH_2-CH_2}{\diagup\!\!\!\!\diagdown}} \Bigg]^{2+} \qquad (19)$$

31   + ROH

The 10^6–10^7 rate enhancement over the uncoordinated ester, as well as the rate ratio $k_{OH}/k_{H_2O} = 10^{11}$, was shown to result from positive $\Delta S^\ddagger$ contributions. These studies have now been extended to the chelated β-alanine ester complex $[Co(en)_2(\beta\text{-alaOPr}^i)]^{3+}$ containing a six-membered chelate ring.[106] Hydrolysis occurs by solvolytic attack at the activated carbonyl center with C—OPr^i

bond fission to give chelated $[Co(en)_2(\beta\text{-alaO})]^{2+}$. The rate law is

$$k_{obs} = k_{H_2O} + (k^I[OH^-] + k^{II}[OH^-]^2)/(1 + K[OH^-])$$

This rate law is interpreted in terms of rate-determining H_2O and OH^- addition to the chelated ester at low and high pH respectively, and rate-determining proton abstraction from an addition intermediate in the pH range 7–10. Catalysis by HPO_4^{2-}, imidazole, N-methylimidazole, ethyl glycinate, and HCO_3^- occurs.

Hay *et al.*[43] have characterized a number of *mer*-[Co(dien)(dipeptideOR)X]$^{2+}$ complexes of the general type (32) with R′ = H, Et; X = NO_2^-,

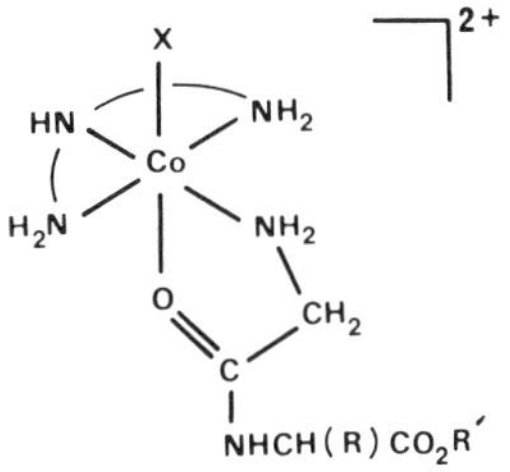

32

and studies the base hydrolysis of the peptide bond in the carbonyl-bonded glycyl peptides. Rate constants for peptide bond hydrolysis fall within the range 0.67 to 0.88 $M^{-1}s^{-1}$ at 25 °C and $I = 0.1\ M$ (Table 7.6). Base hydrolysis of the complexed peptide is *ca.* 2×10^4 times faster than for the uncomplexed peptide at 25 °C. Table 7.7 summarizes the data for the base hydrolysis of ester, amide, and peptide bonds in various cobalt(III) complexes. Generally, rate enhancements are 10^4- to 10^6-fold when comparisons are made with the base hydrolysis rates for the free ligands.

Table 7.6. Rate Constants k_{OH} for Base Hydrolysis of the Peptide Bond in the mer-[Co(dien)(dipeptide/tripeptide)NO₂]$^{n+}$ Complexes at 25 °C and $I = 0.1\ M^a$

Complex[b]	k_{OH} ($M^{-1}s^{-1}$)
$[Co(dien)(glyglyOEt)NO_2]^{2+}$	0.065 ± 0.05
$[Co(dien)(glyglyO)NO_2]^+$	0.86 ± 0.05
$[Co(dien)(glyglyglyO)NO_2]^+$	0.66 ± 0.05
$[Co(dien)(gly-\alpha\text{-alaOEt})NO_2]^{2+}$	0.72 ± 0.05

[a] Data of R. W. Hay, V. M. C. Reid, and D. P. Piplani, *Transition Met. Chem.*, **11**, 302 (1986).
[b] All of the complexes have a *trans*-(O,NO_2) configuration.

Table 7.7. Rate Constants k_{OH} for the Base
Hydrolysis of Ester, Amide, and Peptide
Bonds in Various Cobalt(III) Complexes
$(25\ °C,\ I = 1.0\ M)^a$

Complex	k_{OH} $(M^{-1}s^{-1})$
$[Co(en)_2(glyOPr^i)]^{3+}$	1.5×10^6
$[Co(en)_2(glyNH_2)]^{3+}$	25 ± 1
$[Co(en)_2(glyNHMe)]^{3+}$	1.6 ± 0.2
$[Co(en)_2(glyNMe_2)]^{3+}$	1.1 ± 0.2
$[Co(en)_2(glyglyO)]^{2+}$	2.6 ± 0.2
$\beta_2\text{-}[Co(trien)(glyNHMe)]^{3+}$	2 ± 1
$\beta_2\text{-}[Co(trien)(glyglyOMe)]^{3+}$	5 ± 1
$\beta_2\text{-}[Co(trien)(glyglyOPr^i)]^{3+}$	3 ± 1
$\beta_2\text{-}[Co(trien)(glyglyO)]^{2+}$	3 ± 1

^{a} Data from D. A. Buckingham, C. E. Davis, D. M. Foster, and
A. M. Sargeson, J. Am. Chem. Soc., **92**, 5571 (1970); N. E.
Dixon and A. M. Sargeson, in: Zinc Enzymes (T. G. Spiro,
ed.), Wiley, New York (1983).

7.13.2. Hydration of Alkenes

The hydration of alkenes is a sluggish reaction and stringent conditions
are often required, such as high concentrations of either acid or base to cause
the reaction to occur. However, enzymes such as aconitase are known which
can effect rapid hydration of alkenes in near-neutral conditions at 25–39 °C.
Sargeson et al.[107] have studied the hydration of coordinated carboxyalkenes
by using bis(ethylenediamine)cobalt(III) complexes with either methyl maleate
(**33**) or ethyl fumarate (**34**) coordinated cis to an aqua ligand. Deprotonation

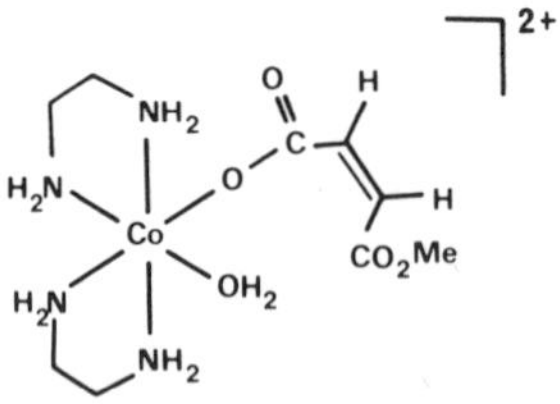

33 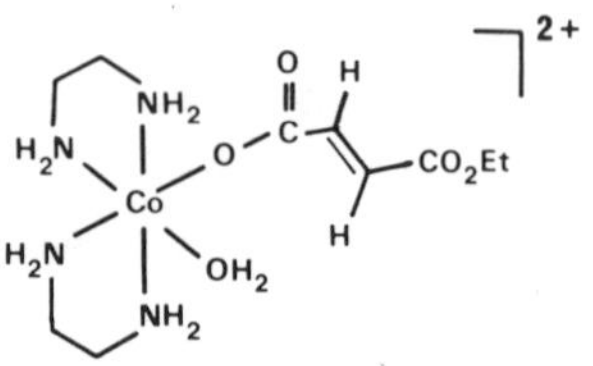   34

of the aqua ligand (pK_a 7.14) gives the $[Co(en)_2(OH)(methylmaleato)]^+$
cation, which undergoes a rapid intramolecular cyclization ($k = 3.4 \times 10^{-2}\ s^{-1}$
at pH 8.05, $I = 1.0\ M$, 25 °C) to give cis-$[Co(en)_2(OH)(methyl\ fumarato)]^+$
and two diastereoisomers of the $[Co(en)_2(methyl\ malato)]^+$ ion. The
intramolecular nature of the cyclization reaction and the exclusive formation

of five-membered rings in the chelated malate product has been established by ^{18}O-tracer experiments and X-ray crystallographic analysis of one of the diastereoisomers. The reaction proceeds by attack of coordinated hydroxide at the alkene as shown in equation (20).

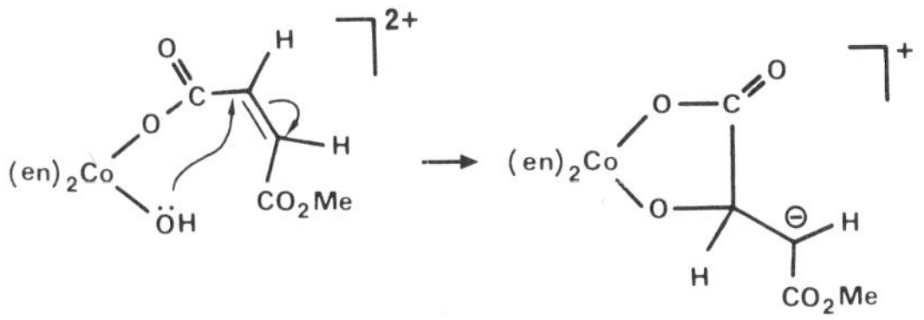

$$(20)$$

7.13.3. Hydrolysis of Phosphates

Metal-ion-promoted hydrolysis of ATP and related polyphosphates and phosphate esters is of broad interest, especially in regard to the roles of metal ions in biological phosphoryl and nucleotidyl transfer. For model studies in this area some advantages are found using cobalt(III) complexes.[108] The complex $[Co(1,3\text{-pn})_2(OH)(OH_2)]^{2+}$ has been observed to promote ATP hydrolysis at rates as high as those reported for any aqua metal ion.[109] Much higher rate enhancements have recently been reported[110] using the $[Co(trpn)\text{-}(OH)(OH_2)]^{2+}$ complex $[trpn = N(CH_2CH_2NH_2)_3 = 3,3',3''\text{-triaminotripro-}$ pylamine]. Rate accelerations in excess of 10^6-fold for ATP hydrolysis occur with this complex and, in addition, rate saturation effects are observed.

7.13.4. Ligand Oxidation

The kinetics of the oxidation of the cysteinsulfenato-bis(ethyl-enediamine)cobalt(III) ion (35) by periodate has been studied in aqueous

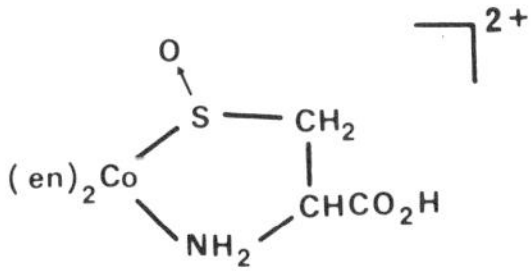

35

solvents containing up to 40% *t*-BuOH and *i*-PrOH.[111] The solvent effects on the oxidation at different pH values are interpreted in terms of solvation of the initial and transition states. Lawrance[112] has shown that O-bonded urea in the cation $[Co(NH_3)_5OC(NH_2)_2]^{3+}$ is oxidized by chlorine and HOCl in aqueous acid. Two sequential steps are observed with both oxidants. Rate constants for both steps show a first-order dependence on [oxidant] with k_1/k_2 always <20. Electronic and vibrational spectra of the intermediate is indicative of a change from an O-bonded to an N-bonded urea. However, a simple isomerization cannot be involved due to the dependence on [oxidant].

7.13.5. Miscellaneous

Coordinated hexafluoroacetylacetonate has been shown[113] to react with hydroxide as in equation (21). Reaction volumes ($\Delta V°$) at infinite dilution

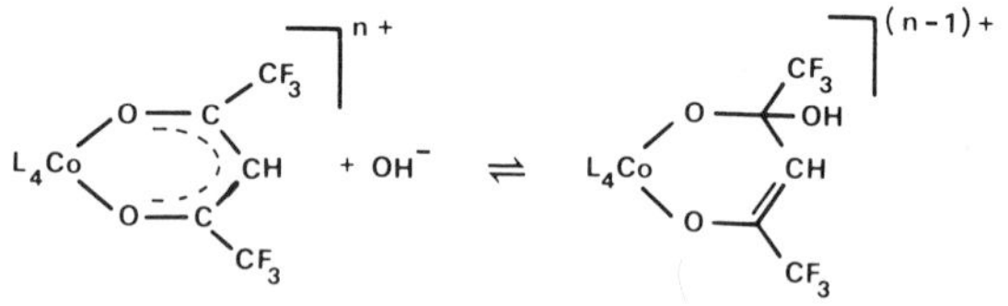

$$\tag{21}$$

have now been determined at 25 °C for $L_4 = (NH_3)_4$, $(en)_2$, α-trien, β-trien, $f(N)$-i-dtma, α-edda, and β-edda. The resulting $\Delta V°$ values support the equilibrium proposed.

The selenide (selenoether) complexes $[Co(\beta\text{-dik})_2(CH_2SeCH_2CH_2\text{-}NH_2)]^+$ (β-dik = acac, Cl-acac, Me-acac) have been prepared and separated into two racemic pairs of the diastereoisomers[114] [$\Delta(R)\Lambda(S)$ and $\Delta(S)\Lambda(R)$ isomers]. The complexes epimerize in solution by inversion at the selenium atom, and the rates have been monitored by HPLC in the temperature range 30.3 to 55.0 °C. The $\Delta G^{\ddagger}_{298}$ values for inversion decrease in the order β-dik = Cl-acac (108 kJ mol^{-1}) > acac (106) > Me-acac (100).

Chapter 8

Substitution Reactions of Inert-Metal Complexes— Coordination Numbers 6 and Above: Other Inert Centers

8.1. Introduction; Groups V to VII

In this first section we shall deal with a few references which are relevant to several of the metal centers covered in this chapter—and indeed often relevant to other centers too. We also deal with a varied selection of references to substitution at Group V, VI, and VII centers. Several cases are marginal, in that it is not obvious whether to classify them as fast inert or slow labile situations. Indeed, sometimes substitution at "inert" centers is appreciably faster than at some centers normally labeled "labile," and *vice versa*. A recent ^{113}Cd NMR study, for instance, revealed remarkably slow ligand exchange at Cd^{2+}.[1] The references incorporated in this section are of interest either in relation to basic questions of substitution mechanisms or to applied chemistry. Further marginal "inert"/"labile" reactions will appear in Section 8.2.4 [iron(III)] and Section 8.7.1 [nickel(I), (II), (III); copper(II)].

8.1.1. General

Earlier volumes of this series have mentioned relatively rapid loss of coordinated trifluoromethylsulfonate from a range of transition-metal centers.

Kinetic data for aquation and for base hydrolysis of these complexes have now been gathered together and discussed in the general context of the chemistry of this rather reluctant ligand.[2]

References on photocalorimetry and on X-ray diffraction studies of solid salts provide important fundamental information on the centrally important $[MX(NH_3)_5]^{2+}$ series of complexes. For X = Cl, comparison of new crystal-structure determinations for M = Ir(III), Os(III), and Cr(III) with results of a redetermination for M = Ru(III) and earlier data for M = Co(III) gives a quantitative picture of relative bond lengths. These give an indication of relatively greater importance of π-bonding at d^3 and d^5 ions than at the d^6 ions.[3] Photocalorimetry, which uses photochemical initiation but actually measures the normal enthalpy change, has been utilized to assess bond strengths in $[MX(MH_3)_5]^{2+}$, M = Rh or Ir, X = Cl, Br, or I, and $[Ru(NH_3)_6]^{3+}$. From these and earlier investigations on analogous cobalt and chromium complexes, the order of decreasing difference in bond strengths between M(III)—NH$_3$ and M(III)—OH$_2$ appears to be[4]:

$$Ir(III) > Rh(III) \simeq Ru(II) > Co(III) > Cr(III)$$

8.1.2. Vanadium

Reaction of the glycine–vanadium(IV) complex $[VO(gly)_2]$ with oxalate proceeds in two steps, the first in the stopped-flow time scale, the second rather slower. The point of interest here is that both steps, giving $[VO(gly)(ox)]^-$ and $[VO(ox)_2]^{2-}$ respectively, are zero-order in oxalate. The implication that both steps are dissociative is out of line with the main body of evidence relating to substitution at vanadium(IV), leading the present authors to suggest that the mechanism of each step involves rate-determining associative attack by water followed by interchange with the incoming oxalate.[5]

8.1.3. Molybdenum

A little more kinetic information has been published on the isomerization of dinuclear molybdenum complexes of the diphosphines (1) to (4) (cf. p. 236 of Volume 4). The slowest $\alpha \rightarrow \beta (5 \rightarrow 6)$ conversion, for the complex containing the diphosphine 1, has a half-life of about one day, in dichloromethane solution. $\alpha \rightarrow \beta$ conversions for other similar bis-diphosphine complexes take place tens or hundreds of times more quickly (lower activation energies)[6]; the reverse $\beta \rightarrow \alpha$ conversion is about ten times slower.[7] For $\{Mo_2Cl_4[(S)\text{-}1,3\text{-}bis\text{-}(diphenylphosphino)butane]_2\}$, rate constants for $\alpha \rightarrow \beta$ and $\beta \rightarrow \alpha$ depend differently on solvent, making the equilibrium constant markedly dependent on the medium.[6] Added diphosphine has no effect on rate constants, ruling out a mechanism involving phosphine dissociation.

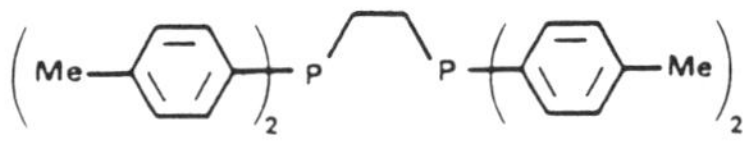

1    2

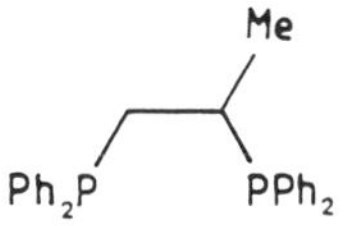

3 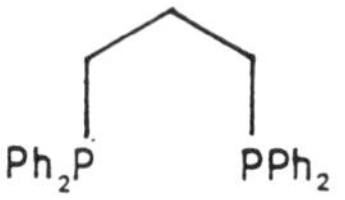   4 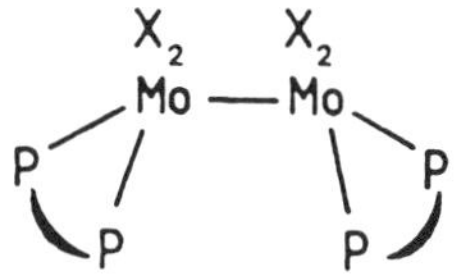   5

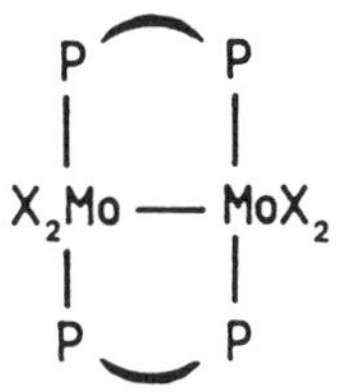

6

8.1.4. Technetium

In recent months there has been much less interest in kinetics and mechanisms of substitution at technetium in its numerous oxidation states. Interest has rather been concentrated in redox and disproportionation processes. Several workers have sought the elusive 6+ oxidation state. Evidence has been obtained for a transient chloro-oxo-Tc(VI) species in reduction of pertechnetate with thionyl chloride or phosphorus oxychloride.[8] Nitrido-anions [TcNX$_4$]$^-$, with X = Cl or a Cl/Br mixture, are less unstable, but the halide ligands are still fairly labile.[9,10] A kinetic study of ligand exchange at these nitridotechnetium(VI) halide anions is in progress.[10] Technetium(V) complexes are generally based on the TcO^{3+} unit. In octahedral derivatives, the ligand *trans* to the oxygen is often quite labile, the other four ligands less so—the oxide ligand is, as so often in oxocations, extremely reluctant to leave. Factors affecting the lability of the ligand *trans* to oxide have been discussed, particularly in relation to kinetic control of synthesis of ternary bidentate-terdentate ligand complexes [TcO(LL)(LLL)]. Chain length is a useful variable in this context.[11] The half-lives for the rearrangement reactions of oxotechnetium(V) products of tin(II) reduction of pertechnetate in the presence of hedp (hydroxyethylidene diphosphonate, 7) are several minutes.[12] The replacement of coordinated cystine by cysteine at technetium(V) is the final step in the three-stage kinetics of reduction of pertechnetate by cysteine.[13]

Some information on pH effects on hydrolysis rates of technetium(IV)-1,7-bis(2-pyridyl)-2,6-diazaheptane (dptn, 8) species is available in a report on tin(II) chloride reduction of pertechnetate in the presence of this ligand; rate constants are derived for base hydrolysis of cationic and uncharged TcO_2–dptn species.[14] Investigations of relevance to mechanisms of substitution at technetium complexes reported in recent years can be tracked down through two general reviews of progress in technetium chemistry.[15]

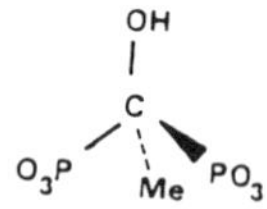

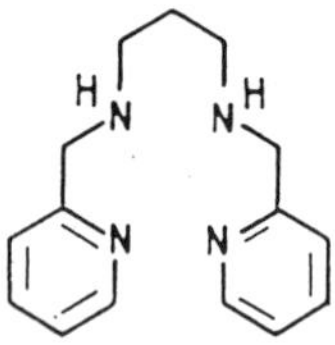

8.1.5. Rhenium

Exchange of labeled oxygen with the $[ReO_2(py)_4]^+$ cation is about ten times faster than at its ethane-1,2-diamine analogue $[ReO_2(en)_2]^+$. Retardation at low pH is attributed to reversible loss of pyridine and its subsequent protonation. The activation enthalpy of $137 \, kJ \, mol^{-1}$, in neutral solution, reflects the strength of the rhenium–oxygen bonds in the *trans*-ReO_2 oxocation unit (cf. Tc^{3+} in Section 8.1.5. above).[16] Photochemical reaction of the 2,6-dimethylphenylisocyanide complex of rhenium(I), $[Re(CNxyl)_6]^+$, with halide occurs by a dissociative process involving a reactive d-d excited state.[17]

8.2. Iron

8.2.1. Pentacyanoferrates(II)

Rate constants, with enthalpies and entropies of activation where available, are listed for dissociation and formation of pentacyanoferrate ions in Table 8.1.[18–21] Rate constants for the nicotinamide (**9**) and 4-picolylamine (**10**) complexes are as expected, but the kinetic data for the protonated and binuclear 4-picolylamine complexes are surprising. The protonated ligand would be expected to dissociate markedly more slowly than 4-picolylamine itself, rather than slightly more rapidly, while binuclear species of this type generally dissociate more rapidly than their mononuclear analogues. The tris-bipyrazine-ruthenium(II) cation (bipyrazine, bpz = **11**) can add up to six $[Fe(CN)_5]^{3-}$ units; kinetic data in Table 8.1 refer only to addition and dissociation of the first of these. The second and subsequent $[Fe(CN)_5]^{3-}$ moieties add successively more slowly, as the charge product becomes less favorable. By the time the sixth $[Fe(CN)_5]^{3-}$ is added, the formation rate constant is down to $0.3 \, M^{-1} \, s^{-1}$. Such slow reaction is hardly surprising in view of the $z_A z_B$ value of 39! Charge product is barely relevant to the reverse dissociation

Table 8.1. *Kinetic Parameters for Dissociation of $[Fe(CN)_5L]^{n-}$ Anions and of Their Formation from $[Fe(CN)_5(OH_2)]^{3-}$, in Aqueous Solution at 298.2 K unless Otherwise Stated*

		Dissociation		Formation
L	$10^3 k$ (s^{-1})	$\Delta H^{\ddagger}$ (kJ mol^{-1})	$\Delta S^{\ddagger}$ (J K^{-1} mol^{-1})	k_f (M^{-1} s^{-1})
Nicotinamide (9) *a*	1.75	107	+104	
b	4.33	89	+25	
c	0.47	107	+137	
4-Picolylamine (10)	0.78			363
4-Picolylammonium	0.97			
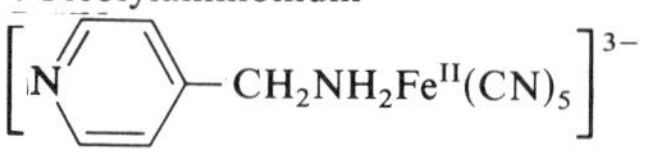 $\left[\text{N} \bigcirc - CH_2NH_2Fe^{II}(CN)_5 \right]^{3-}$	0.63			
$[Ru(bpz)_3]^{2+}$ (bpz = 11)	0.46			35,000

a In water.
b In aqueous acetonitrile, x(MeCN) = 0.249.
c In aqueous glycerol, x(glycerol) = 0.217.

process, and indeed there is only a very small range of rate constants for loss of the sixth, fifth, ... $[Fe(CN)_5]^{3+}$.[21]

It may be taken that the above formation reactions have I_d (Eigen-Wilkins) mechanisms, but it is not always clear whether the various authors have distinguished between D and I_d mechanisms for ligand loss from $[Fe(CN)_5L]^{n-}$. In recent years it has become apparent that the evidence for the operation of the limiting D mechanism in some of the earlier studies of pentacyanoferrates(II) and pentacyanocobaltates(III) was less unequivocal than then believed. It has now been shown that the limiting D mechanism does indeed operate for reaction of $[Co(CN)_5(N_3)]^{3-}$ with thiocyanate in acidic solution.[22] Whatever the precise mechanism, bond strength will have a predominant influence on ligand dissociation rate constants. The relation between spectra, bond strengths, and rate constants for dissociation has been reviewed for pentacyanoferrates(II), in relation to bioinorganic systems.[23] The relation between bonding (σ- and π-) and dissociation reactivity has also been explored for the specific case of the *N*-methylpyrazinium cation radical (12) complex of pentacyanoferrate(II).[24]

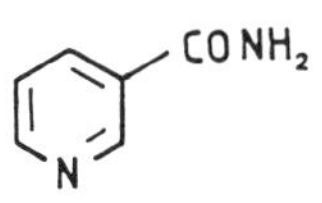

9 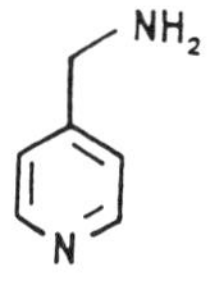   10 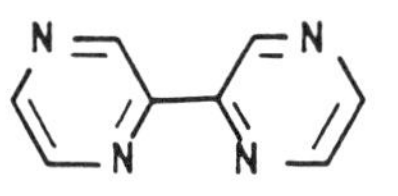   11 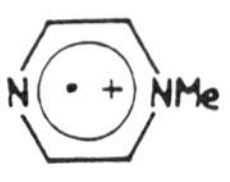   12

Dissociation of nicotinamide (9) from its $[Fe(CN)_5(nic)]^{3-}$ derivative has been studied in a variety of binary aqueous solvent mixtures, with cosolvents methanol, t-butyl alcohol, glycol, glycerol, acetone, and acetonitrile. Attempts to establish a Grunwald–Winstein correlation revealed that not only the magnitude but also the sign of the slope (m) of the appropriate plot depended on the cosolvent. Correlation with G^E values of the respective solvent mixtures was also explored. Solvation of the leaving ligand plays a key role in determining reactivity. Nicotinamide is predominantly hydrophobic; the solvatochromism of $[Fe(CN)_5(nic)]^{3-}$ is consistent with preferential solvation by the organic cosolvents.[18] A report of greatly increased rates of air oxidation of cyanoferrate(II) complexes on going from water as solvent to aqueous DMSO or DMF counsels appropriate precautions if dissociation kinetics are studied in such binary solvent media.[25]

The photochemistry of the $[Fe(CN)_5(NO)]^{2-}$ anion (nitroprusside) continues to exert its fascination. In deaerated water, irradiation at 313 or 365 nm results in photoaquation and photooxidation, the products being $[Fe(CN)_5(OH_2)]^{n-}$, $n = 2$ and 3.[26] In nonaqueous solvents such as methanol, glycol, dimethylformamide, and dimethyl sulfoxide, ligand replacement is accompanied by photooxidation.[27] The loss of the NO ligand in these reactions is characterized by activation volumes between +5 and +9 $cm^3\ mol^{-1}$. This suggests a D mechanism; photosubstitution can compete with photooxidation when the solvent is as nucleophilic as water or pyridine.[28] In pulse radiolysis of nitroprusside, ligand dissociation follows on from intramolecular electron transfer.[29] The other aspect of nitroprusside chemistry that continues to flourish is that of the kinetics and mechanisms of reactions at the coordinated nitrosyl ligand. Reaction with nitrite has been probed through ^{13}C and ^{15}N NMR spectroscopy; both NO_2^- and OH^- attack at the NO ligand.[30] Similarly, there is parallel attack by hydroxide in the reaction with secondary amines. Here the latter give a nitrosamine which can react with the $[Fe(CN)_5(NHR_2)]^{3-}$ product.[31] Kinetics of reaction with the carbanions from $CH_2(CN)_2$, $EtCH(CN)_2$, and $CH_2(CO_2Me)_2$ have also been established.[32] Similar reactions of analogous pentacyanoferrates(III) and pentacyanoruthenates(II) are mentioned below (Sections 8.2.4 and 8.3).

8.2.2. Iron(II)–Diimine Complexes

Dissociation (aquation) of $[Fe(bipy)_3]^{2+}$, $[Fe(phen)_3]^{2+}$, and their ligand-substituted derivatives is still under scrutiny and discussion. Ion-pairing effects on aquation rate constants for $[Fe(phen)_3]^{2+}$ in the presence of arenesulfonates have, with the help of ancillary NMR evidence, been explained in terms of stacking interactions between the phenanthroline ligands and the arene moieties of the anions tucked into the intraligand pockets of the complex.[33] Arrhenius plots for $[Fe(phen)_3]^{2+}$ and its less symmetrical 2-methyl derivative have been used to illustrate heirarchical levels in "Molecular System

Organization".[34] Racemization of $[Fe(phen)_3]^{2+}$ is faster than dissociation. This has long been attributed to parellel intramolecular and dissociative pathways for racemization. This contrasts with, for instance, $[Ni(phen)_3]^{2+}$, where equality of rate constants for racemization and dissociation indicate a purely dissociative mechanism for the former. It has now been shown that the tris(4,7-diphenyl-1,10-phenanthroline-disulfonate)iron(II) cation undergoes racemization and dissociation at the same rate, in contrast to its unsubstituted parent. Activation parameters for the common rate-determining bond-breaking step are $\Delta H^{\ddagger} = 130 \text{ kJ mol}^{-1}$ and $\Delta S^{\ddagger} = +90 \text{ J K}^{-1} \text{ mol}^{-1}$.[35]

The authors of the papers cited in the preceding paragraph accept the well-established mechanisms for aquation of these much-studied complexes. Two recent papers call established dogma into question. A discussion of the kinetics of replacement of edta by phen suggests, oddly, that slow attachment of the third phen ligand to the Fe(II) can be attributed to steric factors rather than to the normally invoked high-spin → low-spin change.[36] More controversial is Lahiri's claim[37] that both Krumholz[38] and Baxendale and George[39] in their classic kinetic studies of the $Fe^{2+}/2,2'$-bipyridyl system contrived to get the rate law for $[Fe(bipy)_3]^{2+}$ formation wrong. However, Lahiri's rate law indicates that the rate-limiting step in the formation of this complex must be the formation of the mono-bipyridyl complex, a conclusion which flies in the face of a great deal of chemical evidence and intuition. It is further claimed[37] that both Basolo and Sykes proposed[40] an erroneous mechanism for the dissociation of $[Fe(bipy)_3]^{2+}$ in acidic aqueous solution.

The mechanism of base hydrolysis of low-spin iron(II)–diimine complexes has been a matter of lively controversy for many years now (cf. earlier Volumes in this Series). As this question remains unresolved it is a trifle worrying to find an experiment on hydroxide attack at the tris(4,7-diphenyl-1,10-phenanthroline)iron(II) cation recommended for student use.[41] Admittedly it is presented simply as an exercise in two-stage kinetics, with any student seeking an explanation of the kinetic pattern having to refer back to earlier papers.[42] The whole topic of covalent hydration and related phenomena in the chemistry of heteroaromatic diimine metal complexes has again been reviewed,[43] this time with particularly clear reference being made to the irreversible ligand changes which appear always to follow nucleophilic attack by groups such as hydroxide or cyanide at coordinated, activated diimine ligands. The fundamental factors underlying such activation have been rehearsed.[44] One should bear in mind that the role of covalent hydration in organic reaction mechanisms is still under scrutiny, as for instance in a recent study of aqueous bromination of 2- and 4-pyrimidinones (2-pyrimidinone = 13).[45] There may be reservations over the direct attack of hydroxide ion at unactivated coordinated diimines, but there is plenty of evidence for such attack by the hydroxyl radical. Recently described cases include reaction with the $[Fe(phen)_3]^{2+}$ [46] and $[Ru(bipy)_3]^{2+}$ [47] cations, and with the cobalt(II), cobalt(III), and copper(II) complexes of the terpyridyl-like macrocyclic ligand

(14).[48] Three recent publications deal with [Fe(phen)$_2$(CN)$_2$], the product of nucleophilic attack of cyanide (somewhere) at the [Fe(phen)$_3$]$^{2+}$ cation.

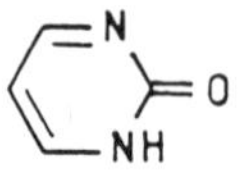

13   14

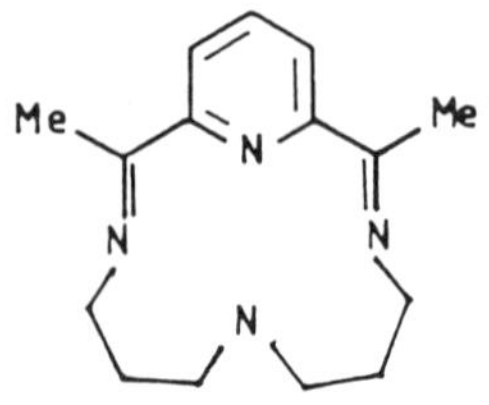

Spectra and circular dichroism of [Fe(phen)$_2$(CN)$_2$] and a range of its Lewis acid adducts establish configurations and indicate configurational inversion in cyanide attack at [Fe(phen)$_3$]$^{2+}$.[49] It is interesting that the dihydrate of [Fe(phen)$_2$(CN)$_2$] is remarkably reluctant to lose water of crystallization, as though this were bonded particularly and unusually strongly (covalent hydration?).[50] Unfortunately it proved impossible to establish the structure of this hydrate, or of its close relation [Fe^{III}(phen)$_2$(CN)$_2$]NO$_3$$xH_2$O, by X-ray diffraction.[51] These ternary iron(II)–diimine–cyanide complexes tend to be remarkably inert to substitution; a little qualitative information on the extreme slowness of solvolysis of [Fe(phen)$_2$(CN)$_2$] in binary aqueous solvent mixtures has recently been published.[52]

Alkyl substituents in the 2,2'-bipyridyl ligands of the [Fe(bipy)$_3$]$^{2+}$ cation have only a small effect on activation volumes for aquation. These increase from +14.8 cm^3 mol^{-1} to +15.5, +16.9, and +17.4 for the 4,4'-dimethyl, 4,4'-diethyl, and 5,5'-dimethyl derivatives, respectively, in 0.01 M HCl.[53] It is slightly surprising that methyl substitution in bipy ligands has the opposite effect from methyl substitution in phen ligands—$\Delta V^{\ddagger}$ values for aquation of [Fe(phen)$_3$]$^{2+}$ and for [Fe(4,7-Me$_2$phen)$_3$]$^{2+}$ are +15.4 and +11.6 cm^3 mol^{-1}, respectively.[54]

It is possible to incorporate three 2,2'-bipyridyl units into an encapsulating ligand. The ligand (15) has two rather large capping groups, suggesting a large cavity in the free ligand. However, the incorporation of an Fe^{2+} ion pulls the diimine groups toward itself, to give a coordination environment around the Fe^{2+} that, from the visible absorption spectrum, must resemble [Fe(bipy)$_3$]$^{2+}$ very closely. As the ligand (15) is more flexible than most encapsulating ligands, it can form its Fe^{2+} complex relatively quickly—formation is complete within an hour. But the complex once formed is extremely inert, undergoing dissociation several orders of magnitude more slowly than [Fe(bipy)$_3$]$^{2+}$.[55] The kinetics and mechanism of formation of the similar encapsulated complex (16) have been investigated, but this time from ligand components cyclohexane-1,2-dione-dioxime (nioxime), borate, and Fe^{2+}. The rate law is rather simple

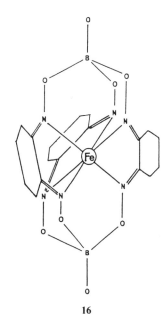

15   **16**

for such a complicated sequence[56]:

$$+d[\mathbf{16}]/dt = k[Fe^{2+}][nioxH_2]^3[H_3BO_3][H^+]^{-1}$$

The very low activation energy, reported to be $2\,kcal\,mol^{-1}$, must reflect a combination of various enthalpies of activation and of reaction for several steps. Two ternary diimine–iron(II) complexes appear in the next section, since their substitution reactions involve the nondiimine groups.

Kinetic studies of tris-diimine–iron(II) complexes in binary aqueous solvent media have dealt with cosolvents methanol, dioxan, and N,N-dimethylformamide. For the tris(4,7-diphenyl-1,10-phenanthroline-disulfonate)iron(II) cation, racemization becomes faster, but dissociation slower, as the methanol content increases. The racemization reactivity trend parallels that for $[Fe(phen)_3]^{2+}$; a reluctance to dissociate in methanol-rich mixtures probably arises from the presence of the six strongly hydrophilic sulfonate groups on the periphery of the complex.[35] Hidden within a mass of transfer chemical potentials for a range of simple and complex ions are initial state–transition state analyses of solvent composition effects on hydroxide attack at the tris-diimine complexes of the hydrophobic ligand (**17**) and the small and hardly hydrophobic ligand gmi, (**18**). Effects of single-ion assumptions on such

17   **18**

analyses are explored for $[Fe(gmi)_3]^{2+}$ and $[Fe(phen)_3]^{2+}$.[57] This analysis, for bimolecular hydroxide attack, complements the analogous treatment of dissociative substitution at cobalt(III) in the first Volume of this series of Reports.[58] Moving on from methanol–water mixtures to methanol itself, it has been found that the activation volume for solvolysis (dissociation) of $[Fe(phen)_3]^{2+}$ is almost the same in methanol ($+14.1$ cm^3 mol^{-1}) as in water ($+15.4$ cm^3 mol^{-1}).[53] Returning to solvation effects in the initial and transition states, the slightly reduced polarity of the transition state for $[Fe(phen)_3]^{2+}$ solvolysis, and consequent modest decrease in solvation, appears to offer a satisfactory rationalization of the observed variation in rate constant with solvent composition in aqueous dioxan. Such an explanation is compatible with the linearity of a logarithm of rate constant versus E_T plot. The relation between rate constants and excess Gibbs function (G^E) ties in with this solvation picture and the normally accepted picture of effects of dioxan on water structure.[59] Finally, in this section on mixed aqueous solvents, variation in formation rate constant with mole fraction dimethylformamide can be related to changes in composition of solvated iron(II) species $[Fe(OH_2)_k(DMF)_{6-k}]^{2+}$. Sparsity of results on pH effects on rate constants for the reverse solvolysis precludes a full understanding of the factors involved.[37]

Several publications deal with micellar effects on aquation and base hydrolysis of $[Fe(phen)_3]^{2+}$ and its ligand-substituted derivatives. Sodium dodecyl sulfate has very little effect on rate constants, for aquation or for hydroxide attack at $[Fe(phen)_3]^{2+}$,[60] or for aquation of $[Fe(3,4,7,8-Me_4phen)_3]^{2+}$.[61] The small hydrophobic and electrostatic interactions between micelles and these complexes will operate similarly on the initial and transition states for dissociative aquation. For base hydrolysis, negligible micelle–hydroxide interactions will again result in very limited effects on rate constants. By way of contrast, the positively charged micellar agents cetyltrimethylammonium bromide, chloride, hydroxide, and cyanide produce marked accelerations in rates for hydroxide or for cyanide attack at $[Fe(4,7-Ph_2phen)_3]^{2+}$ and at $[Fe(3,4,7,8-Me_4phen)_3]^{2+}$. Reactivity differences between these two complexes can be attributed to their different binding constants to the micelles.[62] Cationic micellar agents also greatly accelerate hydroxide attack at ligand-sulfonated derivatives of the $[Fe(4,7-Ph_2phen)_3]^{2+}$ cation.[63] Here the cationic micelles have a high affinity for the now negatively charged iron(II) complex as well as the hydroxide. Kinetics of hydroxide attack at $[Fe(phen)_3]^{2+}$ and its $4,7-Me_2$ and $3,4,7,8-Me_4$ derivatives have also been studied in microemulsions.[64] Biphasic kinetics are attributed to adsorption at the interphase preceding attack by the hydroxide. This semiphysical scheme contrasts with the two-stage mechanism involving preliminary attack of hydroxide (or cyanide) at the coordinated ligand proposed for some of the micelle-mediated reactions mentioned above. Choice between the two mechanisms is based on the observed kinetic consequences of varying the hydrophobicity and the electron-withdrawing powers of ligand substituents.

8.2.3. Other Low-Spin Iron(II) Complexes

The complex $[Fe(LLLL)(MeCN)_2]^{2+}$, where LLLL = the macrocyclic bis-diimine ligand (**19**), reacts with thiocyanate or with imidazole by a mechanism close to the dissociative limit. This is indicated by activation volumes of $+10.1$ and $+13.6$ cm^3 mol^{-1}, respectively. These values are considerably more positive than the activation volume for MeCN exchange at $[Fe(MeCN)_6]^{2+}$, which must therefore be deemed to take place by a mechanism more of the I_d type.[65] There continues to be considerable effort to establish a full picture of substitution reactivities in bis-dimethylglyoxime and related complexes of formula *trans*-$[Fe(dioximate)_2LL']$. For more than 50 complexes of this type derived from benzoquinone dioxime and from naphthoquinone dioxime (**20**), with

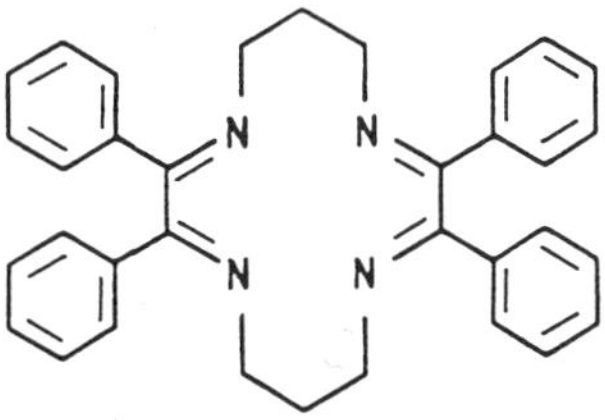

19

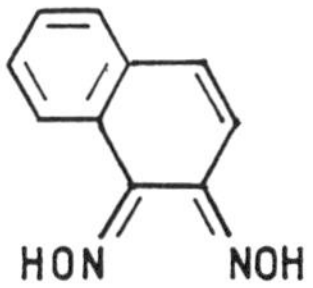

20

L and L' taken variously from nitrogen and phosphorus bases, carbon monoxide, and tosylmethylisocyanide, kinetic parameters for substitution and charge-transfer spectra taken together are compatible with dissociative activation in which π-bonding plays a key part.[66] A slightly more limited investigation of the parent dimethylglyoxime series, a mere 24 complexes this time, leads to a similar *trans*-labilizing order of ligands, again with π-acceptor ligands being strongly delabilizing.[67] The complex $[Fe(dmgH)_2(imidH)_2]$ proves difficult to study, as it is prone to undergo autoxidation unless a large excess of imidazole is present.[68] The importance of π-bonding in determining *trans*-labilization in a series of complexes $[Fe(ppdme)LL']$, where ppdme = protoporphyrin IX dimethyl ester, has been compared with the situation in the dioxime complexes mentioned above.[69] Thermodynamics and kinetics of interaction between iron(II)–"basket handle porphyrins" and uncharged nitrogen bases,[70] tetrahydrofuran,[71] chloride (in a selection of solvents),[72] and dioxygen[73] have been examined, with particular respect to the role of solvation and how this may be affected by the nature, ester-linked or amide-linked, of the "handle." Rate constants for the reaction of dinuclear bis[porphinatoiron(II)] complexes with carbon monoxide have been compared with data for reaction of carbon monoxide with myoglobin and with hemoglobin.[74] Kinetic data for the reverse carbon monoxide loss have also been obtained, as have data for loss of carbon monoxide and of dioxygen from synthetic heme complexes.[75] Finally, in this paragraph on porphyrins and related ligands it may be noted that kinetic data have been presented on analogous nickel(II) systems.[76]

Linking the previous paragraph to this one is an NMR study of the formation of linear polymers from reactions of the phthalocyanines of iron(II) and of ruthenium(II) with the 1,4-diisocyanobenzenes (21), X = Cl or CN. Dimeric and trimeric intermediates were characterized.[77] Photochemical replacement of methyl isocyanide by acetonitrile gives $[Fe(CNMe)_5(NCMe)]^{2+}$ and $[Fe(CNMe)_4(NCMe)_2]^{2+}$ from $[Fe(CNMe)_6]^{2+}$. If bipy is present, $[Fe(CNMe)_4(bipy)]^{2+}$ is formed,[78] which reacts with hydrazine in acetonitrile solution to give the dicarbene complex (22).[79] Both the dicarbene and the bipy complexes react under photochemical conditions with acetonitrile with the replacement of CNMe by NCMe.[79] A succinct overview of this group's work on these isocyanide reactions is given at the end of the former paper.[78]

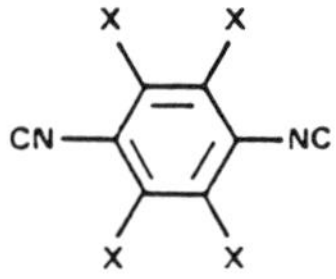

21 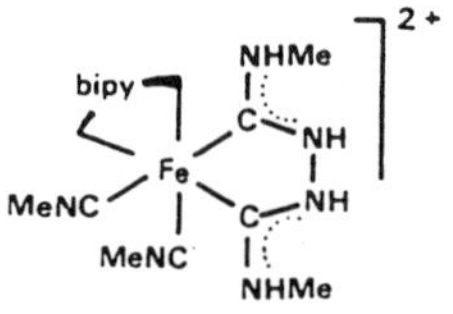   22 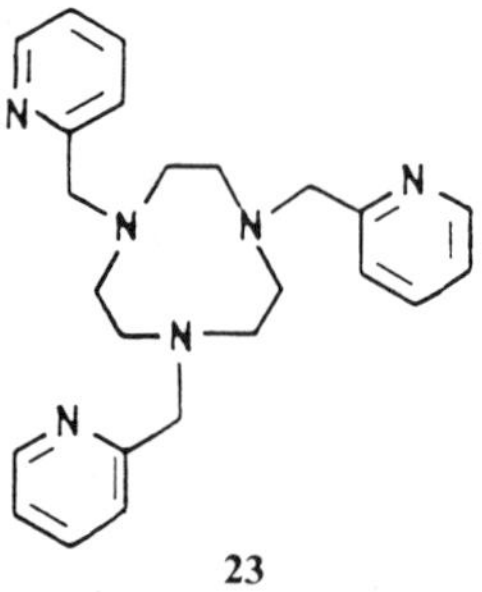   23

In contrast to the dissociative processes discussed in the preceding paragraphs, racemization of $[Fe(tptcn)]^{2+}$, where tptcn = the tris-pendant ligand (23), takes place by an intramolecular mechanism. The rate constant is unusually fast for racemization at low-spin iron(II), presumably because this complex seems to be near the low-spin/high-spin crossover. A fairly high activation energy has been ascribed to the complicated concerted ligand movement required in what is believed to be a Bailar twist.[80]

8.2.4. *Iron(III) Complexes*

Both hexacyanoferrate(II) and hexacyanoferrate(III) form bis-adducts with $[Fe(LLLL)]^+$, where $LLLLH_2$ = the macrocyclic ligand (24). These adducts are thermally stable, but on irradiation of the hexacyanoferrate(II) adduct cyanide transfer occurs, giving $[Fe(LLLL)(CN)_2]^-$.[81] Photolysis products and mechanisms have been fairly well established for hexacyanoferrate(II), but less is known about photolysis of its iron(III) analogue. Recent work has shown that the $[Fe(CN)_5(OH_2)]^{3-}$ and $[Fe(CN)_5(OH)]^{4-}$ produced by photolysis of $[Fe(CN)_6]^{3-}$ in aqueous media react thermally to give dinuclear species.[82] Reaction of iron(III)–edta derivatives with cyanide leads first to hexacyanoferrate(III), but then, thanks to reduction by liberated edta-type

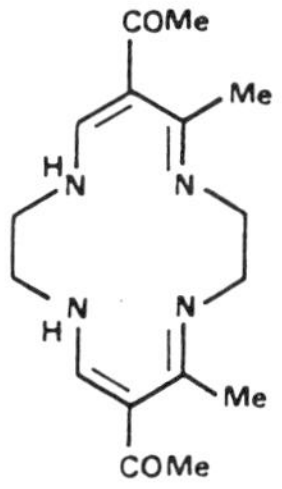

24

ligand, to hexacyanoferrate(II). Kinetics and mechanisms have been described for the complexes derived from hedta (N-hydroxyethylethylenediaminetriacetate), dtpa (diethylenetriaminepentaacetate),[83] pdta (diaminopropanetetraacetate),[84] and triethylenetetraminehexaacetate.[85] Kinetic studies of iron(III)–porphyrin species have included those entailing reaction with cyanide,[86] imidazoles,[87] and amines.[88] For the first of these, and the third in benzene solution, the addition of the second ligand is rate-determining; for the second it is possible to detect transient [Fe(pp)(imid)Cl] intermediates at −78 °C. The oxygen-transfer catalyst *meso*-[tetrakis(2,6-dimethylphenyl)porphinato]iron(III) operates via preequilibrium substitution at the iron(III) prior to rate-determining oxygen transfer.[89] Kinetic parameters have been reported for the removal of iron(III) from the two nonequivalent sites in transferrin by the sulfonatocatechol ligand 3,4-licams. The results here[90] parallel those of an earlier study of the removal of iron(III) from transferrin by edta.[91]

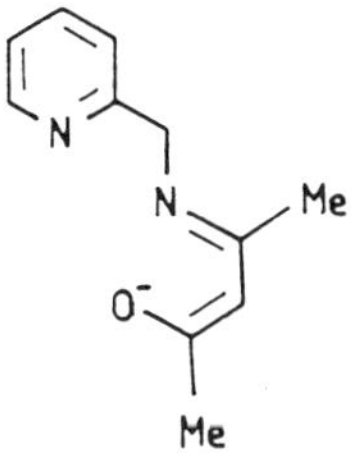

25

Information on intramolecular motions associated with $S = \frac{1}{2}, \frac{5}{2}$ interconversion in the near-crossover iron(III) complex [Fe(acpa)$_2$]$^+$, acpa = **25**, is of interest in its own right and is also relevant to analogous processes in similar iron(II) species (cf. Section 8.2.2).[92] Barriers for $C_{2h} \rightleftharpoons C_{2v}$ interconversions have been estimated for [Fe$_2$(SMe)$_2$(NO)$_4$] and [Fe$_2$(SeMe)$_2$(NO)$_4$] in nine solvents. Barriers are some 10 kJ mol^{-1} higher for the selenium compound.[93] These results can be compared with those for inversion at sulfur and selenium coordinated to such centers as platinum(IV) (Section 8.7.2). The lability of the nitrosyl groups in the mixed valence analogue [Fe$_2$S$_2$(NO)$_4$]$^-$ ("Roussin's red anion"), in [Fe$_4$S$_3$(NO)$_7$]$^-$ ("Roussin's black anion"), and in [Fe(NO)-

($S_2CNMe_2)_2$] has been established, in the last-named case through the rapidity of exchange with added nitrite.[94]

8.3. Ruthenium

The large corpus of knowledge on kinetics of substitution at pentacyanoferrate(II) anions is at last beginning to be complemented by parallel studies of pentacyanoruthenate(II) analogues. The usual strongly curved plots of k_{obs} against incoming ligand concentration have been plotted for replacement of a range of Group V bases L from $[Ru^{II}(CN)_5L]^{n-}$ anions. Rate constants cover only a small range, from 2.4×10^{-5} to $7.3 \times 10^{-5}\,s^{-1}$ for L = pyrazine and N-methylpyrazinium, respectively, with various pyridines leaving at intermediate rates. These rate constants are, of course, considerably less than those for iron(II) analogues under analogous conditions.[95] Rate constants for formation from $[Ru(CN)_5(OH_2)]^{3-}$ have also been determined, for a range of Group V entering ligands[95-97] and for DMSO.[97] Some qualitative comments on the lability of water in the $[Ru(C_6H_6)(OH_2)_3]^{2+}$ cation[98] and on photoaquation of the benzene in the $[Ru(C_6H_6)(NH_3)_3]^{2+}$ cation[99] have been made. Kinetics of replacement of coordinated water in the tetraamine complexes *trans*-$[Ru(NH_3)_4(P\{OR\}_3)(OH_2)]^{2+}$, by isonicotinamide[100] and by thiatriazolate, (**26**),[101] have been determined. There is a steady but small increase in rate as the group R increases in size from methyl up to *n*-butyl.[100] Replacement of trialkyl phosphite by sulfite leads to much faster, by bisulfite (HSO_3^-) to rather slower, anation.[101] The reverse aquation reactions are believed to go by dissociative interchange, as indeed are the formation reactions.[100,101] Kinetics of reaction of the $[Ru(NH_3)_5(OH_2)]^{2+}$ cation electrostatically bound into a Nafion (polyelectrolyte) film with various pyridine derivatives have been determined.[102,103] Interpretation of the observed modest rate enhancements is not simple, but should yield useful insights into details of the binding of complexes to modified electrodes, a key but little investigated feature of electron-transfer kinetic studies at such electrodes.[103]

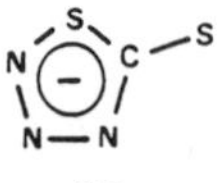

26 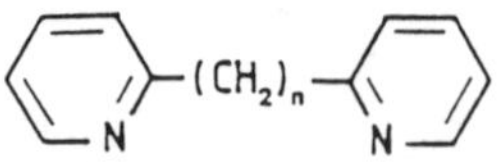   27 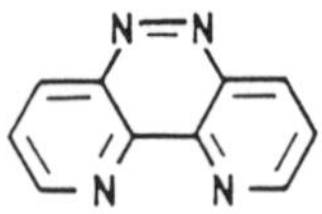   28

There has been the customary activity in relation to substitution at ruthenium(II)–diimine complexes, generally in connection with the photochemistry and photophysics of ions of the $[Ru(bipy)_3]^{2+}$ family. Laser photolysis of the bipyrazine (**11**) complex $[Ru(bipz)_3]^{2+}$ in chloride-containing acetonitrile readily gives $[Ru(bipz)_2(MeCN)Cl]^+$.[104] Ring-opening and ring-closure kinetics have been studied under thermal and photochemical conditions for cations $[Ru(bipy)_2(LL)]^{2+}$, where LL = methylene-bridged bis-pyridine ligands (**27**).[105] Other mixed ligand tris-diimine complexes have been developed in efforts to find complexes which are photoredox active but resistant to photosolvolysis. Ligands examined included methyl and carboxylate ester derivatives of bipy, and taphen (**28**).[106] Complexes containing 2,2'-bipyridyl-4,4'-carboxylate, and ternary complexes of the $[Ru(bipy)_2(CN)_2]$ type, are interesting in that protonation or metalation at coordinated ligands may significantly affect kinetic properties.[107] Other coordinaated ligand reactions include NMR studies of sites and reactivities for H/D exchange at $[Ru(terpy)_2]^{2+}$,[108] and interactions of the various noncoordinated nitrogens of $[Ru(bipz)_3]^{2+}$ with $[Fe(CN)_5]^{3-}$ (cf. Section 8.2.1 above).[21] Kinetic studies involving substitution of nondiimine ligands in ternary complexes of this type include those of thermal and of photochemical replacement of water and chloride ligands in the $[Ru(bipy)_2(CO)(OH_2)]^{2+}$ and $[Ru(bipy)_2(CO)Cl]^+$,[109] and of some ligand replacement reactions of ruthenium in dubious and high oxidation states. Electrochemical reduction of *cis*-$[Ru(bipy)_2Cl_2]$ and of *cis*-$[Ru(bipy)_2LCl]^+$ (L = py, PR_3, CO, etc.) gives complexes formally of ruthenium(I). However, the observed rapid substitution of chloride in these "Ru^I" complexes is more satisfactorily viewed as strong labilization by a bipy$^-$ radical anion ligand coordinated to Ru(II).[110] There is, as ever for NO complexes, ambiguity as to the oxidation state of the metal in $[Ru(NO)-(SEt)_2Br_3]$; in chloroform there is some photodissociation of the NO ligand, which then becomes involved in oxidation of the coordinaated SEt_2, to sulfoxide and then to sulfone.[111] Facile attack by hydroxide at NO coordinated to ruthenium(II) can give oxo-ruthenium(IV) complexes. One such product, *trans*-$[RuClO(py)_4]^+$, undergoes rapid substitution in methanol solution,[112] apparently through an initial Ru(IV) → Ru(III) reduction step. The methoxy-ruthenium(III) product hydrolyzes very slowly, but other alkoxy-ruthenium(III) complexes of this type hydrolyze quickly.[113]

Kinetic studies on ternary dichlororuthenium(II) and (III) species have focused on chelating diphosphine ligands, where the novel feature is the formation of μ-chlorodiruthenium(II) species in certain sterically crowded situations,[114] and on cyclam complexes [see rhodium(III) below].[115] Reaction of ruthenium(II)-phthalocyanine with tetramethyl and tetrachloro-benzene-1,4-diisocyanides to give linear dimers and polymers represents a slightly different type of process [cf. iron(II), Section 8.2.3].[77] Kinetic parameters are available, from NMR and electrochemical techniques, for intramolecular isomerization (N1 $\rightleftharpoons$ N2) for the benzotriazole complexes $[Ru(NH_3)_5(bzt)]^{n+}$, $n = 2$ or 3, bzt = **29**. Rate constants are considerably smaller for the ruthenium(II) complexes; activation entropies are fairly negative

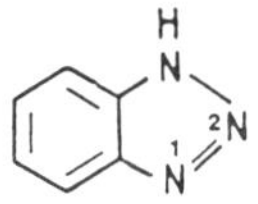

29

for the ruthenium(III) rearrangements.[116] The rate constant for isomerization of the N-urea to the O-urea linkage isomer of $[Ru(NH_3)_5(urea)]^{3+}$ is $9.1 \times 10^{-3} \text{ s}^{-1}$ at 25°C.[117]

8.4. Osmium

$[Os(NH_3)_5(\eta^2\text{-acetone})]^{2+}$, prepared from $[Os(NH_3)_5(CF_3SO_3)]^+$, is a remarkably inert species[118]; $[Ru(NH_3)_5(acetone)]^{2+}$ has been known for some time to be labile.[119] The half-life for solvolysis of the osmium complex in acetonitrile is about 1 week, while exchange with d^6-acetone is negligible in one day. However, such substitutions can be greatly accelerated by the introduction of a trace of osmium(III).[118] Solvolysis, isomerization, and oxidation state changes have been studied for a series of 2,2'-bipyridyl complexes of osmium in oxidation states II through VI. For Os(II) and Os(III) the *trans* forms, $[Os(bipy)_2(OH_2)(OH)]^+$ and $[Os(bipy)_2(OH)_2]^+$ respectively, isomerize to *cis*. For Os(VI), specifically for $[Os(bipy)_2(O)_2]^{2+}$, the *cis* isomer isomerizes to the *trans* in hot acetonitrile, but undergoes solvolysis, probably through a one-end-off mechanism for the leaving bipy, in minutes.[120] The lability of various oxidation states of osmium has also been investigated for a series of μ-oxodiosmium-tetrakis(2,2'-bipyridyl) complexes; the osmiums also have an aqua-, hydroxo-, or oxo-ligand each, depending on the osmium oxidation state. The diosmium(II) compound readily hydrolyzes to $[Os(bipy)_2)_2]^{2+}$, with the rate constant increasing as the pH increases, while the diosmium(V) complex undergoes considerable bipy loss as it hydrolyzes to mononuclear osmium(V) products.[121] Following studies of the acidity of various ligand protons in diimine–ruthenium(II) complexes and of kinetics of H/D exchange for such complexes,[108,122] rate constants for H/D exchange have been obtained, in dimethyl sulfoxide–water media, for $[Os(bipy)_3]^{2+}$. A simple second-order rate law applies, and rate constants increase with pH in line with the increasing chemical potential of OH^- (OD^-). The order of reactivity of ligand protons in this respect is[123]: $3,3' \gg 5,5' > 6,6' > 4,4'$. High reactivity of the 3,3'-protons corresponds to their high acidity.[108,123]

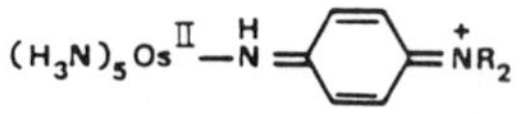

$$(H_3N)_5Os^{II}\!-\!\overset{H}{N}=\!\!\!\langle\!\!\!\rangle\!\!\!=\overset{+}{N}R_2$$

30

Recent investigations of intramolecular processes of osmium complexes include the *cis* $\rightleftharpoons$ *trans* isomerizations mentioned above,[120] stereochemical nonrigidity of $[OsH_3(PPh_3)_4]^+$,[124] and rotation about the C=N bonds in the benzoquininonediimine complex (30).[125] Unusually precise activation parameters ($\Delta H^\ddagger = 5.3 \text{ kcal mol}^{-1}$, $\Delta S^\ddagger = -16 \text{ cal deg}^{-1} \text{ mol}^{-1}$) could be obtained for the hydridophosphine complex as the signals which coalesced were separated by > 4000 Hz at low temperatures. The remarkably low barrier to the C=N bond rotation, a mere 52 kJ mol^{-1} (12 kcal mol^{-1}), was attributed to important contributions from a resonance structure formally containing osmium(IV) and C—N single bonds.

8.5. Rhodium

The debate on the geometry, square pyramidal or trigonal bipyramidal, for five-coordinate transient intermediates in photosubstitution reactions of octahedral rhodium(III) and iridium(III) continues.[126] However, study of pressure effects on photoaquation of cis-$[Rh(bipy)_2Cl_2]^+$ suggests that there may be significant associative character, though the markedly negative activation volume, -9.7 cm^3 mol^{-1}, may also be accounted for by a large solvation contribution in an essentially dissociative mechanism.[127]

Scrambling of ammonia ligands during photoaquation of $[Rh(NH_3)_5Cl]^{2+}$ has been studied through ^{15}N labeling. The ammonia lost from the rhodium comes in equal amounts from the cis and trans positions, indicating that axial ammonia is four times more efficiently labilized than equatorial. This may be contrasted with the specifically equatorial photolabilization demonstrated for $[Rh(NH_3)_5(CN)]^{2+}$.[128] Photoaquation involves parallel loss of ammonia and of X^- from $[Rh(NH_3)_5X]^{2+}$ cations, thermal aquation generally results in loss of X^- only, while addition of mercury(II) accelerates loss of halide and similar ligands greatly. It is generally assumed that mercury(II)- and silver(I)-catalyzed aquation involves short-lived $[Rh—X—M]^{n+}$ intermediates. An investigation of Ag^+-catalyzed aquation of $[Rh(NH_3)_5I]^{2+}$ shows that more than one Rh—X—Ag intermediate must be involved. Species containing such poly-nuclear units as $[RhIAg_2]^{4+}$ and $[RhIAg_3]^{5+}$ are suggested.[129] Other rhodium(III) complexes whose photochemical kinetics have been described include $[Rh(CN)_6]^{3-}$ [130] and several porphyrin derivatives. The octaethylpor-phyrin complex $[Rh(oep)Me]$ loses its methyl ligand on irradiation in degassed benzene, giving a Rh(II)-oep dimer.[131] The tetraphenylporphyrin complex $[Rh(tpp)Cl]$ takes up pyridine in ethanol but loses it, with high quantum yield, on irradiation.[132] $[Rh(oep)(PPh_3)Cl]$ undergoes thermal substitution by triphenyl- or tri-n-butyl-phosphine to give bis-phosphine products, but if it is irradiated in the presence of potential ligands (L) CO, THF, or MeCN, then PPh_3 is lost and $[Rh(oep)LCl]$ formed.[133] Tetrakis(p-trimethylammonium-phenyl)porphyrin labilizes waters also coordinated to rhodium(III) greatly, as shown in a kinetic study of reactions of $[Rh(tapp)(OH_2)_2]^{5+}$ with Cl^-, Br^-, I^-, NCS^-, and $[Co(CN)_6]^{3-}$. Reactivities are similar to those for the analogous tetrasulfonato complex, but around 500 times faster than those for $[Rh(NH_3)_5(OH_2)]^{3+}$. Dissociative anation is proposed for this $[Rh(tapp)-(OH_2)_2]^{5+}$ cation.[134]

Kinetic parameters (k, $\Delta H^\ddagger$, $\Delta S^\ddagger$) have been determined for aquation of $[Rh(NH_3)_5(propionate)]^{2+}$ [135] and for formation of $[Rh(NH_3)_5$-

(malonate)]$^{+}$ [136] in aqueous solution, and for the formation of ([H$_3$N)$_5$-Rh(NC)Co(CN)$_5$] from [Rh(NH$_3$)$_5$(OH$_2$)][Co(CN)$_6$] in the solid state.[137] Rhodium(III)–cobalt(III) reactivity comparisons suggest predominantly associative activation in the first two cases, but dissociative in the third.

A limiting D mechanism operates for pyridine anation of trans-[Rh(dmgH)$_2$Me(OH$_2$)][138] and for anation of [Rh(dmgH)$_2$R(OH$_2$)], where R = CH$_2$CH$_3$, CH$_2$CF$_3$, or CH$_2$Cl.[139] For solvolysis of [Rh(dmgH)$_2$(tu)X], X = Cl, Br, or I, in dipolar protic and aprotic solvents, the so-called "mutual exchange mechanism" appears to have a predominantly associative nature,[140] in contrast with the primarily dissociative character for the analogous cobalt(III) complexes made from nioxime rather than dimethylglyoxime.[141] Rate constants have been determined for proton exchange and for base hydrolysis of cis- and trans-[Rh(cyclam)$_2$Cl$_2$]$^{+}$ (cyclam = 31); rate constants for solvolysis and for reprotonation of the amido-conjugate base were derived. The lower reactivity of these rhodium(III) complexes in comparison with their cobalt(III) analogues derives from the much lower value of the quotient k(solvolysis)/k(reprotonation) for the conjugate bases of the rhodium complexes.[115] Base hydrolysis of the trimethyl phosphate complex [Rh(NH$_3$)$_5$(tmp)]$^{3+}$ results both in hydrolysis of the coordinated tmp and some rhodium–ligand bond cleavage. With respect to the former pathway, the [Rh(NH$_3$)$_5$]$^{3+}$ moiety gives a rate enhancement of about 400 times to tmp hydrolysis.[142]

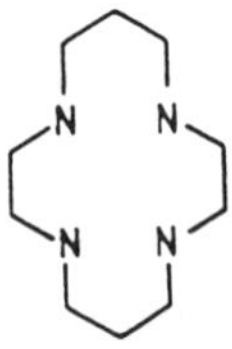

31 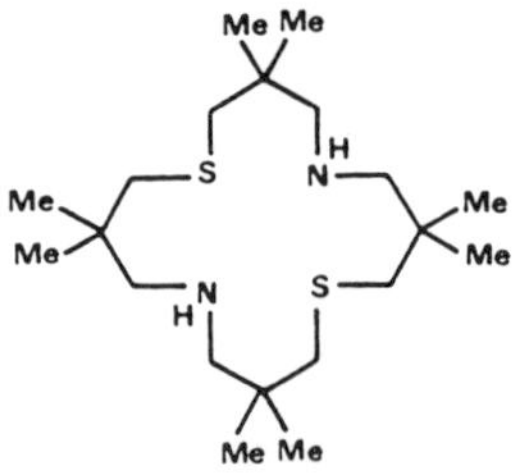   32 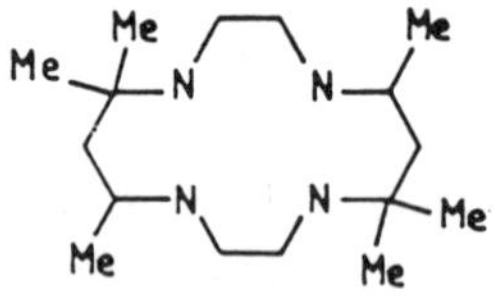   33

The rate constant for water exchange at [RhCl$_2$(OH$_2$)(PMe$_2$Ph)$_3$]$^{+}$ is, at 7×10^4 s^{-1}, some 10^{10} faster than at [Rh(NH$_3$)$_5$(OH$_2$)]$^{3+}$; a D mechanism is suggested,[143] with square-pyramidal geometry favored for the intermediate.[126] Rate constants for water exchange and for isomerization under photochemical conditions have been determined for cis- and trans-[Rh(NH$_3$)$_4$(OH$_2$)$_2$]$^{3+}$; the results are placed in the context of other aqua-ammine–rhodium(III) complexes.[144] Photoracemization of cis-L-[Rh(en)$_2$(OH)$_2$]$^{+}$ and of cis-L-[Rh(en)$_2$(OH)(OH$_2$)]$^{2+}$ has been compared with that of trans analogues. [Rh(en)$_2$(OH)]$^{2+}$ is believed to be the intermediate in each case.[145] Interconversion of configurational isomers of rhodium(III) complexes of macrocycle (32) is slow on the NMR time-scale.[146] The ligand tet-a (33) is hexamethylcyclam (cyclam = 31, see above); the six methyl substituents prevent the formation of a stable cis-[Rh(tet-a)(OH)$_2$]$^+$ cation. The photochemistry of the stable trans-isomer has been investigated to see whether it shows any steric effects from the six methyl substituents. In fact its photobehavior closely parallels that of its cyclam analogue (both are very inert to photoaquation as they contain no easily breakable Rh—N bond); both complexes exhibit unusual photoproperties.[147]

The kinetics of chelate ring opening and closing in the forward and reverse reactions (X = Br, N$_3$, NO$_2$):

$$trans\text{-}[Rh(LL)_2Br_2]^+ + X^- \rightleftharpoons mer\text{-}[Rh(LL)LBr_2X]$$

are very similar to those established some time ago for chloroanalogues; bromide has a slightly greater cis-labilizing effect in these rhodium(III) systems.[148] Rate constants and activation enthalpies for rupturing the

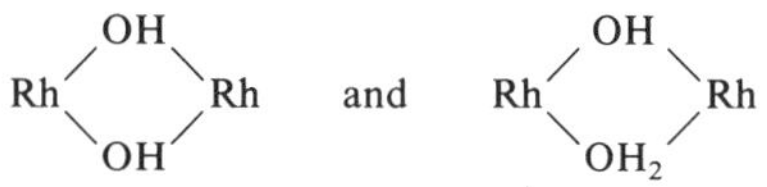

$$\text{Rh} \underset{OH}{\overset{OH}{<>}} \text{Rh} \quad \text{and} \quad \text{Rh} \underset{OH_2}{\overset{OH}{<>}} \text{Rh}$$

rings in (34) and its μ-hydroxo-μ-aqua-analogue have been measured,[149] and a mechanism proposed for cleavage of the rhodium–rhodium bond in compounds [Rh$_2$(CF$_3$CO$_2$)$_4$L$_2$], where L = a phosphine or phosphite ligand.[150]

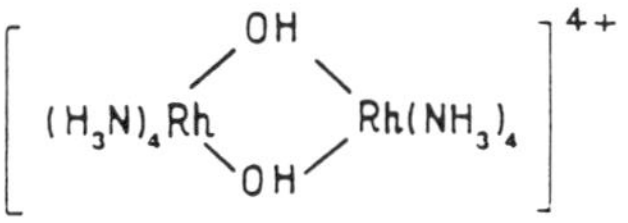

$$\left[(H_3N)_4Rh \underset{OH}{\overset{OH}{<>}} Rh(NH_3)_4 \right]^{4+}$$

34

8.6. Iridium

Comparisons of reactivities and mechanisms with rhodium(III) analogues, and often with cobalt(III) too, are provided for photosubstitution at [Ir(CN)$_6$]$^{3-}$,[130] water exchange at [IrCl$_2$(OH$_2$)(PMe$_2$Ph)$_3$]$^+$,[143] and base hydrolysis of coordinated trimethyl phosphate.[142] A contrast between iridium(III) and ruthenium(III) is provided by the reactions of mixed ligand dimethyl sulfoxide and diethyl sulfoxide complexes with dioxygen. While the latter is catalytically oxidized to the sulfone in the Ru(III)–NO–DESO complex (see above)[111], reaction of the Ir(III)–Cl–DMSO complex is that of DMSO and Cl$^-$ loss.[151]

Mechanisms of replacement of coordinated water in cations of the [M(NH$_3$)$_5$(OH$_2$)]$^{3+}$ type in solution and in the solid state have been compared and contrasted, for Cr(III), Co(III), Rh(III), as well as for Ir(III).[152] Recent Ir(III) examples include replacement by halides[153] and by [Co(CN)$_6$]$^{3-}$ [137]. Kinetics of replacement of water have also been studied for Delépine's anion, [Ir$_3$(μ$_3$-N)(μ-SO$_4$)$_6$(OH$_2$)$_3$]$^{4-}$. For reaction with chloride, bromide, or azide, the kinetic pattern is that of a single first-order step ([incoming ligand] ≫ [Ir₃anion]). This is consistent with equivalence of the three iridium atoms in this mixed valence species [though its Mössbauer spectrum at 4 K indicates localized Ir(III)Ir(IV)Ir(IV)], or with rapid replacement of the second and third water molecules. The more complicated kinetics of reaction with thiocyanate may be due to the involvement of both $-NCS$ and $-SCN$ species, as suggested on the basis of the difference between the kinetic behavior of thiocyanate and of azide,[154] or to the rate constants for replacement of the three waters being comparable but not equal. The great difficulty in satisfactorily fitting experimental absorbance data to the kinetic equations for three consecutive first-order processes[155] precludes an unequivocal answer in this Delépine system. Mechanisms and stereochemistry of halide exchange for various combinations of chloride and bromide exchange at binary and ternary hexahalide anions have been compared for iridium(III) and (IV). Stereoselectivity is more marked for iridium(III).[156] Iodide dissociation is rate-determining in the rearrangement of *sec*-alkyl to *n*-alkyl derivatives for the irdium(III) derivatives $[IrI_2R(CO)(PR'_3)_2]$.[157]

There has been singularly little activity on the photochemical substitution front, except for the demonstration of relatively long-lived transients, resulting from Ir—N bond cleavage in a photoexcited state, in pulsed laser excitation of $[Ir(bipy)_3]^{3+}$ in its normal tris-N,N-ligated and in its bis-N,N/mono-C,N-ligated forms.[158]

8.7. Nickel, Palladium, and Platinum

8.7.1. Nickel

Of the various oxidation states of nickel which are stable, or which can be stabilized by appropriate ligands, in solution, only nickel(IV) (low-spin d^6) is intrinsically inert to substitution. However, other oxidation states can, in appropriate complexes, have fairly long half-lives with respect to substitution. Both mononuclear and binuclear nickel(II)–hexamethylenediaminetetra-acetate complexes react with cyanide at "conventional" rates,[159] while the half-life for the nickel(I) complex of the macrocycle (35) (tetra-*N*-methyl-33) is well over 90 hours.[160] The lability of tetraaza-macrocyclic ligand complexes is very dependent on ring size, as well as on the number and position of methyl substituents. The ring-size effect has been demonstrated for nickel(II) and for copper(II), with [14]aneN₄ and [16]aneN₅ complexes dissociating extremely slowly,[161,162] but [17]aneN₄ and [18]aneN₄ complexes undergoing substitution on the stopped-flow time-scale.[162] Replacement of coordinated water in the nickel(III) complexes $[Ni(LLLL)(OH_2)_2]^{3+}$, LLLL = C-*meso*-5,12-dimethylcyclam (36) or its diethyl analogue, by chloride or bromide is relatively slow for a d^7 center, though still in the stopped-flow time-scale.[163]

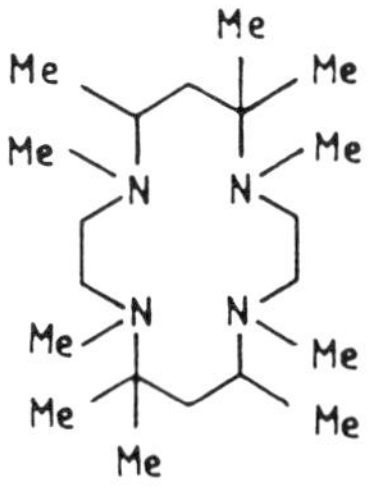

35 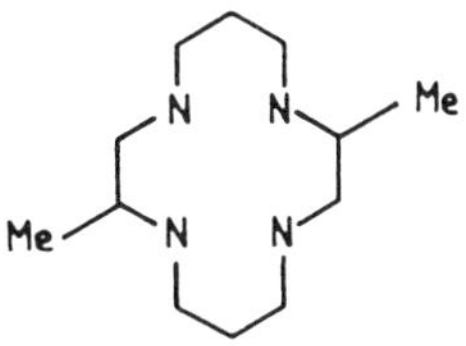   36

8.7.2. Ligand Replacement at Palladium(IV) and Platinum(IV)

Substitution at palladium(IV) is generally fast. Thus it takes place at stopped-flow rates for replacement of chloride by dimethylglyoximate in several chloropalladium(IV) species (in acetone solution), despite an activation energy as high as 133 kJ mol^{-1} for the compound written as [Pd(Ph$_2$PPhF)$_2$Cl$_4$], which is presumably [Pd(PPh$_2${C$_6$H$_4$F})$_2$Cl$_4$].[164] Different mechanisms are claimed for photoaquation of *trans*-[Pt(NH$_3$)$_4$X$_2$]$^{2+}$, depending on whether X = Cl or Br—homolysis and heterolysis respectively.[165] The photoreaction sequence for [Pt(NO$_2$)$_x$X$_{(6-x)}$]$^{2-}$ anions involves isomerization previous to photoredox.[166] Reactions of tetracyanoplatinate(IV) complexes [Pt(CN)$_4$XY]$^{n-}$ are often complicated by substitution and redox steps,[167] while interpretation of the reactivity of Pt—Pt bonded species such as [Pt$_2$(μ-PO$_4$H)$_4$(OH$_2$)$_2$]$^{2-}$ with respect to halide anation is clouded by controversy as to the oxidation state (3 + or 4 +) of the platinum.[168]

8.7.3. Intramolecular Processes

Studies of kinetics of inversion at sulfur or selenium coordinated to platinum(IV)[169] continue with unabated vigor, indeed with fresh impetus in the shape of the first paper in a series to be devoted to complexes containing terdentate ligands of the type MeE(CH$_2$)$_n$E'(CH$_2$)$_n$EMe, with E, E' − S and/or Se.[170] Details of the various intramolecular processes have been unraveled for the diplatinum species (37) with X = Cl, Br, or I, Y = Z = Me, and E = S or Se,[171] and also with X = Cl or Br, Y = H, Z = SMe, and E = S.[172] Effects of ring size have been probed for inversion at selenium in the rings (38) to (41) coordinated to palladium(IV).[173]

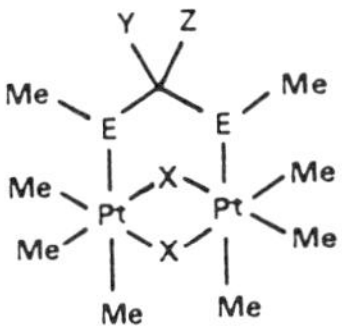

37    38    39 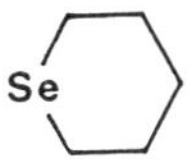   40 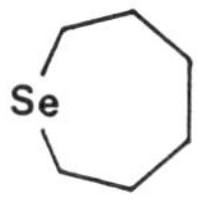   41

Comparative discussions in this area can be increasingly widely based as kinetic information on inversion at coordinated sulfur or selenium becomes available for a growing variety of environments. Recent examples include those of inversion at sulfur bound to platinum(II)[174] and to $[M(CO)_5]$ moieties,[175] at selenium bound to cobalt(III),[176] at sulfur and selenium in nitrosyliron(III) complexes (see Section 8.2.4),[93] and, in a coordinated ligand rather than directly bonded to a metal center, in the "bridge reversal inversion"

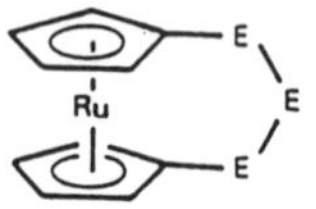

42 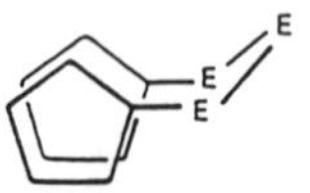   43    44

of compounds (42), E = S or Se (i.e. for the process 43 ⇌ 44).[177] Activation energies are rather similar for these ring inversions to those for inversion at sulfur or selenium directly bonded to a metal, and it is interesting to see analogous effects of ring size on inversion barriers. These decrease as ring size increases, whether this increase be due to the presence of the larger selenium atoms in (42) or to increasing numbers of carbon atoms as in the sequence of ligands (38) to (41).

Chapter 9
Substitution Reactions of Labile Metal Complexes

9.1. General

In the review period ligand substitution processes have been studied for labile metal centers ranging from $Li^{+(1)}$ to UO_2^{2+}.[2-5] The first-row transition metal ions account for approximately half of the reported studies, with nickel(II)[6-26] being particularly prominent probably as a consequence of its reactions being at the slower end of the ligand substitution time scale and hence readily experimentally accessible. By comparison with the complexation of the transition metal ions, in which directional bonding is particularly important, the complexation of alkali metal ions by crown ethers, cryptands, antibiotics, and similar ligands is more dependent on metal ion size and ligand steric characteristics, and is also an area of substantial mechanistic study.[1,27-34] Many of the systems studied have biological implications. The rapid reaction methodology employed ranges from stopped-flow and NMR spectroscopic methods to the less used ultrasonic,[20,35] pressure-jump,[36,37] pulse radiolysis pH-jump,[38] and electric field-jump[32] techniques. Activation volumes determined by a variety of variable-pressure methods[9-11,39-44] continue to prove a valuable aid to mechanistic interpretation.

9.2. Complex Formation Involving Unsubstituted Metal Ions: Unidentate Ligand Substitution and Solvent Exchange

9.2.1. Bivalent Ions

The most ubiquitous ligand substitution process, solvent exchange, continues to be investigated as exemplified by an ^{17}O NMR study of acetic acid exchange on $[M(HOAc)_6]^{2+}$ which yields, for Mn^{2+}, $k_{ex} = 1.6 \pm 0.1 \times 10^7$, $\Delta H^{\ddagger} = 29 \pm 2$, $\Delta S^{\ddagger} = -10 \pm 7$; for Fe^{2+}, $k_{ex} = 5 \pm 1 \times 10^6$, $\Delta H^{\ddagger} = 33 \pm 5$, $\Delta S^{\ddagger} = -8 \pm 20$; for Co^{2+}, $k_{ex} = 1.3 \pm 0.2 \times 10^6$, $\Delta H^{\ddagger} = 37 \pm 3$, $\Delta S^{\ddagger} = -6 \pm 12$; and for Ni^{2+}, $k_{ex} = 2 \pm 1 \times 10^5$, $\Delta H^{\ddagger} = 40 \pm 5$, $\Delta S^{\ddagger} = -10 \pm 20$, where k_{ex} is in s^{-1} at 298.2 K, and the units of $\Delta H^{\ddagger}$ and $\Delta S^{\ddagger}$ are kJ mol^{-1} and J K^{-1} mol^{-1}, respectively.[6] For Cu^{2+}, $k_{ex} \sim 10^7 \, s^{-1}$ at 248.2 K. The variation in lability of $[M(HOAc)_6]^{2+}$ with the nature of M is similar to that observed in other solvents. (Coordinating anions change the lability of coordinated HOAc.[7,45]) The effect of solvent size on lability has been examined in a ^1H NMR study[8] of diethylformamide exchange on $[Co(def)_6]^{2+}$ and $[Ni(def)_6]^{2+}$ $[10^{-3}k_{ex}(298.2 \text{ K}) = 120 \pm 16$ and $1.15 \pm 0.08 \, s^{-1}$; $\Delta H^{\ddagger} = 51.4 \pm 2.0$ and 74.2 ± 1.1 kJ mol^{-1}; and $\Delta S^{\ddagger} = 24.8 \pm 7.0$ and 62.6 ± 3.4 J K^{-1} mol^{-1}, respectively]; and in a ^{13}C NMR study of dimethylacetamide exchange on $[Co(dma)_6]^{2+}$ and $[Ni(dma)_6]^{2+}$ $[10^{-6}k_{ex}(298.2 \text{ K}) = 23.1 \pm 3.5$ and $3.52 \pm 0.70 \, s^{-1}$; $\Delta H^{\ddagger} = 32.7 \pm 0.8$ and 40.4 ± 1.8 kJ mol^{-1}; and $\Delta S^{\ddagger} = 5.8 \pm 3.0$ and 16.0 ± 10.0 J K^{-1} mol^{-1}, respectively]. In each case a D mechanism is assigned. The labilities of the def systems are similar to those of the analogous dmf systems, while those of the dma systems are substantially greater, and this is considered to result predominantly from the increased crowding at the metal center caused by the acetyl moiety of dma.

The high lability of copper(II) has resulted in a paucity of data for its monodentate ligand substitution processes. This situation has been improved through a ^{17}O NMR study of methanol exchange on $[Cu(MeOH)_6]^{2+}$ $[k_{ex}(298.2 \text{ K}) = 3.1 \pm 0.5 \times 10^7 \, s^{-1}$, $\Delta H^{\ddagger} = 17.2 \pm 0.8$ kJ mol^{-1}, $\Delta S^{\ddagger} = -44.0 \pm 4.0$ J K^{-1} mol^{-1}, and $\Delta V^{\ddagger} = 8.3 \pm 0.4$ cm^3 mol^{-1}].[40] The positive $\Delta V^{\ddagger}$ value is interpreted in terms of a d activation mode, and the negative $\Delta S^{\ddagger}$ is thought to be a consequence of the effect of the tetragonal distortion of $[Cu(MeOH)_6]^{2+}$ on either of, or both of, first and second coordination-sphere interactions. The assignment of a d activation mode to $[Cu(MeOH)_6]^{2+}$ is consistent with the trend for the activation mode for $[M(MeOH)_6]^{2+}$ to change from a to d as M varies in the sequence Mn, Fe, Co, Ni. A theoretical study of the Jahn–Teller distortions of Cu(II) complexes has been published in which it is also proposed that $[Pt(H_2O)_4]^{2+}$ and its Pd(II) analogue may be viewed as tetragonally distorted hexaaqua ions in which the axial waters are the more distant from the metal center, and that exchange of water between the axial sites and two *trans* equatorial sites occurs simultaneously through the ε_g Jahn–Teller active mode characterizing d^8 low-spin species.[46]

A stopped-flow spectrophotometric study of reaction (1) yields $k_f(298.2 \text{ K}) = 1.14 \pm 0.11 \times 10^6$ dm^3 mol^{-1} s^{-1}; $\Delta H^{\ddagger} = 17.3 \pm 1.7$ kJ mol^{-1}, and

$$[Pd(H_2O)_4]^{2+} + I^- \underset{k_b}{\overset{k_f}{\rightleftharpoons}} [PdI(H_2O)_3]^+ + H_2O \qquad (1)$$

$\Delta S^{\ddagger} = -71 \pm 5$ J K^{-1} mol^{-1}; and $k_b(298.2 \text{ K}) = 0.92 \pm 0.18 \, s^{-1}$, $\Delta H^{\ddagger} = 45 \pm 3$, and $\Delta S^{\ddagger} = -95 \pm 6$ J K^{-1} mol^{-1}.[47] An I_a mechanism is proposed for these reactions, and also for the analogous reactions of the Pt(II) system ($k_f = 7.7 \pm$ 0.4 dm^3 mol^{-1} s^{-1} and $k_b = 8.0 \pm 0.7 \times 10^{-5}$ s^{-1} at 298.2 K). It is concluded that bond breaking is less important for the Pd(II) ligand substitution processes than is the case for Pt(II). A pressure-jump study of sulfate substitution on the tetrahedral solvatomers, $[Be(H_2O)_i(glycol)_{4-i}]^{2+}$, shows that the substitution rate depends only on the nature of the solvent molecule replaced and not the composition of the solvatomer.[36]

9.2.2. Trivalent Ions

The high acidity of $[Fe(H_2O)_6]^{3+}$ results in the existence of $[Fe(H_2O)_5(OH)]^{2+}$ at low pH, which both complicates ligand substitution studies of the former species, and adds mechanistic interest as $[Fe(H_2O)_6]^{3+}$ undergoes water exchange through an a activation mode while a d activation mode operates for $[Fe(H_2O)_5(OH)]^{2+}$.[48,49] Studies of the substitution of $[Fe(H_2O)_6]^{3+}$ and $[Fe(H_2O)_5(OH)]^{2+}$ by NCS$^-$ have yielded a confusing range of $\Delta V^{\ddagger}$ values. A high-pressure temperature-jump spectrophotometric reinvestigation of this system indicates that the contribution of the $[Fe(H_2O)_6]^{3+}/NCS^-$ path, by comparison with the $[Fe(H_2O)_5(OH)]^{2+}/NCS^-$ path, to the overall substitution rate was too small for reliable determination in some of the earlier studies, and that the data treatment was inappropriate.[41] It is now found that the $[Fe(H_2O)_5(OH)]^{2+}/NCS^-$ substitution rate has a significant ionic-strength dependence. The new $\Delta V^{\ddagger}(298.2$ K) values are $4.3 \pm$ 0.6 cm^3 mol^{-1} at 1.0 mol dm^{-3} ionic strength for $[Fe(H_2O)_6]^{3+}/NCS^-$, and 5.4 to 16.5 cm^3 mol^{-1} over the ionic strength range 0.1 to 1.5 mol dm^{-3} for $[Fe(H_2O)_5(OH)]^{2+}/NCS^-$. A study of the formation of $[Fe(H_2O)_5(H_2PO_4)]^{2+}$ at pH < 2 supports the assignment of a and d activation modes for ligand substitution on $[Fe(H_2O)_6]^{3+}$ and $[Fe(H_2O)_5(OH)]^{2+}$, respectively.[50] The variation of k_f for ligand substitution on $[Ti(H_2O)_6]^{3+}$ over more than three orders of magnitude (Table 9.1), and the general increase of k_f with the basicity of the substituting ligand, is strong evidence for the operation of an I_a mechanism.[51] This deduction is supported by the observation that $\Delta V^{\ddagger} = -12.1 \pm 0.4$ cm^3 mol^{-1} for water exchange on $[Ti(H_2O)_6]^{3+}$ $[k_{ex}(298.2$ K$) = 1.81 \pm 0.03 \times 10^5$ s^{-1}, $\Delta H^{\ddagger} = 43.4 \pm 0.7$ kJ mol^{-1}, and $\Delta S^{\ddagger} = 1.2 \pm$ 2.2 J K^{-1} mol^{-1}].[52]

The high lability of the lanthanide(III) aqua ions is demonstrated by a pulse radiolysis pH-jump study of the complexation of La(III) and Gd(III) by methyl red (o-carboxybenzene-azodimethyl-aniline) which yields $k_f = 3.7 \pm 0.3 \times 10^7$ and $3.5 \pm 0.2 \times 10^7$ dm^3 mol^{-1} s^{-1}, respectively, and $k_b = 6.2 \pm 3.4 \times 10^3$ and $5.6 \pm 1.3 \times 10^3$ s^{-1} in aqueous 0.5 mol dm^{-3} t-butanol and 0.1 mol dm^{-3} NaClO$_4$ at $ca.$ 293 K.[38] By comparison with the aqua ions, six-coordinate $[Tm(tmu)_6]^{3+}$ is less labile as indicated by the parameters for tetramethylurea (tmu) exchange: $k_{ex}(298.2$ K$) = 145 \pm 1$ s^{-1}, $\Delta H^{\ddagger} = 29.3 \pm 0.3$ kJ mol^{-1}, and $\Delta S^{\ddagger} = -105 \pm 1$ J K^{-1} mol^{-1}.[53] These data are

Table 9.1. Formation Rate Constants for Ligand Substitution on
$[Ti(H_2O)_6]^{3+}$ (Reference 51)

Ligand	$k_f(dm^3 mol^{-1} s^{-1})$	Conditions[a]
$ClCH_2COOH$	0.7×10^3	288.2, 0.5
CH_3COOH	1×10^3	288.2, 0.5
NCS^-	8×10^3	281.2–282.2, 1.5
Cl_2CHCOO^-	1.1×10^5	288.2, 0.5
$ClCH_2COO^-$	2.1×10^5	288.2, 0.5
Methylmalonate(1-)	3.2×10^5	288.2, 0.5
$HC_2O_4^-$	3.9×10^5	283.2, 1.0
Malonate(1-)	4.2×10^5	288.2, 0.5
CH_3COO^-	1.8×10^6	288.2, 0.5

[a] Temperature in K, ionic strength in mol dm^{-3}.

consistent with a trend, observed for $[M(tmu)_6]^{3+}$, for k_{ex}(298.2 K) to increase, $\Delta H^{\ddagger}$ to decrease, and $\Delta S^{\ddagger}$ to become more negative as the ionic radius of M (=Sc, Lu, Tm, and Y) increases. Ligand exchange on $[Tm(tmu)_6]^{3+}$ {which is 2×10^5 times less labile than $[Tm(dmf)_8]^{3+}$} is thought to proceed through a d activation mode.

9.3. Complex Formation Involving Unsubstituted Metal Ions: Multidentate Ligand Substitution

9.3.1. Uni- and Bivalent Ions

9.3.1.1. Crown Ethers, Cryptands, Antibiotics, and Similar Ligands

The conformational properties of crown ethers, cryptands, antibiotics, and similar ligands are important in determining their complexation mechanisms, and in consequence theoretical studies of their conformation and complexation are of considerable interest. Conformational analysis shows that cryptand C111 (=4,10,15-trioxa-1,7–diazabicyclo[5.5.5]eicosane) is a rigid molecule in which the endo-endo conformation is the most stable, while the more flexible C222 (=4,7,13,16,21,24-hexaoxa-1,10-diazabicyclo[8.8.8]-eicosane) undergoes facile exo-endo conversion, and exists in exo-exo, exo-endo, and two endo-endo conformations all of which are of similar strain energy.[54] (The lone pair of electrons of tetrahedral nitrogen are respectively oriented toward and away from the cryptand cavity in the endo and exo conformations.) While there is no reported experimental determination of exo-/endo- conformational equilibria (and solid-state X-ray studies show both C222 and its alkali metal ion cryptates to exist in the endo-endo conformation[55,56]) it is proposed that the rate of cryptate formation in solution may be limited by the existence of C222 in several conformations, some of which are kinetically inert. This contrasts with an ultrasonic study[57] which indicates that facile equilibria involving three C222 conformations are unlikely to be rate limiting for cryptate formation; and also with protonation studies of C222 which are interpreted in terms of the *endo-endo* conformation being predominant.[58] (A molecular mechanical study of C222 cryptates of Na^+, K^+, Rb^+, Cs^+, and Ca^{2+} has been reported,[59] as has a similar study of the complexation of alkali metal ions by anisole spherands.[60]) A ^{23}Na NMR study of the decomplexation of $Na.C211^+$ (C211 = 4,7,13,18-tetraoxa-1,10-diazabicyclo[8.5.5]eicosane) yields: $k_b(335 \text{ K}) = 1053.6 \pm 4.1, 832.7 \pm 5.0$, and $554.8 \pm 3.2 \text{ s}^{-1}$; $\Delta H^{\ddagger} = 67.2 \pm 0.3, 69.5 \pm 0.4$, and $83.5 \pm 0.5 \text{ kJ mol}^{-1}$; and $\Delta S^{\ddagger} = 12.6 \pm 0.7, 17.4 \pm 1.2$, and $55.9 \pm 1.2 \text{ J K}^{-1} \text{ mol}^{-1}$ in water, dmso, and dmf, respectively.[27] The decomplexation mechanism is considered to be unimolecular. A similar ^{23}Na NMR study of the decomplexation (also thought to be unimolecular) of $Na·C21C_5^+$ (C21C$_5$ = 4,7,13-trioxa-1,10-diazabicyclo[8.5.5]eicosane) in acetonitrile, propylene carbonate, acetone, methanol, dmf, and pyridine respectively yields: $k_b(298.2 \text{ K}) = 84.8 \pm 1.6, 19.4 \pm 0.5, 878 \pm 6, 1800 \pm 50, 28\,800 \pm 300$, and $93.5 \pm 0.5 \text{ s}^{-1}$; $\Delta H^{\ddagger} = 57.9 \pm 0.7, 70.3 \pm 0.5, 54.4 \pm 0.4, 44.9 \pm 0.1, 40 \pm 0.1$, and $62.8 \pm 0.2 \text{ kJ mol}^{-1}$; and $\Delta S^{\ddagger} = -13.8 \pm 2.1, 15.3 \pm 1.4, -6.1 \pm 1.2, -31.9 \pm 0.4, -25.3 \pm 0.5$, and $3.3 \pm 0.5 \text{ J K}^{-1} \text{ mol}^{-1}$.[28,61] The substantially greater lability of $Na·C21C_5^+$ by comparison with $Na·C211^+$ is attributed to C21C$_5$ possessing one less oxygen binding group than C211, and the variation of the $Na·C21C_5^+$ decomplexation parameters with the nature of the solvent is attributed to variations in the electron-donating ability and steric characteristics of the solvents. The importance of the cryptand and the solvent in determining mechanism is indicated by a ^7Li NMR study, which shows that in acetonitrile, propylene carbonate, and acetone unimolecular decomplexation is the dominant path for Li^+ exchange on $Li·C222^+$, while for $Li·C221^+$ in acetonitrile and propylene carbonate direct displacement of complexed Li^+ by solvated Li^+ is the major exchange path.[1]

While studies of alkali and alkaline earth cryptates have been extensive, much less is known about the complexation of the heavy metal cations. This situation has been improved through a study of the Ag^+ and Pb^{2+} cryptates in dmso characterized by the data in Table 9.2.[62] The stabilities of the Ag^+ and Pb^{2+} cryptates are substantially greater than those of the alkali and alkaline earth ions of similar size as a consequence of the greater covalent interactions in the heavy metal cryptates. For the Ag^+ cryptates the greater stabilities are primarily attributable to their lower decomplexation rates by comparison to those of corresponding alkali metal ion cryptates, while the greater stabilities of the Pb^{2+} cryptates are largely a consequence of their greater formation rates by comparison with those of the corresponding alkaline earth cryptates. The stabilities and labilities of the Ag^+ and Pb^{2+} cryptates exhibit a systematic metal ion/cryptand size dependence in a similar manner to the alkali and alkaline earth cryptates.

Table 9.2. *Formation and Decomplexation Rate Constants, and Stability Constants for Ag^+ and Pb^{2+} Cryptates in Dimethyl Sulfoxide at 298.2 K (Reference 62)*

$$M^{m+} + \text{cryptand} \underset{k_b}{\overset{k_f}{\rightleftharpoons}} [\text{M·cryptand}]^{m+} \qquad K_s = k_f/k_b$$

Cation	Cryptand	k_f $(dm^3\,mol^{-1}\,s^{-1})$	k_b (s^{-1})	log K_s
Ag^+	C211	5.2×10^5	0.36	6.17
	C221	2.0×10^6	5.6×10^{-4}	9.55
	C222	2.9×10^6	0.10	7.27
	$C2_B22$	1.4×10^6	0.15	6.95
	$C2_B2_B2$	1.7×10^6	0.30	6.75
Pb^{2+}	C211	2.4×10^3	0.50	3.68
	C221	1.5×10^5	6.5×10^{-4}	8.37
	C222	2.1×10^5	1.25×10^{-2}	7.23
	$C2_B22$	8.0×10^4	4.0×10^{-2}	6.30
	$C2_B2_B2$	2.2×10^4	0.13	5.40

The influence of solvent on the complexation of alkali metal ions by crown ethers can be substantial as a ^{23}Na NMR study shows.[29] The exchange of Na^+ on $Na.18C6^+$ (18C6 = 18-crown-6) proceeds through a unimolecular mechanism in methanol [k_b(298 K) = $3.65 \pm 0.07 \times 10^4\,s^{-1}$, $\Delta H^{\ddagger} = 38.1 \pm 1.3\,kJ\,mol^{-1}$, and $\Delta S^{\ddagger} = -30 \pm 4\,J\,K^{-1}\,mol^{-1}$], 60/40 mol% tetrahydrofuran/methanol, and 80/20 mol% tetrahydrofuran/propylene carbonate; while in propylene carbonate [k_b(298 K) = $1.3 \pm 0.07 \times 10^5\,dm^3\,mol^{-1}\,s^{-1}$, $\Delta H^{\ddagger} = 16.7 \pm 2.1$, and $\Delta S^{\ddagger} = -90 \pm 7\,J\,K^{-1}\,mol^{-1}$] and 40/60 mol% tetrahydrofuran/propylene carbonate a bimolecular exchange mechanism, in which Na^+ in $Na·18C6^+$ is displaced by a second Na^+, predominates. It is concluded that a unimolecular mechanism will predominate in solvents of high Gutmann donor number (D_N) and high dielectric constant, a bimolecular mechanism will predominate in solvents of low D_N and high dielectric constant, and in solvents of low or high donor number and low dielectric constant the extent of ion pairing will determine which mechanism predominates. Uni- and bimolecular Na^+ exchange paths are also observed for $Na·DB24C8^+$ (DB24C8 = dibenzo-24-crown-8) in nitromethane.[30] At $NaBPh_4$ and DB24C8 concentrations $<10^{-3}\,mol\,dm^{-3}$ the unimolecular path dominates ($\Delta G^{\ddagger} \sim$ 65 kJ mol^{-1} at 300 K), while at higher concentrations the bimolecular path dominates [k_{ex}(295 K) = $2.8 \pm 0.3 \times 10^5\,dm^3\,mol^{-1}\,s^{-1}$, $\Delta H^{\ddagger} = 31 \pm 3\,kJ\,mol^{-1}$, and $\Delta S^{\ddagger} = -32 \pm 10\,J\,K^{-1}\,mol^{-1}$]. The transition state for the bimolecular path is envisaged as a quasi-planar structure in which two groups of four oxygens face opposite sides of the molecular plane, each group complexing one Na^+. A ^{133}Cs NMR study of Cs^+ exchange between the solvated state and either $Cs·DB21C7^+$ or $Cs·DB24C8^+$ in acetone and methanol indicates that a bimolecular mechanism, in which an incoming Cs^+ displaces Cs^+ from the crown ether complex, is the predominant exchange path.[31] In acetone the parameters for Cs^+ exchange on $Cs \cdot DB21C7^+$ and $Cs \cdot DB24C8^+$ are, respectively: $10^{-4}k_{ex}(220\ K) = 0.94 \pm 0.3$ and $7.4 \pm 1.0\ dm^3\ mol^{-1}\ s^{-1}$, $\Delta H^{\ddagger} = 33.9 \pm 2.1$ and $28.0 \pm 1.3\ kJ\ mol^{-1}$, and $\Delta S^{\ddagger} = -11.3 \pm 10.5$ and $-20.9 \pm 6.3\ J\ K^{-1}\ mol^{-1}$. In methanol the parameters for Cs^+ exchange on $Cs \cdot DB21C7^+$ and $Cs \cdot DB24C8^+$ are, respectively: $10^{-4}k_{ex}(220\ K) = 2.7 \pm 0.4$ and $5.6 \pm 2.0\ dm^3\ mol^{-1}\ s^{-1}$, $\Delta H^{\ddagger} = 25.1 \pm 1.3$ and $12.1 \pm 2.1\ kJ\ mol^{-1}$, and $\Delta S^{\ddagger} = -43.9 \pm 6.3$ and $-97.1 \pm 10.5\ J\ K^{-1}\ mol^{-1}$. The lower lability of $Cs \cdot DB21C7^+$, by comparison with $Cs \cdot DB24C8^+$, is attributed to the closer fit of Cs^+ into the cavity formed by DB21C7. There is a tendency for bimolecular alkali metal ion exchange in crown ether complexes to become more important as the ion varies from Na^+ to Cs^+.

An ultrasonic relaxation study shows the complexation of Ag^+ and Tl^+ by 18C6 in dmf to be consistent with the Eigen–Winkler mechanism:

$$M^+ + 18C6 \underset{k_{-1}}{\overset{k_1}{\rightleftharpoons}} M^+\cdots18C6 \underset{k_{-2}}{\overset{k_2}{\rightleftharpoons}} [M\cdot18C6]^+ \underset{k_{-3}}{\overset{k_3}{\rightleftharpoons}} [M\cdot18C6']^+ \qquad (2)$$

in which $M^+\cdots18C6$ is an encounter complex, and $[M\cdot18C6]^+$ and $[M\cdot18C6']^+$ represent different conformations of the final complex.[35] For Ag^+ and Tl^+ respectively at 298.2 K, $k_2 = 2.1 \times 10^7$ and $1.8 \times 10^8\ s^{-1}$; $k_{-2} = 2.3 \times 10^8$ and $9.8 \times 10^6\ s^{-1}$; $k_{-3} = 1.8 \times 10^6$ and $2.1 \times 10^6\ s^{-1}$; $\Delta H^{\ddagger}_{-2} = 23.4$ and $20.5\ kJ\ mol^{-1}$; and $\Delta S^{\ddagger}_{-2} = -6.7$ and $-42.3\ J\ K^{-1}\ mol^{-1}$. The desolvation step (k_2) of the Tl^+ system possesses a similar rate to that for K^+ $(k_2 = 1.8 \times 10^8\ s^{-1})$ and is faster than that of Ag^+ consistent with solvent being more tightly bound to the smaller Ag^+. However, $\tilde{k}_{-3}$ for Ag^+ and Tl^+ are similar, which probably indicates the dominance of conformational changes in determining the rate of this reaction step. In contrast to the labile Ag^+ and Tl^+ 18C6 systems, the complexation of $Hg(CN)_2$ by 18C6 and other crown ethers is slow, and this has been attributed to 18C6 assuming a high-energy conformation with six ether oxygens arranged in an equatorial plane about the CN–Hg–CN axis.[63] A Monte Carlo modeling study of 18C6 shows that the intrinsically most stable C_i conformer, which is the least suited to cation complexing, is the least stable form in water, while the D_{3d} conformer is more strongly hydrated and should be the most stable form in water, and has a preformed cavity suitable for cation binding.[64]

A study of the effects of the ionizable side arms of modified crown ethers on Na^+ complexation shows that complexation by *sym*-dibenzo-16-crown-5-oxyacetate is diffusion controlled $[k_f(298.2\ K) = 1.9 \pm 1.0 \times 10^{10}\ dm^3\ mol^{-1}\ s^{-1}$ in 99% w/w methanol–water] suggesting an interaction between Na^+ and the oxyacetate side arm prior to complexation by the polyether ring, similar to some antibiotic complexation processes.[32] Complexation of Na^+ by the antibiotic monensin (HMon) in ethanol occurs through paths (3) and (4),

$$Na^+ + HMon \underset{k_{-1}}{\overset{k_1}{\rightleftharpoons}} NaHMon^+ \qquad K_1 \qquad (3)$$

$$\text{Na}^+ + \text{Mon}^- \underset{k_{-2}}{\overset{k_2}{\rightleftharpoons}} \text{NaMon} \qquad K_2 \tag{4}$$

characterized by $k_1 = 4.8 \times 10^6 \, \text{dm}^3 \, \text{mol}^{-1} \, \text{s}^{-1}$, $k_{-1} = 810 \, \text{s}^{-1}$, $k_2 = 1.1 \times 10^9 \, \text{dm}^3 \, \text{mol}^{-1} \, \text{s}^{-1}$, and $k_{-2} = 2.2 \, \text{s}^{-1}$ at 298.2 K ($\Delta H_1^{\ddagger} = 13.8 \, \text{kJ mol}^{-1}$, $\Delta S_1^{\ddagger} = -71.0 \, \text{J K}^{-1} \, \text{mol}^{-1}$, $\Delta H_{-1}^{\ddagger} = 35.2 \, \text{kJ mol}^{-1}$, $\Delta S_{-1}^{\ddagger} = -70.8 \, \text{J K}^{-1} \, \text{mol}^{-1}$).[33] The decreased stability of NaHMon$^+$ (log $K_1 = 3.77$) compared to that of NaMon (log $K_2 = 8.8$) is due to protonation causing a 200-fold decrease in the formation rate constant and a 400-fold increase in the decomplexation rate constant for NaHMon$^+$ by comparison with the corresponding rate constants for NaMon.

9.3.1.2. Other Ligands

Ligand substitution on bivalent first-row transition ions continues to be an area of active study. Thus it is reported[9] that for the substitution of 2,2'-bipyridine and 2,2',6',2''-terpyridine on Mn^{2+}, Fe^{2+}, Co^{2+}, and Ni^{2+}, $\Delta V^{\ddagger}$ ranges from -3.4 to $6.7 \, \text{cm}^3 \, \text{mol}^{-1}$ from Mn^{2+} to Ni^{2+} in a trend which parallels that observed for water exchange,[65] and which is consistent with an I_a mechanism operating for Mn^{2+}, and an I_d mechanism operating for Fe^{2+}, Co^{2+}, and Ni^{2+}. Several variable-pressure pressure-jump studies[10,11] of the substitution of $[\text{Ni}(\text{H}_2\text{O})_6]^{2+}$ by carboxylate ligands (L-L^{n-}) are interpreted through equations (5)–(7). From a consideration of the overall volume of

$$[\text{Ni}(\text{H}_2\text{O})_6]^{2+} + \text{L-L}^{n-} \underset{\text{fast}}{\overset{K_0}{\rightleftharpoons}} [\text{Ni}(\text{H}_2\text{O})_6]^{2+} \cdots \text{L-L}^{n-} \qquad \Delta V_0^{\circ} \tag{5}$$

$$\xrightarrow{k_1} [(\text{H}_2\text{O})_5\text{Ni-L-L}]^{(2-n)+} + \text{H}_2\text{O} \qquad \Delta V_1^{\ddagger} \tag{6}$$

$$\xrightarrow{k_2} \left[(\text{H}_2\text{O})_4\text{Ni} \overset{\text{L}}{\underset{\text{L}}{\diagdown}} \right]^{(2-n)+} + 2\text{H}_2\text{O} \qquad \Delta V_2^{\ddagger} \tag{7}$$

reaction, the observed volume of activation of complexation, the calculated volume of formation of the encounter complex (ΔV_0°), and other data, it is deduced that succinate acts as a monodentate ligand while malonate, tartronate, malate, and tartrate act as bidentate ligands, and that $\Delta V_1^{\ddagger}$ characterizing the rate-determining first-bond formation equals 7.3, 7.5, 8.0, 6.6, and 6.5 cm^3 mol^{-1}, respectively, consistent with the operation of an I_d mechanism. In contrast, ring closure is rate determining for L-L^{n-} = glycolate and lactate, for which $\Delta V_2^{\ddagger} = 11.6$ and $10.4 \, \text{cm}^3 \, \text{mol}^{-1}$, respectively. The difference in the rate-determining step is attributed to a greater steric strain existing in the five-membered rings of the glycolate and lactate chelates by comparison with the six-membered rings formed by malonate, tartronate, malate, and tartrate. Such variations in the position of the rate-determining step in the coordination sequence is quite common. Thus the rate-determining steps in the complexation of $[Ni(H_2O)_6]^{2+}$ by 2-aminophenol and 2-aminophenol-4-sulfonate $[k_f(298.2 \text{ K}) = 2.7 \times 10^3$ and $2.9 \times 10^3 \text{ dm}^3 \text{ mol}^{-1} \text{ s}^{-1}$, respectively] are considered to be first bond formation by the hydroxy group in each case.[12] In contrast, ring closure is the slow step in the complexation of $[Ni(H_2O)_6]^{2+}$ by LH (7-hydroxy-8-(2-pyridylazo)-2-naphthol-2-sulphonic acid), which exists predominantly as a quinone hydrazone tautomer), characterized by k_{LH} in

$$k_f = k_{LH} + k_{OH}[OH^-] \tag{8}$$

expression (8) for the observed second-order complexation rate constant k_f. The slowness of this complexation ($k_{LH} = 2.7 \times 10^2 \text{ dm}^3 \text{ mol}^{-1} \text{ s}^{-1}$ at 298.2 K) is thought to be a consequence of ring closure involving the transfer of a strongly hydrogen-bonded proton from LH to the solvent.[13] Complexation involving the deprotonated ligand (L^-) is characterized by $k_{OH}(298.2 \text{ K}) = 1.5 \times 10^9 \text{ dm}^{-6} \text{ mol}^{-2} \text{ s}^{-1} = k_L K_w / K_a$, where K_a is the acid dissociation constant of LH such that $k_L = 1.1 \times 10^6 \text{ dm}^3 \text{ mol}^{-1} \text{ s}^{-1}$. The value of k_L is larger than can be accounted for by an I_d mechanism and suggests the operation of an internal conjugate base mechanism. The substitution of a second LH occurs *ca* 300 times more rapidly than the first LH substitution, and it appears that this cannot be solely attributed to an increase in the lability of water in $[Ni(L)(H_2O)_4]^+$, and an increase in the encounter complex stability through a stacking interaction is postulated as a cause of the rapid second substitution. A related study[14] indicates that rate enhancements for the entry of a second ligand caused by coordinated aromatic ligands are greater with the ionized form of the entering ligand than with the protonated form. It also appears that the coordinated ligand influences the rate of ring closure for the second ligand. The complexation of $[Ni(H_2O)_6]^{2+}$ by histamine proceeds through parallel paths corresponding to complexation by histamine deprotonated at either the imidazole N-1 or N-3, and characterized by first-bond formation-rate constants of 2.3×10^3 and $1.4 \times 10^3 \text{ dm}^3 \text{ mol}^{-1} \text{ s}^{-1}$, respectively, at 298.2 K.[15] The N-3 complex leads directly to the bidentate histamine complex, but the N-1 complex cannot form a bidentate complex and dissociates to form the N-3 complex. This results in biphasic kinetic behavior.

The complexation of Cu^{2+} by hexadentate N,N'-bis(β-carbamoylethyl)ethylenediamine, N,N'-bis(β-carbamoylethyl)trimethylenediamine, N,N'-bis(β-carbamoylethyl-1,2-propylenediamine, and N,N'-bis(β-carbamoylethyl)-2-hydroxytrimethylenediamine is characterized by $10^{-8} k_f(298.2 \text{ K}) = 3.35, 1.48, 2.23,$ and $7.80 \text{ dm}^3 \text{ mol}^{-1} \text{ s}^{-1}$ in aqueous 0.01 mol dm^{-3} $NaClO_4$, and the rate-determining step is assigned to the formation of the first $Cu-N$ bond through an I_d mechanism.[66] For the corresponding Ni^{2+} systems $10^{-5} k_f = 1.58, 0.955, 1.41,$ and $3.02 \text{ dm}^3 \text{ mol}^{-1} \text{ s}^{-1}$, and the rate-determining step is assigned to first $Ni-N$ bond formation through an internal conjugate base mechanism.[16] Complexation by the monoprotonated ligands is much slower in both systems, and the rate-determining step is thought to be proton loss. Several studies of multidentate ligand substitution on Ni(II),

Table 9.3. *Formation Rate Constants for Hydroxycuprate(II) Species Reacting with Tetraamine Ligands at 298.2 K (References 67 and 68)*

Ligand	$k_f(Cu(OH)_3^-)$ $(dm^3\,mol^{-1}\,s^{-1})$	$k_f(Cu(OH)_4^{2-})$ $(dm^3\,mol^{-1}\,s^{-1})$
2,3,2-tet	$1.0 \pm 0.7 \times 10^7$	$4.3 \pm 0.2 \times 10^6$
Et$_2$-2,3,2-tet	$3.0 \pm 0.3 \times 10^6$	$2.9 \pm 0.6 \times 10^5$
Me$_4$trien	$4.1 \pm 0.6 \times 10^6$	$4.2 \pm 1.3 \times 10^5$
Me$_6$trien	$3.4 \pm 0.4 \times 10^5$	$<10^4$
Cyclam	$2.7 \pm 0.4 \times 10^6$	$3.8 \pm 0.9 \times 10^4$
Me$_2$cyclam	5.6×10^5	0.9×10^4
(N-Me)$_4$cyclam	$3.1 \pm 0.4 \times 10^3$	<10
tet-b	3.1×10^4	1.1×10^2
C2$_N$1$_O$1$_O$	$6.6 \pm 0.3 \times 10^4$	$3.8 \pm 0.7 \times 10^3$

Co(II), and Cu(II) in mixed solvent systems have been reported.[17-19] The rate-determining step for the complexation of Cu(II) by 1,1,10,10-tetramethyl-1,4,7,10-tetraazadecane (Me$_4$trien), 1,1,4,7,10,10-hexamethyl-1,4,7,10-tetraazadecane (Me$_6$trien), 1,4,8,11-tetramethyl-1,4,8,11-tetraazacyclotetradecane [(N-Me)$_4$cyclam], and 1,4,7,10-tetraaza-13,18-dioxabicyclo[8.5.5]eicosane (C2$_N$1$_O$1$_O$) in 0.075-0.40 aqueous NaOH varies with the substituting ligand; and the reactivity of these ligands with Cu(OH)$_3^-$ is 10 to >300 greater than with [Cu(OH)$_4$]$^{2-}$ (Table 9.3).[67,68] For Me$_4$trien reacting with either [Cu(OH)$_3$]$^-$ or [Cu(OH)$_4$]$^{2-}$ and Me$_6$trien reacting with [Cu(OH)$_3$]$^-$ the rate-determining step appears to be Jahn–Teller inversion after formation of the first Cu—N bond. The rate-determining step apparently shifts to second-bond formation for the reaction of Me$_6$trien with [Cu(OH)$_4$]$^{2-}$. For cyclic (N-Me)$_4$cyclam and C2$_N$1$_O$1$_O$, second-bond formation or multiple desolvation is proposed as the rate-determining step with [Cu(OH)$_3$]$^-$ and [Cu(OH)$_4$]$^{2-}$. Decomplexation of 1,4,7,10,13-pentaazacyclopentadecane-copper(II), 1,4,7,10,13-pentaazacyclohexadecanecopper(II), and 1,4,7,11,14-pentaazacycloheptadecanecopper(II) is characterized by a second-order dependence on [H$^+$] (thought to indicate biprotonation of the coordinated ligand), and rate constants (298.2 K) equal 0.049, 4.8, and 1200 dm^6 mol^{-2} s^{-1}, respectively, which reflect a weakening of Cu—N bonds as the ring size increases.[69] Similarly, the decomplexation of 1,5,8,12-tetraazacyclooctadecanecopper(II) and 1,5,8,12-tetraazacycloheptadecanenickel(II) is much greater than those of the analogous 1,4,8,11-tetraazacyclotetradecane complexes in aqueous acid solution.[70]

The influence of the reaction medium on the rates of ligand substitution has been demonstrated by a stopped-flow study which shows that k_f(298 K) for substitution of Cu^{2+} by the keto form of 1-phenylbutane-1,3-dione is increased from 12 dm^3 mol^{-1} s^{-1} in water to 217 dm^{-3} mol^{-1} s^{-1} in the electrical double layer around sodium dodecylsulfate micelles where Cu^{2+} ions are

Table 9.4. *Parameters for the Substitution of 4-Isopropyltropolone on* $[M(solvent)_6]^{3+}$
(Reference 42)

M	$k_f(298.2 \text{ K})$ $(\text{dm}^3 \text{ mol}^{-1} \text{s}^{-1})$	$\Delta H_f^{\ddagger}$ (kJ mol^{-1})	$\Delta S_f^{\ddagger}$ (J K mol^{-1})	$\Delta V_f^{\ddagger}$ $(\text{cm}^3 \text{ mol}^{-1})$
		In dmso		
Al	0.22	82.3 ± 1.5	18.5 ± 5.0	$12.2 \pm 1.0 \,(313.2 \text{ K})$
Ga	37	74.2 ± 1.6	33.8 ± 5.4	$10.6 \pm 0.6 \,(308.2 \text{ K})$
In	1.9×10^4	36.7 ± 0.8	-39.8 ± 2.5	$-0.1 \pm 0.6 \,(308.2 \text{ K})$
		In dmf		
In	3.5×10^5	36.1 ± 2.4	-17.7 ± 8.1	$0.3 \pm 0.3 \,(289.2 \text{ K})$

concentrated.[71] Several similar studies of kinetic effects of micelles have been reported.[72-75]

9.3.2. Trivalent Ions

A high-pressure stopped-flow study[42] of the substitution of bidentate 4-isopropyltropolone (Hipt) on $[M(solvent)_6]^{3+}$, where M = Al, Ga and In, in dmso and dmf yields the rate law (9) and the parameters in Table 9.4. These data are consistent with the operation of I_d (M = Al and Ga) and I_a (M = In) mechanisms (10) in which the rate-determining step (which does not involve

$$d[M(\text{ipt})^{2+}]/dt = k_f[M^{3+}][\text{Hipt}] \tag{9}$$

$$M^{3+} + \text{Hipt} \underset{}{\overset{K_0}{\rightleftharpoons}} M^{3+}\cdots\text{Hipt} \overset{k_i}{\longrightarrow} M(\text{ipt})^{2+} + H^+ \tag{10}$$

deprotonation of Hipt and ring closure) is characterized by $k_i \, (=k_f/K_0)$ such that $\Delta V_f^{\ddagger} = \Delta V_0^{\circ} + \Delta V_i^{\ddagger}$. From these data and those for Hipta substitution on Fe^{3+} in water, dmso, and dmf[43] it is deduced that as the solvent molecular bulk decreases, so does the degree of d character in the ligand substitution process, even to the extent of causing a mechanistic change from I_a for solvent exchange on $[Fe(\text{dmso})_6]^{3+}$ to I_d for the substitution of Hipta on this species. The substitution of 4-(2-pyridylazo)resorcinol on Ga(III) and In(III) in water and aqueous solvent mixtures shows the usual increase in the rapidity of ligand substitution in the sequence: $M^{3+} < M(OH)^{2+} < M(OH)_2^+$ for M = Ga or In.[76] In a similar study of the substitution of 1-(2-pyridylazo)-2-naphthol on In(III) in water and aqueous methanol mixtures the greater reactivity of $In(OH)^{2+}$ over In^{3+} is also observed and a activation modes are assigned to their ligand substitution processes.[77] A conductivity pressure-jump study of a mixed solvent system indicates that water in $[Al(H_2O)_5(HCONH_2)]^{3+}$ is substituted ten times more rapidly than in $[Al(H_2O)_6]^{3+}$, and that the formamide ligand is less labile than the aqua ligands.[37] Sulfate substitution apparently occurs through an I mechanism in which SO_4^{2-} resides on a trigonal face of the octahedron in the encounter complex prior to the departure of one of the three solvent molecules delineating the opposite trigonal face.

The formation of the vanadium(III) species $[V(cysteine)_2]^-$ (in which cysteine may be tridentate, or may be bidentate with the remaining two coordination sites occupied by monodentate ligands) occurs through a fast step followed by a slower step ($E_a = 41 \pm 4 \, kJ \, mol^{-1}$ and $A = 1.4 \times 10^{10}$ at pH 8.8) attributed to the substitution of the first and second cysteine, respectively.[78] Under similar conditions the vanadium(II) species $[V(cysteine)_3]^{4-}$ forms in several steps the slowest ($E_a = 41 \pm 4 \, kJ \, mol^{-1}$ and $A = 1.4 \times 10^7$ at pH 8.2) of which is attributed to coordination of the third cysteine.

9.4. Complex Formation Involving Substituted Metal Ions: Ligand Substitution and Solvent Exchange

9.4.1. Uni- and Bivalent Ions

Square-planar ligand substitution reactions[79] are often characterized by equation (11), but in the case of the Rh(I) reactions in methanol shown in equation (12) the solvent dependent term, k_s, is absent as anticipated for the

$$k_f = k_s + k_x[X] \tag{11}$$

$$[Rh(\beta\text{-diketonate})(octadiene)] + phen \xrightarrow{k_f} [Rh(phen)(octadiene)]^+ + \beta\text{-diketonate}^- \tag{12}$$

displacement of a bidentate ligand by solvent, and the variation of the entering ligand (X = phen) dependent term, k_x, for a range of β-diketones is encompassed by $k_x(298.2 \, K) = 29$ and $2.76 \times 10^5 \, dm^3 \, mol^{-1} \, s^{-1}$, respectively, for acetylacetonate (acac$^-$) and hexafluoroacetylacetonate.[80] This variation appears to reflect the tendency of electron-withdrawing substituents of the β-diketonate to labilize the Rh(I) complex toward ligand substitution through an A mechanism. The substitution of $[Rh(acac)(CO)_2]$ by triphenylphosphite (P) in benzene also proceeds through an A mechanism, and produces $[Rh(acac)(CO)P]$ in a fast step, and $[Rh(acac)(P)_2]$ and the displacement of acac$^-$ in successively slower steps.[81] An unusual example of ligand substitution on a univalent ion is the addition reaction of five-coordinate $(2,2',2''$-nitrilotriethoxy)nitrosylvanadate with cyanide according to equation (13),

$$[V(NO)\{N(C_2H_4O)_3\}]^- + CN^- \underset{k_b}{\overset{k_f}{\rightleftharpoons}} [V(NO)(CN)\{N(C_2H_4O)_3\}]^{2-} \tag{13}$$

where $k_f(293.2 \, K) = 1.12 \, dm^3 \, mol^{-1} \, s^{-1}$, $\Delta H_f^{\ddagger} = 60 \pm 4 \, kJ \, mol^{-1}$, and $\Delta S_f^{\ddagger} = -38 \pm 12 \, J \, K^{-1} \, mol^{-1}$; and $k_b(293.2 \, K) = 0.61 \, s^{-1}$, $\Delta H_b^{\ddagger} = 75 \pm 4 \, kJ \, mol^{-1}$, and $\Delta S_b^{\ddagger} = 8 \pm 12 \, J \, K^{-1} \, mol^{-1}$ at pH 10 in aqueous $2.0 \, mol \, dm^{-3}$ NaClO$_4$ solution.[82]

While the majority of nickel(II) mechanistic studies have been concerned with six-coordinate species, lower-coordination-number species continue to attract attention. Thus an ultrasonic relaxation study[20] of the square-planar/octahedral equilibrium (14) existing in aqueous solutions

$$[\text{Ni}(\text{Me}_4\text{cyclam})]^{2+} + 2\text{H}_2\text{O} \underset{k_b}{\overset{k_f}{\rightleftharpoons}} trans\text{-}[\text{Ni}(\text{Me}_4\text{cyclam})(\text{H}_2\text{O})_2]^{2+} \quad (14)$$

of [[1R,4S,8S,11R]-1,4,8,11-tetraazacyclotetradecane]nickel(II) yields k_f(298.2 K) = $3.10 \times 10^7 \text{ s}^{-1}$, $\Delta H_f^{\ddagger} = 10.2 \pm 0.1 \text{ kJ mol}^{-1}$, and $\Delta S_f^{\ddagger} = -67.2 \pm 0.1 \text{ J K}^{-1} \text{ mol}^{-1}$; and k_b(298.2 K) = $2.85 \times 10^7 \text{ s}^{-1}$, $\Delta H_b^{\ddagger} = 39.0 \pm 0.1 \text{ kJ mol}^{-1}$, and $\Delta S_b^{\ddagger} = 28.4 \pm 0.4 \text{ J K}^{-1} \text{ mol}^{-1}$; and the square-planar/octahedral interconversion apparently occurs through a reactive five-coordinate intermediate. Five-coordinate nickel(II) complexes are not necessarily very labile as is demonstrated in the case of def exchange on five-coordinate (diethylformamide)[2,2',2''-tri(dimethyamino)triethylamine]nickel(II) [k_{ex}(298.2 K) = $944 \pm 42 \text{ s}^{-1}$, $\Delta H^{\ddagger} = 23.1 \pm 0.7 \text{ kJ mol}^{-1}$, and $\Delta S^{\ddagger} = -111 \pm 2 \text{ J K}^{-1} \text{ mol}^{-1}$].[21] A facile equilibrium between square-planar and tetrahedral [NiA$_2$] (A$^-$ = N-alkylsalicylaldimate, N,N'-dialkyl-2-aminotropone iminate, or N-alkylbezoylacetone iminate) exists in acetone solution, but substitution of A by N,N'-disalicyleneethylenediamine in acetone occurs through an A mechanism involving the square-planar complex alone.[22] In the presence of pyridine [NiA$_2$] is in facile equilibrium with octahedral [NiA$_2$(pyridine)$_2$], but here also substitution proceeds through the square-planar species only. In methanol solution the substitution of A$^-$ by acetylacetone (Hacac) is characterized by rate law (15) in which k_s characterizes the attack of methanol at the donor

$$\text{rate} = (k_s + k_{\text{Hacac}}[\text{acacH}])[\text{NiA}_2] \quad (15)$$

oxygen of coordinated A$^-$, and k_{Hacac} the substitution of A$^-$ by Hacac through an A mechanism. (A similar rate law is observed for [CuA$_2$].[83]) The conversion of the square-planar N,N'-bis(2-aminoethyl)oxaldiamidenickel(II) complex into trien or edta^{4-} nickel(II) complexes in aqueous solution occurs through paths in which the proton-independent and proton-catalyzed release of the oxaldiamide are rate determining, and a third path in which trien or edta^{4-} is involved in the rate-determining step.[23]

The relationship between complex lability and stability has been investigated for the ligand–ligand replacement reactions in equation (16), where

$$\text{CuL}^{(2-n)+} + \text{cyclamH}^+ \xrightarrow{k_f} \text{Cu(cyclam)}^{2+} + \text{L}^{n-} + \text{H}^+ \quad (16)$$

L^{n-} = glycolate (18, 2.36), malonate (5.3, 5.81), succinate (56, 2.60), picolinate (7.1, 8.60), glycinate (5.2, 8.07), iminodiacetate (3.7, 10.55), nitrilotriacetate (8.5×10^{-2}, 11.50), N-(2-hydroxyethyl)ethylenediaminetriacetate (4.3×10^{-6}, 17.40), 1,2-diaminoethane (7.1, 10.76), 1,3-diaminopropane (3.9, 9.98), diethylenetriamine (4.5×10^{-4}, 16.00), and N,N-bis(3-aminopropylamine) ($9.1 \times$

10^{-5}, 14.20), and $10^{-6}k_r$ dm^3 mol^{-1} s^{-1} and log K_{CuL} (K_{CuL} = the stability constant of CuL$^{(2-n)+}$) at 298.2 K, respectively, appear in parentheses.[84] The relationship between k_r and K_{CuL} is given by equation (17), where k_0 = 1.1×10^7 dm^3 mol^{-1} s^{-1}, and $K_0 = 2.3 \times 10^{10}$ dm^3 mol^{-1}, at 298.2 K in aqueous

$$\log k_r = \log[k_0/(1 + K_{CuL}/K_0)] \tag{17}$$

KNO$_3$ (I = 1 mol dm^{-3}). Thus for the CuL$^{(2-n)+}$ of low K_{CuL} lability is independent of stability, while for those of high K_{CuL} lability is inversely dependent on stability.

The influence of nonleaving ligands on leaving-ligand lability is a recurrent theme in ligand substitution studies. The variation of k_f with the nature of X^{x+} (where X^{x+} is one of a range of nitrogen heterocycles) in reaction (18)

$$[\text{Ru(CN)}_5(\text{H}_2\text{O})]^{3-} + \text{X}^{x+} \underset{k_b}{\overset{k_f}{\rightleftharpoons}} [\text{Ru(CN)}_5\text{X}]^{(3-x)-} + \text{H}_2\text{O} \tag{18}$$

is consistent with the operation of an I_d mechanism, and k_{ex}(298.2 K) = 10 ± 5 s^{-1} has been derived for water exchange on low-spin $[\text{Ru(CN)}_5(\text{H}_2\text{O})]^{3-}$ in 0.10 mol dm^{-3} NaCl, which compares with k_{ex} = 0.014 s^{-1} for low-spin $[\text{Ru(H}_2\text{O)}_6]^{2+}$.[85] This difference is largely attributed to the lower surface charge density on the metal center of $[\text{Ru(CN)}_5(\text{H}_2\text{O})]^{3-}$ labilizing coordinated water. Nevertheless, a comparison with low-spin $[\text{Fe(CN)}_5(\text{H}_2\text{O})]^{3-}$ and high-spin $[\text{Fe(H}_2\text{O)}_6]^{2+}$, for which k_{ex} = 300 and 4.4×10^6 s^{-1}, respectively, indicates that changes in spin state can have substantial effects on lability. A ^{1}H NMR study[86] shows that the lifetime of dmf (τ) in *trans*-$[\text{Mg(15C5)(dmf)}_2]^{2+}$ varies according to equation (19), which typifies an I mechanism, and in which

$$1/\tau = k_i K_0[\text{dmf}]/(1 + K_0[\text{dmf}]) \tag{19}$$

[dmf] is the concentration of free dmf [k_i(313.2 K) = 24.7 ± 2 s^{-1}, $\Delta H^\ddagger$ = 81 ± 44 kJ mol^{-1}, $\Delta S^\ddagger$ = 41 ± 5 J K^{-1} mol^{-1}, and K_0 = 7.8 ± 2 dm^3 mol^{-1} in CD$_3$NO$_2$]. The lability of dmf in *trans*-$[\text{Mg(15C5)(dmf)}_2]^{2+}$ is 10^4 times less than in $[\text{Mg(dmf)}_6]^{2+}$.[87] Substantial variations in lability occur with change in the exchanging ligand as shown by a variable-pressure ^{1}H NMR study of L (=nitriles, ethers, amides, and —PO donor ligands) exchange on SbCl$_5$L in CH$_2$Cl$_2$ and CHCl$_2$CHCl$_2$, which is characterized by a first-order rate law and $\Delta G^\ddagger$(298.2 K) ranging from 37 to 94 kJ mol^{-1}.[44] It is concluded that a D mechanism operates for these systems which are characterized by $\Delta V^\ddagger$ in the range 18.2 to 30.0 cm^3 mol^{-1}.

Ternary complex formation is frequently studied as an aid to understanding biological systems. A very detailed ^{13}C and ^{31}P NMR study of the Mn^{2+}/atp^{4-}/glycine system in water/glycerol solution shows that the binary complexes $[\text{Mn(atp)}]^{2-}$, $[\text{Mn(atp)}_2]^{6-}$, $[\text{Mn(atp)}_3]^{10-}$, and the ternary complex $[\text{Mn(atp)(glycine)}]^{6-}$ are formed (where atp^{4-} is adenosine 5'-triphosphate, and coordinated water is omitted).[88] It appears that only one atp^{4-} binds as a tridentate ligand through phosphate oxygens and that subsequent atp^{4-}

binding occurs through a stacking interaction of its adenosine moiety with that of the first atp^{4-}. Exchange of the first atp^{4-} is characterized by $k_{ex}(298.2\text{ K}) = 2.82 \pm 0.05 \times 10^4\text{ s}^{-1}$, $\Delta H^{\ddagger} = 56.8 \pm 0.5\text{ kJ mol}^{-1}$, and $\Delta S^{\ddagger} = 31 \pm 2\text{ J K}^{-1}\text{ mol}^{-1}$. Exchange of glycine (probably bound through oxygen only) in $[Mn(atp)(glycine)]^{6-}$ is characterized by $k_{ex}(298.2\text{ K}) = 8.2 \pm 1.7 \times 10^6\text{ s}^{-1}$, $\Delta H^{\ddagger} = 30.0 \pm 0.7\text{ kJ mol}^{-1}$, and $\Delta S^{\ddagger} = -12 \pm 3\text{ J K}^{-1}\text{ mol}^{-1}$. A ^{31}P NMR study of six-coordinate $[M(trien)(PO_4)]^{(m-3)+}$ yields $10^{-3}k_{ex}(298.2\text{ K}) = 15.2$, 5.0, 35, and 2.8 s^{-1} (when $M^{m+} = Cu^{2+}$, Ni^{2+}, Fe^{3+}, and Mn^{2+}, respectively) for phosphate exchange in aqueous solution at pH 11.6.[24] It is deduced that the rapidity of phosphate exchange precludes a similar process from being rate determining in the decomposition of acetyl phosphate in similar complexes. The ternary complexes $[Ni(atpH)(bipy)(H_2O)]^-$, $[Ni(atp)(bipy)(H_2O)]^{2-}$, and $[Ni(atp)(bipy)(OH)]^{3-}$ are characterized by $10^{-3}k_f(288.2\text{ K}) = 1.9$, 4.3, and $220\text{ dm}^3\text{ mol}^{-1}\text{ s}^{-1}$, respectively, for substitution of bipy (2,2′-bipyridyl) on $[Ni(atpH)(H_2O)_3]^-$, $[Ni(atp)(H_2O)_3]^{2-}$, and $[Ni(atp)(OH)(H_2O)_2]^{3-}$ in aqueous 0.1 mol dm^{-3} KCl.[25] The variation of k_f is mainly attributable to labilization of coordinated water by the OH^- ligand, and to a lesser extent by atp^{4-} and $atpH^{3-}$. Ring closure appears to be the rate-determining step in the substitution of the aqua ligands of $[Ni(nta)(H_2O)_2]^-$ (nta^{3-} = nitrilotriacetate) by glycine, α-alanine, L-phenylalanine, L-valine, and β-alanine and is characterized by $k_f(308.2\text{ K}) = 3.96$, 3.54, 3.36, 3.28, 2.60 s^{-1} respectively under pseudo-first-order conditions in aqueous 0.1 mol dm^{-3} KNO_3.[26]

Scheme 1

$$\mathrm{Ca(edta)^{2-}} \underset{\text{fast}}{\overset{+\mathrm{Tb^{3+}},\,K_{\mathrm{Tb}}}{\rightleftharpoons}} \mathrm{Ca(edta)Tb^{+}} \xrightarrow{k_1} \mathrm{Ca^{2+} + Tb(edta)^{-}}$$

$$\mathrm{Ca(edta)^{2-}} \;\; \underset{K_{\mathrm{H}}}{\overset{+\mathrm{H^+}}{\downarrow}} \; \underset{\text{fast}}{\overset{-\mathrm{H^+}}{\uparrow}} \;\; \mathrm{Ca(Hedta)^{-}}$$

$$\mathrm{Ca(Hedta)^{-}} \xrightarrow{k_2} \mathrm{Ca^{2+} + Hedta^{3-}} \xrightarrow{+\mathrm{Tb^{3+}},\,k_3} \mathrm{Ca^{2+} + H^{+} + Tb(edta)^{-}}$$

The kinetics of $edta^{4-}$ exchange between Ca^{2+} and Tb^{3+} in the pH range 5.0–6.0 are consistent with the two-path Scheme 1.[89] This predicts a variation of the observed first-order rate constant (k_{obs}) for the displacement of complexed Ca^{2+} according to equation (20), and when $K_{Tb}[Tb^{3+}] \gg 1$, $K_H[H^+]$,

$$k_{obs} = (k_1 K_{Tb}[Tb^{3+}] + k_2 K_H[H^+])/(1 + K_H[H^+] + K_{Tb}[Tb^{3+}]) \qquad (20)$$

an inverse dependence of k_{obs} on $[Tb^{3+}]$ as found experimentally. This indicates that the dissociation of Ca^{2+} from binuclear $Ca(edta)Tb^+$ is slower ($k_1 \sim 18\text{ s}^{-1}$ at 298 K) than from $Ca(edta)^-$. Below pH 5.0 there is evidence for a third path for the displacement of Ca^{2+}. This study was carried out using a stopped-flow spectrophotometer to monitor the increase in luminescence as Tm^{3+} was bound by $edta^{4-}$, which results from the displacement of coordinated water whose OH vibrational manifold provides a nonradiative path for the deexcitation of Tb^{3+}, thereby reducing its lifetime and quantum yield. A similar technique has been employed to study the exchange of metal ions in protein systems.[90] The fluorescent indicator quin 2 has been used to monitor the release of Ca^{2+} from calmodulin and its tryptic fragments, which occurs through two processes characterized by rate constants equal to 24 ± 6.0 and $240 \pm 50 \, s^{-1}$ at 298.2 K in aqueous $0.1 \, mol \, dm^{-3} \, KCl.$[91]

9.4.2. Trivalent Ions and Ions of Higher Valency

Gold(III) is unusual among trivalent metal ions in forming square-planar metal complexes with monodentate ligands.[92] Ligand substitution occurs through an *a* activation mode, and is characterized by a rate law similar to equation (11). In the most recent study it is found that mono-substitution of $[Au(NH_3)_4]^{3+}$ by I^- is characterized by: $k_x(298.2 \, K) = 1.52 \pm 0.03 \times 10^3 \, dm^3 \, mol^{-1} \, s^{-1}$, $\Delta H^{\ddagger} = 54.6 \pm 1.1 \, kJ \, mol^{-1}$, and $\Delta S^{\ddagger} = -1 \pm 4 \, J \, K^{-1} \, mol^{-1}$, and k_s is negligible at pH 2.3 in aqueous $1.0 \, mol \, dm^{-3}$ $NaClO_4.$[93] The effect of bound ligands on the complexation of the octahedral $GaA^{(3-a)+}$ complexes (where $a-$ is the charge of the anionic ligand) is exemplified in the sequence of reactions shown in Scheme 2, which were studied

Scheme 2

$$Ga(OH)_2^+ + H_2L^{2-} \xrightarrow{k_{Ga(OH)_2}} GaL^-$$

$$\Big\| K_{Ga(OH)_2}$$

$$GaOH^{2+} + H_2L^{2-} \xrightarrow{k_{GaOH}} GaL^- + H^+$$

$$\Big\| K_{GaOH}$$

$$Ga^{3+} + H_2L^{2-} \xrightarrow{k_{Ga}} GaL^- + 2H^+$$

$$\Big\| K_{GaNCS}$$

$$GaNCS^{2+} + H_2L^{2-} \xrightarrow{k_{GaNCS}} Ga(NCS)L^{2-} \text{ (and/or } GaL^- + NCS^-) + 2H^+$$

(where H_2L^{2-} = 4,5-dihydroxy-1,3-benzenedisulfonate and coordinated water is not shown)

---

by stopped-flow spectrophotometry.[94] At 298.2 K the observed second-order rate constants k_{Ga}, k_{GaOH}, $k_{Ga(OH)_2}$, and k_{GaNCS} are equal to 16.9, 2.29×10^3, 2.34×10^4, and $7.94 \times 10^3 \, dm^3 \, mol^{-1} \, s^{-1}$, respectively, in aqueous $0.5 \, mol \, dm^{-3}$ $NaClO_4$. Similar studies have been carried out when the bound ligand A^{a-} is oxalate and the substituting ligand is 2-hydroxy-2,4,6-cycloheptatrien-1-one. The ligand substitution mechanism is considered to be I_d and it is found that the bound ligands increase the lability of bound water by comparison with that of water in $[Ga(OH_2)_6]^{3+}$. The logarithm of the difference between the water exchange rate constant for a single coordinated water on the $GaA^{(3-a)}$ complexes and $[Ga(OH_2)_6]^{3+}$, $R(A)$, shows an approximately linear variation with the electron-donor constant, $E(A)$, of A^{a-}. The slope $R(A)/E(A) = \gamma = 1.1$ for $Ga(III)$ which compares with $\gamma = 0.47, 0.40$, and 0.30 for $Co(II)$, $Ni(II)$, and $Zn(II)$. This implies that as the hardness of the metal center increases, so does the lability of coordinated water show increased sensitivity to the electron-donating ability of the other ligands.

The complexation of $Fe(III)$ by hydroxamic acid based ligands has been an area of substantial study, because such ligands resemble the siderophores which mediate biological $Fe(III)$ transport. A stopped-flow and rapid-scan UV/visible spectroscopic study of the equilibria in equations (21)–(23) (where

$$[FeA_3] + H^+ \underset{k_{-1}}{\overset{k_1}{\rightleftharpoons}} [FeA_2]^+ + HA \tag{21}$$

$$[FeA_2]^+ + H^+ \underset{k_{-2}}{\overset{k_2}{\rightleftharpoons}} [FeA]^{2+} + HA \tag{22}$$

$$[FeA]^{2+} + H^+ \underset{k_{-3}}{\overset{k_3}{\rightleftharpoons}} Fe^{3+} + HA \tag{23}$$

$A^- = $ bidentate CH_3CONHO^- and the $Fe(III)$ complexes are six-coordinate, but coordinated water is not shown) in aqueous solution at 209.2 K yields: $k_1 = 1.0 \pm 0.02 \times 10^5$, $k_{-1} = 1.7 \pm 0.07 \times 10^3$, $k_2 = 1.4 \pm 0.03 \times 10^3$, $k_{-2} = 1.6 \pm 0.6 \times 10^3$, $k_3 = 0.06$, and $k_{-3} = 4.8 \text{ dm}^3 \text{ mol}^{-1} \text{ s}^{-1}$ over a range of concentration conditions which do not appear to cause major variations.[95] The values of k_{-1} and k_{-2} are very similar to $k_f \sim 2 \times 10^3 \text{ dm}^3 \text{ mol}^{-1} \text{ s}^{-1}$ for HA substitution on $[Fe(H_2O)_5OH]^{2+}$, which proceeds through an I_d mechanism, and it appears that coordinated hydroxamate has a labilizing effect similar to hydroxide and that substitution of coordinated water in $[FeA_2]^+$ and $[FeA]^{2+}$ proceeds through an I_d mechanism. A related study[96] finds that the formation of $[Fe(H_2O)_4(4\text{-}CH_3OC_6H_4C(X)N(O)H)]^{2+}$ (X = O, S) proceeds through parallel paths involving $[Fe(H_2O)_6]^{3+}$ and $[Fe(H_2O)_5OH]^{2+}$, where I_a mechanisms are considered to operate in both cases, although in the latter case the a character is probably a consequence of hydrogen bonding between the hydroxo ligand and the entering ligand. Initial bond formation for both paths occurs through the $\rangle C{=}O$ or $\rangle C{=}S$ donor atoms. A stopped-flow spectrophotometric study of imidazole substitution on the $Fe(III)$ centers of cytochrome c and cytochrome c with $Ru(III)(NH_3)_5$ bound to the His-33 residue (in 1.0 mol dm^{-3} NaCl aqueous solution at pH 6.9) yields: $k_1(298.2 \text{ K}) = 28 \pm 2 \text{ s}^{-1}$, $\Delta H^\ddagger = 58.2 \pm 2.1 \text{ kJ mol}^{-1}$, and $\Delta S^\ddagger = -21 \pm 4 \text{ J K}^{-1} \text{mol}^{-1}$; and $k_{-2}(298.2 \text{ K}) = 1.5 \pm 0.1 \text{ s}^{-1}$, $\Delta H^\ddagger = 46.9 \pm 2.1 \text{ kJ mol}^{-1}$, and $\Delta S^\ddagger = -84 \pm 8 \text{ J K}^{-1} \text{mol}^{-1}$, for imidazole (Im) substitution shown in Scheme 3 in which Met-80 and His-18 are cytochrome protein residues.[97]

Scheme 3

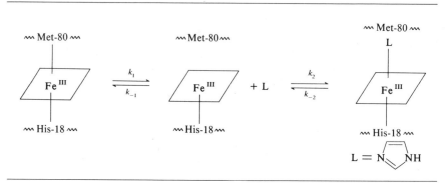

$$L = N\!\!\diagup\!\!\diagdown NH$$

The chemistry of vitamin B_{12} continues to stimulate interest in Co(III) model systems as exemplified by a ^1H NMR study of 3,5-lutidine exchange on [Co(saloph)(3,5-lutidine)R] [where saloph is the dianion of disalicylidine-o-phenylenediamine, which occupies the four axial coordination sites of Co(III), and R is an alkyl group *trans* to 3,5-lutidine].[98] When R = CH_2CN, CH_2CF_3, and CH_3, respectively, k_b(298.2 K) = 2.25 ± 0.27 × 10^4, 4.35 ± 0.21 × 10^5, and 2.75 ± 0.35 × 10^8 s^{-1} in CDCl$_3$ in the first case and in CD$_2$Cl$_2$ in the last two cases for 3,5-lutidine dissociation, which proceeds through a five-coordinate intermediate. This variation of k_b illustrates the variation of the *trans* labilizing effect in these species which are *ca* 10 times more labile than the corresponding [Co(DH)$_2$(3,5-lutidine)R] (where DH is the monoanion of dimethylglyoxime), which in turn demonstrates a substantial variation in the *cis* labilizing effect. There appears to be a synergy between the *trans* and *cis* effects. Ligand (L) exchange has also been investigated in the [Co{(DO)-(DOH)pn}(L)R]$^+$ [where (DO)(DOH)pn is the mono anion of $N^2,N^{2'}$-propanediylbis(2,3-butanedione-2-imine-3-oxime] species for P and N donor ligands, but their lability is considerably less than for [Co(saloph)(3,5-lutidine)R].[99,100] Rapid axial ligand substitution has been studied for reaction (24) where L is one of three tetraaza-tetraimine macrocyclic ligands and

$$[CoL(CH_3CN)_2]^{3+} + 2X^- \longrightarrow [CoLX_2]^+ + 2CH_3CN \qquad (24)$$

$X^- = Cl^-$ or Br$^-$.[101] The substitution proceeds through 1:1 and 2:1 encounter complexes, with the slower substitution of the second CH_3CN being characterized by rate constants (X = Br$^-$) varying from 0.873–48.5 s^{-1} at 303.2 K as L becomes more electron withdrawing. Ligand substitution is considered to proceed through an I_a mechanism, which is unusual for Co(III). The rapid formation of O-bonded sulfito complexes from *trans*-[Co(en)$_2$(H$_2$O)$_2$]$^{3+}$ and SO$_2$ in water evidently proceeds through substitution on the oxygen of a hydroxo ligand rather than on Co(III).[102]

Ligand substitution studies on low-spin d^7 Ni(III) are of considerable interest as the size of this ion is similar to that of low-spin Co(III), which is one of the few trivalent transition-metal ions to exhibit d character in its ligand substitution processes. For reaction (25) in aqueous perchloric acid it is found

$$[\text{NiL}(\text{H}_2\text{O})_2]^{3+} + \text{X}^- \underset{k_b}{\overset{k_f}{\rightleftharpoons}} [\text{NiL}(\text{H}_2\text{O})\text{X}]^{2+} + \text{H}_2\text{O} \tag{25}$$

that when L = [α]-C-*meso*-5,12-dimethylcyclam, k_f(298.2 K) = 2180 ± 66 and 889 ± 31 dm^3 mol^{-1} s^{-1} for X$^-$ = Cl$^-$ and Br$^-$, respectively; when L = C-*meso*-5,12-diethylcyclam, k_f(298.2 K) = 3100 ± 400 and 2800 ± 200 dm^3 mol^{-1} s^{-1} for X$^-$ = Cl$^-$ and Br$^-$, respectively; and when L = C-*rac*-5,12-dimethylcyclam, k_f(286.6 K) = 14 000 ± 2000 dm^3 mol^{-1} s^{-1} when X$^-$ = Cl$^-$.[103] The variation of k_f with the nature of X$^-$ for a given L is small, which suggests the operation of a d activation mode. Reaction (25) is followed by a second reaction thought to be the elimination of H$_2$O to produce a five-coordinate species.

The decomplexation of 1,7-diaza-4,10,13-trioxacyclopentadecane-N,N'-diacetate (L^{2-}, which has seven donor atoms) complexes of a range of lanthanide(III) ions has been studied by stopped-flow methods using Cu^{2+} as a scavenger for free L^{2-}.[104,105] A multipath decomplexation mechanism is proposed in Scheme 4 (from which coordinated water is omitted) in which

Scheme 4

LnL^{+*} and LnHL^{2+*} represent species in which L^{2-} and HL$^-$ are partially decomplexed. There is no decomplexation-rate dependence on Cu^{2+} concentration, which indicates Cu^{2+} is not involved in the rate-determining step for decomplexation. (This mechanism is similar to those proposed for the decomplexation of other Ln complexes of multidentate ligands.[106]) When Ln = La, Pr, and Eu the observed first-order decomplexation rate constant is expressed by equation (26), and when Ln = Tb, Er, Yb, and Lu it is expressed by equation (27), both of which incorporate rate constants from Scheme 4.

$$k_{obs} = k_d + k_H[H^+] + k_{ac}[\text{acetate}] \tag{26}$$

$$k_{obs} = k_d + (K_H k_{lim}[H^+])/(1 + K_H[H^+]) \tag{27}$$

The acetate-catalyzed (k_{ac}) decomplexation of the lighter Ln complexes is probably a result of the coordination of acetate and the consequent decrease of complex charge rendering the complex more amenable to protonation. The absence of an acetate term in equation (27) for the heavier Ln complexes is attributed to differences in solution structures and the tighter binding of coordinated water molecules by comparison to those thought to be displaced in the lighter Ln complexes. It appears that the heavier Ln complexes are characterized by a relatively high protonation constant ($K_H = k_H/k_{-H}$) such that, in the [H$^+$] range 8.4×10^{-6}-2.5×10^{-4} mol dm^{-3}, a limiting k_{lim} is observed for LnHL^{2+}, while in the case of the lighter Ln complexes a linear dependence on [H$^+$] characterized by k_H is observed. There is an approximate inverse relationship between both log k_d and log k_H and the logarithm of the stability constant of LnL$^+$. A selection of values for the parameters of equations (26) and (27) appears in Table 9.5. Polyamino polycarboxalate ligands labilize Cr(III) and other trivalent metal complexes toward ligand substitution[107] as in reaction (28), where both Co(III) and Cr(III) are six coordinate and edta^{4-}

$$[(NH_3)_5Co\{(edta)Cr(H_2O)\}]^{2+} + X^- \underset{k_b}{\overset{k_f}{\rightleftharpoons}} [(NH_3)_5Co\{(edta)CrX\}]^+ + H_2O \tag{28}$$

forms one and five bonds to the first and second metal centers, respectively.[108] When X$^-$ = OAc$^-$, N$_3^-$, and NCS$^-$, $k_f = 44.5 \pm 1.1$, 676 ± 15, and 133 ± 1 dm^3 mol^{-1} s^{-1}, and $k_b = 2.1 \pm 0.4$, 7.25 ± 0.08, and 4.2 ± 0.1 s^{-1}, respectively,

Table 9.5. *Parameters for the Decomplexation of LnL$^+$ at 298.2 K in Aqueous 0.1 mol dm^{-3} LiClO$_4$ (Reference 104)*[a]

Ln	$10^3 k_d'$ (s^{-1})	k_H (dm^3 mol^{-1} s^{-1})	$10^{-3} K_H$ (dm^3 mol^{-1})	$10^2 k_{lim}$ (s^{-1})	$10^2 k_{acetate}$ (dm^3 mol^{-1} s^{-1})
La	46.9 ± 1.5	718 ± 11			218 ± 4
Pr	4.14 ± 0.20	81.9 ± 1.5			
Eu	1.76 ± 0.06	31.1 ± 0.4			8.09 ± 0.23
Tb	2.18 ± 0.11		2.96 ± 0.54	1.46 ± 0.20	
Er	3.32 ± 0.13		6.64 ± 0.85	1.04 ± 0.12	
Yb	8.76 ± 0.18		9.39 ± 0.89	1.50 ± 0.13	
Lu	10.8 ± 0.6		12.2 ± 1.1	4.11 ± 0.36	

[a] $k_d' = k_d + k_{ac}[5 \times 10^{-3}]$ for La, Pr, and Eu; $k_d' = k_d$ for Tb, Er, Yb, and Lu.

in aqueous $1.00 \, \text{mol} \, \text{dm}^{-3}$ $NaClO_4$ at 298.2 K; and the k_f values are 10^6–10^8 greater than the corresponding values observed for $[Cr(H_2O)_6]^{3+}$ and $[Cr(NH_3)_5H_2O]^{3+}$. An a activation mode is suggested for the ligand substitution processes, and the nucleophilic assistance of the pendant group appears to be more important than its steric effect on the ligand substitution rate.

A ^1H and ^{19}F NMR study of the exchange of bidentate 2-thenoyltrifluoroacetonate (tta$^-$) on $[Th(tta)_4]$ in CD_3CN shows that the rate of tta$^-$ exchange is linearly dependent on the concentration of Htta in the enol form.[109] Exchange apparently proceeds through a nine-coordinate intermediate, $[Th(tta)_4(Htta_{enol})]$, in which the proton of the entering enol transfers to a tta$^-$ ligand which then rapidly dissociates to complete the ligand exchange $[k_{ex}(303.2 \, K) = 153 \pm 29 \, \text{kg} \, \text{mol}^{-1} \, \text{s}^{-1}$, $\Delta H^\ddagger = 43.9 \pm 0.5 \, \text{kJ} \, \text{mol}^{-1}$, $\Delta S^\ddagger = -58 \pm 1.5 \, \text{J} \, K^{-1} \, \text{mol}^{-1}]$. The exchange of tta$^-$ is retarded by dmso, probably as a consequence of the formation of $[Th(tta)_4(dmso)]$ which suppresses the formation of the $[Th(tta)_4(Htta_{enol})]$ intermediate. The exchange of dmso on $[Th(tta)_4(dmso)]$ in $(CD_3)_2CO$ is interpreted in terms of an I mechanism $[k_i(197.2 \, K) = 580 \pm 20 \, s^{-1}, K_0 = 3.3 \, \text{kg} \, \text{mol}^{-1}, \Delta H^\ddagger = 25.1 \pm 2.5 \, \text{kJ} \, \text{mol}^{-1}$, and $\Delta S^\ddagger = -65 \pm 13 \, \text{J} \, K^{-1} \, \text{mol}^{-1}]$. Similar mechanisms are proposed for the exchange of acetylacetonate, acac$^-$, on $[Th(acac)_4]$,[110] $[U(acac)_4]$,[111] and $[UO_2(acac)_2L]^{2+}$ (L = dmso, dmf, or trimethylphosphate) and the substitution of acac$^-$ by dibenzoylmethanate in the latter species.[2,3] The observed lability toward acac$^-$ exchange: $[Th(acac)_4] \geq [U(acac)_4] > [Zr(acac)_4] \approx [Hf(acac)_4] > [Ti(acac)_3]^+$, is explicable in terms of variation in ionic radii and electronic configurations.[110]

The relevance of dioxygen metalloporphyrin complexes to the understanding of biological oxygen-binding systems[112] has stimulated an interest in the reaction of metalloporphyrins with hydrogen peroxide. Thus the observed rate constant (k_{obs}) for hydrogen peroxide substitution on 5,10,15,20-(4-N-methylpyridiniumyl)porphineoxotitanium(IV), $[TiO(tmpyp)]^{4+}$, to produce $[Ti(O_2)$-$(tmpyp)]^{4+}$ is found to be of the form in equation (29) where, at 298.2 K,

$$k_{obs} = k + k_H[H^+] \qquad (29)$$

$k = 2.86 \times 10^{-2} \, \text{dm}^3 \, \text{mol}^{-1} \, \text{s}^{-1}$, $k_H = 51.7 \, \text{dm}^6 \, \text{mol}^{-2} \, \text{s}^{-1}$, $\Delta H^\ddagger$ and $\Delta H_H^\ddagger = 40.1 \pm 1.0$ and $38.5 \pm 1.0 \, \text{kJ} \, \text{mol}^{-1}$; $\Delta S^\ddagger$ and $\Delta S_H^\ddagger = -140 \pm 3$ and $-83 \pm 3 \, \text{J} \, K^{-1} \, \text{mol}^{-1}$; and $\Delta V^\ddagger$ and $\Delta V_H^\ddagger = -18.6 \pm 0.3$ and $-3.9 \pm 0.2 \, \text{cm}^3 \, \text{mol}^{-1}$ in aqueous $1.0 \, \text{mol} \, \text{dm}^{-3}$ ionic strength solution ($HNO_3/NaNO_3$).[113] The rate-determining step is the coordination of hydrogen peroxide through an I_a mechanism, and the k_H term arises from a reaction path in which the oxo group of $[TiO(tmpyp)]^{4+}$ is protonated. A similar study of hydrogen peroxide substitution on $[MoO(tmpyp)H_2O]^{5+}$ {which undergoes protolysis prior to the rate-determining step to produce $[MoO(tmpyp)(O_2)]^{3+}$} yields $k(298.2 \, K) = 1.05 \times 10^3 \, \text{dm}^3 \, \text{mol}^{-1} \, \text{s}^{-1}$, $\Delta H^\ddagger = 37 \pm 1 \, \text{kJ} \, \text{mol}^{-1}$, $\Delta S^\ddagger = -63 \pm 3 \, \text{J} \, K^{-1} \, \text{mol}^{-1}$, and $\Delta V^\ddagger = -0.2 \pm 0.3 \, \text{cm}^3 \, \text{mol}^{-1}$.[114] In this case an I mechanism operates.

Other reaction paths involving monomeric and dimeric species are also encountered in this Mo(V) system.

The equilibria between the six-coordinate V(IV) species in equation (30) [where the L^{n-} studied are shown in structures (1), (2), and (3) with $X^- = NCS^-$

$$[VO(L)(H_2O)]^{(2-n)+} + X^{-1} \underset{k_b}{\overset{k_f}{\rightleftharpoons}} [VO(L)(X)]^{(1-n)+} + H_2O \qquad (30)$$

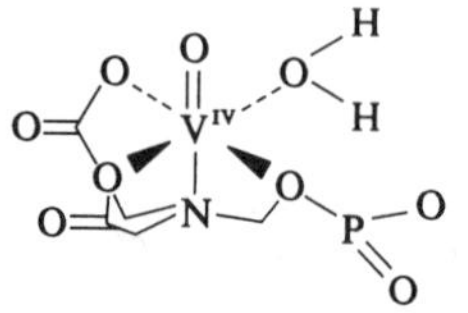

$[VO(pida)(H_2O)]^{2-}$	$[VO(nta)(H_2O)]^-$	$[VO(pmida)(H_2O)]$
1	**2**	**3**

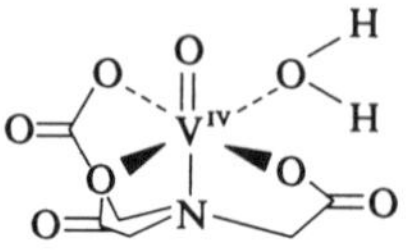

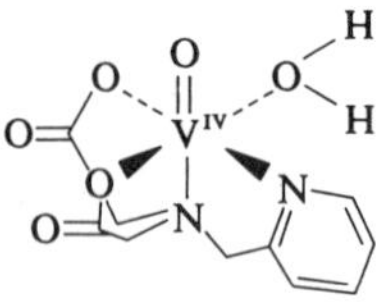

or N_3^-] are characterized by the parameters in Table 9.6.[115] The k_f characterizing substitution by NCS^- and N_3^- for all three complexes are substantially different, consistent with the operation of an I_a mechanism. The large $\Delta H_f^{\ddagger}$ and positive $\Delta S_f^{\ddagger}$ characterizing $[VO(pida)(H_2O)]^{2-}$ are attributed to the effects of hydrogen bonding between the phosphonate group and the aqua ligand on the substitution process. The displacement of glycine from $[VO(gly)_2H_2O]$ by oxalate to form $[VO(ox)_2H_2O]^{2-}$ (H_2O occupies an axial site in both complexes) occurs in two steps.[116] The rate-determining formation of a six-coordinate intermediate, in which the $-NH_2$ group of the leaving glycinate is displaced by water, is characterized by $k_1(298.2\ K) = 11.3\ s^{-1}$. This water and the monodentate glycinate are then displaced by the first oxalate in faster steps. The displacement of the second glycinate occurs through a similar series of reactions and is characterized by $k_2(298.2\ K) = 29.7\ s^{-1}$. In the pH range 6–10 $trans$-$[WO_2(CN)_4]^{4-}$ protonates to form $trans$-$[WO(H_2O)(CN)_4]^{2-}$, which undergoes substitution by N_3^- to form $[WO(N_3)(CN)_4]^{3-}$ characterized by $k_f = 4.2 \pm 1\ dm^3\ mol^{-1}\ s^{-1}$, $k_b = 0.20 \pm 0.06\ s^{-1}$, and $\Delta V_f^{\ddagger} = 10.6 \pm 0.05\ cm^3\ mol^{-1}$, considered consistent with the operation of a d activation mode.[117]

In one of the few reported kinetic studies of the complexation of a polyatomic cation by crown ethers, it is found that a 1:1 complex forms between UO_2^{2+} and 18C6, and 1:1 and 1:2 complexes are produced by UO_2^{2+} and diaza-18C6 (1,7,10,16-tetraoxa-4,13-diazacyclooctadecane) in propylene carbonate.[4] In the presence of excess 18C6 concentration three consecutive steps are observed in the formation of $UO_2(18C6)^{2+}$(inclusive), and are assigned to the reactions shown in Scheme 5. In the first step rapid preequilibria produce two encounter complexes $UO_2^{2+}\cdots L$ and $L\cdots UO_2^{2+}\cdots L$ ($L = 18C6$), the first of which leads to the formation of UO_2L^{2+}(external) in which a single bond to UO_2^{2+} is formed and partial UO_2^{2+} desolvation occurs. The second step produces UO_2L^{2+}(exclusive) in which further bonds are formed between UO_2^{2+} and L, and further desolvation occurs. The third step involves complete

Table 9.6. *Kinetic Data for Ligand Substitution on* $[VO(L)(H_2O)]^{(2-n)+}$ *(Reference 115)*

Complex	X^-	$k_f(298.2 \text{ K})$ (dm^3 mol^{-1} s^{-1})	$\Delta H_f^\ddagger$ (kJ mol^{-1})	$\Delta S_f^\ddagger$ (J K^{-1} mol^{-1})	$k_b(298.2 \text{ K})$ (s^{-1})	$\Delta H_b^\ddagger$ (kJ mol^{-1})	$\Delta S_b^\ddagger$ (J K^{-1} mol^{-1})
[VO(pida)(H$_2$O)]$^{2-}$	NCS$^-$	0.004	123	128	0.028	66	−54
[VO(nta)(H$_2$O)]$^-$	NCS$^-$	0.62	41	−122	0.170	69	−25
[VO(pmida)(H$_2$O)]	NCS$^-$	0.26	49	−90	0.061	61	−61
[VO(pida)(H$_2$O)]$^{2-}$	N$_3^-$	0.056	88	31	0.084	62	−53
[VO(nta)(H$_2$O)]$^-$	N$_3^-$	4.10	54	−52	0.480	53	−75
[VO(pmida)(H$_2$O)]	N$_3^-$	3.70	47	−75	0.130	64	−47

Scheme 5

$$UO_2^{2+} + 2L \overset{K_1}{\rightleftharpoons} UO_2^{2+}\cdots L + L \overset{K_1'}{\rightleftharpoons} L\cdots UO_2^{2+}\cdots L$$

$$\downarrow k_1$$

$$UO_2L^{2+} + L \overset{K_2}{\rightleftharpoons} L\cdots UO_2L^{2+}$$
$$\text{(external)} \qquad\qquad \text{(external)}$$

$$\downarrow k_2$$

$$UO_2L^{2+} \overset{k_3}{\underset{k_{-3}}{\rightleftharpoons}} UO_2L^{2+}$$
$$\text{(exclusive)} \qquad \text{(inclusive)}$$

$(k_1 = 930 \pm 50 \text{ s}^{-1}$, $K_1 = 145 \pm 10 \text{ dm}^3 \text{ mol}^{-1}$, $K_1' = 420 \pm 20 \text{ dm}^3 \text{ mol}^{-1}$, $k_2 = 18 \pm 2 \text{ s}^{-1}$, $K_2 = 90 \pm 10 \text{ dm}^3 \text{ mol}^{-1}$, and $k_3 + k_{-3} = 0.022 \pm 0.002 \text{ s}^{-1}$ in propylene carbonate at 298.2 K)

desolvation of UO_2^{2+} as it enters the center of the cavity formed by L to produce UO_2L^{2+}(inclusive). Neither of the encounter complexes $L\cdots UO_2^{2+}\cdots L$ or $L\cdots UO_2L^{2+}$(external) leads to further reaction. Four steps are observed for the formation of $[UO_2(\text{diaza-18C6})]^{2+}$(inclusive) and $[UO_2(\text{diaza-18C6})_2]^{2+}$(biexternal) (where in the latter species each diaza-18C6 is bonded to UO_2^{2+} through a single nitrogen) in the presence of excess diaza-18C6 concentration, and the proposed mechanism is similar to that shown in Scheme 5 with the addition of another step for the formation of $[UO_2(\text{diaza-18C6})_2]^{2+}$(biexternal). The simpler substitution of salicylate on $UO_2(\text{OH})^+$ in water apparently proceeds through a fast first-bond formation between a carboxylate oxygen and uranium(VI), which is followed by a slower ring-closure step $[k(298.2 \text{ K}) = 52.6 \pm 0.02 \text{ s}^{-1}$, $\Delta H^{\ddagger} = 52.2 \pm 0.20 \text{ kJ mol}^{-1}$, and $\Delta S^{\ddagger} = -37.7 \pm 0.45 \text{ J K}^{-1} \text{ mol}^{-1}]$ involving the condensation of the salicylate phenolic group and the OH^- ligand to eliminate a water molecule.[5] For salicylate and substituted salicylates, k decreases as the acidity of the carboxylic acid group of the salicylic acids increases. The kinetic aspects of the potential use of multidentate ligands in the extraction of UO_2^{2+} from sea water have been considered in two studies.[118,119]

Part 3

Reactions of Organometallic Compounds

Chapter 10
Substitution and Insertion Reactions

10.1. Substitution Reactions

10.1.1. Carbon Monoxide Displacement in Mononuclear Complexes

Rest and co-workers report the photochemically induced carbonyl dissociation of σ-allyl complexes of manganese and tungsten in frozen gas matrices at approximately 12 K.[1] For $[Mn(\eta^1\text{-}C_3H_5)(CO)_5]$ the loss of the carbonyl ligand is accompanied by a $\sigma \rightarrow \pi$ rearrangement of the allyl ligand to produce $[Mn(\eta^3\text{-}C_3H_5)(CO)_4]$. The tetracarbonyl species loses another carbonyl ligand reversibly to produce $[Mn(\eta^3\text{-}C_3H_5)(CO)_3]$. In the tungsten complex $[W(\eta^1\text{-}C_3H_5)(\eta^5\text{-}C_5H_5)(CO)_3]$ loss of a carbonyl ligand and $\sigma \rightarrow \pi$ rearrangement occur in two steps. Isolation of the 16-electron intermediate $[W(\eta^1\text{-}C_3H_5)(\eta^5C_5H_5)(CO)_2]$ shows that the loss of the carbonyl ligand must occur before the $\sigma \rightarrow \pi$ allyl rearrangement. The authors suggest that CO dissociation and $\sigma \rightarrow \pi$ rearrangement in general proceed in solution by two-step reactions even when the intermediate stages cannot be observed. The reaction for the tungsten case is shown in equation (1).

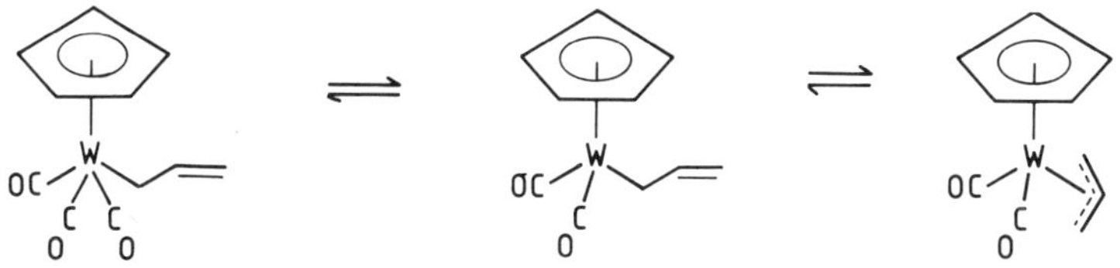

$$\tag{1}$$

The reactivity of $[(\eta^5\text{-}C_9H_7)V(CO)_4]$ with $P(n\text{-Bu})_3$ and $P(OEt)_3$ to form the phosphine-substituted compound was studied by Basolo and co-workers.[2]

They found that the reaction mechanism is CO dissociative (S_N1) with kinetic parameters of $k(100\,°C) = 3.7 \times 10^{-4}\,s^{-1}$, $\Delta H^{\ddagger} = 31.9 \pm 0.3\,kcal\,mol^{-1}$, and $\Delta S^{\ddagger} = 10.5 \pm 0.9\,cal\,mol^{-1}\,K^{-1}$ for the reaction with $P(n\text{-Bu})_3$ in decalin. Comparison of the rate of substitution of the indenyl complex to that of the cyclopentadienyl shows a rate enhancement for the indenyl over the cyclopentadienyl ligand of 13-fold at $100\,°C$ (for $[(\eta^5\text{-}C_5H_5)V(CO)_4]$ substitution, $\Delta H^{\ddagger} = 35.5\,kcal\,mol^{-1}$ and $\Delta S^{\ddagger} = 14.6\,cal\,mol^{-1}\,K^{-1}$). This is not as large an effect from the indenyl ligand as in the associative mechanism, where a rate enhancement of 10^8-fold over the Cp-substituted complex is observed.[3] The rate enhancement is believed to arise from stabilization of the transition state by interaction of the six-membered aromatic ring of the indenyl ligand with the metal center, thus compensating for the loss of M—CO bonding. A similar increase in rate is seen by substituting carbocyclic ligands with $\eta^5\text{-}N$-heterocycle ligands, such as in reaction of the $\eta^5\text{-}N$-heterocycle manganese tricarbonyls with $P(OEt)_3$, $P(n\text{-Bu})_3$, PPh_3, or $P(Cy)_3$.[4] The CO substitution reactions are second order overall, first order in metal complex and first order in nucleophile for the compounds $[(\eta^5\text{-pyrrolyl})Mn(CO)_3]$, $[(\eta^5\text{-indolyl})Mn(CO)_3]$, and $[(\eta^5\text{-1-pyrindinyl})Mn(CO)_3]$. Kinetic parameters for the pyrrolyl and indolyl species are listed in Table 10.1. The plots of the kinetic data (k_{obs} against ligand concentration) are linear and pass through the origin, which suggests the reactions are entirely second order with no evidence for a dissociative pathway (no nonzero intercept). The indolyl compound was thermally unstable, so no kinetic data are reported. The mechanism for the reaction

Table 10.1. Kinetic Data for the Substitution Reactions of $[(\eta^5\text{-}C_4H_4N)Mn(CO)_3]$ and of $[(\eta^5\text{-}C_8H_6N)Mn(CO)_3]$ in Decalin (Reference 4)

Complex	L	T (°C)	k ($M^{-1}s^{-1}$)	$\Delta H^{\ddagger}$ (kcal mol^{-1})	$\Delta S^{\ddagger}$ (eu)
$[(\eta^5\text{-}C_4H_4N)Mn(CO)_3]$	$P(n\text{-Bu})_3$	130	3.88×10^{-4}	15.5 ± 1.6	-37.4 ± 3.9
		142	7.35×10^{-4}		
		151	10.1×10^{-4}		
	$P(OEt)_3$	129	0.43×10^{-4}	22.7 ± 0.1	-22.6 ± 2.4
		143	1.19×10^{-4}		
		151	2.04×10^{-4}		
	PPh_3	130	0.19×10^{-4}		
	$P(Cy)_3$	130	0.38×10^{-4}		
$[(\eta^5\text{-}C_8H_6N)Mn(CO)_3]$	$P(n\text{-Bu})_3$	120	1.54×10^{-3}	14.9 ± 0.9	-34.0 ± 2.3
		130	2.70×10^{-3}		
		140	4.14×10^{-3}		
	$P(OEt)_3$	110	0.429×10^{-3}	14.5 ± 1.3	-36.5 ± 3.2
		120	0.826×10^{-3}		
		130	1.14×10^{-3}		
		140	1.96×10^{-3}		

of the pyrrolyl complex is shown in equation (2). The facile CO substitution of these N-heterocycles is in contrast to the reported inertness of the carbocyclic

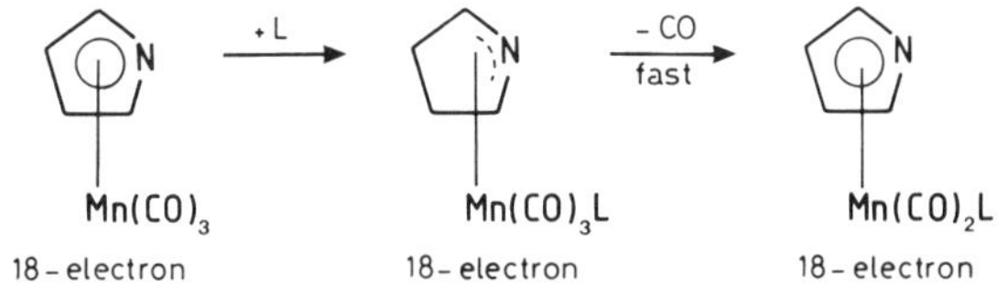

$$\underset{\substack{\text{Mn(CO)}_3 \\ \text{18-electron}}}{} \xrightarrow{+L} \underset{\substack{\text{Mn(CO)}_3 L \\ \text{18-electron}}}{} \xrightarrow[\text{fast}]{-CO} \underset{\substack{\text{Mn(CO)}_2 L \\ \text{18-electron}}}{} \tag{2}$$

analogues.[5] This greater reactivity is attributed to greater electronegativity of nitrogen compared to carbon which enhances the localization of electron density on the N-heterocyclic ligand. The pyrindinyl complex undergoes substitution at a rate 40 times faster than the corresponding indenyl derivative due to the localization of a pair of electrons on the pyrindinyl ligand which allows for the full aromaticity of the benzene ring as shown in equation (3).

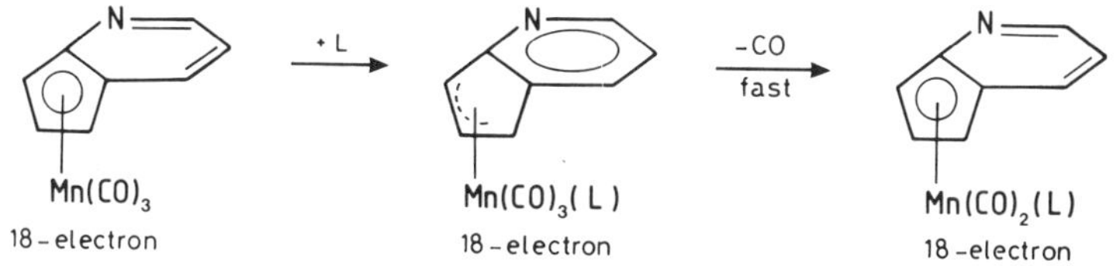

$$\underset{\substack{\text{Mn(CO)}_3 \\ \text{18-electron}}}{} \xrightarrow{+L} \underset{\substack{\text{Mn(CO)}_3(L) \\ \text{18-electron}}}{} \xrightarrow[\text{fast}]{-CO} \underset{\substack{\text{Mn(CO)}_2(L) \\ \text{18-electron}}}{} \tag{3}$$

The photochemical CO displacement in the complexes $[(\eta^5\text{-Cp})\text{-Fe(CO)}_2(\eta^1\text{-Cp})]$ (1), $[(\eta^5\text{-Cp})\text{Fe(CO)}_2(\eta^1\text{-indenyl})]$ (2), or $[(\eta^5\text{-indenyl})\text{-Fe(CO)}_2(\eta^1\text{-indenyl})]$ (3) leads to the generation of an η^3 species from the η^1 species.[6] The subsequent thermal conversion of the η^3 species to the η^5 species and the loss of the second CO ligand produce the sandwich compounds $[\text{Fe}(\eta^5\text{-Cp})_2]$, $[\text{Fe}(\eta^5\text{-Cp})(\eta^5\text{-indenyl})]$, or $[\text{Fe}(\eta^5\text{-indenyl})_2]$, respectively. The formation of the sandwich compounds is thus a two-step process as shown in equations (4) and (5) for species (1). Equation (4) appears to occur as one

$$[(\eta^5\text{-Cp})\text{Fe(CO)}_2(\eta^1\text{-Cp})] \xrightarrow{h\nu} [(\eta^5\text{-Cp})\text{Fe(CO)}(\eta^3\text{-Cp})] + CO \tag{4}$$

$$[(\eta^5\text{-Cp})\text{Fe(CO)}(\eta^3\text{-Cp})] \xrightarrow{\Delta} [\text{Fe}(\eta^5\text{-Cp})_2] + CO \tag{5}$$

step with concomitant CO loss and $\eta^1 \to \eta^3$ ring slippage occurring, as no 16-electron intermediate is observed even at temperatures as low as 40 K. The kinetics of the thermal reaction shown in equation (5) are also reported, because this permits investigation of what would ordinarily be regarded as a fast step in a reaction sequence. The monocarbonyl species derived from (1)–(3) exist as two isomers which are labeled *endo* and *exo*, as shown below for species (1). The *endo* species undergoes thermal isomerization to the *exo* species and CO loss to form the sandwich compound as shown in equation (6). The rate constant for this first order reaction is $k_1 = 1.66 \times 10^{-2}\,\text{s}^{-1}$ at

$$\underset{endo}{[(\eta^5\text{-Cp})\text{Fe(CO)}(\eta^3\text{-Cp})]} \xrightarrow[k_1]{\Delta} \underset{exo}{[(\eta^5\text{-Cp})\text{Fe(CO)}(\eta^3\text{-Cp})]} + [(\eta^5\text{-Cp})_2\text{Fe}] \tag{6}$$

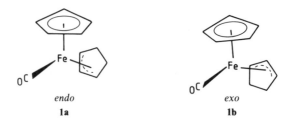

endo   exo
1a   1b

166 K. The *exo* isomer produces only the sandwich compound and does so at a much slower rate with $k_2 = 1.37 \times 10^{-4} \, s^{-1}$ at 166 K as shown in equation (7). The kinetic parameters for the first-order reaction have been measured

$$[(\eta^5\text{-Cp})Fe(CO)(\eta^3\text{-Cp})] \xrightarrow[k_2]{\Delta} [Fe(\eta^5\text{-Cp})_2] + 2CO \qquad (7)$$

for the *exo* isomers of species (1) and (2) at 298 K. For (1), $E_a = 13.9 \, \text{kcal mol}^{-1}$, $\Delta H^\ddagger = 12.9 \, \text{kcal mol}^{-1}$, and $\Delta S^\ddagger = 1.9 \, \text{cal mol}^{-1} \, K^{-1}$. For (2), $E_a = 21.7 \, \text{kcal mol}^{-1}$, $\Delta H^\ddagger = 21.0 \, \text{kcal mol}^{-1}$, and $\Delta S^\ddagger = 0.5 \, \text{cal mol}^{-1} \, K^{-1}$. It is interesting to note the large difference in the activation energies for (1) (η^3-Cp) and (2) (η^3-indenyl). This suggests that the dissociative loss of CO does not precede $\eta^3 \rightarrow \eta^5$ conversion, but instead the sandwich complexes are formed by a concerted $\eta^3 \rightarrow \eta^5$ conversion/CO loss pathway from the monocarbonyls produced photochemically from (1)–(3).

The carbonyl substitution of bis(η^5-Cp) dicarbonyl compounds of the titanium triad metals shows a marked difference in the kinetics and mechanism of reaction within the triad.[7] The mechanism of substitution with titanium complexes is dissociative, with the reaction being first order in metal complex and zero order in entering nucleophile. Zr and Hf compounds are associative with the rate being first order in both metal complex and entering nucleophile. The proposed associative mechanism involves $\eta^5 \rightarrow \eta^3 \rightarrow \eta^5$ ring slippage to maintain the 18-electron count at the metal center. Table 10.2 lists the kinetic parameters for these reactions. The differences in mechanism probably arise from steric considerations caused by the smaller size of titanium. Evidence for steric effects being important is seen in comparing the equilibrium constants for substitution for a series of phosphorous ligands, and comparing the cone angle to the equilibrium constant. Table 10.3 shows that the equilibrium constants are larger for phosphorous ligands having smaller cone angles.

The indenyl ligand effect is seen for CO exchange in the titanium compounds. The indenyl compound undergoes carbonyl exchange 11-fold faster than the Cp compound, while the Cp* (Cp* = η^5-pentamethyl Cp) derivative is only $\frac{1}{4}$ as fast as Cp.

Another type of CO displacement occurs through intramolecular nucleophilic attack on a coordinated carbonyl ligand.[8] The substitution of $[Ru(CO)_3(PPh_3)_2]$ by two molecules of 2,2'-dipyridyldisulfide goes through intermediate (4), which has a structure indicative of incipient intramolecular

Table 10.2. Kinetic Parameters for Substitution Reactions of $[(\eta^5\text{-ring})_2M(CO)_2]$ with CO or PR_3 in Decalin (Reference 7)

$[(\eta^5\text{-ring})_2M(CO)_2]$	$\Delta H^\ddagger$ (kcal mol^{-1})	$\Delta S^\ddagger$ (eu)	$k\ s^{-1\,a}$ (at 45 °C)	$k_{rel}{}^b$
$[Cp_2Ti(^{13}CO)_2]^c$	26.9 ± 0.4	11.7 ± 1.1	8.14 × 10^{-4f}	1.0
$[Cp_2Ti(CO)_2]^d$	27.9 ± 1.8	15.0 ± 5.6	8.78 × 10^{-4f}	1.08
$[Cp_2^*Ti(^{13}CO)_2]^c$	28.5 ± 0.5	14.1 ± 1.5	2.15 × 10^{-4f}	0.26
$[Ind_2Ti(^{13}CO)_2]^c$	24.1 ± 1.6	7.9 ± 5.4	9.11 × 10^{-3f}	11.2
$[Cp_2Zr(CO)_2]^e$	12.0 ± 0.4	−31.2 ± 1.4	6.22 × 10^{-3g}	7.64
$[Cp_2HfCO)_2]^c$	12.6 ± 0.8	−31.0 ± 2.6	4.22 × 10^{-1h}	3.01
$[Cp_2Hf(CO)_2]^d$	15.2 ± 0.6	−31.0 ± 1.2	3.91 × 10^{-5g}	0.048

^{a} First- or second-order rate constant as marked.
^{b} Relative reaction rates for $[PR_3] = 1\ M$ or for CO pressure of 1 atm (5.8 × 10^{-3} M at 45 °C).
^{c} Reactions with CO at 1 atm.
^{d} Reactions with PMe$_2$Ph.
^{e} Reactions with P(OEt)$_3$.
^{f} First-order rate constant.
^{g} Second-order rate constant (s^{-1} M^{-1}).

nucleophilic attack on a coordinated carbonyl group as shown in equation (8). The N–C distance in (4) indicated by the dotted line is 2.782(7), which

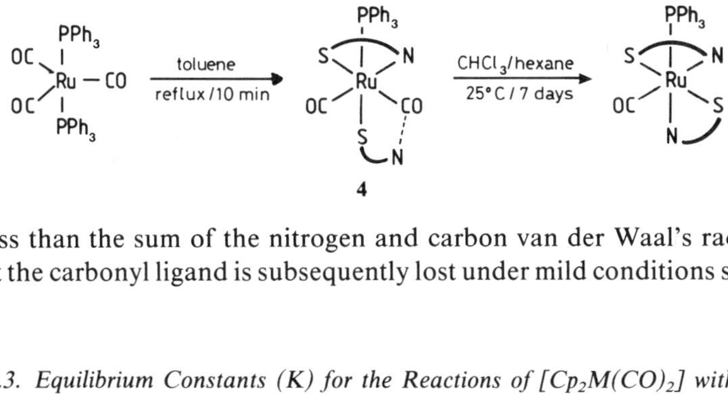

(8)

is 0.5 less than the sum of the nitrogen and carbon van der Waal's radii. The fact that the carbonyl ligand is subsequently lost under mild conditions suggests

Table 10.3. Equilibrium Constants (K) for the Reactions of $[Cp_2M(CO)_2]$ with PR_3 in Decalin at 30.0 °C (Reference 7)

$[Cp_2M(CO)_2]$	PR_3	Cone angle	ΔHNP	K
$[Cp_2Ti(CO)_2]$	PMe$_3$	118	114	1.9 ± 0.2 × 10^{-3}
$[Cp_2Ti(CO)_2]$	PMe$_2$Ph	122	281	2.7 ± 0.3 × 10^{-4}
$[Cp_2Ti(CO)_2]$	PMePh$_2$	136	424	6.7 ± 0.2 × 10^{-5}
$[Cp_2Ti(CO)_2]$	PPh$_3$	145	573	7 ± 2 × 10^{-7}
$[Cp_2Ti(CO)_2]$	P(n-Bu)$_3$	132	131	1.7 ± 0.2 × 10^{-6}
$[Cp_2Ti(CO)_2]$	P(OEt)$_3$	109	520	2.9 ± 0.3 × 10^{-4}
$[Cp_2Ti(CO)_2]$	P(O-i-Pr)$_3$	130	500	2.8 ± 0.4 × 10^{-6}
$[C_2Zr(CO)_2]$	P(OEt)$_3$	109	520	3.3 ± 0.1 × 10^{-3}

Table 10.4. *Associative Rate Constants for Substitution*
into $[Mn(CO)_5]$ *(Reference 10)*

Ligand	Cone angle	$k(M^{-1}s^{-1})$
$P(n\text{-}Bu)_3$	132	$1.0 \pm 0.1 \times 10^9$
$P(i\text{-}Pr)_3$	160	$6.7 \pm 0.7 \times 10^7$
$P(O\text{-}i\text{-}Pr)_3$	130	$3.1 \pm 0.3 \times 10^7$
PPh_3	145	$1.7 \pm 0.2 \times 10^7$
$AsPh_3$	~140	$6.5 \pm 0.8 \times 10^4$

that the short distance is due to incipient nucleophilic attack rather than molecular packing interactions or stereochemical constraints. The infrared spectrum of (4) has the higher of the two $\nu(CO)$ values at 2035 cm^{-1}, which is in the range for the carbonyl ligands susceptible to nucleophilic attack.

Electron-transfer processes may also be used to activate the nucleophilic substitution of a carbonyl ligand.[9] Such is the case for substitution of $[(N\text{-}N)Fe(CO)_3]$ (N-N = 2,3-diazanorbornene) by $P(OMe)_3$, PPh_3, or $P(OPh)_3$. Reduction of the iron complex produces the radical anion, which undergoes nucleophilic substitution more quickly than the neutral species. This provides the first example of electron-induced nucleophilic substitution observed in a mononuclear complex under reduction conditions. Interest in substitution of radical species also extends to those generated photochemically.[10] Substitution of $[Mn(CO)_5]$ radicals by PPh_3 or $AsPh_3$ in hexane has been shown to occur by an associative mechanism. Competition studies between CCl_4 and PPh_3 or CH_2Br_2 and $AsPh_3$ were used to establish the mechanism. The second-order rate constants are listed in Table 10.4. The transition state or intermediate must be a 19-electron species with a σ-acyl-like carbonyl ligand.

Competition experiments have also been used to study the substitution reactions of the $[CpW(CO)_3]$ radical with phosphines and phosphites.[11] The only metal containing products from the reaction of $[CpW(CO)_3H]$ with Ph_3C^+ in the presence of Ph_3CCl and PR_3 are $[CpW(CO)_3Cl]$ and $[CpW(CO)_2(PR_3)Cl]$. The order of reactivity of the phosphines is $PBu_3 > P(OPh)_3 > PPh_3$, and substitution is significantly inhibited by CO. An associative mechanism is proposed on the basis of these results. A 19-electron species plays a role, either as an intermediate or as a transition state. The authors suggest the 19-electron species is stabilized by either a two-center–three-electron interaction or by delocalization of an electron on the Cp ligand.

Two competing mechanistic pathways are observed for the slow CO exchange in the 17-electron complexes $[(C_5H_7)_2VCO]$, $[Cp(C_5H_7)VCO]$, and $[Cp(2,4\text{-}C_7H_{11})VCO]$.[12] This is in contrast to the rapid associative exchange observed[13] for $[Cp_2VCO]$ and $[Cp_2^*VCO]$ as shown by their relative rates in Table 10.5. CO exchange in $[(C_5H_7)VCO]$ involves both a ligand-independent

Table 10.5. *Comparison of Rates of CO Exchange at 60.0 °C in [η^5-L_2V^{13}CO]*
(References 12 and 13)

L_2	ν_{CO} (cm^{-1})	k_1 (s^{-1})	k_2 (M^{-1}s^{-1})	Rel. rate
$(C_5H_7)_2$	1959	8.1×10^{-6}	3.8×10^{-3}	1
$Cp(C_5H_7)$	1938	2.7×10^{-4}	5.7×10^{-3}	10
$Cp(2,4\text{-}C_7H_{11})$	1935	3.1×10^{-4}	1.3×10^{-3}	13
Cp_2	1881	$\sim 10^{-4}$	~ 800	1.5×10^5
Cp_2^*	1842	$\sim 10^{-4}$	~ 260	5×10^4

dissociative mechanism ($\Delta H_1^{\ddagger} = 28.1 \pm 0.4$ kcal mol^{-1} $\Delta S_1^{\ddagger} = 2 \pm$ 1 cal mol^{-1} K^{-1}) and a ligand-dependent pathway ($\Delta H_2^{\ddagger} = 22.7 \pm$ 0.4 kcal mol^{-1}, $\Delta S_2^{\ddagger} = -2 \pm 1$ cal mol^{-1} K^{-1}). The possibilities proposed for the ligand-dependent path include a dissociative interchange (I_d) mechanism, an energetically unfavorable $\eta^5 \rightarrow \eta^3$ pentadienyl ring slippage, or a u-shaped $\rightarrow$ sickle-shaped ligand rearrangement. Both steric and electronic factors are used to explain the differences in reactivities between the Cp and pentadienyl complexes. Intramolecular site exchange of carbonyl ligands in [HCr(CO)$_5$]$^-$ has been studied by variable-temperature ^{13}C NMR.[14] An Arrhenius plot yielded an activation energy of 18.5 ± 0.5 kcal mol^{-1}, $\Delta S^{\ddagger} = 14.0 \pm 2$ eu, and $\Delta G_{298}^{\ddagger} = 13.8 \pm 0.5$ kcal mol^{-1}. The two mechanisms proposed are shown in Scheme 1: **A** is a hydride migration mechanism, and **B** is a trigonal twist mechanism, with or without substantial metal–ligand bond breaking in the transition state. The second mechanism is favored.

Two reviews published this year include substitution reactions. The first paper focuses on the carbonyl derivatives of titanium, zirconium, and haf-

Scheme 1

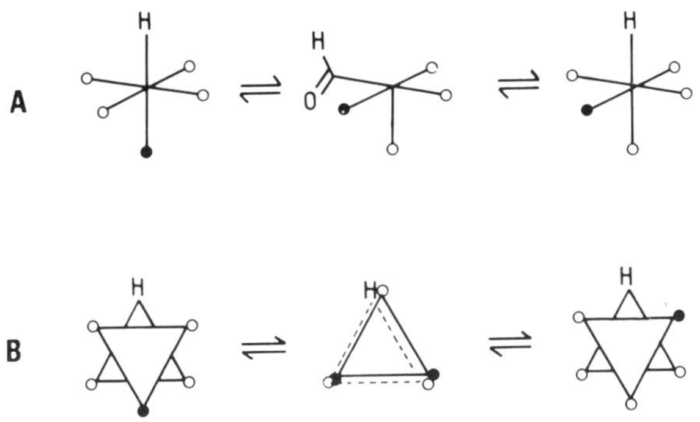

nium,[15] while the second covers ligand substitution reactions with Group V and VI donor ligands.[16]

10.1.2. Arene and Diene Displacement in Mononuclear Complexes

The first kinetic investigation of an organometallic selenocarbonyl complex has been reported by Butler and Ismail.[17] The arene substitution reactions of the chalcocarbonyl complexes $[(\eta\text{-arene})Cr(CO)_2(CX)]$ [X = Se, arene = Bz; X = S, arene = Bz, PhMe, $PhNMe_2$, $PhCO_2Me$, $o\text{-}C_6H_4Me_2$, $p\text{-}C_6H_4Me_2$, $m\text{-}C_6H_4(CO_2Me)Me$, $p\text{-}C_6H_4(CO_2Me)_2$, $p\text{-}C_6H_4(OMe)_2$, $1,3,5\text{-}C_6H_3Me_3$] by tertiary phosphite $[L = P(OMe)_3, P(OPh)_3, P(n\text{-}BuO)_3, P(OC_6H_{11})_3]$ in methyl cyclohexane to form $mer\text{-}[Cr(CO)_2(CX)L_3]$ are reported. The reactions are first order in both metal complex and entering ligand. The reactions of the seleno- and thiocarbonyl derivatives are significantly faster than those of the corresponding tricarbonyl ligand. The explanation for this is the better π-acceptor abilities of CS and CSe compared to that of CO which leads to a weaker metal–arene bond.

Arene exchange and substitution are compared in $[(\text{napthalene})\text{-}Cr(CO)_3]$.[18] The exchange and substitution reactions follow different rate laws in n-nonane. Exchange with toluene is first order in metal complex and is independent of toluene concentration. The fact that this process has an activation entropy of almost zero ($\Delta H^{\ddagger} = 31.5$ kcal mol^{-1}, $\Delta S^{\ddagger} = -0.3$ cal mol^{-1} K^{-1}) suggests that the rate-determining step involves very little interaction between the incoming arene and the metal complex. The substitution reaction with $P(OMe)_3$ to give $fac\text{-}\{Cr(CO)_3[P(OMe)_3]_3\}$, on the other hand, displays a first-order dependence on both the metal complex and the ligand. The rate is faster, and the large activation energy is negative ($\Delta H^{\ddagger} = 14.4$ kcal mol^{-1}, $\Delta S^{\ddagger} = -23.9$ cal mol^{-1} K^{-1}). This suggests a concerted mechanism in which substantial metal–ligand interaction occurs in the rate-determining step. The arene ring in $[\eta\text{-}C_7H_7W(CO)_3]^+$ can be displaced by triphenylphosphine to yield $[(PPh_3)_3W(CO)_3]$.[19] Kinetic studies suggest the mechanism involves the rapid formation of a π-complex, followed by the rate-limiting attack by a second PPh_3 ligand at the metal (Scheme 2).

An η^3 slipped-ring metal radical has been characterized crystallographically.[20] $[(\eta^3\text{-indenyl})(\eta^5\text{-indenyl})V(CO)_2]$ is the first complex of a first-row metal radical characterized structurally. These results suggest that the 17-electron systems Cp_2VCO and $(\eta^5\text{-}C_5Me_5)VCO$, which substitute CO by a bimolecular process, may involve a 16-electron transition state or intermediate that contains a slipped ring. Substitution of the arene ring in $[\eta^5\text{-}C_5H_5)Ru\text{-}(\eta^6\text{-arene})]PF_6$ and $[(\eta^5\text{-}C_5Me_5)Ru(\eta^6\text{-arene})]PF_6$ complexes (arene = naphthalene, anthracene, pyrene, chrysene, azulene) by acetonitrile is found to occur by an associative mechanism.[21] This is supported by the rate law, which is first order in both metal complex and nucleophile, and by the kinetic parameters for the reaction of $[(\eta^5\text{-}C_5Me_5)Ru(\eta^6\text{-anthracene})]^+$ with CH_3CN

Scheme 3

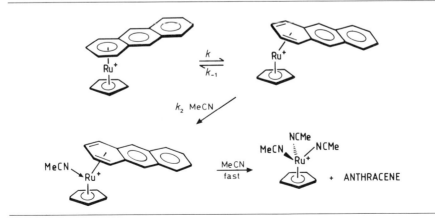

with values $\Delta H^{\ddagger} = +14.9 \pm 0.3 \text{ kcal mol}^{-1}$ and $\Delta S^{\ddagger} = -13.3 \pm 0.9$ eu. Two associative mechanisms are proposed. Scheme 3 includes a preequilibrium between an η^6-arene metal complex and an η^4-arene species, while Scheme 4 involves direct nucleophilic attack on the metal center of the η^6 complex.

Scheme 3

Scheme 4

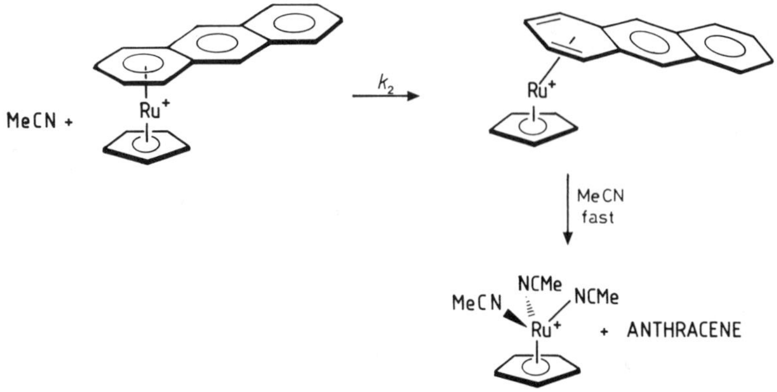

The propenyl substituent of *trans*-propenylbenzene can participate in both an intramolecular and an intermolecular fashion in arene exchange[22] in (*trans*-propenylbenzene)tricarbonylchromium with benzene involving intermediates (**5**) and (**6**), respectively. The rates of intramolecular exchange

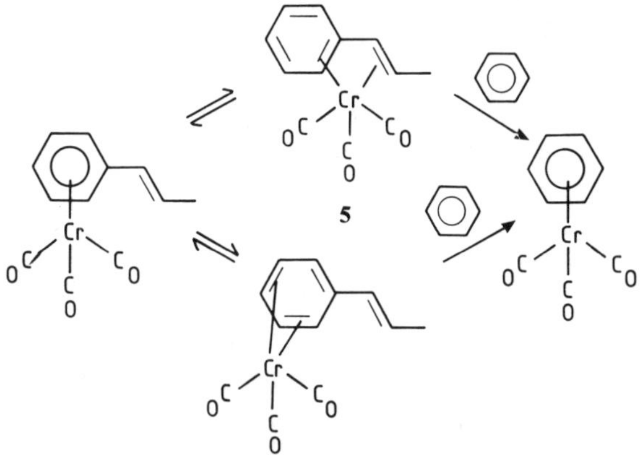

involving coordination through an alkene and the effects of added donor catalysts are reported.

A review of metal–hydrocarbon π-complexes published recently includes substitution reactions.[23]

10.1.3. Displacement of Miscellaneous Ligands

The enthalpies of ligand exchange of [L_nMo(CO)$_3$] with substituted arenes, sodium cyclopentadiene, nitriles, isocyanides, phosphines, and other ligands are reported.[24] These enthalpies span a range of over 50 kcal mol^{-1} and can be used to predict the position of thermodynamic equilibrium. The relative order of stabilization for two-electron donors is THF < cyanide ~ phospineoxide < PCl_3 < pyridine < arsines < phosphines ~ phosphites ~ isocyanides ~ carbon monoxide. The relative stabilities for arenes are benzene < toluene < *p*-xylene < *m*-xylene < *o*-xylene < mesitylene < (trimethylsilyl)-benzene < hexamethylbenzene < tetraphenylborate < dimethylaniline < *p*-bis(dimethylamino)benzene, which span a range of 7.5 kcal mol^{-1}. This is the same order observed for the $[(arene)Cr(CO)_3]$ complexes.[25]

At 25 °C, $[Et_4N][HFe(CO)_4]$ reacts with methyl iodide in CH_3CN to eliminate methane and produce the unstable $[Fe(CO)_4(NCCH_3)]$.[26] The second-order reaction, first order in iron and in CH_3I, has a rate constant of $1.7 \pm 0.1 \times 10^{-2} M^{-1} s^{-1}$. The activation parameters, $E_a = 11.2 \pm 1.0$ kcal mol^{-1}, $\Delta H^{\ddagger} = 10.6 \pm 1.0$ kcal mol^{-1}, and $\Delta S^{\ddagger} \sim -33$ cal mol^{-1} K^{-1}, are consistent with a bimolecular process. Mechanistic probes were used to determine if the reaction occurs by electron transfer, hydride transfer, or attack by the metal center on CH_3I. The results of these studies show that the reaction occurs by an S_N2 nucleophilic displacement reaction with some electron-transfer component. The rate-determining step is proposed as a nucleophilic substitution.

The exchange reactions of $[(Me_2CO)_2Ir(X)_2(PPh_3)_2]$ with free acetone when X = H or D have been studied by Crabtree and Habib.[27] The k_H/k_D ratio of 1.13 supports the proposed dissociative mechanism. This exchange is compared to the substitution of pyridine for Cl$^-$ in *trans*-$[PtHCl(PR_3)_2]$.[28] The mechanism in this case is associative and is validated by the k_H/k_D value of 1.4.

The thermolysis of $[(\eta^5\text{-}C_5Me_5)(PMe_3)Ir(Cy)(H)]$ in benzene produces cyclohexane and $[(\eta^5\text{-}C_5Me_5)(PMe_3)Ir(Ph)(H)]$.[29] Kinetic studies reveal that the reaction is first order in metal complex and is inhibited by cylohexane. The activation parameters are $\Delta H^{\ddagger} = 35.6 \pm 0.5$ kcal mol^{-1} and $\Delta S^{\ddagger} = 10 \pm 2$ eu. The mechanism is a simple reductive elimination of one hydrocarbon followed by oxidative addition of the other as shown in Scheme 5. Further mechanistic studies suggest the initial reductive elimination involves a cyclohexane/$[(\eta^5\text{-}C_5Me_5)(PMe_3)Ir]$ σ complex as an intermediate. Evidence for this proposal includes an inverse isotope effect of 0.7 for reductive elimination of cyclohexane and H/D scrambling between the hydride and the α-cyclohexyl position of $[(\eta^5\text{-}C_5Me_5)(PMe_3Ir(Cy)(H)]$ (see Scheme 6).

Dobson and co-workers have published a series of papers on the displacement of chelating ligands coordinated through sulfur donor atoms from $[(chelate)M(CO)_4]$ (M = Cr, Mo, W) complexes by alkyl and aryl phosphines and phosphites.[30-34] Both an associative and a dissociative mechanism have been proposed. In the dissociative path, the chelate ring opens to afford a five-coordinate intermediate, which can undergo competitive ring closure or attack of L as in equation(9). The associative path is probably an interchange

Scheme 5

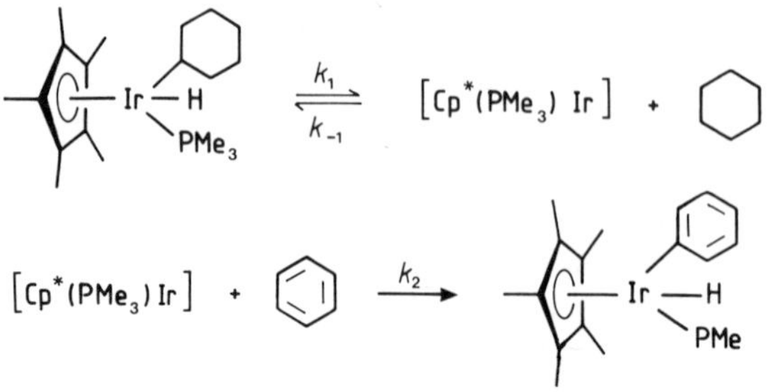

Scheme 6

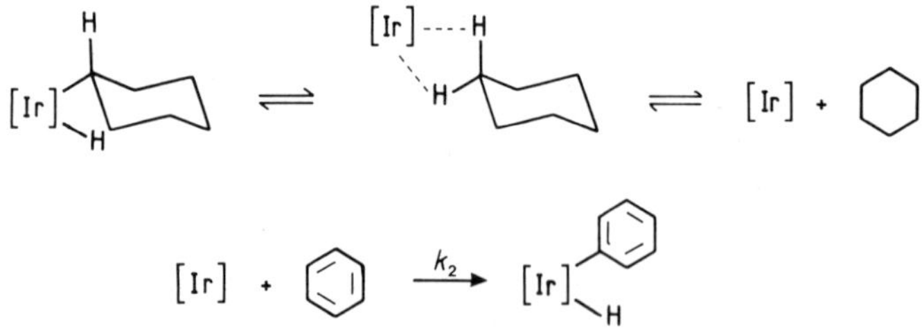

Table 10.6. Activation Parameters for the Reaction of [(chelate)Cr(CO)$_4$] with P(OEt)$_3$ (Reference 30)

S⌒S	$\Delta H^{\ddagger}$ (kJ mol^{-1})	$\Delta S^{\ddagger}$ (J K^{-1} mol^{-1})	$\Delta V^{\ddagger}$ (cm^3 mol^{-1})
dto	117 ± 3	56 ± 10	+14.7 ± 0.7 (55 °C)
BTE	92 ± 1	−7 ± 4	+14.0 ± 0.6 (45 °C)

process which involves some initial interaction between L and the metal complex as indicated in equation (10).

The use of normal kinetic parameters has not been successful in determining the mechanism in these reactions. Both possible pathways result in similar second-order rate expressions, since k_2 is the rate-determining step for the dissociative mechanism. The conventional kinetic parameters ($\Delta H^{\ddagger}$ and $\Delta S^{\ddagger}$)

have led to equivocal mechanistic assignments.[30] The volume of activation is found to be a much more sensitive measure of the mechanism. Table 10.6 lists the results of activation parameters for the reaction of [(chelate) $Cr(CO)_4$] with $P(OEt)_3$ in 1,2-dichloroethane with chelate equal to 3,6-dithiaoctane (dto) or 1,2-bis(tert-butylthio)ethane (BTE). Even though the $\Delta S^{\ddagger}$ values vary, the $\Delta V^{\ddagger}$ values are very similar. The large positive values of the activation volume rule out the possibility of an associative mechanism, and are in good agreement with the values expected for a dissociative mechanism.

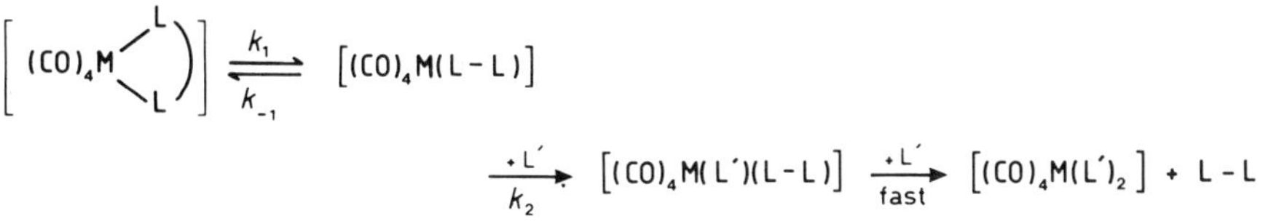

$$\left[(CO)_4M \overset{L}{\underset{L}{\diagup}} \right] \underset{k_{-1}}{\overset{k_1}{\rightleftharpoons}} \left[(CO)_4M(L-L) \right]$$

$$\overset{+L'}{\underset{k_2}{\longrightarrow}} \left[(CO)_4M(L')(L-L) \right] \overset{+L'}{\underset{fast}{\longrightarrow}} \left[(CO)_4M(L')_2 \right] + L-L$$

(9)

$$\left[(CO)_4M \overset{L}{\underset{L}{\diagup}} \right] + L' \longrightarrow \left[(CO)_4M \overset{L'}{\underset{L}{\diagdown}} \right]$$

$$\longrightarrow \left[(CO)_4M(L')(L-L) \right] \overset{+L'}{\underset{fast}{\longrightarrow}} \left[(CO)_4M(L')_2 \right] + L-L$$

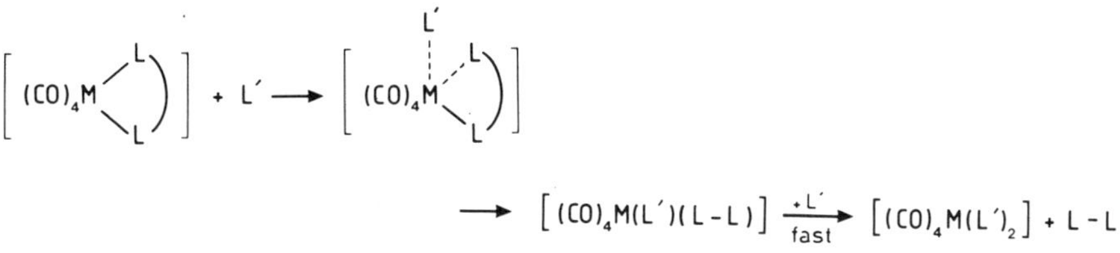

(10)

Chelating ligands with a rigid backbone are expected to inhibit dissociative ring replacement, since ring opening is more severely restricted than for a flexible ring backbone.[31] However, the displacement of the rigid chelating ligand bmtb [4-methyl-1,2-bis(methylthio)benzene] from [(bmtb)$Cr(CO)_4$] by L [L = $P(OPr^i)_3$, $P(OEt)_3$] occurs by a dissociative mechanism. The reaction obeys a second-order rate law with activation parameters, when L = $P(OPr^i)_3$, in 1,2-dichloromethane (DCE) of $\Delta H^{\ddagger} = 20.8\ kcal\ mol^{-1}$ and $\Delta S^{\ddagger} =$

Table 10.7. Activation Parameters for Reaction of Five-Coordinate Intermediates from Pulsed-Laser Flash Photolysis of [(chelate)$W(CO)_4$] (References 32 and 33)

Solvent	$\Delta H^{\ddagger}_{-1}$ (kcal mol^{-1})	$\Delta S^{\ddagger}_{-1}$ (eu)	$\Delta H^{\ddagger}_{2}$ (kcal mol^{-1})	$\Delta S^{\ddagger}_{2}$ (eu)
	[(η^1-DTO)$W(CO)_4$]			
DCE	9.8 ± 0.1	-2.9 ± 0.5		
CB	8.8 ± 0.5	-3.5 ± 1.6	8.1 ± 3.3	-9.1 ± 11.3
	[(η^1-DTN)$W(CO)_4$]			
DCE	11.9 ± 0.4	1.2 ± 1.2		
CB	11.5 ± 0.3	2.6 ± 1.5	6.6 ± 0.8	-12.9 ± 2.1

Table 10.8. Activation Parameters for Axial Ligand Substitution Reactions of
[FeN₄TL] Complexes in Toluene (Reference 35)

T	L	$\Delta H^{\ddagger}$ (kcal mol^{-1})	$\Delta S^{\ddagger}$ (cal deg^{-1} mol^{-1})
MeIm	MeIm	25.7 ± 0.8	17 ± 3
MeIm	PBu₃	33 ± 1	25 ± 4
MeIm	P(OBu)₃	29.0 ± 0.2	19 ± 1
MeIm	CO	26.8 ± 0.5	7 ± 2
py	CO	28 ± 1	14 ± 4
P(OBu)₃	MeIm	30.9 ± 0.4	19 ± 2
PBu₃	MeIm	29.5 ± 0.1	21 ± 2
BzNC	py	28.4 ± 0.7	9 ± 2
BzNC	MeIm	31 ± 1	12 ± 3
TMIC	py	31.2 ± 0.2	16 ± 1

-8.2 cal mol^{-1} K^{-1} and, in chlorobenzene, $\Delta H^{\ddagger} = 20.9$ kcal mol^{-1} and $\Delta S^{\ddagger} = -4.0$ cal mol^{-1} K^{-1}.

A dissociative mechanism is also observed for pulsed-laser flash photolysis of [(chelate)M(CO)₄] complexes (M = Cr, Mo, W).[32,33] The intermediates produced photochemically are the same five-coordinate species as are produced thermally. The activation parameters and rate consants are listed in Table 10.7 for the reactions of [(DTO)W(CO)₄] and [(DTN)W(CO)₄] with phosphines; DTO = 2,2,7,7-tetramethyl-3,6-dithiaocatne, DTN = 2,2,8,8-tetramethyl-3,7-dithianonane.

The displacement of chelating ligands coordinated through nitrogen has also been reported.[34] Dissociative axial ligand substitution in ferrous dimethylglyoxime complexes, [Fe(N)₄LT], has been reported by Stynes and Chen.[35] The activation parameters are listed in Table 10.8.

10.1.4. Displacement Reactions in Metal Dimers

Fast time-resolved IR spectroscopy has been used to provide structural information about metal carbonyl intermediates.[36–46] The technique consists of kinetic measurements made at one IR wavelength for each UV flash, with the wavelength changed between UV flashes. Data accumulated across the spectral region of interest are used to construct spectra corresponding to any time delay after the flash. This permits observation of species produced on 1.5–3 μs for solutions and 30–100 ns for gas-phase samples. Three general types of reaction have been studied since the first paper was published in 1982; (1) gas-phase photochemistry, (2) photochemistry of [Cr(CO)₆] in solution, and (3) photolysis of dinuclear metal carbonyls. The gas photochemistry of both [Fe(CO)₅] and [Cr(CO)₆] has been studied. The kinetics of substitution

Table 10.9. Rate Constants and Activation Parameters for the Reaction of
$[Cp_2Fe_2(\mu\text{-}CO)_3]$ with MeCN, $P(Bu^n)_3$, and PPh_3 in Cyclohexane
(Reference 45)

Ligand	$10^{-5} k$ $(s^{-1} dm^3 mol^{-1})$	$\Delta H^{\ddagger}$ $(kJ\,mol^{-1})$	$\Delta S^{\ddagger}$ $(J\,mol^{-1}\,K^{-1})$
MeCN	7.6 (24.0 °C)	24.4	−50.0
PBu_3^n	1.25 (24.6 °C)	28.1	−52.3
PPh_3	2.1 (24.7 °C)	23.3	−64.8

reactions in solution of $[Cr(CO)_5(solvent)]$ and $[Cr(CO)_6]$ with CO, H_2O,[37] H_2,[38,39] and N_2[39,40] have been reported. The activation energy for substitution with CO or H_2O was $5.3 \pm 1.2\,kcal\,mol^{-1}$, but the rate of substitution with H_2 was much faster than with CO.

The intermediates in the photolysis of the dinuclear metal carbonyls $[Mn_2(CO)_{10}]$, $[MnRe(CO)_{10}]$, and $[Re_2(CO)_{10}]$ have been identified. For $[Mn_2(CO)_{10}]$, both matrix isolation[41] and time-resolved IR spectroscopy[42] have shown that the primary photoproducts are $[Mn(CO)_5]$ and $[Mn_2(CO)_9]$. Both techniques suggest that $[Mn_2(CO)_9]$ has an asymmetric π-bonded CO bridge. A simlar species, $[MnRe(CO)_9]$, with bridging CO group is seen during photolysis of $[MnRe(CO)_{10}]$.[43] However, the photolysis of $[Re_2(CO)_{10}]$ produces $[Re_2(CO)_9]$ with no bridging CO group. The structure and photochemical reactions of $[Fe_2(CO)_8]$ have also been reported.[44] The study of the photosubstitution of $[(\eta\text{-}Cp)Fe(CO)_2]_2$ with time-resolved IR spectroscopy has led to the identification of the principle intermediate as $[(\eta\text{-}Cp)_2Fe_2(\mu\text{-}CO)_3]$.[45] The activation parameters for the reaction of this intermediate with MeCN, $P(n\text{-}Bu)_3$, and PPh_3 in cyclohexane solution listed in Table 10.9 are consistent with an associative bimolecular reaction mechanism.

When the Cp ligand is replaced by $\eta\text{-}C_5Me_5$ (Cp*), the primary photoproduct is the radical $[Cp^*Fe(CO)_2]^{\cdot}$ and the minor product is $[(\eta\text{-}Cp^*)_2Fe_2(\mu\text{-}CO)_3]$.[46] The radicals recombine to produce both the cis and trans isomers of $[Cp^*Fe(CO)_2]_2$. The cis isomer undergoes intramolecular isomerization to produce the more stable trans species by a first-order process with activation energy of $16.2\,kcal\,mol^{-1}$.

$[(OC)_4Mo(\mu\text{-}PEt_2)_2Mo(CO)_4]$ undergoes reversible carbonyl substitution with phosphine ligands in decalin to give $[(OC)_4Mo(\mu\text{-}PEt_2)_2Mo(CO)_3L]$ by the mechanism shown in Scheme 7.[47] Both the forward and the reverse reactions occur by a dissociative mechanism involving the reactive intermediate $[(OC)_4Mo(PEt_2)_2(CO)_3]$, which contains a coordinatively unsaturated six-coordinate molybdenum center. The value of the competition ratio, k_{-1}/k_2 varies considerably with change of phosphine from 4.38 for L = $P(OEt)_3$ to 2.09×10^5 for L = $PPh_2(C_6H_{11})$ due to steric hindrance of the incoming group. A plot of the difference in activation entropy $\Delta S_{-1}^{\ddagger} - \Delta S_2^{\ddagger}$ against Tolman's

Scheme 7

$$[(OC)_4Mo(\mu\text{-}PEt_2)_2Mo(CO)_4] \underset{k_{-1}}{\overset{k_1}{\rightleftharpoons}}$$

$$[(OC)_4Mo(\mu\text{-}PEt_2)_2Mo(CO)_3] + CO$$

$$[(OC)_4Mo(2\mu\text{-}PEt_2)_2Mo(CO)_3] + L \underset{k_{-2}}{\overset{k_2}{\rightleftharpoons}}]$$

$$[(OC)_4Mo(\mu\text{-}PEt_2)_2Mo(CO)_3L]$$

$$k_{obs} = k_f + k_r =$$

$$\frac{k_1}{1+(k_{-1}[CO]/k_2[L])} + \frac{k_{-2}}{1+(k_2[L]/k_{-1}[CO])}$$

cone angle shows that the back reaction predominates for large phosphines. The rate, entropy, and enthalpy change for each step have been measured. The reaction mechanism is outlined in detail in the plot of the energy level versus reaction coordinate displayed in Figure 10.1.

The addition of CO to triply bonded $[R_1CpMo(CO)_2Mo(CO)_2CpR_2]$ yields singly bonded $[R_1CpMo(CO)_3Mo(CO)_3Cp_2]$.[48] Crossover experiments with R_1 and R_2 equal to H or Me show that no crossover products are obtained. This excludes a mechanism involving bond homolysis, and shows that the reaction occurs by the sequential addition of two molecules of CO. The

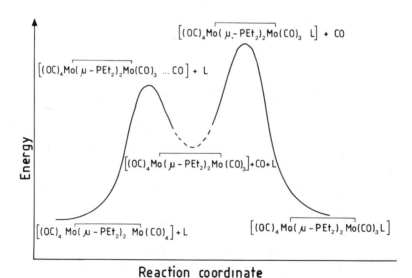

Reaction coordinate

Figure 10.1.   Reaction coordinate for substitution of CO by L in

$$[(CO)\overline{Mo(\mu\text{-}PEt_2)_2Mo}(CO)_4].$$

Scheme 8

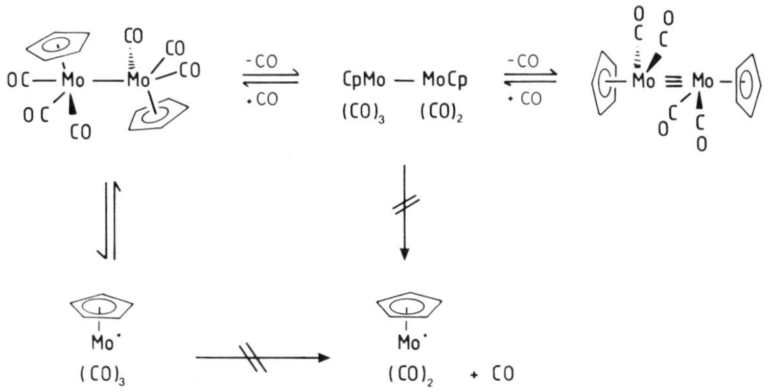

principle of microscopic reversibility states that the loss of CO from the singly bonded species must occur by the same mechanism, not one involving bond homolysis (Scheme 8).

Tyler and co-workers have studied the photochemical disproportionation of $[Cp_2Mo_2(CO)_6]$ with halide ligands to produce $[CpMo(CO)_3X]$ and $[CpMo(CO)_3]^-$.[49] They have shown that 19-electron intermediates are reasonable intermediate species. The probable intermediate is $[CpMo(CO)_3X]^-$. Evidence for 19-electron intermediates in the photochemical disproportionation reactions of $[Mn_2(CO)_{12}]$ and $[Mn_2(CO)_6(depe)_2]$ with nucleophiles is presented.[50] The reaction proceeds by photoinitiated Mn—Mn metal bond cleavage followed by attack by the nucleophile. Good electron donors, such as amines and oxygen–donor ligands, are required for disproportionation products to be obtained. Poorer electron donors, such as phosphines, give only substitution products. The inter- and intramolecular carbon monoxide exchange has been studied in manganese–rhenium decarbonyl.[51] The intermolecular exchange reaction between unlabeled $[MnRe(CO)_{10}]$ with ^{13}CO in decalin probably involves a CO dissociation mechanism, as the rate is similar to that reported for substitution with PPh_3. The intramolecular migration of the label to the Mn atom is a first-order reaction with $k = 0.47 \pm 0.09 \times 10^{-4}\,s^{-1}$ at 65 °C, $\Delta H^{\ddagger} = 12.7 \pm 1.4\,cal\,mol^{-1}$, and $\Delta S^{\ddagger} = -41 \pm 6\,cal\,mol^{-1}\,K^{-1}$. The large negative entropy is consistent with a pairwise CO exchange mechanism as shown in Scheme 9. The photochemical reactions of manganese–rhenium diimine carbonyl complexes are also reported.[52] The photolysis of $[(CO)_5MM'(CO)_3(\alpha\text{-diimine})]$ (M, M' = Mn, Re) in 2-MeTHF at temperatures varying from 133 to 230 K causes homolysis of the metal–metal bond. In the presence of $P(n\text{-Bu})_3$, several substituted complexes such as $[M(CO)_5]^-[M'(CO)_3(\alpha\text{-diimine})(P(n\text{-Bu})_3)]^+$ were isolated. These ions were formed by thermal disproportionation of a photosubstituted product

Scheme 9

$[(CO)_5MM'(CO)_2(\alpha\text{-diimine})(2\text{-Me-THF})]$ rather than by direct heterolysis of the metal–metal bond.

The heterobimetallic species (**7**) undergoes CO displacement with nucleophilic attack on the η^2-acetyl carbon by the molybdenum of the $[CpMo(CO)_3]^-$ fragment of (**7**).[53] This forms an acetyl/oxycarbene bridge in (**8**) as shown in Scheme 10. This reaction has a first-order dependence on

Scheme 10

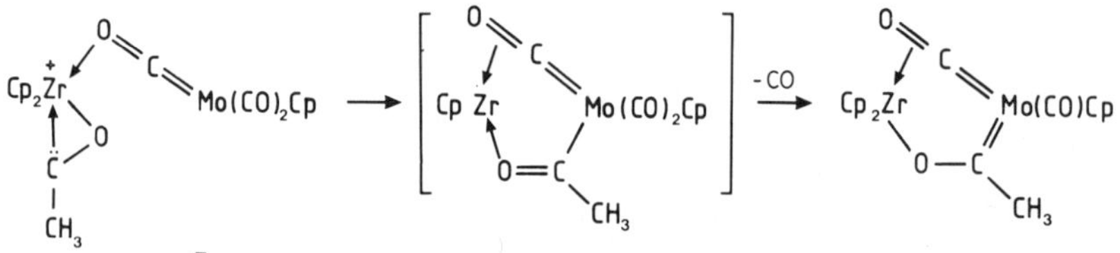

(**7**) and is independent of the CO pressure and of the presence of added (**8**), and has activation parameters of $\Delta H^{\ddagger} = 17.7 \pm 0.7\,\text{kcal mol}^{-1}$, $\Delta S^{\ddagger} = -21 \pm 2$ eu, and $k_{obs}(25\,°C) = 2.0 \pm 0.2 \times 10^{-5}\,\text{s}^{-1}$. This mechanism may proceed through an O-outside intermediate such as (**9**), in which nucleophilic attack is sterically feasible as shown in equation (11).

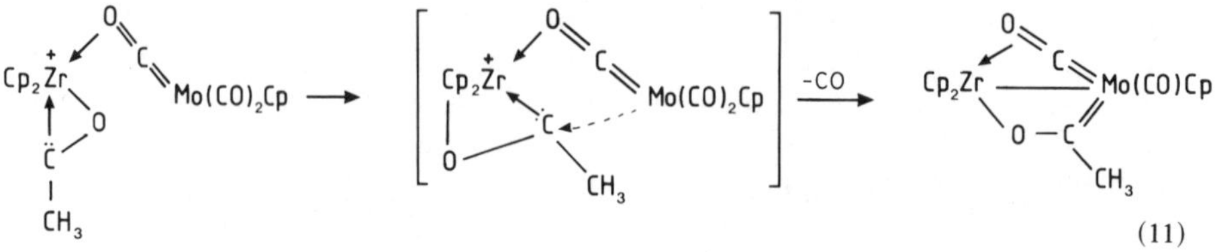

$$(11)$$

The terminal halide in dinuclear platinum(I) complexes $[Pt_2Cl_2\text{-}(\mu\text{-dppm})_2]^+$, abbreviated (Cl--Cl) (where dppm = $Ph_2PCH_2PPh_2$), undergoes substitution with PPh_3.[54] In CH_2Cl_2 at 25 °C the second-order reaction produces only the monosubstituted product (Cl--PPh$_3$) with a rate constant of $15.2 \pm 0.7 \, M^{-1} \, s^{-1}$. The disubstituted product (PPh$_3$--PPh$_3$) forms only in a more polar solvent, such as methanol. The reverse reaction, (PPh$_3$--PPh$_3$), reacts with halide ions (X = Cl, Br$^-$, I$^-$) to yield (X--PPh$_3$). Scheme 11 outlines the mechanism for the reactions with and without added salt. The reaction occurs quickly in the presence of inert perchlorate salts through a mechanism involving rate-limiting loss of PPh$_3$ and then addition of X$^-$ (**A**). In the absence of added salt, the reaction occurs in three distinct kinetic steps (**B**). Stopped-flow kinetics has been used to measure the rate of each step for all three halide ions.

Scheme 11

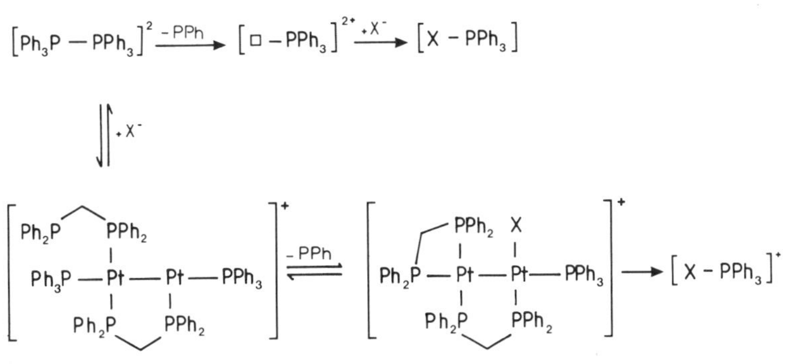

Carbon monoxide adds reversibly to $[HPt(\mu\text{-}PP)_2PtCO]^+$, (H--CO) (PP = Et$_2PCH_2PEt_2$ or Ph$_2$PCH$_2$PPh$_2$), to produce $[HPt(\mu\text{-}PP)_2Pt(CO)_2]^+$ and $[H(CO)Pt(\mu\text{-}PP)_2Pt(CO)_2]^+$.[55] This reaction involves disproportionation of Pt(I) to give a Pt(0)–Pt(II) species. This is in contrast to the symmetrical complex $[Pt_2Cl_2(\mu\text{-}dppm)_2]$ which adds CO to give either $[Pt_2Cl_2(\mu\text{-}CO)(\mu\text{-}dppm)_2]$ or $[Pt_2Cl(CO)(\mu\text{-}dppm)_2]Cl$ while retaining equal oxidation states for the two platinum centers. Substitution in (H--CO) is much faster than in other diplatinum(I) complexes due to the easy cleavage of the Pt—Pt bond in (H--CO).

10.1.5. *Substitution Reactions of Polynuclear Metal Complexes*

The kinetics of acetonitrile substitution reactions of $[Os_3(CO)_{12-n}(CH_3CN)_n]$ ($n = 1, 2$) with phosphorus or arsenic donor ligands in *p*-xylene or toluene have been described.[56] The reaction occurs by way of a dissociative process as is indicated in equations (12) and (13). When L = PPh$_3$, the k_2/k_{-1}

ratio was found to be 0.103 ± 0.002, i.e., the intermediate $[Os_3(CO)_{11}]$ is more reactive toward CH_3CN than PPh_3 by a factor of 10.

$$[Os_3(CO)_{11}(CH_3CN)] \overset{k_1}{\underset{k_{-1}}{\rightleftharpoons}} [Os_3(CO)_{11}] + CH_3CN \qquad (12)$$

$$[Os_3(CO)_{11}] + L \overset{k_2}{\longrightarrow} [Os_3(CO)_{11}L] \qquad (13)$$

The activation parameters for CH_3CN loss from the osmium center were determined to be $\Delta H^\ddagger = 26.9$ kcal mol^{-1} and $\Delta S^\ddagger = 22$ eu. The analogous parameters for CH_3CN loss in $[Os_3(CO)_{10}(CH_3CN)_2]$ were 21.8 kcal mol^{-1} and 6.9 eu, respectively. An intermediate with a bridging CO structure (10) was proposed in an effort to explain these results, i.e., the reaction has some intramolecular associative character leading to a lower value of $\Delta H^\ddagger$ and a more unfavorable value of $\Delta S^\ddagger$.

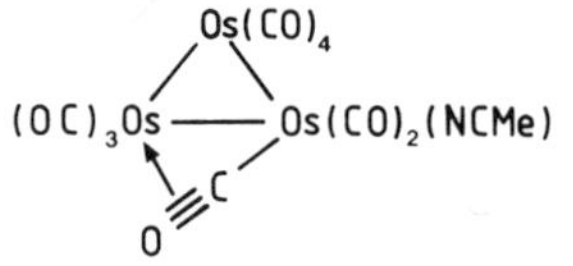

10

Tri-n-butylphosphine substituted metalcarbonyl clusters, $[M_3(CO)_{12-n^-}(PnBu_3)_n]$ ($M = Os$, $n = 0, 1$; $M = Ru$, $n = 1, 2$), have been reported to react with $PnBu_3$ according to equation (14).[57,58] The product distribution was

$$k_{obs} = k_1 + k_2[PnBu_3] \qquad (14)$$

found to depend on both the phosphine concentration and the reaction temperature. That is, the pathway governed by k_1 led only to substitution at the intact cluster, while the $k_2[PnBu_3]$ term led to cluster fragmentation with formation of mononuclear products. The bimolecular fragmentation pathway was given the mechanistic designation F_N2 and is strongly dependent on the strength of the M—M bond. The values of $\Delta H^\ddagger$ for sequential loss of carbon monoxide indicate that introduction of the first $PnBu_3$ ligand stabilizes the cluster toward CO dissociation, but a second one labilizes the cluster toward further substitution. A summary of the kinetic parameters for these substitution processes is provided in Table 10.10.

There have been two recent reports on the photofragmentation and photosubstitution kinetics of $[M_3(CO)_{12}]$ ($M = Ru$ and Os) clusters. Poë and Sekhar have studied the 436 nm photochemical kinetics of reactions of $[Os_3(CO)_{12}]$ with 1-octene and $P(OEt)_3$ in benzene.[59] The absorption of 436 nm radiation does not in itself cause CO dissociation or cluster fragmentation. Instead, both reactions are thought to proceed via a common photogenerated intermediate which is suggested to be a chemically reactive isomeric form of $[Os_3(CO)_{12}]$. This intermediate (11) is proposed to be an electronically excited state of

Table 10.10. Kinetic Parameters for Reactions of $[M_3(CO)_{12-n}(PnBu_3)]$ with $PnBu_3$ (References 57 and 58)

	$n = 0$		$n = 1$		$n = 2$	
	Ru	Os	Ru	Os	Ru	Os
$\Delta H_2^{\ddagger}$ (kcal mol^{-1})	12.1 ± 0.2	20.6 ± 0.2	7.9 ± 0.3	15.0 ± 1.4	17.3 ± 2.2	—
$\Delta S_2^{\ddagger}$ (cal k^{-1} mol^{-1})	-27.5 ± 0.6	-22.7 ± 0.6	-38.7 ± 0.9	-35.0 ± 4.3	-19 ± 7	—
$10^5 k_2(100\,^{\circ}\text{C})$ (l mol^{-1} s^{-1})	6×10^4	6	5×10^4	25	4×10^3	—
$\Delta H_1^{\ddagger}$ (kcal mol^{-1})	31.8 ± 0.2	32.9 ± 0.3	22.1 ± 0.6	39.3 ± 1.2	26.5 ± 1.1	34.7 ± 0.2
$\Delta S_1^{\ddagger}$ (cal K^{-1} mol^{-1})	20.2 ± 0.6	7.6 ± 0.9	-1.1 ± 1.8	23.1 ± 3.1	11.4 ± 3.4	20.2 ± 0.6
$10^5 k_1(100\,^{\circ}\text{C})$ (s^{-1})	5×10^3	2	5×10^4	1	6×10^4	0.5

$$\begin{array}{c} Os(CO)_4 \\ (OC)_4Os \overset{\diagup}{\underset{\diagdown}{}} Os(CO)_3 \\ \underset{\diagdown}{} \overset{\diagup}{} \\ C \\ \| \\ O \end{array}$$

11

$[Os_3(CO)_{12}]$ which is characterized by the presence of a vacant coordination site on one Os atom. Reaction of (**11**) with 1-octene led to fragmentation of the cluster with formation of $[Os(CO)_4(1\text{-octene})]$ and $[Os_2(CO)_8]$ by way of the intermediate $[Os_3(CO)_{12}(1\text{-octene})]$. On the other hand, reaction of (**11**) with $P(OEt)_3$ led to CO substitution with production of $[Os_3(CO)_{11}P(OEt)_3]$. Similarly, Ford and co-workers[60] have carried out an extensive investigation of the medium, ligand, and wavelength effects on quantum yields and flash photolysis kinetics for the photofragmentation and photosubstitution reactions of $[Ru_3(CO)_{12}]$ and its phosphorus donor ligand substituted derivatives, $[Ru_3(CO)_{12-n}L_n]$ {L = $P(OMe)_3$, PPh_3, $P(p\text{-tolyl})_3$, or $P[O(o\text{-tolyl})]_3$}. The photofragmentation $[Ru_3(CO)_{12}] + 3L \rightarrow 3[Ru(CO)_4L]$ was found to arise from the lowest-energy excited state of $[Ru_3(CO)_{12}]$, the structure of which was analogous to that of species (**11**). A second intermediate, $[Ru_3(CO)_{12}L]$, results upon reaction with L. Kinetic flash-photolysis experiments show the lifetime of this intermediate to be markedly dependent on the nature of L. Photosubstitution reactions leading to $[Ru_3(CO)_{11}L]$ derivatives were proposed to occur by way of the $[Ru_3(CO)_{11}]$ fragment afforded from higher-energy excited states. The reactivity of $[Ru_3(CO)_{11}]$ with various L was established by flash photolysis to be CO > $P(OMe)_3$ > PPh_3.

Carbon monoxide substitution in the anionic hydride, $[HRu_3(CO)_{11}]^-$, with PPh_3 in tetrahydrofuran solution has been investigated and found to be reversible with an equilibrium constant $K = 0.23$ at 25 °C for the reaction as depicted in equation (15).[61] The reaction proceeds by way of dissociative

$$[HRu_3(CO)_{11}]^- + PPh_3 \underset{k_{-1}}{\overset{k_1}{\rightleftharpoons}} [HRu_3(CO)_{10}PPh_3]^- + CO \qquad (15)$$

pathway in either direction with $k_1 = 2.1\ \mathrm{s}^{-1}$ at 25 °C ($\Delta H^{\ddagger} = 16.0\ \mathrm{kcal\ mol}^{-1}$, $\Delta S^{\ddagger} = -1.9$ eu) and $k_{-1} = 0.46\ \mathrm{s}^{-1}$ at 25 °C ($\Delta H^{\ddagger} = 18.2\ \mathrm{kcal\ mol}^{-1}$, $\Delta S^{\ddagger} = 1.0$ eu). Consistent with a dissociative mechanism, the rate of reaction of $[HRu_3(CO)_{11}]^-$ with $n\text{-}Bu_3P$ was the same as that noted for PPh_3. Dissociative CO loss in the neutral $[Ru_3(CO)_{12}]$ derivative occurs with a $\Delta H^{\ddagger}$ value some 10 kcal mol^{-1} higher. This gross difference in reactivity is attributed to a stabilized transition state or intermediate, perhaps of type (**12**) or (**13**).

Further studies on ligand substitution reactions of $[HRu_3(\mu\text{-}CX)(CO)_{10}]$ (X = OMe or NMe_2) with arsines and phosphines have been reported.[62] Kinetic measurements for the reactions of $AsPh_3$ with $[HRu_3(\mu\text{-}CNMe_2)\text{-}(CO)_{10}]$ and $AsPh_3$, PPh_3, and $P(OMe)_3$ with $[HRu_3(\mu\text{-}COMe)(CO)_{10}]$ were determined. The rate laws and activation parameters were found to be con-

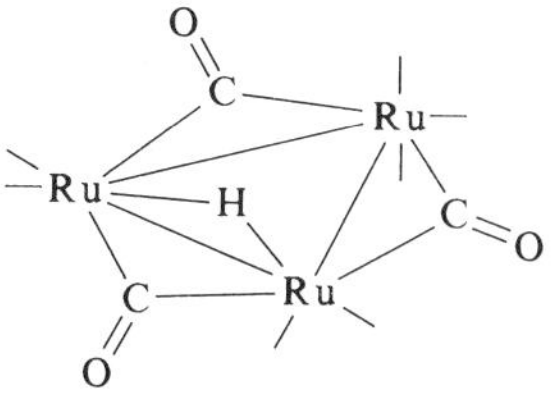

12 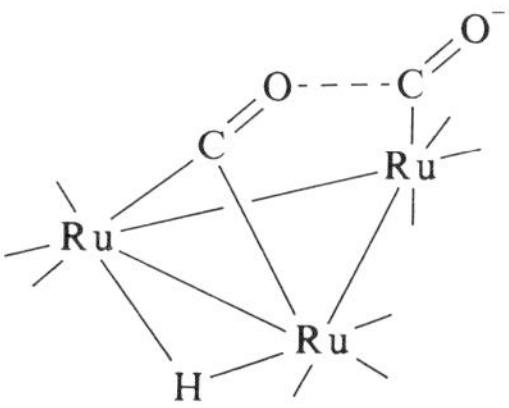   **13**

sistent with a CO dissociation mechanism, with CO dissociation being slower in the $[HRu_3(\mu\text{-}CNMe_2)(CO)_{10}]$ derivative as compared with its OMe analogue (X = OMe, $\Delta H^\ddagger = 26.6$ kcal mol^{-1}, $\Delta S^\ddagger = 8$ eu; X = NMe$_2$, $\Delta H^\ddagger = 26.9$ kcal mol^{-1}, $\Delta S^\ddagger = 3$ eu). Activation parameters for the reverse reaction, i.e., dissociation of AsPh$_3$ from $[HRu_3(\mu\text{-}CNMe_2)(CO)_9(AsPh_3)]$ in the presence of CO, were found to be $\Delta H^\ddagger = 25.4$ kcal mol^{-1} and $\Delta S^\ddagger = 2$ eu. The proposed mechanism for ligand substitution involving these $[HRu_3(\mu\text{-}CX)\text{-}(CO)_{10}]$ derivatives is described in Scheme 12 in which the CX ligand changes from a $\mu\text{-}\eta^1$ three-electron donor ligand to a $\mu_3\text{-}\eta^2$ five-electron donor in the transition state. The implications for this mechanism on cluster hydrogenation processes were discussed.

Phosphine ligand substitution in $[Pt_3(CO)_3L_3]$ derivatives with phosphines (L') has been found[63] to readily occur at ambient temperature as in equation

Scheme 12

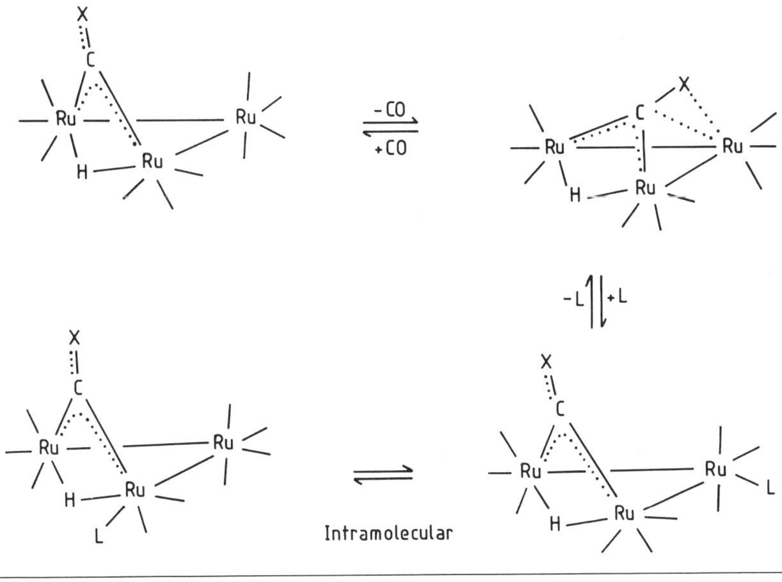

Intramolecular

(16). The substitution process was proposed to take place by an associative process, however, no quantitative rate data were provided. Consistent with

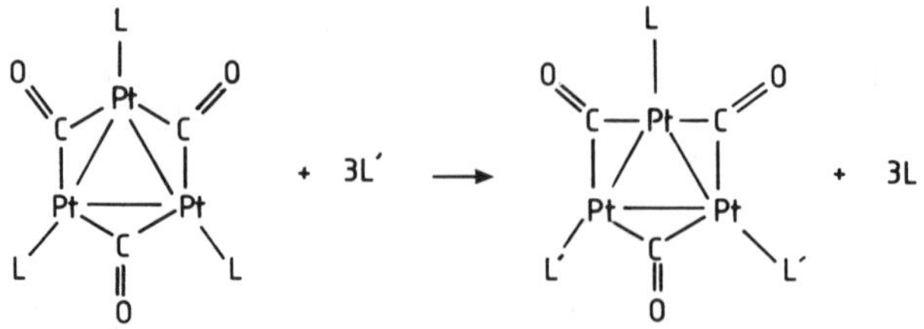

$$\text{(16)}$$

this mechanism when an equimolar mixture of $[Pt_3(^{12}CO)_3(P\text{-}t\text{-}Bu_2Ph)_3]$ and $[Pt_3(^{13}CO)_3(Pt\text{-}Bu_2Ph)_3]$ was reacted with a ninefold excess of PCy_3, the initial products were $[Pt_3(^{12}CO)_3(PCy_3)_3]$ and $[Pt_3(^{13}CO)_3(PCy_3)_3]$, with ^{13}CO scrambling occurring at a much slower rate. The presence of less bulky phosphine ligands in these trinuclear clusters, such as PPh_3, was observed to result in cluster fragmentation of the trimers to monomers.

Ligand substitution reactions in the closed tetrahedral clusters $[RPFe_3(CO)_{10-n}L_n]$ (R = alkyl, aryl; L = phosphite, isonitrile; n = 0, 1, 2) to provide in a stepwise manner $[RPFe_3(CO)_{10-n}L_{n+1}]$ and $[RPFe_3(CO)_{10-n}L_{n+2}]$ have been investigated.[64] Elimination of CO or L from these M—M bond-opened species results in ligand substitution in the closed tetrahedral derivatives as outlined in Scheme 13. Hence, ligand substitution occurs not by a dissociative process, but by a series of addition–elimination steps involving bond-opened intermediates. Reactions of $[Ir_4(CO)_{12}]$ with $P(OEt)_3$, $P(OCH_2)_3CEt$, and PCy_3 have revealed these processes to be first order in both metal cluster and L concentrations,[69] with the activation parameters as

Scheme 13

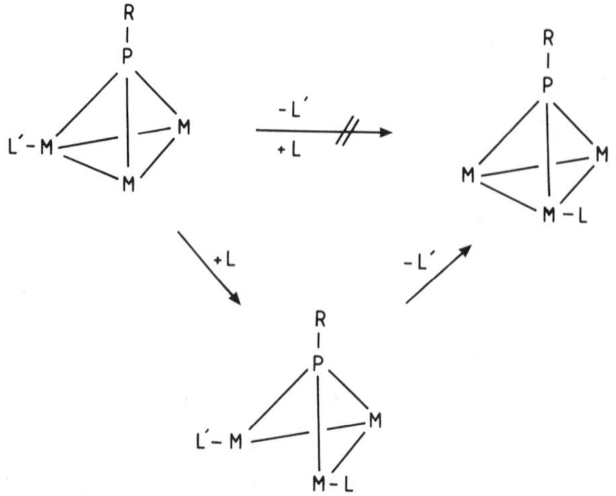

Table 10.11. *Activation Parameters for Associative*
Reactions of [Ir₄(CO)₁₂] *(Reference 69)*

Ligand	$\Delta H_2^{\ddagger}$ (kcal mol^{-1})	$\Delta S_2^{\ddagger}$ (cal K^{-1} mol^{-1})
P-n-Bu$_3$	11.67 ± 0.38	-30.2 ± 1.3
PCy$_3$	16.5 ± 1.6	-26 ± 4
etbb	14.08 ± 0.53	-21.4 ± 1.6
P(OEt)$_3$	18.2 ± 1.1	-10 ± 4
PPh$_3$	20.5 ± 0.4	-16 ± 1
P(OPh)$_3$	17.3 ± 0.2	-25 ± 1

depicted in Table 10.11. Incorporation of data from the literature for the PPh$_3$ and P(OPh)$_3$ derivatives[70] allowed for these data to be expressed in terms of a linear free-energy relationship involving electronic effects (ΔHNP) and steric effects (Tolman's cone angles).

The successive one-electron reduction of the bicapped triiron cluster [Fe$_3$(μ_3-PPh)$_2$(CO)$_9$] to afford the anion radical [Fe$_3$(PPh)$_2$(CO)$_9$]$^-$ and diamagnetic [Fe$_3$(PPh)$_2$(CO)$_9$]$^{2-}$ has been described by Ohst and Kochi.[71] Although thermal ligand substitution reactions of [Fe$_3$(μ_3-PPh)$_2$(CO)$_9$] with P(OMe)$_3$ have been shown to occur slowly and with low selectivity,[72] the anion radical undergoes rapid CO ligand substitution with phosphine ligands.[73] The rate-limiting rearrangement of [Fe$_3$(PPh)$_2$(CO)$_9$]$^-$ by a slippage of a phosphinidene capping ligand from μ_3 to μ_2 coordination represents the key step in the electron-transfer catalysis of ligand substitution on this triiron cluster derivative (Scheme 14).

Scheme 14

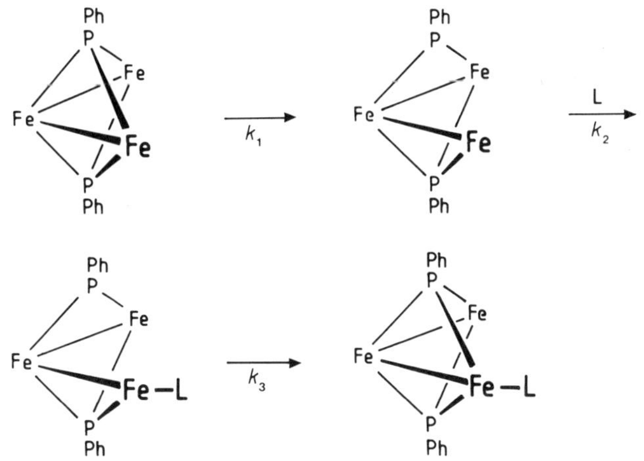

The tetracobalt carbonyl cluster (**14**) has been found to readily undergo CO ligand substitution reactions with $P(OMe)_3$ to afford sequentially mono, bis, tris, and tetrakis derivatives, where the phosphite ligands are each bound to separate cobalt centers as demonstrated by X-ray crystallography.[74] The kinetics and activation parameters indicate that mono- and bis-substitution occur primarily by an associative pathway, while tris- and tetrakis-substitution take place via a dissociative mechansim. Other phosphine ligands, such as PPh_3, react in a much less facile manner with (**14**). On the other hand, the

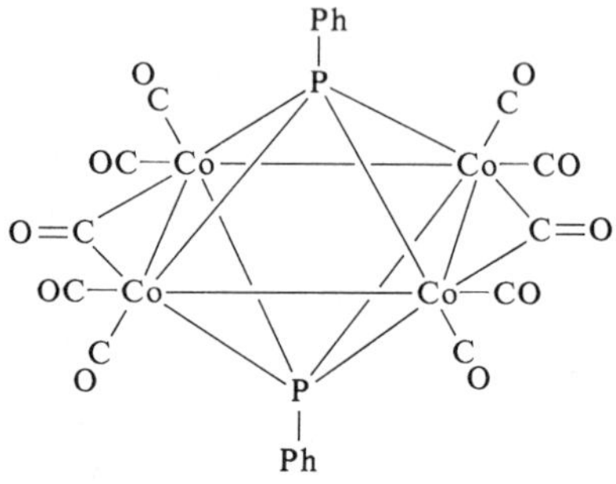

14

anion radical $[CO_4(CO)_{10}(PPh_3)_2]^-$ was observed to be substitutionally quite labile. The kinetic lability of the anion radical has allowed for the selective synthesis of a variety of mono-substituted derivatives.[75] A dissociative

Scheme 15

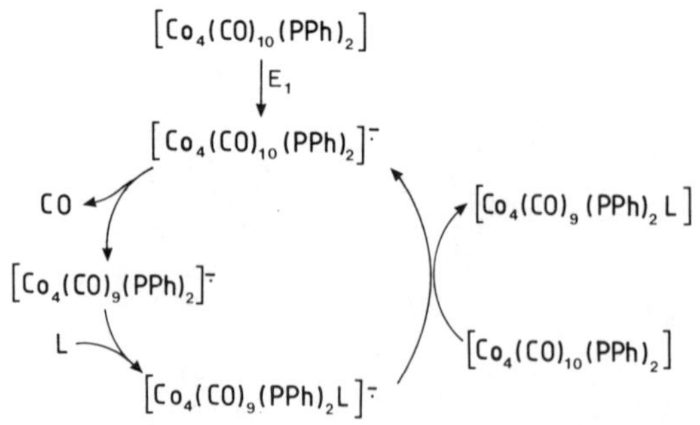

mechanism is proposed for the electrocatalytic substitution process (Scheme 15), where E_i represents the electrode or chemical potential necessary to reduce a portion of the tetracobalt species to the radical anion.

10.2. Insertion Reactions

10.2.1. Carbon Monoxide Insertion

Several papers focus on carbonylation of metal carbon bonds in square-planar complexes. Previous studies have covered compounds of the type $[MXR(CO)L]$ and $[MXRL_2]$. Current studies have expanded to include $[MR(CO)L_2]$ and studies of the second type, where L_2 is bidentate. Trans-$[MeIr(CO)(PPh_3)_2]$ has been shown[76] to undergo carbonylation by a two-step mechanism in which a carbonyl ligand first adds to produce an 18-electron five-coordinate species, then the 18-electron acyl complex is formed as in equation (17). Studies on $[PtClPh(PY)]$ have looked at carbonylation as a

$$\text{trans-}[CH_3Ir(CO)(PPh_3)_2] \xrightarrow[CO]{-78\,°C} [CH_3Ir(CO)_2(PPh_3)_2]$$

$$\xrightarrow{\text{room temp, CO}} [CH_3C(O)Ir(CO)_2(PPh_3)_2] \qquad (17)$$

function of the donor atom Y and the size of the chelate ring.[77] Dissociation of L followed by organic group migration has been demonstrated as the mechanism for carbonylation. The reaction occurs slowly when PY = dppp, PC_2S, or PC_2N and does not take place when PY = dppe or appe.

Ab initio studies show that the carbonyl insertion reaction of $[M(CH_3)(H)-(CO)(PPh_3)]$ (where M = Pt, Pd) takes place via methyl migration rather than carbonyl migration.[78,79] Calculation of the optimized transition-state structure indicates a three-center transition state in which the methyl group has moved half way toward the carbonyl ligand, while the carbonyl group moves only slightly toward the methyl group. The differences in the Pd and Pt complexes, the effects on substitution on R, and the trans effect are also discussed. The carbonylation of metal oxygen bonds in square-planar complexes has also been studied.[80] Carbonylation of $[(DPPE)PtCH_3(OCH_3)]$ is first order in metal complex and first order in CO. Low temperature ^{13}C NMR has been used to observe a coordinate intermediate produced by coordination of a CO ligand. The acyl is then formed by a rate-limiting inner-sphere insertion of CO into an available metal–oxygen bond. Crossover experiments rule out the involvement of dissociated species during the reaction. For the iridium case, trans-$[RIr(CO)(PPh_3)_2]$ (where R = Ph, n-Pr, t-Bu, and Me), carbonylation has been shown to occur by alkoxide displacement of CO to form the intermediate $[Ir(CO)_3(PPh_3)_2][OR]$ followed by alkoxide attack on a carbonyl ligand of the cation.[81] For the case with R = C_6F_5, carbonylation does not occur.[82] The reason suggested for this is that the alkoxide anion formed in the proposed mechanism would not be of sufficient nucleophilicity to attack the cation and form the carboalkoxy species.

Steric influences on the rate of CO insertion into the molybdenum–carbon bond have been examined in the phosphine substitution of a series of $[(\eta^5\text{-}Cp^*)(CO)_3MoCH_2Ph]$ compounds.[83] The scope of steric factors has

Table 10.12. Rate Constants for the Reaction of [Cp(CO)₃MoCH₂Ph] Systems with Nucleophiles in Acetonitrile at 30 °C (Reference 83)

Cp	L	$10^4 k_1$ (s^{-1})	$10^3 k_{-1}/k_2$ $(mol\,l^{-1})$	$10^3 k_3$ $(l\,mol^{-1}\,s^{-1})$
C_5H_5	PMe_2Ph	3.4	4.6	
C_5H_4Me	PMe_2Ph	4.7	4.3	
C_5Me_5	PMe_2Ph	9.2	140	
C_5H_5	BuNC	2.7	5.2	3.3
C_5H_4Me	BuNC	5.3	23	5.2
C_5Me_5	BuNC	7.9	160	11
C_5H_5	CyNC	2.8	28	
C_5Me_5	CyNC	12	250	

been measured by introducing substitution on the cyclopentadienyl ligand and by varying the phosphine ligand. Specifically, carbon monoxide insertion has been followed for reaction of $[(\eta^5\text{-}Cp^*)(CO)_3MoCH_2Ph]$ with tertiary phosphines and alkyl isocyanides in polar solvents to produce the substituted acyl complex, where Cp* is cyclopentadienyl, methylcyclopentadienyl, or pentamethylcyclodienyl. The insertion is a two-step process which involves formation of a solvent (S) stabilized intermediate followed by reaction with the nucleophile (L) to produce the acyl product as in equation (18). The rate constant k_1 was found to increase with increasing cyclopentadianyl substitution for reaction with both phosphines and alkyl isocyanide as shown in Table 10.12. The values of k_{-1}/k_2 reflect changes in k_{-1} which correspond to the k_1 values and differences in k_2 values. The k_2 term decreases sharply for the pentamethyl complex compared to the cyclopentadienyl and methylcyclopentadienyl systems, due to much greater congestion in the pentamethyl complex.

The currently accepted mechanism for carbonylation of Mn– or Mo–carbon bonds in compounds of the type $[L_nM(CO)R]$ involves interaction of the solvent as shown in equation (18). Halpern suggests that the role of the

$$[(CO)MR] \underset{k_{-1}-S}{\overset{k_1 S}{\rightleftharpoons}} [SM(COR)] \underset{k_{-2}-L}{\overset{k_2 L}{\rightleftharpoons}} [LM(COR)] + S \qquad (18)$$

solvent molecule with the unstable acyl intermediate is to catalyze its formation rather that to stabilize it.[84] This is proposed in a study of the reaction between $[p\text{-}CH_3OC_6H_4CH_2Mn(CO)_5]$ and $cis\text{-}[HMn(CO)_4(PMe_2Ph)]$ which provides evidence for a dissociative trapping mechanism. Steric factors have also been found to play an important role in the octahedral iron(II) case.[85] Table 10.13 shows that for $fac\text{-}[(diars)Fe(CO)_2(L)Me]^+$, the rate is highly dependent on steric factors, such as the cone angle of L, while being fairly insensitive to electronic effects. The first-order rate constant shows an overall increase of a factor of 270 as L varies from ETPB (cone angle 101°) to Ph_2PMe (cone angle

Table 10.13. Rate Constants for Carbonylation of $[(AsAs)Fe(CO)_2L(Me)]^+$ at 290 K (Reference 85)

	ETPB	P(OMe)$_2$	PhP(OMe)$_2$	PMe$_3$	PhPMe$_2$	Ph$_2$P(OMe)	Ph$_2$PMe
$10^4 k_{obs}$ (s^{-1})	0.30 ± 0.05	0.96 ± 0.1	2.1 ± 0.4	5.9 ± 1.0	9.1 ± 0.6	10.7 ± 0.7	80 ± 8
Cone angle	101	107	115	118	122	132	136
ν(CO)	2042.3	2027.0	2028.0	2018.8	2016.6	2021.9	2016.9
	1990.4	1974.2	1975.8	1964.9	1964.2	1970.4	1964
Force constant (mdyne Å^{-1})	16.42	16.16	16.18	16.04	16.00	16.09	16.00
Electronic parameter	2086.8	2079.5	2075.8	2064.1	2065.3	2072	2067

Table 10.14. *Comparative Data at 79 °C for the*
Carbonylation of cis-[CH$_3$Mn(CO)$_4$L]
(Reference 87)

L	k_f (s^{-1})	K_{eq} (M^{-1})
CO	1.9×10^{-5}	120
P(OMe)$_3$	2.3×10^{-5}	140
P(OPh)$_3$	3.6×10^{-5}	170
P(OPh)$_3$^{a}	3.5×10^{-5}	170
PBu$_3$	4.5×10^{-4}	180

^{a} 1/2 atmosphere CO.

136°). This suggests that the rate-determining step is unimolecular with no inter- or intramolecular nucleophilic participation.

Electronic effects rather than steric factors are important as is seen in the isoelectronic and isostructural *fac*-[(diars)Mn(CO)$_3$Me] complex, which has a much slower rate of CO insertion.[86] Similar results are reported[87] for the carbonylation of *cis*-[MeMn(CO)$_4$L] [L = CO, P(OPh)$_3$, P(OMe)$_3$, and PBu$_3$]. Table 10.14 shows that the rate constants and activation parameters vary only slightly with L. The transition state for methyl migration probably does not contain a coordinatively unsaturated metal center. The entering ligand also contributes little, as reflected by the similar rates and activation parameters for the reaction with 1/2 atmosphere CO, with 1 atmosphere CO, and with 1 atmosphere H$_2$.

CO insertion into manganese–carbon bonds is an intermediate step when the silylated phosphine (C$_6$H$_5$)$_2$PSi(CH$_3$)$_3$ is reacted with [(CO)$_5$MnR] (R = ryl, alkyl).[88] The driving force of the reaction is the production of an α-(silyloxy) alkyl complex [(CO)$_4$MnC(R)(OSi(CH$_3$)$_3$P(C$_6$H$_5$)$_2$] as shown in equation (19). The half-life for the disappearance of [(CO)$_5$MnMe] in this reaction is 2.2 h, which is similar to the half-life of 2.5 h reported for the carbonylation reaction with PPh$_3$.

$$\left[(CO)_5Mn-R \right] + (C_6H_5)_2PSi(CH_3)_3 \longrightarrow \left[\begin{array}{c} O \\ \parallel \\ (CO)_4Mn-C-R \\ | \\ (C_6H_5)_2PSi(CH_3)_3 \end{array} \right] \longrightarrow \left[\begin{array}{c} OSi(CH_3)_3 \\ | \\ (CO)_4Mn-C-R \\ | \diagup \\ (C_6H_5)_2P \end{array} \right]$$

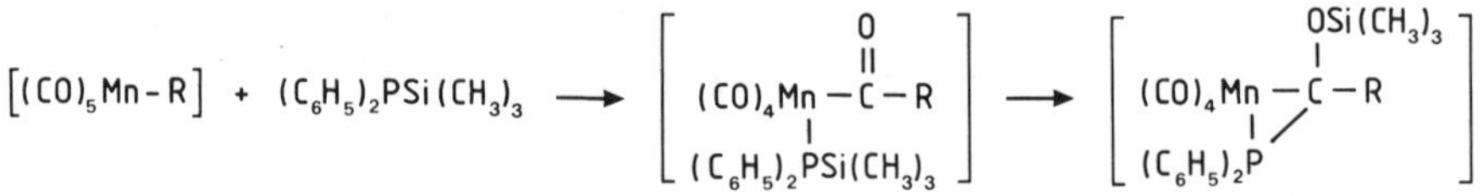

(19)

The heats of reaction of [RMo(CO)$_3$C$_5$H$_5$] (R = Me, Et) with phosphines to produce the phosphine-substituted acyl complex [RC(O)Mo(CO)$_2$PR$_3$-(C$_5$H$_5$)] have been reported.[89] This reaction consists of two parts, phosphine substitution and carbonyl insertion. The enthalpy of phosphine substitution in [HMo(CO)$_3$C$_5$H$_5$] is used to calculate the enthalpy of carbonyl insertion as shown in Table 10.15. The more basic phosphines such as PMe$_3$ are more

Table 10.15. Enthalpies of the Reaction $[RMo(CO)_3Cp] + PR \rightarrow [RC(O)Mo(CO)_2$-$(PR_3)Cp]$ with Calculated Enthalpies of CO Insertion and Infrared Data (Reference 89)

| | R = Me | | | R = Et | | |
Phosphine	ΔH_{react} (kcal mol^{-1})	$\Delta H_{CO ins}$ (kcal mol^{-1})	Acyl str (cm^{-1})	ΔH_{react} (kcal mol^{-1})	$\Delta H_{CO ins}$ (kcal mol^{-1})	Acyl str (cm^{-1})
PMe$_3$	19.9	14.8	1619	22.6	17.5	1618
P-n-Bu$_3$	19.1	14.1	1621	21.8	16.8	1619
PMe$_2$Ph	17.3	13.5	1621	19.7	15.9	1620
PMePh$_2$	14.4	12.8	1624	18.1	16.5	1623
PPh$_3$				15.4	16.3	1624
P(OMe)$_3$	17.4	12.8	1632	19.8	15.2	1630

favored for the carbonyl insertion reaction by about 2 kcal mol^{-1} over P(OMe)$_3$. Insertion into the Mo—Et bond is about 3 kcal mol^{-1} more favorable than into the Mo—Me bond.

10.2.2. Olefin Insertion

Facile olefin insertion reactions into the Rh—Rh and Rh—H bonds of [Rh$_2$(OEP)$_2$] and [(OEP)RhH] have been observed; OEP = octaethylporphyrin.[90] These reactions proceed through a free radical chain mechanism. Insertion into the dimer occurs readily at ambient temperatures in C$_6$D$_6$ according to the rate law and mechanism shown in Scheme 16 with $\Delta H_1^\circ = -11.2 \pm 2.5$ kcal mol^{-1}, $\Delta S_1^\circ = -15 \pm 8$ cal mol^{-1} K^{-1}, and $k_3^{obs} = 1.35 \pm 0.1 \times 10^2$ $M^{-3/2}$ s^{-1} at 30 °C. Two factors are suggested to account for this unusual free radical chain process: (1) unusually strong (OEP)Rh—C bonds, which contribute to the driving force of the chain propagaing steps, and (2) the absence of axial ligands in [Rh$_2$(OEP)$_2$] which makes the homolytic displacement steps possible.

A theoretical study of olefin insertions into Ti—C and Ti—H bonds has been completed.[91] These calculations suggest that the Ti—C bond and the C—C or H—C bond are formed concertedly with two sets of paired localized

Scheme 16

$$-d[Rh_2(OEP)_2]/dt = k_3^{obs}[Rh_2(OEP)_2]^{3/2}[PhCH=CH_2]$$

initiation/termination: $[Rh_2(OEP)_2] \rightleftharpoons 2[(OEP)Rh]^\bullet$

propagation: $[(OEP)Rh]^\bullet + PhCH=CH_2 \rightleftharpoons [(OEP)RhCH_2\dot{C}HPh]$

$[(OEP)RhCH_2\dot{C}HPh] + [Rh_2(OEP)_2] \rightarrow [(OEP)RhCH_2CH(Ph)Rh(OEP)] + [(OEP)Rh]^\bullet$

orbitals playing the major roles. The mechanism involves a four-centered transition state as shown for the insertion of $CH_2 = CH_2$ into $CH_3T : Cl_2^+$.

10.2.3. Carbon Dioxide Insertion

A series of papers on CO_2 insertion into tungsten-alkyl and -aryl bonds have been published by Darensbourg and co-workers.[92-94] The reaction of *cis*-$[RW(CO)_4L]^-$ [R = Me, Et, and Ph; L = CO, PMe$_3$, and P(OMe)$_3$] to provide metallocarboxylates is shown to obey second-order kinetics, first order in anionic metal substrate and first order in carbon dioxide. When R = Me and L = P(OMe)$_3$, the activation parameters are $\Delta H^\ddagger = 10.2 \pm 0.2$ kcal mol^{-1}, $\Delta S^\ddagger = -43.3 \pm 0.9$ cal mol^{-1} K^{-1}, and $\Delta G^\ddagger$ at 25 °C = 23.1 ± 0.5 kcal mol^{-1}. A concerted reaction pathway involving an attack of the nucleophilic $[RW(CO)_5]^-$ ion at the electrophilic carbon or carbon oxygen center of CO_2 is proposed. This proposal is supported by the dependence of the rate of these reactions on carbon monoxide and the nature of the metal center, the activation parameters, and the observed retention of configuration at the α-carbon. Unequivocal demonstration of the α-carbon stereochemistry in the carbon dioxide insertion reaction has been accomplished by the conversion of *threo*-$[W(CO)_5CHDCHDPh]^-$ with CO_2 to *threo*-$[W(CO)_5O_2CCHDCHDPh]^-$, indicating that a high degree of retention of configuration at the carbon was involved.[93] The kinetic measurements are consistent with the mechanism in equation (20). Several factors have been found to affect the rate of CO_2 insertion

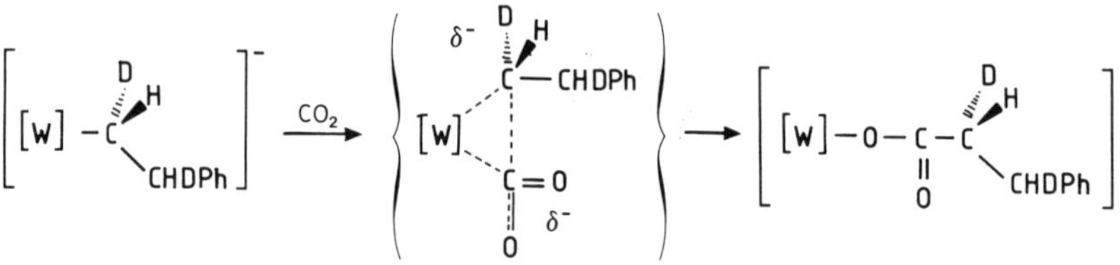

$$\qquad\qquad\qquad\qquad\qquad\qquad\qquad\qquad\qquad\qquad\qquad\qquad (20)$$

in the tungsten alkyl anions. Replacement of a carbonyl ligand by a more electron-donating phosphine or phosphite ligand results in a significant enhancement of the rate of CO_2 insertion. For example, the rate of CO_2 insertion into *cis*-$[CH_3W(CO)_4PMe_3]^-$ is 243 times greater than that in $[CH_3W(CO)_5]^-$. The rate of CO_2 insertion is also enhanced in the presence of Na^+ or $Na(18\text{-crown-}6)^+$ over that carried out with the noninteracting PPN$^+$ [bis(triphenylphosphine)iminium cation] or Na-kryptofix-221$^+$.[94] The latter rate accelerations are attributed to a cation stabilization of the incipient carboxylate during the carbon dioxide insertion process.

A kinetic investigation of carbon dioxide insertion into the Re—H bond in *fac*-$[Re(chelate)(CO)_3H]$, chelate = 2,2'-bipyridine (bipy) or 4,4'-X$_2$-2,2'-bipy, has appeared.[95] The reaction follows second-order kinetics, first order in metal complex and first order in CO_2. The reaction kinetics were consistent with a simple second-order rate law shown in equation (21). The bimolecular rate constant is $1.97 \pm 0.07 \times 10^{-4} \, M^{-1} \, s^{-1}$ with activation parameters $\Delta H^{\ddagger} = 12.8 \pm 0.9 \, \text{kcal mol}^{-1}$ and $\Delta S^{\ddagger} = -33.0 \pm 6.9$ eu. A dissociative mechanism has

$$-d[\text{Re}]/dt = k_i[\text{Re}][\text{CO}_2] \tag{21}$$

been ruled out, since the kinetics are clearly second order and the addition of trapping agents such as PPh_3 has no effect on the insertion rate. Instead, the hydrogen is transferred directly to CO_2, essentially as a hydride ion via a highly polar, charge-separated state. The intermediate or transition state in this associative hydride-transfer process is linear (**15**) or cyclic (**16**) in form.

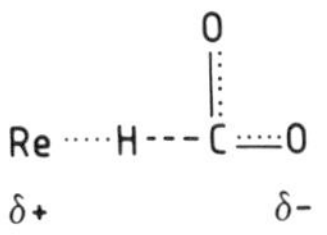

15 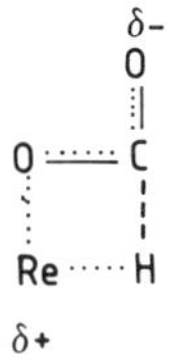   **16**

Evidence for this charge-separated state includes a large solvent effect and a substantial inverse $(k_{\text{H}}/k_{\text{D}})$ isotope effect.

The complex $[\text{Cp*Ir(H)(OEt)PPh}_3]$ has been found to undergo CS_2 insertion into the Ir—OEt bond to form $[\text{Cp*Ir(H)(PPh}_3)(\text{SC(S)OEt})]$ while not inserting CO or C_2H_4. This first-order reaction is independent of CS_2 concentration. Three possible intermediates are proposed, with (**18**) or (**19**) being favored[96] over (**17**).

17   **18** 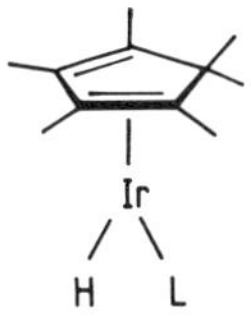   **19**

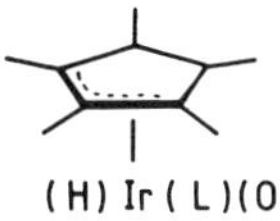

Chapter 11

Metal–Alkyl and Metal–Hydride Bond Formation and Fission; Oxidative Addition and Reductive Elimination

11.1. Introduction

During the reporting period (mid-1985 through 1986) there was considerable activity in several areas relevant to this chapter. Research dealing with C—H bond activation and molecular hydrogen complexes (H—H activation) has been particularly interesting (and fashionable). A variety of reviews have appeared. Especially welcome is Volume 2 of *The Chemistry of the Metal-Carbon Bond*,[1] which contains chapters on heterolytic, homolytic, and electrochemical cleavage of M—C bonds as well as an excellent chapter on oxidative addition and reductive elimination. A comprehensive survey of the organometallic chemistry of alkanes (C—H activation) has appeared.[2] A review of metal carbonyl clusters contains sections on C—H activation and oxidative addition.[3] Also noteworthy are reviews dealing with various aspects of transition metal alkyl,[4-6] alkenyl[7] and alkynyl[7] complexes, oxidative addition to metal carbonyl clusters[8] and square-planar complexes,[9] and free radical mechanisms in organometallic chemistry.[10]

11.2.   Dihydrogen Complexes

Interest in η^2-dihydrogen complexes increased dramatically in 1984 when Kubas *et al.*[11] reported the first fully characterized examples of stable isolable dihydrogen complexes, namely $[M(CO)_3(PR_3)_2(H_2)]$ (M = Mo, W; R = *i*-Pr, Cy). Since then, solid-state proton NMR has been used to determine the H—H bond length in $[W(CO)_3(PR_3)_2(H_2)]$ as 0.890 ± 0.006 Å.[12] The H_2 in $[M(CO)_3(PR_3)_2(H_2)]$ is labile and readily replaced by other ligands.[13] In the presence of D_2 the H_2 ligand is eventually equilibrated to give H_2, D_2, *and* HD species, implying cleavage of the H—H bond. This occurs even in the solid state, so dissociation of CO or PR_3 is not required for the reaction. IR and NMR data show[13,14] that $[M(CO)_3(PR_3)_2(H_2)]$ in solution exists in equilibrium with the dihydride as shown in equation (1). This shows that in

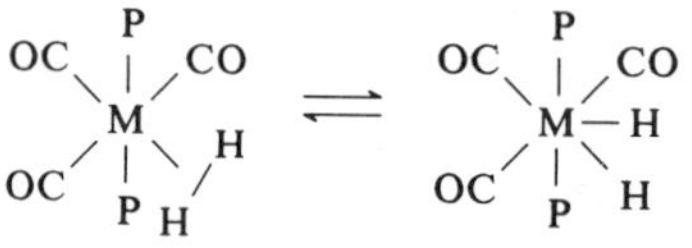

$$
\begin{array}{c}
\text{OC} \quad \overset{\displaystyle P}{\underset{\displaystyle P}{\overset{|}{\underset{|}{\text{M}}}}} \diagdown\text{CO} \\
\text{OC} \qquad \qquad \text{H}
\end{array}
\quad \rightleftharpoons \quad
\begin{array}{c}
\text{OC} \quad \overset{\displaystyle P}{\underset{\displaystyle P}{\overset{|}{\underset{|}{\text{M}}}}}\text{—H} \diagdown\text{CO} \\
\text{OC} \qquad \qquad \text{H}
\end{array}
\qquad (1)
$$

these systems the η^2-H_2 complex is not far in energy from the oxidative addition product. X-ray structures of $[W(CO)_3(PR_3)_2]$ (R = *i*-Pr,Cy) show[15] an agostic C···H···W interaction, which in these cases constitutes rare examples of such an interaction involving alkyl groups. Hence the 16-electron $[W(CO)_3(PR_3)_2]$ complexes possess the ability to activate C—H and H—H bonds. In other words, the arrested H—H oxidative addition in $[M(CO)_3(PR_3)_2(H_2)]$ is analogous to arrested C—H oxidative addition present in agostic C···H···M complexes. A variety of new dihydrogen complexes have been reported. UV photolysis of $[M(CO)_6]$ (M = Cr,Mo,W) in Xe at −70 °C gives $[M(CO)_5(H_2)]$ and *cis*-$[M(CO)_4(H_2)_2]$ (for M = Cr and W).[16] The thermal stability of $[M(CO)_5(H_2)]$ is W > Cr ≫ Mo (least stable). Under D_2, $[Cr(CO)_5(H_2)]$ forms $[Cr(CO)_5(D_2)]$ but *not* $[Cr(CO)_5(HD)]$. However, in the presence of $[Cr(CO)_4(H_2)_2]$ and D_2, thermal H_2/D_2 exchange does produce $[Cr(CO)_5(HD)]$ from $[Cr(CO)_5(H_2)]$, possibly via the intramolecular exchange shown in Scheme 1. The existence of $[M(CO)_5(H_2)]$ shows that the bulky tertiary phosphine ligands present in the original η^2-H_2 complexes of Kubas *et al.*[11] are not essential for stability. In practical terms, however, the synthesis

---

Scheme 1

$$[Cr(CO)_4(H_2)_2] \overset{D_2}{\rightleftharpoons} [Cr(CO)_4(H_2)(D_2)]$$

$$[Cr(CO)_4(HD)_2] \overset{D_2}{\rightleftharpoons} [Cr(CO)_4(D_2)_2] + HD$$

$$[Cr(CO)_5(H_2)] \overset{HD}{\rightleftharpoons} [Cr(CO)_5(HD)]$$

---

of η^2-H_2 complexes requires a 16-electron configuration and for thermal reactions near room temperature this is conveniently achieved by the use of bulky phosphine ligands; with smaller phosphines the $[M(CO)_3(PR_3)_2]$ precursors would be unavailable. UV photolysis at low temperatures has been used to prepare several other dihydrogen complexes: $[Fe(CO)(NO)_2(H_2)]$,[17] $[Co(CO)_3(NO)(H_2)]$,[17] and $[CpM(CO)_2(H)(H_2)]$ (M = Mo,W).[18,19]

Dihydrogen complexes have also been made by protonation of metal hydrides. HBF_4 addition to $[M(dppe)_2(H)_2]$ (M = Fe,Ru) yields *trans*-$[M(dppe)_2(H_2)(H)]BF_4$.[20] The iron complex (**1**) is stable to H_2 loss up to 50 °C; the ruthenium analogue is less stable. At temperatures greater than -20 °C the Fe—H hydride in (**1**) exchanges with the two equivalent hydrogens in Fe—H_2. Protonation of $[CpRu(PPh_3)(CNBu^t)(H)]$ yields $[CpRu(PPh_3)-(CNBu^t)(H_2)]PF_6$, which is stable at room temperature under H_2.[21] Similarly, protonation of the polyhydride $[Ir(PCy_3)_2(H)_5]$ gives the bis-dihydrogen complex $[Ir(PCy_3)_2(H)_2(H_2)_2]^+$, which liberates H_2 when treated with MeCN to give the known $[Ir(PCy_3)_2(MeCN)_2(H)_2]^+$.[22] Replacement of the water ligand in (**2**) by H_2 generates (**3**), abbreviated as $[Ir(bq)L_2(H_2)(H)]^+$; the indicated

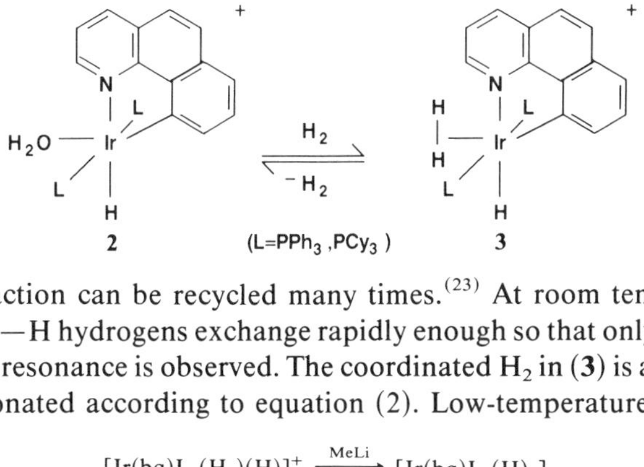

2   (L=PPh$_3$,PCy$_3$)   3

reversible reaction can be recycled many times.[23] At room temperature the Ir—H_2 and Ir—H hydrogens exchange rapidly enough so that only an averaged proton NMR resonance is observed. The coordinated H_2 in (**3**) is acidic enough to be deprotonated according to equation (2). Low-temperature experiments

$$[Ir(bq)L_2(H_2)(H)]^+ \underset{H^+}{\overset{MeLi}{\rightleftharpoons}} [Ir(bq)L_2(H)_2] \qquad (2)$$

with the deuterium analogues show that the deprotonation is from the H_2 and not the hydride ligand in (**3**). Exchange of D for H in (**2**) occurs readily; this is suggested[23] to take place via an intermediate dihydrogen complex as shown in Scheme 2. The measurement of proton spin-lattice relaxation times (T_1) is an effective way to distinguish between M—H and M—H_2 coordination, the latter linkage leading to much faster relaxation because of the short H-H

Scheme 2

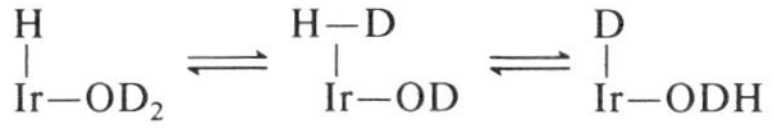

distance and the resultant strong dipole–dipole coupling.[13,22,23] An extended Huckel study[24] of the interaction of d^6 ML_5 fragments with H_2 to give $[ML_5(H_2)]$ or $[ML_5(H)_2]$ leads to the conclusion that η^2-H_2 formation is exothermic with the M—H_2 bond strength being greater for L = H^- compared to L = CO. Not surprisingly, the ease of conversion of $[ML_5(H_2)]$ to the d^4 $[ML_5(H)_2]$ complex is related to the ability of the ML_5 moiety to populate the σ^* H_2 orbital. Formation of the dihydride is made more difficult by replacing a metal with high-lying d orbitals (W) by one with low-lying d orbitals (Cr, Fe), and by replacing a σ-donor L with a π-acceptor L. It is also suggested that the η^2-H_2 structure is favored when the *trans* ligand is CO rather than a purely σ-donor.

11.3. Metal Hydride Complexes

The pK_a values of mononuclear carbonyl hydrides have been measured in acetonitrile and are given in Table 11.1.[25] Acidities range from strong for $[HCo(CO)_4]$, which has a pK_a similar to that of HCl in MeCN, to weak for $[CpW(CO)_2(PMe_3)H]$. The substitution of C_5Me_5 for C_5H_5 is seen to raise the pK_a by at least several units. Acidity seems to decrease down a triad (Cr, Mo, W; Mn, Re), although the nickel triad complexes $[HM(P(OMe)_3)_4]^+$ are an exception.[26] It also appears that acidity generally increases left to right

Table 11.1. Thermodynamic Acidities of
Carbonyl Hydrides in Acetonitrile at 25 °C[a]

Hydride	pK_a
$HCo(CO)_4$	8.3
$HCo(CO)_3P(OPh)_3$	11.3
$H_2Fe(CO)_4$	11.4
$CpCr(CO)_3H$	13.3
$CpMo(CO)_3H$	13.9
$HMn(CO)_5$	15.1
$HCo(CO)_3PPh_3$	15.4
$CpW(CO)_3H$	16.1
$Cp^*Mo(CO)_3H$	17.1
$H_2Ru(CO)_4$	18.7
$CpFe(CO)_2H$	19.4
$CpRu(CO)_2H$	20.2
$H_2Os(CO)_4$	20.8
$HRe(CO)_5$	21.1
$Cp^*Fe(CO)_2H$	26.3
$CpW(CO)_2(PMe_3)H$	26.6

[a] Data from Reference 25. Cp^* is C_5Me_5.

across a transition series, being especially pronounced for the first row. The reaction of metal carbonyl hydrides with $[Re(C(O)Et)(CO)_4(MeCN)]$ forms EtCHO.[27] The rate constants for a series of hydrides are the inverse of the kinetic acidities, suggesting that the hydrides act as nucleophiles. The coordinated MeCN must dissociate prior to reaction, showing that the site of attack is the metal and not the acyl carbon. The rate of deprotonation of $[HMo(\widehat{PP})_2]^+$ ($\widehat{PP}$ = dppe, dmpe, depe) is very slow, taking days with some amine bases.[28] The rate is greatly enhanced by the addition of hard anions like Cl^-. A detailed kinetic study shows that steric effects due to the chelating phosphine ligands are large. It is proposed that the small anions act to penetrate the complex, accept the proton, and transfer it to the larger amine base.

Theoretical homolytic bond dissociation energies for MH^+ molecules (M = Ca through Zn)[29] are in good agreement with experiment.[30] Reductive elimination of H_2 from cis-$[PtH_2(PMe_3)_2]$ according to equation (3) follows

$$cis\text{-}[PtH_2(PMe_3)_2] \xrightarrow{k} [Pt(PMe_3)_4] + Pt_{(s)} + H_2 \qquad (3)$$

a simple first-order rate law.[31] HD is not produced from a mixture of the protio and deutero complexes, thus ruling out a dinuclear mechanism. The dependence of the activation parameters on solvent are given in Table 11.2, which shows that as the coordinating power of the solvent decreases (at approximately constant polarity), the rate increases modestly while $\Delta H^{\ddagger}$ and $\Delta S^{\ddagger}$ undergo large compensatory changes. It is thought that the transition state is a late one which permits coordination by two solvent molecules to the metal as the H_2 departs. As steric factors render the solvent less able to coordinate in the transition state, the Pt—H bond cleavage is less assisted, leading to a greater $\Delta H^{\ddagger}$; for the same reason $\Delta S^{\ddagger}$ becomes less negative.

The thermal reductive elimination of H_2 from $[H_2IrCl(CO)(PPh_3)_2]$ and $[H_2IrCl(PPh_3)_3]$ is slow. However, flash photolysis produces the common intermediate $[H_2IrCl(PPh_3)_2]$, which rapidly eliminates H_2.[32,33] Interestingly, flash photolysis of $[H_2RhCl(PPh_3)_3]$ leads only to concerted hydrogen elimination without prior ligand dissociation.[33] Chronoamperometry was used to study the kinetics of hydrogen elimination from protonated [1.1]ferrocenophanes.[34] Thermolysis and photolysis of dinuclear zirconocene hydride complexes produces H_2 along with a paramagnetic zirconocene

Table 11.2. Kinetic Data at 21 °C for the Reductive Elimination of Hydrogen from cis-[PtH₂(PMe₃)₂] in Several Solvents

	THF	2,5-Me₂THF	2,2,5,5-Me₄THF
$10^4 k$ (s^{-1})	7.3	18	30
$\Delta H^{\ddagger}$ $(kJ\,mol^{-1})$	39	58	83
$\Delta S^{\ddagger}$ $(J\,K^{-1}\,mol^{-1})$	−171	−96	−8

hydride species.[35] The mechanism is a complex one and involves the hydrogens on the cyclopentadienyl ligands.

Careful measurements of the (weak) kinetic isotope effect for H_2 and D_2 oxidative addition to $[Ir(CO)Cl(PPh_3)_2]$ have been compared to model calculations with the conclusion that the transition state is triangular with reactant-like character and involves substantial hydrogen tunneling.[36] Oxidative addition of H_2 to $[Ir(CO)Br(chiraphos)]$, which contains the optically active (2S,3S)-bis(diphenylphosphino)butane, forms disastereomers (**4a**) and (**4b**) stereoselectively and with a kinetic differentiation corresponding to 1.7 kJ mol^{-1} in $\Delta G^{\ddagger}$.[37] Upon warming from -25 °C to room temperature, the diastereomers convert to the thermodynamic ones, (**5a**) and (**5b**), with a kinetic differentiation of 2.1 kJ mol^{-1}; the thermodynamic differentiation between (**5a**) and (**5b**) corresponds to only 0.16 kJ mol^{-1}. In the proposed mechanism shown in Scheme 3, (**4a**) and (**4b**) equilibrate via reductive elimination to give (**6**),

Scheme 3

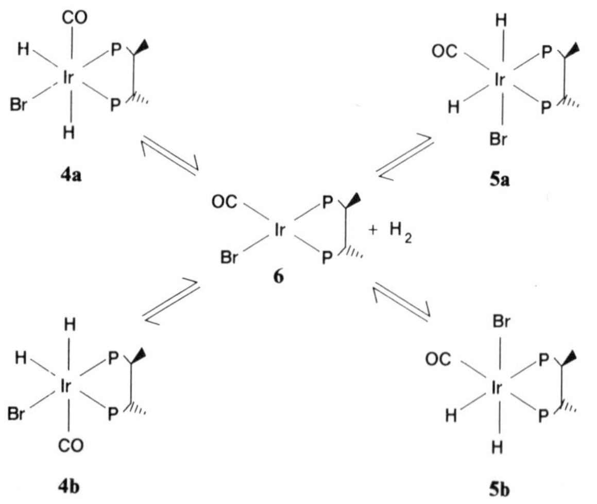

which more slowly adds H_2 to give (**5a**) and (**5b**). Similar chemistry obtains with Ph_3SiH, except that the kinetic discrimination is more pronounced. Oxidative addition of R_3SiH to $[IrX(CO)(dppe)]$ (X = Br, CN) proceeds stereoselectively under kinetic control to give initially the isomer having H *trans* to CO and Si *trans* to P.[38] In contrast, the *cis* oxidative addition of HY (Y = Br, I) gives initially the isomer having H *trans* to X and Y *trans* to P. The experimental conditions and observed stereoselectivity argue strongly for a concerted mechanism. The conclusion then is that H—Y approaches with its axis parallel to P—Ir—X, while R_3SiH has its axis parallel to OC—Ir—P. It is suggested that this difference arises because the silane approach to the Ir center is nucleophilic while that of HY is electrophilic.

11.4. C—H Activation

A number of reviews dealing with C—H activation have recently appeared. The most comprehensive one is by Crabtree.[2] Others include ones by Graham,[39] Deem,[40] Green and O'Hare,[41] Ephritikhine,[42] and Schwartz.[43] An extended Huckel MO analysis of H—H and C—H activation by d^0 metal fragments RML_n supports[44] the previously proposed four-center four-electron transition state. The only requirement for this type of reaction is that the metal complex have an empty MO able to interact with the filled σ bonding orbital in the C—H or H—H bond being cleaved. The methane exchange in reaction (4) is a good example of this type of reaction.[45] In the

$$(Cp^*)_2LuCH_3 + {}^{13}CH_4 \rightarrow [(Cp^*)_2Lu^{13}CH_3] + CH_4 \qquad (4)$$

gas phase Ru^+, Rh^+, and Pd^+ ions dehydrogenate alkanes via a 1,2-mechanism, in contrast to the 1,4-mechanism followed by Co^+ and Ni^+ ions.[46] The carbenes $FeCH_2^+$ and $CoCH_2^+$ also insert into C—H bonds of alkanes in the gas phase.[47]

The three-center, two-electron interaction between a vacant metal orbital and an α- or β-C—H bond has recently been recognized as being not uncommon. Such *agostic* bonding constitutes a mechanism whereby an electron-deficient metal center can increase its electron density and at the same time weaken or "activate" the C—H bond. The agostic interaction can therefore be seen as a prelude to oxidative addition, a fact that accounts for much of the current interest in C···H···M bonds. By measuring the shift in $E°$ for the oxidation of $[Cp_2WRR']$ as R and R' are varied, it was concluded that a weak three-center, three-electron agostic interaction exists in the 17-electron radical cations when R (R') is methyl or ethyl.[48] This may explain why hydrogen atom abstraction from $[Cp_2WRR']^+$ occurs at the α carbon. Thermolysis of (7) generates (8) and (9), which possess an activated aliphatic C—H bond.[49] Deprotonation of (8) and (9) occurs readily to give the diene (10). UV photoelectron spectroscopy has been used to probe the electron structure of (cyclohexenyl)manganese tricarbonyl, which is one of the first complexes to

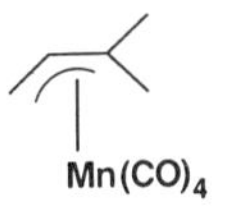

$Mn(CO)_4$	$(OC)_3Mn$···H, CH_2	CH_2, H···$Mn(CO)_3$	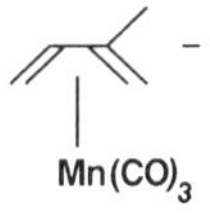 $Mn(CO)_3$
7	8	9	10

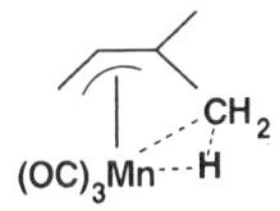

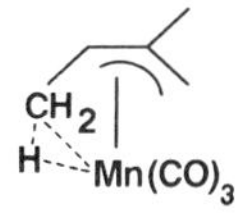

be recognized to exhibit agostic bonding.[50] Other examples of agostic interactions as well as relevant molecular orbital calculations have been published.[51-55] Intramolecular C—H activation (cyclometallation) of aliphatic CH bonds in 2,6-di-*ter*-butylphenol (ArOH) on Ti(IV) and Zr(IV) metal centers proceeds according to reaction (5).[56] Kinetic measurements

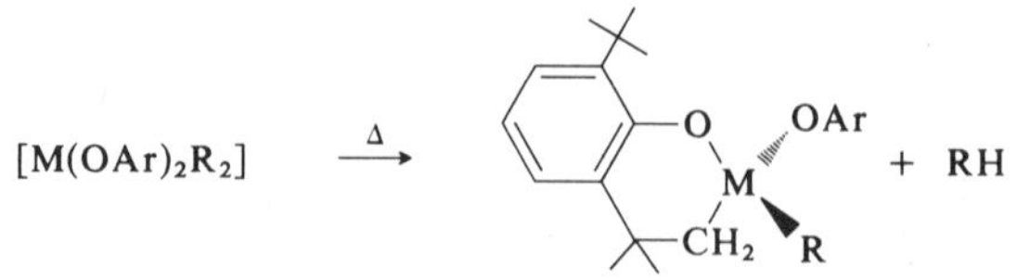

$$[M(OAr)_2R_2] \xrightarrow{\Delta} \qquad + \ RH \qquad (5)$$

(M=Ti, Zr; R=CH$_2$Ph, CH$_2$SiMe$_3$)

(R = CH$_2$Ph) show the cyclometallation step to be unimolecular with $\Delta H^{\ddagger}$ (kJ mol^{-1}) = 96 (Ti) and 90 (Zr); $\Delta S^{\ddagger}$ (J K^{-1} mol^{-1}) = -54 (Ti) and -79 (Zr). A multicenter transition state similar to that discussed above[44] is proposed. The complex [Ta(OAr)$_2$Me$_3$] undergoes two successive thermal C—H insertions shown in Scheme 4.[57] The formation of (12) and (13) from (11) is believed to occur via the usual four-center, four-electron transition state expected with d^0 systems. Values of $\Delta H^{\ddagger}$ for (11) → (12) and (12) → (13) are similar to that for reaction (5) but $\Delta S^{\ddagger}$ is much less negative in the Ta(V) cases. This may be because the mobility of the *tert*-butyl groups is more restricted in (11) as compared to the four-coordinate Ti and Zr complexes.

Scheme 4

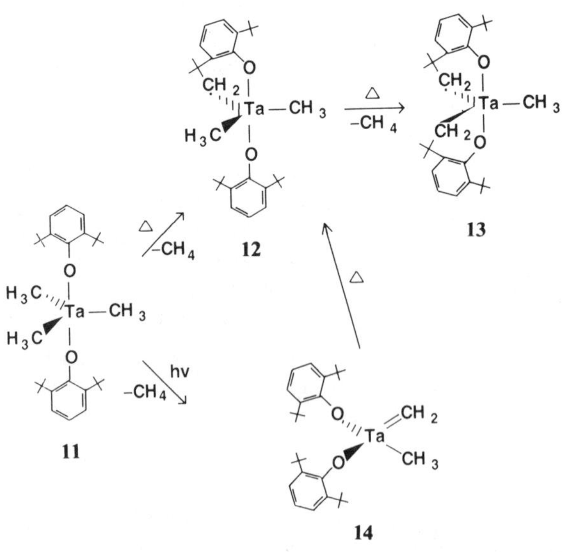

Photolysis of (11) produces the methylidene complex (14) which undergoes thermal C—H activation to give (12). Interestingly, $\Delta H^{\ddagger}$ for (14) → (12) is only about half that for (11) → (12) while $\Delta S^{\ddagger}$ is much more negative. The smaller $\Delta H^{\ddagger}$ probably arises from the need to break only a π bond in (14), while a (presumably) stronger σ bond must be cleaved in (11). A more negative $\Delta S^{\ddagger}$ is also reasonable for a variety of reasons.[57]

The reaction of various alkanes with $[Mn(CO)_3]^-$ in the gas phase shows that the reactivity order for oxidative addition of aliphatic bonds to $[Mn(CO)_3]^-$ is primary < secondary < tertiary.[58] It is suggested that the difficulty of adding tertiary CH bonds in the condensed phase is due to the steric bulk of the complex used rather than any intrinsic reactivity differences. Irradiation of $[CpRe(PPh_3)_2(H)_2]$ in C_6D_6 or THF-d_8 solvent catalyzes H/D scrambling between the solvent and a variety of added alkanes, including methane.[59] Oxygen does not affect the process, suggesting that radicals are not involved. A mechanism is favored that involves photochemical loss of PPh_3, oxidative addition of solvent followed by reductive elimination to generate $[CpRe(PPh_3)(H)(D)]$, which then transfers deuterium to the alkane by another oxidative addition–reductive elimination cycle. Competition studies showed that aromatic and primary aliphatic C—H bonds have similar reactivities in this system. Although irradiation of $[(C_6H_6)Re(PPh_3)_2H]$ dissociates PPh_3, as with $[CpRe(PPh_3)_2(H)_2]$, H/D exchange does not occur for alkanes in C_6D_6 solvent.[60] H/D exchange does occur with arenes, but by a mechanism that does not directly involve the hydride ligand. Irradiation of $[CpRe(PMe_3)_3]$ produces $[CpRe(PMe_3)_2]$ (15) which oxidatively adds primary, cyclopropyl, methane, aromatic, and vinyl hydrogens, but *not* secondary and tertiary CH bonds.[61,62] Cyclometallation through a PMe_3 ligand competes with alkane activation. The inability of (15) to activate the CH bonds of cyclohexane contrasts with previously studied iridium and rhodium complexes and means that cyclohexane is a convenient solvent for other CH oxidative additions, e.g., methane. The C—H insertion products with (15) are stable enough to be isolated and characterized, but all reductively eliminate RH upon warming. Although the absolute stabilities of the rhenium insertion products are lower than analogous iridium ones, the relative stabilities parallel each other with the half-lives for reductive elimination of $[CpRe(PMe_3)_2(R)(H)]$ following the order $C_6H_5 > C_3H_5 \approx C_2H_3 > CH_3 > PMe_2CH_2 > n\text{-}C_6H_{13}$. It is not clear whether kinetics or thermodynamics is the controlling factor, but the high selectivity of (15) with respect to intermolecular C—H bond activation seems to follow the thermodynamic stability of the product: aryl > vinyl > methyl > primary alkyl ≫ secondary alkyl. The competitive cyclometallation of (15) may be due to steric crowding, which at the same time may also play a role in the high selectivities in intermolecular activation. The rhenium complexes $[Cp^*Re(CO)(PMe_3)_2]$,[61] $[Cp^*Re(CO)_2(PMe_3)]$,[61] and $[Cp^*Re(CO)(PMe_3)(N_2)]$[63] when irradiated generate coordinatively unsaturated species analogous to (15) that also insert into CH bonds. Activation of the CH bonds in alkylbenzenes[64] and cycloheptatriene[65] by rhenium complexes has also been reported.

A number of examples of CH activation have appeared with iron triad complexes. Photolysis of $[Fe(dmpe)_2(H)_2]$ produces $[Fe(dmpe)_2]$, which inserts into the $sp^2 C—H$ bonds of alkenes and arenes.[66] The analogous $[Fe(depe)_2]$ can be generated both photochemically and thermally. It reacts with ethylene to give a mixture of vinyl hydride and η^2-ethylene complexes, which upon warming converts entirely to the π-complex.[67] Recooling the reaction mixture to the temperature originally used in the thermal reaction does not regenerate the vinyl hydride complex, proving that the C—H insertion does *not* proceed via prior formation of the π-complex. (This applies to iridium chemistry as well, *vide infra*.) A quantitative analysis showed that $[Fe(depe)_2]$ reacts with ethylene to form the vinyl hydride 23 times faster than to form the π-complex. Photolytic dehydrogenation of $[M(P(CH_2CH_2CH_2PMe_2)_3)(H)_2]$ in benzene gives a cyclometallated product for M = Fe, but the ruthenium complex inserts into benzene to give the phenyl hydride.[68] This metal dependent switch from intra- to intermolecular CH activation is suggested to be due to the differing sizes of the central metal. Thermolysis of *cis*-$[Os(PMe_3)_4(H)R]$ (16), (R = Me, CH_2CMe_3, CH_2SiMe_3) activates the C—H bond in benzene to give a quantitative yield of $[Os(PMe_3)_4(H)Ph]$.[69,70] The reaction proceeds via initial dissociation of PMe_3 to afford $[Os(PMe_3)_3(H)R]$, which rapidly and reversibly undergoes cyclometallation when R = CH_2CMe_3 and CH_2SiMe_3. Oxidative addition of C_6H_6 to $[Os(PMe_3)_3(H)R]$ followed by reductive elimination of RH and coordination of PMe_3 completes the reaction. The alkyl group in (16) exerts a strong influence on the rate of the initial PMe_3 dissociation. The complexes (16) do not activate free alkanes. The addition of excess PMe_3 to $[(arene)Os(PMe_3)C_2H_4]$ in C_6D_6 results in C—H activation that is intramolecular when the arene is benzene and is intermolecular when it is *p*-cymene, see reactions (6) and (7).[71] Photolysis of (17) induces the interesting cyclometallation reaction shown in Scheme 5.[72]

$$[(benzene)Os(PMe_3)C_2H_4] + PMe_3 \xrightarrow{C_6D_6} [Os(PMe_3)_4(H)Ph] \qquad (6)$$

$$[(p\text{-cymene})Os(PMe_3)C_2H_4] + PMe_3 \xrightarrow{C_6D_6} [Os(PMe_3)_4(D)Ph] \qquad (7)$$

Scheme 5

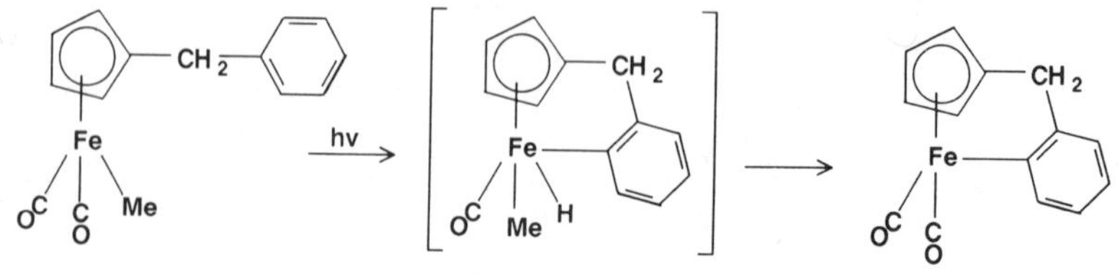

17

Not surprisingly, much of the research concerning C—H activation has involved rhodium and iridium complexes. The metalloporphyrins $[M(OEP)]_2$ (M = Rh,Ir; OEP = octaethylporphyrin dianion) react thermally with toluene and other alkylbenzenes at the benzylic C—H according to equation (8).[73,74]

$$[M(OEP)]_2 + PhCH_3 \rightarrow [(OEP)MCH_2Ph] + [(OEP)MH] \qquad (8)$$

The mechanism probably involves homolytic dissociation to form [M(OEP)] radicals that attack the weakest (i.e., benzylic) C—H bonds. Ethylene reacts with $[Cp^*Ir(PMe_3)]$ to give both vinyl hydride and η^2-ethylene complexes.[75] The π-complex is the thermodynamic product and therefore cannot be an intermediate in the C—H insertion reaction. A kinetic and thermodynamic study of alkane exchange reactions starting with $[Cp^*Ir(PMe_3)(Cy)(H)]$ (18) led to a number of important observations.[76] Reaction of (18) in benzene gives $[Cp^*Ir(PMe_3)(Ph)(H)]$ by a mechanism consisting primarily of reductive elimination of C_6H_{12} and oxidative addition of benzene (Scheme 6). An inverse

Scheme 6

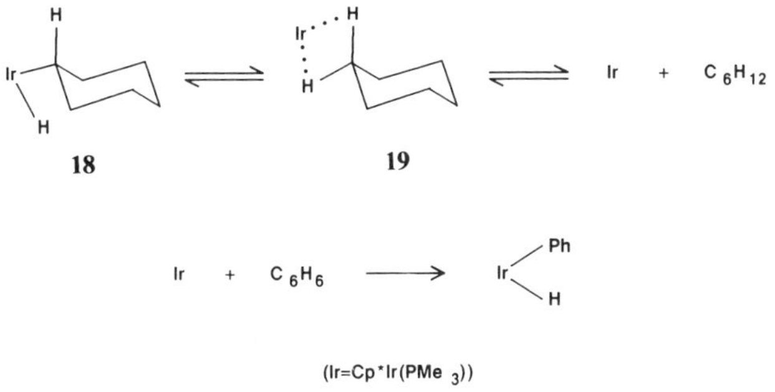

$(Ir=Cp^*Ir(PMe_3))$

kinetic isotope effect (k_H/k_D) for the reductive elimination of C_6H_{12} and H/D scrambling between the hydride and α-cyclohexyl positions in (18) constitute evidence for the agostic C···H···M interaction shown as (19) in Scheme 6. The qualitative order of iridium–carbon dissociation energies in $[Cp^*Ir(PMe_3)$-(R)(H)]$ complexes was found to be: R = neopentyl < cyclohexyl < primary 2,3-dimethylbutyl < primary pentyl $\ll$ phenyl. When generated at low temperatures in alkanes, $[Cp^*Rh(PMe_3)]$ gives products arising from insertion into primary C—H bonds.[77] There is evidence that secondary C—H bonds are also activated, but rapid intramolecular rearrangement occurs to the primary products. Kinetics and isotope labeling suggest that this rearrangement occurs through the intermediacy of η^2-C—H alkane complexes. The rate-determining step in arene oxidative addition to $[Cp^*Rh(PMe_3)]$ is coordination of a double bond; the second step (oxidative addition) occurs with $k_H/k_D = 1.4$.[78] In agreement with other studies, the isotope effect for reductive elimination of arene from $[Cp^*Rh(PMe_3)(aryl)(H)]$ points to an agostic C···H···M interaction in the transition state. Irradiation of $[CpRh(CO)C_2H_4]$ in low-temperature matrices and in solution produces [CpRhL] species that activate the C—H bond in methane and the Si—H bond in $HSiEt_3$.[79] Similarly, photolysis of $[CpIr(C_2H_4)_2]$ leads to the vinyl hydride $[CpIr(C_2H_4)(C_2H_3)(H)]$ which undergoes subsequent photolysis to the vinylidene complex, $[CpIr(CCH_2)(H)_2]$.[80] The thermal reaction of $[IrCl(PR_3)_2]$ (R = *i*-Pr) with benzene at 80 °C gives the phenyl hydride complex.[81] The dinuclear complex (20) reacts with C_6D_6 at 45 °C to give (21) via a reductive elimination–oxidative addition cycle.[82] The interesting feature of this reaction is the much lower temperature required compared to the mononuclear $[Cp^*Ir(PMe_3)(Ph)(H)]$. It seems that the presence of a second metal center can greatly influence the C—H activation energetics. In addition to the examples just cited, several other cases of C—H activation by rhodium and iridium complexes have been published.[83–89]

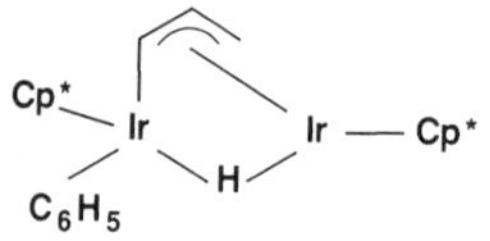

20 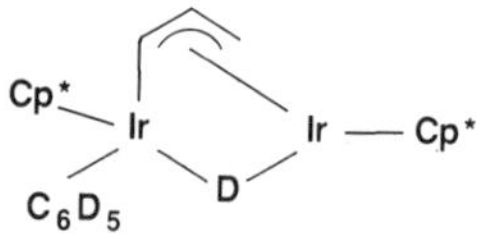   **21**

Several relevant papers have appeared dealing with C—H addition by platinum and palladium complexes. Thermolysis of $[Pt(cy_2PCH_2CH_2Pcy_2)(CH_2CMe_3)(H)]$ reductively eliminates CMe_4.[90] The coordinatively unsaturated $[Pt(cy_2PCH_2CH_2Pcy_2)]$ thus formed inserts into the C—H bonds of a variety of hydrocarbons, including arenes, acetylenes, and cyclopentane. Electrochemical reduction of *cis*-$[PtCl_2(PEt_3)_2]$ generates $[Pt(PEt_3)_2]$, which abstracts a β-proton from the tetrabutylammonium ion present in the bulk electrolyte.[91] The details of C—H activation by heterogeneous platinum catalysts have been investigated.[92] A kinetic study[93] of the *ortho*-palladation of *N*,*N*-dimethylbenzylamines shows that the intramolecular activation of C—H bonds in arenes by Pd(II) is kinetically favored over intermolecular activation by a factor of at least 360 mol dm^{-3}. Thermal cyclometallation of $[(Cp^*)_2Th(CH_2CMe_3)_2]$ in cyclohexane proceeds unimolecularly according to the mechanism illustrated in Scheme 7.[94] The data support the four-center, four-electron transition state expected for d^0, f^0 systems. Complex (22) undergoes ring-opening C—H activation reactions with a variety of RH molecules to give $[(Cp^*)_2Th(CH_2CMe_3)(R)]$.[95] Relative rates are in the order; RH = $SnMe_4 > SiMe_4 >$ cyclopropane $\approx PMe_3 >$ benzene $> CH_4 \geq C_2H_6 \gg$ cyclohexane. Again, a four-center transition state is postulated. A complete ther-

Scheme 7

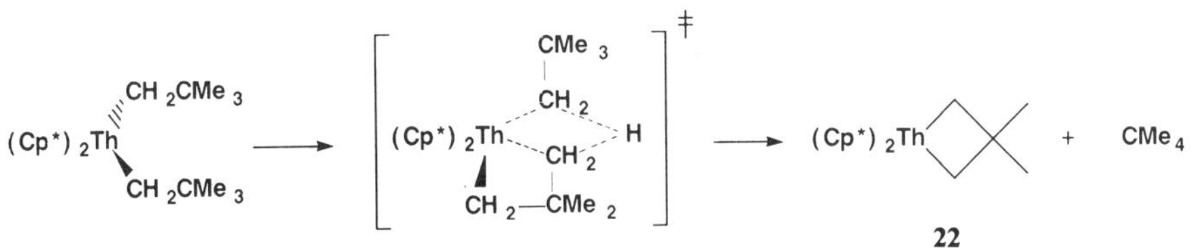

modynamic and kinetic analysis for the reversible reaction (9) has been published.[96]

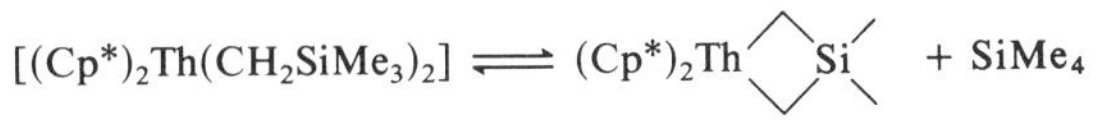

$$[(Cp^*)_2Th(CH_2SiMe_3)_2] \rightleftharpoons (Cp^*)_2Th\diamond Si + SiMe_4 \qquad (9)$$

11.5. Oxidative Addition and Reductive Elimination Involving Two Metal Centers

A brief account has been published concerning oxidative addition and reductive elimination at dinuclear centers containing the (Mo—Mo) or (W—W) fragment.[97] The photochemical reaction of $[Mn_2(CO)_8(PBu_3)_2]$ and $PhCH_2Cl$ goes cleanly according to equation (10).[98] While this reaction is

$$[Mn_2(CO)_8(PBu_3)] + PhCH_2Cl \rightarrow [Mn(CO)_4(PBu_3)Cl] + [Mn(CO)_4(PBu_3)(CH_2Ph)] \quad (10)$$

formally an oxidative addition across the Mn—Mn bond, kinetic data indicate that mononuclear intermediates are the reactive species as shown in Scheme 8. Reaction (11) for M = Co ($n = 4$) and M = Mn ($n = 5$) occurs via rate-

$$[(OC)_nMCH_2COOEt] + HM(CO)_n \rightarrow MeCOOEt + [M_2(CO)_n] \quad (11)$$

determining combination of $HMn(CO)_n$ and $[(OC)_{n-1}MCH_2COOEt]$.[99] The rate with M = Mn is only about one order of magnitude less than with cobalt.

Scheme 8

$$[Mn_2(CO)_8(PR_3)_2] \xrightarrow{h\nu} [Mn_2(CO)_7(PR_3)_2] + CO$$

$$[Mn_2(CO)_7(PR_3)_2] \rightarrow [Mn(CO)_3PR_3] + [Mn(CO)_4PR_3]$$

$$[Mn(CO)_3PR_3] + R'X \rightarrow [Mn(CO)_3(PR_3)(R')X]$$

$$[Mn(CO)_3(PR_3)(R')X] + [Mn(CO)_4PR_3] \rightarrow [Mn(CO)_3(PR_3)R'] + [Mn(CO)_4(PR_3)X]$$

$$[Mn(CO)_3(PR_3)R'] + CO \rightarrow [Mn(CO)_4(PR_3)R']$$

The formation of aldehydes from transition metal hydrides and alkyl carbonyl complexes is a general reaction. An investigation of a number of these reactions indicates that the rate-determining step is generally formation of an acyl intermediate.[100] Direct observation of the solvated acyl intermediate was possible in some cases. Reactions (12) and (13) occur by radical chain mechanisms with the chain carriers $[Mn(CO)_5]$ and $[Os(CO)_4H]$ arising from hydrogen

$$[Au(PPh_3)Me] + [HMn(CO)_5] \rightarrow [(PPh_3)AuMn(CO)_5] + CH_4 \qquad (12)$$

$$2[Au(PPh_3)Me] + [H_2Os(CO)_4] \rightarrow [(Ph_3PAu)_2Os(CO)_4] + 2CH_4 \qquad (13)$$

atom abstraction from the metal hydrides by initiators.[101] Alkane elimination from $cis\text{-}[Os(CO)_4(H)R]$ is proposed to occur by rate-limiting formation of the acyl hydride, which then reacts rapidly with $[Os(CO)_4(H)R]$ or nucleophile PR_3 to eliminate RH as shown in Scheme 9.[102] It was noted that simple intramolecular reductive elimination from any $cis\text{-}[Os(CO)_4(R)(R')]$ complex (including R = R' = H) has never been observed. This reflects the high energy and possibly the spin state of $[Os(CO)_4]$.

Scheme 9

$$[Os(CO)_4(H)R] \xrightarrow{k} [Os(CO)_3(COR)(H)]$$

$$[Os(CO)_3(COR)(H)] + [Os(CO)_4(H)(R)] \rightarrow [(OC)_4(H)OsOs(R)(CO)_4] + RH$$

$$[Os(CO)_3(COR)(H)] + PR_3 \rightarrow [Os(CO)_4PR_3] + RH$$

Acetyl chloride adds to one Rh(I) center in (23) to give (24).[103] When $X = Cl^-$, the addition is reversible. Increasing the electron density on the

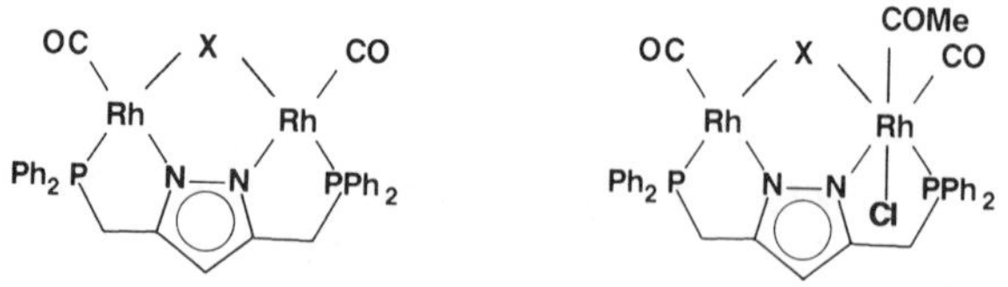

23

24

metal(s) by changing X to PPh_2^- renders the addition irreversible. IR and NMR data show that (24) is best formulated as a Rh(I)–Rh(III) complex. When $X = PPh_2^-$, (24) is sufficiently electron rich that a second MeCOCl can be added (reversibly) to the Rh(I). The iridium analogue of (23) undergoes the same reactions with the distinction that all MeCOCl oxidative additions are irreversible. The porphyrin dimer $[Rh_2(OEP)_2]$ oxidatively adds a variety of molecules XY ($PhCH{=}CH_2$, $PhCH_2Br$, $HSnBu_3$, 9,10–dihydroanthracene)

Scheme 10

$[Rh_2(OEP)_2] \rightarrow 2[Rh(OEP)]$

$[Rh(OEP)] + PhCH=CH_2 \rightarrow [Rh(OEP)CH_2CHPh]$

$[Rh(OEP)CH_2CHPh] + [Rh_2(OEP)_2] \rightarrow [Rh(OEP)CH_2CH(Ph)Rh(OEP)] + [Rh(OEP)]$

to give $[Rh(OEP)X]$ and $[Rh(OEP)Y]$.[104] With styrene the product is $[Rh(OEP)CH_2CH(Ph)Rh(OEP)]$. A free radical mechanism obtains for these reactions, which is illustrated in Scheme 10.

The oxidative addition of MeI to (**25**) and (**26**) was studied in order to compare the reactivity of mononuclear and binuclear complexes.[105] The steric and electronic differences between (**25**) and (**26**) should be fairly small,

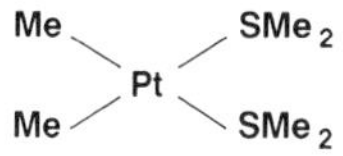

<table>
<tr><td>

Me, SMe$_2$
   Pt
Me, SMe$_2$

25

</td><td>

Me   Me$_2$ Me
     S
 Pt   Pt
Me   S   Me
    Me$_2$

26

</td></tr>
</table>

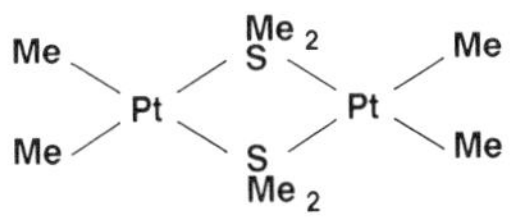

allowing any effects, such as cooperativity in (**26**), to be probed. The mechanism of MeI addition to (**25**) is S_N2 and in MeCN the cationic intermediate $[Pt(SMe)_2(Me)_3(NCMe)]^+$ was observed. A molecule of MeI adds to each metal center in (**26**) with addition to the first being rate determining. Having two metal atoms close together does not help the oxidative addition in this case since (**26**) reacts *ca* 20 times slower than (**25**). Reductive elimination of methane from (**27**) occurs smoothly in spite of the fact that a dinuclear 1,2-elimination is symmetry forbidden.[106] NMR data and other considerations

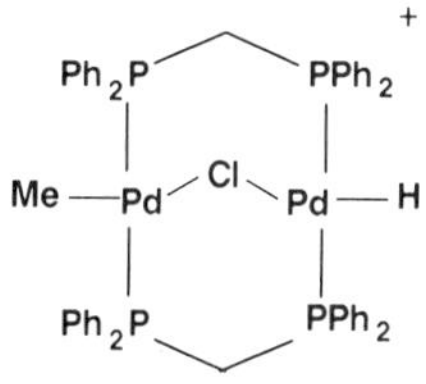

<table>
<tr><td>

Ph$_2$P   PPh$_2$
           +
Me—Pd Cl Pd—H
Ph$_2$P   PPh$_2$

27

</td><td>

Ph$_2$P   PPh$_2$
           +
Me—Pd H Pd—Cl
Ph$_2$P   PPh$_2$

28

</td></tr>
</table>

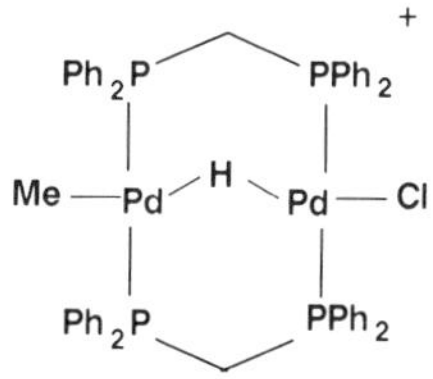

are consistent with interconversions like (**27**) → (**28**), so that in fact the reductive elimination involves only one metal center (**28**). Thermolysis of (**29**) gives the clean reductive elimination of propane.[107] The mechanism of this interesting reaction has not been established. The oxidative addition of alkyl halides (RX) to (**30**) produces the Au(II) dimer (**31**).[108] A kinetic study suggests that the initial step is one-electron transfer from (**30**) to the alkyl halide (R'X).

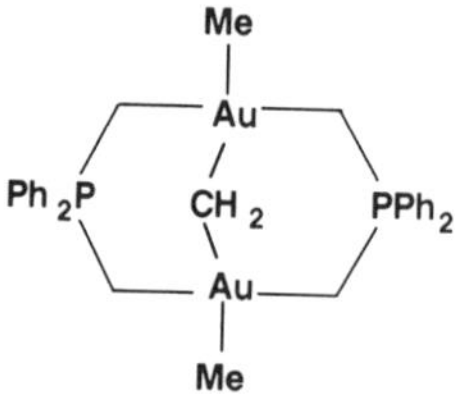

29

30

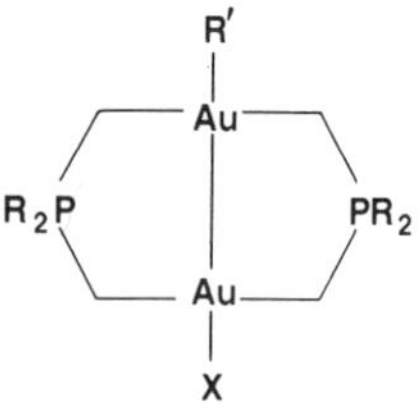

31

11.6. Other Reactions

11.6.1. Titanium

MO calculations on $[TiCl_2(C_2H_4)]$ have been used as a basis for probing the factors responsible for the two limiting ethylene bonding modes, namely, a π-complex versus a metallacyclopropane structure.[109] In another study stepwise bond dissociation enthalpies $D_1(M-L)$ and $D_2(M-L)$ in $[Cp_2TiL_2]$ ($L = Cl^-$, CO, Me^-, SMe^-) were estimated using extended Huckel calculations and thermodynamic data.[110] Ligand to metal π donation results in larger differences in $D_2 - D_1$ than σ donation.

11.6.2. Chromium

Cr^{2+} reacts with iodoacetamide to give $[Cr(OH_2)_5I]^{2+}$ and generate the CH_2CONH_2 radical that reacts further to yield $[Cr(OH_2)_5CH_2CO(NH_2)]^{2+}$ and $[Cr(OH_2)_5OC(NH_2)CH_3]^{3+}$ in a $3:2$ ratio.[111] In general, the reaction of Cr^{2+} with ICH_2Y is faster for more electron-withdrawing Y groups. In the acid-induced decomposition of σ-bonded (pyridylmethyl)chromium(III) complexes, $[Cr(CH_2C_5H_4N)L_n]$ ($L = $ dien, trien, etc.), aquation occurs before $Cr-C$ bond fission, which is proposed to occur homolytically.[112] The radicals CH_2OH, $CHMeOH$, and CMe_2OH react with $[Cr(edta)]^{2-}$ and $[Cr(nta)]^-$ to give α-hydroxyalkyl chromium(III) complexes.[113] The hydrolysis rates of these complexes suggest that solvent water and not a water ligand acts as the electrophile.

11.6.3. Manganese and Rhenium

The complex $[(arene)Mn(CO)_2Me]$ can be made by treatment of the corresponding iodide with MeLi.[114] A general route to $[(arene)Mn(CO)_2R]$ ($R = $ Me, Et, i-Pr, CH_2Ph) involves reduction of $[(arene)Mn(CO)_2I]$ to give $[(arene)Mn(CO)_2]^-$, followed by alkylation.[115] A series of papers concerning the synthesis and reactivity of manganese and rhenium α-hydroxyalkyl complexes have been published.[116–118]

11.6.4. Iron, Ruthenium, and Osmium

When electrochemically reduced, iron(III) porphyrins can be conveniently alkylated at the metal by reaction with alkyl halides.[119] Oxidative addition of halogens, HCl, and alkyl halides (RX) to $[Ru(CO)_2(triphos)]$ {triphos = 1,1,1-tris[(diphenylphosphino)methyl]ethane} yields $[Ru(CO)_2(Y)(triphos)]^+$ (Y = R,X,H).[120] Acetyl chloride adds to give the unstable acetyl product (Y = COMe), which eliminates ketene to give the hydride (Y = H). This unusual instability of an acetyl complex is attributed to the high *trans* influence of the phosphine ligand. $[RuCl(H)(PPh_3)_3]$ reacts with alkynes to give the *ortho*-metalated complex $[RuCl(PPh_2(C_6H_4))PPh_3)_2]$.[121] The exchange between the hydride ligand and the aryl *ortho* hydrogens is rapid and reversible. The electrophilic cleavage of the Os—Me bond in $[Cp^*Os(CO)(PMe_2Ph)Me]$ occurs upon addition of Br_2 or $HgBr_2$ to yield $[Cp^*Os(CO)(PMe_2Ph)Br]$.[122] The complexes $[Cp^*Os(CO)(PMe_2Ph)(Me)(Y)]Br$ (Y = Br,HgBr) are isolable intermediates in these reactions. Preliminary kinetic data suggest that the reaction with Y = Br occurs via nucleophilic attack by Br^- on the cation rather than by intramolecular reductive elimination steps.

11.6.5. Cobalt, Rhodium, and Iridium

Acyl cobalt carbonyls, $[Co(CO)_4C(O)R]$, react with H_2 to yield RCHO.[123] The mechanism involves initial dissociation of CO followed by oxidative addition of H_2 and finally reductive elimination. There has been considerable research dealing with vitamin B_{12} and associated model complexes. Substitution of coordinated water by $CH(CN)_2^-$ in aquocobalamin (B_{12a}) and diaquocobinamide at 25 °C has $\log K_{eq} = 7.4$ and 11.5, respectively.[124] The dealkylation of alkylcobalamin by $[PtCl_6^{2-}]$ in acidic solution is suggested to occur by rate-determining one-electron transfer to form a $[R\text{-}B_{12}^+,PtCl_6^{3-}]$ ion pair that collapses to products.[125] Similarly, the halide-assisted demethylation of Me-B_{12} by iodine in aqueous solution involves electron transfer followed by heterolytic Co—C bond cleavage to give B_{12r} according to Scheme 11.[126] Cleavage of the Co—C bond in adenosylcobalamin (Ado-B_{12}) proceeds via both heterolytically to yield B_{12a} and homolytically to yield B_{12r}.[127] At pH 4.0 heterolysis dominates and at pH 7.0

Scheme 11

$$Me\text{-}B_{12} + I_2 \rightarrow Me\text{-}B_{12}^+ + I_2^-$$

$$Me\text{-}B_{12}^+ + X^- \rightarrow B_{12r} + MeX$$

$$B_{12r} + I_2 + H_2O \rightarrow H_2O\text{-}B_{12}^+ + I_2^-$$

$$2I_2^- \rightarrow I_2 + 2I^-$$

homolysis is the major pathway. The base-on Co—C bond dissociation energy of Ado-B_{12} in water is reported as $131 \pm 5 \, kJ \, mol^{-1}$.

Electrochemical reduction of alkylcobaloximes, $[RCo(DH)_2(OH_2)]$, in aprotic solvents at a dropping mercury electrode produces a Co(II) radical anion that undergoes homolytic Co—C fission to yield the Co(I) cobaloxime.[128] $Co^{II}(salen)$ complexes can be converted into $RCo^{II}(salen)$ derivatives by hydrazines $RNHNH_2$ under oxidative conditions (O_2 or *tert*-BuOOH).[129] Initially a Co(III)–superoxo or –butylperoxo complex is formed that then reacts with the hydrazine to produce radicals (R) that alkylate the cobalt center. The rearrangement of 5-hexenylcobalt(III) Schiff base complexes to cyclopentylmethyl analogues is triggered by chemical or electrochemical oxidation.[130] The mechanism is a radical chain process with the oxidation initiating homolytic fission to yield the 5-hexenyl radical that rapidly isomerizes to the cyclopentylmethyl radical. The latter attacks the Co(III) reactant (S_H2) to give product and 5-hexenyl radical (the chain carrier). The kinetic chain lengths are high. Aliphatic free radicals (R) react with $[R'Co(DH)_2(OH_2)]$ to form RR' and $[Co(DH)_2(OH_2)]$.[131] Based on kinetic results with R' groups of varying steric bulk, it is proposed that the radicals attack the nitrogen end of the N=C bond of the macrocycle *cis* to the Co—R' bond, followed by reductive elimination of RR'. The effect of the *trans* ligand on the Co—C bond length in $[RCo(DH)_2L]$ and related complexes has been investigated.[132-134] Also noteworthy is a study of the alkali-induced decomposition of $[(HOCH_2CH_2)Co(DH)_2py]$ to form $[Cp(DH)_2py]^-$ and acetaldehyde.[135]

The oxidative addition of MeI to $Li[RhI_2(CO)_2]$ in acetic acid is promoted by the presence of LiOAc and LiI.[136] It is postulated that the promoter anion serves to coordinate to the reactant and thereby increase its nucleophilicity toward MeI. The addition of MeI to $[Rh(acac)(CO)(PPh_3)]$ proceeds through the ionic intermediate $[Rh(acac)(CO)(PPH_3)(Me)]I$ to give $[Rh(acac)-(COMe)(PPh_3)I]$, which rearranges to give the final product, $[Rh(acac)(CO)(PPh_3)(Me)I]$.[137] Refluxing (32) in benzene leads to intramolecular oxidative addition of a C—Cl bond to give (33).[138] Complexes

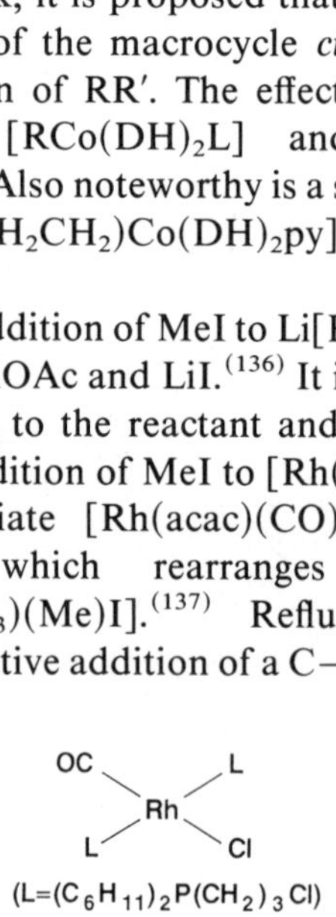

$(L=(C_6H_{11})_2P(CH_2)_3Cl)$

$(R=C_6H_{11})$

32 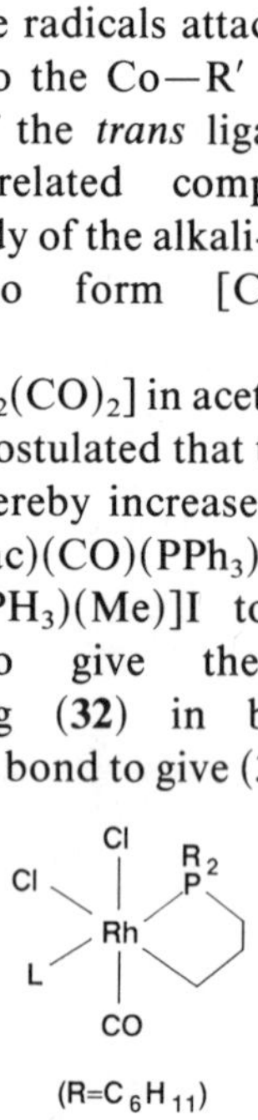   33

(34) and (35) undergo dissociation of PMe$_3$ to yield as intermediates (36) and (37), both of which have the option of reductive elimination of methanol or β-elimination of formaldehyde.[139] Scheme 12 summarizes the chemistry. The ratio of MeOH to CH_2O formed is 0.15 for (36) and 13 for (37). The overall

Scheme 12

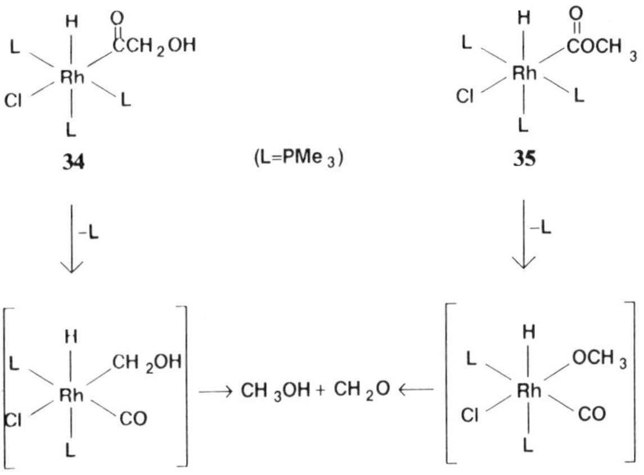

rates of loss of (**34**) and (**35**) are almost identical, indicative of rate-limiting PMe$_3$ dissociation. The conclusion is that methanol elimination from a methoxy hydride complex is preferred over methanol elimination from a hydroxymethyl hydride complex. Put another way, β-elimination is easier than methanol formation for the former and conversely for the latter. Thermal or photolytic generation of [Cp*Rh(PMe$_3$)] in cyclopropane at $-60\,^\circ$C yields the C—H insertion product, [Cp*Rh(PMe$_3$)(cyclopropyl)(H)], which rearranges at $-20\,^\circ$C via intramolecular migration of the Cp*Rh(PMe$_3$) unit to the α-carbon–carbon bond of the cyclopropyl ring to form the C—C insertion product [Cp*Rh(PMe$_3$)(CH$_2$CH$_2$CH$_2$)].[140] The reaction of formaldehyde with [Rh(OEP)H] gives the hydroxymethyl complex, [Rh(OEP)CH$_2$OH], which self-condenses to a metallo ether, [Rh(OEP)CH$_2$OCH$_2$Rh(OEP)], and also undergoes intramolecular reductive elimination of methanol according to reaction (14).[141] Somewhat similar chemistry obtains for the [Rh(TPP)]

$$[\text{Rh(OEP)CH}_2\text{OH}] + [\text{Rh(OEP)H}] \rightarrow [\text{Rh}_2(\text{OEP})_2] + \text{MeOH} \qquad (14)$$

analogues (TPP = tetraphenylporphyrin anion).[142] The electrochemical reduction of [Rh(TPP)(NHMe$_2$)$_2$]$^+$ in CH$_2$Cl$_2$ generates [Rh^{II}(TPP)], which can dimerize or attack the solvent to afford [Rh(TPP)CH$_2$Cl].[143] Alkyl exchange between (**38**) and (**39**) according to reaction (15) follows the reactivity

$$(\mathbf{38}) + (\mathbf{39}) \rightarrow [\text{Rh(BPDOBF}_2)(\text{Me})\text{I}] + [\text{Rh(PPDOBF}_2)] \qquad (15)$$

sequence R = benzyl > methyl ≫ *n*-butyl and is interpreted as a nucleophilic attack by Rh(I) (**38**) at carbon.[144] This behavior is entirely consistent with the known reactivity of (**38**) toward alkyl halides.[145,146] For the oxidative

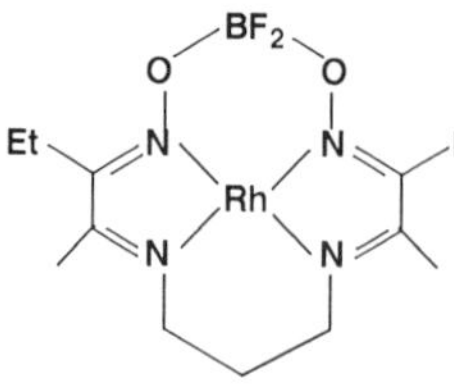

38

$[Rh(BPDOBF_2)]$

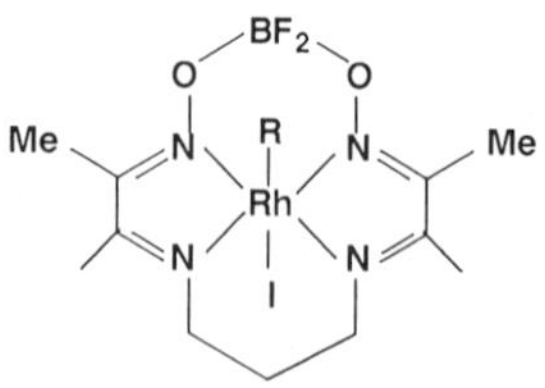

39

$[Rh(PPDOBF_2)(R)I]$

addition of (**38**) to RX, the reactivity orders are X = I > Br > OTs > Cl; R = methyl > primary > secondary > tertiary (no reaction). A simple S_N2 mechanism probably holds except when the electrophile is sterically congested and has a very polarizable leaving group (e.g., 2-iodopropane) in which case a one-electron (radical) pathway is taken.

A rare example of oxidative addition of water was recently discovered and is illustrated in Scheme 13.[147] The hydroxy hydride complex (**40**) does

Scheme 13

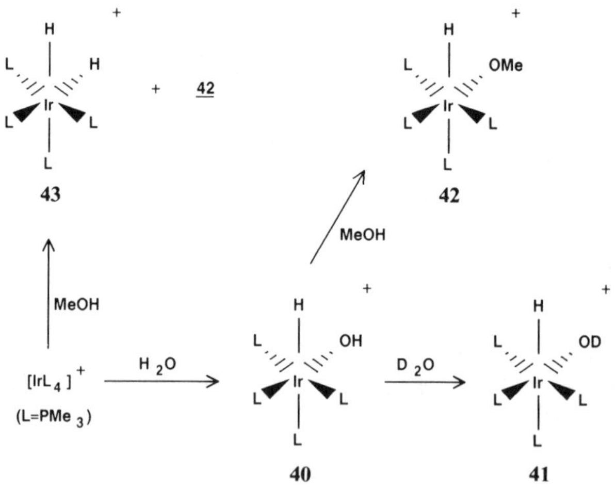

not eliminate water even at 100 °C and is only weakly basic. Addition of D_2O to (**40**) affords (**41**), showing that a reductive elimination–oxidative addition sequence is not occurring. In methanol complex (**40**) converts to the methoxy hydride (**42**), but methanol addition to $[IrL_4]^+$ yields the dihydride (**43**) as well as (**42**). This suggests that the oxidative addition of methanol to $[IrL_4]^+$

involves initial formation of $[\text{IrL}_3(\text{H})(\text{OMe})]^+$ which then β-eliminates and coordinates another L. The ethoxy hydride complex $[\text{Cp*Ir}(\text{PPh}_3)(\text{OEt})\text{H}]$ has a chemistry substantially different from that of early or higher valent metal alkoxides or of the analogous alkyl hydride complexes $[\text{Cp*Ir}(\text{PPh}_3)(\text{R})\text{H}]$.[148] Oxidative addition of a variety of small molecules to *trans*-$[\text{Ir}(\text{CO})(\text{PPh}_3)_2(\text{Me})]$[149] and of functionalized alkyl halides to $[\text{IrCl}(\text{CO})(\text{PR}_3)_2]$[150] have been described. Rearrangement of *sec*-alkyliridium(III) complexes $[\text{Ir}(\text{X})(\text{CO})(\text{I})(\text{R})\text{L}_2]$ (X = Cl,I; L = PMe_3, PMe_2Ph) proceeds cleanly by a mechanism involving rate-determining iodide dissociation *trans* to the alkyl ligand.[151] The reaction is completed by β-elimination, hydride insertion, and stereospecific return of the iodide. Similar chemistry obtains for rearrangements of $[\text{Ir}(\text{X})(\text{CO})(\text{Me})(\text{R})\text{L}_2]$.[152]

11.6.6. Nickel, Palladium, and Platinum

Exchange between Ar and Ar' in *trans*-$[\text{Ni}(\text{Ar})(\text{SAr}')(\text{PEt}_3)_2]$ and in *cis*-$[\text{Ni}(\text{Ar})(\text{SAr}')(\text{dmpe})]$ takes place via reversible reductive elimination of ArSAr' followed by oxidative addition.[153] The alkylnickel(II) complexes $[\text{RNi}(\text{tmc})]^+$ (tmc = 1,4,8,11-tetramethyl-1,4,8,11-tetraazacyclotetradecane) are readily made by the reaction of $[\text{Ni}(\text{tmc})]^+$ and primary alkyl halides.[154] The rate-determining step in this reaction is one-electron transfer from $[\text{Ni}(\text{tmc})]^+$ to RX to generate R$^\bullet$, which is trapped by $[\text{Ni}(\text{tmc})]^+$. Alkyl halides R'X react with $[\text{RNi}(\text{tmc})]^+$ by electron transfer to yield alkylnickel(III) species and alkyl halide radical anions, both of which rapidly eliminate alkyl radicals that react to produce R_2, RH, and R(−H).[155] The hydrolysis of $[\text{RNi}(\text{tmc})]^+$ in alkaline aqueous solution yields RH and $[\text{Ni}(\text{tmc})(\text{OH})]^+$.[156] The hydrocyanation of ethylene as catalyzed by $[\text{NiL}_2(\text{C}_2\text{H}_4)]$ (L = P(O-*o*-tolyl)$_3$) at −40 °C proceeds through the intermediate $[\text{NiL}(\text{CN})(\text{C}_2\text{H}_5)(\text{C}_2\text{H}_4)]$ (**44**) which reacts with excess L to eliminate $\text{CH}_3\text{CH}_2\text{CN}$.[157] Kinetic data pertaining to the reductive elimination step suggest that (**44**) must first coordinate a second L before losing $\text{CH}_3\text{CH}_2\text{CN}$. Hence this is a rare example of associative reductive elimination.

Several theoretical MO studies have been published concerning oxidative addition and reductive elimination at palladium and platinum centers.[158,159] Strongly luminescent orthometalated Pt(IV) complexes, $[\text{Pt}(\text{Phpy})_2(\text{CH}_2\text{Cl})\text{Cl}]$ and $[\text{Pt}(\text{Thpy})_2(\text{CH}_2\text{Cl})\text{Cl}]$, can be prepared in high yield by photolysis of (**45**) and (**46**), respectively.[160] The oxidative addition

45   46

of allylic phenyl sulfides according to reaction (16) is kinetically retarded to a much greater extent when $C(\gamma)$ bears a substituent as compared to $C(\alpha)$, indicating that a crucial step in the reaction is attack by Pd on $C(\gamma)$.[161] As shown in reaction (17), iodine adds to (47) to yield a η^1-I_2 adduct, which

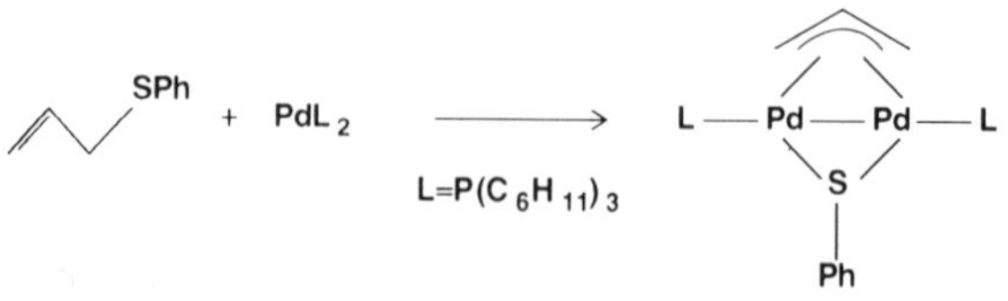

$$\text{SPh-allyl} + \text{PdL}_2 \longrightarrow \text{L—Pd—Pd—L} \quad (16)$$

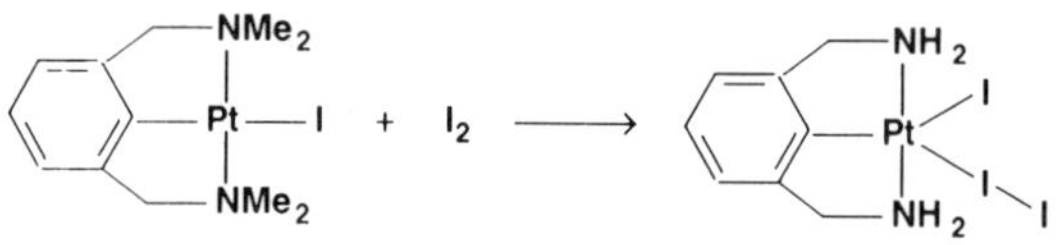

$$L = P(C_6H_{11})_3$$

$$(47) + I_2 \longrightarrow (48) \quad (17)$$

does not undergo subsequent oxidative addition as might be expected.[162] It is proposed that (48) serves as a model for the initial stage of oxidative addition of halogens to metal complexes. The reaction of $[\text{PtMe}_2(\text{NN})]$ (NN = 1,10-phenanthroline or bipyridine) with i-PrBr gives $[\text{PtMe}_2(\text{Br})(i\text{-Pr})(\text{NN})]$ by a S_N2 mechanism.[163] In contrast i-PrI reacts by a free radical chain mechanism with equation (18) being the initiation step. Under an inert atmosphere the product is $[\text{PtMe}_2(\text{I})(i\text{-Pr})(\text{NN})]$, but under oxygen the i-Pr radicals

$$[\text{PtMe}_2(\text{NN})] + i\text{-PrI} \rightarrow [\text{PtMe}_2(\text{I})(\text{NN})] + i\text{-Pr}^\bullet \quad (18)$$

are converted to i-PrO$_2$, which then react to give $[\text{PtMe}_2(\text{I})(i\text{-PrO}_2)(\text{NN})]$. The formation of such alkylperoxo complexes constitutes a useful test for a free radical mechanism of oxidative addition. The i-Pr radicals produced in equation (18) may also be trapped by alkenes (CH$_2$=CHX), leading to the formation of $[\text{PtMe}_2(\text{CHXCH}_2\text{Pr})(\text{I})(\text{NN})]$ in high yield.[164] A kinetic study of the oxidative addition of MeI to mononuclear and binuclear Pt(II) complexes having similar ligand sets shows that the mononuclear ones react somewhat more rapidly.[165] Although reductive elimination from four-coordinate Pd(II) frequently requires predissociation of a ligand, this is apparently not the case for reaction (19).[166] Excess PPh$_3$ retards the reaction but some

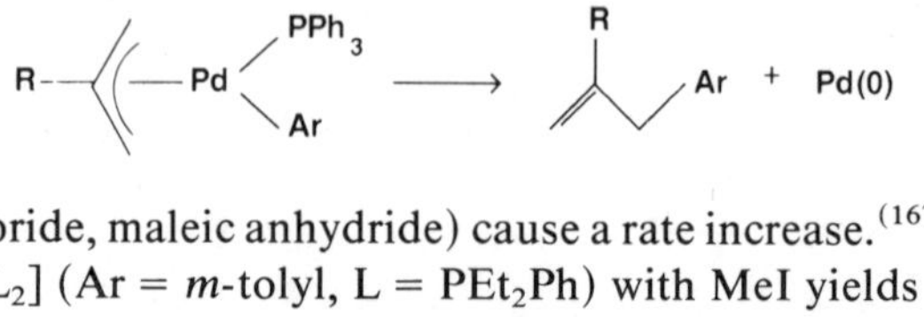

$$R \text{—} \text{Pd}(\text{PPh}_3)(\text{Ar}) \longrightarrow R\text{—Ar} + \text{Pd(0)} \quad (19)$$

olefins (allyl chloride, maleic anhydride) cause a rate increase.[167] The reaction of *trans*-$[\text{PdAr}_2\text{L}_2]$ (Ar = m-tolyl, L = PEt$_2$Ph) with MeI yields m-xylene and 3,3-bitolyl.[168] The addition of *trans*-$[\text{PdMe}(\text{I})\text{L}_2]$ increases the reaction rate and the selectivity for m-xylene. The kinetic data suggest that the reaction

Scheme 14

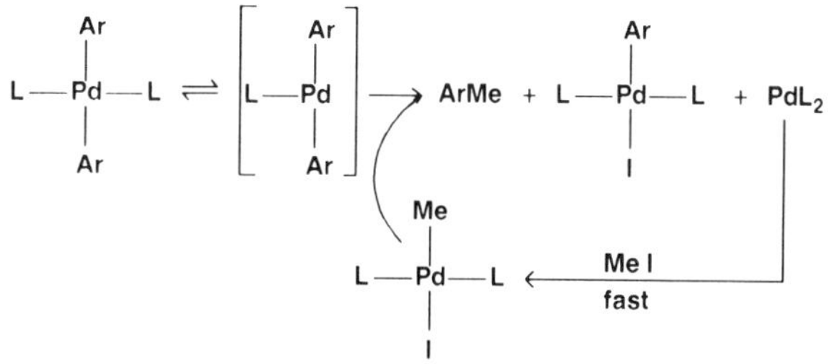

does not proceed through Pd(IV) intermediates but rather the key step is the intermolecular reaction with *trans*-[PdMe(I)L$_2$] as illustrated in Scheme 14.

The complex *trans*-[Pd(PPh$_3$)$_2$(CH$_2$Ph)Cl] (**49**) is stable at 85 °C in benzene.[169] However, the addition of phosphine sponge, [Pd(NCPh)$_2$Cl$_2$], leads to quantitative formation of PhCH$_2$Cl, probably via intermediate formation of the (isolated) [Pd(PPh$_3$)(CH$_2$Ph)Cl]$_2$ dimer. Abstraction of chloride from (**49**) with AgBF$_4$ in C$_6$D$_6$ gives immediate formation of C$_6$D$_5$CH$_2$C$_6$H$_5$. A free radical pathway for this latter reaction was ruled out, leaving the conclusion that it constitutes the first example of electrophilic alkylation of an arene by a transition metal alkyl complex. Further experiments indicated that the alkylation requires prior coordination of the arene to the metal. When coordination of the arene is blocked, Pd—CH$_2$Ph bond homolysis occurs. Protonolysis of *trans*-[PtCl(CH$_2$CMe$_3$)(PEt$_3$)$_2$)] by HX in aqueous methanol occurs by simple electrophilic attack at the Pt—C bond.[170] The thermal decomposition of *trans*-[PtCl(CH$_2$CMe$_3$)L$_2$] (L = P(cyclopentyl)$_3$) to give *trans*-[PtCl(H)L$_2$] and 1,1-dimethylcyclopropane is proposed to follow the mechanism in Scheme 15.[171]

Scheme 15

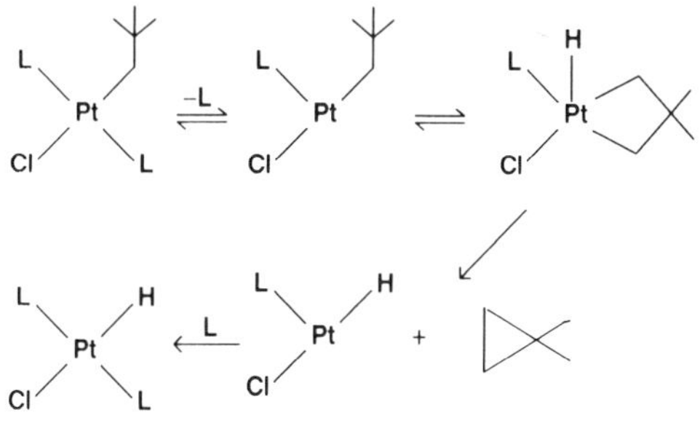

Chapter 12

Reactivity of Coordinated Hydrocarbons

12.1. Introduction

As in previous volumes, this chapter reviews kinetic and mechanistic studies on the stoichiometric reactions of coordinated hydrocarbons with nucleophiles and electrophiles. Concomitant with a significant increase in activity in this area has been a shift in emphasis from (π-polyene) metal complexes to studies of the reactivity patterns of η^1- to η^3-ligands such as carbenes, alkenes, and enyl groups. The regio- and stereospecificity of such processes have also attracted increasing attention, in particular the use of chiral complexes in asymmetric synthesis. Related catalytic processes are discussed elsewhere (Chapter 14), as are intramolecular ligand rearrangements (Chapter 13) and insertions (Chapter 10).

12.2. Nucleophilic Addition and Substitution

12.2.1. Carbon Monoxide Ligands

Nucleophilic attack on carbonyl ligands is mainly discussed elsewhere (Chapter 14), and is generally only mentioned here when attack at carbon monoxide occurs competitively with reaction at a coordinated hydrocarbon. Of particular interest, however, are a review of the Fischer–Tropsch synthesis[1] and theoretical[2-4] and mechanistic[5,6] studies of the reduction of metal carbonyls involving formyl species.

12.2.2. Addition at Carbene Ligands

Studies in recent years have shown the iron moiety $[CpFe(CO)(PPh_3)]$ to be a very versatile chiral auxiliary, permitting high stereochemical control during reactions at attached carbon ligands such as acyls and carbenes.[7] A recent example is the hydride reduction of the alkoxycarbene complex (1) in THF, which proceeds with complete stereoselectivity to give (2) as a single diastereomer according to equation (1). The remarkable stereoselectivity can

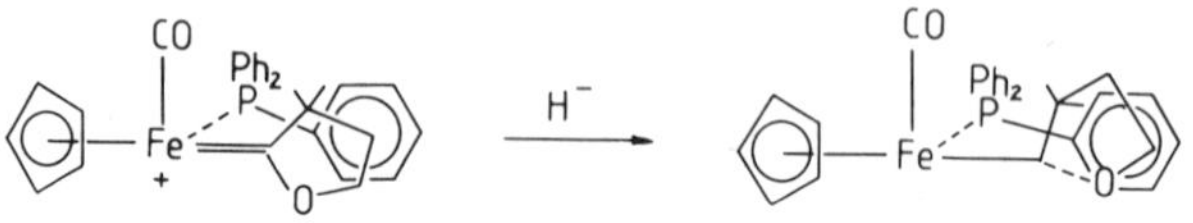

$$\text{(1)}$$

be explained in terms of a conformational analysis carried out for model $[CpFe(CO)(PPhH_2)(CH_2R)]$ species.[8] This analysis, which assumes a pseudooctahedral geometry, predicts the most stable conformer of (1) to be that in which the bulky quaternary carbon centre lies in the most open site between the Cp and CO ligands (see Newman projection 3). This forces the

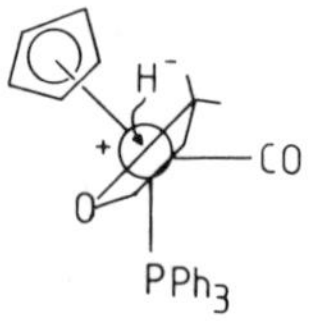

acyl oxygen *anti* to the CO ligand, and approach by the hydride nucleophile can only occur toward the face of the carbene opposite the blocking phenyl group of the PPh_3 in (1). Exploration of asymmetric induction by the related ruthenium moiety $[CpRu\{Ph_2PCH(Me)CH(Me)PPh_2\}]$, in which chiral centres are located on both the metal and the diphosphine ligand, has also begun.[9] For example, stereospecific attack by methanol occurs on the alkylidene ligands of (4), R = Me, Ph, to give the methoxycarbene complexes (5) shown in equation (2). Retention of configuration at ruthenium was confirmed from

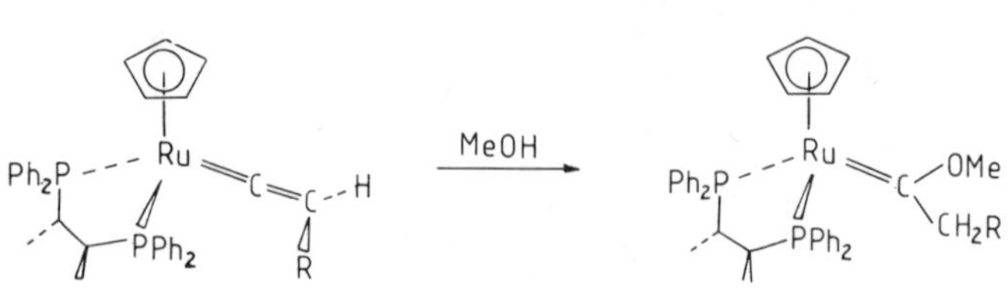

$$\text{(2)}$$

X-ray crystal structures of (4), R = Me and (5), R = Ph. The reaction between the chromium carbene complex $[(CO)_5Cr\{C(OMe)Ph\}]$ and alkynes ($RC\equiv CR'$) provides a useful route to substituted hydronaphthoquinones. A detailed study, employing a wide range of sterically hindered, electron-rich and electron-withdrawing substitutents (R,R'), has shown that regioselectivity is dominated by the steric effects of the alkyne substituents, and that electronic effects of R,R' are unimportant.[10]

Electrophilic metal carbenes have been implicated as intermediates in a variety of reactions. For example, the role of metal carbene species in $[Rh_2(OAc)_4]$-catalyzed cyclopropanation of alkenes is generally accepted to involve metal-hydrido species, which undergo subsequent olefin and carbonyl insertions as in equation (3). Significantly, an alternative route involving the initial formation of carbene (6) and ketene (7) intermediates has been demonstrated in the stoichiometric reaction of $[W(CO)_3Cl(AsPh_3)_2]$ with norbornene and ethanol to give the ester (8) as in Scheme 1.[12] The same intermediates

Scheme 1

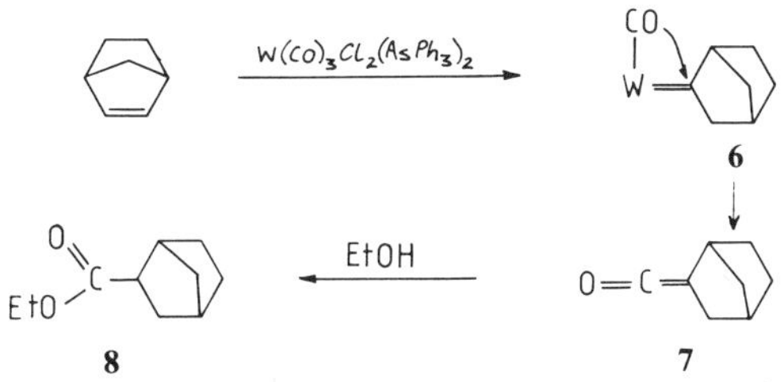

were proposed for the related hydroformylation of norbornene in the presence of $[W(CO)_3Cl_2(AsPh_3)_2]$.

$$(CO)M-H \xrightarrow{\text{>c=c<}} (CO)M-\overset{|}{\underset{|}{C}}-\overset{|}{\underset{|}{C}}-H \rightleftharpoons M-\overset{|}{\underset{\|}{C}}-\overset{|}{\underset{|}{C}}-\overset{|}{\underset{|}{C}}-H \rightarrow \text{Products} \qquad (3)$$

Protonation of the methoxymethyl complex $[CpFe(CO)_2(CH_2OMe)]$ generates the methyl complex $[CpFe(CO)_2(CH_3)]$.[13] A mechanistic study[14] of the analogous reaction with $[CpFe(CO)(PPh_3)(CH_2OMe)]$ reveals the initial formation of the carbene complex $[CpFe(CO)(PPh_3)(=CH_2)]^+$. This highly electrophilic carbene abstracts hydride from the starting complex to give equimolar amounts of the methyl product $[CpFe(CO)(PPh_3)(CH_3)]$ and $[CpFe(CO)(PPh_3)\{=CH(OMe)\}]^+$.

12.2.3. Addition at Acyl Ligands

A series of elegant papers by Davies *et al.* have demonstrated the ability of the chiral auxiliary $[CpFe(CO)(PPh_3)]$ to achieve high stereoselectivities in a wide range of carbon–carbon bond-forming reactions at coordinated acyl ligands. Highly diastereoselective tandem Michael additions of nucleophiles (RLi) and subsequent electrophilic alkylations (R'X) to α,β-unsaturated acyl complexes such as (9) have been reported.[14-16] For example, reaction (4) with R = R' = Me generates the α,β-dimethylbutanoyl complex (11) as a single diastereomer (d.e. > 100:1). The stereochemical control in these reactions was rationalized in terms of preferential formation of E-enolates (10) in the *anti* conformations (acyl O to CO), with approach of both the nucleophile (RLi) and electrophile (R'X) exclusively at the unhindered face of the double bond, the other face being completely shielded by the PPh$_3$ ligand as in equation (4). These predictions were confirmed by an X-ray analysis of (9), and a

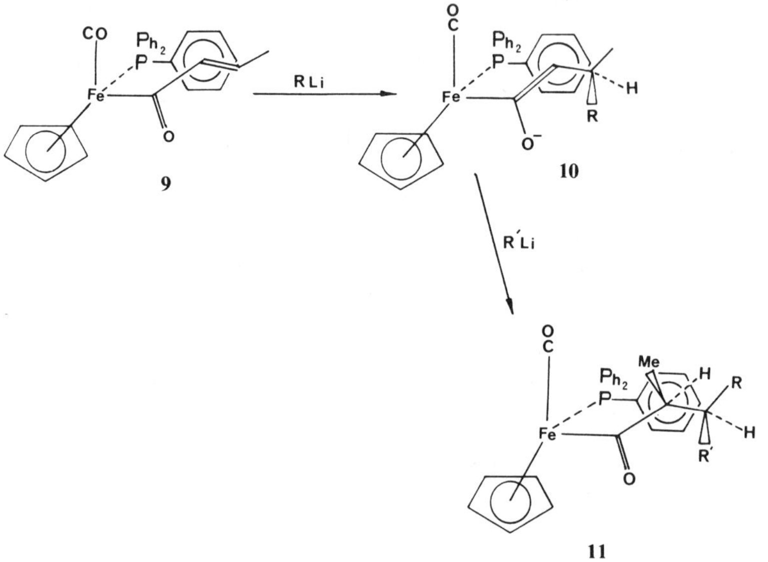

(4)

detailed conformational analysis for the acyl complex [CpFe(CO)(PPh$_3$)-(COCH$_3$)] based on extended Huckel and *ab initio* SCF MO calculations.[17,18] These analyses show the conformational preference of the acyl oxygen to be *anti* periplanar to CO due to steric interactions between the acyl ligand and two of the phenyl groups of the PPh$_3$. The analysis further predicts that increased size of the electrophile should enhance observed stereoselectivity. This was borne out in practice. For example, alkylation of the enolate (10) by the large methyl tosylate instead of MeI increased diastereoselectivity markedly.[18]

   The extension of this methodology to the stereoselective synthesis of β-lactams has been explored, for example, via Michael addition of LiNHCH$_2$Ph to (9) and electrophilic quenching with MeI or MeOH.[15,16] Repetition with optically pure substrates such as (S)-E-(9) provided enantiomerically pure β-lactams. The use of chiral [CpFe(CO)(PPh$_3$)(COR)] complexes in asymmetric synthesis will be further facilitated by the assignment of the absolute configuration of (R)-[CpFe(CO)(PPh$_3$){COCH$_2$CH$_2$O-(R)-menthyl}] via X-ray analysis.[19] Methods have also been devised for assessing the optical purity of such acyl complexes. Other developments include the stereoselective synthesis of quaternary carbon centres[20] and the use of α,β-unsaturated acyl complexes, such as (9), as chiral dienophile equivalents in asymmetric Diels–Alder cycloaddition reactions.[21]

12.2.4. Addition at Alkylidyne Ligands

Facile alkylidyne–alkyne coupling has been shown to occur in the reactions of the μ_3-alkylidyne clusters [H$_3$Ru$_3$(μ_3-CX)(CO)$_9$] (X = OMe, Me, Ph, or CH$_2$CH$_2$CMe$_3$) with a wide range of alkynes RC$_2$R as shown in Scheme 2.[22] Subsequent mild hydrogenation of the 1,3-dimetalloallyl clusters [HRu$_3$-η^3-XCCRCR)(CO)$_9$] (12) yields alkylidynes of longer chain length (13). This unprecedented reaction sequence suggests a new mechanism

Scheme 2

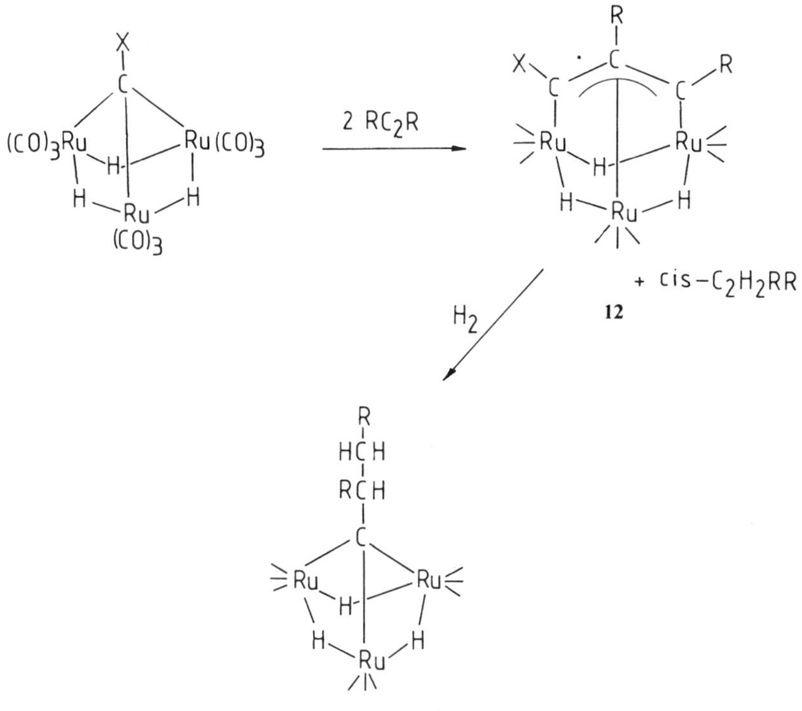

for Fischer–Tropsch chain growth in CO hydrogenation. Analogous reactions occur with $[HM_3(\mu\text{-COMe})(CO)_{10}]$ (M = Ru, Os).

12.2.5. Addition at η^2-Alkenes

A new reactivity parameter, k_{CO}^*, has been proposed for predicting the susceptibility of π-ethene ligands to nucleophilic addition.[23] This parameter represents the Cotton–Kraihanzel C—O stretching force constant of a hypothetical compound in with a CO ligand replaces the ethene. The k_{CO}^* values appear useful in predicting large qualitative differences in reactivity, such as setting lower limits for reactions of π-ethene ligands with different nucleophiles. However, it is not known whether k_{CO}^* is related to kinetic or thermodynamic factors, due to the current lack of rate and equilibrium data for such additions.

The oxidation of allyl alcohol (14) in aqueous $PdCl_4^{2-}$ gives β-hydroxypropanol (15) and hydroxyacetone (16) as major products (*ca* 55%).[24] Deuterium isotope distribution studies using $CD_2{=}CHCH_2OH$ and $CH_2{=}CHCD_2OH$ as substrates confirm that (15) and (16) are formed by the hydroxypalladium-hydride shift mechanism found for other acyclic alkenes as in Scheme 3. The deuterium isotope effect, k_H/k_D, for the hydride shift in

Scheme 3

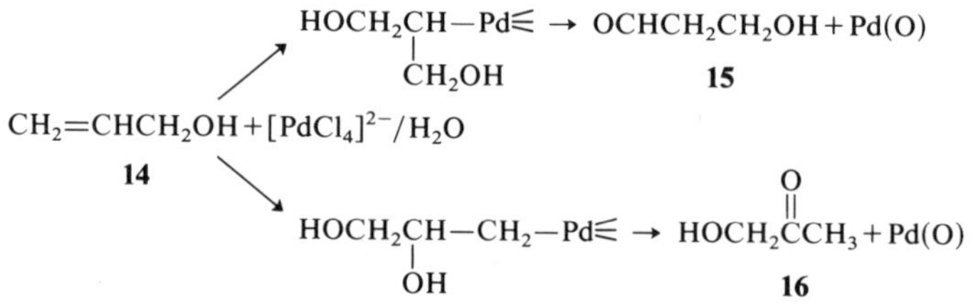

$$CH_2{=}CHCH_2OH + [PdCl_4]^{2-}/H_2O$$

14

$$\underset{\overset{|}{CH_2OH}}{HOCH_2CH}{-}Pd{\leqslant} \;\rightarrow\; OCHCH_2CH_2OH + Pd(O)$$

15

$$\underset{\overset{|}{OH}}{HOCH_2CH}{-}CH_2{-}Pd{\leqslant} \;\rightarrow\; HOCH_2\overset{\overset{O}{\|}}{C}CH_3 + Pd(O)$$

16

the formation of (15) is 1.9, which is very close to that (1.8) found previously for oxidation of 1,2-ethene-d_2. The labeling experiments also confirm that the other major product, acrolein (30%), is formed via direct hydride extraction from the alcohol carbon by Pd(II), a mechanism established for saturated alcohol oxidation.

The mechanism of oxidation of terminal alkenes by $[PdCl(NO_2)(CH_3CN)_2]$ in acetic acid has been reexamined.[25] Using (*E*)-1-deuterio-1-decene as substrate, products (17) and (18) were formed in approximately equal amounts as in equation (5). Their stereochemistry (threo/erythro ratio 88/12) is inconsistent with a previously[26] proposed mechanism requiring the

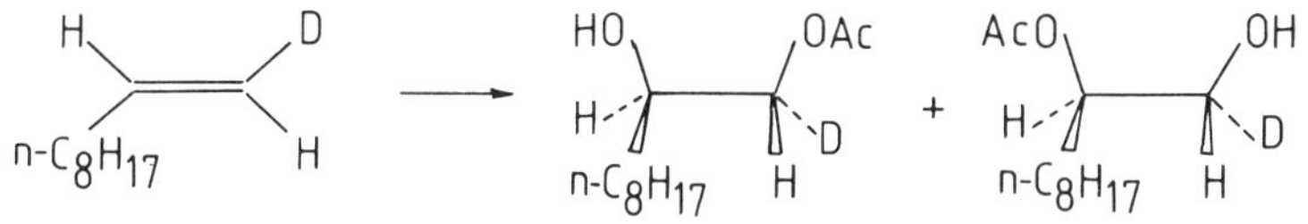

$$(5)$$

palladium–carbon bond to be cleaved with retention of configuration at carbon. A new mechanism is postulated involving *trans* acetoxypalladation, followed by an oxidative cleavage of the Pd—C bond with inversion.

A theoretical analysis of Pd^{2+}-catalyzed nucleophilic addition to alkenes, employing the concept of paired interacting orbitals, favors *trans* addition for OH^-, but *cis* attack by H^-.[27] These predictions agree with experimental observations. A particularly interesting development for future mechanistic studies in the use of FAB mass spectrometry to identify intermediates in organometallic reactions.[28] For example, adducts formed by addition of aryl palladium across the enol ether carbon–carbon double bond have been detected in the Pd-mediated coupling of aryl mercuric acetates with cyclic enol ethers. Addition of carbanions and morpholine to both (Z)- and (E)-isomers of the η^2-crotylacetate complex (**19**), R = H, R′ = Me; or R = Me, R′ = H, has been

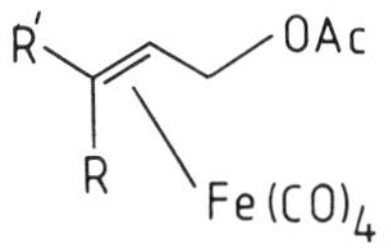

19

shown to be mildly regiospecific, with a preference for attack at the less substituted end of the C=C double bond.[29] The reactions are stereospecific, e.g. (Z)-(**19**) → (Z)-adducts. Optically-active metal–alkene complexes, in which the alkene itself is a center of asymmetry, should provide unique substrates for asymmetric synthesis. The preparation of such a complex, viz. the dihydrodioxin complex **20** in equation (6), is therefore of some interest.[30] It adds a broad range of nucleophiles, yielding a single optically-active adduct (**21**) in each case, suggesting strong stereoelectronic control of the addition.

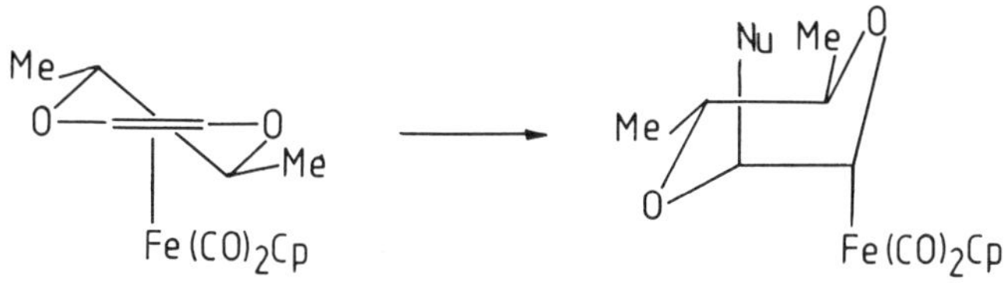

$$(6)$$

Although not involving addition at an alkene, the Pd^{2+}-induced ring opening of vinylcyclopropane in the presence of nucleophiles (Cl^-, MeO^-, AcO^-) may be conveniently mentioned here. Chloropalladation of (+)-2-

carene (**22**) in benzene with $[PdCl_2(CH_3CN)_2]$ occurs with retention of configuration at carbon, to give the (π-allyl) complex (**23**), X = Cl as in Scheme 4.[31] In contrast, for chloropalladation in chloroform/ethanol the Pd—C bond formation takes place with inversion, giving (**24**). Oxy-palladations in methanol or acetic acid solvents give products of type (**23**), X = MeO or AcO.

Scheme 4

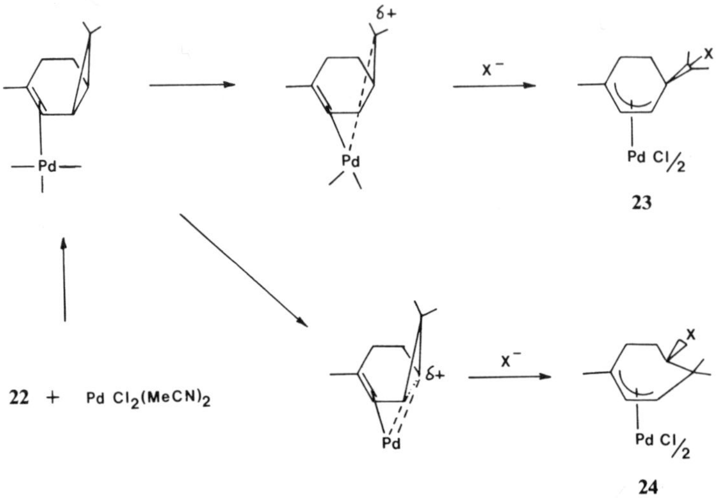

12.2.6. Addition at η^2-Alkynes

Reaction of the alkyne complexes $[CpFe(CO)\{P(OPh)_3\}(MeC{\equiv}CR)^+$ (R = Me, Ph, CO_2Me) with the nucleophiles R_3N, R_2N^-, RO^-, and RLi reveals a variety of reaction paths depending on substrate, nucleophile, and conditions.[32] Reaction with MeLi is particularly indiscriminate, showing nucleophilic attack at each of the Cp, alkyne, and iron centers. These reactions clearly violate the empirical rule of Davies, Green, and Mingos[33] that nucleophiles will react preferentially at even over odd polyenes. The cause of this marked difference in coordinated alkynes compared with alkenes was attributed to the presence of an additional π-⊥-orbital on the alkyne.[32]

12.2.7. Addition at η^3-Enyls

Previous studies of nucleophilic attack on (π-allyl) palladium complexes have revealed two distinct stereochemical pathways: (1) nucleophiles such as

Scheme 5

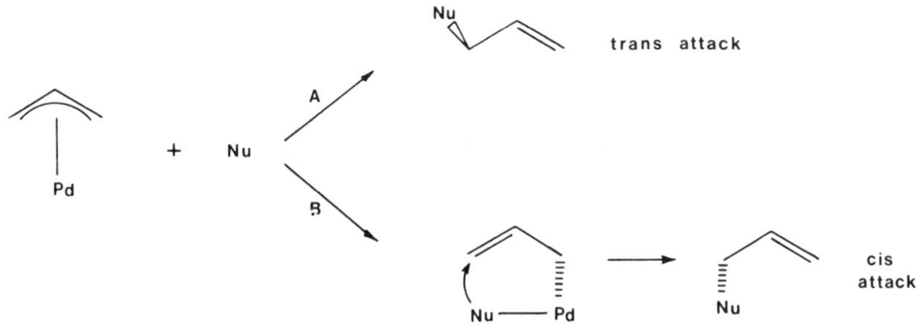

H⁻, Me⁻, Ar⁻, vinyl, and allyl add *cis* via a migration from the metal to the allyl ligand (path A in Scheme 5), (2) nucleophiles such as carbanions, amines, amides, alcohols, and chloride prefer to add *trans* (path B).[34] It has now been shown that carboxylates (in the presence of *p*-benzoquinone) can be directed to either *cis* or *trans* attack, depending on the ligand environment.[35] In the presence of chloride ions (LiCl), acetate adds to $[(\pi\text{-allyl})\text{PdCl}]_2$ complexes via path A exclusively, while if Cl⁻ is removed only path B operates. In the latter case, prior coordination of AcO⁻ to the Pd is followed by *cis* migration to the allyl ligand as in equation (7). The role of chloride ligands

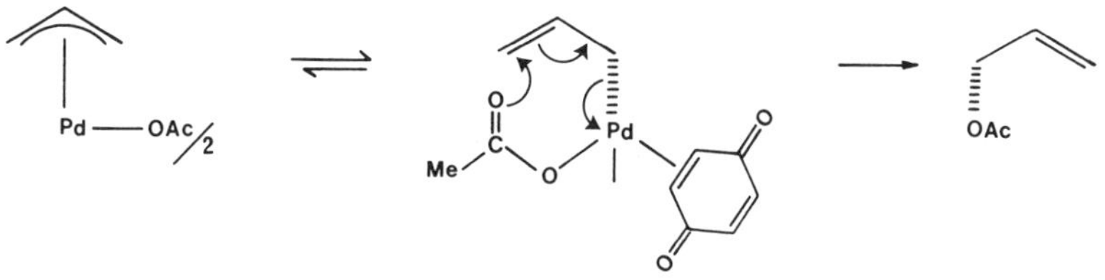

(7)

in these dual stereoselective reactions is apparently to prevent coordination of the acetate nucleophile. The donor–acceptor properties of auxiliary ligands (L) have also been shown to influence strongly the regiospecificity of amine addition to $(\eta^3$-geranyl)– and $(\eta^3$-neryl)–palladium complexes, $[(\eta^3\text{-allyl})\text{PdL}_2]^+$.[36] Donor ligands L, such as pyridine and TMEDA, direct the addition to the less substituted terminus of the π-allyl ligand, while a relative acceptor such as PPh₃ directs to the more substituted terminus. The variation in amination rates with the nature of L (PPh₃ ≫ bipy > TMEDA) can also be rationalized in terms of the relative donor–acceptor properties of these auxiliary ligands.

The regio- and stereochemistry of carbanion addition to related $[\text{CpMo(NO)(CO)}(\eta^3\text{-allyl})]^+$ cations (**25**) has also been examined.[37] Stereoselectivity is determined primarily by the preference of the allyl ligand in (**25**) for either the *exo*-(**25a**) or *endo*-(**25b**) conformation, with highest

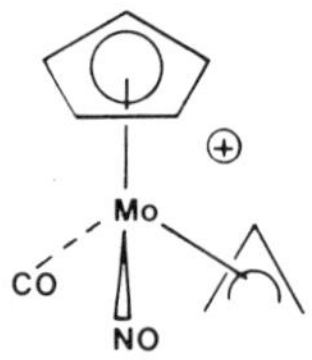

25a 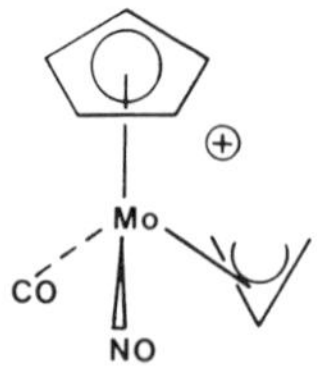   25b

stereoselectivities observed in the former instance. In this *exo* conformation, nucleophilic attack at the allyl terminus *cis* to the NO ligand is preferred, with the nucleophile approaching the allyl face external (*anti*) to the metal. The ability of certain carbanions to catalyze *endo–exo* interconversion can lead to a high degree of stereoselectivity in carbanion addition. Similar studies have been made of addition by heteronuclear anions (Nuc = H^-, MeO^-, PhS^-) to type (25) cations.[38] In each case, attack occurs at the terminus of the coordinated allyl group. In the cases of H^- and MeO^-, no spectroscopic evidence was found for Mo—Nuc or Mo—CO·Nuc species, supporting direct addition of the nucleophile to the *anti* face of the allyl ligand. With [CpMo(NO)-(CO)(η^3-cyclooctenyl)]$^+$ as substrate, addition of PhS^- to the prochiral allylic ligand is diastereoselective, producing only one stereoisomer (26) in high yield. Its structure was confirmed by an X-ray analysis. Of considerable interest to such stereochemical studies is the report that *exo-* and *endo*-conformational isomers of [CpMo(CO)(L)(η^3-allyl)]$^+$ (L = CO or NO) species can be differentiated by ^{95}Mo NMR spectroscopy.[39] *Exo*-isomers of a given allyl complex yield resonances upfield of their corresponding *endo*-isomers.

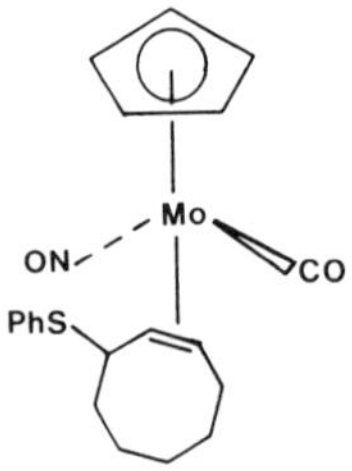

26 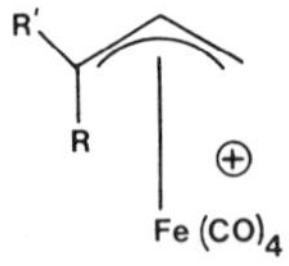   27

The regioselectivity of carbanion addition to [(η^3-crotyl)Fe(CO)$_4$]$^+$ complexes (27) is similar to that seen above with related Pd(II) substrates, with preferential attack at the less substituted allyl terminus.[29] A novel observation is the higher regioselectivity shown by the *anti* isomer (27a), R = Me, R′ = H, compared with the *syn* analogue (27b), R = H, R′ = Me.

12.2.8. Addition at η^4-Dienes

A novel explanation has been put forward for why cyclic dienes in aqueous acetone do not undergo hydroxypalladation as readily as do acyclic dienes and mono alkenes.[40] Weak coordination of the second C=C group to the Pd(II) is assumed, giving either square-pyramidal (**28**) or trigonal-bipyramidal (**29, 30**) complexes. It is further assumed that H_2O/OH^- attack is inter-molecular, and occurs readily only when the target C=C bond is perpendicular to the X–Pd–X axis. Thus, attack is impossible in structures **28**, **29**, and **30** where the dihedral angle is zero, but facile for more flexible acyclic dienes and mono-enes.

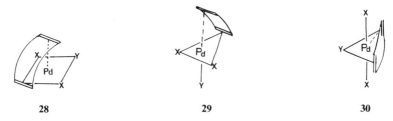

| 28 | 29 | 30 |

Nucleophilic attack on a coordinated C=C bond in preformed $[Pd(\eta^4\text{-}diene)Cl_2]$ complexes, where the two C=C bonds are indeed approximately perpendicular to the $PdCl_2$ plane,[41] is of course well known to be facile. In a recent example, attack of MeO^- on the cycloocta-1,4-diene palladium complexes (**31**), R = H or Me, was investigated.[41] X-ray structural data for substrates (**31**) establish C(5) as the most distant alkene carbon from the Pd in both cases, which would (according to a recent theoretical analysis[42]) target it as the site for kinetically-controlled nucleophilic attack. The isolated product in each case has, however, the $1,4,5\text{-}\eta^3$-cyclooctenyl structure (**32**), R = H or Me, rather than the predicted $1,3,4\text{-}\eta^3$-arrangement.

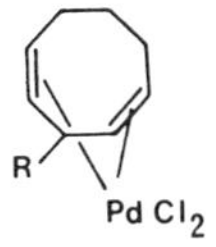

31

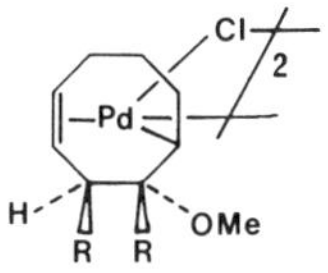

32

Kinetic and spectroscopic studies of the reactions of cyclohexylamine with the complexes $[MX_2(\eta^4\text{-cycloocta-1,5-diene})]$ (M = Pd, X = Cl, Br; M = Pt, X = Br) in acetone reveal the rate law (8).[43] This behavior was rationalized in terms of the mechanism in Scheme 6, involving reversible formation of the zwitterion (**33**) followed by amine-assisted deprotonation to give product (**34**). Assuming a steady-state concentration for intermediate (**33**), and $k_{-1} \gg k_2[\text{amine}]$, k_{obs} is given by expression (9). For X = Br, the Pd(II) complex is

$$\text{Rate} = k[\text{M}][\text{amine}]^2 = k_{obs}[\text{M}] \tag{8}$$

$$k_{obs} = K_1 k_2[\text{amine}]^2 \tag{9}$$

about 70 times more reactive than its Pt(II) analogue. This provides the first quantitative comparison of the relative abilities of Pd(II) and Pt(II) to activate

Scheme 6

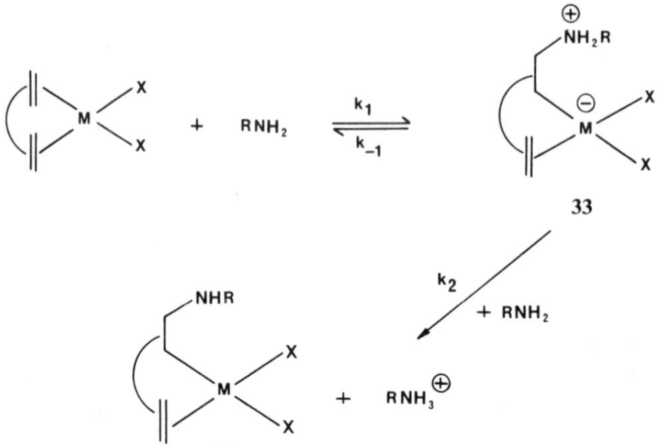

33

34

alkenes toward nucleophilic attack. The order Pd(II) ≫ Pt(II) may arise from less effective π-back bonding in the Pd case. A range of nucleophiles, including H^-, py, PhMgBr, and PMe_3, attack $[Co(CO)_3(\eta^4\text{-}1,3\text{-butadiene})]^+$ regiospecifically at the diene C-1.[44] This behavior parallels that reported for $[CpMo(CO)_2(\text{diene})]^+$ systems, but constrast with the C-2 selectivity (kinetic) of neutral isoelectronic $[Fe(CO)_3(\eta^4\text{-diene})]$ species. The differing regioselectivity may arise from the charge differences between the systems.

12.2.9. Addition at η^5-Dienyls

Surprisingly, the nucleophiles MeLi, PhNHLi, and Ph_2NLi react with the complex $[CpFe(CO)\{P(OPh)_3\}(MeC\equiv CPh)]^+$ mainly at the Cp ring to give $[(\eta^5\text{-}C_5H_4\cdot Nuc)Fe(CO)\{P(OPh)_3\}(\eta^1\text{-MeC}=CHPh)]$ products.[32] These reactions violate an empirical rule[33] stating that nucleophiles will react preferentially at even over odd polyenes. $[Fe(CO)_3(1\text{-}5\text{-}\eta\text{-dienyl})]^+$ complexes (**35**) continue to attract considerable attention. Kinetic studies of the reversible reactions (10) (dienyl = C_6H_7, 2-$MeOC_6H_6$, or C_7H_9) in MeCN reveal the rate law (11).[45] The observed rate trend $C_6H_7 > 2\text{-MeOC}_6H_6 > C_7H_9$ (46:6:1 at

$$[Fe(CO)_3(1\text{-}5\text{-}\eta\text{-dienyl})]^+ + 2\text{-Etpy} \underset{k_{-1}}{\overset{k_1}{\rightleftharpoons}} [Fe(CO)_3(1\text{-}4\text{-}\eta\text{-diene}\cdot 2\text{-Etpy})]^+ \qquad (10)$$

$$k_{obs} = k_1[2\text{-Etpy}] + k_{-1} \qquad (11)$$

0 °C) and the low $\Delta H_1^\ddagger$ and large negative $\Delta S_1^\ddagger$ values are consistent with direct addition (k_1) of 2-Etpy to the dienyl rings of cations (**35**). Comparison with

earlier data[46] on related systems shows that the steric retardation caused by the introduction of blocking groups in the 2-position of the pyridine nucleophile decreases in the order 2-Et > 2-Me > H (16:5:1) for the parent $[Fe(CO)_3(C_6H_7)]^+$ cation.

The major initial products from the reaction of $[Fe(CO)_3(1\text{-}5\text{-}\eta\text{-}C_6H_7)]^+$ with I^- in nitromethane or acetone solvents have been shown to be the ring adduct $[Fe(CO)_3(1\text{-}4\text{-}\eta\text{-}IC_6H_7)]$, and the novel acyliodide complex $[Fe(CO)_2(COI)(1\text{-}5\text{-}\eta\text{-}C_6H_7)]^+$ (36) arising from I^- attack at a CO ligand.[47] Similar reactions occur with the related cations (35), dienyl = 2-MeOC$_6$H$_6$ or C$_7$H$_9$. Exposure of (36) to light results in its rapid disappearance, which may account for the failure to detect (36) in earlier studies. The carbonyl-displaced product $[Fe(CO)_2I(1\text{-}5\text{-}\eta\text{-}C_6H_7)]^+$ only becomes a significant product after longer reaction periods. These results are at variance with a theoretical study[48] of nucleophilic attack on type (35) cations (dienyl = C$_6$H$_7$ or C$_7$H$_9$), which predicts initial attack by I^- at the iron center. However, the prediction[48] of initial attack by OH^- at the CO ligand rather than the dienyl ring in high-polarity solvents agrees with experimental observations. This theoretical analysis, which calculates interaction energies using frontier and near-frontier orbitals on the interacting species, also suggests the formation of metal and carbonyl attack intermediates in the reactions of (35) with SH^- and NCS^-. Self-consistent charge and configuration (SCCC) molecular-orbital calculations have also been carried out on the dienyl cations (35), dienyl = C$_7$H$_9$ and C$_8$H$_{11}$.[49] In both complexes the ring adopts a nonplanar geometry, with the aliphatic part bent away from the metal. The most stable conformations in both cases have the hydrogen atoms of the aliphatic group as staggered as possible. Chiral discrimination occurs in the reaction of the cycloheptadienyl cation (37) with the sulfoximine-stabilized enolate (38) shown in equation (12).[50] Enantiomeric excesses of up to 50% are obtained depending on the base and solvent employed in generating the nucleophile (38).

$$37 \qquad\qquad 38 \qquad\qquad\qquad\qquad\qquad\qquad (12)$$

Kinetic studies of the addition of phosphorus and nitrogen nucleophiles to the rings in cyclohexadienyl and cycloheptadienyl complexes of the type $[Mn(CO)(NO)L(1\text{-}5\text{-}\eta\text{-}dienyl)]^+$ (L = CO, PR$_3$), such as that in equation (13), reveal the general rate law (14).[51,52] The Mn(CO)(NO)L$^+$ moiety is shown to be electronically equivalent to the Fe(CO)$_2$L$^+$ unit for activating a dienyl ring. The Brønsted slopes, α, of about 0.5 found for the addition of triarylphosphines to several of these cations establish the importance of nucleophile basicity in these reactions. As expected, the presence of a substituent (Me, Ph) at the C(6)-saturated carbon in $[Mn(CO)(NO)L(1-5-\eta-6-RC_6H_6)]^+$ causes a very large steric retardation of k (factor of 300–5000), consistent with *exo*-addition to the cyclohexadienyl ring. However, for the related cycloheptadienyl substrates (**39**) no steric retardation is caused by a C_6H_5 substitutent (R) at C(6), since nucleophilic additon can now occur at C(1) rather than C(5) (equation 13). *Exo*-addition of phosphorus nuclophiles in equation (13) was

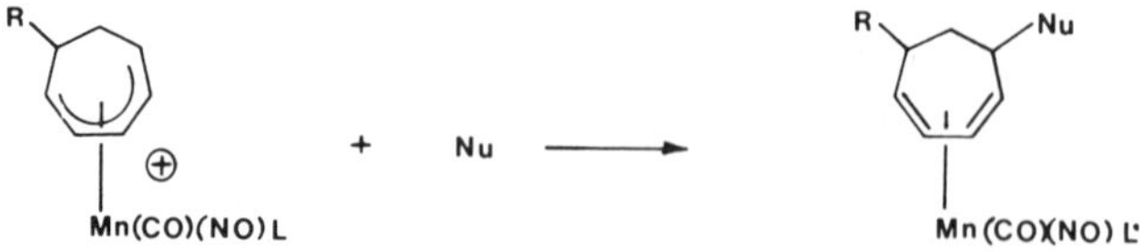

$$k_{obs} = k[Nu] \tag{14}$$

established by NMR spectroscopy using double deuteriation experiments.[53] However, with H^- as nucleophile, stereoselective *endo*-addition to the cycloheptadienyl ring of (**39**), R = H, L = CO, occurs,[53] as had been shown previously for related $[Mn(CO)(NO)L(cyclohexadienyl)]^+$ cations. Also significant in this respect is the conversion of the *exo*-phosphonate complex (**40**) to its *endo*-isomer via deprotonation with *n*-BuLi and addition of water as in equation (15).[54] This providesone of the very few examples of *exo* ⇌ *endo*

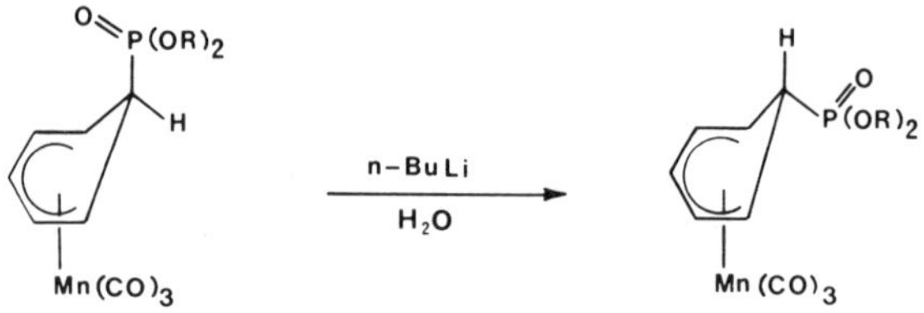

isomerization in a coordinated cyclic ring. It may be rationalized in terms of deprotonation to give a p-orbital at C(6) perpendicular to the {C(1), C(6), C(5)} plane, followed by protonation (from water) in an *exo*-fashion because of steric hindrance by the $Mn(CO)_3$ moiety.

12.2.10. Attack at η^6-Arenes

Although no kinetic data are available, several recent synthetic papers allow semiquantitative conclusions to be drawn concerning the substitution reactions of metal-coordinated haloarenes. The rate of chloride substitution in reactions (16) decreases in the order Cl > COOH > NMe$_2$ (*ca* 700:70:1) in keeping with the electron-withdrawing/donating properties of the *para*-X substituent.[55] Coordination of *ortho*-, *meta*-, or *para*-ClC$_6$H$_4$CF$_3$ to a neutral Cr(arene) moiety has been found to enhance the rate of chloride substitution by thiophenoxide.[56] This contrasts with previous observations that susceptibil-

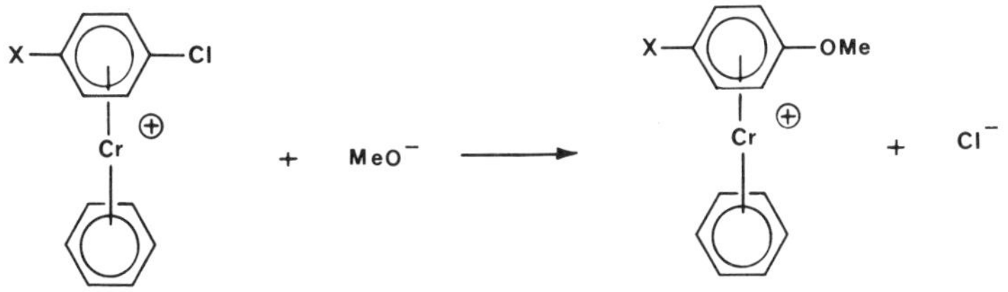

$$X \text{-} \bigcirc \text{-} Cl \quad \overset{\oplus}{Cr} \quad + \quad MeO^- \quad \longrightarrow \quad X \text{-} \bigcirc \text{-} OMe \quad \overset{\oplus}{Cr} \quad + \quad Cl^- \qquad (16)$$

ity to nucleophilic substitution is decreased in bis(fluoroarene)chromium(0) complexes compared to free fluoroarenes. The paramount importance of a positive charge in activating coordinated arenes is seen by the observation that neutral $[FeCp(\eta^6\text{-}C_6H_5X)]$ (X = F, Cl) complexes are inert to halogen displacement by alkoxides and amines, in contrast to the related $[FeCp(\eta^6\text{-}C_6H_5X)]^+$ cations which undergo facile nucleophilic substitution.[57]

Facile migration of alkyl groups from manganese to the benzene ring occurs upon reacting complexes $[Mn(CO)_2R(\eta^6\text{-}C_6H_6)]$ (41; R = Me, Et, Pr^i) with PPh$_3$, yielding *endo*-$[Mn(CO)_2(PPh_3)(\eta^5\text{-}R\cdot C_6H_6)]$ species (42) as in equation (17).[58] Migration rates vary in the order Me > Et > Pr^i (480:20:1).

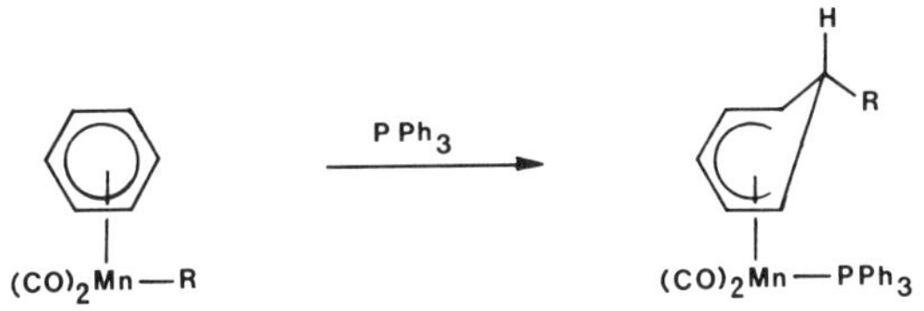

$$\bigcirc \quad \overset{PPh_3}{\longrightarrow} \quad \bigcirc \qquad (17)$$

$(CO)_2Mn\text{-}R \qquad\qquad (CO)_2Mn\text{-}PPh_3$

41   42

Competitive with reaction (17) is alkyl migration to a carbonyl ligand to give acyl species $[Mn(CO)(COR)(\eta^6\text{-}C_6H_6)]$ in a "dead-end" preequilibrium. Direct migration of R to the benzene ring in equation (17) is not favored, since the analogous hydride species $[Mn(CO)_2D(\eta^6\text{-}C_6H_6)]$ not only undergoes much slower migration, but also yields an *exo*-$[Mn(CO)_2(PPh_3)(\eta^5\text{-}C_6H_6D)]$ product. Instead, the intermediacy of η^4-arene species of the type (43) is proposed. In the gas-phase reaction of H$^-$ with $[Cr(CO)_3(\eta^6\text{-}C_6H_6)]$

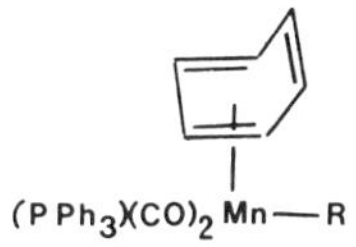

$(PPh_3)(CO)_2Mn\text{-}R$

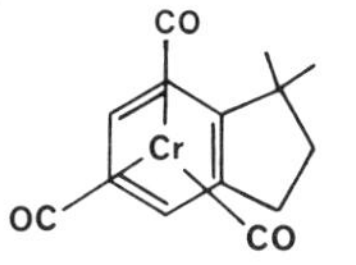

43   44

leading to $HCr(CO)_3^-$, initial hydride ion attack at the arene ligand is confirmed by experiments with D$^-$. These give $[HCr(CO)_3]^-$ and $[DCr(CO)_3]^-$ in the ratio of 7:1.[59] If alternative pathways involving initial D$^-$ attack at the metal or a CO ligand were involved, then only $[DCr(CO)_3]^-$ would be produced.

Several papers have discussed the regioselectivity of nucleophilic attack on the arene ring of $[Cr(CO)_3(\eta^6\text{-}arene)]$ complexes.[60-63] For example, the

1,1-dimethylindane complex (**44**) (where steric effects cause the adoption of the preferred conformation shown) is attacked by carbanion nucleophiles preferentially at carbon atoms eclipsed by a Cr—CO bond.[60]

12.2.11. Addition at η^7-Tropylium Ligands

Extended Hückel energies calculated for adducts arising from attack by alkoxides at various sites in $[M(CO)_3(\eta^7\text{-}C_7H_7)]^+$ (**45**; M = Cr, Mo, W) complexes show that the ring *exo*-addition product is the most stable in each case.[64] This correlates with the general isolation of the 7-*exo* ring adduct as the thermodynamically stable product. However, calculated charges, frontier electron densities, and interaction energies predict M > C(O) > C(ring) as the favored sites of attack by RO⁻. These latter predictions are borne out by low-temperature spectroscopic studies of the reactions of cations (**45**), M = Mo or W, with MeO⁻ in CH_2Cl_2.[65] Initial formation of metal-adducts (**46**) is observed, followed by the sequential appearance of carbomethoxy (**47**) and ring (**48**) adducts. However, with the chromium cation (**45**) only the 7-*exo*-alkoxy ring adduct is observed. These results, together with subsequent kinetic studies,[66] may be rationalized in terms of Scheme 7. Two steps are observed by stopped-flow spectroscopy for M = Mo or W, which are assumed to involve the rapid preequilibrium formation (k_1) of either (**46**) or (**47**), followed by slower ring addition (k_2). The latter attack of MeO⁻ on the tropylium ring

Scheme 7

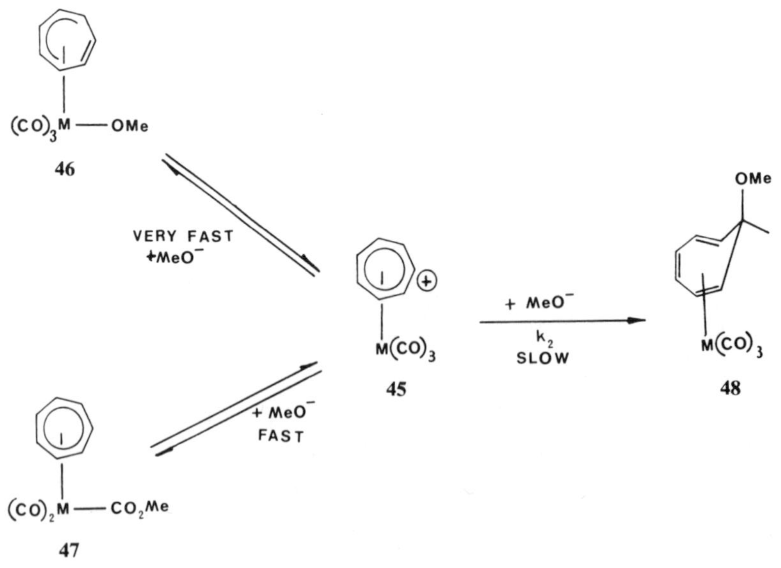

shows a strong metal dependence, k_2 decreasing down the series $Cr > Mo > W$ (50:10:1). On the other hand, attack (k_1) at the metal (or CO) occurs at about the same rate for both Mo and W. Failure to observe preequilibrium intermediates such as (46) and (47) in the Cr case arises because their rate of formation is comparable with that of direct attack of MeO^- on the ring.

12.3. Electrophilic Attack

Mechanistic studies of electrophilic attack on coordinated π-hydrocarbons have grown significantly. However, except for two kinetic studies reported below, emphasis has concentrated on stereochemical aspects.

12.3.1. Attack at η^1-Hydrocarbons

Kinetic measurements on the protonolysis of the σ-bonded 2-norbornanone complex (49) to give (50) reveal the rate law (18).[67] This result,

$$k_{obs} = k[H^+] \qquad (18)$$

together with the large kinetic isotope effect $(k_H/k_D = 4)$, indicates rate-determining attack of a proton on the substrate. A three-center transition state of the type (51) is proposed as shown in Scheme 8.

Scheme 8

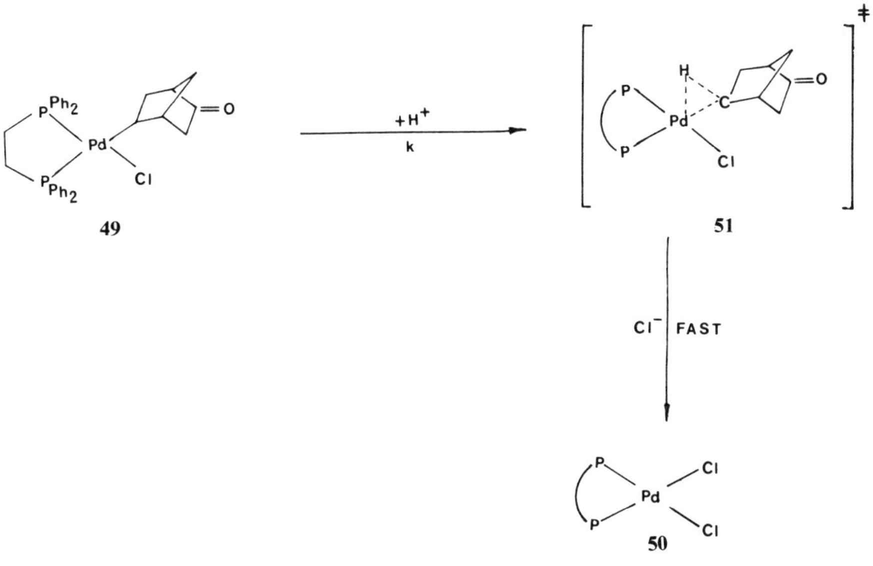

The rhenium-substituted ylide (**52**), prepared by deprotonation of $[CpRe(NO)(PPh_3)(CH_2P^+Ar_3)]^+$ (Ar = p-MeC$_6$H$_4$), is methylated stereospecifically to give (*SS, RR*)-(**53**) according to equation (19).[68] This methylation occurs in the opposite stereochemical sense to that recently found with

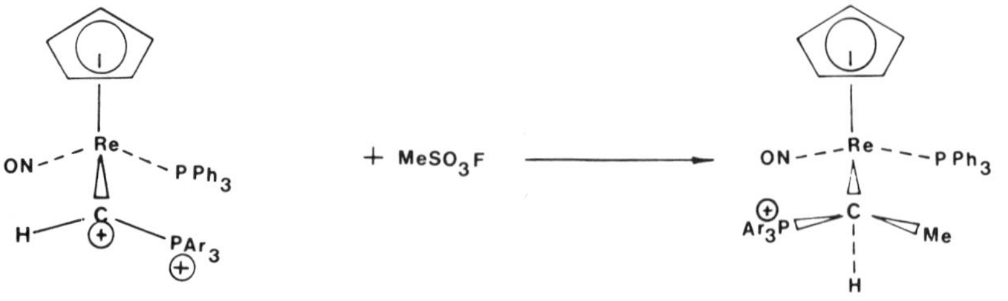

52   **53**

the rhenium-substituted carbanion $[CpRe(NO)(PPh_3)(:CHCN)]^-$, indicating that the factors controlling alkylation stereospecificity are different for anionic and neutral Re-C$_\alpha$ nucleophiles. High diastereoselectivity has also been observed in the alkylation (CH$_3$I, PhCH$_2$Br) of α-methoxyvinyl complexes (Z)-$[CpRe(NO)(PPh_3)\{C(OMe)=CHR\}]$ (R = Me, CH$_2$Ph).[69]   The stereochemistry of these reactions, and probable transition-state geometries, were established by an X-ray crystal structure of the alkylation product (SR,RS)-$[CpRe(NO)(PPh_3)\{COCH(Me)CH_2Ph\}]$.

The reaction of Ph$_3$C$^+$ with alkyl rhenium complexes $[CpRe(NO)(PPh_3)\-(CH_2R)]$ (R = Ph, *n*-alkyl) causes stereospecific α-hydride abstraction to give alkylidene complexes $[CpRe(NO)(PPh_3)(=CHR)]^+$. A theoretical analysis of the energy-reaction coordinate diagrams for these transformations has been modified[70] in the light of new information on the relative stabilities of the Re-C$_\alpha$ rotamers of the starting compounds.

12.3.2. Attack on η^4-η^6-Hydrocarbons

Owing to steric reasons, the trityl cation Ph$_3$C$^+$ can generally only abstract *exo*-hydrogens from metal-coordinated cyclohexa-1,3-dienes. However, abstraction of an *endo*-hydrogen atom from (**54**) has now been achieved with Ph$_3$C$^+$ via an electron transfer path as in equation (20).[71] The EPR spectrum

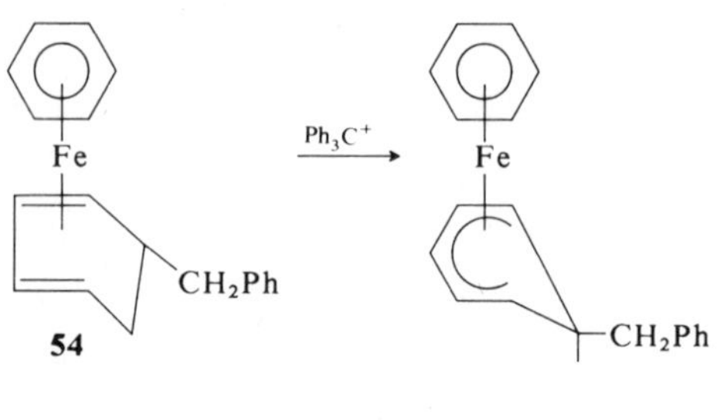

54

of the reaction mixture in CH_2Cl_2 shows signals due to the 17-electron Fe(I) intermediate (54^+) and $Ph_3C^.$ when recorded at $-140\,°C$.

Treatment of $[(\eta^5\text{-}C_5H_5)Re(NO)(PPh_3)(COR)$ $(R = Me, Bz, Ph)$ with $LiNPr_2^i$ and then MeI produces $[(\eta^5\text{-}C_5H_4COR)Re(NO)(PPh_3)(CH_3)].^{(72)}$ This contrasts with similar reactions on iron acyl complexes $[CpFe(CO)(PPh_3)-(COR)]$, where the initial formation of enolate anions led to elaboration at the acyl ligands (see Section 12.2.3). With the rhenium complexes, deuterium-labeling experiments show initial deprotonation of the Cp ring to give (55), followed by rapid migration of the acyl ligand to the Cp ring shown in equation (21). Crossover experiments confirm that the rearrangement (55) → (56) is

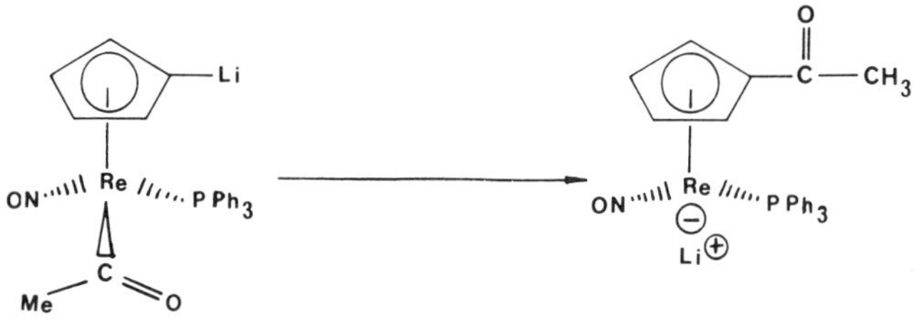

$$(21)$$

55   **56**

intramolecular. Electrophilic attack (acylation, lithiation) on $[Cr(CO)_3(\eta^6\text{-}1,1\text{-}$dimethylindane)]$, which has the preferred conformation (44), occurs preferentially at carbon atoms staggered with respect to Cr—CO bonds.$^{(60)}$ This contrasts with nucleophilic attack on (44) which targets arene carbon atoms in eclipsed positions.

12.3.3. Reactions at Side Chains

The t-butanoyl substitutent of (57) rearranges to a 3-methylbutan-2-oyl group in strong acids such as CF_3SO_3H and H_2SO_4 as in equation (22).$^{(73)}$

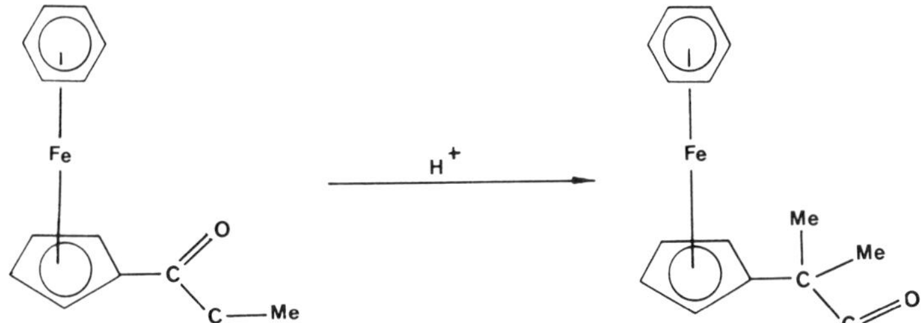

$$(22)$$

57

Comparison with free t-butylphenyl ketone shows that coordination to the $CpFe(arene)^+$ moiety causes a $10^4\text{-}10^5$-fold increase in the rate of the rearrangement.

Chapter 13

Rearrangements, Intramolecular Exchange, and Isomerizations of Organometallic Compounds

13.1. Mononuclear Compounds

13.1.1. Isomerizations and Ligand-Site Exchange

In $[(\eta\text{-}C_5Me_5)ReH_6]$, two types of hydrides are observed[1] at $-80\,°C$. Exchange between IrH and IrH_2 is observed in $[IrH(H_2)(7,8\text{-benzoquinoli-}nate)(PR_3)_2]^+$. It was proposed that the exchange involves a classical $[IrH_3(7,8\text{-}benzoquinolinate)(PR_3)_2]^+$ transition state.[2] The equilibrium between cis and trans isomers of $[PtH_2(PMe_3)_2]$ has been investigated by ^{31}P NMR spectroscopy.[3] A ^1H NMR study of (1) has shown that there is a very low energy barrier to isomerization by pseudorotation at silicon,[4] while (2) has a high

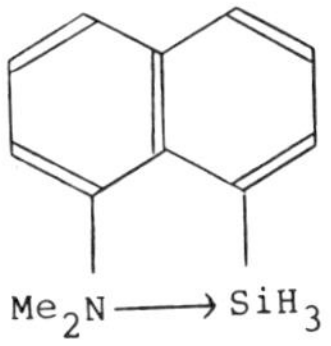

Me$_2$N$\longrightarrow$SiH$_3$

1

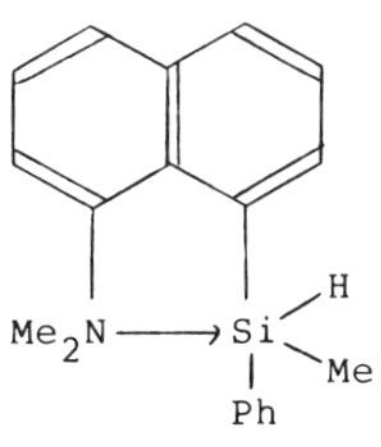

Me$_2$N$\longrightarrow$Si$\begin{smallmatrix}H\\Me\end{smallmatrix}$

Ph

2

activation energy.[5] Variable-temperature ^1H and ^{19}F NMR spectroscopy has been used to investigate[6] pseudorotation in $[Ph_2SiF_3]^-$, $[MePhSiF_3]^-$, $[PhSiF_4]^-$, and $[Ph_3SiF_2]^-$. Pseudorotation has been observed[7] in Bu_3^nPO adducts of Ph_2SnX_2.

The ^1H NMR spectrum shows the equilibrium in Scheme 1. At room temperature, there is only one cyclopentadienyl signal due to this equilibrium.

Scheme 1

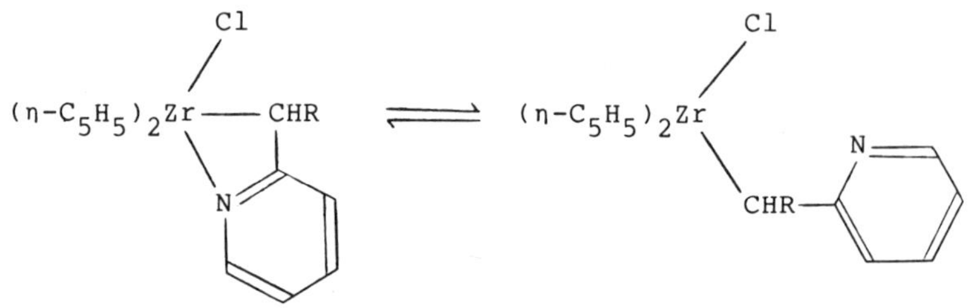

This literature analysis appears to be oversimplified due to the presence of a chiral centre in the $CH(SiMe_3)R$ group which requires racemization at zirconium to equilibrate the cyclopentadienyl groups.[8] $[(\eta\text{-}C_5Me_5)\text{-}Ta(SiMe_3)_3Me_3]$ is fluxional with the methyl groups exchanging with $\Delta G^\ddagger$ of 12.5 ± 0.5 kcal mol^{-1}.[9] ^{19}F NMR spectroscopy has been used to determine the rate of *cis/trans* isomerism in $[R_2Fe(CO)_4]$, $R = C_2F_5$, C_3F_7, C_4F_9, C_6F_{13}.[10] The ^1H NMR spectrum of $[IrH(CO)(Ph_2PCH_2)_3CMe]$ shows a quartet at room temperature, but an AA'X pattern at low temperature due to scrambling.[11]

A permutational analysis has been given of isomerism and rearrangements of *cis* bis(chelate) complexes of the type $M(AB)_2XY$, where AB is a bidentate chelate with dissimilar coordinating atoms. It was concluded that even in the absence of signal assignment, use of two-dimensional NMR spectroscopy gives the number of modes of rearrangements and the type of exchange pattern responsible for the observed isomerizations in the $M(AB)_2XY$ systems.[12] This treatment was then applied to $Sn(PhCOCHCOMe)PhCl$. It was concluded that the rearrangements could be explained in terms of the Bailar twist, but were inconsistent with Ray–Dutt twists or square-pyramidal intermediates.[13] By the use of two-dimensional ^{31}P NMR spectroscopy the trigonal twist mechanism has also been demonstrated[14] to occur in $[Cr(CO)_{5-n}(CS)\text{-}\{P(OR)_3\}_n]$. Variable-temperature ^1H, ^{13}C, and ^{31}P NMR studies on $[CrH(CO)_2(Me_2PCH_2CH_2PMe_2)_2]$ have shown exchange of the phosphorus nuclei causing the two phosphorus signals to become equivalent, and the hydride to change from a triplet of triplets to a quintet.[15] ^{13}C-enriched [W(CO)$_3$(S$_2$CNC$_4$H$_4$)$_2$] shows two fluxional processes, the first exchanging only two carbonyl groups, and the second exchanging the third group.[16]

Ligand exchange in [Cr(CO)$_2$(PMe$_3$)$_2$(η^4-diene)] and [Cr(CO)$_2$-{P(OMe)$_3$}$_2$(η^4-diene)] has been studied using ^{31}P NMR spectroscopy and interpreted in terms of a trigonal twist. The mobility of the ligand spheres depends on the position of the alkyl groups of the diene ligands. A methyl group in the 1-E position increases the barrier of the ligand movement by *ca* 16 kJ mol^{-1}, but in the 2-position the barrier is lowered by *ca* 5 kJ mol^{-1}.[17] [MoH(CH=CMeCOOR)(Me$_2$PCH$_2$CH$_2$PMe$_2$)$_2$] is stereochemically nonrigid according to ^{31}P NMR spectroscopy, which shows four signals at low temperature but only one at room temperature.[18] According to ^{13}C and ^{31}P NMR spectroscopy, the fluxionality of [TaX(CO)$_3$(PMe$_3$)$_3$] at room temperature is frozen out on cooling.[19] *Cis-trans* carbonyl exchange has been observed for [HCr(CO)$_4$(^{13}CO)]$^-$ and $\Delta G^{\ddagger} = 13.8 \pm 0.5$ kcal mol^{-1}, $\Delta H^{\ddagger} = 18.0 \pm 0.5$, and $\Delta S^{\ddagger} = 14.0 \pm 2$ eu.[20] $\Delta G^{\ddagger}$ has been determined for CO exchange in [CpMo(CO)$_3$H] using ^{13}C spectroscopy.[21]

IR spectroscopy has revealed that in solution many complexes of the type [M(CO)$_4$L] (M = Fe, Ru, Os; L = P, As, Sb) exhibit axial-equatorial isomerism. The tendency to give the less common equatorial isomer is Ru > Os > Fe; Sb > As > P. The order of ligands was rationalized in terms of the σ-donor ability of the element. In agreement with theoretical predictions, weaker donors prefer the equatorial site.[22] The interconversion of two isomers of [WBr$_2$(CO)$_2$(PMe$_2$Ph)$_3$] is frozen out at $-90\,°C$.[23] The activation energy for [Mo(OC)$_2$(PPh$_3$)$_2$] in 6,6,6,6-(OC)$_2$(Ph$_3$P)$_2$-6-MoB$_9$H$_{13}$ has been determined.[24] The value of $\Delta G^{\ddagger}$ for the enantiomerization of Be(RN-Pr^i-salicylaldiminate)$_2$ has been determined using ^{1}H NMR spectroscopy.[25] ^{13}C NMR spectroscopy has been used to determine the rate of racemization of [Fe{(2-C$_5$H$_4$NCH$_2$)-1,4,7-triazacyclononane}]$^{2+}$.[26] NMR spectroscopy has also been used to study[27] tetrahedral-square planar equilibrium in [{Et$_2$P(S)NR}$_2$Ni]. ^{19}F and ^{31}P NMR spectroscopy has been used to demonstrate[28] a slow equilibrium between

$$[(\overline{NMe_2CH_2C_6H_4})\overline{Pd}\{C(CF_3)=C(CF_3)C_6H_4CH_2\overline{N}Me_2\}(PMe_2Ph)]$$

and

$$[(\overline{N}Me_2CH_2C_6H_4)\overline{Pd}\{C(CF_3)=C(CF_3)C_6H_4CH_2NMe_2\}(PMe_2Ph)].$$

^{31}P NMR spectroscopy has revealed[29] an equilibrium between η^1 and η^2 bonding of the (mesityl)P=CPh$_2$ ligand in [(Ph$_3$P)$_2$Pt(mesitylP=CPh$_2$)]. A study of the temperature dependence of the ^{1}H NMR spectra of tetrahedral Zn{2,6-(Pr^{i}N=CH)$_2$-4-CH$_3$C$_6$H$_2$O}$_2$ has indicated the presence of two parallel dynamic processes, leading to the inversion of the tetrahedral configuration of the complex. At <258 to 283 K, the rapid interconversion of the enantiomers occurs without bond cleavage in the chelate ring. At higher temperatures, the interconversion involves rupture of the Zn—O or Zn—N bond of one ligand and recoordination. For the analogous beryllium complex, interconversion of the enantiomers involves only bond cleavage followed by recoordination.[30]

Variable-temperature ^{1}H and ^{31}P NMR spectroscopy of $[Ir\{P(O)(OMe)_2\}-\{P(OMe)_3\}_4]$ shows exchange by the Berry mechanism.[31] ^{1}H and ^{31}P NMR spectroscopy has also been used to investigate PMe$_2$Ph exchange in trans,mer-$[MCl_2(OH_2)(PMe_2Ph)_3]^+$, M = Rh, Ir. The exchange occurs by dissociation of water followed by pseudorotation with the rate for the rhodium complex being faster than for the iridium complex.[32] The temperature dependence of the ^{1}H and ^{13}C NMR spectra of $[Cu(Ph_2PCH_2CH_2PEt_2)_2]^+$ has been interpreted in terms of inversion at the tetrahedral Cu(I) center via an intramolecular ring-opening mechanism. A similar dynamic ligand exchange was reported for the analogous complexes of Ag(I) and Au(I). The rate of inversion of the metal center increases in the order Cu(I) < Au(I) < Ag(I).[33]

$\Delta G^{\ddagger}$ has been determined for the metal centered rearrangement of $[M(S_2CNMe_2)_4]^+$ and $[M(OSCNMe_2)_4]^+$, M = Nb, Ta, and is 10–12 kcal mol^{-1} for the dithiocarbamate complexes and 14–15 kcal mol^{-1} for the monothiocarbamate complexes.[34] A dynamic process equilibrates the two dithiocarbamate alkyl substituents on the NMR time scale in $[W(S)(PhC_2Ph)-(S_2CNR_2)_2]$ with an activation barrier of 15.9 kcal mol^{-1}. This process is promoted by dechelation of the η^2-thiocarboxamide sulfur donor to generate a fluxional five-coordinate intermediate.[35]

13.1.2. Simple Ligand Rotations about the Metal–Ligand Bond and Related Rocking or Flipping Motions

Line shape analysis of the ^{1}H NMR spectrum of 1,1'-diphenyluranocene as a function of temperature has given the activation parameters $\Delta H^{\ddagger} = 4.4 \pm 0.3$ kcal mol^{-1} and $\Delta S^{\ddagger} = -4.7 \pm 1.3$ eu for the rotation of the phenyl group about the bond to the eight-membered ring.[36] Variable-temperature NMR measurements on $[\{\eta\text{-}(Me_2SiH)_6C_6\}Cr(CO)_3]$ have yielded a Me$_2$SiH—C rotational barrier of 14.2 kcal mol^{-1}, which can be compared with 15.7 kcal mol^{-1}, calculated using empirical force-field calculations for the free arene. The calculations indicate that rotation takes place by a stepwise mechanism rather than by correlated dynamic gearing of all six groups.[37] ^{1}H and ^{13}C NMR spectroscopy has been used to determine the barrier for the rotation of aryl groups in $[(\eta\text{-arene})_2Re_2(\mu\text{-}H)_2(\mu\text{-}CHAr)]$ as 55.1, aryl = C$_6$H$_5$, 64.8, aryl = 4-C$_6$H$_4$Me, and 78.0 kJ mol^{-1}, aryl = 3,5-Me$_2$C$_6$H$_3$.[38]

The activation energy for methyl group rotation has been determined,[39] along with that for molecular tumbling, for Me$_2$SiHCl, MeSiHCl$_2$, and SiHCl$_3$, from spin-lattice relaxation measurements on ^{13}C, ^{29}Si, and ^{35}Cl. The temperature dependence of T_1 of (3) has been determined and the rotational barriers of the methyl groups determined as 5.6, 11, and 16 kJ mol^{-1}, the first being due to the geminal pair on silicon.[40] ^{1}H and ^{13}C NMR spectroscopy has been used to determine[41] the activation energy for C—N bond rotation in $[(\eta^5\text{-}C_5H_4Me)Mn(CO)_2\{CPh(NMe_2)\}]$. The activation energy for B—NMe$_2$ rotation in $Ph(NMe_2)(NHMMe_3)$, M=C, Si, has been determined using variable-temperature ^{13}C NMR spectroscopy.[42] The activation energy for B—N rotation in $Me_2EtSnCMe=CEtBEtNC_4H_4$ has also been determined.[43] the results of a variable-temperature ^{13}C NMR study of $Ph(RE)B\overline{N}CHMeCH_2CH_2CH_2$, E = O, S, show that the p_π-p_π bonding between B and O is ca 3 kcal mol^{-1} stronger than between boron and sulfur.[44] In $[(R^1R^2N)_2B]^+$, the barrier to rotation of the B—N bonds is ca 82 kJ mol^{-1}.[45] ^{1}H NMR spectroscopy has been used to determine[46] the activation energy for NMe$_2$ rotation in $[(Me_2N)_2(Ph_3Sn)Mo\equiv Mo(SnPh_3)(NMe_2)_2]$.

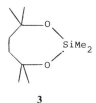

3

The ^{1}H NMR spectrum of $[(\eta\text{-}C_5H_5)_2MPh(CH=PPh_3)]$, M = Ti, Zr, shows restricted rotation about the M—CH=PPh$_3$ with $\Delta G^\ddagger$ of 12.0 ± 0.2 kcal mol^{-1} (Ti) and 8.6 ± 0.3 (Zr), suggesting stronger π interaction for titanium.[47] The ^{1}H and ^{13}C NMR spectra of $[WO\{O(\eta\text{-}C_5H_4)Fe(\eta\text{-}C_5H_5)\}\{(\eta\text{-}C_5H_4)Fe(\eta\text{-}C_5H_5)\}_3]$ show hindered rotation of the W—C bonds with $\Delta G^\ddagger$ = 62.5 ± 0.5 kJ mol^{-1}.[48] The temperature-dependent ^{13}C NMR spectrum of $[\{R^1C(N_2)\}_2Pd(PR^2_3)_2]$, R^1 = CO$_2$Et, C(O)Me, Ph, 4-tol, Pr^i, Bu^t; R^2 = Et, Bu, Ph, shows restricted rotation about the Pd—C bond.[49] ^{31}P NMR spectroscopy has been used to investigate restricted rotation about the Pt—CH(SO$_2$CF$_3$)$_2$ bond in $trans\text{-}[Pt(PPh_3)_2(trans\text{-}PhCH=CHPh)\{CH(SO_2CF_3)_2\}]$ and $\Delta H^\ddagger$ of 11.6 ± 0.7 kcal mol^{-1} and $\Delta S^\ddagger$ of 1.5 ± 3 eu were determined.[50] SnMe$_2$\{C(1-naphthyl)=CMe$_2$\}$_2$ exists as a mixture of rotamers (as in Scheme 2), which interconvert with $\Delta G^\ddagger$ of 109.75 ± 0.84 kJ mol^{-1}.[51] The temperature dependence of the ^{19}F NMR spectrum of $[CpFe(=CF_2)(CO)(PPh_3)]^+$ has been attributed to CF$_2$ rotation.[52] ^{1}H and ^{13}C NMR spectroscopy has been used to determine the activation energy for Ru=C bond rotation in $[(\eta\text{-}C_5H_5)(dppe)Ru=CH_2]^+$ as 10.9 kcal mol^{-1}.[53] Variable-temperature ^{1}H NMR spectra of $[(\eta\text{-}C_5H_5)(Bu^n_3P)(OC)R\overline{u}=CCH=CHCH=CHCH=\overset{\cdot}{C}H]^+$ show restricted rotation about the Ru=C bond with $\Delta G^\ddagger$ of 8.5 kcal mol^{-1}.[54]

Solid-state ^{2}H NMR spectra of $[W(CO)_3(PPr^i_3)_2(\eta\text{-}D_2)]$ show rapid rotation of the D$_2$ about the metal-D$_2$ axis. IR and variable-temperature ^{1}H and ^{31}P NMR spectra of solutions of $[W(CO)_3(PPr^i_3)_2(\eta\text{-}H_2)]$ have revealed an equilibrium with the seven-coordinate $[WH_2CO)_3(PPr^{13})_2]$.[55,56] $Trans\text{-}[Fe(\eta^2\text{-}H_2)(H)(dppe)_2]^+$ shows two fluxional processes. The low-energy process, with $\Delta G^\ddagger$ = ca 8.3 ± 0.7 kcal mol^{-1}, is attributed to H$_2$ rotation. The

Scheme 2

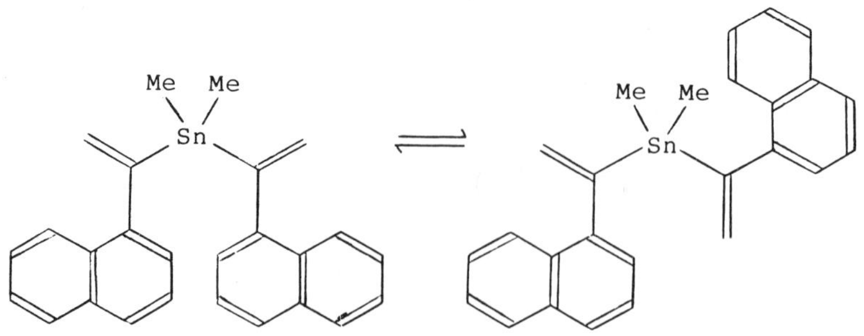

second process exchanges terminal hydride and η^2-H_2 ligands with $\Delta H^{\ddagger} =$ 13.9 ± 0.7 kcal mol^{-1}, $\Delta S^{\ddagger} = -1 \pm 3$ eu.[57] Hydrogen exchange between the hydrides and η^2-H_2 ligands in $[IrH_2(H_2)_2(PCy_3)_2]^+$ can be frozen out at low temperatures. The T_1 of the coordinated H_2 is very short.[58] Line broadening observed in the ^1H NMR spectrum of $[(Cy_2PCH_2CH_2PCy_2)PtH_2]$ has been taken as evidence for hydride exchange via an η^2-H_2 intermediate, but the changes could be due to viscosity changes.[59]

Dynamic and saturation transfer ^1H spectroscopy has been used to determine the rate of rotation of the Co—N bond in some 2-aminopyridine complexes of vitamin B_{12} coenzyme.[60] ^{13}C and ^{31}P NMR spectroscopy has been used to demonstrate restricted rotation about the P—M and P—Bu^t bonds in *trans*-$[(Bu_2^tPR)_2M(CO)X]$, M=Rh, Ir. The restricted rotation about the P—M bond results in the presence of three rotamers.[61,62] Similarly, there is restricted rotation of the diene in $[(\eta^1, \eta^5$-$C_5H_4CH_2CH_2)Mo(CO)_2(\eta^2$-diene)]$ with $\Delta G^{\ddagger}_{280}$ of 62.0 ± 0.2 kJ mol^{-1}, diene = butadiene; 66.2 ± 0.2 kJ mol^{-1}, diene = 1,3-pentadiene; and 61.0 ± 0.2 kJ mol^{-1}, diene = isoprene.[63] The barrier of the rotation of olefin ligands in $[(OC)_3L^1L^2W(\eta$-olefin)]$ has been determined from dynamic NMR spectroscopy.[64] Dynamic NMR studies of $[M(CO)-(MeC_2Me)L_2X_2]$, M = Mo, W, have revealed metal–alkyne rotational barriers ranging from 9 to 13 kcal mol^{-1}.[65] The activation energy for alkyne rotation in $[PhCH=W(PMe_3)_2(PHC_2H)Cl_2]$ has been determined as 50.6 ± 2.9 kJ mol^{-1}.[66] According to ^1H and ^{31}P NMR studies, there is alkyne propeller rotation in $[WBr_2(CO)L_2(R^1C\equiv CR^2)]$.[67] ^{19}F NMR spectroscopy has been used to investigate alkyne and S–W rotation in $[(\eta$-$C_5H_5)W(SC_6F_5)(PR_3)$-$(\eta$-$CF_3C_2CF_3)]$.[68] Olefin rotation occurs in $[(\eta$-$C_5H_5)Co\{P(O)(OEt)_2\}_3Rh$-$(\eta$-$C_2H_4)_2]$ with $\Delta G^{\ddagger}$ of *ca* 54 kJ mol^{-1}.[69] EHMO calculations have been performed on ligand rotations in $[(\eta$-indenyl)Rh(η-$C_2H_4)_2]$. The ethylene rotation barrier is 10.5 kcal mol^{-1}, while the indenyl rotation has $\Delta G^{\ddagger}$ of 8.4 ± 0.5 kcal mol^{-1}.[70] The NMR spectrum of (4) shows restricted olefin rotation.[71]

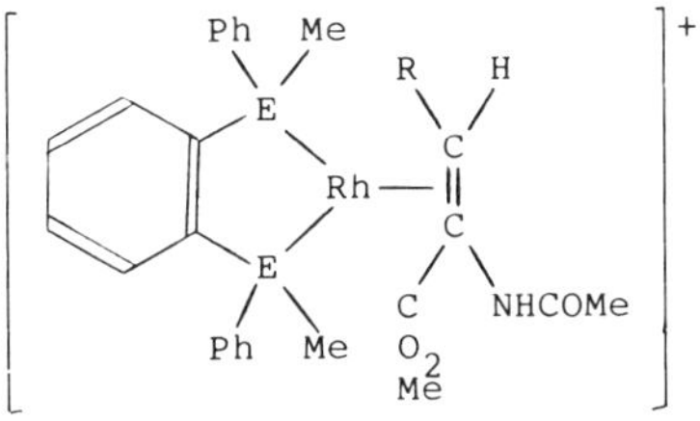

4

Restricted rotation of the CO_2 ligand in $[Mo(CO)_4(CO_2)_2]$ has been investigated by ^{13}C and ^{31}P NMR spectroscopy.[72] ^1H and ^{13}C NMR spectroscopy has been used to determine the activation energy of S_2O in $[(\eta\text{-}C_5Me_5)Re(CO)_2(\eta^2\text{-}S_3O)]$ as 57 kJ mol^{-1}.[73] The dynamic behavior of the C_3 ring in $[(\eta\text{-}C_5H_5)Mo(\eta^3\text{-}C_3Ph_3)(CO)_2]$ has been examined.[74] $[(\eta\text{-}C_5H_5)Mo(CO)_2(\eta^3\text{-pentadienyl})]$ has two isomers which interconvert with $\Delta G^{\ddagger}$ of 18.2 ± 0.3 kcal mol^{-1}.[75] Spin saturation transfer ^1H NMR spectroscopy has shown that the fluxionality of $[FeL(NO)_2(\eta\text{-allyl})]^+$, L = phosphorus ligand, involves slow allyl rotation at room temperature.[76] In solution, the *anti* isomers of $[(\eta^3\text{-}2,4\text{-}Me_2\text{-}C_5H_5)Rh(PR_3)_2]$ undergo a fluxional process which exchanges the ends of the η^3-2,4-dimethylpentadienyl ligand via 18-electron $[(\eta^5\text{-}Me_2\text{-}C_5H_5)Rh(PR_3)_2]$ intermediates. The free energy of activation for the exchange process increases with the cone angle of the PR_3 ligand. At higher temperatures, equilibration of *anti* and *syn* isomers was observed, probably via intermediates with sickle-shaped dienyl ligands.[77] Restricted rotation and isomerization of some allylic palladium complexes have been studied by ^1H NMR spectroscopy.[78] The ^{13}C NMR spectrum of (**5**) shows complete averaging with $\Delta G^{\ddagger}$ of 21.4 kcal mol^{-1}, $\Delta H^{\ddagger}$ of 20.9 kcal mol^{-1}, and $\Delta S^{\ddagger}$ of -1.3 eu. This was claimed to be the first example of stereochemical nonrigid behavior of a polycyclic conjugated hydrocarbon–metal complex.[79]

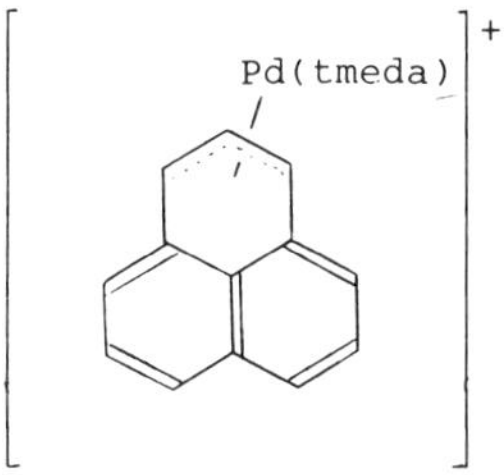

5 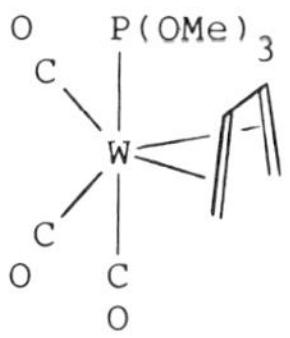   **6** 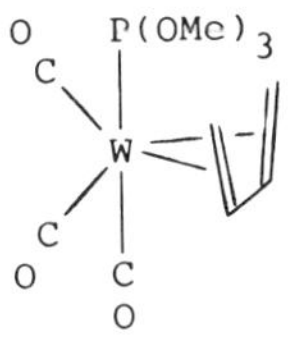   **7**

The activation energy for the rotation of the $s\text{-}cis\text{-}\eta^4$-butadiene ligand in $[(\eta\text{-}C_5Me_5)_2Th(\eta\text{-}C_4H_6)]$ has been determined.[80,81] The activation energy for the conversion of $[(s\text{-}cis\text{-}\eta^4\text{-butadiene})Zr(\eta\text{-}C_5H_4Bu^t)_2]$ to $[(s\text{-}trans\text{-}\eta^4\text{-}$butadiene$)Zr(\eta\text{-}C_5H_4Bu^t)_2]$ has been determined by using ^1H NMR spectroscopy to follow the rearrangement.[82] An NMR study of (**6**) and (**7**) has shown that they interconvert with $\Delta G^{\ddagger}_{200}$ of 36.2 ± 0.2 kJ mol^{-1}. Carbonyl scrambling occurs with $\Delta G_{250}^{\ddagger}$ of 50.6 ± 0.2 kJ mol^{-1}.[83] ^{1}H NMR spectroscopy shows that diene rotation in $[(\eta\text{-}C_5H_5)Me(OC)Mo(\eta\text{-}C_5H_4)]$ has $\Delta G^{\ddagger}$ of 64.1 ± 2 kJ mol^{-1}.[84] The variable-temperature ^{1}H NMR spectra of $[M(C_5H_4CPh_2)\text{-}(cod)]^+$, M = Rh, Ir, show evidence for diphenylfulvene ligand rotation about the ML$_2$ axis.[85]

The activation energy for η^5-dienyl rotation has been determined for $[(\eta\text{-}C_5H_7)Mn\{PhP(CH_2CH_2PEt_2)_2\}]$ as 18.3 ± 0.5 kcal mol^{-1}, for $[(\eta\text{-}2,4\text{-}Me_2\text{-}C_5H_5)Mn\{PhP(CH_2CH_2PEt_2)_2\}]$ as 17.3 ± 0.2 kcal mol^{-1}, for $[(\eta\text{-}C_5H_7)Mn\{MeC(CH_2PMe_2)_3\}]$ as 11.4 ± 0.6 kcal mol^{-1}, and for $[(\eta\text{-}2,4\text{-}Me_2\text{-}C_5H_5)Mn\{MeC(CH_2PMe_2)_3\}]$ as 10.9 ± 0.2 kcal mol^{-1}.[86] There is restricted rotation about the Re-$(\eta^5\text{-dienyl})$ bond in $[ReH_2(PMe_2Ph)(\eta^5\text{-}C_8H_{11})]$ according to ^{13}C and ^{31}P NMR spectroscopy.[87] Similarly, there is restricted rotation of the pentadienyl group in $[(\eta\text{-}C_5H_7)Re(PMe_2Ph)_3]$ with $\Delta G^{\ddagger} = 16.4$ kcal mol^{-1} and in $[(\eta\text{-}2,4\text{-}Me_2\text{-}C_5H_5)Re(PMe_2Ph)_3]$ with $\Delta G^{\ddagger} = 13.7$ kcal mol^{-1}.[88] The ^{1}H and ^{13}C NMR spectra of (8) shows restricted rotation of the rings and the activation energy was determined.[89] The ^{1}H NMR spectrum of $[(\eta^5\text{-}C_7H_9)Fe(CO)_2Me]$ shows dienyl rotation with $\Delta G^{\ddagger}$ of 13.1 kcal mol^{-1}.[90] Restricted indenyl rotation is observed in $[(\eta\text{-}C_5H_5)_2Zr(\mu\text{-}PPh_2)_2Rh(\eta\text{-indenyl})]$ with an activation energy of 14 to 15 kcal mol^{-1}.[91] The rotation of the azaborazole ligand in (9), L = CO or L$_2$ = cod, is frozen out at low temperatures, resulting in the cyclo-octadiene olefinic signals splitting from two to four signals.[92]

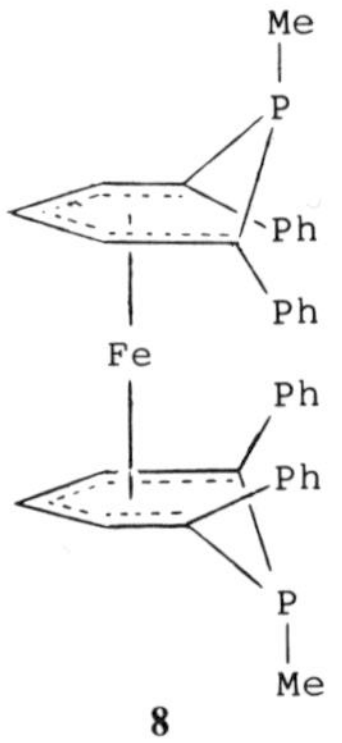

8

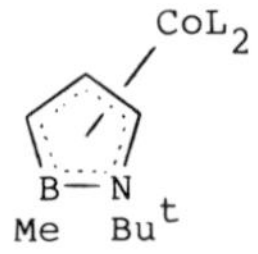

9

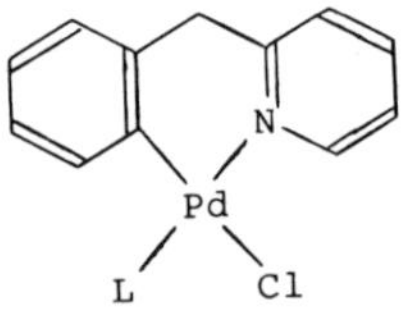

10

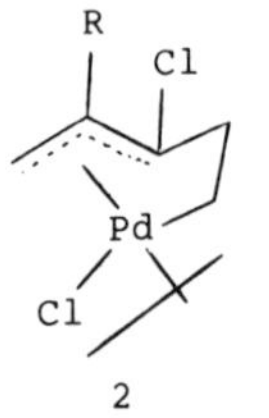

11

The dynamics of the inversion of the six-membered ring in (10), L = substituted pyridine, has been studied by ^{1}H NMR spectroscopy and $\Delta G^{\ddagger}$ was determined as ca 57 kJ mol^{-1}. The activation energy is virtually insensitive to the substituent on the pyridine ring.[93] Dynamic NMR experiments on (11) have led to the conclusion that very fast chloropalladation/dechloropalladation occurs in solution in such a way that the carbon alternately getting and losing chlorine dynamically inverts its configuration.[94]

The variable-temperature ^{31}P-$\{^1$H$\}$ NMR spectrum of $[(OC)_5MoPPh_2\text{-}NSNPh_2PMo(CO)_5]$ provides evidence for a rapid cis,trans–trans,cis interconversion with an energy barrier of 9.4 ± 0.2 kcal mol^{-1}.[95] Dynamic ^{1}H and ^{13}C NMR spectroscopy has been used to study the fluxionality of some ferrocene and ruthenocene cryptands due to restricted rotation about CO$-$N bonds.[96,97]

13.1.3. Migration of Metal Atoms between Different Ligand Sites

The activation energy has been determined for hydride exchange between the hydride and η^2-H$_2$ in $[(\eta$-C$_5$H$_5)$Ir$(\eta^2$-H$_2)$(H)(PMe$_3)]^+$.[98] A line shape analysis of the ^{6}Li and ^{13}C Li exchange broadened spectra of [Bu^{t}Li]$_4$ has yielded activation parameters for the fluxional exchange and possible mechanisms for fluxional exchange were discussed.[99] ^{6}Li NMR spectroscopy has been successful in demonstrating ^{6}Li-^{6}Li exchange in 3,4-dilithio-2,5-Me$_2$-2,4-hexadiene.[100] ^{1}H and ^{13}C NMR spectroscopy has been used to determine the ratio of axial and equatorial isomers in the substituted cyclohexanes, [(*cis*-1-R-C$_6$H$_{10}$-4-C$_6$H$_5$)Cr(CO)$_3$].[101] The observation of partially relaxed coupling to ^{9}Be in the ^{13}C NMR spectrum of beryllocene in solution has resulted in an estimation of 10^{10} s^{-1} for the rate of molecular inversion.[102] Low-temperature ^{1}H and ^{13}C NMR spectroscopy has been used to show that the allyl is σ-bonded in (C$_5$H$_5$)Mg(2-Me-allyl). The activation energy for metal [1,3]-shifts was determined.[103]

The ^{13}C NMR spectrum of [ReMe$(\eta^1$-C$_5$H$_5)$(NO)(PMe$_3)_3$] is a singlet due to fluxionality of the cyclopentadienyl group. The equilibrium constant with [(C$_5$H$_5)$ReMe(NO)(PMe$_3)$] was determined.[104] The sequence of mobilities for the cyclopentadienyl mercury complexes is Hg$(\eta^1$-C$_5$H$_5)_2$ > Hg$(\eta^1$-C$_5$H$_5)$Cl $\sim$ Hg$(\eta^1$-C$_5$Cl$_5)$Ph > Hg$(\eta^1$-C$_5$Me$_5)$.[105] "Fluxional η^1-cyclopentadienyl compounds of main group elements" have been reviewed.[106] "Fluxionality of polyene and polyenyl metal complexes" has also been reviewed.[107]

In (12), the iron migrates between double bonds in the allene with $E_a = 13.9$ kcal mol^{-1}.[108] (13) has a circumambulatory rearrangement in which both C(8) and the Fe(CO)$_3$ moieties migrate.[109] ^{1}H, ^{13}C, and ^{1}H-NOESY NMR spectroscopy has been used to investigate role exchange of the benzene rings in [$(\eta^6$-C$_6$H$_6)(\eta^4$-C$_6$H$_6)$Os] with $\Delta G^{\ddagger}$ of 80 ± 5 kJ mol^{-1}. No exchange within the η^4 ring was found.[110] [$(\eta$-C$_5$Me$_5)$Co$\{\eta^4$-C$_6$(CO$_2$Me)$_6\}$] is fluxional due to the motion of the cobalt on the arene ring with a barrier of 53 kJ mol^{-1}.[111] The kinetics of the interconversion of [$(\eta$-(C$_5$Me$_5)$Co(1,2,5,6-η^4-C$_8$H$_8)$] and [$(\eta$-C$_5$Me$_5)$Co(1-4-η^4-C$_8$H$_8)$] has been determined by ^{1}H NMR spectroscopy.[112]

Magnetization transfer has been used to determine the individual rates of metal migration between the sites in [Cr$(\eta$-C$_8$H$_8)$(CO)$_3$]. It was shown that the predominant mechanism is a [1,3]-shift with a significant [1,2]-shift also occurring.[113] ^{1}H NMR spectroscopy has been used to determine the activation energy for Cr(CO)$_3$ migration between arene rings in (14). The nature of L has virtually no effect.[114]

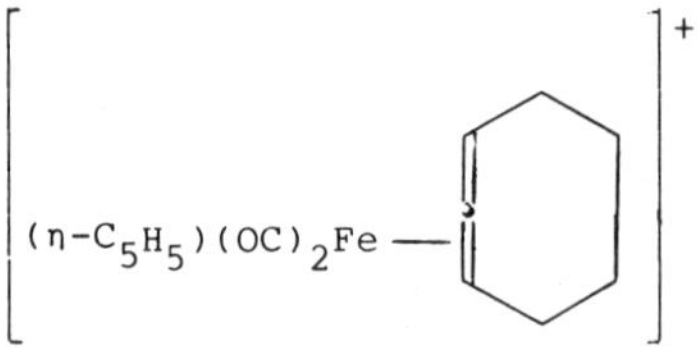

12

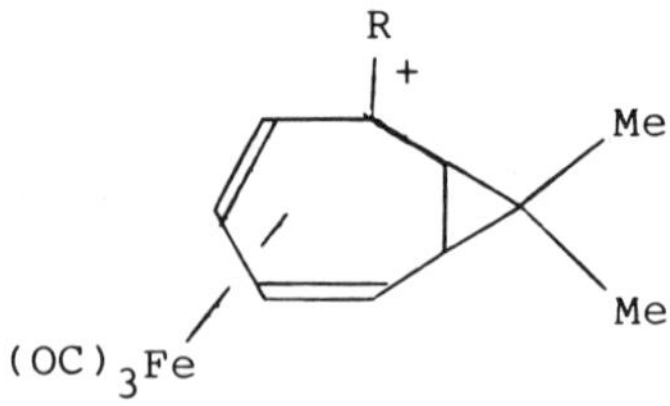

13

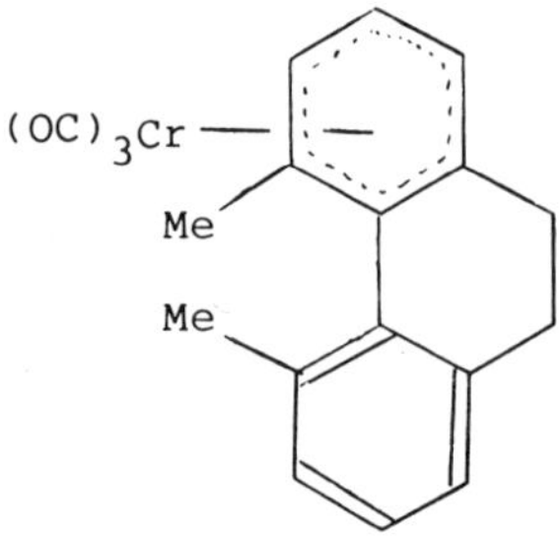

14

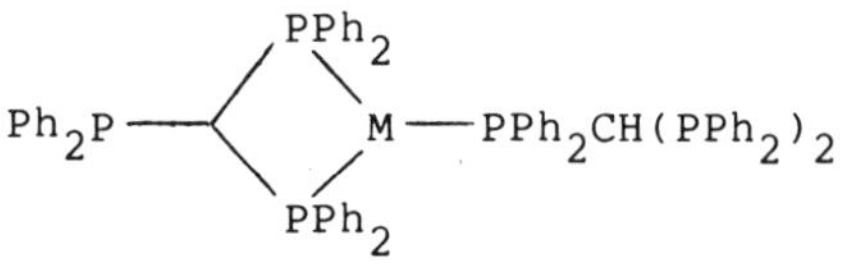

15

The fluxional exchange in Scheme 3 has been studied by ^{1}H NMR spectroscopy.[115] The activation energy for phosphorus exchange and rotation of the P—C bond in (15), M = Ge, Sn Pb,[116] and Sn{C(PMe$_2$)$_3$}$_2$,[117] has been determined. [PtCl(PEt$_3$){(SPPh$_2$)$_3$C}] contains an S,S-bonded chelate with dynamic stereochemistry controlled by one inert and one labile Pt—S bond. The value of $\Delta G^{\ddagger}$ has been determined as 67.1 kJ mol^{-1}.[118] ^{1}H and ^{13}C NMR spectra are consistent with the sulfur atoms undergoing rapid exchange in Me$_n$GeX$_{4-n-m}$(S$_2$CNMe$_2$)$_m$.[119]

Scheme 3

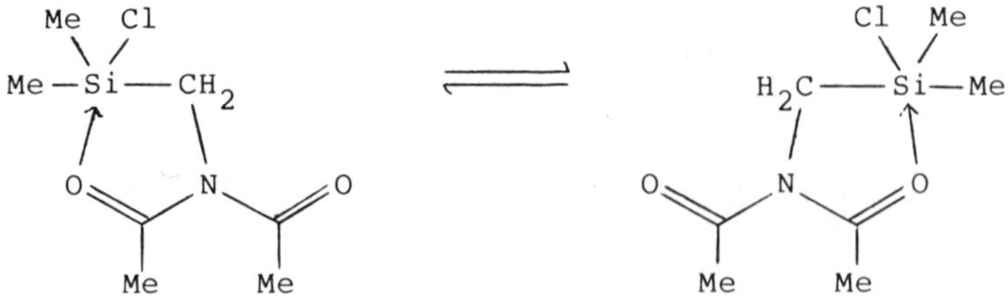

13.1.4. Agostic Bonding and Hydrogen Atom Migrations

[Cr(CO)$_2$L(C$_7$H$_{10}$)] has an agostic hydrogen interaction and $\Delta G^{\ddagger}$ was determined for exchange across the ring and hydrogen transfer.[120] The ^{1}H

NMR spectrum of $[(\eta\text{-}C_3H_5)_3MoCHMe_2]$ shows agostic interaction to a methyl hydrogen atom. At $-120\,°C$, a separate agostic hydrogen is observed, but it is averaged above $-50\,°C$.[121] (**16**) undergoes two rearrangement processes. The low-energy process involves exchange of the three protons of the bridging methyl group, $\Delta G^{\ddagger} = 8.9\,\text{kcal mol}^{-1}$, involving a 16-electron intermediate. The high-energy process involves hydride transfer across the allyl via an 18-electron diene hydride intermediate with $\Delta G^{\ddagger} = 17.3\,\text{kcal mol}^{-1}$.[122] Methyl group rotation in (**17**) and (**18**) has activation energies of 43.5 ± 1 and $41.7 \pm 1\,\text{kJ mol}^{-1}$ respectively due to agostic interactions.[123] In $[(\eta^5\text{-}C_6H_7)ReH_2(PPh_3)_2]$, ^1H NMR spectroscopy has demonstrated restricted rotation of the cyclohexadienyl ligand and exchange of hydride between the metal and the ring.[124] Dynamic NMR spectroscopy has been used to demonstrate hydrogen exchange between the hydride and olefin in $[(\eta^5\text{-}C_5Me_5)CoH(\eta\text{-}C_2H_4)\{P(OMe)_3\}]$. The hydride is agostic as demonstrated by the observation of $^1J(^{13}C, ^1H) = 61\,\text{Hz}$.[125]

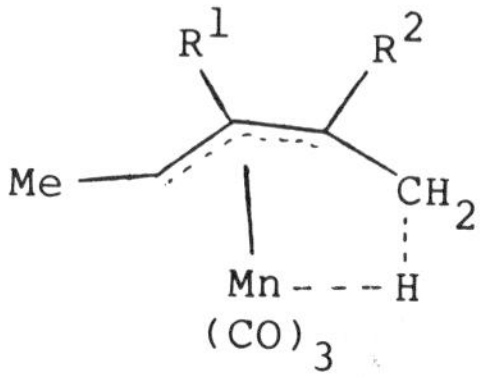

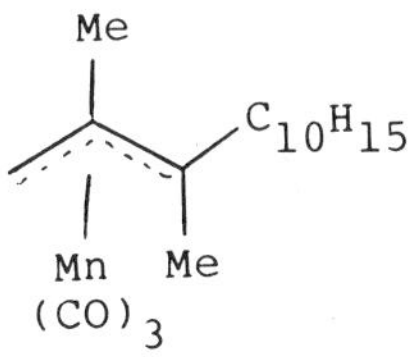

The ^1H NMR spectrum of $[(\eta\text{-}C_5H_5)_2Nb(BH_4)]$ shows terminal-bridge hydride exchange with $\Delta G^{\ddagger}$ of $16.4 \pm 0.4\,\text{kcal mol}^{-1}$.[126] Below $-20\,°C$, ^1H and ^{31}P NMR spectra indicate a rigid stereochemistry for $[FeH(BH_4)\{(Ph_2CH_2)_3CMe\}]$, but on raising the temperature, hydride scrambling occurs, and the coordinated phosphorus atoms become magnetically equivalent.[127]

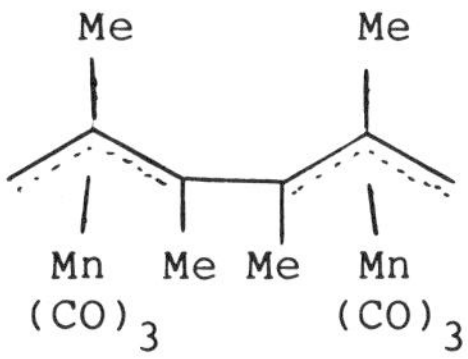

13.1.5. *Internal Ligand Rearrangements*

$[Hg(\eta^1\text{-}dppm)_2]^{2+}$ is fluxional on the NMR time scale as a result of fast intramolecular displacement of the coordinated phosphorus by the free phosphorus atom in the η^1-dppm ligands. At temperatures above 253 K, intermolecular ligand exchange sets in.[128] Variable-temperature ^1H NMR studies of (**19**) have indicated that the twist-boat ThS_5 ring is fluxional with $\Delta G^{\ddagger}$ of

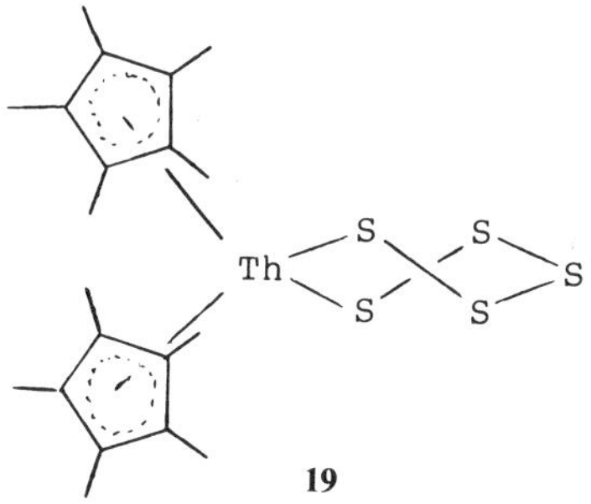

19

57.4(5) kJ mol^{-1}.[129] Variable-temperature ^{1}H NMR spectra of [(η-C$_5$H$_5$)$_2$M(SC$_6$H$_4$NH)], M = Ti, Zr, Hf, have been used to determine $\Delta G^{\ddagger}$ for ring inversion.[130] Similarly, the activation energies for ring inversion in [(η-C$_5$H$_5$)$_2$M(E$_2$C$_6$H$_4$)], M = Ti, Zr, Hf; E = S, Se, Te, have been determined.[131,132]

Variable-temperature ^{1}H NMR spectra of [M(CO)$_5$(Me$_3$SiCH$_2$EECH$_2$-SiMe$_3$)][133] and [M(CO)$_5$(Me$_2$CCH$_2$EECH$_2$)],[134] M = Cr, Mo, W; E = S, Se, have established the occurrence of two distinct internal dynamic phenomena. The coordinated sulfur or selenium atoms undergo facile pyramidal inversion, while above ambient a novel [1,2]-metal shift between adjacent pairs of chalcogen atoms of the ligands occurs. Cis- and trans-isomers of [Cr(CO)$_4${(MeS)$_2$CHCH(SMe)$_2$}] have been identified using NOESY. The activation energy for sulfur inversion was determined.[133]

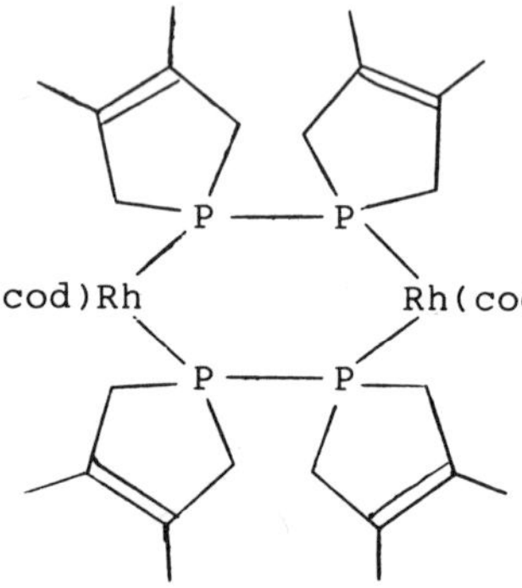

20

^{1}H and ^{13}C NMR spectroscopy has shown ring inversion for (**20**) with $\Delta G^{\ddagger}$ of 14.0 ± 0.2 kcal mol^{-1}.[136] The barrier for thioether inversion has been determined in cis-[Pd{RSCH$_2$C(CF$_3$)$_2$O}$_2$] as 73 to 74 kJ mol^{-1}, in trans-[Pd{RSCH$_2$C(CF$_3$)$_2$O}$_2$] as 67 to 69 kJ mol^{-1}, and in cis-[Pt{RSCH$_2$C(CF$_3$)$_2$O}$_2$] as 88 to 90 kJ mol^{-1}.[137] Similar investigations have also been carried out on [M{RSCH$_2$C(CF$_3$)$_2$O}(PR$_3$)$_2$]$^+$ and [M{RSCH$_2$C(CF$_3$)$_2$O}Cl(PPh$_3$)], M = Pd, Pt.[138] The stereodynamics of trans-[PdX$_2$(SeCH$_2$CMe$_2$CH$_2$)$_2$] have been studied by variable-temperature ^{1}H NMR spectroscopy. The conformational changes of the ligand rings are always

Scheme 4

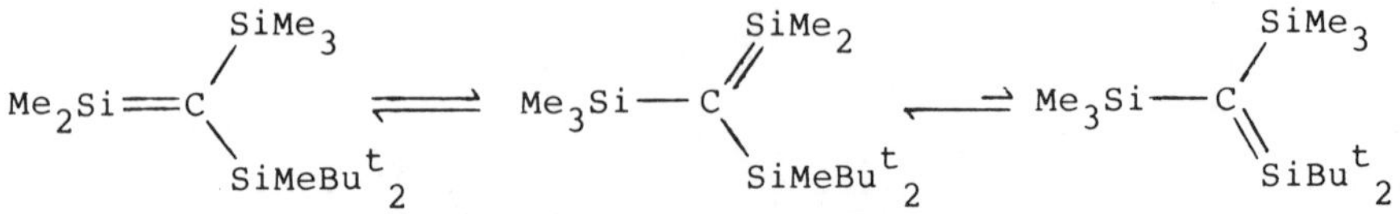

fast, but the inversion at the coordinated selenium is much slower, with $\Delta G^{\ddagger}$ of 66 to 78 kJ mol^{-1}.[139] A general method for evaluating rate constants in complex exchange networks from two-dimensional NOESY NMR spectra has been proposed and applied to ^{195}Pt in [PtXMe$_3$(MeSCH$_2$CH$_2$SMe)].[140] The ^{1}H and ^{13}C NMR spectrum of Me$_2$Si=C(SiMe$_3$)(SiMeBu$_2^t$) indicates[141] a rapid intramolecular methyl exchange, according to Scheme 4.

13.2. Dinuclear Compounds

13.2.1. Rotation about the Metal–Metal Direction

The ^{1}H NMR spectrum of Mo$_2$(CH$_2$Bu^t)$_2$(OCH$_2$Bu^t)$_4$ is temperature dependent due to the CH$_2$ group becoming inequivalent because of restricted rotation about the Mo≡Mo bond with a barrier of 12 ± 0.5 kcal mol^{-1}.[142] Variable-temperature ^{1}H NMR spectroscopy, including magnetization transfer and two-dimensional NMR spectroscopy, has been used to investigate rotation of the molybdenum–molybdenum bond in 1,2-Mo$_2$(OAr)$_2$(NMe$_2$)$_4$ and related compounds.[143] [Re$_2$(CO)$_8${μ-Ph$_2$P(CH$_2$)$_n$PPh$_2$}] adopts a staggered conformation due to restricted rotation about the Re—Re bond with $\Delta G^{\ddagger}$ of 12.3, $n = 2$, 11.0, $n = 3$, and 9.6 kcal mol^{-1}, $n = 4$.[144] [{W(CO)$_3$}$_2$-μ-{(η^5-C$_5$H$_4$)$_2$SiMe$_2$}] shows *anti–gauche* exchange with $\Delta G^{\ddagger}$ of 67.8 kcal mol^{-1}.[145]

13.2.2. Carbonyl Ligand Migrations

The activation energy for carbonyl exchange in (21), involving bridged intermediates, has been determined on the ^{13}CO-enriched compound using line shape analysis.[146] Variable-temperature ^{13}C NMR studies have shown bridge–terminal carbonyl exchange in [($\eta^{3,4}$-C$_7$H$_7$)Fe(CO)$_2$(μ-CO)Rh-(dppe)].[147] The reaction of [Rh(octaethylporphyrin)]$_2$ with CO yields two isomers, one with a terminal carbonyl, and the other with a bridging carbonyl.[148]

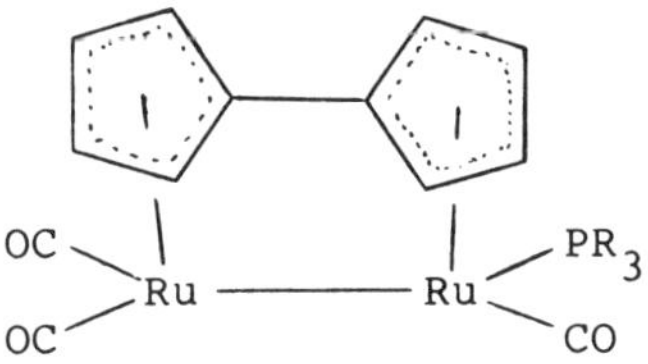

21

13.2.3. Migration of Other Ligands between Metal Atoms

The variable-temperature ^{1}H and ^{13}C NMR studies of (22), M = Fe, Ru, show that a fluxional process of the dienone ligand leads to exchange of σ- and π-bonds at the two metal centers.[149]

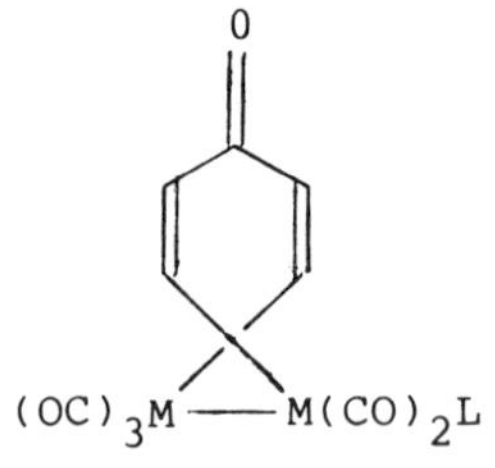

22

13.2.4. Other Exchange Reactions

Two isomeric $[Re_2(CO)_8(\mu\text{-}\eta^{2,1}\text{-}C_7H_7)]$ complexes show temperature-dependent ^1H NMR spectra due to hindered ligand movements of the cycloheptatrienyl bridges with $\Delta G^{\ddagger}$ of 39.9 ± 2 and 48.0 ± 2 kJ mol^{-1}.[150] A ^{13}C NMR study of $[(OC)_3FeCR^1SR^2SFe(CO)_3]$ has shown that the carbonyl fluxionality around each metal is independent.[151] A variable-temperature ^1H NMR investigation of (23) has shown two types of C_6Me_6 and CH groups at low temperature and one at room temperature. Interconversion occurs with $\Delta G^{\ddagger}$ of 10.3 ± 1.2 kcal mol^{-1}.[152] *Cis–trans* isomerism of $[(\mu\text{-}CH_2)(\mu\text{-}CO)\{(\eta\text{-}C_5H_5)\text{-}Fe(CO)_2\}_2]$ has been studied by ^1H NMR spectroscopy in solvents of different polarities and rate constants determined.[153]

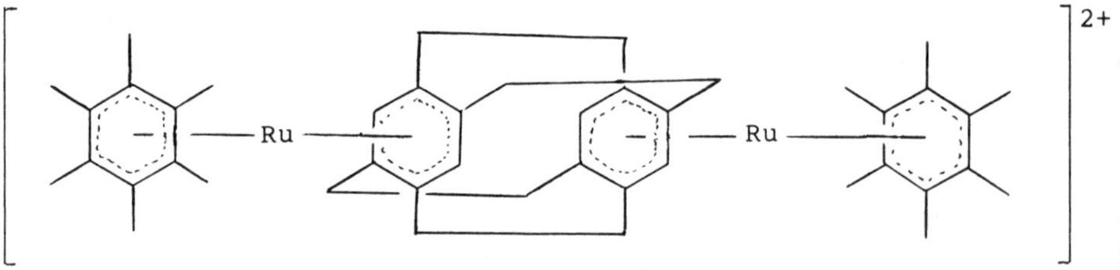

23

PEt$_3$ exchange in $[MnPt(\mu\text{-}H)(\mu\text{-}CO)(CO)_4(PEt_3)_2(Mn\text{-}Pt)]$ has been observed by ^1H NMR spectroscopy.[154] ^1H NMR spectroscopy has shown dynamic behavior for $[\{(Me_2PCH_2PMe_2)(PMe_3)Co\}_2PMe_2]$ with the CH$_2$ group being a singlet at room temperature and AB at low temperature.[155] The ^{31}P NMR spectra of $[(Cy_3P)Rh(\mu\text{-}CO)_2Rh(CO)(PCy_3)]$ have indicated an intramolecular geminal PCy$_3$ exchange process on Rh(PCy$_3$)$_2$.[156] The ^{31}P NMR spectrum of $[Rh_2\{(Me_2PCH_2)_2PMe\}_2(CO)_2]^{2+}$ is temperature dependent, with three phosphorus environments at low temperature and two at high.[157] ^{13}C and ^{31}P NMR spectroscopy has been used to investigate the dynamic behavior of $[\{PhC(S)SMe\}Fe_2(CO)_5\{P(OMe)_3\}]$. There are three isomers at $-90\,°C$ which interconvert on heating.[158] The X-ray crystal structure of $[Pt_2(\mu\text{-}dppm)_2Cl(PPh_3)]^+$ indicates considerable tetrahedral distortion of the normally square-planar environment around its PPh$_3$ ligated platinum atom. These distortions relieve repulsions between the phenyl groups on the PPh$_3$ and those on the adjacent dppm ligand, but cause a significant retardation of the rates of axial/equatorial interchange of the substituents on the Pt$_2$(μ-dppm)$_2$ ring.[159]

Rapid exchange of bridging and terminal (Me$_3$SiCH$_2$)$_2$As groups in Ga$_2$\{μ-As(CH$_2$SiMe$_3$)$_2$\}$_2$\{As(CH$_2$SiMe$_3$)$_2$\}$_2$Cl$_2$ has been observed.[160]

[(PtXMe$_3$)$_2$(ĖCH$_2$CMe$_2$CH$_2$Ė)], E = S, Se, show pyramidal inversions of the E atom pairs and scrambling of the PtMe groups. The value of $\Delta G^{\ddagger}$ for pyramidal inversion is 66 to 75 kJ mol^{-1}.[161] Variable-temperature ^{1}H NMR studies of [(PtXMe$_3$)$_2$\{HC(SMe)$_3$\}] have shown inversion of pyramidal sulfur atoms, a 1,3-metal pivoting motion, and a scrambling of Pt–Me environments by a PtMe$_3$ rotation mechanism. The latter two motions are concerted and involve a common transition state.[162] ^{119}Sn two-dimensional NOESY NMR spectroscopy has been used to demonstrate unambiguously that, in CH$_2$\{PhSn(SCH$_2$CH$_2$)$_2$NMe\}$_2$, isomerization occurs at the tin center in an uncorrelated way.[163] Dynamic ^{1}H NMR spectroscopy has been used to show isomerism in [Al(OPr^i)(acac)$_2$]$_2$ between *meso* and *d,l* isomers and restricted rotation of the bridging isopropoxide groups at low temperatures.[164]

13.3. Cluster Compounds

13.3.1. Rearrangements Involving the Relative Motion of Metal Atoms in a Cluster

The activation parameters for the interconversion of two isomers of [(μ-H)M$_3$(μ-CNMe$_2$)(CO)$_9$L], M = Ru, Os, have been determined by following the conversion by ^{1}H NMR spectroscopy.[165] The activation energy for intramolecular metal-core rearrangements in [M$_2$Ru$_4$(μ_3-H)(CO)$_{12}$L$_2$], M = Cu, Ag, has been determined from ^{31}P NMR spectroscopy.[166] The ^{13}C NMR spectrum of [(Me$_3$P)(OC)$_4$OsOs$_3$(CO)$_{11}$], (24), at $-115\,°$C, is consistent with the solid-state structure, but by $-89\,°$C, all but two signals have collapsed into the base line. An unprecedented mechanism involving breaking the longest Os—Os bond was proposed to rationalize this fluxional behavior.[167] ^{1}H NMR spectroscopy has shown that [Os$_4$(CO)$_{11}$(μ_3-S)(HC$_2$CO$_2$Me)(μ-H)$_2$] is dynamic with exchange between two isomers.[168] Variable-temperature ^{1}H

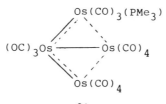

24

NMR studies of $[PtOs_3(CO)_9(PMe_2Ph)_2(\mu_3\text{-}S)_2]$ have shown that it undergoes a rapid rearrangement of the cluster framework with $\Delta G^{\ddagger} = 7.5$ kcal mol^{-1}. A process in which the platinum atom interchanges bridging sites between the two Os—Os bonds was proposed.[169] ^{1}H and ^{11}B NMR spectroscopy has been used to demonstrate an equilibrium between 2- and 3-MeB$_5$H$_{10}$. The temperature dependence of the equilibrium constant was determined.[170]

13.3.2. Localized CO Exchanges

^{13}C magnetization transfer measurements on $[Re_3(CO)_{10}H_3]^{2-}$, **(25)**, have shown four dynamic processes: (1) local carbonyl exchange on Re^1, (2) rotation

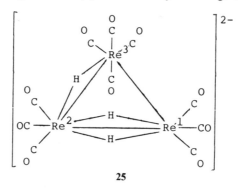

25

of the Re$(\mu\text{-}H)_2$Re fragment on ReH(CO)$_4$, (3) hydride transfer between the singly bridged and unbridged edges, and (4) local carbonyl exchange on Re^3.[171] Variable-temperature ^{13}C NMR spectra of $[HFeRu_3N(CO)_{12}]$ shows localized carbonyl exchange on iron which freezes out at -95 °C and on Ru^1 which freezes out at -70 °C.[172] The ^{13}C NMR spectrum of $[Ru_3(\mu\text{-}H)(\mu_3\text{-}MeC\!\!=\!\!CCH\!\!=\!\!NMe_2)(CO)_9]$ shows localized dynamic exchange of the axial and equatorial carbonyl ligands at the ruthenium atoms and rotation of the μ_3-alkyne, leading to a single ^{13}CO signal at 90 °C.[173] The ^{13}C NMR spectrum of $[HOs_3(CO)_{10}Et]$ indicates a dynamic process which creates an average plane of symmetry perpendicular to the triangle. At temperatures above -20 °C, it converts to $[H_2Os_3(CO)_{10}(CHMe)]$ and rate and equilibrium constants were determined.[174] The dynamics of $[(\mu\text{-}H)Os_3(CO)_9(\mu\text{-}HC_2H)\text{-}(\mu\text{-}SPr^i)]$ have been investigated by ^{1}H and ^{13}C NMR spectroscopy, and shown to involve localized carbonyl and hydride exchange.[175] The ^{13}C NMR spectrum of $[Fe_2Rh(\mu\text{-}H)(\mu_3\text{-}COMe)(CO)_7(\eta\text{-}C_5H_5)]$ shows terminal carbonyl exchange.[176]

13.3.3. Delocalized CO Exchanges

Variable-temperature ^{13}C NMR spectra show that $[(Me_3P)\text{-}(OC)_4OsOs_3(CO)_{11}]$ is stereochemically nonrigid, involving all the carbonyl ligands.[177] The problem of the detection of the expected four carbonyl signals for $[Co_4(CO)_{12}]$ has finally been solved. The previous difficulties were attributed to ^{13}C relaxation of the second kind caused by $^1J(^{59}Co, ^{13}C)$. The spectra were analyzed in terms of carbonyl exchange.[178] The dynamics of $[Co_4(CO)_8(L\text{-}L)_2]$, L-L = dppm, dmpm, $Ph_2PCH_2PMe_2$, have been determined by ^{13}C and ^{31}P NMR spectroscopy. All eight carbonyls exchange at the same rate, indicating a concerted mechanism for the exchange process.[179] Variable-temperature ^{13}C NMR studies of $[Co_4(CO)_9(\eta^6\text{-triptycene})]$ and $[Co_4(CO)_9(\eta^6\text{-mesitylene})]$ have yielded activation parameters for both carbonyl exchange and triptycene rotation. Unfortunately, the authors appear to have been misled by the temperature-dependent ^{59}Co quadrupolar broadening and the activation parameters appear to be grossly in error.[180] the ^{13}C and ^{31}P NMR spectra of $[Co_2Rh_2(CO)_{11}(PEt_3)]$ show extensive CO exchange at room temperature, but it is static at $-70\,°C$.[181]

In solution, $[Rh_3\{\mu\text{-}(Ph_2PC_2)PPh\}_2(\mu\text{-CO})(CO)_2Br_2]^+$ undergoes rapid bridge-terminal carbonyl exchange, which leaves one of the terminal carbonyl groups unaffected.[182] ^1H, ^2H, ^{13}C, and ^{103}Rh NMR spectroscopy has been used to investigate ligand migrations in $[Rh_{13}(CO)_{24}H_n]^{(5-n)-}$ and $[Rh_{14}(CO)_{25}H_n]^{(4-n)-}$.[183]

13.3.4. Other Exchange Reactions

The M−H−B and B−H−B hydride exchange in $[(\eta\text{-}C_5Me_5)Rh\text{-}(B_{10}H_{11}Cl)(PMe_2Ph)]$ occurs with $\Delta G^{\ddagger}$ of 33 kJ mol^{-1}.[184] The ^1H NMR spectrum of (**26**) shows two cyclopentadienyl and two methyl signals which exchange with $\Delta G^{\ddagger}$ of 60 kJ mol^{-1}, M = Mo, and 67 kJ mol^{-1}, M = W.[185] In (**27**), the NMR spectrum shows two cyclopentadienyl and two R groups at low temperature, which exchange with $\Delta G^{\ddagger}_{241}$ of 6.3 ± 0.1 kcal mol^{-1}.[186]

The ^1H NMR spectrum of $[(\mu\text{-H})(\eta\text{-}C_5H_5)MoOs_3(CO)_{11}]$ shows hydride exchange.[187] The ^1H NMR spectrum of $[H_2Fe_2Ru_2(CO)_{13}]$ shows one hydride

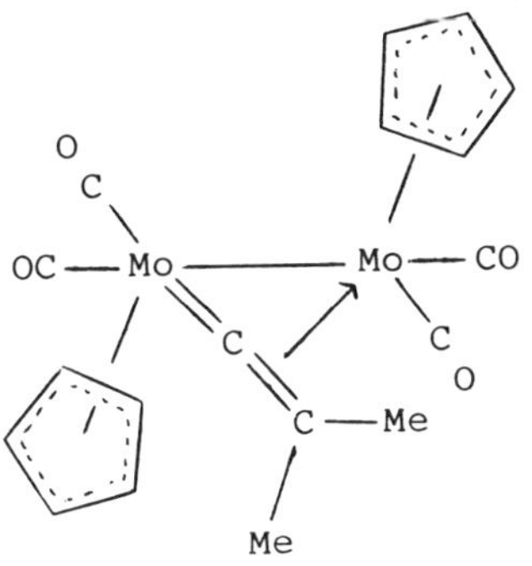

26

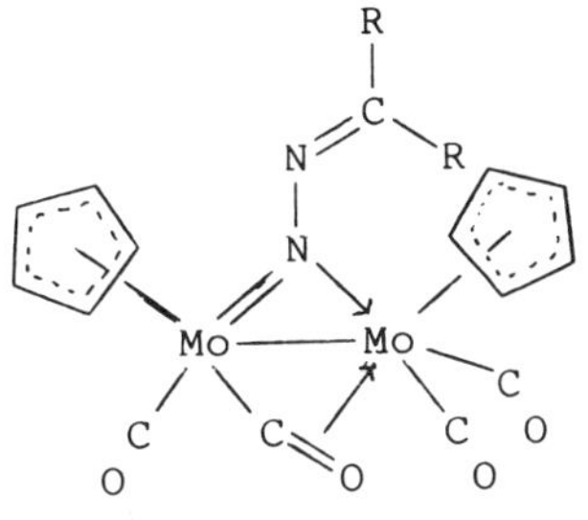

27

signal at room temperature and two at low temperatures.[188] The variable-temperature ^1H NMR spectra of $[(\eta\text{-}C_5H_5)NiCoM(CO)_6(MeC_2Ph)]$, M = Fe, Ru, Os, show restricted rotation of the alkyne.[189] Variable-temperaure ^{13}C NMR studies on $[CoFe_2(\mu\text{-}H)(\mu_3\text{-}COMe)(CO)_7(\eta\text{-}C_5H_5)]$, $[Fe_2Ni(\mu_3\text{-}COMe)(\mu_3\text{-}CO)_6(\eta\text{-}C_5H_5)]$, and $[FeNi(\mu\text{-}CO)_2(CO)(\eta\text{-}C_5H_5)_2]$ have indicated that conformational equilibria exist in solution, with two distinct species observable in the IR spectrum of $[CoFe_2(\mu\text{-}H)(\mu_3\text{-}COMe)(CO)_7(\eta\text{-}C_5H_5)]$.[190]

^1H and ^{13}C NMR magnetization transfer measurements have been used to show that the inequivalent methylene hydrogen atoms in $[Os_3(CO)_{11}(\mu\text{-}CH_2)]$ undergo slow exchange. There is also bridge–terminal exchange.[191] In $[Os_3(CO)_9(\mu\text{-}dppm)(\eta^1\text{-}dppm)]$, the μ-dppm gives two ^{31}P NMR signals at low temperature, which exchange on warming.[192] The lowest-energy dynamic process observed for $[H_2Os_3(CO)_9(MeC_2Me)]$ in the ^{13}C NMR spectrum is consistent with an oscillatory motion of the alkyne and of the hydride bridging the Os—Os edge, coupled with the edge hopping of the second hydride.[193] the activation energy for alkyne rotation in $[RuCo_2(CO)_9(R^1C_2R^2)]$ has been determined using ^1H NMR spectroscopy.[194] The ^1H NMR spectrum of $[Ir_4(CO)_{10}^{}(1,2,5,6\text{-}\eta^4\text{-}C_8H_8)]$ shows four cyclo-octatetraene signals at low temperatures, and two on warming, due to ligand scrambling on the cluster.[195]

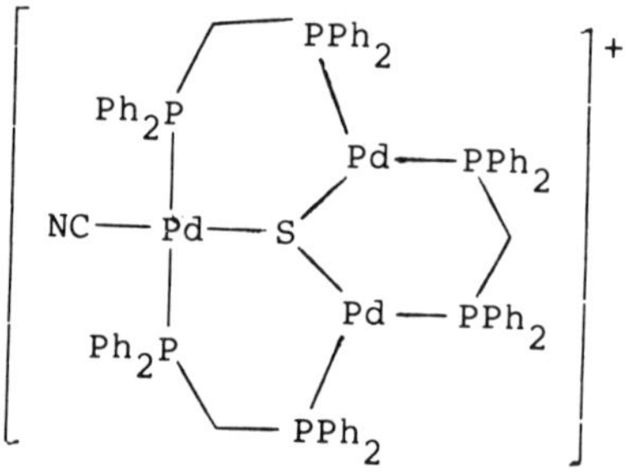

28

The ^1H NMR spectrum of (28) shows a dynamic process, where the CH_2 group of the unique dppm is AB at low temperature and a singlet at high temperature.[196] Variable-temperature ^1H and ^{31}P NMR studies have shown two isomers for $[M_2Ru_2(\mu_3\text{-}H)_2\{\mu\text{-}Ph_2As(CH_2)_nPPh_2\}(CO)_{12}]$, M = Cu, Ag, in solution at low temperature.[197] The ^{31}P NMR spectrum shows one signal for $[Au_3(PPh_3)_6(CNPr^i)_2]^{3+}$ at 228 K, but several signals at lower temperatures.[198] ^{31}P NMR spectroscopy has shown that $[Au_5ReH_4(PPh_3)_7]^{2+}$ is fluxional, with PPh_3 exchange on the gold atoms.[199]

Chapter 14

Homogeneous Catalysis of Organic Reactions by Complexes of Metal Ions

14.1. Reactions Involving Carbon Monoxide

14.1.1. Hydroformylations and Hydrocarboxylations

14.1.1.1. Rhodium

The hydroformylation reaction and related reactions of CO have recently been reviewed.[1] A number of mechanistic studies have appeared using 1-hexene as the substrate. The hydroformylation of 1-hexene using $[Rh(SBu^t)(CO)(P(OMe)_3)]_2$ selectively yields heptanal and 2-methylhexanal.[2] In order to gain mechanistic information, gas uptake data have been utilized in a kinetic analysis which indicates first-order dependence on the substrate and catalyst. In another report it was also found that addition of t-BuSH to $[RhH(CO)(PPh_3)]$ generates a much more active dinuclear hydroformylation catalyst: $[Rh_2(\mu\text{-}S\text{-}t\text{-}Bu)_2(CO)_2PPh_3)_2]$, **1**.[3] Addition of excess ligand to (**1**) is necessary to prevent catalyst deactivation. The selective hydroformylation reaction of 1-hexene has also been achieved by various μ-pyrazolato dirhodium(I) complexes $[Rh_2(M-P_3)_2(CO)_2L_2]$.[4] The highest conversion rates were observed for $L = P(OPh)_3$; these rates decreased when the basicity of the ligand was increased. In another study with 1-hexene, small amounts of triphenylphosphite increase the n/iso ratio; IR examination of the reaction mixture revealed that $[HRh(CO)\{P(OPh)_3\}_2]$ was the active form of the catalyst.[5] The reaction of $[Rh(acac)\{P(OPh)_3\}_2]$ with CO, H_2, and olefins has also been investigated using UV-vis, IR, and NMR techniques.[6] In the presence of H_2 and free phoshite [Rh(acac)P$_2$] produces HRhP$_4$, which isomerizes olefins. Addition of CO to HRhP$_4$ produces a good hydroformylation catalyst. No evidence of any hydride complexes could be found in the absence of free phosphite. The bimetallic early–late transition metal complex [Cp$_2$Zr(CH$_2$PPh$_2$)$_2$RhH(PPh$_3$)] was found to be more active and selective than the precursor [RhH(PPh$_3$)$_4$].[7] The increase in catalytic activity may be due to the increase in polarity of the Rh—H bond in the presence of the Zr moiety which is evidenced by comparing the chemical shift of the hydride region in the proton NMR.

The hydroformylation of 1,3-dienes may lead to useful products, such as adipic aldehyde or pentenals. Bidentate phosphine ligands bridged by two carbon atoms strongly promote the rhodium-catalyzed hydroformylation of 1,3-dienes; butadiene was converted to pentanal with 90% selectivity and less than 1% branched products.[8] The high linearity compared to the hydroformylation of 1-alkenes is believed to be due to the difference in stability of the products of insertion of CO into the intermediate η^3-butenyl complex. The η^1-allyl is formed with a large preference at the C1 position leading to the linear product. A rhodium hydroformylation catalyst system has also been used to prepare highly functionalized organics, such as nortropidines-nitrogen containing cyclic olefins.[9] A higher regiospecific route to alpha methylarylpropionaldehyde was developed by hydroformylating styrene with a rhodium α,α-TREDIP ligand system.[10] A 62:1 *iso* regioselectivity was found in the hydroformylation of styrene.

14.1.1.2. Cobalt

The cleavage of acylcobalt carbonyls by [HCo(CO)$_4$] or H_2 is regarded as key step in the hydroformylation reaction. The reductive cleavage of *n*-butyryltetracarbonylcobalt, 2, and isobutyrylcobalt tetracarbonyl, 3, by H_2 and [HCo(CO)$_4$] were studied in order to obtain information about the aldehyde formation step.[11] Isobutyraldehyde and *n*-butyraldehyde were formed in quantitative yields from the reaction of (3) and (2), respectively, using either [HCo(CO)$_4$] or H_2 under a CO pressure of 0.5–3 bar. From the results of the kinetic measurements at 25 °C, reaction rates were found to be first-order in (2) and (3), approximately first-order in [HCo(CO)$_4$] or H_2, and approximately negative first-order in CO. The inhibiting effect of CO on the reactions of (2) and (3) with [HCo(CO)$_4$] or H_2 suggest a tricarbonyl intermediate as shown in Scheme 1. The equilibrium lies far on the side of the tetracarbonyls from the infrared analysis. The k_3/k_4 values extrapolated to the temperatures necessary for the catalytic process indicate that H_2 is responsible for aldehyde formation from acylcobalt carbonyls under industrial conditions. It has also been reported that the reaction between [HCo(CO)$_4$] and ethyl acrylate under kinetic conditions results in the formation of ethyl propionate, ethyl-2-formylpropionate, and [1-(ethoxycarbonyl)ethyl]propionylcobalt tetracar-

Scheme 1

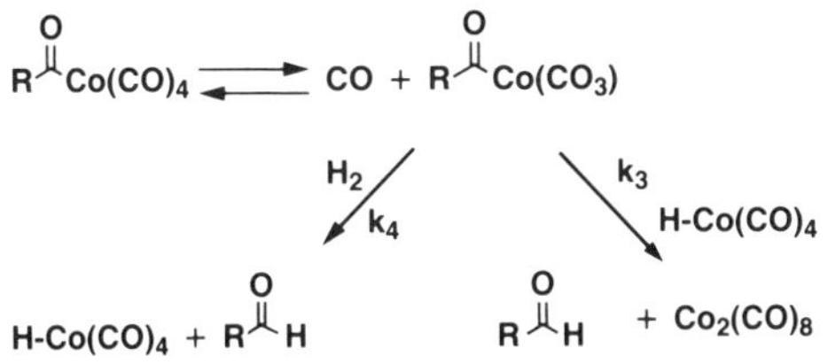

$$R\overset{\overset{\displaystyle O}{\|}}{}Co(CO)_4 \rightleftharpoons CO + R\overset{\overset{\displaystyle O}{\|}}{}Co(CO_3)$$

$$\swarrow \begin{smallmatrix}H_2\\k_4\end{smallmatrix} \qquad \begin{smallmatrix}k_3\\H\text{-}Co(CO)_4\end{smallmatrix}\searrow$$

$$H\text{-}Co(CO)_4 + R\overset{\overset{\displaystyle O}{\|}}{}H \qquad\qquad R\overset{\overset{\displaystyle O}{\|}}{}H + Co_2(CO)_8$$

bonyl.[12] This latter complex is transformed at higher temperatures into [3-(ethoxycarbonyl)propionyl]cobalt tetracarbonyl, which is the precursor of ethyl 3-formylpropionate. The branched product is the kinetically controlled product and the straight-chain compound is the thermodynamically controlled product. Also, the bimetallic complex $[Co_2(CO)_8Ru_3(CO)_{12}]$ showed high catalytic activity for the hydroformylation and hydroesterification of olefins, when compared to either the Co or Ru catalyst alone.[13] The synergistic effect is speculated to be due to a dinuclear reductive elimination between a cobalt-acyl and a ruthenium hydride.

It has also shown that phosphorous–carbon bond cleavage is a mode of catalyst deactivation during the hydroformylation catalyzed by triarylphosphine-substituted cobalt carbonyl species.[14] In another study it was found that phosphorous–carbon bond cleavage does not appear to be free radical in nature but appears to be influenced electronically by the substituents present on the aryl ring.[15] Electron-donating groups effect a fivefold lower rate than those with lower electron-donating ability. The reaction appears to fit the traditional requirements of an oxidative addition in both its electronic response and product distribution.

14.1.1.3. Platinum

Platinum complexes are highly active and selective catalysts for the hydroformylation of olefins. A ^{31}P NMR study on the reaction of *trans*-[PtCl(CO-*n*-hexyl)L$_2$] with SnCl$_2$ revealed two species in equilibrium with the starting platinum–acyl complex: one is *trans*-[Pt(SnCl$_3$)(CO-*n*-hexyl)L$_2$] and the other is a complex that contains Pt and Sn but no Pt—Sn bond.[16] Under ambient conditions this system reacts with H$_2$ to produce *n*-heptanal and a complex mixture of Pt hydrides. This same mixture of complex hydrides was obtained by reaction of *trans*-[PtHClL$_2$] and SnCl$_2$.[17] Among the many hydrides formed from the above reaction, only *trans*-[PtH(SnCl$_3$)L$_2$] rapidly inserts ethylene at $-80\,^\circ$C to yield *cis*-[PtEt(SnCl$_3$)L$_2$]. At $-10\,^\circ$C, *cis*-[PtEt(SnCl$_3$)L$_2$] irreversibly rearranges to the *trans*-isomer. Species containing the Pt—Sn moiety are involved in the insertion of the olefin, insertion of CO into the

Pt—C bond, and in the hydrogenolysis of the acyl intermediate. The asymmetric hydroformylation of the three isomeric linear butenes using [(R,R-Diop)Pt(SnCl$_3$)Cl] has been reinvestigated.[18] The chirality of the 2-methylbutanal arising from 1-butene is opposite to that arising from Z and E 2-butene. The results reported in this paper show that the regioselectivity and asymmetric induction are determined during or before the formation of the alkylplatinum intermediate.

Platinum catalysts containing diphenylphosphine oxide ligands, such as [Pt(H)(Ph$_2$PO)(Ph$_2$POH)(PPh$_3$)], catalyze the hydroformylation of hept-1-ene and hept-2-ene yielding products of high linearity (90% and 60%, respectively).[19] The hydroformylation activity could be followed stepwise by treating the catalyst with ethylene to give the platinum–ethyl intermediate. Carbon monoxide addition to the Pt–ethyl complex afforded the Pt–acyl complex at a faster rate than the ethylene insertion step. When [Pt(H)(Ph$_2$PO)(Ph$_2$HPO)-(PPh$_3$)] was treated with excess Ph$_2$HPO, the new complex [Pt(H)(Ph$_2$PO)-(Ph$_2$POH)$_2$] was formed. When ethylene was hydroformylated with [Pt(H)-(Ph$_2$PO)(Ph$_2$POH)$_2$], the priopionyl intermediate was observed by ^{1}H and ^{31}P NMR. This reaction produces both propan-1-ol and pentan-3-one.

14.1.2. Carbonylation of Alcohols and Esters

A recent review on the mechanistic aspects of transition-metal-mediated alcohol carbonylation was published in late 1986.[20] A number of reports have issued concerning the mechanism of the iodide-promoted rhodium-catalyzed carbonylation of linear and secondary alcohols; most of these investigations are covered in the review on alcohol carbonylations.

The results of a comprehensive kinetic, spectroscopic, and analytical study into the rhodium-catalyzed carbonylation of primary alcohols (R = Me, Et, and n-Pr) demonstrate that the reaction rate is first-order in both [Rh] and added [HI] and independent of CO pressure.[21] In all cases the only rhodium-containing species observed in the infrared spectra, obtained from solutions at operating conditions, was [RhI$_2$(CO)$_2$]$^-$. Carbonylation of EtOH-d$_5$ showed no kinetic isotope effect. The relative rate data and activation parameters obtained from these alcohols (R = Me, Et, and n-Pr) suggest that the oxidative addition of the corresponding alkyl iodide to the rhodium(I) center is nucleophilic in nature and the relative rate data compare well with the rates expected for the S_N2 displacement for organic halides. The spectroscopic and kinetic results are all consistent with the rate-determining S_N2 displacement of iodide from the corresponding alkyl iodides by [RhI$_2$(CO)$_2$]$^-$. These results are consistent with the previously reported mechanism for the methanol system. It was also discovered in the carbonylation of n-propanol, that the amount of isobutyric acid, but not the overall reaction rate, varied inversely with CO pressure. To explain this observation, an additional pathway, occurring after the rate-determining nucleophilic-type oxidative addition step, was proposed:

the initially formed alkyldicarbonylrhodium(III) species dissociates CO and allows for formation of an hydrido–olefin complex by β-hydride elimination. This mechanism was further verified by studying[22] the carbonylation of ^{13}C labeled ethanol as shown in Scheme 2. Both studies indicate that the initially

Scheme 2

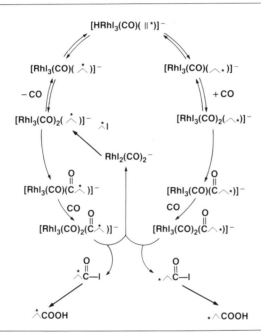

generated alkyl-dicarbonyl rhodium(III) species is inherently unstable and that this instablity can be relieved by loss of CO to generate a five-coordinate complex. The presence of isomerization in both the ethanol and *n*-propanol study indicate that loss of CO can compete effectively with rearrangement to the acyl-dicarbonyl rhodium(III) complex. In a study by Dake,[23] the specific rates and kinetic features of the carbonylation of *n*-propanol and *n*-butanol were reported. On comparison of the rates of *n*-alcohols under identical conditions it was observed that the rates varied in the following order: methanol > ethanol > *n*-propanol > *n*-butanol.

   In contrast to the findings of Hjortkjaer and Jorgenson,[24] Dekleva and Forster[25] found an inhibitory effect of increased CO pressure on the rates of the carbonylation of isopropanol using the Rh/I catalyst system. The carbonylation of isopropanol has a measurable kinetic isotope effect and a more complicated kinetic behavior when compared to the carbonylation of linear primary alcohols. By examining the products (*iso-* vs. *n*-butyric acids) both in the absence and presence of added propylene as a function of pressure, it was concluded that a hydrocarboxylation pathway was responsible for the predominant production of *n*-butyric acid. From studies with specific deuterium labeled isopropanols, a second route involving an electron-transfer mechanism is suggested to explain the retention of deuterium at rates too fast to be attributed to a nucleophilic displacement reaction. The rate effects of lithium iodide on the rhodium-catalyzed acetic anhydride synthesis cannot be accounted for by a methanol carbonylation cycle.[26] The mechanism proposed is a modified methanol carbonylation cycle which possesses two regimes of chemistry. At high lithium concentration, the rate and activation parameters are consistent with the methanol carbonylation reaction. At low Li concentration, a new rate-determining step involving the consumption of acetyl iodide by lithium acetate is operative. In another study, iodide and acetate anions were also shown to be promoters in the oxidative addition of MeI to Rh(I) carbonyl complexes.[27]

14.1.3. Homologation Reactions Involving Synthesis Gas

There is much interest in the use of ruthenium in the catalyzed synthesis of oxygenated compounds from CO/H_2. While the mechanism of this homologation reaction has been investigated, the actual catalytic species has not been identified.[28] A recent review highlights the contributions in this area.[29] The reaction of ethyl orthoformate with CO/H_2 in the presence of ruthenium carbonyl iodide has been studied.[29] Results of experiments with $CO + D_2$ suggest a stepwise process of hydrogenation and carbonylation. Roper found that homologated acids were generated more rapidly when esters were used as starting materials than when acetic acid was used.[30] Water, which is formed in the homologation reaction of acetic acid, may inhibit the activation of the acyl moiety necessary for the reaction sequence.[31] Addition of Lewis acid promoters to the $[Ru(CO)_3I_3]^-$ catalyst system increased the selectivity of ethyl acetate from methyl acetate at the expense of hydrocarbon formation.[32] The role played by the Lewis acids in the selectivity may be related to the acceleration of the alkyl migration–carbonyl insertion step. In another paper, it was demonstrated that there are four major pathways occurring simultaneously in the iodide/Ru promoted homologation of methyl esters.[33] These are homolgation of the methyl ester, carbonylation to acetic acid, reduction to alcohols, and hydrogenolysis to methane. Quantitative measurements of several of these pathways have been obtained as the promoter, the concentration of the promoter, and the partial pressures of hydrogen and carbon monoxide are varied. Because of the direct relationship between the overall rate to all products and the hydrogen pressure, it is suggested that the active species results from the hydrogenation of the ruthenium iodocarbonyl complex. An acetyl–Ru complex has been implicated as the common intermediate for both the homologation reaction and the carbonylation pathway. A homogeneous Co—Ru/iodide catalyst system has been used to catalyze the homologation of ethyl esters.[34] The synergistic effect of the ruthenium cocatalyst is believed to arise from the formation of the olefin derived from β-hydrogen elimination of the alkyl moiety. This olefin is then hydroformylated/hydrogenated to the higher alcohol. The homologation of methanol has also been studied with cobalt/iodide catalyst systems. The selectivity to acetaldehyde has been increased by the addition of α,γ-bis(diphenylphosphino)alkanes.[35] The reaction of the complexes $[CH_3C(O)M(CO)_xL]$ (M = Co, Mn) with $HM'R_3$ (M = Si, Sn) has been found to be consistent with a pathway involving initial CO dissociation from $[CH_3C(O)M(CO)_xL]$, oxidative addition of the M'—H bond, and reductive elimination of acetaldehyde.[36] For $HSnR_3$ the rate-determining step is CO dissociation and for $HSiR_3$ the rate-determining step is oxidative addition.

14.1.4. Carbonylation of Organic Halides

There have been many examples of Pd, Pt, and Rh catalyzed carbonylations of organic halides and pseudohalides. The product-forming step in the Pd-catalyzed methoxycarbonylation of benzyl chloride was determined by using trimethylphoshine in the alkoxycarbonylation reaction.[37] Complex (**4**) and (**5**) can be isolated in the reaction sequence shown in Scheme 3. When

Scheme 3

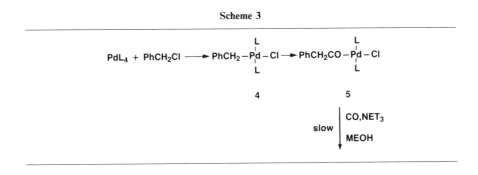

(**5**) is partially decomposed under an atomosphere of ^{13}CO there is no incorporation of ^{13}CO into the recovered complex or in the product ester, demonstrating the irreversibility of the CO insertion step.

A new methodology for making carbon–cobalt bonds has resulted in the development of a new catalytic reaction of alkyl acetates and lactones with hydrosilane, CO, and $[Co_2(CO)_8]$.[38] This provides a novel method for the introduction of the siloxymethylidene group into a carbon bearing an acetoxy group as in equation (1). The suggested key catalytic species is $[R_3SiCo(CO)_4]$.

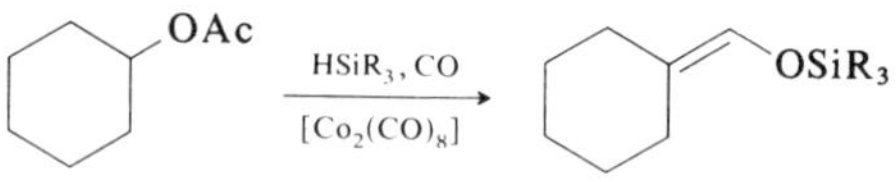

$$\text{(cyclohexyl-OAc)} \xrightarrow[\text{[Co}_2\text{(CO)}_8]}{\text{HSiR}_3,\text{ CO}} \text{(cyclohexenyl-OSiR}_3) \qquad (1)$$

14.1.5. Decarbonylations

More mechanistic insight into the decarbonylation reaction of acyl chlorides by Rh(I) complexes was obtained by studying the reaction of ^{13}CO, $[RhCl(CO)(PPh_3)_2]$ 6, and aliphatic acid chlorides.[39] It was found that (6) catalyzes the incorporation of ^{13}CO into a wide variety of acid chlorides including aryl, alkyl, and vinylic acid chlorides. Although free CO is involved in the labeling, the CO exchange does not proceed by decarbonylation/carbonylation sequences involving aryl or alkyl chlorides or olefins. The kinetically preferred paths for the organometallic intermediates involve equilibration and reductive elimination to give acid chlorides and (6). The barriers to the formation of decarbonylation products (alkyl chlorides, olefins) are greater than the barriers to reductive elimination of acyl chlorides. In another study complexes $[C_5Me_5MMe_2(Me_2SO)]$ (M = Rh and Ir) react with aldehydes to form alkyl- and aryl-carbonyl complexes.[40] From the experimental results it is suggested that the key intermediate/transition state involves a M(V) intermediate.

The palladium-catalyzed decarbonylation of acyl cyanides produces the corresponding nitriles in excellent yields.[41] When an acyl cyanide has a β-hydrogen, alkenes are formed. The decarbonylation of tricyclic bridgehead acid chlorides using a Pd-catalyst also produced tricyclic olefins.[42]

14.1.6. Water Gas Shift Reaction

The water gas shift (WGS) reaction in alkaline media has been suggested to involve (1) a nucleophilic attack of $-OH$ or H_2O on CO coordinated to a metal center, giving a hydroxycarbonyl hydride, (2) a thermal decarboxylation of the hydroxycarbonyl complex to afford CO_2 and a metal hydride, and (3) H_2 evolution by reaction of the hydride with a proton or water. A systematic isolation of all possible intermediates in the WGS reaction catalyzed by $[Ru(bipy)_2(CO)Cl]^+$ has been reported (Scheme 4).[43] The above reaction pathway is strongly supported by the isolation or characterization of all the intermediates.

A study of the catalytic activity of mixed metal clusters of Fe and Co group metals under basic WGS conditions demonstrated a synergistic effect in the Fe/Ru mixed metal systems.[44] A much weaker effect was observed in the Fe/Ir system. The reduction of nitrobenzenes to the corresponding anilines is

Scheme 4

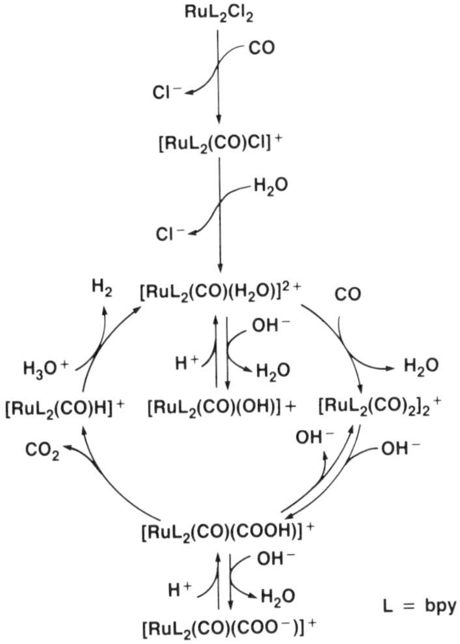

$$RuL_2Cl_2$$

CO

Cl^-

$$[RuL_2(CO)Cl]^+$$

H_2O

Cl^-

H_2 $\quad [RuL_2(CO)(H_2O)]^{2+}$ $\quad$ CO

OH^-

H_3O^+ $\qquad H^+ \searrow H_2O$ $\qquad H_2O$

$$[RuL_2(CO)H]^+ \quad [RuL_2(CO)(OH)]^+ \quad [RuL_2(CO)_2]_2{}^+$$

CO_2 $\qquad OH^- \quad OH^-$

$$[RuL_2(CO)(COOH)]^+$$

OH^-

$H^+ \searrow H_2O$ $\qquad$ L = bpy

$$[RuL_2(CO)(COO^-)]^+$$

best achieved under WGS conditions in basic medium and polar solvents.[45] For catalytic reactions involving monohydrides such as $[RuHCl(CO)(PPh)_3]$, electron-withdrawing substituents on the aromatic ring lower the rate of reduction of the nitro group while electron-donating groups increase the rate of reduction.[45]

14.2. Hydrogenations

During this period many reports dealing with the use of homogeneous hydrogenation catalysts have appeared. Since many of these lack mechanistic detail or possess mechanisms common to other well-known systems and hence offer no new insight, they are not discussed here. An interesting class of compounds that has been documented in recent years by X-ray studies comprises the so-called nonclassical hydride or molecular hydrogen complex which exhibits η^2-H_2 coordination.[46-48] Such complexes are potential candidates for intermediates along the reaction coordinate to dihydride formation—a key step in dihydrogen activation for catalytic hydrogenations. In fact, evidence has been presented for the dissociation of the H—H bond in

$[M(CO)_3(PR_3)_2(\eta^2\text{-}H_2)]$ ($M = W, Mo$) to the dihydride complexes.[49] Low-temperature matrix (liquid Xe) IR studies reveal that dihydrogen adducts of Fe,[50] Co,[50] and Ru[51] exist, as well as $[M(CO)_5(H_2)]$ ($M = Cr, Mo, W$)[51] and *cis*-$[Cr(CO)_4(H_2)_2]$.[51] Interestingly, $[Cr(CO)_5(H\text{—}D)]$ is formed in the D_2/H_2 exchange of *cis*-$[Cr(CO)_4(H_2)_2]$ and the mechanism proposed is reminiscent of olefin metathesis in that D_2/H_2 exchange to yield *cis*-$[Cr(CO)_4(D_2)(H_2)]$ is followed by an intramolecular formation of HD.[51] Discrete Fe and Ru[52] complexes of the type $[M(H_2)H(PP)_2]^+$ are very stable and their X-ray structure clearly reveals the symmetrical $\eta^2\text{-}H_2$ ligand *trans* to the hydride. Polyhydride complexes of Ir,[53,59] Fe,[55] Ru,[55] and Os[55] have been shown to contain the $\eta^2\text{-}(H_2)$ ligand[53,55] (e.g., $[IrH_2(H_2)_2P_2]^+$). A ^1H NMR method based on the extremely short T_1 value of a $\eta^2\text{-}(H_2)$ compared to a normal terminal hydride has been described as well.[53,54] Theoretical studies (extended Hückel) reveal that the H_2 approach leading to the η^2 structure is exothermic and is particularly favored when the incoming H_2 molecule approaches the metal *trans* to a Π-accepting ligand.[56]

14.2.1. Alkene Hydrogenations

Directing effects in catalytic homogeneous hydrogenations have been exploited by a number of workers in recent years for obtaining stereochemical control over the hydrogen addition to functionalized olefins. The presence of a ligating group on an olefin has been shown to direct the attack of hydrogen addition from the face of the molecule containing the directing group. Some spectacular selectivities have been obtained with the $[Ir(COD)(PCy_3)\text{-}$(pyridine)]PF_6 catalyst in dichloromethane.[57] While there are many subtle effects to rationalize, it is significant that directing effects have only been observed with catalysts capable of attaining the 12-electron configuration, while 14-electron catalyst fragments, such as "$RuHCl(PPh_3)_2$"[58] and "$RhCl(PPh_3)_2$," do not give directing effects. The Ir cationic catalysts employed by Crabtree *et al.*[57] appear to bind the functional groups in the order $CONH_2 < OH < C{=}O < CO_2R < OMe$ and that these functional groups bind more tightly than the olefin. This explains the observation that the product must not bind to the catalyst too strongly or it can act as a poison. In model studies using norbornenol, only one hydride-containing complex, (7), could be observed by NMR,[58] but at higher temperatures this stable compound only reductively eliminates H_2; no hydrogenation occurs. This suggests that

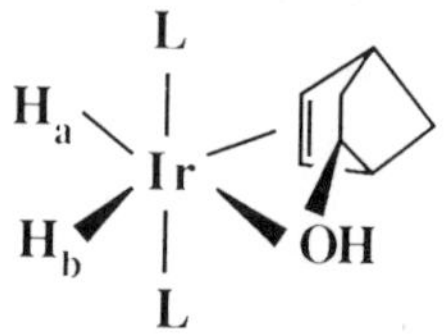

7

the hydrogenation intermediate contains phosphines which are *cis*, in analogy to the Rh asymmetric hydrogenation mechanism. The *endo* face specific hydrogenation of *endo*-6-methylenebicyclo[2.2.2]octan-2-ol with D_2 and Ir catalysts was studied by ^{13}C NMR.[58] The isotope (H/D) redistributions observed in the products are extensive and are interpreted in terms of an intramolecular mechanism in which product formation occurs by breakdown of an alkyliridium trihydride.

The use of metal clusters as homogeneous catalysts is interesting from the standpoint that such polynuclear species represent soluble analogues of heterogeneous catalysts. In fact, there are few examples of actual polynuclear catalysts, because it is usually dissociation of a coordinatively unsaturated mononuclear species that gives rise to catalytic activity. A series of tetranuclear $[(Cp)NiOs_3(\mu\text{-}H)_3(CO)_9]$ and $[(Cp)NiOs_3(\mu\text{-}H)_3(CO)_8L]$ clusters catalyze the homogeneous 1,2-hydrogenation of dienes and nonconjugated dienes are isomerized.[59] Arguments are presented for the intact cluster functioning as catalyst; such arguments include that these clusters exhibit both facile H/D exchange and excellent site and product selectivities. Complexes of structure (**8**) were detected spectroscopically and their reaction with H_2 or D_2 yield hydrogenation product plus starting cluster. The mechanism proposed involved formation of a metal (Os) alkyl bond via hydride migration. The vacant site on the cluster is then capable of activating H_2. That these clusters do not react with alkynes to form open-form (butterfly) complexes suggests that the cluster remains in the *closo*-form during catalytic hydrogenation. The so-called anion promoted clusters $[M_3(CO)_{11}X]^-$ (M = Ru[60a] or Os[60b] and X = N_3^- or NCO^-) have also been shown to be hydrogenation catalysts, although Ru is more active than the Os system. Utilization of this reactivity difference made it possible to intercept intermediates in the Os system. The anionic Os cluster reacts with H_2 to produce $[H_2Os_3(NCO)(CO)_{10}]^-$ which will react with an α,β-unsaturated ester, such as maleic anhydride, to yield an Os carbonyl cluster containing a terminally bound alkyl ligand. This is the first example of such a species. A crystal structure of $[HOs_3(NCO)(\sigma\text{-succinoyl})(CO)_9]^-$ (**9**) was reported (hydride not located). This hydride, σ-alkyl cluster does

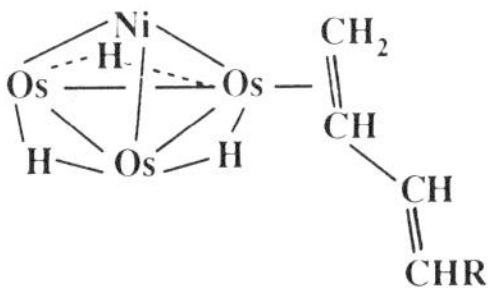

8

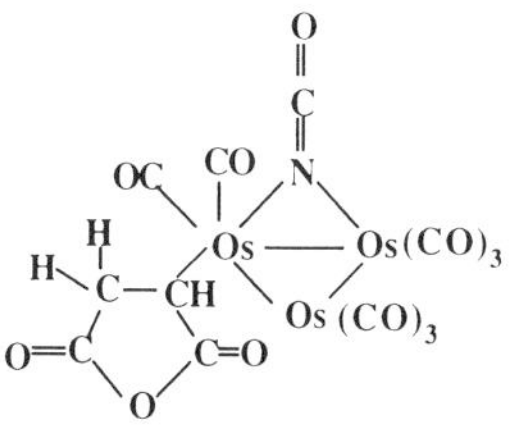

9

reductively eliminate slowly (3 h, 75 °C under CO) to yield the starting anionic cluster and succinic anhydride.[60b]

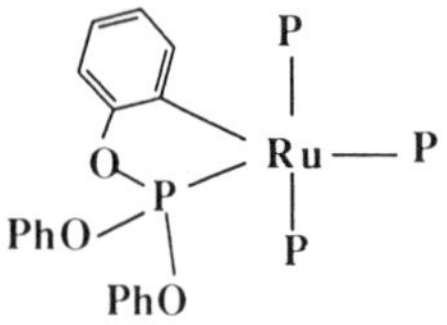

10 $\quad$ [orthometalated Ru complex with $P(OPh)_3$ ligands, PhO, Ru—P]

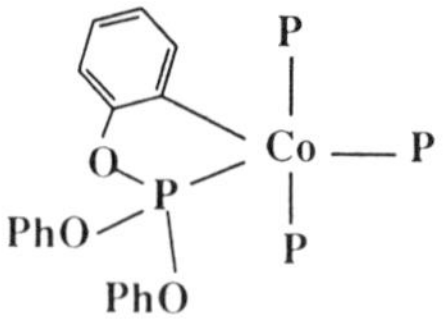

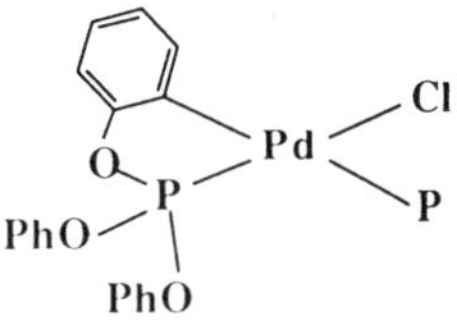

11 $\quad$ [orthometalated Co complex with $P(OPh)_3$ ligands, PhO, Co—P]

12 $\quad$ [orthometalated Pd complex with $P(OPh)_3$ ligands, PhO, Pd—Cl, P]

In an interesting comparison study it has been shown that the complexes $[HClRu(P(OPh)_3)_4]$, $[HCo(P(OPh)_3)_4]$, and $[Cl_2Pd(P(OPh)_3)_2]$ are not hydrogenation catalysts, but that their orthometalated analogues (**10**), (**11**), and (**12**) are. While *in situ* NMR studies are described, no intermediates are observed. Interestingly, upon exposure to trace O_2 levels phosphine oxide appears and catalytic activity increases, but this is due to colloid formation which also results in loss of selectivity and greater isomerization.[61] It is interesting to note that the complexes $[RuCl_2(Ph_2P(CH_2)_mPPh_2)_2]$ have been shown to be efficient olefin hydrogenation catalysts in the presence of NH_4PF_6. It is proposed that loss of halide from Ru (driven by ion-pairing in polar solvents) generates vacant coordination sites for catalysis.[62]

14.2.2. Arene and Alkyne Hydrogenations

Orthometaled Ru complexes have been found to be excellent arene and alkyne hydrogenation catalysts. The anionic ruthenium hydride $[RuH_2(PPh_3)_2(PPh_2C_6H_4)]^-$, **13**, has been studied in relation to the catalytic hydrogenation of anthracene.[63] Through the elegant use of independent syntheses and *in situ* NMR studies, much of the coordination chemistry and mechanism involved in the anthracene system has been elucidated. The anion (**13**) reacts with H_2 to yield *fac*-$[RuH_3(PPh_3)_3]^-$, **14**, which reacts with anthracene to yield $[RuH(PPh_3)_2(\eta^4\text{-anthracene})]^-$, **15**, and 1,2,3,4-tetrahydroanthracene as in equation (2). This reaction exhibits the same rate law as

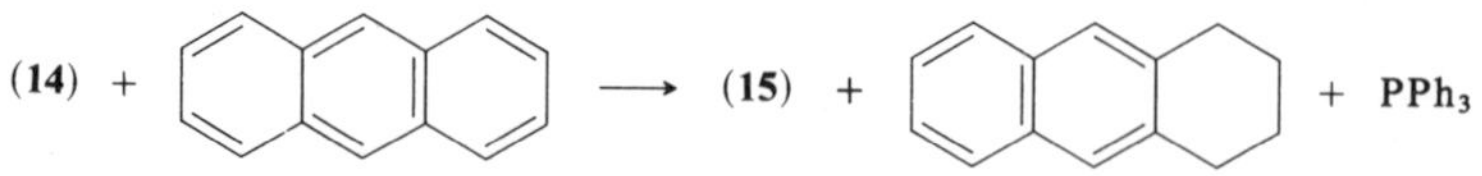

$$(\mathbf{14}) + \text{[anthracene]} \longrightarrow (\mathbf{15}) + \text{[tetrahydroanthracene]} + PPh_3 \qquad (2)$$

the isotope exchange of (**14**) with D_2, implying that both reactions proceed through a common unimolecular rate-determining step, the reductive elimination of H_2, to form a common intermediate $[RuH(PPh_3)_3]^-$ [an isomer of (**13**)] as shown in equation (3).

$$(\mathbf{13}) \xrightarrow{-H_2} [RuH(PPh_3)_3]^- \xrightarrow{D_2} [RuHD_2(PPh_3)_3]^- \qquad (3)$$

Another complex $[RuH_5(PPh_3)_2]^-$, **16**, forms when (**15**) reacts with H_2, also generating tetrahydroanthracene. The $[RuHCl(PPh_3)_3]$ complex has also been shown to be an active hydrogenation catalyst for polynuclear heteroaromatic compounds, such as quinoline to 1,2,3,4-tetrahydroquinoline.[64]

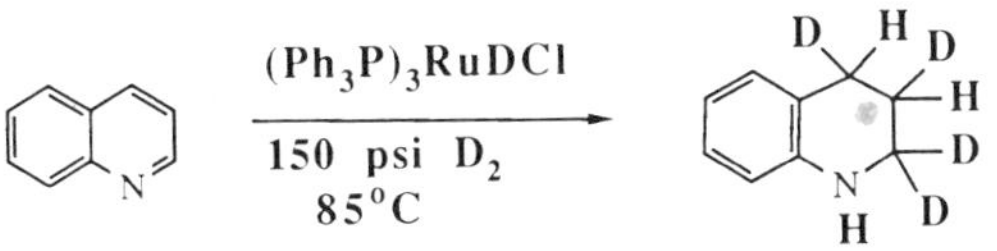

$$\text{quinoline} \xrightarrow[\substack{150 \text{ psi } D_2 \\ 85^\circ C}]{(Ph_3P)_3RuDCl} \text{product} \qquad (4)$$

Deuterium gas was used in the hydrogenations and the pattern of D incorporation was studied via ^{1}H NMR as a function of time [equation (4)]. Scheme 5, shown here in abbreviated form, accounts for the unusual D incorporation and, as noted, requires an orthometalation step for D incorporation at the 8-position.

Scheme 5

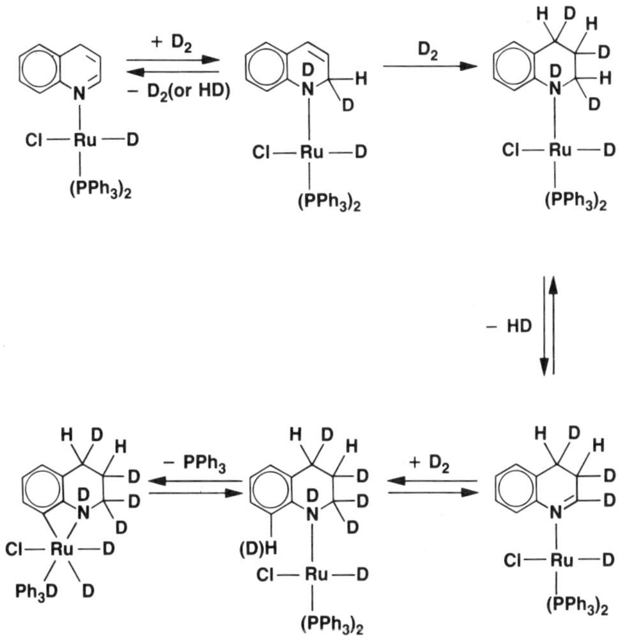

The study of the catalytic hydrogenation of alkynes using the same catalyst [RuHCl(PPh$_3$)$_3$] has shown that the orthometalated analogue of the above-mentioned (**13**), the complex **17**, [RuCl(PPh$_3$)$_2$(P(C$_6$H$_5$)$_2$(C$_6$H$_4$)], is produced when alkynes are reacted with [RuHCl(PPh$_3$)$_3$].[65] Further, reaction of (**17**) with D$_2$ established that a rapid equilibrium exchange occurs between hydride and *ortho*-arylphosphine sites. Isotope exchange during actual catalytic alkyne hydrogenation conditions reveals that (**17**) is not an intermediate in the hydrogenation owing to the lack of deuterium incorporation into the *ortho*-aryl sites. The complex {RuH[1,4-bis(diphenylphosphinobutane)$_2$]}PF$_6$ has been shown to be an excellent nonisomerizing alkyne and olefin hydrogenation catalyst.[66] The neutral dihydride [RuH$_2$(dppb)$_2$], **18**, is not an intermediate in these systems, but in the presence of base this dihydride forms in quantitative yield, implicating the known Ru(IV) dihydride [RuH$_3$(dppb)$_2$]$^+$ as an intermediate since this is known to deprotonate with base to yield (**18**).[66] A number of dimeric rhodium complexes containing the bridging bis(diphenylphosphino)methane ligand are known to be efficient alkyne hydrogenation catalysts. A recent NMR study of the hydrogenation reaction of [Rh$_2$H$_2$(CO)$_2$(dppm)$_2$] with alkynes reveals that a chemically induced dynamic nuclear polarization (CIDNP) occurs.[67] This complex reacts with phenylacetylene under H$_2$ to yield the complex (**19**) and styrene. Deuterium incorporation studies also

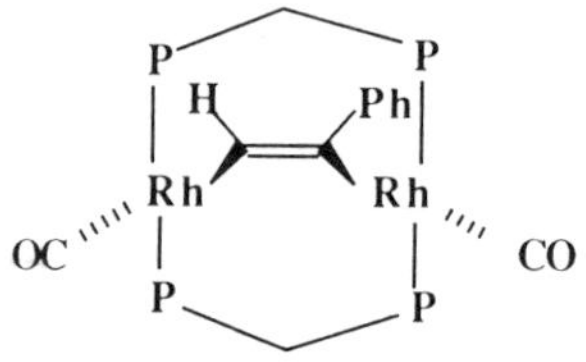

P   P
| H   Ph|
Rh'   Rh
OC | CO
P   P

19

reveal that *cis*-addition of D$_2$ occurs. Evidence was presented[67] which strongly supports the conclusion that the radical pair responsible for the CIDNP must be a metal-centered biradical which transfers its H atoms to the alkyne *cis* in an essentially concerted manner [equation (5)].

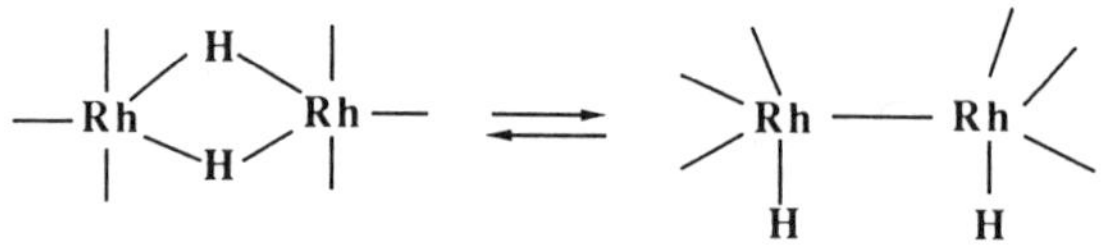

$$-\text{Rh} \underset{\text{H}}{\overset{\text{H}}{\diamond}} \text{Rh}- \rightleftharpoons \text{Rh}----\text{Rh} \qquad (5)$$
$$\qquad\qquad\qquad\qquad \text{H} \quad\; \text{H}$$

   In another study of the dppm dimer catalysts, Ir was substituted for Rh since it was felt that with Ir more stable intermediates could result and thus be isolable. Although these systems are not catalytic themselves, several potentially relevant complexes have been identified. The dimer [IrCl(CO)(Dppm)]$_2$, **20**, reacts with H$_2$ to yield the *cis*-dihydride, (**21**). This complex reacts with alkyne to yield a stable isolable complex in which alkyne insertion into only one of the Ir—H bonds occurs to yield a σ-alkenyl complex which does not

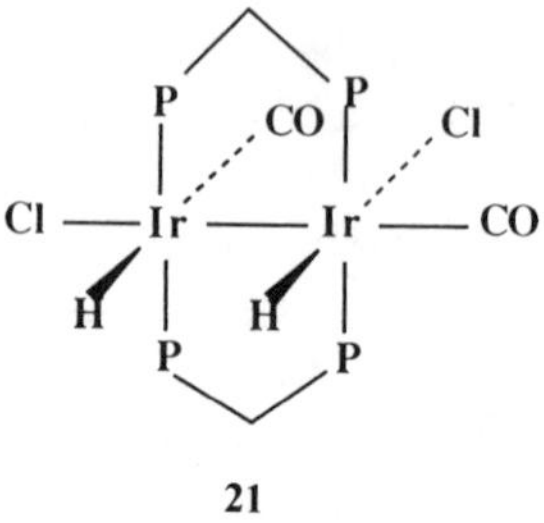

P   CO P   Cl
| . | /
Cl —Ir ——— Ir —CO
   H | H |
      P   P

21

undergo further reaction between the remaining hydride and the coordinated metalated olefin.[68]

14.2.3. Other Functional Group Hydrogenations

The hydrogenation of esters to alcohols is catalyzed by $[H_4Ru_4(CO)_8(PR_3)_4]$ clusters, but recent work has revealed that mononuclear and dinuclear species are present in these systems and appear to be lead to catalysis not the cluster itself.[69] A less traditional homogeneous hydrogenation catalyst, $[(MeCp)_2Mo_2(S)_2(SH)_2]$, **22**, has been reported which will efficiently reduce $C=N$ bonds in imines, isothiocyanates, and isocyanates. In addition, (**22**) functions as a catalyst for reduction of arylnitro compounds to anilines. Further hydroxylamines are reduced and nitrosobenzene to 1,2-diphenylhydrazine. These complexes are of great interest, since the site of hydrogen activation is proposed to be the bridging sulfide ligands. The reaction of the dimer (**22**) with azobenzene [equation (6)] produces diphenylhydrazine and

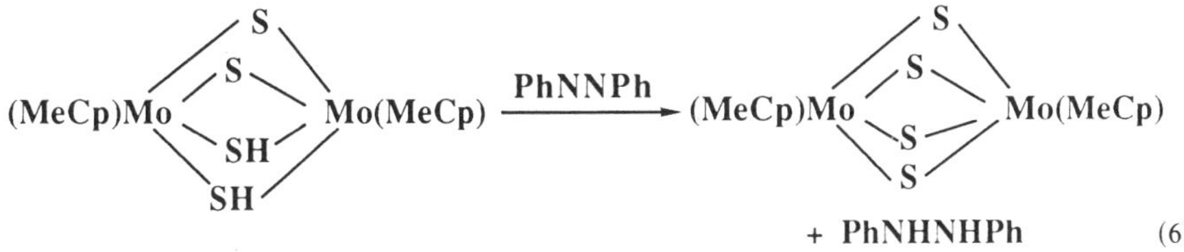

$$
(MeCp)Mo\underset{\underset{SH}{\overset{S}{\diagup}}{\overset{S}{\diagdown}}}{\diagup}Mo(MeCp) \xrightarrow{PhNNPh} (MeCp)Mo\underset{\underset{S}{\overset{S}{\diagup}}{\overset{S}{\diagdown}}}{\diagup}Mo(MeCp) + PhNHNHPh \quad (6)
$$

22

the new dimer tetra-μ-sulfido[bis(MeCp)]di(molybdenum), **23**, which when treated with H_2 regenerates (**22**).[70] The same complex (**22**) has also been reported by separate workers to catalyze SO_2 reduction to S_8 in the first example of a homogeneous hydrogenation of SO_2.[71]

Imines have been reduced catalytically by both $HCo(CO)_4$ (in which an imine-adduct is observed)[72] and $RhCl(PPh_3)$ or $[Rh(diene)(PR_3)_2]PF_6$ type complexes.[73] Nitriles were reduced catalytically by both permethyl-scandocene and -zirconocene hydrides.[74] The reactions are very similar to aluminium hydride reductions of nitriles, i.e., initial adducts are observed. The reactions are driven by the high Lewis acidity of these early transition metal complex hydrides which polarizes the $-C\equiv N$ bond. The stepwise hydrogeneration of nitriles was also observed with this system and it is noteworthy that the $Sc-N$ bond undergoes facile hydrogenolysis.

14.2.4. Transfer Hydrogenations

Transfer of hydrogen from donor molecules, such as primary or secondary alcohols, to acceptor molecules can be catalyzed by transition metal complexes and can in fact often be used to advantage over conventional hydrogenations, e.g., selective reduction of α,β-unsaturated aldehydes or ketones to allylic alcohols. Typically, catalysts effective for homogeneous hydrogenations based on such metals as Rh, Ir, Ru, etc. are most often those used for transfer hydrogenation. Cationic rhodium(I) complexes, [Rh(NBD)-(L)$_2$] [where L = P(p-X-C$_6$H$_4$)$_3$ with X = Cl, F, H, CH$_3$], have been used as catalysts for efficient hydrogen transfer from isopropanol solvent to a variety of ketones.[75] Kinetic studies using acetophenone revealed that the reaction is first-order in substrate and second-order in [Rh], indicating perhaps that a binuclear species may be catalytically active. The reaction requires alkoxide to displace the coordinated diene, water poisons the reaction, and the more electron-withdrawing phosphines increase the rate. This prompted the authors to propose that mono-hydride Rh species form via deprotonation of intermediate dihydride species.[75] The first example of a nonphotochemical catalyzed dehydrogenation of methanol was reported.[76] In this study Ru complexes of the type [Ru$_2$(OAc)$_4$Cl] + PPh$_3$ or [Ru(OAc)Cl(PEtPh$_2$)$_3$], **24**, were used. Complex (**24**) was shown to possess a *mer*-phosphine arrangement in methanol and, at 30 °C in the presence of added HOAc, catalyzed the conversion of CH$_3$OH to CO + hydrogen. In a lengthy study probing the effect that the structure possessed by Rh, Ru, Ir, and Os catalysts has on rate, it was found that the highest activity for methanol reduction of cyclohexanone was exhibited by [RuCl$_2$(PPh$_3$)$_3$].[77] The reduction of 4-t-butylcyclohexanone by this system gives predominantly the *trans*-cyclohexanol. Monitoring the reactions by IR reveals that catalytic chemistry does not begin until an as yet unidentified ruthenium carbonyl complex forms.

14.2.5. Catalytic Reduction of Carbon Monoxide

In the homogeneous catalyzed hydrogenation of CO, a hydroxymethyl, HMCH$_2$OH, intermediate has been suggested. By comparing the reactivity of (**25**) and (**26**) for their ability to form methanol, it was concluded that methanol formation via an alkoxy hydride intermediate, (**25**), is preferred over methanol from an hydroxymethyl hydride complex, (**26**), which favors beta hydride elimination to yield formaldehyde.[78]

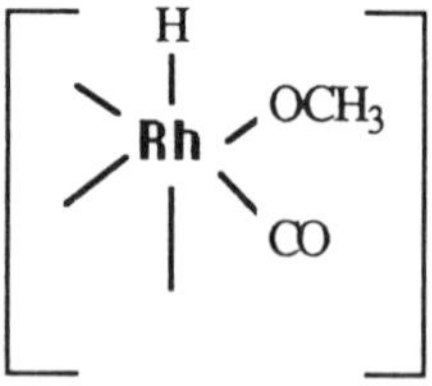

25 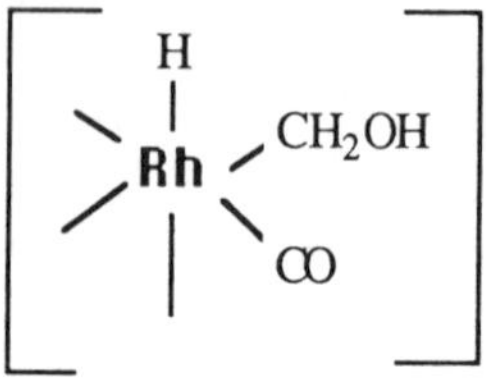   26

Model studies on the complex, *trans*-[Ru(CHO)(CO)(P-P)$_2$]$^+$, suggest that the catalyzed CO reduction occurs by successive intermolecular additions of H$^-$ and H$^+$ to a coordinated CO.[79] It was also found that the ratio of ethylene glycol to methanol is increased when small amounts of Rh are added

to an iodide/Ru catalyst.[80] From a study in whch the the total metal/I ratio
is held constant and the Ru-to-Rh ratio is varied, it was demonstrated that
rhodium effects selectivity; the increase in the selectivity to ethylene glycol is
due to an increased rate of ethylene glycol production and a decrease in
methanol productivity. The results are consistent with the transfer of a hydride
ligand from [HRu(CO)$_4$]$^-$ to a carbonyl ligand on a Rh(III) complex as in
equation (7) with a < b. Bulky alkylphosphines enhance the activity and

$$Ru-H^- + Rh-CO \rightarrow Rh-CHO \overset{_a \nearrow \quad CH_3OH}{\underset{_b \searrow \quad HOCH_2CH_2OH}{}} \tag{7}$$

stability of the rhodium carbonyl catalyst in the synthesis of ethylene glycol
from CO/H$_2$; high-pressure IR studies indicate that [HRh(CO)$_3$L] plays an
important role in product formation.[81] During the hydrogenation of CO by
an iridium catalyst, triarylphosphines enhanced the selectivity and activity for
ethylene glycol formation while trialkylphosphines promoted methanol formation.[82] It has also been reported that [HRu(CO)$_4$]$^-$ can be generated by the
reaction of [Ru$_3$(CO)$_{12}$], bis(triphenylphosphine)iminium acetate, and CO/H$_2$
in N-methylpyrrolidone[83]; methanol is the major product of this reaction
at 23 °C.

14.3. Oxidations

14.3.1. Molecular Oxygen Oxidations

The utilization of molecular oxygen as a selective oxidizing agent of
organic molecules has tremendous economic potential. A large quantity
of work has appeared in this field reflecting the continuous interest in the
use of this abundant oxidant for selective transformations. The major hurdle
to the effective use of oxygen as a selective oxidant for oxygen atom transfer
chemistry is the need for coreductants (such as alcohols, ascorbate, phosphines,
etc.) in addition to the substrate. The advent of catalytic systems which can
promote oxygen reactivity in a dioxygenase sense (as opposed to the well-
known monooxygenase mode of reactivity) is necessary to bypass the economic
shortcomings inherent in the use of coreductant systems. During this period
new dioxygenase systems have been reported. The first of these is a trans-
[Ru(VI)(dioxo)tetramesitylporphyrinato)] complex, 27, which catalyzes olefin
epoxidations under ambient conditions.[84] The reactions are zero-order in
oxygen and olefin concentrations and possess the stoichiometry of 2 mole
epoxide produced per mole O$_2$ consumed. It is also known that the corresponding Ru(II) complex reacts with O$_2$ to yield (27). This complex is also a
competent olefin epoxidation agent itself yielding epoxide plus the Ru(II)
complex. In this system a mechanism has been proposed in which complex

(27) is the active oxidant producing a transient Ru(IV)-monooxo intermediate which disproportionates to yield trans-[Ru(O)$_2$(TMP)] [i.e., (27)] and the [Ru(II)(TMP)] complex.[84]

Thioethers were reported to be catalytically oxidized selectively to sulfoxides utilizing molecular oxygen by ceric ammonium nitrate.[85] This system also possesses a reaction stoichiometry consistent with dioxygenase-like activity. The reactions are zero-order in oxygen, but first-order in Ce(IV) and thioether concentration. A mechanism involving Ce(IV) initiation of radical cation chemistry as in equations (8)–(11) is proposed and supported by kinetic arguments and mechanistic studies.

$$Ce(IV) + SR_2 \xrightarrow{k_1} Ce(III) + R_2S^+ \tag{8}$$

$$R_2S^+ + {}^3O_2 \xrightleftharpoons{k_2} R_2\overset{+}{S}-OO^{\cdot} \tag{9}$$

$$R_2\overset{+}{S}-OO^{\cdot} + Ce(III) \xrightarrow{k_3} Ce(IV) + R_2\overset{+}{S}OO^- \tag{10}$$

$$R_2\overset{+}{S}-OO^- + SR_2 \xrightarrow{k_4} 2\,R_2S \rightarrow O \tag{11}$$

Mechanistic studies continue to appear dealing with the use of palladium(II) nitro complexes as catalysts for molecular oxygen oxidations of terminal olefins to form ketones and epoxides (as their glycol monoacetates in acetic acid). These systems also are important, since they too do not require coreductants to activate molecular oxygen. The complex [Pd(CO)(NO$_2$)-(CH$_3$CN)$_2$], 28, is such a catalyst and has been the subject of detailed studies designed to elucidate the mechanism of oxygen atom transfer from the coordinated nitro group to the olefin. The stoichiometric oxidation of terminal alkenes with (28) was studied in acetic acid and found to yield primarily mixtures of 2-acetoxy-1-alkanols and 1-acetoxy-2-alkanols. These products contain an ^{18}O label (from ^{18}O labeled nitro) in the acetoxy group. To account for these results a mechanism was suggested in which an acetoxy palladation followed by acetyl migration to an oxygen of the coordinated NO$_2$ with subsequent acetoxy migration to the Pd center and reductive elimination occurs.[86] This mechanism predicts an overall trans-addition of OH and AcO across the olefin double bond. The actual stereochemistry of this step was shown to involve nearly total inversion at the Pd—C bond.[87] This rules out the mechanism given by Mares et al.[86] and supports a mechanism involving an intermediate cyclic five-membered ring carbocation. The same complex (28) was also reported to catalyze the oxygen oxidation of norbornene.[88] In this study a cyclic intermediate (29) was proposed. This Pd catalyst and other related Pd complexes are known to be competent catalysts for the conversion of terminal olefins to ketones. Recently, though, it was reported that complex (28) with CuCl$_2$ (cocatalyst) will catalyze the oxygen oxidation of terminal olefins to aldehydes in t-butanol.[89] It was proposed that cycloaddition of nitro-palladium to an olefin leads to oxygen atom transfer via a five-membered ring oxazapalladacyclopentane. Subsequent β-elimination leads to the observed aldehyde product

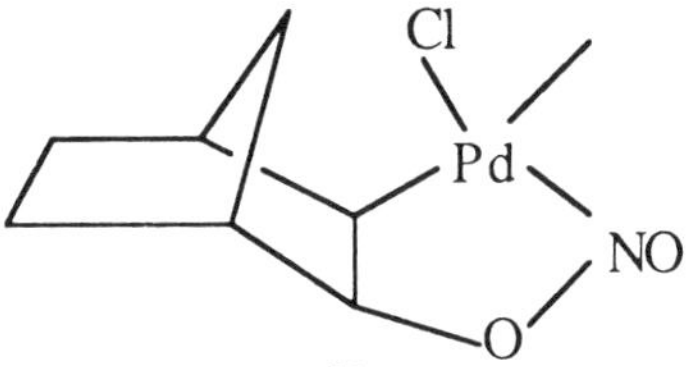

29

and reoxidation of the resultant Pd-nitrosyl with oxygen completes the catalytic cycle.

The rhodium complex **30**, [RhCl(PPh$_3$)$_3$(O$_2$)], has been reported to be a catalyst for the oxygenation of 1,4-cyclooctadiene yielding cyclooctane-1,4-dione in dry benzene under ambient conditions. This represents a most interesting activation of dioxygen since, as was shown by ^{18}O labeling studies, the ketonic oxygens in the product diketone each arise from one Rh—O$_2$ complex.[90] The kinetic studies were consistent with a pathway involving (1) prior oxygen interaction with rhodium, (2) a fast preequilibrium reaction displacing PPh$_3$ yielding a [Rh^{III}Cl(PPh$_3$)$_2$(O$_2$)(diene)] complex, **31**, and (3) the subsequent intramolecular attack, rearrangements, and reductive elimination yields diketone. As in the case of a monoolefin oxygenation with the same complex, a dioxarhodacyclopentane intermediate is invoked. In contrast to O—O bond scission resulting in ketone plus an oxorhodium species,[91,92] intramolecular attack of the peroxidic oxygen on the second olefin occurs as in Scheme 6.[90]

Scheme 6

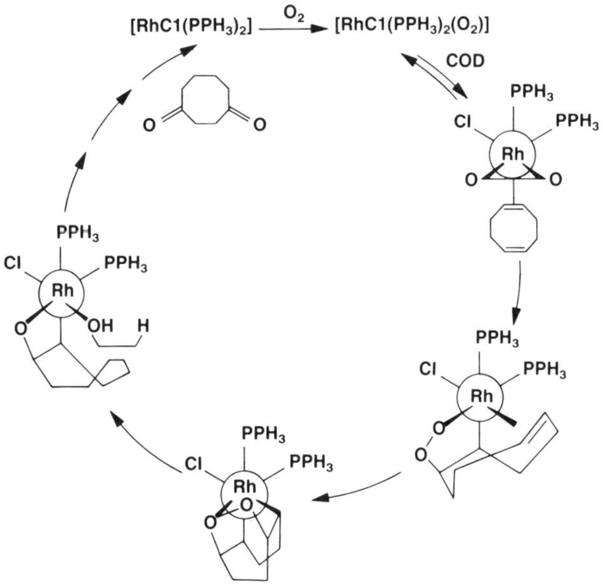

Many examples of free-radical or one-electron autoxidation processes have been published this period. The cobalt-catalyzed autoxidation of alkyl aromatic compounds to the corresponding arylcarboxylic acids is commercially significant technology. The mechanism of this reaction involves a radical cation intermediate shown in equations (12) and (13). Bromide ion is known to have

$$\text{ArCH}_3 + \text{Co(III)} \rightarrow \text{ArCH}_3^+ + \text{Co(II)} \qquad (12)$$

$$\text{ArCH}_3^+ \rightarrow \text{ArCH}_2^{\cdot} + \text{H}^+ \qquad (13)$$

a promotional effect in these systems and its role has been the subject of numerous reports. In acetic acid, xylene oxidations were studied using a Co/pyridine/Br$^-$ catalyst system.[93] Cobalt clusters of structure (32) were actually isolated from catalytic autoxidation reactions and have been shown to be competent oxidants of arylaromatics, i.e., converting toluene quantitatively to benzyl bromide in the presence of added bromide and in the presence of O$_2$ autocatalytically to benzoic acid. The mechanism is presumed to proceed by oxidation of Br$^-$ to Br$^{\cdot}$ by the cluster. The bromine radical is a chain initiator and abstracts a hydrogen atom to yield a benzylic radical which reacts with O$_2$ to yield organic peroxide derived oxidation products. These clusters are also significant in that higher steady-state concentrations of Co(III) result due to their lower oxidation potential, i.e., their oxidation potential is sufficient to oxidize bromide, but not xylene. This bromine radical pathway for autoxidation has been implicated by other workers as well.[94]

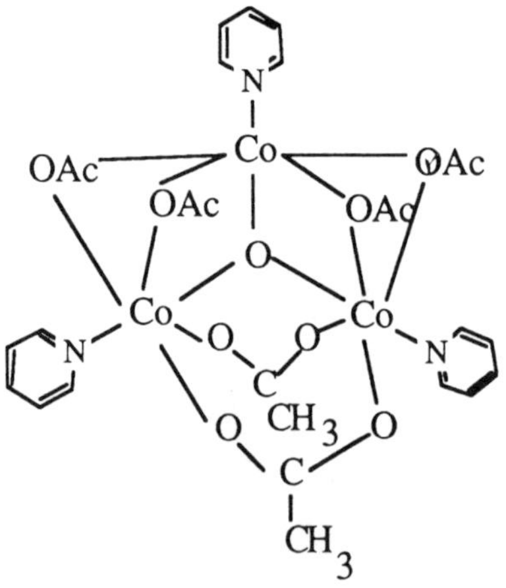

32

The mechanism of the oxygen oxidation of benzene has been studied using ^{18}O tracer techniques when catalyzed by copper ions[95] and with Fenton's reagent.[96] In the copper system, the results were in agreement with a Fenton-type mechanism shown in equations (14)–(18) in which the phenol and hydroquinone products arise from hydroxycyclohexadienyl radical and its oxygenated adduct, respectively.[95] A number of mechanistic studies have been

$$2Cu^+ + O_2 + 2H^+ \rightarrow 2Cu^{2+} + H_2O_2 \tag{14}$$

$$Cu^+ + H_2O_2 + H^+ \rightarrow Cu^{2+} + H_2O + HO^{\bullet} \tag{15}$$

$$HO + \text{(benzene ring)} \longrightarrow \text{(cyclohexadienyl, }H, OH) \tag{16}$$

$$\text{(cyclohexadienyl }H, OH) + Cu^{2+} \longrightarrow \text{(phenol)}-OH + Cu^+ \tag{17}$$

$$\text{(cyclohexadienyl }H, OH) + O_2 \longrightarrow \text{(cyclohexadiene }H, OH, OO^{\bullet}) \tag{18}$$

published in which catechols have been oxidized with oxygen catalytically in the presence of various metal complexes. Catalytic oxidation of 3,5-di-t-butyl catechol in the presence of $FeCl_3$, bipyridine, and pyridine with molecular oxygen has been shown via ^{18}O-tracer studies and by isolation of intermediates to proceed by stepwise oxygen insertion into a $1:1$ catecholate Fe(III) complex.[97] Both an intradiol pathway (proceeding through a cyclic seven-membered ring anhydride) and an extradiol pathway (proceeding via cleavage) were identified as leading to the cyclic five-membered ring lactone products. In contrast to the iron system the molecular oxygen oxidation of 3,5-di-t-butylcatechol catalyzed by a $Cu^I(Cl)$(pyridine) complex yields predominantly o-quinone products.[98] Based on kinetic studies (second-order dependence on catalyst concentration, first-order oxygen dependence, and zero-order in substrate) and on the observation of a rate-determining step involving formation of a dimeric μ-dioxoCu(II) complex, a mechanism was proposed implicating Cu(I) and Cu(III) species. The $[Co^{II}(acac)_2]$ catalyzed oxygen oxidation of 3,5-di-t-butylcatechol has also been studied in detail and, as in the Cu case, o-quinone is the major product.[99] Based on kinetic and spectroscopic studies a ternary complex of 3,5-di-t-butylcatechol, cobalt, and oxygen reacts intramolecularly (in the rate-determining step) to yield quinone and hydrogen peroxide. The intermediacy of 3,5-t-butyl-o-benzosemiquinone in the metal-catalyzed oxygen oxidation of 3,5-di-t-butylcatechol has been probed in tracer studies using a cobalt catalyst.[100] Cobalt complexes enhance the rate of oxidative cleavage of the semiquinone radical and labeling studies are consistent with intermolecular scrambling, in contrast to the chemistry observed elsewhere.[100] The cobalt acetate catalyzed autoxidation of acetals has also been studied and a free radical chain pathway proposed in which initiation is catalyzed by homolysis of the CH bond as in equation (19) and catalyst regenerated by peroxide decomposition [equation (20)].[101]

$$RCH(OR')_2 + Co^{3+} \rightarrow R\dot{C}(OR')_2 + Co^{2+} + H^+ \tag{19}$$

$$R''OOH + Co^{2+} \rightarrow R''O^{\bullet} + Co^{3+} + OH^- \tag{20}$$

14.3.2. Oxidations Involving Other Oxidants

The catalytic use of such an oxidant as *t*-BuOOH to carry out selective epoxidations of olefins (especially the propylene-Halcon process) is very important commercially. These oxidations, utilizing d^0 metal complexes (Mo^{VI}, V^V, Ti^{IV}) as catalysts, are very selective and can be stereoselective. While it is generally believed that these epoxidations proceed via a heterolytic pathway, there remains discussion as to whether (1) nucleophilic attack of olefin on a coordinated electrophilic peroxidic oxygen occurs or (2) an olefin coordinates then inserts to form a five-membered ring dioxametalacyclopentane which decomposes to yield alkoxide plus epoxide. Vanadium complexes of the type shown as (33) have been synthesized which will catalyze the selective

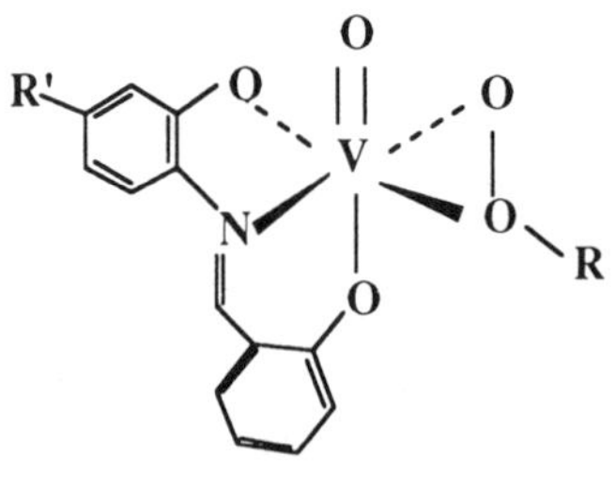

33

oxidation of olefins to epoxides.[102] These vandyl complexes are the first example of well-defined metal peroxide complexes which perform Halcon chemistry, i.e., no allylic oxidation (e.g., *cis*-olefin to *cis*-epoxide), inhibited by water and alchols, rates faster in nondonor solvents, rates increase as the nucleophilic character of the olefin. Kinetic studies support a mechanism in which an olefin peroxide complex forms via reaction of olefin with (33). The rate-determining step is the decomposition of this complex to yield epoxide and metal alkoxide complex which reacts with *t*-BuOOH to regenerate (33) plus *t*-BuOH. More strongly coordinating olefins inhibit the reaction. All these facts are most consistent with a dioxametallacyclopentane intermediate.[102]

With M(III)(Salen) (M = Cr or Mn) catalytic expoxidation of olefins can be accomplished using iodosylbenzene or pyridine *N*-oxide (M = Cr)[103] or *t*-BuOOH (and PhIO) with Mn(III).[104] Olefin epoxidation in such systems involves formation of an oxo-metal(V) species with a rate-limiting attack on the olefin by this electrophilic agent. In another system utilizing the Pt-group transition metal, ruthenium, it has been observed that hypohalite can be used to cleanly oxidize olefin to epoxides in the presence of a catalytic amount of the $[Ru(IV)(bipyridyl)(terpyridine)oxo]^{2+}$ complex.[105] The catalytic cycle apparently involves a two-electron Ru(II)/Ru(IV) redox cycle in which a concerted O-atom transfer occurs, although an oxametalacyclobutane intermediate cannot be ruled out. Hypohalite has also been studied as a catalytic oxidant using [Mn(III)(tetraphenylporphyrinato)(acetate)] as a catalyst.[106]

These studies are also consistent with a [Mn(V)(oxo)(TTP)] intermediate, which reacts with olefin directly to yield epoxide.

The catalyzed hydrogen peroxide epoxidation of olefins is potentially an important reaction. In this same chapter in Volume 4 of this series two reports were cited in which such chemistry was described. During this period full papers have appeared describing these systems in greater detail. A [Pt(Diphos)-(CF₃)(OH)] complex catalyzes the H_2O_2 epoxidation of olefins via external attack of HO_2^- on the coordinated alkene of a $[Pt(Diphos)(CF_3)(olefin)]^+$ complex forming a dioxaplatinacyclopentane intermediate.[107] In the second system non-Fenton chemistry can be realized using anhydrous hydrogen peroxide with Fe^{2+} in acetonitrile.[108] In this Fe(II) system arguments are presented for a mechanism involving no oxidation-state changes for the Fe(II) center during a catalytic epoxidation cycle; thus, the authors do not invoke an iron(IV)(oxo) species as an intermediate.

14.3.3. Biomimetic Oxidations

In recent years, tremendous gains in our understanding of the mode of action of the hemoprotein-based cytochrome P-450 monooxygenase enzymes have been made by studying the chemistry of model iron(porphyrinato) complexes. Ample evidence has accumulated using strong oxo-atom transfer oxidants which suggests that an $[iron(IV)(oxo)(porphyrinato \text{ radical cation})]^+$ complex is the species responsible for oxidizing substrates. The use of various metal porphyrin complexes as catalysts for oxidations of hydrocarbons using such strong oxidants as PhIO, R_3NO, ROOH, OCl^-, etc. has been reviewed.[109] The olefin oxidation reaction has been the subject of numerous studies using $[Fe(IV)(oxo)(porphyrinato)]^+$ catalyst systems. In one such study, kinetic evidence for an intermediate $[(olefin)iron(oxo)(porphyrinato)]^+$ complex was detected in a catalytic olefin epoxidation system.[110] In a more recent study using the $[Fe(IV)(oxo)(tetramesitylporphyrinato)]^+$ complex, the catalytic cyclooctene epoxidation reaction was studied at $-42\,°C$. At this low temperature the spectral signal of a unique species appears.[111] The spectrum of this material decays in a first-order fashion to produce the $[Fe(III)(TMP)]^+$ complex and cyclooctene oxide. These results showed unambiguously that an [iron(oxo)(olefin)(porphyrinato)] complex is an intermediate in these iron porphyrin P450-like olefin epoxidation systems. In order to gain a better insight into the nature of the reaction pathway of the olefin with the $[Fe(oxo)(porphyrinato)]^+$ complex [e.g., can this process be characterized by (1) direct oxygen atom transfer, (2) free-radical addition following by fast ring closure, (3) electrophilic addition followed by fast ring closure, (4) metalaoxacyclobutane formation, or (5) electron transfer to yield olefin radical cation], this reaction step has been probed using a series of similar metaloporphyrins of varying electronegativity.[112] With these systems the catalytic epoxidation of norbornene was studied using perfluoroiodosylbenzene. High amounts of endo-2,3-epoxynorbornane were formed and the ratios of endo/exo product were also independent of the oxidant used (other iodosylbenzenes). These results are consistent with the reaction of olefin with an oxidizing species {such as $[Fe(O)(Por)]^+$} followed by direct electron transfer to yield olefin radical cation.[112] These conclusions are also consistent with the observations of a catalytic epoxidation study using deuterium-labeled styrenes.[113] In model studies designed to investigate the role of axial ligands in [Fe(III)(porphyrinato)L] complexes on the catalytic epoxidation of cyclohexene with PhIO, the rates and yields of cyclohexene oxide were increased with more electron-withdrawing substituents.[114] Given that the rate-determining step is the oxygen atom transfer step from an $[Fe(IV)(oxo)(porphyrinato)]^+$ complex to olefin, a reduction in the electron density will enhance the electrophilicity or oxidizing power of the iron-oxo center with the consequence that the catalyst becomes more active. This trend is also observed when more electron-withdrawing groups are substituted on the porphyrin ring itself. Additional support for a carbocation (radical cation) intermediate in P450-like oxidations is presented in a study of *trans*-cyclooctene oxidation with PhIO catalyzed by [Fe(III)Cl(TPP)]. Skeletal rearrangement products were observed which are consistent with the rearrangement of an electron-transfer generated carbocation intermediate.[115]

The details of the formation of the $[Fe(IV)(oxo)(porphyrinato)]^+$ species from dioxygen in living systems are largely unexplored. The preceding discussion has dealt only with the so-called peroxide shunt pathway in which the actual high-energy oxidant is synthesized from an [Fe(III)(porphyrinato)] complex plus a potent oxo atom transfer reagent. Using the [Fe(III)(OH)-(tetramesitylporphyrinato)]$^+$ complex in place of the chloro analogue, it was found that p-NO_2-$C_6H_4CO_3H$ generates the [Fe(III)(TMP)(peroxybenzoate)] intermediate which decays in the first-order fasion to yield the [(oxo)iron(IV)porphyrin radical cation] complex $[Fe(IV)(O)(TMP)]^+$.[116] This peracid complex is relevant to the catalytic P450 chemistry, since an [Fe(III)(peroxide)] complex is a proposed intermediate in these systems. The synthesis of such a model complex and the finding that it converts to the $[Fe(IV)(oxo)(porphyrinato)]^+$ oxidation level via an acid-catalyzed heterolytic cleavage is very significant.

14.4. Metathesis Reactions

The Sixth International Symposium on Olefin Metathesis was held during this period and a special double issue of *Journal of Molecular Catalysis*[117] was devoted to this subject. Contributions to the volume covered metathesis polymerizations, as well as basic studies of homogeneous and heterogeneous catalysts. The following have been published: a review covering recent advances in tungsten–carbene complexes as applied to metathesis,[118] a short review of the Shell Higher Olefin Process,[119] and a book.[120]

While it has been demonstrated that olefin metathesis reactions are propagated by metal–carbene complexes and that many such complexes exist, few have been effective metathesis catalysts themselves. Consequently, detailed mechanistic insight is often lacking due to the complexity of such systems, e.g., additives of a Lewis acid character required to generate *in situ* catalytic systems. A recent communication describes the synthesis and reactivity of the first example of a well-defined complex that is itself a highly active, neutral metathesis catalyst (no added Lewis acid required for activity).[121] This is a very active catalyst for *cis*-pentene metathesis (T.N. at 25 °C is at 1000/min), and ^{1}H NMR studies reveal that the major species in solution is a single isomer of $[W(=CHEt)(=NR)(OR_f)_2]$, (**34**). One of the most significant advances in olefin matathesis would be the development of catalysts capable of promoting metathesis of functionalized olefins. Certainly in systems where catalytic activity depends on added Lewis acids, this will be a problem since the substrates contain basic functionality. Complex (**34**) will catalyze the metathesis of methyl oleate, but this system gradually deactivates with the buildup of the W(VI) oxo complex $[W(=O)(=NR)(OR_f)_2]$. This result suggests that the deactivation pathway involves metathesis of the carbonyl (C=O) double bond. Another neutral tungsten complex, $[W(CO)_3Cl_2(AsPh_3)_2]$, has been utilized in the metathesis polymerization of norbornene.[122] The mechanism for the formation of a carbene intermediate involves prior olefin coordination followed by a 2,3-hydrogen shift.

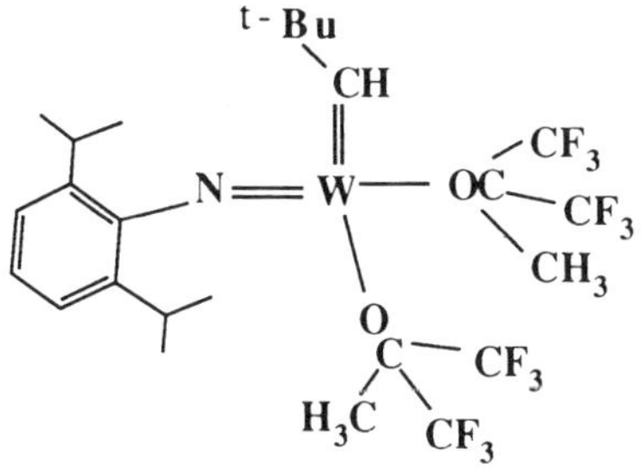

34

The viability of the carbene-forming step in metathesis has precedent in a study of the protonation of tungsten alkylidene complexes. Addition of a carboxylic acid to $[W(\equiv CEt)(OCMe_3)_3]$ yields an unstable complex, $[W(=CHEt)(OCMe_3)_2(O_2CR)_2]$, which rearranges to yield $[W(propylene)-(OCMe_3)_2(O_2CR)_2]$, a reaction which is second-order in tungsten.[123] The *in situ* generation of active olefin metathesis catalysts from $[WCl_6]$ using an alkyne initiator has been studied in the cyclopentene system and evidence was presented which supports the notion that metal carbenes form from the alkyne and initiate olefin metathesis.[124] The ring-opening polymerization of norbornene was studied with the $[Cp_2Ti=CH_2]$ catalyst (generated *in situ*).[125] Bis(cyclopentadienyl)titanacyclobutanes form from the reaction with these catalysts and on thermolysis α-substituted carbenes are formed, which in the presence of added norbornene generates polymer. The polymerization proceeds via this metathesis mechanism with the rate-limiting step the cleavage of a titanacyclobutane affording the high-energy carbene intermediate.

14.5. Isomerization Reactions

A review on the mechanism and catalyst involved in the asymmetric isomerization of functionalized olefins was issued in 1985.[126] A detailed study of the 1-butene isomerization by $[(\eta^6\text{-arene})NiR_2]$ was found to be first order in both 1-butene and Ni complex.[127] The ratio of *cis*-2-butene to the *trans* isomer is dependent on solvent, suggesting a role of solvent in the catalytic cycle. A mechanism is proposed which involves the intermediacy of nickel hydride species generated by the insertion of 1-butene into the Ni—R bond followed by β-hydride elimination. The isomerization of 1,5-cyclooctadiene to 1,4-COD and 1,3-COD catalyzed by $[(R\text{-}Cp)_2TiCl_2]/R'MgX$ systems was recently studied.[128] A Ti/Mg ratio from 1:2 to 1:10 was necessary for an effective isomerization catalyst. Also, hydride complexes of Rh(I) were used to study the isomerization of (Z)-dimethyl-butenedioate to (E)-dimethyl-butenedioate in benzene at 25 °C.[129] The reaction catalyzed by $[RhH(PPh_3)_4]$ is first-order in Z-isomer and in catalyst. Deuterium-labeling studies are consistent with an addition–elimination mechanism of isomerization involving a σ-alkyl rhodium complex as an intermediate.

A regiocontrolled synthesis of allylsilanes by Rh(I)- or Ir(I)-catalyzed isomerization of olefins as in equation (21) has been reported.[130] The results suggest that the trimethylsilyl group can become the auxiliary to bring about the energy difference among the regio isomers of the possible olefin products.

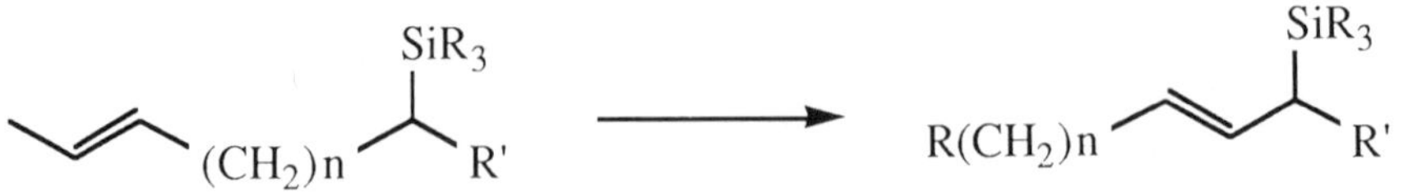

$$(21)$$

14.6. Oligomerizations of Olefins and Alkynes

A comprehensive study of the transition-metal-catalyzed dimerization of ethylene and propylene was the subject of a recent review.[131] Alkyl and hydrido–metal complexes are believed to be key intermediates in such metal-catalyzed transformations of olefins as di-, oligo-, or polymerization. Olefin complexes of organometals as models for organometallic catalysts were also the subject of a recent review.[132] Reduction of $RuCl_3 \cdot H_2O$ with $Zn/MeOH$ in the presence of methyl acrylate generates a catalytic system for the dimerization of alkyl acrylates to hexendioates.[133] Selectivity and rate are influenced by ligands, solvent, reducing agent, and Brønsted acidity. It has also been shown that zero-valent ruthenium complexes such as $[(\eta_6\text{-}C_6H_6)\text{-}(CH_2CHCO_2CH_2)_2Ru]$, **35**, catalyze the dimerization of acrylates as in equation (22) in the absence of additives at 140 °C.[134] Kinetic studies using (**35**) as the

$$2\,CH_2{=}CHCO_2CH_3 \;\rightarrow\; CH_3O_2CCH{=}CHCH_2CH_2CO_2CH_3 \qquad (22)$$

$$\text{MA} \qquad\qquad\qquad \text{DHD}$$

catalyst precursor reveal that the rate of dimerization obeys the rate law: $d[DHD]/dt = k[Ru]^{0.5}[MA]$.

Kinetic studies indicate that oligomerization of methyl acrylate arises from decay of the dimerization catalyst to a new catalytic species which promotes oligomerization. It was also reported that dimethyl phthalate was catalytically dimerized in the presence of 1,10-phenanthroline–palladium complex and Cu(II) to give exclusively 3,3',4,4'-tetramethyl biphenyltetracarboxylate, (**36**).[135] The Cu(II) salt appears to stabilize the Pd intermediate.

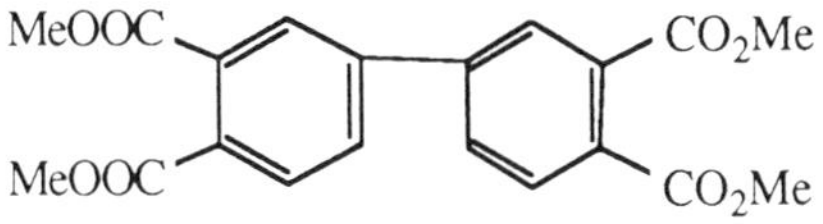

36

The Cu(II) concentration was independent of the rate of formation of (**36**) at the initial stage of the reaction. The 2,3,3',4'-isomer, (**37**), is formed as in equation (23) when the Pd(II)/Cu(II)/2,4-pentanedione system is used. The

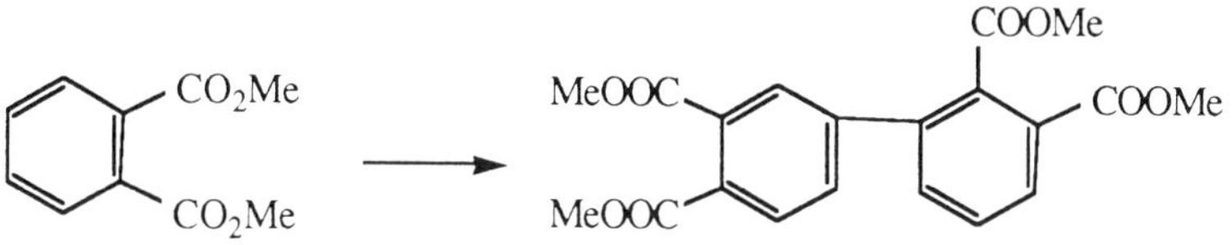

$$(23)$$

37

mechanism of the Pd(II)-catalyzed dimerization of methyl acrylate with methyl methacrylate has been investigated using deuterated methyl methacrylate, $D_2C{=}CCH_3COOCH_3$.[136] Statistical scrambling of deuterium within the allyl system using $[PdCl_2(PhCN)_2]/AgBF_4$ indicated a participation of a Π allylic structure. Transfer of D from d_2-methylmethacrylate to methyl acrylate occurs in codimerization experiments. These observations have been interpreted as an oxidative addition of Pd(II) into the vinyl C—H bond forming a Pd(IV) intermediate in a fast reversible reaction.

The mechanism of olefin polymerization by soluble Ziegler–Natta catalysts has been the object of continuous study. It has been proposed that a cationic complex (38) is the active species, although (38) has never been directly observed or isolated in the Ziegler–Natta systems. In a recent study complex (39) was synthesized, characterized, and shown to polymerize ethylene in the absence of an aluminum cocatalyst.[137] The observation of ethylene polymerization by (39) supports the original Long–Breslow–Newburg mechanism that cationic complex (38) is the active species.

$$Cp_2Ti^+ \quad AlR_{n-1}Cl^-_{5-n} \qquad [Cp_2Zr(CH_3)(THF)][BPh_4]$$
$$\backslash R$$

$$\textbf{38} \qquad\qquad\qquad\qquad \textbf{39}$$

The ring-opening polymerization of cyclic olefins is an important application of the metathesis reaction. Polymers of 1-methylbicyclo[2.2.1]hept-2-ene have been synthesized using twenty-four different metathesis catalysts.[138] Arguments are presented that the fully biased (head to tail) polymers proceed largely through a metal–carbene species (40), in which the last-added monomeric unit has its methyl group adjacent to the metal center. It has also been reported that titanacyclobutanes derived from strained cyclic olefins are useful catalysts for the metathetical polymerization of norbornene and also permit a detailed mechanistic study of ring-opening polymerization.[139] Bis(cyclopentadienyl)titanacyclobutanes (41) and (42) have been prepared from the reaction of norbornene with precursors of "$Cp_2Ti=CH_2$" and $Cp_2Ti=CHC(CH_3)_2CHCH_2$, respectively.

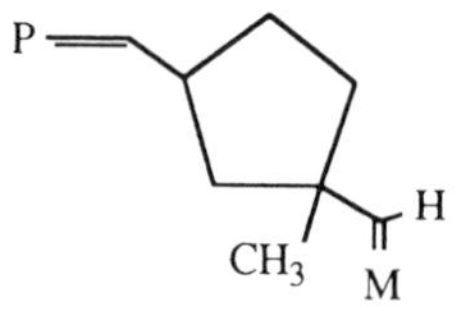

$$\textbf{40} \qquad\qquad\qquad \textbf{41} \qquad\qquad\qquad \textbf{42}$$

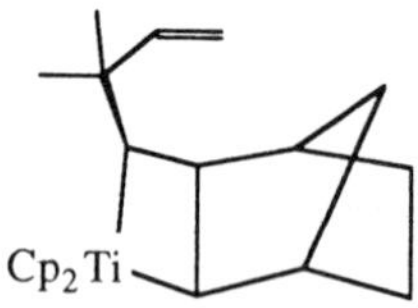

The kinetics of both polymerizations are first order in norbornene with $\Delta G^{\ddagger} = 24$ kcal mol^{-1} and $\Delta S^{\ddagger} = 9$ eu. The experimental evidence suggests the polymerizations catalyzed by either (41) or (42) proceed by the same mechanism except for differences in the initiation step. Scheme 7 shows the mechanism for the polymerization catalyzed by (41). Simultaneous incorporation of norbornene-d^2 into the polymer by all active sites confirms that this sytem is best described as a living polymer. This system is the only living polymerization system for the ring-opening polymerization of cyclic olefins reported to date.

The catalytic dimerization and telomerization of 1,3-dienes have been studied by numerous authors. The reaction of η^3-allylbis(triarylphosphite)nickel(II) complexes with butadiene was monitored by ^{31}P NMR spectroscopy[140]; both the anti- and syn-configurations could be identified. The

Scheme 7

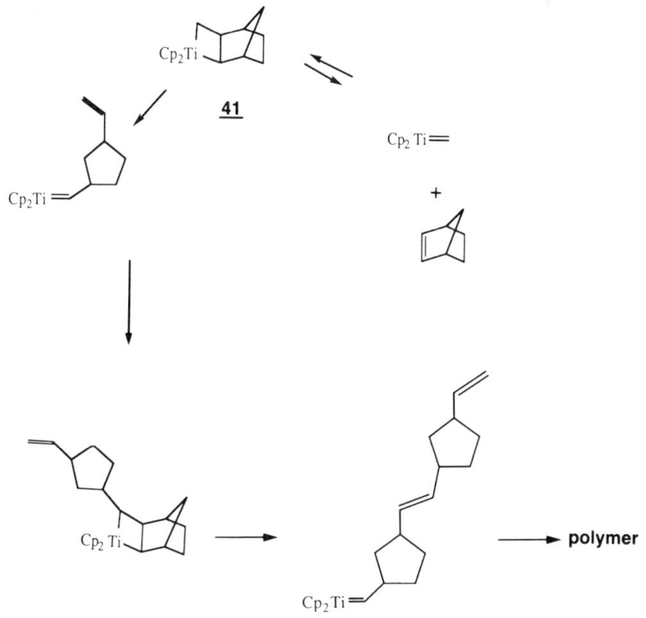

anti–syn isomerization, the higher reactivity of the thermodynamically more stable *syn*-form, and the formation of the *anti*-structure as a result of each individual butadiene insertion step were also demonstrated. It was also found that cationic palladium complexes, such as $[Pd(dppe)(py)_2][BF_4]_2$, are active catalysts for the telomerizations of isoprene with diethylamine to give *N,N*-diethyl(dimethyl-2,7-octadienyl)amines.[141] Experiments with *N*-deuteriodiethylamine revealed that the hydrogen of the nucleophile adds selectively to the C-6 of the octadienyl chain. In another report, evidence supporting the postulate that species such as (**43**) in equation (24) are intermediates in the Pd-catalyzed dimerization and telomerization of butadiene was obtained by studying the stoichiometric reaction of these species with nucleophilies.[142]

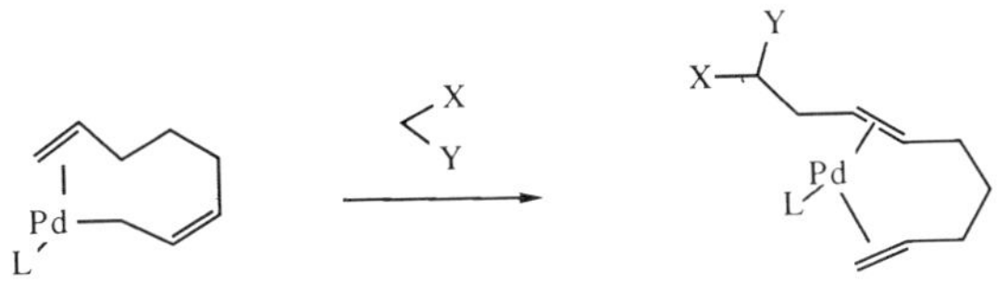

$$(24)$$

The influence of substituted phosphorous heterocycles such as chiral 1,3,2-oxazaphospholidines in the Ni-catalyzed butadiene oligomerization was investigated.[143] Although no optical induction was observed, P-methyl-substituted 1,3,2-oxazaphospholidine and 1,3,2-diazaphopholidine-modified Ni catalysts in toluene dimerize butadiene to give predominantly linear products, such as 1,3,6-octatriene. Both the methyl-substituted phosphorous group and the five-membered-ring heterocycle containing a nitrogen atom next to the phosphorous group appear to be essential requirements for preference of linear dimerization over cyclic dimerization.

Current interest in the physical properties of polyalkynes has focused attention on the mechanisms of the metal-catalyzed polymerization. Complexes of the type (44) undergo reactions with alkynes to give oligomerization products,[144] which result from insertion of the incoming alkyne into the M=C bond of the η^2-vinyl complex as in equation (25). This transformation of (44)

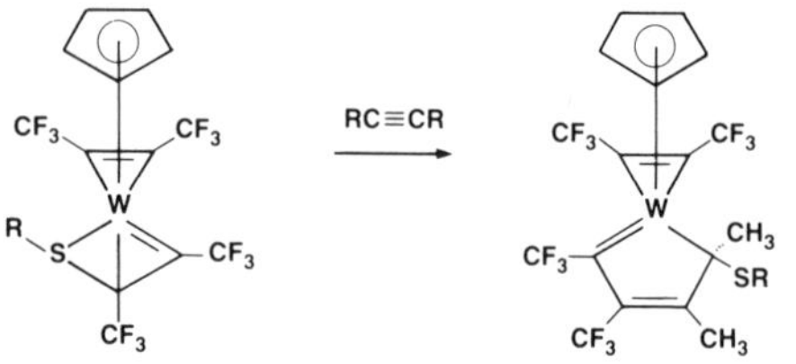

$$(25)$$

44

into alkyne-inserted products provides a model for alkyne oligomerization involving η^2-vinyl intermediates.

The comparative rates of cyclotrimerization of 3-hexyne and dimethyl-acetylenedicarboxylate, DMAD, with the catalysts [$(\eta^5$-C$_5$R$_5$)Rh(COD)] and [$(^5\eta$-C$_5$H$_4$PPh$_3$)RhCO$_2$]$^+$ have been studied.[145] In these reactions the rate of cyclotrimerization of DMAD decreases as the π-acceptor strength of the cyclopentadienyl ligand increases; this trend is reversed for hex-3-yne.

14.7. Carbon–Carbon Bond Forming Reactions

Many examples of catalytic carbon–carbon bond forming reactions have been published during this period. Reviews on enantioselective catalysis,[146] nucleophillic attack on transition metal organometallics,[147] and the catalytic chemistry of organopalladium and organonickel complexes have appeared.[148] A book entitled *Asymmetric Catalysis* devotes a chapter to chiral carbon–carbon bond forming reactions.[149]

14.7.1. Allylic Alkylations

Catalytic asymmetric allylic alkylation is a direct and powerful method of making carbon–carbon bonds. The basic mechanism of the palladium-catalyzed reaction has been solved for aryl-substituted chiral allyls and a detailed description of this mechanism can be found in a 1986 book entitled *Asymmetric Catalysis*.[149] In this book, Bosnich points out that allylic alkylation is an example of reactant control of asymmetric synthesis in which the major diastereomer gives the major product enantiomer. There are two reports of the use of functional groups to enhance stereoselectivity by intramolecular proximity effects. In the first example high optical yields (80–100%) were obtained in the palladium-catalyzed allylations of a chiral enamine which contained a chiral group at the alpha position of the ester carbonyl, i.e., the allylation of chiral enamines derived from (S)-proline allyl ester. This novel method provides a simple and highly efficient entry to optically active α-allyl ketones. The high optical yields are most likely due to an intramolecular allylation via a cyclic transition state chelated with the palladium catalyst.[150] In the second example, modification of optically active ferrocenylphosphine ligands with hydroxyl groups that would selectively attract soft carbon nucleophiles led to high optical yields of 1,3-disubstituted allyl acetates (92% ee).[151] It is believed that the hydroxyl groups on the ligand, which are located outside the π-allyl of the (π-allyl)palladium intermediate, interact attractively with the soft carbon nucleophiles by hydrogen bonding and this interaction is responsible for the high stereoselectivity. Asymmetric induction studies have also provided mechanistic information in the nickel and palladium cross-coupling reaction between allylic compounds and organometallic reagents. Results from asymmetric induction and deuterium studies in the Pd-catalyzed coupling of allyl acetates and phenylzinc chloride support Schwartz's[152] and Negishi's[153] postulated mechanism: at some stage a four-coordinated neutral allylic complex containing a single phosphine ligand is involved.[154] In another study of the reaction of phenyl halides, using *sec*-butyl-magnesium halides and chiral nickel catalysts, the chiraphos ligand always gave the best optical yield.[155] It is suggested that the use of chiral ligands with C_2 symmetry such as chiraphos reduces the number of possible diastereomeric reaction intermediates. Further studies are needed to solve the nickel-catalyzed allylic alkylation reaction.

Platinum complexes offer a good way of studying the mechanism of allylic alkylations due to similarity to palladium and the ability of Pt allyls to form stable olefin complexes. A ^{13}C NMR study of a series of reactions involving dimethyl sodium malonate together with ^{13}C-labeled η^3-butenyl-[$R,R,$-4,5-bis(diphenylphosphinomethyl)-2,2-dimethyl-1,3-dioxolane-p,p']-platinum tetrafluoroborate, **45**, and unlabeled substrate demonstrated that the platinum allyl is a true intermediate in the catalytic allylic alkylation.[156] Comparison of the stoichiometric reactions of [(η^3-crotyl)Fe(CO)$_4$]BF$_4$ and [(η^2-crotylacetate)Fe(CO)$_4$] and nucleophiles with the catalytic reaction of crotylacetate, a nucleophile, and [Fe$_2$(CO)$_9$] demonstrates that neither the η^2 or η^3 complex are required intermediates in these catalytic systems. Studies of the interaction of sodium dimethylmalonate with [Fe$_2$(CO)$_9$] have provided the evidence of Fe$_4$(CO)$_4$DMM ions in the catalytic reaction.[157]

14.7.2. Coupling Reactions

The palladium-catalyzed reaction of vinyl triflates with organostannanes in the presence of LiCl offers a mild and general method of making carbon–carbon bonds.[158] 4-*tert*-butylcyclohexyl triflate couples in high yield with vinyltributylstannate in the presence of 2 mol% of tetrakis(triphenylphosphine)palladium(0) and 3 equivalents of LiCl to give 1-vinyl-4-*tert*-butycyclohexene; this reaction is first-order. *Trans*-(4-*t*-butylcyclohexyl-1-enyl)-chlorobis(triphenylphosphine)palladium(II), **46**, can be isolated from the above reaction, and LiCl is essential for this reaction. NMR studies and kinetic results are consistent with a catalytic cycle in which oxidative addition of the vinyl triflate to the Pd(0) complex yields complex (**46**) and lithium triflate.[158] The mechanism of the coupling of aryl chlorides with a catalytic amount of a nickel salt and excess of a reducing metal has been investigated.[159] Mechanistic studies suggest that both the Ni species and zinc metal (the reducing metal) are involved in the rate-limiting step. Cross-coupling experiments have implicated an arylnickel species [speculated to be a Ni(I) aryl] which preferentially reacts with aryl chloride to yield a diarylnickel chloride intermediate which, in turn, reductively eliminates to a biphenyl.[159]

14.7.3. C—H Activation Reactions

One of the goals of the intense research effort in C—H bond activation is the development of catalytic carbon–carbon bond forming reactions. Recently some new catalytic carbon–carbon bond forming reactions via C—H activation have been reported; these alkylation reactions involve both aromatic and benzylic C—H activation.

Catalytic *ortho*-ethylation of phenol has been accomplished by the insertion of ethylene into ruthenium complexes containing *ortho*-metalated triphenyl phosphite linkages; the ruthenium complex together with phenoxide catalyzes the selective *ortho* alkylation of phenol.[160] About 15 turnovers of phenol per mol of ruthenium can be obtained at 200 °C. The rate-determining step in this reaction appears to be substitution of ethylene for triphenylphosphite in complex (**47**). A proposed mechanism involves insertion of ethylene into the *ortho*-metalated Ru—C bond, followed by another *ortho*-metalation

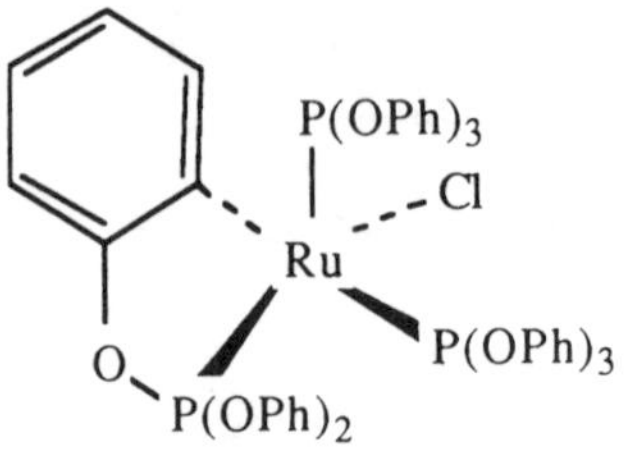

47

step to produce a di-*ortho*-metalated complex, which undergoes a reductive elimination to produce a bound ethylated triphenylphosphite. The ethylation reaction is completely inhibited by triphenylphosphite.

A new catalytic route to indoles was discovered by the reaction of 2,6-xylyl isocyanide in the presence of 20 mol% of [Ru(DMPE)$_2$(napthyl)H], **48**.[161] This complex is the first reported homogeneous metal complex to undergo reversible activation of arene and certain activated aliphatic C—H bonds. The catalytic route to indoles is thought to go through a six-membered metallocycle ring as shown in Scheme 8. The conversion of complex (**49**) to *trans*-(**50**) and the isomerization of *trans*-(**50**) to *cis*-(**50**) prior to indole formation can be observed by ^{1}H NMR.

Scheme 8

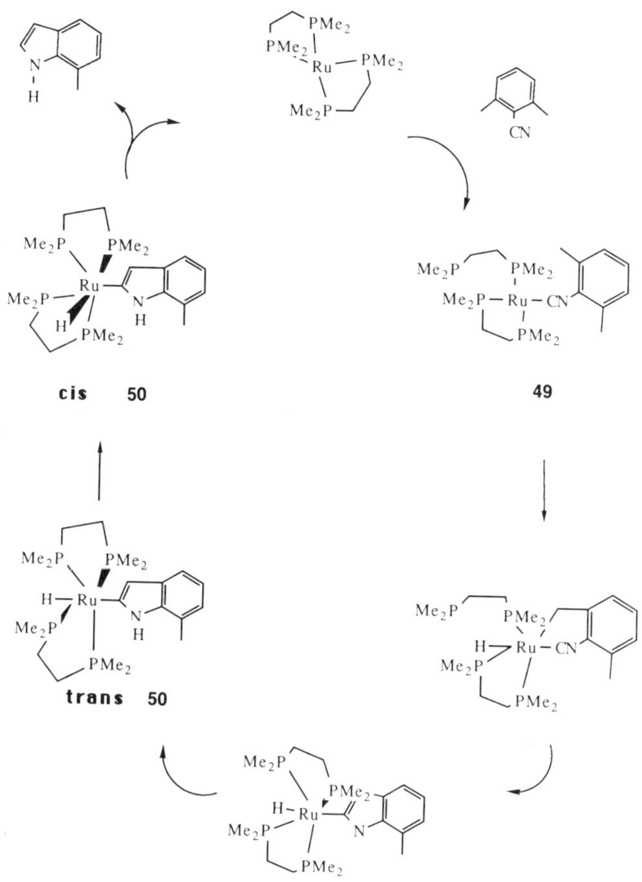

14.7.4. Cycloaddition Reactions

The first example of a nickel-catalyzed intramolecular [4 + 4] cycloaddition reaction has been reported.[162] This reaction, which has previously been used in the intermolecular synthesis of four-, six-, eight-, and twelve-membered rings from 1,3-dienes, has now been used for the synthesis of polycycles by an efficient direct intramolecular [4 + 4] cycloaddition reaction. Several interesting mechanistic conclusions follow from this study. First, catalyst variation influences the efficiency and product-type selectivity of this reaction. Second, dienes connected by three atoms give *cis*-fused rings while those separated by four atoms give *trans*-fused ring products. Both results are in accord with a mechanism that involves preferential formation and reaction of the more stable tetraene-nickel and bis-π-allyl complexes. The high stereoinduction (99:1) in the reaction of tetraene (**51**) suggests that the ester group directs chemoselective and facial-selective coordination of the catalyst to the proximate diene as in equation (26).

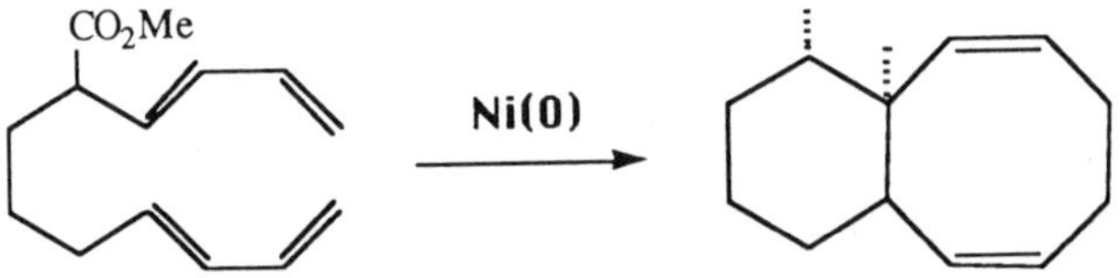

$$CO_2Me \xrightarrow{\text{Ni(0)}} \qquad (26)$$

51

14.7.5. Claisen Rearrangements

The Cope and Claisen rearrangements are of considerable synthetic utilty. Recently, the mechanism of the metal-catalyzed allyl imidate rearrangement was investigated with Pd(0), Pd(II), Rh(I), and Ir(I) complexes.[163] The Pd(II) catalyst is characterized by exclusive [3,3] regioselectivity, while the Pd(0), Rh(I), and Ir(I) catalyst generally give both [3,3] and [1,3] rearrangements. After a series of experiments using chiral substrates and substrates with specific deuterium labels, the mechanism of these reactions were elucidated. The Pd(II) reaction is proposed to proceed via a cyclic carbonium ion intermediate, while the Pd(0) catalysis is a form of catalytic allylation involving oxidative addition followed by nucleophilic attack on the π-allyl intermediate. In another report the metal-catalyzed Claisen rearrangement of allylic thionobenzoates was investigated.[164] The Pd(II) rearrangement of the allylic thionobenzoates parallels the study with the allyl imidates. The Pd(II) catalyst may be the most useful synthetically, because of the high turnover rate and greater steric control implicit in the cyclization mechanism. The Pd(II) catalyst produces exclusively [3,3] regioselectivity.

14.7.6. Aldol Condensations

Catalytic asymmetric aldol reactions were accomplished in high enantio-selectivity (>90% ee) and high *trans* selectivity (>97%) using a chiral ferro-cenylphosphine-gold(I) catalyzed reaction of secondary aldehydes with an isocyanoacetate [equation (27)].[165] The length of the amino side-chain on

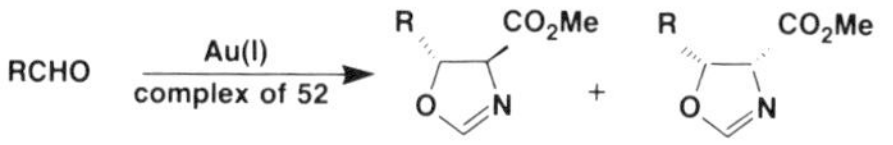

$$RCHO \xrightarrow[\text{complex of 52}]{Au(I)} \quad \begin{array}{c} R \quad CO_2Me \\ \diagdown \diagup \\ O \diagdown N \end{array} \quad + \quad \begin{array}{c} R \quad CO_2Me \\ \diagdown \diagup \\ O \diagdown N \end{array} \tag{27}$$

(**52**) appears to be critical in obtaining high optical yields. Gold is also essential for high selectivity; silver or copper analogues are less active. This may be due to the stronger affinity of gold(I) to complex with phosphorous ligands.

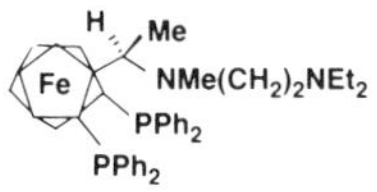

52

A new aldol-type carbon–carbon bond forming reaction was obtained by the reaction of vinyl ketones with aldehydes using a catalytic amount of [HRh(PPh$_3$)$_4$].[166] It is postulated that reaction of the Rh hydride with the vinyl ketone results in the formation of a rhodium(I) enolate, which reacts with an aldehyde to form the coupled product as in equation (28).

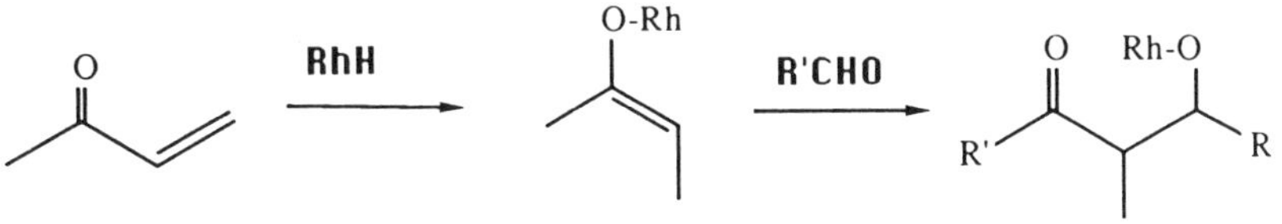

$$\tag{28}$$

14.8. Hydrosilylations

The hydrosilylation of ketones to yield the corresponding silyl ether alcohols is an important reaction, especially as this has been demonstrated to be an effective route to chiral alcohols via a catalytic asymmetric hydrosilyla-tion. This approach has been extended to oximes using *in situ* generated catalysts derived from [Rh(cod)Cl]$_2$ and optically active phosphines.[167] The resultant optically active silyl amines can be hydrolyzed to chiral amines. The mechanism proceeds by formation of *syn–anti* silyloxime isomers as detected by ^{1}H NMR (Scheme 9). The silylamine is also observed by ^{1}H NMR and is proposed to form by silylation of the coordinated nitrene.

Another interesting application of hydrosilylation of ketones is the Pd-catalyzed reduction of α,β-unsaturated ketones to yield the saturated ketone.[168] The catalyst comprises a soluble Pd source, e.g., [Pd(PPh$_3$)$_4$], and a Lewis acid such as ZnCl$_2$. An intermediate Pd(0) olefin complex is formed

Scheme 9

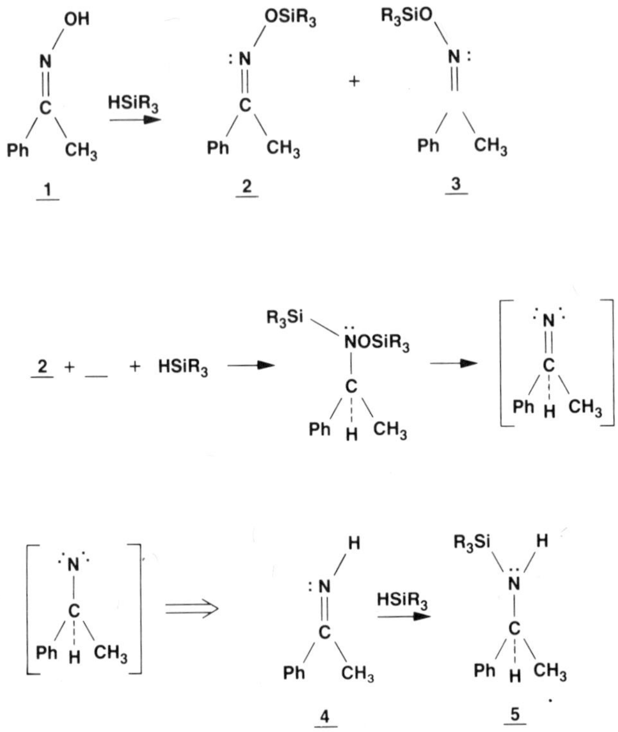

in a rapid equilibrium as demonstrated by spin saturation transfer in the NMR spectrum of $[Pd(PPh_3)_4]$ and benzalacetone. A silyl enol intermediate is observed and is proposed to be formed via a route involving 1,4-addition followed by reductive elimination. The Lewis acid catalyzed hydrolysis of the silyl enol ether yields the ketone product. The hydrosilylation of olefins and alkynes can be accomplished catalytically using $[Rh(CO)_2Cl]_2$ and diazadiene ligands. In stoichiometric control studies complexes of type (**53**) are formed,[169] and these are effective catalysts.

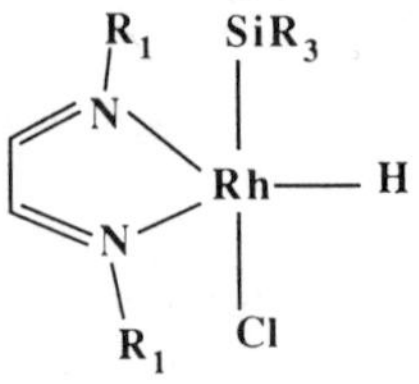

53

Addition of an olefin or alkyne to this coordinatively unsaturated Rh(III) species, followed by insertion into the Rh—H bond, yields a vinyl or alkyl species which can reductively eliminate to yield the observed (for internal alkynes) *cis*-addition product. Iridium complexes have also been cited for hydrosilylation of olefins and a recent study on the oxidative addition of hydrosilanes to [IrBr(CO)(dppe)]-type complexes reveals that addition proceeds stereoselectively under kinetic control with only one of four possible *cis*-addition products formed, structure (**54**), although this is not the thermodynamic product.[170] It is proposed that this product results from a nucleophilic

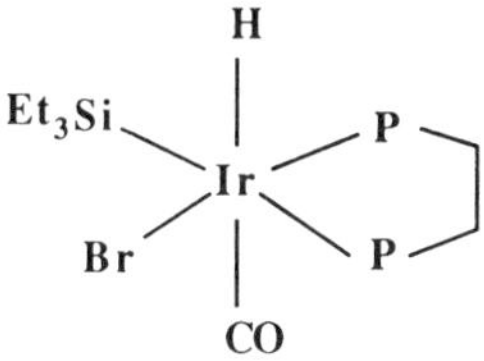

54

silane addition. This study has obvious implications for Rh-catalyzed hydrosilylations, especially asymmetric hydrosilylations. Finally, a recent study has shown that the Pt-catalyzed hydrosilylation of olefins using [Pt(COD)$_2$] and HSiEt$_3$ is not a true homogeneous system but proceeds by silane reduction of the COD ligands to generate Pt colloids, which are the active hydrosilylation catalysts.[171] The authors conclude that in cases where the reduction of both ligand and metal are possible, colloid catalysis should be considered.

14.9. Hydrocyanations

The addition of HCN to unsaturated compounds is an important reaction with wide applicability in organic synthesis. The production of such important monomers as adiponitrile for the polymer industry via the nickel-catalyzed HCN addition to butadiene (DuPont) constitutes the most significant application. The mechanism of the HCN addition to ethylene catalyzed by [Ni(P(*o*-tolyl)$_3$)$_4$] was studied via ^{31}P, ^{1}H, and ^{13}C NMR at low temperatures.[172] In the absence of HCN only signals due to [(CH$_2$CH$_2$)Ni(L)$_2$] are evident, but addition of HCN generates quantitatively a new species whose structure was identified by ^{13}C, ^{31}P, and ^{1}H using labeled H^{13}CN and ^{13}CH$_2$CH$_2$ to be a mixture of the three possible isomers [(C$_2$H$_4$)NiL(Et)(CN)]. Magnetization transfer techniques were applied to the study of the phosphine exchange process and evidence suggests that an unobserved five-coordinate species [(C$_2$H$_4$)NiL$_2$(Et)(CN)] is formed. This intermediate then reductively eliminates propanenitrile and regenerates [(C$_2$H$_4$)NiL$_2$] as in Scheme 10. At higher concentrations of L additional nonproductive equilibria involving Ni species [(C$_2$H$_4$)NiL$_3$], [NiL$_4$], etc. are present. This work defines a well-characterized

Scheme 10

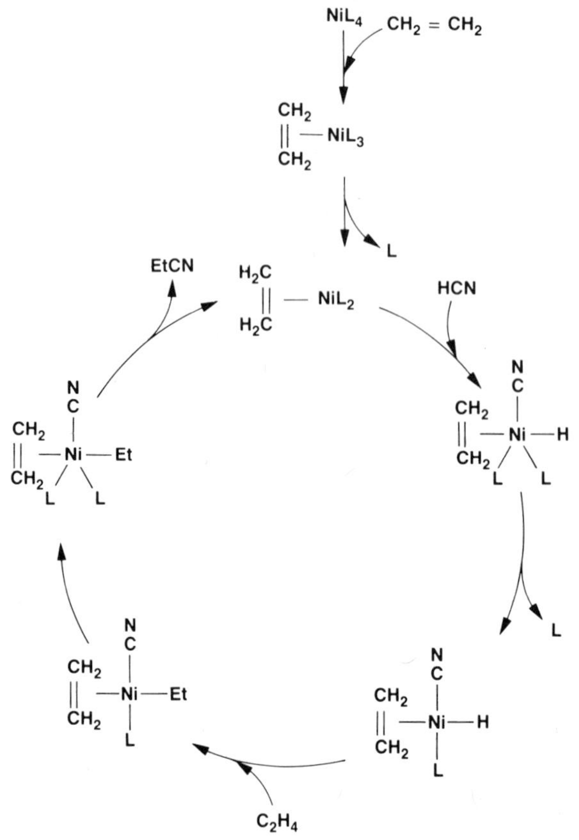

associative pathway for reductive elimination. The stereochemistry of the Ni-catalyzed addition of HCN to 3,3-dimethyl-1-butene and 1,3-cyclo-hexadiene has been investigated using DCN.[173] The formation of *erythro*-2,3-dideuterio-4,4-dimethylpentanenitrile from DCN addition to (*E*)-1-deuterio-2,3-dimethyl-1-butene shows that DCN adds *cis* to the double bond. Hydrogen–deuterium exchange was observed between olefin and DCN and a large deuterium isotope effect (about 6.8) was noted. These observations were rationalized in terms of the mechanism shown in Scheme 11 involving a *cis*-addition of a Ni hydride to a double bond followed by attack of a coordinated cyanide at the thus formed σ-bound hydrocarbon ligand. The observation that DCN addition to 1,3-cyclohexadiene occurs both (1,2) and (1,4) with *cis*-stereochemistry is consistent with this mechanistic picture, except that an intermediate (π-allyl)-nickel cyanide complex produces the observed products in equimolar amounts. The use of Cu catalysts for hydrocyanation

Scheme 11

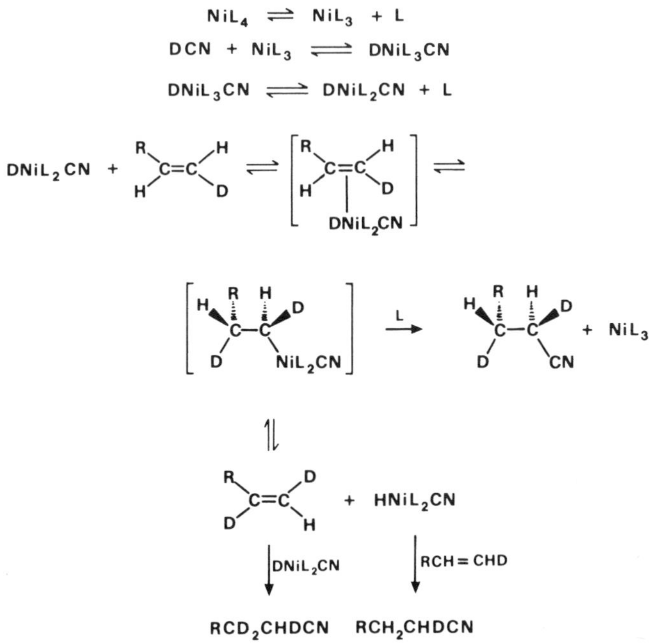

$$NiL_4 \rightleftharpoons NiL_3 + L$$

$$DCN + NiL_3 \rightleftharpoons DNiL_3CN$$

$$DNiL_3CN \rightleftharpoons DNiL_2CN + L$$

is known, but has been less well studied. Aspects of this Cu-catalyzed reaction bearing upon the mechanism have been reported using Cu(I)Br.[174] The reaction is believed to proceed through the intermediacy of Cu(π-allyl) species. Finally, a review of the hydrocyanation of butadiene from a DuPont perspective has been published.[175]

14.10. *C—H Activation*

The use of low-valent transition metals to activate aliphatic and aromatic C—H bonds has been the topic of many recent reviews.[176] It has been reported that $[Ir(i\text{-}Pr_3P)_2H_5]$ is an efficient catalyst for the selective activation of the *trans* vinyl C—H bonds of 3,3-dimethylbutene.[177] A deuterium-exchange study reveals a remarkable selectivity of the *trans* vinyl C—H bond, and the postulated mechanism in equation (29). These results suggest that vinyl C—H activation may be more common than previously recognized. The complex

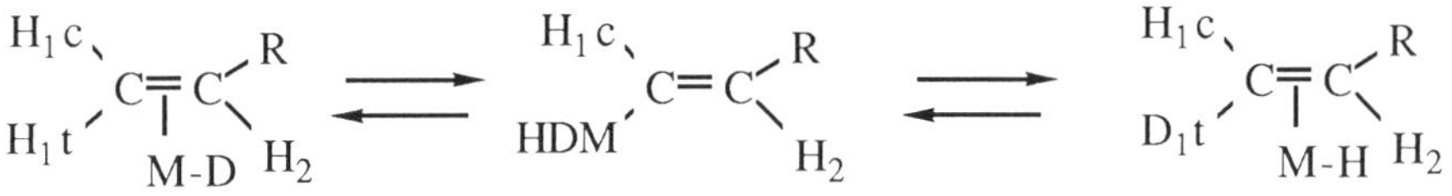

$$(29)$$

$[(\eta^6\text{-}C_6H_6)Re(PPh_3)H]$ (55), formed by photochemical loss of phosphine, is capable of catalytic H/D exchange between benzene-d_6 and other arenes in solution as well as with *meta* and *para* hydrogens of the coordinated phosphine ligands.[178] Labeling studies show that a simple oxidative addition mechanism is not operating in these exchanges. This system is unique in that all other arene H/D exchange catalysts also exchange deuterium into the metal hydride position.

The kinetic and equilibrium isotope effect involved in arene C—H activation by the intermediate $[(C_5Me_5)Rh(PMe_3)]$ confirm[179] the intermediacy of an η^2-arene intermediate (55) as in equation (30). The reductive elimination

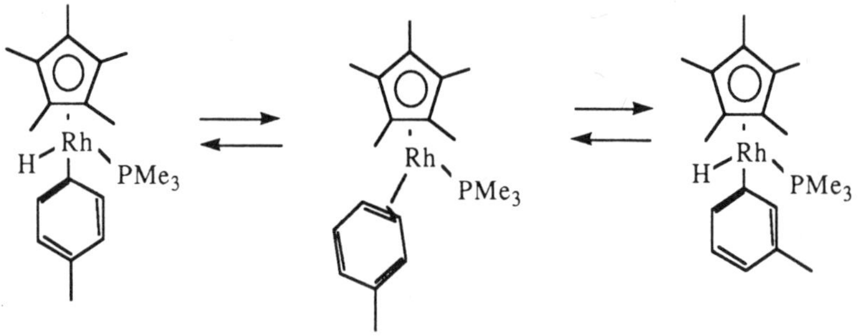

(30)

55

of benzene was found to have an inverse isotope effect and can be accommodated by a transition state for reductive elimination in which substantial C—H bonding is present.

There have been two reports on the catalytic activation of methane using soluble transition-metal complexes. The intermediate complex $[(Cp)Re(PPh_3)H_2]$, **56**, which is formed by photochemical loss of phosphine, is capable of catalyzing H/D exchange of benzene, THF, and alkanes including methane.[180] The catalytic H/D exchange with $[(Cp)Re(PPh_3)H_2]$ represents C—H activation involving an intermediate that is neither electrophilic or nucleophilic and probably utilizes a Re(III)/Re(V) couple. It was also reported that the soluble iridium pentahydride $[(i\text{-}Pr)_2IrH_5]$ (activated by neohexene) catalyzes H–D exchange between benzene-d_6 and CH_4 under mild conditions.[181] The conditions required for catalytic H–D exchange between methane and C_6D_6 are more vigorous than those used in the deuteration of *trans*-vinylic hydrogens. Thermodynamic and kinetic aspects of the intramolecular cyclometalation process and its intermolecular microscopic reverse were obtained by studying[182] reaction (31). It was found that the

$$[Cp_2'Th(CH_2SiMe_3)_2] \rightleftharpoons [Cp_2'\overline{ThCH_2SiMe_2CH_2}] + SiMe_4 \qquad (31)$$

intramolecularly metalated product is thermodynamically more stable than the corresponding intermolecular product because $T\Delta S > \Delta H > 0$.

Part 4

Compilations of
Numerical Data

Chapter 15

Volumes of Activation for Inorganic and Organometallic Reactions: A Tabulated Compilation

15.1. Introduction

As in Volume 4 of this series,[1] this chapter is devoted to a compilation of activation and reaction volume data for inorganic and organometallic reactions published during the period July 1985 to December 1986. The application of high-pressure techniques in the elucidation of reaction mechanisms of coordination compounds has received increasing support from various groups. The additional kinetic parameter of pressure not only adds a further dimension to mechanistic studies, but also completes the comprehension of reaction kinetics which must be accounted for by the suggested reaction mechanism. The past 18 months were highlighted by a number of important contributions at international conferences,[2] a Nato Advanced Study Institute on Advances in High Pressure Studies of Chemical and Biochemical Systems,[3] a few review articles,[4-8] and a monograph entitled *Inorganic High Pressure Chemistry: Kinetics and Mechanisms*[9] that covers all the work performed on inorganic, organometallic, and bioinorganic systems up to the end of 1985. Especially noteworthy is the increased role played by high-pressure techniques in synthetic inorganic chemistry,[6] in biochemistry,[3,9] and in the study of photochemical and photophysical processes.[9,10]

A number of developments in closely related areas should be mentioned in a general way. Recently developed techniques and instrumentation include: EXAFS measurements under high pressure,[11] effect of high pressure on electron mobility,[12] high-pressure FT–IR studies of catalytic surface species,[13] rotational reorientation dynamics at high pressure,[14] micellar microfluidities at high pressure,[15] high-pressure Raman spectroscopy,[16] and the application of high-pressure absorption spectroscopy (UV–VIS and IR) in polymerizing systems.[17] The application of high pressure in the tuning of electronic spectra in the liquid and solid state has received attention from various groups.[3,18-21] In a similar way the study of biochemical systems under pressure has revealed interesting effects.[3,9,22-26] A number of theoretical aspects have received attention: Gibson's concept of an excess pressure was applied to an analysis of the properties of solutions containing both salts and neutral solutes in terms of partial and apparent molar volumes,[27] a discussion of activation parameters for chemical reactions in solution,[28] analysis of volumes of activation in terms of isochoric and isobaric behavior,[29] and theoretical calculation of $\Delta V^{\ddagger}$ for outer-sphere electron-transfer reactions.[30] High-pressure techniques were also applied to study the mechanism of thermal transformations of organosilicon compounds.[31] Whalley *et al.* developed two techniques to study chemical kinetics at extremely high pressures of up to 10 000 MPa, including a carbide-anvil apparatus and a piston-cylinder apparatus.[32] Yoshimura and Nakahara[33-36] have undertaken a series of studies to investigate the additivity rule of partial molal volumes.

Of particular significance is the high-pressure kinetic work performed in an effort to modify the transition-state theory.[3,9] The mechanistic interpretation of activation parameters ($\Delta H^{\ddagger}$, $\Delta S^{\ddagger}$, and $\Delta V^{\ddagger}$) is usually based on a simplified version of the transition-state theory in which it is assumed that the transmission coefficient is independent of temperature and pressure and close to unity. Stochastic models have been introduced to account for the deviation of the transmission coefficient from unity, the original ideas being introduced by Kramers.[37] Jonas *et al.*[38,39] used the pressure dependence of the transmission coefficient to define a collisional contribution toward the experimentally determined $\Delta V^{\ddagger}$, and so account for coupling effects with the medium (solvent). Troe and co-workers[40-44] follow an alternative approach in which they realize that simplified transition-state theory treatment neglects the transport aspect of the process. The transport contribution toward $\Delta V^{\ddagger}$ is defined as $RT(\partial \ln \eta / \partial P)_T$ (where η is the viscosity of the medium), and accounts for the dynamical friction of the medium. It may be concluded that the described modifications are especially of significance in cases where the reaction exhibits a strong dependence on the nature of the solvent and the viscosity of the medium. The mentioned modifications are first efforts to account for the observed effects and further developments will surely be forthcoming.[45]

15.2. Data in Tabular Form

The data in the following tables have been arranged as in the past.[1] A similar sequence of reaction types, and ordering according to atomic number of the central metal atom, have been adopted. In contrast to the previous tabulation, some reaction volume data are included for cases where no $\Delta V^{\ddagger}$ values are available. This is mainly done to illustrate the magnitude of the overall volume changes that occur and that could be expected for a certain type of reaction, assisting the interpretation of $\Delta V^{\ddagger}$ data for related systems. The methods employed to determine $\Delta \bar{V}$ are: (a) from the pressure dependence of the equilibrium constant; (b) from dilatometric or partial molar volume measurements; (c) from kinetic data. Other general remarks are: $\Delta V^{\ddagger}$ data are quoted at ambient pressure—in case of significant curvature in the ln k versus P plots the compressibility coefficient $(\Delta \beta^{\ddagger})$ is also given; no. of data refers to the different pressures at which kinetic measurements (usually more than one experiment) were performed, with the maximum applied pressure being quoted in the fourth column; concentration is given in mol liter^{-1} (M) or mol kg^{-1} (m); a list of abbreviations precedes the tabulated data.

Abbreviations used in the Tables

acac	acetylacetone
bipy	2,2'bipyridine
BTE	1,2-bis(tert-butylthio)ethane
def	N,N-diethylformamide
dien	diethylenetriamine
dto	3,6-dithiaoctane
edta	ethylenediaminetetraacetate
en	ethylenediamine
Hipt	4-isopropyltropalone
Im	imidazole
Me$_6$tren	2,2'-2''-tris(dimethylamino)triethylamine
phen	1,10-phenanthroline
py	pyridine
terpy	2,2':6',2''-terpyridine
2,3,2-tet	1,3-bis-(2'-aminoethylamino)propane
tmpyp	5,10,15,20-tetrakis(4-N-methylpyridiniumyl)porphine
tn	trimethylenediamine
tptdt	2,3,9,10-tetraphenyl-1,4,8,11-tetraaza-1,3,8,10-cyclo-tetradecatetraene
tpyp	5,10,15,20-tetra-4-pyridylporphine
trien	triethylenetetramine

Table 15.1. Volumes of Activation

Reaction	Solvent	T (°C)	P (kbar)	No. of data	$\Delta V^{\ddagger}$ (cm³ mol⁻¹)	$\Delta \beta^{\ddagger}$ (cm³ mol⁻¹ kbar⁻¹)	$\Delta \bar{V}$ (cm³ mol⁻¹)	Method	Ref.	Remarks
Solvent exchange										
$[Al(acac)_3] + acac^-$	Hacac	25	2.4	7	+10.2 ± 1.8				45	spontaneous reaction
$[Co(H_2O)_6]^{2+} + H_2O$	H_2O	83	2.4	14	+5.0 ± 0.4				46	H₂O catalyzed reaction
		72	2.4	14	+5.5 ± 0.6					
					+5.2 ± 0.7					
$[Co(Me_6tren)def]^{2+} + def$	def	92	2.5	13	−1.3 ± 0.2				47	
$[Cu(Me_6tren)def]^{2+} + def$	def	92	2.0	11	+5.3 ± 0.3				47	
$[Cu(MeOH)_6]^{2+} + MeOH$	MeOH	−62	2.0	11	+8.3 ± 0.4				48	
Ligand exchange										
$[SbCl_5 \cdot L] + L^* \rightleftharpoons [SbCl_5 \cdot L^*] + L$	CH_2Cl_2		2.0	ca. 10					49	
$L = Cl_2(Me_2N)PO$		1			+23.0 ± 0.6					
Me_2O		0			+27.2 ± 1.4	+5.9 ± 1.0				
Me_2CO		−9			+28.1 ± 2.0	+8.1 ± 1.9				
Et_2O		−10			+30.0 ± 1.5	+3.8 ± 1.4				
Me_3CCN		−37			+18.2 ± 0.9	+3.2 ± 0.6				
$MeCN$		−19			+24.7 ± 1.7	+4.1 ± 1.8				
Ligand substitution										
$Al^{3+} Hipt \rightarrow [Al(ipt)]^{2+} + H^+$	Me_2SO	40	1.2	6	+12.2 ± 1.0				50	$\mu = 0.19\ m$
$[TiO(tmpyp)]^{4+} + H_2O_2 \rightarrow [Ti(O_2)(tmpyp)]^{4+} + H_2O$	H_2O	25	1.2	7	−18.6 ± 0.3				51	$\mu = 1.0\ M$
$[Cr(CO)_4(S{\frown}S)] + 2P(OEt)_3 \rightarrow$ $[Cr(CO)_4(P(OEt)_3)_2] + S{\frown}S$	$(CH_2Cl)_2$				−3.9 ± 0.2				52	spontaneous reaction
$S{\frown}S = dto$		55	1.5	5	+14.7 ± 0.7					
BTE		45	1.5	5	+14.0 ± 0.6					
$trans\text{-}[Cr(en)_2Br_2]^+ + H_2O \rightarrow$ $trans\text{-}[Cr(en)_2(H_2O)Br]^{2+} + Br^-$	H_2O	15	2.0	5	+2.1				53	acid-catalyzed reaction
		20			+2.2					
		25			+2.2					
		30			+2.3					

Reaction	Solvent	T (°C)	P	n	ΔV‡	ΔV		Ref	Conditions
$trans\text{-}[Cr(tn)_2Br_2]^+ + H_2O \rightarrow$ $trans\text{-}[Cr(tn)_2(H_2O)Br]^{2+} + Br^-$	H_2O	15, 20, 25, 30	2.0	5	+1.4, +1.7, +1.8, +1.9			53	
$[Mn(H_2O)_6]^{2+} + terpy \rightarrow$ $[Mn(terpy)(H_2O)_3]^{2+} + 3H_2O$	H_2O	25	1.0	5	-1.3 ± 0.3, -3.4 ± 0.7			54	$\mu = 0.1\ M$, [L] ≫ [M]; $\mu = 0.1\ M$, [M] ≫ [L]
$[Mn(terpy)(H_2O)_3]^{2+} + terpy \rightarrow$ $[Mn(H_2O)_6]^{2+} + terpy$	H_2O	25	1.0	5	-7.7 ± 2.2			54	$\mu = 0.1\ M$
$[Mn(H_2O)_6]^{2+} + edta^{4-} \rightarrow$ $[Mn(edta)]^{2-} + 6H_2O$	H_2O	25				+32.8	b	55	$\mu \rightarrow 0$
$[Mn(H_2O)_6]^{2+} + Hedta^{3-} \rightarrow$ $[Mn(edtaH)(H_2O)]^- + 5H_2O$	H_2O	25				+25.2	b	55	$\mu \rightarrow 0$
$[Fe(H_2O)_6]^{3+} + SCN^- \rightarrow$ $[Fe(H_2O)_5SCN]^{2+} + H_2O$	H_2O	25	1.2	4	$+4.3 \pm 0.6$	$+8.2 \pm 0.4$	a	56	$\mu = 1.0\ M$
$[Fe(H_2O)_5OH]^{2+} + SCN^- \rightarrow$ $[Fe(H_2O)_4(OH)SCN]^+ + H_2O$	H_2O	25	1.2	4	$+5.4 \pm 1.5$	$+17.0 \pm 0.9$	a	56	$\mu = 0.1\ M$
$[Fe(H_2O)_6]^{2+} + terpy \rightarrow$ $[Fe(terpy)(H_2O)_3]^{2+} + 3H_2O$	H_2O	25	1.0	5	$+3.4 \pm 0.6$, $+3.7 \pm 0.8$			54	$\mu = 0.1\ M$, [L] ≫ [M]; $\mu = 0.1\ M$, [M] ≫ [L]
$[Fe(tptdt)(MeCN)_2]^{2+} + Im \rightarrow$ $[Fe(tptdt)(Im)(MeCN)]^{2+} + MeCN$	MeCN	25	2.0	9	$+13.7 \pm 0.9$			57	
$[Fe(tptdt)(MeCN)_2]^{2+} + SCN^- \rightarrow$ $[Fe(tptdt)(SCN)(MeCN)]^+ + MeCN$	MeCN	25	2.0	9	$+10.1 \pm 0.2$			57	
$[Co(H_2O)_6]^{2+} + bipy \rightarrow$ $[Co(bipy)(H_2O)_4]^{2+} + 2H_2O$	H_2O	25	1.0	5	$+4.3 \pm 1.0$, $+7.5 \pm 1.4$			54	$\mu = 0.1\ M$, [L] ≫ [M]; $\mu = 0.1\ M$, [M] ≫ [L]
$[Co(H_2O)_6]^{2+} + terpy \rightarrow$ $[Co(terpy)(H_2O)_3]^{2+} + 3H_2O$	H_2O	25	1.0	5	$+4.5 \pm 0.8$, $+3.8 \pm 0.8$			54	$\mu = 0.1\ M$, [L] ≫ [M]; $\mu = 0.1\ M$, [M] ≫ [L]
$[Co(terpy)_2]^{2+}$ (low spin) $\rightleftharpoons$ $[Co(terpy)_2]^{2+}$ (high spin)	H_2O	25	1.4	5		$+10.1 \pm 0.4$	a	58	
$[Co(H_2O)_6]^{2+} + edta^{4-} \rightarrow$ $[Co(edta)]^{2-} + 6H_2O$	H_2O	25				+44.1	b	55	$\mu \rightarrow 0$
$[Co(H_2O)_6]^{2+} + Hedta^{3-} \rightarrow$ $[Co(edtaH)(H_2O)]^- + 5H_2O$	H_2O	25				+34.4	b	55	$\mu \rightarrow 0$
$[Co(edta)]^{2-} + H^+ \rightarrow [Co(edtaH)]^-$	H_2O	25				$+3.5 \pm 0.2$	b	59	
$[Co(NH_2Me)_5L]^{3+} + H_2O \rightarrow$ $[Co(NH_2Me)_5H_2O]^{3+} + L$ $L = Me_2SO$	H_2O	23	1.7	4	$+5.9 \pm 0.2$			60	0.1 M $HClO_4$
DMF		38			$+6.3 \pm 0.3$				
MeCN		45			$+6.0 \pm 0.5$				

(continued)

Table 15.1 (*continued*)

Reaction	Solvent	T (°C)	P (kbar)	No. of data	$\Delta V^{\ddagger}$ (cm³ mol⁻¹)	$\Delta \beta^{\ddagger}$ (cm³ mol⁻¹ kbar⁻¹)	$\Delta \bar{V}$ (cm³ mol⁻¹)	Method	Ref.	Remarks
trans-[Co(AA)₂Cl₂]⁺ + H₂O →	H₂O								61	
trans-[Co(AA)(H₂O)Cl]²⁺ + Cl⁻										
AA = en		25	2.0	5	+0.8					
N-Eten		25	1.5	4	+0.3					
N-Meen		40	1.5	4	+6.0					
tn		5	1.5	4	+7.3					
cis-[Co(en)₂(NO₂)Cl]⁺ + H₂O →	H₂O	35	1.4	5	−2.9 ± 0.3				62	0.1 M HClO₄
cis-[Co(en)₂(NO₂)H₂O]²⁺ + Cl⁻										
trans-[Co(en)₂(CN)Cl]⁺ + H₂O →	H₂O	35	1.4	5	−2.0 ± 0.4				62	0.1 M HClO₄
trans-[Co(en)₂(CN)H₂O]²⁺ + Cl⁻										
cis-α-[Co(trien)Cl₂]⁺ + H₂O →	H₂O	27	1.4	5	−5.0 ± 0.4				62	0.1 M HClO₄
cis-α-[Co(trien)(Cl)H₂O]²⁺ + Cl⁻										
cis-β-[Co(trien)Cl₂]⁺ + H₂O →	H₂O	25	1.4	5	−2.0 ± 0.6				62	0.1 M HClO₄
cis-β-[Co(trien)(Cl)H₂O]²⁺ + Cl⁻										
cis-[Co(NH₃)₄(H₂O)X]²⁺ + H₂O →	H₂O								63	
cis-[Co(NH₃)₄(H₂O)₂]³⁺ + X⁻										
X = Cl		55	2.0	5	−4.3 ± 0.2		−11.2	b		$\mu = 0.5\ M$
							−12.1	b		$\mu = 0.03\ M$
Br		43	2.0	5	−3.5 ± 0.6		−12.4	b		$\mu = 0.5\ M$
							−11.8	b		$\mu = 0.04\ M$
NO₃		37	1.6	5	−2.1 ± 0.1		−9.7	b		$\mu = 0.5\ M$
							−9.0	b		$\mu = 0.01\ M$
cis-[Co(NH₃)₄(OH)X]⁺ + H₂O →	H₂O								63	
cis-[Co(NH₃)₄(H₂O)OH]²⁺ + X⁻										
X = Cl		55	2.0	5	+3.3 ± 0.3					
Br		43	2.0	5	+4.9 ± 0.7					
NO₃		37	1.6	5	+5.4 ± 0.5					
[Co(NH₃)₄(NH₂X)Cl]²⁺ + OH⁻ →	H₂O	25	1.0						64	$\mu = 0.013\ M$
[Co(NH₃)₄(NH₂X)OH]²⁺ + Cl⁻										
X = cis, Me					+29.4 ± 0.4		+9.8	b		
trans, Me					+28.6 ± 1.3		+10.4	b		
trans, Et					+28.3 ± 1.4		+10.1	b		
cis, n-Pr					+26.4 ± 1.5		+11.4	b		

trans, n-Pr →					+29.9 ± 1.2				
trans, n-Bu					+28.7 ± 0.7				
trans, i-Bu					+28.5 ± 1.2				
$[Ni(H_2O)_6]^{2+}$ + bipy →	H_2O	25	1.0	5	+5.5 ± 0.3	+11.9	b	54	$\mu = 0.1\ M$, [L] ≫ [M]
$[Ni(bipy)(H_2O)_4]^{2+}$ + $2H_2O$					+5.1 ± 0.4	+12.0	b	54	$\mu = 0.1\ M$, [M] ≫ [L]
$[Ni(H_2O)_6]^{2+}$ + terpy →	H_2O	25	1.0	5	+6.7 ± 0.2			54	$\mu = 0.1\ M$, [L] ≫ [M]
$[Ni(terpy)(H_2O)_3]^{2+}$ + $3H_2O$					+4.5 ± 0.6			54	$\mu = 0.1\ M$, [M] ≫ [L]
Ni^{2+} + L^{2-} → [NiL]	H_2O	20	1.0	6				65	$\mu \to 0$
L = malate					+13.8 ± 0.6	+17.8 ± 0.6	a		
tartrate					+13.7 ± 1.2	+17.8 ± 0.5	a		
[NiL] → Ni^{2+} + L^{2-}								65	calculated
L = malate					−4.0 ± 0.5				
tartrate					−4.1 ± 0.7				
$[Ni(H_2O)_6]^{2+}$ + $edta^{4-}$ → $[Ni(edta)]^{2-}$ + $6H_2O$	H_2O	25				+44.4	b	55	$\mu \to 0$
$[Ni(H_2O)_6]^{2+}$ + $Hedta^{3-}$ → $[Ni(edtaH)(H_2O)]^-$ + $5H_2O$	H_2O	25				+33.7	b	55	$\mu \to 0$
$[Ni(edta)]^{2-}$ + H^+ → $[Ni(edtaH)]^-$	H_2O	25				+3.3 ± 0.3	b	59	$\mu \to 0$
$[Ni(2,3,2\text{-tet})]^{2+}$ (high spin) ⇌ $[Ni(2,3,2\text{-tet})]^{2+}$ (low spin)	H_2O	57				+3.1 ± 0.3	b	46	$\mu = 0.25\ M$
		62				+2.5 ± 0.1	b		$\mu = 4.2\ M$
		45				+2.1 ± 0.1	b		$\mu = 4.2\ M$
$[Cu(H_2O)_6]^{2+}$ + $edta^{4-}$ → $[Cu(edta)]^{2-}$ + $6H_2O$	H_2O	25				+45.8	b	55	$\mu \to 0$
$[Cu(H_2O)_6]^{2+}$ + $Hedta^{3-}$ → $[Cu(edtaH)(H_2O)]^-$ + $5H_2O$	H_2O	25				+36.5	b	55	$\mu \to 0$
$[Cu(edta)]^{2-}$ + H^+ → $[Cu(edtaH)]^-$	H_2O	25				+4.8 ± 0.2	b	59	$\mu \to 0$
$[Cu(edtaH)]^-$ + H^+ → $[Cu(edtaH_2)]$	H_2O	25				+8 ± 1	b	59	$\mu \to 0$
$[Zn(H_2O)_6]^{2+}$ + $edta^{4-}$ → $[Zn(edta)]^{2-}$ + $6H_2O$	H_2O	25				+44.4	b	55	$\mu \to 0$
$[Zn(H_2O)_6]^{2+}$ + $Hedta^{3-}$ → $[Zn(edtaH)(H_2O)]^-$ + $5H_2O$	H_2O	25				+33.6	b	55	$\mu \to 0$
Ga^{3+} + Hipt → $[Ga(ipt)]^{2+}$ + H^+	Me_2SO	35	1.2	7	+10.6 ± 0.6			50, 66	$\mu = 0.27\ m$
$[Mo(CO)_5py]$ + phen → $[Mo(CO)_4(phen)]$ + CO + py	PhMe	50	1.5	4	+0.5 ± 0.5			67	
$[Mo(CO)_5(4\text{-Mepy})]$ + bipy → $[Mo(CO)_4(bipy)]$ + CO + 4-Mepy	PhMe	40	1.0	3	+0.9 ± 0.5			67	
$[Mo(CO)_5(4\text{-Mepy})]$ + phen → $[Mo(CO)_4(phen)]$ + CO + 4-Mepy	PhMe	50	1.0	3	−0.3 ± 0.7			67	
	PhMe	40	1.0	3	+0.1 ± 0.5			67	

(continued)

Table 15.1 (continued)

Reaction	Solvent	T (°C)	P (kbar)	No. of data	$\Delta V^{\ddagger}$ (cm³ mol⁻¹)	$\Delta \beta^{\ddagger}$ (cm³ mol⁻¹ kbar⁻¹)	$\Delta \bar{V}$ (cm³ mol⁻¹)	Method	Ref.	Remarks
cis-[Mo(CO)₄(py)₂] + bipy → [Mo(CO)₄(bipy)] + 2py	(CH₂Cl)₂	15	2.0	5	+4.5 ± 0.6				67	
cis-[Mo(CO)₄(py)₂] + phen → [Mo(CO)₄(phen)] + 2py	(CH₂Cl)₂	15	2.0	5	+4.1 ± 0.5				67	
[MoO(tmpyp)H₂O]⁵⁺ + H₂O₂ → [MoO(O₂)(tmpyp)]³⁺ + 2H⁺ + H₂O	H₂O	25	1.8	7	−0.2 ± 0.3				68	$\mu = 1.05\ m$
[MoO(O₂)(tmpyp)]³⁺ + H⁺ + H₂O → [MoO(tmpyp)H₂O]⁵⁺ + H₂O₂	H₂O	25	1.8	7	+5.2 ± 1.6				68	$\mu = 1.05\ m$
Mo₃O₄(H₂O)₉]⁴⁺ + NCS⁻ → [Mo₃O₄(H₂O)₈NCS]³⁺ + H₂O	H₂O	15	1.5	4	+6.1 ± 0.4				69	
		20	1.5	4	+5.5 ± 0.1					
		25	1.5	4	+5.4 ± 0.2					
		30	1.5	4	+4.4 ± 0.2					
[Pd(L)X]^(2−n)+ + H₂O → [Pd(L)H₂O]²⁺ + X^{n−}	H₂O									
L = dien, X^{n−} = CO₃²⁻		25	1.0	5	−9.7 ± 0.4				70	$\mu = 1.0\ M$
Me₄dien CO₃²⁻		25	1.0	5	−8.6 ± 0.4				70	$\mu = 1.0\ M$
Et₄dien CO₃²⁻		25	1.5	7	−6.7 ± 0.5				70	$\mu = 1.0\ M$
L = MeEt₄dien Cl⁻		25	1.7	7	−12.2 ± 0.6				71	$\mu = 0.1\ M$
		40	1.0	5	−13.2 ± 1.0				71	$\mu = 0.1\ M$
		30	1.5	7	−13.5 ± 1.8				71	$\mu = 1.0\ M$
I⁻		30	1.5	7	−8.4 ± 0.6				71	$\mu = 0.1\ M$
Br⁻		25	1.5	7	−11.9 ± 0.5				71	$\mu = 0.1\ M$
py		50	1.5	4	−3.7 ± 0.4				71	$\mu = 0.1\ M$
NH₃		50	1.5	6	−3.0 ± 0.4				71	$\mu = 0.1\ M$
C₂O₄²⁻		14	1.5	6	−10.6 ± 0.2				71	$\mu = 0.1\ M$
Cl⁻		25	1.5	7	−11.6 ± 0.2				71	$\mu = 0.1\ M$
Et₅dien										
[Cd(H₂O)₆]²⁺ + bipy → [Cd(bipy)(H₂O)₄]²⁺ + 2H₂O	H₂O	0	2.0	5	−5.5 ± 1.0		+2.0 ± 0.8	a	72	
[Cd(bipy)(H₂O)₄]²⁺ + 2H₂O → [Cd(H₂O)₆]²⁺ + bipy	H₂O	0	2.0	5	−6.9 ± 1.2				72	
In³⁺ + Hipt → [In(ipt)]²⁺ + H⁺	Me₂SO	35	1.2	6	−0.1 ± 0.6				50	$\mu = 0.05\ m$
	DMF	16	2.0	9	+0.3 ± 0.3				50	$\mu = 0.05\ m$

Reaction	Solvent	T (°C)					Ref.		Conditions
$[W(H_2O)(CN)_4]^{2-} + N_3^- \rightarrow$ $[W(N_3)(CN)_4]^{3-} + H_2O$	H_2O	25	1.0	5	$+10.6 \pm 0.5$		73		$\mu = 1.0\ M$, pH $= 5.8$
Isomerization									
$[Co(en)_2(OS(O)CH_2CH_2NH_2)]^{2+} \rightarrow$ $[Co(en)_2(SO_2CH_2CH_2NH_2)]^{2+}$	H_2O	60	1.5	4	-9.0 ± 0.7		74		$0.01\ M$ $HClO_4$
Electron transfer									
$[Mn(CNR)_6]^+ + [Mn(CNR)_6]^{2+} \rightarrow$	MeCN		2.2	6–10			75		
R = Me		0			-2.4 ± 0.8				
		9			-2.1 ± 0.8				
Et		6			-5.5 ± 0.5				
t-Bu		6			-10.2 ± 0.5				
		25			-13.3 ± 0.9	-4.5 ± 0.6			
C_6H_{11}		6			-17.4 ± 1.3				
		12			-20.2 ± 1.6	-8.2 ± 1.4			
n-Bu		6			-19.7 ± 2.4	-8.1 ± 2.0		76, 77	$\mu = 0.01\ M$
$[Co(terpy)_2]^{2+} + [Co(bipy)_3]^{3+} \rightarrow$ $[Co(terpy)_2]^{3+} + [Co(bipy)_3]^{2+}$	H_2O	25	1.0	6	-9.6 ± 0.9				
	$HCONH_2$	25	1.0	6	-13.8 ± 1.1				
	MeCN	25	1.5	7	-5.1 ± 1.4				
$[Co(NH_3)_5X]^{3+} + [Fe(CN)_6]^{4-} \rightleftharpoons$ $\{Co(NH_3)_5X^{3+}\cdot Fe(CN)_6^{4-}\}$									
X = H_2O	H_2O	25	1.0	5		$+3.5 \pm 0.6$	78	c	$\mu = 0.5\ M$
Me_2SO	H_2O	25	1.0	6		-11 ± 3	79	c	$\mu = 1.0\ M$
$\{Co(NH_3)_5X^{3+}\cdot Fe(CN)_6^{4-}\} \rightarrow$ $Co^{2+} + X + 5NH_3 + [Fe(CN)_6]^{3-}$									
X = H_2O	H_2O	25	1.0	5	$+37.6 \pm 1.2$		78		$\mu = 0.5\ M$
py	H_2O	15	1.2	7	$+30.7 \pm 1.3$		76, 79		$\mu = 1.0\ M$
			1.0	5	$+28.9 \pm 1.4$				
Me_2SO	H_2O	25	1.0	6	$+34.4 \pm 1.1$		76, 79		$\mu = 1.0\ M$
Cl^-	H_2O	35	1.0	5	$+26.1 \pm 1.3$		76, 79		$\mu = 1.0\ M$
$[MoO(L)H_2O]^{5+} + H^+ + H_2O_2 \rightarrow [Mo(O_2)(L)H_2O]^{6+} + H_2O + OH$	H_2O	35	1.2	6			68		
L = tmpyp					$+1.3 \pm 0.3$				
tpyp					$+4.7 \pm 0.4$				

(*continued*)

Table 15.1 (continued)

Reaction	Solvent	T (°C)	P (kbar)	No. of data	$\Delta V^{\ddagger}$ (cm³ mol⁻¹)	$\Delta \beta^{\ddagger}$ (cm³ mol⁻¹ kbar⁻¹)	$\Delta \bar{V}$ (cm³ mol⁻¹)	Method	Ref.	Remarks
Homolysis										
$[Cr(H_2O)_5R]^{3+} + H_2O \rightarrow$ $[Cr(H_2O)_6]^{2+} + \cdot R^+$ $R = CH_2C_5H_4NH$	H_2O	63	3.5	10	$+20.0 \pm 0.9$		$+0.7 \pm 0.6$		80	0.026 M HClO₄
Photochemical										
$[Fe(CN)_5NO]^{2-} + S \rightarrow$ $[Fe(CN)_5S]^{3-} + NO^+$	S	25	2.0	5					81	
$S = H_2O$					$+8.8 \pm 0.4$					$\lambda_{irr} = 436$ nm
					$+7.8 \pm 1.0$					$\lambda_{irr} = 313$ nm
MeOH					$+10.3 \pm 0.6$					$\lambda_{irr} = 436$ nm
					$+13.0 \pm 1.9$					$\lambda_{irr} = 313$ nm
Me₂SO					$+11.1 \pm 0.4$					$\lambda_{irr} = 436$ nm
					$+11.4 \pm 1.6$					$\lambda_{irr} = 405$ nm
					$+14.1 \pm 1.0$					$\lambda_{irr} = 313$ nm
$[Co(en)_2(SO_2CH_2CH_2NH_2)]^{2+} \rightarrow$ $[Co(en)_2(OS(O)CH_2CH_2NH_2)]^{2+}$	H_2O	25	2.0	9	$+6.5 \pm 0.6$				74	0.01 M HClO₄
$[Ru(\eta^6\text{-}C_6H_6)(H_2O)_3]^{2+} + 3H_2O \rightarrow$ $Ru(H_2O)_6]^{2+} + C_6H_6$	H_2O	25	2.0	9	$+1.1 \pm 0.4$				82	
$cis\text{-}[Rh(bipy)_2Cl_2]^+ + H_2O \rightarrow$ $cis\text{-}[Rh(bipy)_2(H_2O)Cl]^{2+} + Cl^-$	H_2O	25	2.0	5	-9.7 ± 0.8				83	pH = 2.7
$trans\text{-}[Pt(CN)_4(N_3)_2]^{2-} \rightarrow$ $[Pt(CN)_4]^{2-} + 3N_2$	H_2O	25	2.0	9	$+8.1 \pm 0.4$				84	
	EtOH	25	2.0	9	$+14.3 \pm 0.9$				84	
Photophysical: emission lifetime										
$[Ru(bipy)_3]^{2+}$	H_2O	2	3.0	6	-1.5				85	
		15			-1.0					
					-1.0					
					-0.4					
		25			-1.0					9.6 M LiCl
		40			$+2.5$					5.0 M urea
		60			$+4.2$					

Complex / Transition	Solvent	T (°C)	pH	ΔV	Ref.	Remarks
				+3.1		
				+6.6		
	D₂O	70		+7.5		9.6 M LiCl 5.0 M urea
		3.5	3.0	0.0	85	
		25		+0.9		
		40		+3.7		
		60		+5.9		
		70		+8.9		
[Ru(bipy)₃]²⁺ CT → GS	H₂O			−2.2 ± 0.2		
	D₂O			−1.0 ± 0.2		
	MeCN		6	−1.7 ± 0.2	85	calculated
CT → LF	H₂O			+9.7 ± 1.0		
	D₂O			+9.1 ± 1.0		
	MeCN			+12.5 ± 1.0		
[Ru(phen)₃]²⁺	H₂O	2		+2.0	85	
		25		+5.6		
		40		+8.0		
		60		+10.6		
		70		+2.9		
	D₂O	3.5		+6.4		
		25		+9.4		
		40		+10.0		
		60		+11.7		
		70				
[Ru(phen)₃]²⁺ CT → GS	H₂O			−2.2 ± 0.2		
	D₂O			−1.1 ± 0.2		
CT → LF	H₂O			+10.4 ± 1.0		
	D₂O			+11.5 ± 1.0	85	calculated
cis-[Rh(bipy)₂Cl₂]⁺	FMA	23	3.3	+0.2 ± 0.2	83	
	MeOH			+0.6 ± 0.6		
	MeCN			+0.9 ± 0.2		
	H₂O		7 to 10	−2.2 ± 0.3		

References

References for Chapter 1

1. *J. Phys. Chem.*, **20**, No. (17), (1986). Special issue edited by A. H. Zewail, J. N. L. Connor, and N. Sutin.
2. R. A. Marcus, *J. Chem. Phys.*, **24**, 966, 979 (1956).
3. Ref. 1, pp. 3454–3460.
4. R. A. Marcus, *J. Phys. Chem.*, **90**, 3460 (1986).
5. Ref. 1, pp. 3467–3656.
6. Ref. 1, pp. 3657–4218.
7. C. J. Schlesener, C. Amatore, and J. K. Kochi, *J. Phys. Chem.*, **90**, 3747 (1986).
8. H. Sumi and R. A. Marcus, *J. Chem. Phys.*, **84**, 4272, 4894 (1986).
9. G. Williams, G. F. Moore, and R. J. P. Williams, *Comments Inorg. Chem.*, **4**, 55 (1985).
10. R. A. Marcus and N. Sutin. *Biochim. Biophys. Acta*, **811**, 265 (1985).
11. T. Guarr and G. McLendon, *Coord. Chem. Rev.*, **68**, 1 (1985).
12. G. L. Closs, L. T. Calcaterra, N. J. Green, K. W. Penfield, and J. R. Miller, *J. Phys. Chem.*, **90**, 3673 (1986).
13. *Photochemistry and Photophysics of Metal Complexes*, Symposium of 1984 International Chemical Congress of the Pacific Basin Societies, Honolulu (P. C. Ford and A. B. P. Lever eds.), *Coord. Chem. Rev.*, **64** (1985).
14. *Photochemistry and Photocatalysis*, NATO ASI Series C, 1985.
15. G. McLendon, T. Guarr, M. McGuire, K. Simolo, S. Strauch, and K. Taylor, *Coord. Chem. Rev.*, **64**, 113 (1985).
16. N. S. Hush, *Coord. Chem. Rev.*, **64**, 135 (1985).
17. E. Pelizetti and E. Pramauro, *J. Chem. Phys.*, **84**, 146; *Chem. Abstr.*, **103**, 76783.
18. A. Vogler, A. H. Osman, and H. Kunkely, *Coord. Chem. Rev.*, **64**, 159 (1985).
19. N. Mataga, *J. Mol. Struct.* (*Theochem*), **135**, 279 (1986).
20. B. S. Brunschwig, S. Ehrenson, and N. Sutin, *J. Phys. Chem.*, **90**, 3657 (1986).
21. R. A. Marcus and N. Sutin, *Comments Inorg. Chem.*, **5**, 119 (1986).
22. A. Haim, *Comments Inorg. Chem.*, **4**, 113 (1985).
23. W. E. Geiger, *Prog. Inorg. Chem.*, **33**, 275 (1985).
24. A. Chakravorty, *Comments Inorg. Chem.*, **4**, 1 (1985).

25. A. P. Murani, in: *Proc. Workshop High-Energy* . . . Los Alamos National Lab., Report LA-10227-C, Vol. 2 (1984); *Chem. Abstr.*, **105**, 071242.
26. C. Gleitzer and J. B. Goodenough, *Struct. Bonding* (Berlin), **61**, 1 (1985).
27. D. N. Hendrickson, S. M. Oh, T.-Y. Dong, T. Kambara, M. J. Cohn, and M. F. Moore, *Comments Inorg. Chem.*, **4**, 329 (1985).
28. D. E. Richardson, *J. Phys. Chem.*, **90**, 3697 (1986).
29. P. M. Guyon, T. R. Govers, and T. Baer, *Z. Phys. D.*, **1**, 89 (1986).

References for Chapter 2

1. R. J. Balahura and A. Johnston, *Inorg. Chem.*, **25**, 652 (1986).
2. B. H. Berrie and J. E. Earley, *Inorg. Chem.*, **23**, 774 (1984).
3. O. Olubuyide, K. Lu, A. O. Oyetunji, and J. E. Earley, *Inorg. Chem.*, **25**, 4798 (1986).
4. R. C. Thompson, *Inorg. Chem.*, **25**, 184 (1986).
5. R. C. Thompson, *Inorg. Chem.*, **24**, 3542 (1985).
6. D. H. Marcartney, *Inorg. Chem.*, **25** 2222 (1986).
7. F. P. Rotzinger, *Inorg. Chem.*, **25**, 4570 (1986).
8. W. C. Kupferschmidt and R. B. Jordan, *Inorg. Chem.*, **24**, 3357 (1985).
9. K. Tsukahara, *Inorg. Chim. Acta*, **120**, 11 (1986).
10. D. H. Macartney and N. Sutin, *Inorg. Chem.*, **24**, 3403 (1985).
11. S. E. Castillo-Blum, D. T. Richens, and A. G. Sykes, *J. Chem. Soc., Chem. Commun.*, 1120 (1986).
12. E. F. Hills, M. Mosner, and A. G. Sykes, *Inorg. Chem.*, **25**, 339 (1986).
13. J. W. Herbert änd D. H. Macartney, *Inorg. Chem.*, **24**, 4398 (1985).
14. R. B. Ali, K. Sarawek, A. Wright, and R. D. Cannon, *Inorg. Chem.*, **22**, 351 (1983).
15. R. N. Bose, V. D. Neff, and E. S. Gould, *Inorg. Chem.*, **25**, 165 (1986).
16. R. N. Bose and E. S. Gould, *Inorg. Chem.*, **24**, 2832 (1985).
17. Y.-T. Fanchiang, R. N. Bose, E. Gelerinter, and E. S. Gould, *Inorg. Chem.*, **24**, 4679 (1985).
18. R. N. Bose and E. S. Gould, *Inorg. Chem.*, **24**, 2645 (1985).
19. R. N. Bose and E. S. Gould, *Inorg. Chem.*, **25**, 94 (1986).
20. R. Viswanathan and V. R. Vijayaraghavan, *Bull. Chem. Soc. Jpn.*, **59**, 3243 (1986).
21. R. Viswanathan and V. R. Vijayaraghavan, *Indian J. Chem.*, **24A**, 866 (1985).
22. A. Bakac, M. E. Brynildson, and J. H. Espenson, *Inorg. Chem.*, **25**, 4108 (1986).
23. N. Ridgewick-Brown and R. D. Cannon, *Inorg. Chem.*, **24**, 2463 (1985).
24. M. Kozik, C. F. Hammer, and L. C. W. Baker, *J. Am. Chem. Soc.*, **108**, 7627 (1986).
25. C. Millan and H. Diebler, *Inorg. Chem.*, **24**, 3729 (1985).
26. H. Diebler and C. Millan, *Polyhedron*, **5**, 539 (1986).
27. M. A. Harmer, D. T. Richens, A. B. Soares, A. D. Thornton, and A. G. Sykes, *Inorg. Chem.*, **20**, 4155 (1981).
28. N. K. Mohanty and R. K. Nanda, *Bull. Chem. Soc. Jpn.*, **58**, 3597 (1985).
29. M. Vincenti, C. Minero, E. Pramauro, and E. Pelizzetti, *Inorg. Chim. Acta*, **110**, 51 (1985).
30. M. Vincenti, E. Pramauro, E. Pelizzetti, S. Diekman, and J. Frahm, *Inorg. Chem.*, **24**, 4533 (1985).
31. P. J. Smolenaers and J. K. Beattie, *Inorg. Chem.*, **25**, 2259 (1986).
32. J. K. Beattie and P. J. Smolenaers, *J. Phys. Chem.*, **90**, 3684 (1986).
33. M.-S. Chan and A. C. Wahl, *J. Phys. Chem.*, **89**, 5829 (1985).
34. U. Fufholz and A. Haim, *J. Phys. Chem.*, **90**, 3686 (1986).
35. J. O. Ehighaokhuo, J. F. Ojo, and O. Olubuyide, *J. Chem. Soc., Dalton Trans.*, 1665 (1985).
36. A. M. Lannon, A. G. Lappin, and M. G. Segal, *J. Chem. Soc., Dalton Trans.*, 619 (1986).
37. G. D. Armstrong, J. D. Sinclair-Day, and A. G. Sykes, *J. Phys. Chem.*, **90**, 3805 (1986).
38. S. Jagannathan and R. C. Patel, *Inorg. Chem.*, **24**, 3634 (1986).

39. K. Wieghardt, M. Kleine-Boymann, B. Nuber, and J. Weiss, *Inorg. Chem.*, **25**, 1309 (1986).
40. J. A. Boeyens, A. G. S. Furhs, R. D. Hancock, and K. Wieghardt, *Inorg. Chem.*, **24**, 2926 (1985).
41. K. Wieghardt, W. Walz, B. Nuber, J. Weiss, A. Ozarowski, H. Stratemeier, and D. Reinen, *Inorg. Chem.* **25**, 1650 (1986).
42. A. McAuley and P. R. Norman, *Isr. J. Chem.*, **25**, 106 (1985).
43. M. G. Fairbank, A. McAuley, P. R. Norman, and O. Olubuyide, *Can. J. Chem.*, **63**, 2983 (1985).
44. M. G. Fairbank, P. R. Norman, and A. McAuley, *Inorg. Chem.*, **24**, 2639 (1985).
45. I. Krack and R. van Eldick, *Inorg. Chem.*, **25**, 1743 (1986).
46. Y. Sasaki, K. Endo, A. Nagasawa, and K. Saito, *Inorg. Chem.*, **25**, 4845 (1986).
47. M. Kanesato, M. Ebihara, Y. Sasaki, and K. Saito, *J. Am. Chem. Soc.*, **105**, 5711 (1983).
48. R. van Eldick and H. Kelm, *Inorg. Chim. Acta*, **73**, 91 (1983).
49. J. F. Ojo, O. Olubuyide, and O. Oyetunji, *Inorg. Chim. Acta*, **119**, 25 (1986)
50. J. F. Endicott and T. Ramasami, *J. Phys. Chem.*, **90**, 3740 (1986).
51. U. Furholz and A. Haim, *Inorg. Chem.*, **24**, 3091 (1985).
52. V. Balzani and F. Scandola, *Inorg. Chem.*, **25**, 4457 (1986).
53. A. Bakac, V. Butkovic, J. H. Espenson, R. Marcec, and M. Orhanovic, *Inorg. Chem.*, **25**, 341 (1986).
54. P. Connolly, J. H. Espenson, and A. Bakac, *Inorg. Chem.*, **25**, 2169 (1986).
55. J. D. Melton, J. H. Espenson, and A. Bakac, *Inorg. Chem.*, **25**, 4104 (1986).
56. Y.-T. Fanchiang, *J. Chem. Soc., Dalton Trans.*, 1375 (1985).
57. R. M. Nielson, J. P. Hunt, H. W. Dodgen, and S. Wherland, *Inorg. Chem.*, **25**, 1964 (1986).
58. P. Braun and R. van Eldick, *J. Chem. Soc., Chem. Commun.*, 1349 (1985).
59. A. M. Kjaer and J. Ulstrup, *Inorg. Chem.*, **25**, 644 (1986).
60. T. G. Braga and A. C. Wahl, *J. Phys. Chem.*, **89**, 5822 (1985).
61. D. M. Triegaardt and A. C. Wahl, *J. Phys. Chem.*, **90**, 1957 (1986).
62. E. Pelizzetti, E. Pramauro, M. J. Blandamer, J. Burgess, and N. Gosal, *Inorg. Chim. Acta*, **102**, 163 (1985).
63. A. I. Carbone, F. P. Cavasino, C. Sbriziolo, and E. Pelizzetti, *J. Phys. Chem.*, **89**, 3578 (1985).
64. D. Borchardt and S. Wherland, *Inorg. Chem.*, **25**, 901 (1986).
65. J. R. Eyler and D. E. Richardson, *J. Am. Chem. Soc.*, **107**, 6130 (1985).
66. A. Shirazi, M. Barbush, S. Ghosh, and D. W. Dixon, *Inorg. Chem.*, **24**, 2495 (1985).
67. R. Langley, P. Hambright, and R. F. X. Williams, *Inorg. Chem.*, **24**, 3716 (1985).
68. R. Langley, P. Hambright, and R. F. X. Williams, *Inorg. Chim. Acta*, **104**, L25 (1985).
69. L. J. Kirschenbaum, E. T. Borrish, and J. D. Rush, *Isr. J. Chem.*, **25**, 159 (1985).
70. E. T. Borrish, L. J. Kirschenbaum, and E. Mentasti, *J. Chem. Soc., Dalton Trans.*, 1789 (1985).
71. B. Wang, Y. Sasaki, K. Okazaki, K. Kanesato, and K. Saito, *Inorg. Chem.*, **25**, 3745 (1986).
72. G. E. Kirvan and D. W. Margerum, *Inorg. Chem.*, **24**, 3245 (1985).
73. A. G. Lappin, D. P. Martone, and P. Osvath, *Inorg. Chem.*, **24**, 4187 (1985).
74. D. P. Martone, P. Osvath, C. Eigenbrot, M. C. M. Laranjeira, R. D. Peacock, and A. G. Lappin, *Inorg. Chem.*, **24**, 4693 (1985).
75. P. Osvath and A. G. Lappin, *J. Chem. Soc., Chem. Commun.*, 1056 (1986).
76. Y. Sasaki, K. Meguro, and K. Saito, *Inorg. Chem.*, **25**, 2277 (1986).
77. J. K. Barton, C. V. Kumar, and N. J. Turro, *J. Am. Chem. Soc.*, **108**, 6391 (1986).
78. Z. Khurram, W. Bottcher, and A. Haim, *Inorg. Chem.*, **24** 1966 (1985).
79. D. Sandrini, M. T. Gandolfi, M. Maestri, F. Bolletta, and V. Balzani, *Inorg. Chem.*, **23**, 3017 (1984).
80. I. I. Creaser, L. R. Gahan, R. J. Geue, A. Launionis, P. A. Lay, J. D. Lydon, M. G. McCarthy, A. W.-H. Mau, A. M. Sargeson, and W. H. F. Sasse, *Inorg. Chem.*, **24**, 2671 (1985).
81. N. Sabbatini, S. Dellonte, A. Bonazzi, M. Ciano, and V. Balzani, *Inorg. Chem.*, **25**, 1738 (1986).
82. C. Chiorboli, F. Scandola, and H. Kisch, *J. Phys. Chem.*, **90**, 2211 (1986).
83. D. K. Liu, B. S. Brunschwig, C. Creutz, and N. Sutin, *J. Am. Chem. Soc.*, **108**, 1749 (1986).
84. S. S. Shah and A. W. Maverick, *Inorg. Chem.*, **25**, 1867 (1986).
85. S. Sakaki, G. Koga, F. Sato, and K. Ohkubo, *J. Chem. Soc., Dalton Trans.*, 1959 (1985).

86. G. D. Armstrong, T. Ramasami, and A. G. Sykes, *Inorg. Chem.*, **24**, 3230 (1985).
87. G. D. Armstrong and A. G. Sykes, *Inorg. Chem.*, **25**, 3725 (1986).
88. P. C. Harrington and R. G. Wilkins, *J. Am. Chem. Soc.*, **103**, 1550 (1981).
89. G. D. Armstrong, J. A. Chambers, and A. G. Sykes, *J. Chem. Soc., Dalton Trans.*, 755 (1986).
90. G. D. Armstrong, S. K. Chapman, M. J. Sisley, A. G. Sykes, A. Aitken, N. Osheroff, and E. Margoliash, *Biochemistry*, **25**, 6947 (1986).
91. A. G. Sykes, *Chem. Soc. Rev.*, **14**, 283 (1985).
92. J. McGinnis, J. D. Sinclair-Day, and A. G. Sykes, *J. Chem. Soc., Dalton Trans.*, 2007 (1986).
93. J. McGinnis, J. D. Sinclair-Day, and A. G. Sykes, *J. Chem. Soc., Dalton Trans.*, 2011 (1986).
94. F. A. Armstrong, P. C. Driscoll, H. A. O. Hill, and C. Redfield, *J. Inorg. Biochem.*, **28**, 171 (1986).
95. J. D. Sinclair-Day and A. G. Sykes, *J. Chem. Soc., Dalton Trans.*, 2069 (1986).
96. B. S. Brunschwig, P. J. DeLaive, A. M. English, M. Goldberg, H. B. Gray, S. L. Mayo, and N. Sutin, *Inorg. Chem.*, **24**, 3743 (1985).
97. J. McGinnis, W. J. Ingledew, and A. G. Sykes, *Inorg. Chem.*, **25**, 3730 (1986).
98. S. K. Chapman, W. H. Orme-Johnson, J. McGinnis, J. D. Sinclair-Day, A. G. Sykes, P.-I. Ohlsson, and K.-G. Paul, *J. Chem. Soc., Dalton Trans.*, 2063 (1986).
99. H. B. Gray, *Chem. Soc. Rev.*, **15**, 17 (1986).
100. R. Bechtold, M. B. Gardineer, A. Kazmi, B. van Hedryck, and S. S. Isied, *J. Phys. Chem.*, **90**, 3800 (1986).
101. R. J. Crutchley, W. R. Ellis, and H. B. Gray, *J. Am. Chem. Soc.*, **107**, 5002 (1985).
102. S. E. Peterson-Kennedy, J. L. McGourty, J. A. Kalweit, and B. M. Hoffman, *J. Am. Chem. Soc.*, **108**, 1739 (1986).
103. K. C. Cho, C. M. Che, K. M. Ng, and C. L. Choy, *J. Am. Chem. Soc.*, **108**, 2814 (1986).
104. E. Cheung, K. Taylor, J. A. Kornblatt, A. M. English, G. McLendon, and J. R. Miller, *Proc. Natl. Acad. Sci. U.S.A.*, **83**, 1330 (1986).
105. N. Liang, C. H. Kang., P. S. Ho, E. Margoliash, and B. M. Hoffman, *J. Am. Chem. Soc.*, **108**, 4665 (1986).
106. K. T. Conklin and G. McLendon, *Inorg. Chem.*, **25**, 4804 (1986).

References for Chapter 3

1. J. Straehle, *Comments Inorg. Chem.*, **4**, 295 (1985).
2. E. S. Gould, *Acc. Chem. Res.*, **19**, 66 (1986).
3. D. E. Richardson, *Comments Inorg. Chem.*, **3**, 367 (1985).
4. D. N. Hendrickson, S. M. Oh, T. Y. Dong, T. Kombara, M. J. Cohn, and M. F. Moore, *Comments Inorg. Chem.*, **4**, 329 (1985).
5. C. C. Winterbourn, *Environ. Health Perspect.*, **62**, 321 (1985).
6. A. Naqui, B. Chance, and E. Cadenas, *Ann. Rev. Biochem.*, **55**, 137 (1986).
7. I. P. Skibida, *Russ. Chem. Rev.*, **54**, 875 (1985).
8. B. H. J. Bielski, D. E. Cabelli, R. L. Arudi, and A. B. Ross, *J. Phys. Chem. Ref. Data*, **14**, 1041 (1985).
9. V. Balzani, N. Sabbatini, and F. Scandola, *Chem. Rev.*, **86**, 319 (1986).
10. G. J. Kavarnos and N. J. Turro, *Chem. Rev.*, **86**, 401 (1986).
11. J. Halpern, *J. Organomet. Chem.*, **300**, 139 (1986).
12. F. R. Keene, C. J. Creutz, and N. Sutin, *Coord. Chem. Rev.*, **64**, 247 (1985).
13. K. Aoyagi, M. Mukaida, H. Kakihana, and K. Shimizu, *J. Chem. Soc., Dalton Trans.*, 1733 (1985).
14. J. H. Butler and L. I. Gordon, *Inorg. Chem.*, **25**, 4573 (1986).
15. E. Saviuc and A. Calusaru, *Rev. Roum. Chim.*, **31**, 753 (1986); *Chem. Abstr.*, **106**, 39199d.
16. V. L. Goedken, S.-M. Peng, and Y. Park, *J. Am. Chem. Soc.*, **96**, 284 (1974).

17. E. G. Samsel and J. K. Kochi, *Inorg. Chem.*, **25**, 2450 (1986).

18. S. L. Bruhn, A. Bakac, and J. H. Espenson, *Inorg. Chem.*, **25**, 535 (1986).

19. A. K. Gupta, K. S. Gupta, and Y. K. Gupta, *Inorg. Chem.*, **24**, 3670 (1985).

20. W. R. Murphy, Jr., K. Takeuchi, M. H. Barley, and T. J. Meyer, *Inorg. Chem.*, **25**, 1041 (1986).

21. K. Tanaka, M. Honjo, and T. Tanaka, *Inorg. Chem.*, **24**, 2662 (1985).

22. J. D. Melton, A. Bakac, and J. H. Espenson, *Inorg. Chem.*, **25**, 3360 (1986).

23. M. S. Ram and D. M. Stanbury, *Inorg. Chem.*, **24**, 2954 (1985).

24. L. Eberson and F. Radner, *Acta Chem. Scand., Ser. B*, **38**, 861 (1984).

25. A. N. Red'kin and V. A. Smirnov, *Kinet. Katal.*, **26(4)**, 1000 (1985).

26. P. Legzdins, B. Wassink, F. W. B. Einstein, and A. C. Willis, *J. Am. Chem. Soc.*, **108**, 317 (1986).

27. D. Hedden, D. M. Roundhill, and M. Walkinshaw, *Inorg. Chem.*, **24**, 3146 (1985).

28. K. Tsukahara, *Bull. Chem. Soc. Jpn.*, **59**, 1709 (1986).

29. R. J. Balahura, G. Ferguson, B. L. Ruhl, and R. G. Wilkins, *Inorg. Chem.*, **22**, 3990 (1983).

30. F. T. Bonner and B. Ravid, *Inorg. Chem.*, **14**, 558 (1975).

31. M. N. Hughes and P. E. Wimbledon, *J. Chem. Soc., Dalton Trans.*, 703 (1976).

32. M. N. Hughes and P. E. Wimbledon, *J. Chem. Soc., Dalton Trans.*, 1650 (1977).

33. M. J. Akhtar, C. A. Lutz, and F. T. Bonner, *Inorg. Chem.*, **18**, 2369 (1979).

34. M. P. Doyle and S. N. Mahapatro, *J. Am. Chem. Soc.*, **106**, 3678 (1984).

35. D. A. Bazylinski, J. Goretski, and T. C. Hollocher, *J. Am. Chem. Soc.*, **107**, 7986 (1985).

36. D. A. Bazylinski and T. C. Hollocher, *Inorg. Chem.*, **24**, 4285 (1985).

37. M. J. Akhtar, F. T. Bonner, M. N. Hughes, E. J. Humphreys, and C.-S. Lu, *Inorg. Chem.*, **25**, 4635 (1986).

38. G. Bazsa and I. R. Epstein, *J. Phys. Chem.*, **89**, 3050 (1985).

39. K. Wieghardt, M. Woeste, P. S. Roy, and P. Chaudhuri, *J. Am. Chem. Soc.*, **107**, 8276 (1985).

40. M. S. Ram and D. M. Stanbury, *J. Phys. Chem.*, **90**, 3691 (1986).

41. Z. B. Alfassi, W. A. Prütz, and R. H. Schuler, *J. Phys. Chem.*, **90**, 1198 (1986).

42. A. Vogler, C. Quett, A. Paukner, and H. Kunkely, *J. Am. Chem. Soc.*, **108**, 8263 (1986).

43. C. Bartocci, A. Maldotti, V. Carassiti, and O. Traverso, *Inorg. Chim. Acta*, **107**, 5 (1985).

44. C. E. Castro, M. Jamin, W. Yokoyama, and R. Wade, *J. Am. Chem. Soc.*, **108**, 4179 (1986).

45. U. Nickel, B. M. Zhou, and G. Gulden, *Ber. Bunsenges. Phys. Chem.*, **89**, 999 (1985).

46. F. Grases, J. Palou, and E. Amat, *Transition Met. Chem.*, **11**, 253 (1986).

47. T. Ohno, *J. Phys. Chem.*, **89**, 5709 (1985).

48. D. Sandrini, M. Maestri, P. Belser, A. von Zelewsky, and V. Balzani, *J. Phys. Chem.*, **89**, 3675 (1985).

49. G. Jones, II and V. Malba, *J. Org. Chem.*, **50**, 5776 (1985).

50. A. K. Colter, A. G. Parsons, and K. Foohey, *Can. J. Chem.*, **63**, 2237 (1985).

51. E.-I. Ochiai, D. I. Shaffer, D. L. Wampler, and P. D. Schettler, Jr., *Transition Met. Chem.*, **11**, 241 (1986).

52. R. E. Sassoon, Z. Aizenshtat, and J. Rabani, *J. Phys. Chem.*, **89**, 1182 (1985).

53. R. E. Sassoon and J. Rabani, *J. Phys. Chem.*, **89**, 5500 (1985).

54. M. Venturi, Q. G. Mulazzani, M. Ciano, and M. Z. Hoffman, *Inorg. Chem.*, **25**, 4493 (1986).

55. H. Rau, R. Frank, and G. Greiner, *J. Phys. Chem.*, **90**, 2476 (1986).

56. S. Sakaki, G. Koga, and K. Ohkubo, *Inorg. Chem.*, **25**, 2330 (1986).

57. P. A. Metcalf and C. P. Kubiak, *J. Am. Chem. Soc.*, **108**, 4682 (1986).

58. A. Slama-Schwok and J. Rabani, *J. Phys. Chem.*, **90**, 1176 (1986).

59. J. A. Baumann, G. E. Bossard, T. A. George, D. B. Howell, L. M. Koczon, R. K. Lester, and C. M. Noddings, *Inorg. Chem.*, **24**, 3568 (1985).

60. N. Jubran, H. Cohen, and D. Meyerstein, *Isr. J. Chem.*, **25**, 118 (1985).

61. N. Jubran, G. Ginzburg, H. Cohen, Y. Koresh, and D. Meyerstein, *Inorg. Chem.*, **24**, 251 (1985).

62. M. P. Doyle, S. N. Mahapatro, C. M. VanZyl, and M. R. Hester, *J. Am. Chem. Soc.*, **107**, 6136 (1985).

63. J. H. Hall, R. Lopez de la Vega, and W. L. Purcell, *Inorg. Chim. Acta*, **102**, 157 (1985).

64. E. W. Harlan, J. M. Berg, and R. H. Holm, *J. Am. Chem. Soc.*, **108**, 6992 (1986).
65. J. D. Rush and B. H. J. Bielski, *Inorg. Chem.*, **24**, 4282 (1985).
66. P. Ramamurthy and P. Natarajan, *Inorg. Chem.*, **25**, 3554 (1986).
67. E. F. Hills, P. R. Norman, T. Ramasami, D. T. Richens, and A. G. Sykes, *J. Chem. Soc.*, *Dalton Trans.*, 157 (1986).
68. S. Asperger, I. Murati, D. Pavlovic, and A. Sustra, *J. Chem. Soc., Chem. Commun.*, 814 (1986).
69. H. E. Toma and A. C. C. Silva, *Can. J. Chem.*, **64**, 1280 (1986).
70. G. A. Tondreau and R. G. Wilkins, *Inorg. Chem.*, **25**, 2745 (1986).
71. K. Matsumoto and K. Fuwa, *J. Am. Chem. Soc.*, **104**, 897 (1982).
72. K. Matsumoto and T. Watanabe, *J. Am. Chem. Soc.*, **108**, 1308 (1986).
73. J. D. Rush and B. H. J. Bielski, *Inorg. Chem.*, **24**, 3895 (1985).
74. D. R. Prasdad and M. Z. Hoffman, *J. Am. Chem. Soc.*, **108**, 2568 (1986).
75. A. J. Elliot, S. Padamshi, and J. Pika, *Can. J. Chem.*, **64**, 314 (1986).
76. M. R. Leonov, V. A. Il'yushenkov, V. M. Fomin, and N. V. Il'yushenkova, *Radiokhimiya*, **22**, 12 (1986).
77. D. Pavlovic, S. Asperger, and B. Domi, *J. Chem. Soc., Dalton Trans.*, 2535 (1986).
78. E. Spodine, A. M. Atria, and P. Meza, *Transition Met. Chem.*, **11**, 205 (1986).
79. E. Ahmed, S. Ahmed, A. Begum, A. Khair, and A. K. Roy, *Transition Met. Chem.*, **11**, 271 (1986).
80. N. Jubran, G. Ginzburg, H. Cohen, Y. Koresh, and D. Meyerstein, *Inorg. Chem.*, **25**, 1908 (1986).
81. S. Fukuzumi, K. Ishikawa, and T. Tanaka, *Chem. Lett.*, 1 (1986).
82. G. Favero, S. Issa, A. Turco, and U. Vettori, *J. Organometal. Chem.*, **315**, 237 (1986).
83. D. T. Sawyer, G. S. Srivatsa, M. E. Bodini, W. P. Schaefer, and R. M. Wing, *J. Am. Chem. Soc.*, **108**, 936 (1986).
84. C. J. Raleigh and A. E. Martell, *Inorg. Chem.*, **25**, 1190 (1986).
85. B. R. James and R. H. Morris, *Can. J. Chem.*, **64**, 897 (1986).
86. G. Davies, M. A. El-Sayed, A. El-Toukhy, T. R. Gilbert, and K. Nabih, *Inorg. Chem.*, **25**, 1929 (1986).
87. M. A. El-Sayed, A. El-Toukhy, and G. Davies, *Inorg. Chem.*, **24**, 3387 (1985).
88. K. D. Karlin, R. W. Cruse, M. S. Haka, Y. Gultneh, and B. I. Cohen, *Inorg. Chim. Acta*, **125**, L43 (1986).
89. S. M. Nelson, A. Lavery, and M. G. B. Drew, *J. Chem. Soc., Dalton Trans.*, 911 (1986).
90. V. V. Shapovalov, I. D. Ozherel'ev, and D. M. Palade, *Russ. J. Inorg. Chem.*, **30**, 828 (1985).
91. F. M. Ashmawy, R. M. Issa, S. A. Amer, C. A. McAuliffe, and R. V. Parish, *J. Chem. Soc.*, *Dalton Trans.*, 421 (1986).
92. S. Bhaduri, N. Y. Sapre, and A. Basu, *J. Chem. Soc., Chem. Commun.*, 197 (1986).
93. F. C. Frederick and L. T. Taylor, *Polyhedron*, **5**, 887 (1986).
94. G. R. A. Johnson, N. B. Nazhat, and R. A. Saadalla-Nazhat, *J. Chem. Soc., Chem. Commun.*, 407 (1985).
95. W. H. Koppenol, *J. Free Rad. Biol. Med.*, **1**, 281 (1985).
96. J. D. Rush and W. H. Koppenol, *J. Biol. Chem.*, **261**, 6730 (1986).
97. A. F. C. Gorren, H. Dekker, and R. Wever, *Biochim. Biophys. Acta*, **90**, 809 (1985).
98. K. B. Yatsimirskii and E. V. Rybak-Akimova, *Teor. Eksp. Khim.*, **22(3)**, 309 (1986), (CA105: 140678p).
99. G. D. Armstrong and A. G. Sykes, *Inorg. Chem.*, **25**, 3514 (1986).
100. S. Goldstein and G. Czapski, *J. Am. Chem. Soc.*, **108**, 2244 (1986).
101. M. Inamo, S. Funahashi, Y. Ito, Y. Hamada, and M. Tanaka, *Inorg. Chem.*, **24**, 2468 (1985).
102. J. M. Aubry, *J. Am. Chem. Soc.*, **107**, 5844 (1985).
103. S. K. Ghosh and E. S. Gould, *Inorg. Chem.*, **25**, 3357 (1986).
104. R. H. Simoyi, P. De Kepper, I. R. Epstein, and K. Kustin, *Inorg. Chem.*, **25**, 538 (1986).
105. M. F. Zipplies, W. A. Lee, and T. C. Bruice, *J. Am. Chem. Soc.*, **108**, 4433 (1986).
106. D. M. Doley, J. L. Karas, T. F. Jones, C. E. Coté, and S. B. Smith, *Inorg. Chem.*, **25**, 4761 (1986).

107. T. M. Florence, J. L. Stauber, and K. J. Mann, *J. Inorg. Biochem.*, **24**, 243 (1985).
108. H. Sugimoto and D. T. Sawyer, *J. Am. Chem. Soc.*, **107**, 5712 (1985).
109. M. L. Kremer, *Int. J. Chem. Kinet.*, **17**, 1299 (1985).
110. S. Lunák, P. Sedlak, and J. Veprek-Siska, *Collect. Czech. Chem. Commun.*, **51**, 973 (1986).
111. H. Bobadilla, A. Decinti, and L. Gil, *Polyhedron*, **5**, 1429 (1986).
112. N. M. Guindy, M. I. Ismail, Z. A. Gamra, and N. E. Milad, *Oxid. Commun.*, **8**, 119 (1985).
113. E. N. Rizkalla, L. H. J. Lajunen, and G. R. Choppin, *Inorg. Chim. Acta*, **119**, 93 (1986).
114. A. E. Gekhman, N. I. Moiseeva, E. A. Blymberg, and I. I. Moiseev, *Izvad. Akad. Nauk SSSR, Ser. Khim.*, **34**, 2653 (1985).
115. M. Orbán, *J. Am. Chem. Soc.*, **108**, 6893 (1986).
116. A. Bakac and J. H. Espenson, *J. Am. Chem. Soc.*, **108**, 713 (1986).
117. M. S. Ram, A. Bakac, and J. H. Espenson, *Inorg. Chem.*, **25**, 3267 (1986).
118. P. N. Balasubramanian and T. C. Bruice, *J. Am. Chem. Soc.*, **108**, 5495 (1986).
119. L.-C. Yuan and T. C. Bruice, *J. Am. Chem. Soc.*, **108**, 1643 (1986).
120. W. A. Lee and T. C. Bruice, *Inorg. Chem.*, **25**, 131 (1986).
121. Yu. S. Zimin, E. P. Talzi, V. M. Nekipelov, V. D. Chinakov, and K. I. Zamaraev, *React. Kinet. Catal. Lett.*, **29**, 225 (1985).
122. E. S. Schmidt, T. C. Bruice, R. S. Brown, and C. L. Wilkins, *Inorg. Chem.*, **25**, 4799 (1986).
123. A. Rajavelu and S. Vangalur, *Indian J. Chem., Sect. A*, **25A**, 836 (1986); *Chem. Abstr.*, **106**, 39189a.
124. C. K. Ranganathan, T. Ramasami, D. Ramaswamy, and M. Santappa, *Inorg. Chem.*, **25**, 915 (1986).
125. A. K. Basak and A. E. Martell, *Inorg. Chem.*, **25**, 1182 (1986).
126. M. Ebihara, Y. Sasaki, and K. Saito, *Inorg. Chem.*, **24**, 3831 (1985).
127. M. Kikkawa, Y. Sasaki, S. Kawata, Y. Hatakeyama, F. B. Ueno, and K. Saito, *Inorg. Chem.*, **24**, 4096 (1985).
128. J. T. Groves and Y. Watanabe, *Inorg. Chem.*, **25**, 4808 (1986).
129. R. M. Kellett and T. G. Spiro, *Inorg. Chem.*, **24**, 2373 (1985).
130. L. Roecker, W. Kutner, J. A. Gilbert, M. Simmons, R. W. Murray, and T. J. Meyer, *Inorg. Chem.*, **24**, 3784 (1985).
131. D. Kotkar, V. Joshi, and P. K. Ghosch, *Inorg. Chem.*, **25**, 4334 (1986).
132. P. K. Ghosch, B. S. Brunschwig, M. Chou, C. Creutz, and N. Sutin, *J. Am. Chem. Soc.*, **106**, 4772 (1984).
133. P. A. Lay and W. H. F. Sasse, *Inorg. Chem.*, **24**, 4707 (1985).
134. G. S. Srivatsa and D. T. Sawyer, *Inorg. Chem.*, **24**, 1734 (1985).
135. P. Chandayot and Y.-T. Fanchiang, *Inorg. Chem.*, **24**, 3535 (1985).
136. M. P. Heyward and C. F. Wells, *J. Chem. Soc., Dalton Trans.*, 2593 (1986).
137. J. Konstantatos, G. Kalatzis, and E. Vrachnou-Astra, *J. Chem. Soc., Dalton Trans.*, 2461 (1985).
138. F. M. Ashmawy, C. A. McAuliffe, R. V. (Dick) Parish, and J. Tames, *J. Chem. Soc., Dalton Trans.*, 1391 (1985).
139. J. Desilvestro, D. Duonghong, M. Kleijn, and M. Grätzel, *Chimia*, **39**, 102 (1985).
140. C. M. Dicken, F.-L. Lu, M. W. Nee, and T. C. Bruice, *J. Am. Chem. Soc.*, **107**, 5776 (1985).
141. C. M. Dicken, T. C. Woon, and T. C. Bruice, *J. Am. Chem. Soc.*, **108**, 1636 (1986).
142. T. C. Woon, C. M. Dicken, and T. C. Bruice, *J. Am. Chem. Soc.*, **108**, 7990 (1986).
143. L.-C. Yuan, T. S. Calderwood, and T. C. Bruice, *J. Am. Chem. Soc.*, **107**, 8273 (1985).
144. D. E. Katsoulis and M. T. Pope, *J. Chem. Soc., Chem. Commun.*, 1186 (1986).
145. B. D. Gupta and M. Kumar, *Inorg. Chim. Acta*, **113**, 9 (1986).
146. Y.-T. Fanchiang, *Organometallics*, **4**, 1515 (1985).
147. S. Fukuzumi, K. Ishikawa, and T. Tanaka, *Chem. Lett.*, 1801 (1986).
148. A. G. Blackman, D. A. Buckingham, C. R. Clark, and S. Kulkarni, *Aust. J. Chem.*, **39**, 1465 (1986).
149. G. A. Lawrance, *Transition Met. Chem.*, **11**, 396 (1986).

150. L. Drougge and L. I. Elding, *Inorg. Chem.*, **24**, 2292 (1985).

151. C. Sishta, M. Ke, B. R. James, and D. Dolphin, *J. Chem. Soc., Chem. Commun.*, 787 (1986).

152. L. M. Wilson and R. D. Cannon, *Inorg. Chem.*, **24**, 4366 (1985).

153. G. Crisponi, P. Deplano, and E. F. Trogu, *J. Chem. Soc., Dalton Trans.*, 365 (1986).

154. A. Barbati, F. Calderazzo, R. Poli, and P. F. Zanazzi, *J. Chem. Soc., Dalton Trans.*, 2569 (1986).

155. A. Harriman, P. A. Christensen, G. Porter, K. Morehouse, P. Neta, and M.-C. Richoux, *J. Chem. Soc., Faraday Trans. 1*, **82**, 3215 (1986).

156. J. D. Lydon and R. C. Thompson, *Inorg. Chem.*, **25**, 3694 (1986).

157. J. Schlosserová, M. Hronec, and V. Vesely, *J. Chem. Soc., Faraday Trans. 1*, **81**, 2095 (1985).

158. S. Caron, E. Ianovici, P. Lerch, and A. G. Maddock, *Inorg. Chim. Acta*, **109**, 209 (1985).

159. C. D. Hubbard and J. G. Jones, *Inorg. Chim. Acta*, **125**, 71 (1986).

160. L. I. Elding and L. H. Skibsted, *Inorg. Chem.*, **25**, 4084 (1986).

161. F. Pina, M. Maestri, R. Ballardini, Q. G. Mulazzani, M. D'Angelantonio, and V. Balzani, *Inorg. Chem.*, **25**, 4249 (1986).

162. E. F. Hills, C. Sharp, and A. G. Sykes, *Inorg. Chem.*, **25**, 2566 (1986).

163. K. Yokoi, I. Watanabe, and S. Ikeda, *Bull. Chem. Soc. Jpn.*, **58**, 2172 (1985).

164. A. A. Abdeli-Khalek and M. M. Elsemongy, *Inorg. Chim. Acta*, **100**, 219 (1985).

165. J. Benko, O. Vollárová, and D. Tahotná, *Transition Met. Chem.*, **11**, 30 (1986).

166. O. Vollárová, J. Benko, and E. Skalná, *Transition Met. Chem.*, **10**, 401 (1985).

167. P. N. Balasubramanian, J. W. Reed, and E. S. Gould, *Inorg. Chem.*, **24**, 1794 (1985).

168. A. A. Abdel-Khalek, *Transition Met. Chem.*, **11**, 67 (1986).

169. G. Bazsa and I. Fábián, *J. Chem. Soc., Dalton Trans.*, 2675 (1986).

170. M. S. Milyukova, M. N. Litvina, and B. F. Myasoedov, *Radiokhimiya*, **27**, 744 (1985); *Chem. Abstr.*, **104**, 121960n.

171. H. Ohura, Y. Kisaki, and S. Yamasaki, *Nippon Kagaku Kaishi*, **12**, 2326 (1985); *Chem. Abstr.*, **104**, 121955q.

172. L. Adamciková and P. Sevcik, *Collect. Czech. Chem. Commun.*, **50**, 2338 (1985).

173. V. Gáspár, G. Bazsa, and M. T. Beck, *J. Phys. Chem.*, **89**, 5495 (1985).

174. R. J. Field, E. Körös, and R. M. Noyes, *J. Am. Chem. Soc.*, **94**, 8649 (1972).

175. M. Varga, L. Gyorgyi, and E. Körös, *J. Am. Chem. Soc.*, **107**, 4780 (1985).

176. L. Gyorgyi, M. Varga, and E. Körös, *React. Kinet. Catal. Lett.*, **28**, 275 (1985); *Chem. Abstr.*, **103**, 184443w.

177. M. F. Crowley and R. J. Field, *React. Kinet. Catal. Lett.*, **28**, 233 (1985); *Chem. Abstr.*, **103**, 184442v.

178. Z. Nagy-Ungvarai and E. Körös, *React. Kinet. Catal. Lett.*, **27**, 83 (1985); *Chem. Abstr.*, **103** (18), 148200f.

179. D. O. Cooke, *React. Kinet. Catal. Lett.*, **27**, 379 (1985); *Chem. Abstr.*, **103**, 93640s.

180. X. Ji-ide and N. Shi-sheng, *Inorg. Chem.*, **25**, 1264 (1986).

181. G. A. Lawrance and C. B. Ward, *Transition Met. Chem.*, **10**, 258 (1985).

182. C. R. Dennis, J. G. Leipoldt, S. S. Basson, and A. J. Van Wyk, *Inorg. Chem.*, **25**, 1268 (1986).

183. J. Banos, F. Sánchez-Burgos, and M. C. Carmona-Guzmán, *J. Chem. Soc., Dalton Trans.*, 1975 (1985).

184. V. A. Ermakov, *Radiokhimiya*, **27**, 726 (1985); *Chem. Abstr.*, **104**, 57007c.

185. R. C. Thompson, *Inorg. Chem.*, **25**, 184 (1986).

186. R. C. Thompson, *Inorg. Chem.*, **24**, 3542 (1985).

187. A. A. El-Awady and G. M. Harris, *Inorg. Chem.*, **25**, 1323 (1986).

188. V. K. Joshi, R. van Eldik, and G. M. Harris, *Inorg. Chem.*, **25**, 2229 (1986).

189. R. N. Bose, N. Rajasekar, D. M. Thompson, and E. S. Gould, *Inorg. Chem.*, **25**, 3349 (1986).

190. L. I. Simándi, M. Jáky, C. R. Savage, and Z. A. Schelly, *J. Am. Chem. Soc.*, **107**, 4220 (1985).

191. A. Rodriguez, S. Lopez, M. C. Carmona-Guzmán, F. Sánchez, and C. Piazza, *J. Chem. Soc., Dalton Trans.*, 1265 (1986).

192. R. Langley, P. Hambright, K. Alston, and P. Neta, *Inorg. Chem.*, **25**, 114 (1986).

193. D. M. Davies and J. M. Lawther, *J. Chem. Soc., Chem. Commun.*, 385 (1986).

194. C. R. Dennis, J. G. Leipoldt, S. S. Basson, and G. J. Lamprecht, *Polyhedron* **4**, 1621 (1985).
195. P. Chandayot and Y.-T. Fanchiang, *Inorg. Chem.*, **24**, 3532 (1985).
196. A. M. Newton, *Can. J. Chem.*, **64**, 311 (1986).
197. A. R. Murdock, T. Tyree, W. Otterbein, L. Kinney, M. Carreras, J. N. Cooper, and R. C. Elder, *Inorg. Chem.*, **24**, 3674 (1985).
198. K. L. Nash, J. M. Cleveland, J. C. Sullivan, and M. Woods, *Inorg. Chem.*, **25**, 1169 (1986).
199. C.-L. Lee, G. Besenyei, B. R. James, D. A. Nelson, and M. A. Lilga, *J. Chem. Soc., Chem. Commun.*, 1175 (1985).
200. R. L. Bartzatt and J. Carr, *Transition Met. Chem.*, **11**, 116 (1986).
201. T. J. Kemp and M. A. Shand, *Inorg. Chem.*, **25**, 3940 (1986).
202. T. Miyashita and M. Matsuda, *Bull. Chem. Soc. Jpn.*, **58**, 3031 (1985).
203. G. Dasgupta and M. K. Mahanti, *React. Kinet. Catal. Lett.*, **28**, 153 (1985).
204. K. Antal, I. Bányai, and M. T. Beck, *J. Chem. Soc., Dalton Trans.*, 1191 (1985).
205. F. Grases, J. Palou, and E. Amat, *Transition Met. Chem.*, **11**, 253 (1986).
206. P. O'Brien, J. Barrett, and F. Swanson, *Inorg. Chim. Acta*, **108**, L19 (1985).
207. G. C. Pillai and E. S. Gould, *Inorg. Chem.*, **25**, 3353 (1986).
208. M. D. Lilani, G. K. Sharma, and R. Shanker, *Indian J. Chem.*, **25A**, 370 (1986).
209. P. S. R. Murti and K. S. Tripathy, *Indian J. Chem.*, **24A**, 217 (1985).
210. V. A. Antipin, G. S. Parshin, V. P. Kazakov, and S. N. Zagidullin, *React. Kinet. Catal. Lett.*, **27(1)**, 103 (1985); *Chem. Abstr.*, **103**, 93617q.
211. S. K. Ghosh, R. N. Bose, K. Laali, and E. S. Gould, *Inorg. Chem.*, **25**, 4737 (1986).
212. K. M. Inani, P. D. Sharma, and Y. K. Gupta, *J. Chem. Soc., Dalton Trans.*, 2571 (1985).
213. A. K. Gupta, K. S. Gupta, and Y. K. Gupta, *Indian J. Chem.*, **25A**, 653 (1986); *Chem. Abstr.*, **105**, 85833h.
214. B. Gupta, A. K. Gupta, K. S. Gupta, and Y. K. Gupta, *Indian J. Chem.*, **24A**, 927 (1985); *Chem. Abstr.*, **104**, 75880m.
215. A. K. Gupta, K. S. Gupta, and Y. K. Gupta, *Inorg. Chem.*, **24**, 3670 (1985); *Chem. Abstr.*, **103**, 147927m.
216. S. S. Miller, K. Zahir, and A. Haim, *Inorg. Chem.*, **24**, 3980 (1985).
217. J. D. Rush and B. H. J. Bielski, *J. Phys. Chem.*, **89**, 5062 (1985).
218. A. Plonka, J. Mayer, D. Metodiewa, J. L. Gebicki, A. Zgirski, and M. Grabska, *J. Radioanal. Nucl. Chem.*, **101**, 221 (1986).
219. H. Khan, *J. Radioanal. Nucl. Chem.*, **97**, 21 (1986).
220. N. Jubran, D. Meyerstein, J. Koresh, and H. Cohen, *J. Chem. Soc., Dalton Trans.*, 2509 (1986).
221. D. K. Geiger and G. Ferraudi, *Inorg. Chim. Acta*, **117**, 139 (1986).
222. J. Grodkowski, P. Neta, C. J. Schlesener, and J. K. Kochi, *J. Phys. Chem.*, **89**, 4373 (1985).
223. J. D. Rush and B. H. J. Bielski, *J. Am. Chem. Soc.*, **108**, 523 (1986).
224. B. G. Ershov, M. A. Akinshin, A. V. Gordeev, and N. L. Sukhov, *Radiat. Phys. Chem.*, **27**, 91 (1986).
225. A. Harriman, P. A. Christensen, G. Porter, K. Morehouse, P. Neta, and M.-C. Richoux, *J. Chem. Soc., Faraday Trans. 1*, **82**, 3215 (1986).
226. B. J. Parsons, M. Al-Hakim, G. O. Phillips, and A. J. Swallow, *J. Chem. Soc., Faraday Trans. 1*, **82**, 1575 (1986).
227. R. D. Saini and P. K. Bhattacharyya, *Radiat. Phys. Chem.*, **27**, 189 (1986).
228. P. Neta, J. Silverman, V. Markovic, and J. Rabani, *J. Phys. Chem.*, **90**, 703 (1986).
229. S. Baral, C. Lume-Pereira, E. Janata, and A. Henglein, *J. Phys. Chem.*, **89**, 5779 (1985).
230. N. Oshima, H. Suzuki, and Y. Moro-oka, *Inorg. Chem.*, **25**, 3407 (1986).
231. K. Hasegawa, T. Imamura, and M. Fujimoto, *Inorg. Chem.*, **25**, 2154 (1986).
232. T. Ozawa and A. Hanaki, *J. Chem. Soc., Dalton Trans.*, 1513 (1985).
233. D. Weinraub, P. Levy, and M. Faraggi, *Int. J. Radiat. Biol. Relat. Stud. Phys., Chem., Med.*, **50**, 649 (1986); *Chem. Abstr.*, **106**, 46582b.
234. M. Faraggi, P. Peretz, and D. Weinraub, *Int. J. Radiat. Biol. Relat. Stud. Phys., Chem. Med.*, **49**, 951 (1986); *Chem. Abstr.*, **105**, 111109y.

235. P. Connolly and J. H. Espenson, *Inorg. Chem.*, **25**, 2684 (1986).
236. S. Heller, S. Kusserow, W.-H. Böhmer, and K. Madeja, *Z. Anorg. Allg. Chem.*, **536**, 65 (1986).
237. K. J. Brewer, W. R. Murphy, Jr., K. J. Moore, E. C. Eberle, and J. D. Petersen, *Inorg. Chem.*, **25**, 2470 (1986).
238. I. Kovács, F. Ungváry, and L. Marko, *Organometallics*, **5**, 209 (1986).
239. M. Dasgupta and M. K. Mahanti, *Transition Met. Chem.*, **11**, 286 (1986).
240. M. Dasgupta and M. K. Mahanti, *Bull. Soc. Chim. France*, **5**, 711 (1986).
241. M. Dasgupta and M. K. Mahanti, *Afinidad*, **43**, 353 (1986); *Chem. Abstr.*, **105**, 103413c.
242. P. Kita and R. B. Jordan, *Inorg. Chem.*, **25**, 4791 (1986).
243. P. Kita and R. B. Jordan, *Inorg. Chem.*, **24**, 2701 (1985).
244. A. Bakac and J. H. Espenson, *J. Am. Chem. Soc.*, **108**, 719 (1986).
245. L. S. Hegedus and D. H. P. Thompson, *J. Am. Chem. Soc.*, **107**, 5663 (1985).
246. J. P. Collman, J. I. Brauman, and A. M. Madonik, *Organometallics*, **5**, 310 (1986).
247. J. P. Collman, J. I. Brauman, and A. M. Madonik, *Organometallics*, **5**, 218 (1986).
248. J. E. Anderson, C.-L. Yao, and K. M. Kadish, *Inorg. Chem.*, **25**, 719 (1986).
249. J. M. Hanckel, K.-W. Lee, P. Rushman, and T. L. Brown, *Inorg. Chem.*, **25**, 1852 (1986).
250. K.-W. Lee, J. M. Hanckel, and T. L. Brown, *J. Am. Chem. Soc.*, **108**, 2266 (1986).
251. C.-M. Che and W.-M. Lee, *J. Chem. Soc., Chem. Commun.*, 512 (1986).
252. D. M. Roundhill and S. J. Atherton, *Inorg. Chem.*, **25**, 4072 (1986).
253. D. M. Roundhill, *J. Am. Chem. Soc.*, **107**, 4354 (1985).
254. F. Ozawa, M. Fujimori, T. Yamamoto, and A. Yamamoto, *Organometallics*, **5**, 2144 (1986).
255. R. S. Wade and C. E. Castro, *Inorg. Chem.*, **24**, 2862 (1985).
256. J. A. McCleverty, C. W. Ninnes, and I. Wotochowicz, *J. Chem. Soc., Dalton Trans.*, 743 (1986).
257. J. P. Collman, P. J. Brothers, L. McElwee-White, and E. Rose, *J. Am. Chem. Soc.*, **107**, 6110 (1985).
258. B. Kräutler and C. Caderas, *Helv. Chim. Acta*, **67**, 1891 (1984).
259. S. Fukuzumi, K. Ishikawa, and T. Tanaka, *Chem. Lett.*, 1355 (1985).
260. N. A. Cooley, K. A. Watson, S. Fortier, and M. C. Baird, *Organometallics*, **5**, 2563 (1986).
261. E. C. Ashby, W.-Y. Su, and T. N. Pham, *Organometallics*, **4**, 1493 (1985).
262. D. Lexa, J.-M. Saveant, and D. L. Wang, *Organometallics*, **5**, 1428 (1986).
263. M. E. Marmion and K. J. Takeuchi, *J. Am. Chem. Soc.*, **108**, 510 (1986).
264. C. L. Hill and D. A. Bouchard, *J. Am. Chem. Soc.*, **107**, 5148 (1985).
265. A. Nishinaga, T. Kondo, and T. Matsuura, *Chem. Lett.*, 1319 (1985).
266. N. Kitajima, K. Whang, Y. Moro-oka, A. Uchida, and Y. Sasada, *J. Chem. Soc., Chem. Commun.*, 1504 (1986).
267. C. S. Kim and W. S. Lee, *Taehan Hwahakhoe Chi*, **30**, 409 (1986); *Chem. Abstr.*, **105**, 214527n.
268. P. G. Arjunan, J. G. Lakshmanan, N. Chandrasekara, K. Ramalingam, and K. Selvaraj, *J. Chem. Soc., Perkin Trans. 2*, 1183 (1985).
269. O. Bortolini, L. Bragante, F. Di Furia, G. Modena, and L. Cardellini, *Chim. Oggi*, **6**, 69 (1986); *Chem. Abstr.*, **105**, 202021e.
270. C. Bartocci, R. Amadelli, A. Maldotti, and V. Carassiti, *Polyhedron*, **5**, 1297 (1986).
271. P. K. Panda, R. K. Panda, and P. S. R. Murti, *Int. J. Chem. Kinet.*, **19**, 155 (1987).
272. P. A. Bott and K. A. K. Lott, *Inorg. Chim. Acta*, **111**, L33 (1986).
273. E. F. G. Herington, *J. Chem. Soc.*, 2747 (1956).
274. A. C. Pathy and G. P. Panigrahi, *Indian J. Chem.*, **25A**, 147 (1986).
275. D. A. Markovic, D. S. Veselinovic, and V. Obradovic, *Can. J. Chem.*, **64**, 2334 (1986).
276. M. Kohno, *Bull. Chem. Soc. Jpn.*, **59**, 2053 (1986).
277. T. Gunji, M. Hirano, and T. Morimoto, *J. Chem. Soc., Perkin Trans. 2*, 1827 (1985).
278. M. Ignaczak and M. Markiewicz, *Polish J. Chem.*, **59**, 747 (1985).
279. S. Fukuzumi, K. Ishikawa, and T. Tanaka, *Nippon Kagaku Kaishi*, **1**, 62 (1985); *Chem. Abstr.*, **103**, 142144u.
280. H. Keypour, J. Silver, M. T. Wilson, and M. Y. Hamed, *Inorg. Chim. Acta*, **125**, 97 (1986).

281. A. M. Lannon, A. G. Lappin, and M. G. Segal, *J. Chem. Soc., Dalton Trans.*, 619 (1986).
282. O. A. Travina, S. O. Travin, and A. P. Purmal, *Khim. Fiz.*, **5**, 72 (1986); *Chem. Abstr.*, **104**, 116851y.
283. K. Tsukahara, T. Okazawa, H. Takahashi, and Y. Yamamoto, *Inorg. Chem.*, **25**, 4756 (1986).
284. R. F. Johnston and R. A. Holwerda, *Inorg. Chem.*, **24**, 3176 (1985).
285. R. D. Scurlock, D. D. Gilbert, and J. M. DeKorte, *Inorg. Chem.*, **24**, 2393 (1985).
286. J. Labuda and J. Sima, *Inorg. Chim. Acta*, **112**, 59 (1986).
287. F. Grases, C. Genestar, and E. Amat, *Int. J. Chem. Kinet.*, **18**, 899 (1986).
288. P. Connolly, J. H. Espenson, and A. Bakac, *Inorg. Chem.*, **25**, 1070 (1986).
289. H. Cohen and D. Meyerstein, *Inorg. Chem.*, **25**, 1505 (1986).
290. A. Rotman, H. Cohen, and D. Meyerstein, *Inorg. Chem.*, **24**, 4158 (1985).
291. M. Masarwa, H. Cohen, and D. Meyerstein, *Inorg. Chem.*, **25**, 4897 (1986).
292. C. Lume-Pereira, S. Baral, A. Henglein, and E. Janata, *J. Phys. Chem.*, **89**, 5772 (1985).
293. R. C. McHatton, J. H. Espenson, and A. Bakac, *J. Am. Chem. Soc.*, **108**, 5885 (1986).
294. A. Bakac, V. Butković, J. H. Espenson, R. Marcec, and M. Orhanović, *Inorg. Chem.*, **25**, 2562 (1986).
295. A. Bakac, V. Butković, J. H. Espenson, R. Marcec, and M. Orhanović, *Inorg. Chem.*, **25**, 341 (1986).
296. M. C. Richoux, P. Neta, A. Harriman, S. Baral, and P. Hambright, *J. Phys. Chem.*, **90**, 2462 (1986).
297. S. Baral and A. G. Lappin, *J. Chem. Soc., Dalton Trans.*, 2213 (1985).
298. K. D. Whitburn, M. Z. Hoffman, N. V. Brezniak, and M. G. Simic, *Inorg. Chem.*, **25**, 3037 (1986).
299. M. Venturi, Q. G. Mulazzani, M. Ciano, and M. Z. Hoffman, *Inorg. Chem.*, **25**, 4493 (1986).
300. H. Cohen, D. Meyerstein, A. J. Shusterman, and M. Weiss, *J. Chem. Soc., Chem. Commun.*, 424 (1985).
301. P. C. Wilkins and R. G. Wilkins, *Inorg. Chem.*, **25**, 1908 (1986).
302. B. Le Motais, A. M. Koulkes-Pujo, and L. G. Hubert-Pfalzgraf, *Radiat. Phys. Chem.*, **29**, 21 (1987).
303. A.-M. Koulkes-Pujo, B. Le Motais, and L. G. Hubert-Pfalzgraf, *J. Chem. Soc., Dalton Trans.*, 1741 (1986).
304. G. S. Nahor and J. Rabani, *J. Phys. Chem.*, **89**, 4541 (1985).
305. Q. Yao and A. W. Maverick, *J. Am. Chem. Soc.*, **108**, 5364 (1986).
306. D. Brault and P. Neta, *Chem. Phys. Lett.*, **121**, 28 (1985).
307. A.-D. Leu and D. A. Armstrong, *J. Phys. Chem.*, **90**, 1449 (1986).
308. J. Grodkowski and P. Neta, *J. Phys. Chem.*, **88**, 1205 (1986).
309. Z. B. Alfassi, A. Harriman, S. Mosseri, and P. Neta, *Int. J. Chem. Kinet.*, **18**, 1315 (1986).
310. L. J. Johnston, J. Lusztyk, D. D. M. Wayner, A. N. Abeywickreyma, A. L. J. Beckwith, J. C. Scaiano, and K. U. Ingold, *J. Am. Chem. Soc.*, **107**, 4594 (1985).
311. K. Tsukahara and R. G. Wilkins, *Inorg. Chem.*, **24**, 3399 (1985).
312. S. J. Atherton, K. Tsukahara, and R. G. Wilkins, *J. Am. Chem. Soc.*, **108**, 3380 (1986).
313. Y. L. Chow and G. E. Buono-Core, *J. Am. Chem. Soc.*, **108**, 1234 (1986).
314. Y. L. Chow, G. E. Buono-Core, C. W. B. Lee, and J. C. Scaiano, *J. Am. Chem. Soc.*, **108**, 7620 (1986).
315. S. N. Bhattacharyya and P. C. Mandal, *J. Chem. Soc., Faraday Trans. 1*, **81**, 2569 (1985).
316. L. A. Tavadyan, V. A. Mardoyan, and A. B. Nalbandyan, *Arm. Khim. Zh.*, **39**, 137 (1986); *Chem. Abstr.*, **105**, 12911x.
317. A. A. Gridnev, I. M. Belgovskii, and N. S. Enikolopyan, *Dokl. Akad. Nauk SSSR*, **289**, 616 (1986); *Chem. Abstr.*, **105**, 202139z.
318. G. C. Pillai, J. W. Reed, and E. S. Gould, *Inorg. Chem.*, **25**, 4734 (1986).
319. G. C. Pillai and E. S. Gould, *Inorg. Chem.*, **25**, 4740 (1986).
320. D. Katakis, E. Vrachnou-Astra, and J. Konstantatos, *J. Chem. Soc., Dalton Trans.*, 1491 (1986).
321. G. Calvaruso, A. I. Carbone, and F. P. Cavasino, *J. Chem. Soc., Dalton Trans.*, 1683 (1985).

322. G. P. Haight, G. M. Jursich, M. T. Kelso, and P. J. Merrill, *Inorg. Chem.*, **24**, 2740 (1985).
323. L. Roecker and T. J. Meyer, *J. Am. Chem. Soc.*, **108**, 4066 (1986).
324. R. N. Mehrotra, *Bull. Chem. Soc. Jpn.*, **58**, 2389 (1985).
325. N. K. Mohanty and R. K. Nanda, *Bull. Chem. Soc. Jpn.*, **58**, 3597 (1985).
326. A. S. Sarac, *Int. J. Chem. Kinet.*, **17**, 1333 (1985).
327. J. Dziegiec, *Polish J. Chem.*, **59**, 731 (1985).
328. N. M. Gluindy, S. Hanafi, and S. S. Anis, *Thermochim. Acta*, **92**, 751 (1985).
329. J. Dziegiec, *Polish J. Chem.*, **59**, 721 (1985).
330. N. Gupta, P. C. Nigam, and R. M. Naik, *Indian J. Chem.*, **25A(1)**, 39 (1986); *Chem. Abstr.*, **104**, 156587y.
331. H. Yonemura, H. Nishino, and K. Kurosawa, *Bull. Chem. Soc. Jpn.*, **59**, 3153 (1986).
332. M. Ignaczak and G. Andrijewski, *Polish J. Chem.*, **59**, 675 (1985).
333. P. V. Sreenivasulu, M. Adinarayana, B. Sethuram, and T. Navaneeth Rao, *Int. J. Chem. Kinet.*, **17**, 1017 (1985).
334. M. M. Girgis, S. A. El-Shatoury, and Z. H. Kalil, *Can. J. Chem.*, **63**, 3317 (1985).
335. R. Gurumurthy and M. Gopalakrishnan, *Indian J. Chem.*, **25A**, 476 (1986).
336. V. K. Sharma, K. Sharma, A. P. Payasi, and P. S. Tiwari, *Z. Phys. Chemie, Leipzig*, **267**, 821 (1986).
337. V. P. Shilov, *Radiokhimiya*, **27**, 581 (1985).
338. S. K. Saha, M. C. Ghosh, and P. Banerjee, *J. Chem. Soc., Dalton Trans.*, 1301 (1986).
339. K. K. S. Gupta, S. Dey, S. Gupta, and A. Banerjee, *J. Chem. Soc., Perkin Trans. 2*, 1503 (1985).
340. N. D. Valechha and A. Pradhan, *Indian J. Chem.*, **24A**, 773 (1985).
341. Y. Zhang, B. Wenderoth, W. Su, and E. C. Ashby, *J. Organomet. Chem.*, **292**, 29 (1985).
342. A. Vlcek, Jr. *J. Organomet. Chem.*, **306**, 63 (1986).
343. S. Gambarotta, S. Strologo, C. Floriani, A. Chiesi-Villa, and C. Guastini, *J. Am. Chem. Soc.*, **107**, 6278 (1985).
344. W. C. Kupferschmidt and R. B. Jordan, *Inorg. Chem.*, **24**, 3357 (1985).
345. S. E. Creager and R. W. Murray, *Inorg. Chem.*, **24**, 3824 (1985).
346. J.-E. Backvall and A. Heumann, *J. Am. Chem. Soc.*, **108**, 7107 (1986).
347. F. Mares, S. E. Diamond, F. J. Regina, and J. P. Solar, *J. Am. Chem. Soc.*, **107**, 3545 (1985).
348. J. T. Groves and Y. Watanabe, *J. Am. Chem. Soc.*, **108**, 507 (1986).
349. J. F. Perez-Benito and D. G. Lee, *Can. J. Chem.*, **63**, 3545 (1985).
350. A. M. Van Den Bergen, R. L. Elliott, C. J. Lyons, K. P. Mackinnon, and B. O. West, *J. Organomet. Chem.*, **297**, 361 (1985).
351. C. J. Schlesener, C. Amatore, and J. K. Kochi, *J. Phys. Chem.*, **90**, 3747 (1986).
352. R. S. Paonessa, N. C. Thomas, and J. Halpern, *J. Am. Chem. Soc.*, **107**, 4333 (1985).
353. K. J. Del Rossi and B. B. Wayland, *J. Am. Chem. Soc.*, **107**, 7941 (1985).
354. T. Ohno, *J. Phys. Chem.*, **89**, 5709 (1985).
355. T. Morito, M. Hirano, K. Echigoya, and T. Sato, *J. Chem. Soc., Perkin Trans. 2*, 1205 (1986).
356. T. Morimoto, M. Hirano, and T. Koyama, *J. Chem. Soc., Perkin Trans. 2*, 1109 (1985).
357. M. Hirano and T. Morimoto, *J. Chem. Soc., Perkin Trans. 2*, 1105 (1985).
358. P. J. Baricelli and P. C. H. Mitchell, *Inorg. Chim. Acta*, **115**, 163 (1986).
359. C. A. Bignozzi, C. Chiorboli, M. T. Indelli, M. A. Rampi Scandola, G. Varani, and F. Scandola, *J. Am. Chem. Soc.*, **108**, 7872 (1986).
360. I. V. Kozhevnikov, V. I. Kim, E. P. Talzi, and V. N. Sidelnikov, *J. Chem. Soc., Chem. Commun.*, 1392 (1985).

References for Chapter 4

1. M. van Duin, J. A. Peters, A. P. G. Kieboom, and H. van Bekkum, *Tetrahedron*, **41**, 3411 (1985).
2. J. G. Dawber and S. I. E. Green, *J. Chem. Soc., Faraday Trans. 1*, **82**, 3407 (1986).

3. M. Makkee, A. P. G. Kieboom, and H. van Bekkum, *Rec. Trav. Chim. Pays-Bas*, **104**, 230 (1985).
4. M. van Duin, J. A. Peters, A. P. G. Kieboom, and H. van Bekkum, *Rec. Trav. Chim. Pays-Bas*, **104**, 488 (1986).
5. D. E. Minter, C. R. Kelly, and H. C. Kelly, *Inorg. Chem.*, **25**, 3291 (1986).
6. R. Csuk, N. Müller, and H. Sterk, *Z. Naturforsch.*, **40b**, 987 (1985).
7. U. Höbel, H. Nöth, and H. Prigge, *Chem. Ber.*, **119**, 325 (1986).
8. R. H. Cragg, T. J. Miller, and D. O'N. Smith, *J. Organomet. Chem.*, **302**, 19 (1986).
9. C. Brown, R. H. Cragg, T. J. Miller, and D. O'N. Smith, *J. Organomet. Chem.*, **296**, C17 (1985).
10. R. H. Cragg, T. J. Miller, and D. O'N. Smith, *J. Organomet. Chem.*, **291**, 273 (1985).
11. P. Kölle and H. Nöth, *Chem. Ber.*, **119**, 313 (1986).
12. A. Brandl and H. Nöth, *Chem. Ber.*, **118**, 3759 (1985).
13. R. Greatrex, N. N. Greenwood, and C. D. Potter, *J. Chem. Soc., Dalton Trans.*, 81 (1986).
14. M. Colombier, J. Atchekazaï, and H. Mongeot, *Inorg. Chim. Acta*, **115**, 11 (1986).
15. D. Power and T. R. Spalding, *Polyhedron*, **4**, 1329 (1985).
16. H. Horii and S. Taniguchi, *J. Chem. Soc., Chem. Commun.*, 915 (1986).
17. J. A. Baban, V. P. J. Marti, and B. P. Roberts, *J. Chem. Soc., Perkin Trans. 2*, 1723 (1985).
18. J. A. Baban, J. P. Goddard, and B. P. Roberts, *J. Chem. Res. (S)*, 30 (1986).
19. I. G. Green and B. P. Roberts, *J. Chem. Soc., Perkin Trans. 2*, 1597 (1986).
20. V. P. J. Marti and B. P. Roberts, *J. Chem. Soc., Perkin Trans. 2*, 1613 (1986).
21. J. A. Baban, B. P. Roberts, and A. C. H. Tsang, *J. Chem. Res. (S)*, 334 (1986).
22. J. A. Baban, B. P. Roberts, and A. C. H. Tsang, *J. Chem. Soc., Chem. Commun.*, 955 (1985).
23. J. A. Baban and B. P. Roberts, *J. Chem. Soc., Perkin Trans. 2*, 1607 (1986).
24. J. A. Baban, N. J. Goodchild, and B. P. Roberts, *J. Chem. Soc, Perkin Trans. 2*, 157 (1986).
25. H. C. Brown, P. V. Ramachandran, and J. V. N. Vara Prasad, *J. Org. Chem.*, **50**, 5583 (1985).
26. H. C. Brown and U. S. Racherla, *J. Org. Chem.*, **51**, 895 (1986).
27. H. C. Brown, D. Basavaiah, S. U. Kulkarni, H. D. Lee, E. Negishi, and J.-J. Katz, *J. Org. Chem.*, **51**, 5270 (1986).
28. H. C. Brown, T. Imai, and N. G. Bhat, *J. Org. Chem.*, **51**, 5277 (1986).
29. H. C. Brown, H. D. Lee, and S. U. Kulkarni, *J. Org. Chem.*, **51**, 5282 (1986).
30. H. C. Brown, B. Nazer, J. S. Cha, and J. A. Sikorski, *J. Org. Chem.*, **51**, 5264 (1986).
31. H. C. Brown, P. V. Ramachandran, and J. Chandrasekharan, *Organometallics*, **5**, 2138 (1986).
32. B. E. Mann, P. W. Cutts, J. McKenna, J. M. McKenna, and C. M. Spencer, *Angew. Chem., Int. Ed. Engl.*, **25**, 577 (1986).
33. D. J. Nelson and P. J. Cooper, *Tetrahedron Lett.*, **27**, 4693 (1986).
34. M. L. McKee, *Inorg. Chem.*, **25**, 3545 (1986).
35. J. A. Soderquist and M. R. Najafi, *J. Org. Chem.*, **51**, 1330 (1986).
36. H. C. Brown, M. M. Midland, and G. W. Kabalka, *Tetrahedron*, **42**, 5523 (1986).
37. H. C. Brown, C. Snyder, B. C. Subba Rao, and G. Zweifel, *Tetrahedron*, **42**, 5505 (1986).
38. H. C. Brown and C. P. Garg, *Tetrahedron*, **42**, 5511 (1986).
39. H. C. Brown, S. U. Kulkarni, C. Gundu Rao, and V. D. Patil, *Tetrahedron*, **42**, 5515 (1986).
40. H. C. Brown and K. J. Murray, *Tetrahedron*, **42**, 5497 (1986).
41. B. Ng, T. Onak, and K. Fuller, *Inorg. Chem.*, **24**, 4371 (1986).
42. Z. J. Abdou, M. Soltis, B. Ok, G. Siwap, T. Banuelos, W. Nam, and T. Onak, *Inorg. Chem.*, **24**, 2363 (1985).
43. B. Ng, T. Onak, T. Banuelos, F. Gomez, and E. W. DiStefano, *Inorg. Chem.*, **24**, 4091 (1985).
44. Z. J. Abdou, G. Abdou, T. Onak, and S. Lee, *Inorg. Chem.*, **25**, 2678 (1986).
45. B. M. Gimarc and J. J. Ott, *Inorg. Chem.*, **25**, 83 (1986).
46. B. M. Gimarc and J. J. Ott, *Inorg. Chem.*, **25**, 2708 (1986).
47. P. Paneth and M. H. O'Leary, *J. Am. Chem. Soc.*, **107**, 7381 (1985).
48. J-Y. Liang and W. N. Lipscomb, *J. Am. Chem. Soc.*, **108**, 5051 (1986).
49. M. E. Grice, K. Song, and W. J. Chesnavich, *J. Phys. Chem.*, **90**, 3503 (1986).

50. J. Flanagan, D. P. Jones, W. P. Griffiths, A. C. Skapski, and A. P. West, *J. Chem. Soc., Chem. Commun.*, 20 (1986).

51. J. O. Edwards, T. E. Erstfeld, K. M. Ibne-Rasa, G. Levey, and M. Moyer, *Int. J. Chem. Kinet.*, **18**, 165 (1986).

52. C. D. Ritchie and Y. Tang, *J. Org. Chem.*, **51**, 3555 (1986).

53. D. Barth, C. Tondre, and J.-J. Delpuech, *Int. J. Chem. Kinet.*, **18**, 445 (1986).

54. H. Yashiro, J. Takeshita, A. Okuwaki, and T. Okabe, *Bull. Chem. Soc. Jpn.*, **59**, 1331 (1986).

55. T. P. Elgano, V. Ramakrishnan, S. Vancheesan, and J. C. Kuriacose, *Tetrahedron*, **41**, 3837 (1985).

56. Q. G. Mulazzani, M. D'Angelantonio, M. Venturi, M. Z. Hoffman, and M. A. J. Rodgers, *J. Phys. Chem.*, **90**, 5347 (1986).

57. P. S. Surdhar, D. A. Armstrong, and V. Massey, *Can. J. Chem.*, **64**, 67 (1986).

58. A. R. Fratzke and P. J. Reilly, *Int. J. Chem. Kinet.*, **18**, 775 (1986).

59. A. R. Fratzke and P. J. Reilly, *Int. J. Chem. Kinet.*, **18**, 757 (1986).

60. H. Horii, Y. Abe, and S. Taniguchi, *Bull. Chem. Soc. Jpn.*, **59**, 721 (1986).

61. H. Horii, Y. Abe, and S. Taniguchi, *Bull. Chem. Soc. Jpn.*, **58**, 2751 (1985).

62. Y. Chiang and A. J. Kresge, *J. Am. Chem. Soc.*, **107**, 6363 (1985).

63. R. A. McClelland and C. Moreau, *Can. J. Chem.*, **63**, 2673 (1985).

64. S. Jaarinen, J. Niiranen, and J. Koskikallio, *Int. J. Chem. Kinet.*, **17**, 925 (1985).

65. R. A. McClelland, N. Banait, and S. Steenken, *J. Am. Chem. Soc.*, **108**, 7023 (1986).

66. S. Steenken, J. Buschek, and R. A. McClelland, *J. Am. Chem. Soc.*, **108**, 2808 (1986).

67. R. Schneider, U. Grabis, and H. Mayr, *Angew. Chem., Int. Ed. Engl.*, **25**, 89 (1986).

68. R. Schneider and H. Mayr, *Angew. Chem., Int. Ed. Engl.*, **25**, 1016 (1986).

69. H. Mayr, R. Schneider, and U. Grabis, *Angew. Chem., Int. Ed. Engl.*, **25**, 1017 (1986).

70. R. Ta-Shma and W. P. Jencks, *J. Am. Chem. Soc.*, **108**, 8040 (1986).

71. A. R. Stein and E. A. Moffatt, *Can. J. Chem.*, **63**, 3433 (1985).

72. R. Mehnert, O. Brede, and W. Naumann, *Ber. Bunsenges. Phys. Chem.*, **89**, 1031 (1985).

73. S. F. Nelson, D. L. Kapp, R. Akaba, and D. H. Evans, *J. Am. Chem. Soc.*, **108**, 6863 (1986).

74. R. H. Smith, Jr., S. R. Koepke, Y. Tondeur, C. L. Denlinger, and C. J. Michejda, *J. Chem. Soc., Chem. Commun.*, 936 (1985).

75. K. Laali and V. Gold, *J. Org. Chem.*, **51**, 2395 (1986).

76. V. V. Krishnamurthy, G. K. Surya Prakash, P. S. Iyer, and G. A. Olah, *J. Am. Chem. Soc.*, **107**, 5015 (1985).

77. W. Kirmse, J. Rode, and K. Rode, *Chem. Ber.*, **119**, 3672 (1986).

78. W. Kirmse and J. Rode, *Chem. Ber.*, **119**, 3694 (1986).

79. G. A. Olah, G. K. Surya Prakash, R. W. Ellis, and J. A. Olah, *J. Chem. Soc., Chem. Commun.*, 9 (1986).

80. H. Choukroun, D. Brunel, and A. Germain, *J. Chem. Soc., Chem. Commun.*, 6 (1986).

81. G. Carr and D. Whittaker, *J. Chem. Soc., Chem. Commun.*, 1245 (1986).

82. M. Lahti, *Acta Chem. Scand., Ser. A*, **40**, 350 (1986).

83. J. L. Maienschein and P. E. Barry, *Int. J. Chem. Kinet.*, **18**, 739 (1986).

84. T. Minato and S. Yamabe, *J. Am. Chem. Soc.*, **107**, 4621 (1985).

85. D. K. Bohme and A. B. Raksit, *Can. J. Chem.*, **63**, 3007 (1985).

86. E. Kaufmann, P. von R. Schleyer, K. N. Houk, and Y.-D. Wu, *J. Am. Chem. Soc.*, **107**, 5560 (1985).

87. D. D. M. Wayner and D. Griller, *J. Am. Chem. Soc.*, **107**, 7764 (1985).

88. N. J. Bunce, K. U. Ingold, J. P. Landers, J. Lusztyk, and J. C. Scaiano, *J. Am. Chem. Soc.*, **107**, 5464 (1985).

89. J. A. Hawari, S. Davis, P. S. Engel, B. C. Gilbert, and D. Griller, *J. Am. Chem. Soc.*, **107**, 4721 (1985).

90. G. L. Closs and O. D. Redwine, *J. Am. Chem. Soc.*, **107**, 6131 (1985).

91. A. L. J. Beckwith, V. W. Bowry, M. O'Leary, G. Moad, E. Rizzardo, and D. H. Solomon, *J. Chem. Soc., Chem. Commun.*, 1003 (1986).

92. J. A. Franz, N. K. Suleman, and M. S. Alnajjar, *J. Org. Chem.*, **51**, 19 (1986).
93. L. J. Johnston, J. Lusztyk, D. D. M. Wayner, A. N. Abeywickreyma, A. L. Beckwith, J. C. Scaiano, and K. U. Ingold, *J. Am. Chem. Soc.*, **107**, 4594 (1985).
94. C. Chatgilialoglu, *J. Org. Chem.*, **51**, 2871 (1986).
95. L. J. Johnston and K. U. Ingold, *J. Am. Chem. Soc.*, **108**, 2343 (1986).
96. L. J. Johnston and J. C. Scaiano, *J. Am. Chem. Soc.*, **107**, 6368 (1985).
97. A. Bakac and J. H. Espenson, *J. Phys. Chem.*, **90**, 325 (1986).
98. S. W. Benson, *J. Phys. Chem.*, **89**, 4366 (1985).
99. D. Malwitz and J. O. Metzger, *Angew. Chem., Int. Ed. Engl.*, **25**, 762 (1986).
100. D. Malwitz and J. O. Metzger, *Chem. Ber.*, **119**, 3558 (1986).
101. J. O. Metzger, *Angew. Chem., Int. Ed. Engl.*, **25**, 80 (1986).
102. J. A. Baban, J. P. Goddard, and B. P. Roberts, *J. Chem. Soc., Perkin Trans. 2*, 1269 (1986).
103. R. L. Barcus, L. M. Hadel, L. J. Johnston, M. S. Platz, T. G. Savino, and J. C. Scaiano, *J. Am. Chem. Soc.*, **108**, 3928 (1986).
104. H. Tomioka, Y. Ozaki, and Y. Izawa, *Tegrahedron*, **41**, 4987 (1985).
105. H. Tomioka, N. Hayashi, and Y. Izawa, *Chem. Lett.*, 695 (1986).
106. D. Bethell and V. D. Parker, *J. Am. Chem. Soc.*, **108**, 7194 (1986).
107. S. Cheng and M. D. Hawley, *J. Org. Chem.*, **51**, 3799 (1986).
108. T. Campino, J. G. Santos, and F. Ibanez, *J. Chem. Soc., Perkin Trans. 2*, 1021 (1986).
109. L. Griffiths, C. S. Cundy, and R. J. Plaisted, *J. Chem. Soc., Chem. Commun.*, 2265 (1986).
110. S. D. Kinrade and T. W. Swaddle, *J. Chem. Soc., Chem. Commun.*, 120 (1986).
111. S. D. Kinrade and T. W. Swaddle, *J. Am. Chem. Soc.* **108**, 7159 (1986).
112. C. T. G. Knight, J. Kirkpatrick, and E. Oldfield, *J. Chem. Soc., Chem. Commun.*, 66 (1986).
113. I. L. Svensson, S. Sjöberg, and L. O. Öhman, *J. Chem. Soc., Faraday Trans.*, **82**, 3635 (1986).
114. I. Hasegawa, K. Kuroda, and C. Kato, *Bull. Chem. Soc. Jpn.*, **59**, 2279 (1986).
115. I. Artaki, M. Bradley, T. W. Zerda, and J. Jonas, *J. Phys. Chem.* **89**, 4399 (1985).
116. N. Shirai, K. Moriya, and Y. Kawazoe, *Tetrahedron*, **42**, 2211 (1986).
117. K. A. Smith, *J. Org. Chem.*, **51**, 3827 (1986).
118. G. J. Chen and C. Tamborski, *J. Organomet. Chem.*, **293**, 313 (1985).
119. W. Rutz, D. Lange, and H. Kelling, *Z. Anorg. Allg. Chem.*, **528**, 98 (1986).
120. W. Rutz, D. Lange, E. Popowski, and H. Kelling, *Z. Anorg. Allg. Chem.*, **536**, 197 (1986).
121. W. Rutz, D. Lange, E. Popowski, and H. Kelling, *Z. Anorg. Allg. Chem.*, **542**, 217 (1986).
122. K. Rühlmann, J. Brumme, U. Scheim, and H. Grosse-Ruyken, *J. Organomet. Chem.*, **291**, 165 (1985).
123. Z. Lasocki and M. Witekowa, *J. Organomet. Chem.*, **311**, 17 (1986).
124. R. Curci, R. Mello, and L. Troisi, *Tetrahedron*, **42**, 877 (1986).
125. B. C. Chakoumakos and G. V. Gibbs, *J. Phys. Chem.*, **90**, 996 (1986).
126. J. A. Stone, A. C. M. Wojtyniak, and W. Wytenberg, *Can. J. Chem.*, **64**, 575 (1986).
127. E. Lukevics and M. Dzintara, *J. Organomet. Chem.*, **295**, 265 (1985).
128. U. Scheim, H. Grosse-Ruyken, K. Rühlmann, and A. Porzel, *J. Organomet. Chem.*, **293**, 29 (1985).
129. U. Scheim, K. Rühlmann, H. Grosse-Ruyken, and A. Porzel, *J. Organomet. Chem.*, **314**, 39 (1986).
130. A. R. Bassindale and T. Stout, *Tetrahedron Lett.*, **26**, 3403 (1985).
131. A. R. Bassindale and T. Stout, *J. Chem. Soc., Perkin Trans. 2*, 221 (1986).
132. A. R. Bassindale and T. Stout, *J. Chem. Soc., Perkin Trans. 2*, 227 (1986).
133. J. B. Lambert, J. A. McConnell, and W. J. Schulz, Jr., *J. Am. Chem. Soc.*, **108**, 2482 (1986).
134. R. Tacke, M. Link, and H. Zilch, *Chem. Ber.*, **118**, 4637 (1985).
135. R. Damrauer and S. E. Danahey, *Organometallics*, **5**, 1490 (1986).
136. J. Boyer, C. Breliere, R. J. P. Corriu, A. Kpoton, M. Poirier, and G. Royo, *J. Organomet. Chem.*, **311**, C39 (1986).
137. C. Brelière, F. Carré, R. J. P. Corriu, M. Poirier, and G. Royo, *Organometallics*, **5**, 389 (1986).
138. J. D. Nies, J. M. Bellama, and N. Ben-Zvi, *J. Organomet. Chem.*, **296**, 315 (1985).

139. J. Boyer, R. J. P. Corriu, A. Kpoton, M. Mazhar, M. Poirier, and G. Royo, *J. Organomet. Chem.*, **301**, 131 (1986).

140. R. J. P. Corriu, M. Mazhar, M. Poirier, and G. Royo, *J. Organomet. Chem.*, **306**, C5 (1986).

141. A. Boudin, G. Cerveau, C. Chuit, R. J. P. Corriu, and C. Reye, *Angew. Chem. Int. Ed. Engl.*, **25**, 473 (1986).

142. A. Boudin, G. Cerveau, C. Chuit, R. J. P. Corriu, and C. Reye, *Angew. Chem. Int. Ed. Engl.*, **25**, 474 (1986).

143. W. H. Stevenson, III, S. Wilson, J. C. Martin, and W. B. Farnham, *J. Am. Chem. Soc.*, **107**, 6340 (1985).

144. W. H. Stevenson, III and J. C. Martin, *J. Am. Chem. Soc.*, **107**, 6352 (1985).

145. G. Klebe, *J. Organomet. Chem.*, **293**, 147 (1985).

146. Yu. L. Frolov, S. G. Shevchenko, and M. G. Voronkov, *J. Organomet. Chem.*, **292**, 159 (1985).

147. R. Damrauer, V. E. Yost, S. E. Danahey, and B. K. O'Connell, *Organometallics*, **4**, 1779 (1985).

148. D. Yang and D. D. Tanner, *J. Org. Chem.*, **51**, 2267 (1986).

149. J. M. Rozell, Jr. and P. R. Jones, *Organometallics*, **4**, 2206 (1986).

150. H. Oda, M. Sato, Y. Morizawa, K. Oshima, and H. Nozaki, *Tetrahedron*, **41**, 3257 (1985).

151. H. Zilch and R. Tacke, *J. Organomet. Chem.*, **316**, 243 (1986).

152. T. V. RajanBabu, G. S. Reddy, and T. Fukunaga, *J. Am. Chem. Soc.*, **107**, 5473 (1985).

153. L. P. Davis, L. W. Burggraf, M. S. Gordon, and K. K. Daldridge, *J. Am. Chem. Soc.*, **107** 4415 (1985).

154. C. Eaborn and D. E. Reed, *J. Chem. Soc., Perkin Trans. 2*, 1695 (1985).

155. C. Eaborn and D. E. Reed, *J. Chem. Soc., Perkin Trans. 2*, 1687 (1985).

156. C. Eaborn and P. D. Lickiss, *J. Organomet. Chem.*, **294**, 305 (1985).

157. C. Eaborn, P. D. Lickiss, S. T. Najim, and M. N. Romanelli, *J. Chem. Soc., Chem. Commun.*, 1754 (1985).

158. A. I. Al-Wassil, C. Eaborn, and M. N. Romanelli, *J. Chem. Soc., Perkin Trans. 2*, 1363 (1986).

159. C. Eaborn, P. D. Lickiss, S. T. Najim, and M. N. Romanelli, *J. Organomet. Chem.* **315**, C5 (1986).

160. G. A. Ayoko and C. Eaborn, *J. Chem. Soc., Perkin Trans. 2*, 1289 (1986).

161. G. A. Ayoko and C. Eaborn, *J. Chem. Soc., Perkin Trans. 2*, 1357 (1986).

162. G. A. Ayoko and C. Eaborn, *J. Chem. Soc., Chem. Commun.*, 630 (1986).

163. R. I. Damja, C. Eaborn, and W.-C. Sham, *J. Organomet. Chem.*, **291**, 25 (1985).

164. R. I. Damja and C. Eaborn, *J. Organomet. Chem.*, **290**, 267 (1985).

165. N. H. Buttrus, R. I. Damja, C. Eaborn, P. B. Hitchcock, and P. D. Lickiss, *J. Chem. Soc., Chem. Commun.*, 1385 (1985).

166. J. Chmielecka, J. Chojnowski, C. Eaborn, and W. A. Stanczyk, *J. Chem. Soc., Perkin Trans. 2*, 1779 (1985).

167. P. Dembech, G. Seconi, C. Eaborn, J. A. Rodriguez, and J. G. Stamper, *J. Chem. Soc., Perkin Trans. 2*, 197 (1986).

168. F. Effenberger and W. Spiegler, *Chem. Ber.*, **118**, 3872 (1985).

169. J. W. Wilt, *Tetrahedron*, **41**, 3979 (1985).

170. N. Auner, R. Walsh, and J. Westrup, *J. Chem. Soc., Chem. Commun.*, 207 (1986).

171. H. Mayr and R. Polk, *Tetrahedron*, **42**, 4211 (1986).

172. J. Yoshida, T. Murata, and S. Isoe, *Tetrahedron Lett.*, **27**, 3373 (1986).

173. V. J. Shiner, Jr., M. W. Ensinger, and G. S. Kriz, *J. Am. Chem. Soc.*, **108**, 842 (1986).

174. E. R. Davidson and V. J. Shiner, Jr., *J. Am. Chem. Soc.*, **108**, 3135 (1986).

175. M. Kira, H. Yoshida, and H. Sakurai, *J. Am. Chem. Soc.*, **107**, 7767 (1985).

176. M. T. Craw, M. Alberti, M. C. Depew, and J. K. S. Wan., *Bull. Chem. Soc. Jpn.*, **58**, 3675 (1985).

177. T. J. Barton, A. Revis, I. M. T. Davidson, S. Ijadi-Maghsoodi, K. J. Hughes, and M. S. Gordon, *J. Am. Chem. Soc.*, **108**, 4022 (1986).

178. J. G. Martin, M. A. Ring, and H. E. O'Neal, *Organometallics*, **5**, 1228 (1986).

179. I. M. T. Davidson and S. Ijadi-Maghsoodi, *Organometallics*, **5**, 2086 (1986).

180. J. Lusztyk, B. Maillard, and K. U. Ingold, *J. Org. Chem.*, **51**, 2457 (1986).

181. L. Fabry, P. Potzinger, B. Reimann, A. Ritter and H. P. Steenbergen, *Organometallics*, **5**, 1231 (1986).
182. H. Watanabe, M. Kato, E. Tabei, H. Kuwabara, N. Hirai, T. Sato, and Y. Nagai, *J. Chem. Soc., Chem. Commun.*, 1662 (1986).
183. H. Watanabe, K. Yoshizumi, T. Muraoka, M. Kato, Y. Nagai, and T. Sato, *Chem. Lett.* 1683 (1985).
184. Y. Nakadaira, N. Komatsu, and H. Sakurai, *Chem. Lett.* 1781 (1985).
185. T. A. Blinka and R. West, *Organometallics*, **5**, 128 (1986).
186. G. A. Razuvaev, V. V. Semenova, T. N. Brevnova, and A. N. Kornev, *Zh. Obshch. Khim.*, **56**, 884 (1986).
187. G. A. Razuvaev, T. N. Brevnova, V. V. Semenova, A. N. Kornev, M. A. Lopatin, G. V. Belysheva, and A. N. Egorochkin, *Bull. Acad. Sci. USSR, Div. Chem. Sci.*, **34**, 2010 (1985).
188. M. Weidenbruch and K.-L. Thom, *J. Organomet. Chem.*, **308**, 177 (1986).
189. H. B. Yokelson, J. Maxha, D. A. Siegel, and R. West, *J. Am. Chem. Soc.*, **108**, 4239 (1986).
190. N. Auner, I. M. T. Davidson, and S. Ijadi-Maghsoodi, *Organometallics*, **4**, 2210 (1985).
191. N. Auner, I. M. T. Davidson, S. Ijadi-Maghsoodi, and F. T. Lawrence, *Organometallics*, **5**, 431 (1986).
192. R. T. Conlin and Y.-W. Kwak, *Organometallics*, **5**, 1205 (1986).
193. D. S. Rogers, M. A. Ring, and H. E. O'Neal, *Organometallics*, **5**, 1521 (1986).
194. M.-H. Yeh, L. Linder, D. K. Hoffman, and T. J. Barton, *J. Am. Chem. Soc.*, **108**, 7849 (1986).
195. P. P. Gaspar and D. Lei, *Organometallics*, **5**, 1276 (1986).
196. D. Lei and P. P. Gaspar, *Organometallics*, **4**, 1471 (1985).
197. H. Appler, L. W. Gross, B. Mayer, and W. P. Neumann, *J. Organomet. Chem.*, **291**, 9 (1985).
198. S. K. Bains, P. N. Noble, and R. Walsh, *J. Chem. Soc., Faraday Trans. 2*, **82**, 837 (1986).
199. S. Konieczy, P. P. Gaspar, and J. Wormhoudt, *J. Organomet. Chem.*, **307**, 151 (1986).
200. D. S. Rogers, M. A. Ring, and H. E. O'Neal, *Organometallics*, **5**, 1467 (1986).
201. R. Damrauer, C. H. DePuy, I. M. T. Davidson, and K. J. Hughes, *Organometallics*, **5**, 2054 (1986).
202. D. Lei and P. P. Gaspar, *J. Chem. Soc., Chem. Commun.*, 1149 (1985).
203. A. G. Brook, K. D. Safa, P. D. Lickiss, and K. M. Baines, *J. Am. Chem. Soc.*, **107**, 4338 (1985).
204. R. Damrauer, C. H. DePuy, I. M. T. Davidson, and K. J. Hughes, *Organometallics*, **5**, 2050 (1986).
205. A. G. Brook and H.-J. Wessely, *Organometallics*, **4**, 1487 (1985).
206. P. R. Jones, J. M. Rozell, Jr., and B. M. Campbell, *Organometallics*, **4**, 1321 (1985).
207. R. T. Conlin and Y.-K. Kwak, *J. Am. Chem. Soc.*, **108**, 834 (1986).
208. B. H. Boo and P. P. Gaspar, *Organometalllics*, **5**, 698 (1986).
209. I. M. T. Davidson and A. Fenton, *Organometallics*, **4**, 2060 (1985).
210. I. M. T. Davidson, A. Fenton, G. Manuel, and G. Bertrand, *Organometallics*, **4**, 1324 (1985).
211. W. Tumas, K. E. Salomon, and J. I. Brauman, *J. Am. Chem. Soc.*, **108**, 2541 (1986).
212. T. J. Barton and B. L. Groh, *J. Am. Chem. Soc.*, **107**, 7221 (1985).
213. T. Kudo and S. Nagase, *Organometallics*, **5**, 1207 (1986).
214. A. Tachibana, H. Fueno, and T. Yamabe, *J. Am. Chem. Soc.*, **108**, 4346 (1986).
215. L. Linder, A. Revis, and T. J. Barton, *J. Am. Chem. Soc.*, **108**, 2742 (1986).
216. T. J. Barton and G. P. Hussmann, *J. Am. Chem. Soc.*, **107**, 7581 (1985).
217. G. Raabe and J. Michl, *Chem. Rev.*, **85**, 419 (1985).
218. A. G. Brook and K. M. Baines, *Adv. Organomet. Chem.*, **25**, 1 (1986).
219. A. G. Brook, *J. Organomet. Chem.*, **300**, 21 (1986).
220. R. Walsh, *J. Phys. Chem.*, **90**, 389 (1986).
221. C. T. G. Knight, R. J. Kirkpatrick, and E. Oldfield, *J. Am. Chem. Soc.*, **108**, 30 (1986).
222. D. J. Brauer, J. Wilke, and R. Eujen, *J. Organomet. Chem.*, **316**, 261 (1986).
223. W Stanczyk, *J. Organomet. Chem.*, **299**, 15 (1986).
224. T. V. Leshina, V. I. Valyaev, M. B. Taraban, V. I. Maryasova, V. I. Rakhlin, S. Kh. Khangazheev, R. G. Mirskov, and M. G. Voronkov, *J. Organomet. Chem.*, **299**, 271 (1986).

225. S. Kozuka, S. Tamura, T. Yamazaki, S. Yamaguchi, and W. Tagaki, *Bull. Chem. Soc. Jpn.*, **58**, 3277 (1985).
226. P. Riviere, A. Castel, D. Guyot, and J. Satge, *J. Organomet. Chem.*, **290**, C15 (1985).
227. E. Michels and W. P. Neumann, *Tetrahedron Lett.*, **27**, 2455 (1986).
228. P. Jutzi and B. Hampel, *Organometallics*, **5**, 730 (1986).
229. K.-T. Kang, G. Manuel, and P. Weber, *Chem. Lett.*, 1685 (1986).
230. J. Barrau, M. El Amine, G. Rima, and J. Satge, *Can. J. Chem.*, **64**, 615 (1986).
231. D. A. Bazylinski and T. C. Hollocher, *Inorg. Chem.*, **24**, 1285 (1985).
232. C. E. Donald, M. N. Hughes, J. M. Thompson, and F. T. Bonner, *Inorg. Chem.*, **25**, 2676 (1986).
233. M. J. Akhtar, F. T. Bonner, M. N. Hughes, E. J. Humphreys, and C.-S. Lu, *Inorg. Chem.*, **25**, 4635 (1986).
234. N. V. Blough and O. C. Zafiriou, *Inorg. Chem.*, **24**, 3502 (1985).
235. F. T. Bonner and N.-Y. Wang, *Inorg. Chem.*, **25**, 1858 (1986).
236. N.-Y. Wang and F. T. Bonner, *Inorg. Chem.*, **25**, 1863 (1986).
237. D. Littlejohn, K. Y. Hu, and S. G. Chang, *Inorg. Chem.*, **25**, 3131 (1986).
238. D. Littlejohn, A. R. Wizansky, and S. G. Chang, *Inorg. Chem.*, **25**, 4610 (1986).
239. W. C. Nottingham and J. R. Sutter, *Int. J. Chem. Kinet.*, **18**, 1289 (1986).
240. M.-H. Herzog-Cance, A. Potier, and J. Potier, *Can. J. Chem.*, **63**, 1492 (1985).
241. A. Boughriet, A. Coumare, J.-C. Fischer, and M. Wartel, *J. Electroanal. Chem. Interfacial Electrochem.*, **200**, 217 (1986).
242. A. Boughriet, M. Wartel, and J.-C. Fischer, *Can. J. Chem,.* **64**, 5 (1986).
243. A. Boughriet, M. Wartel, J.-C. Fischer, and Y. Auger, *J. Electroanal. Chem. Interfacial Electrochem.*, **186**, 201 (1985).
244. A. Boughriet, M. Wartel, J.-C. Fischer, and C. Bremard, *J. Electroanal. Chem. Interfacial Electrochem.*, **190**, 103 (1985).
245. A. Boughriet, A. Coumare, J.-C. Fischer, M. Wartel, and G. Leman, *J. Electroanal. Chem. Interfacial Electrochem.*, **209**, 323 (1986).
246. A. H. Clemens, J. H. Ridd, and J. P. B. Sandall, *J. Chem. Soc., Perkin Trans. 2*, 1227 (1985).
247. A. H. Clemens, P. Helsby, J. H. Ridd, F. Al-Omran, and J. P. B. Sandall, *J. Chem. Soc., Perkin Trans. 2*, 1217 (1985).
248. P. Barrow, J. V. Bullen, A. Dent, T. Murphy, J. H. Ridd, and O. Sabek, *J. Chem. Soc., Chem. Commun.*, 1649 (1986).
249. M. R. Amin, L. Dekker, D. B. Hibbert, J. H. Ridd, and J. P. B. Sandall, *J. Chem. Soc., Chem. Commun.*, 658 (1986).
250. M. Ali and J. H. Ridd, *J. Chem. Soc., Perkin Trans. 2*, 327 (1986).
251. L. R. Dix and R. B. Moodie, *J. Chem. Soc., Perkin Trans. 2*, 1097 (1986).
252. M. P. Hartshorn, R. J. Martyn, W. T. Robinson, K. H. Sutton, J. Vaughan, and J. M. White, *Aust. J. Chem.*, **38**, 1613 (1985).
253. L. Eberson and F. Radner, *Acta Chem. Scand., Ser. B*, **40**, 71 (1986).
254. B. Zielinska, J. Arey, R. Atkinson, T. Ramdahl, A. M. Winer, and J. N. Pitts, Jr., *J. Am. Chem. Soc.*, **108**, 4126 (1986).
255. H. Tezuka, M. Kato, and Y. Sonehara, *J. Chem. Soc., Perkin Trans. 2*, 1643 (1985).
256. N. J. Bunce, S. R. Cater, and J. M. Willson, *J. Chem. Soc., Perkin Trans. 2*, 2013 (1985).
257. D. S. Ross, C.-L. Gu, G. P. Hum, and R. Malhotra, *Int. J. Chem. Kinet.*, **18**, 1277 (1986).
258. R. E. Huie and P. Neta, *J. Phys. Chem.*, **90**, 1193 (1986).
259. Z. B. Alfassi, R. E. Huie, and P. Neta, *J. Phys. Chem.*, **90**, 4156 (1986).
260. R. G. Coombes, A. W. Diggle, and S. P. Kempsell, *Tetrahedron Lett.*, **27**, 2037 (1986).
261. P. Neta and R. Huie, *J. Phys. Chem.*, **90**, 4644 (1986).
262. L. G. Forni, V. O. Mora-Arellano, J. E. Packer, and R. L. Willson, *J. Chem. Soc., Perkin Trans. 2*, 1 (1986).
263. D. G. Karraker, *Inorg. Chem.*, **24**, 4470 (1985).
264. G. A. Olah, R. Herges, J. D. Felberg, and G. K. Surya Prakash, *J. Am. Chem. Soc.*, **107**, 5282 (1985).

265. G. A. Olah, R. Herges, K. Laali, and G. A. Segal, *J. Am. Chem. Soc.*, **108**, 2054 (1986).
266. A. M. M. Doherty, M. S. Garley, K. R. Howes, and G. Stedman, *J. Chem. Soc., Perkin Trans.* 2, 143 (1986).
267. T. Bryant, D. L. H. Williams, M. H. H. Ali, and G. Stedman, *J. Chem. Soc., Perkin Trans.* 2, 193 (1986).
268. J. Casado, A. Castro, J. R. Leis, M. Mosquera, and M. E. Peña, *J. Chem. Soc., Perkin Trans.* 2, 1859 (1986)
269. J. Casado, A. Castro, E. Iglesias, M. E. Peña, and J. V. Tato, *Can. J. Chem.*, **64**, 133 (1986).
270. A. Castro, J. R. Leis, and M. E. Peña, *J. Chem. Res. (S)*, 216 (1986).
271. T. Bryant and D. L. H. Williams, *J. Chem. Soc., Perkin Trans.* 2, 1083 (1985).
272. A. Castro, E. Iglesias, J. R. Leis, M. E. Peña, J. V. Tato, and D. L. H. Williams, *J. Chem. Soc., Perkin Trans.* 2, 1165 (1986).
273. A. Castro, J. R. Leis, M. E. Peña, and J. V. Tato, *J. Chem. Soc., Perkin Trans.* 2, 117 (1986).
274. M. Isobe, *Bull. Chem. Soc. Jpn.*, **58**, 2844 (1986).
275. J. Casado, A. Castro, F. M. Lorenzo, and F. Meijide, *Monatsh. Chem.*, **117**, 335 (1986).
276. J. Casado, A. Castro, F. M. Lorenzo, F. Meijide, and M. Mosquera, *Bull. Soc. Chim. Fr.*, 597 (1985).
277. E. J. Hart, C.-H. Fischer, and A. Henglein, *J. Phys. Chem.*, **90**, 5989 (1986).
278. E. J. Hart and A. Henglein, *J. Phys. Chem.*, **90**, 5992 (1986).
279. O. Pytela, V. Vala, M. Ludwig, and M. Vecera, *Coll. Czech. Chem. Commun.*, **51**, 347 (1986).
280. P. Svoboda, O. Pytela, and M. Vecera, *Coll. Czech. Chem. Commun.*, **51**, 553 (1986).
281. R. H. Smith, Jr., C. L. Denlinger, R. Kupper, A. F. Mehl, and C. J. Michejda, *J. Am. Chem. Soc.*, **108**, 3726 (1986).
282. M. S. Ram and D. M. Stanbury, *J. Phys. Chem.*, **90**, 3691 (1986).
283. M. S. Ram and D. M. Stanbury, *Inorg. Chem.*, **24**, 4234 (1985).
284. Z. B. Alfassi and R. H. Schuler, *J. Phys. Chem.*, **89**, 3359 (1985).
285. M. Ye, K. P. Madden, R. W. Fessenden, and R. H. Schuler, *J. Phys. Chem.*, **90**, 5397 (1986).
286. Z. B. Alfassi, W. A. Prütz, and R. H. Schuler, *J. Phys. Chem.*, **89**, 1198 (1986).
287. J. C. Brand, B. P. Roberts, and R. Strube, *J. Chem. Soc., Perkin Trans.* 2, 1659 (1985).
288. R. van Eldik, H. Kelm, M. Schmittel, and C. Rüchardt, *J. Org. Chem.*, **50**, 2998 (1985).
289. N. Nishimura, T. Tanaka, and Y. Sueishi, *J. Chem. Soc., Chem. Commun.*, 903 (1985).
290. G. Lundkvist, U. Jonsson, and B. Lindeke, *J. Chem. Soc., Perkin Trans.* 2, 1377 (1986).
291. M. Culcasi, P. Tordo, and G. Gronchi, *J. Phys. Chem.* **90**, 1403 (1986).
292. E. M. Y. Quinga and G. D. Mendenhall, *J. Org. Chem.*, **50**, 2836 (1985).
293. E. M. Y. Quinga and G. D. Mendenhall, *J. Am. Chem. Soc.* **108**, 474 (1986).
294. P. S. Radhakrishnamurti, N. K. Rath, and R. K. Panda, *J. Chem. Soc., Dalton Trans.*, 1189 (1986).
295. S. L. Bruhn, A. Bakac, and J. H. Espenson, *Inorg. Chem.*, **25**, 535 (1986).
296. S. F. Nelsen and J. T. Ippoliti, *J. Org. Chem.*, **51**, 3169 (1986).
297. M. Dietrich, J. Heinze, H. Fischer, and F. A. Neugebauer, *Angew. Chem., Int. Ed. Engl.*, **25**, 1021 (1986).
298. S. F. Nelsen and S. C. Blackstock, *J. Am. Chem. Soc.* **107**, 7189 (1985).
299. S. F. Nelsen, in: *Molecular Structure and Energetics, Studies of Organic Molecules* (J. F. Liebman and A. Greenberg, eds.), Vol. 3, pp. 1–56, VCH Publishers, Deerfield Beach, Florida (1986).
300. J. M. Antelo, F. Arce, J. L. Armesto, A. Garcia-Verdugo, F. J. Penedo, and A. Varela, *Int. J. Chem. Kinet.*, **17**, 1231 (1985).
301. J. M. Antelo, F. Arce, J. Franco, M. J. Forneas, M. E. Sanchez, and A. Verela, *Int. J. Chem. Kinet.*, **18**, 1249 (1986).
302. K. Kumar, R. A. Day, and D. W. Margerum, *Inorg. Chem.*, **25**, 4344 (1986).
303. M. Ferriol, J. Gazet, and R. Ouani, *Bull. Soc. Chim. Fr.*, 507 (1986).
304. B. Yamuna, D. S. Mahadevappa, and H. M. K. Naidu, *Ind. J. Chem., Sect. A*, **25**, 663 (1986).
305. D. D. Tanner, R. Arhart, and C. P. Meintzer, *Tetrahedron*, **41**, 4261 (1985).

306. B. R. Cho, S. K. Namgoong, and R. A. Bartsch, *J. Org. Chem.*, **51**, 1320 (1986).
307. B. R. Cho, J. C. Yoon, and R. A. Bartsch, *J. Org. Chem.*, **50**, 4943 (1985).
308. M. Finkelstein, S. A. Hart, W. M. Moore, and S. D. Ross, *J. Org. Chem.*, **51**, 3548 (1986).
309. M. Finkelstein, W. M. Moore, S. D. Ross, and L. Eberson, *Acta Chem. Scand., Ser. B*, **40**, 402 (1986).
310. L. Eberson, M. Finkelstein, B. Folkesson, G. A. Hutchins, L. Jönsson, R. Larsson, W. M. Moore, and S. D. Ross, *J. Org. Chem.*, **51**, 4400 (1986).
311. L. Eberson, J. E. Barry, M. Finkelstein, W. M. Moore, and S. D. Ross, *Acta Chem. Scand., Ser. B*, **40**, 283 (1986).
312. J. C. Day, N. Govindaraj, D. S. McBain, P. S. Skell, and J. M. Tanko, *J. Org. Chem.*, **51**, 4959 (1986).
313. S. S. Kim, S. Y. Choi, and C. H. Kang, *J. Am. Chem. Soc.*, **107**, 4234 (1985).
314. U. Lüning, D. S. McBain, and P. S. Skell, *J. Org. Chem.*, **51**, 2077 (1986).
315. S. A. Glover, A. Goosen, C. W. McCleland, and J. L. Schoonraad, *J. Chem. Soc., Perkin Trans. 2*, 645 (1986).
316. D. D. Tanner, D. W. Reed, S. L. Tan, C. P. Meintzer, C. Walling, and A. Sopchik, *J. Am. Chem. Soc.*, **107**, 6576 (1985).
317. D. D. Tanner, and C. P. Meintzer, *J. Am. Chem. Soc.*, **107**, 6584 (1985).
318. U. Lüning, S. Seshadri, and P. S. Skell, *J. Org. Chem.*, **51**, 2071 (1986).
319. P. S. Skell, U. Lüning, D. S. McBain, and J. M. Tanko, *J. Am. Chem. Soc.*, **108**, 121 (1986).
320. C. Walling, G. M. El-Taliawi, and A. Sopchik, *J. Org. Chem.*, **51**, 736 (1986).
321. U. Lüning and P. S. Skell, *Tetrahedron*, **41**, 4289 (1985).
322. Y. L. Chow and R. A. Perry, *Can. J. Chem.*, **63**, 2203 (1985).
323. Y. L. Chow, T. W. Mojelsky, L. J. Magdzinski, and M. Tichy, *Can. J. Chem.*, **63**, 2197 (1985).
324. Y. Miura, T. Kunishi, and M. Kinoshita, *J. Org. Chem.*, **50**, 5862 (1985).
325. Y. Miura, Y. Shibata, and M. Kinoshita, *J. Org. Chem.*, **51**, 1239 (1986).
326. M. Negareche, Y. Badrudin, Y. Berchadsky, A. Friedmann, and P. Tordo, *J. Org. Chem.*, **51**, 342 (1986).
327. J. C. Evans, S. K. Jackson, C. C. Rowlands, and M. D. Barratt, *Tetrahedron*, **41**, 5191 (1985).
328. J. C. Evans, S. K. Jackson, C. C. Rowlands, and M. D. Barratt, *Tetrahedron*, **41**, 5195 (1985).
329. K. H. Schmidt, A. Bromberg, and D. Meisel, *J. Phys. Chem.*, **89**, 4352 (1985).
330. J. Mönig, R. Chapman, and K.-D. Asmus, *J. Phys. Chem.*, **89**, 3139 (1985).
331. S. Das and C. von Sonntag, *Z. Naturforsch.*, **41b**, 505 (1986).
332. D. Shaffer and J. Heicklen, *J. Phys. Chem.*, **90**, 4408 (1986).
333. R. W. Alder, M. Bonifacic, and K.-D. Asmus, *J. Chem. Soc., Perkin Trans. 2*, 277 (1986).
334. S. A. Fairhurst, K. M. Johnson, L. H. Sutcliffe, K. F. Preston, A. J. Banister, Z. V. Hauptman, and J. Passmore, *J. Chem. Soc., Dalton Trans.*, 1465 (1986).
335. F. Gerson, J. Knöbel, U. Buser, E. Vogel, and M. Zehnder, *J. Am. Chem. Soc.*, **108**, 3781 (1986).
336. A. Kumar and G. Bhattacharjee, *J. Chem. Soc., Perkin Trans. 2*, 61 (1986).
337. B. Singh, A. K. Samant, B. B. L. Saxena, and M. B. Singh, *Tetrahedron*, **42**, 857 (1986).
338. M. L. Bishnoi and K. K. Banerji, *Tetrahedron*, **41**, 6047 (1985).
339. M. K. Reddy, Ch. S. Reddy, and E. V. Sundaram, *Tetrahedron*, **41**, 3071 (1985).
340. S. Mittal, V. Sharma, and K. K. Banerji, *J. Chem. Res (S)*, 264 (1986).
341. M. L. Bishnoi, S. C. Negi, and K. K. Banerji, *Ind. J. Chem., Sect. A*, **25**, 660 (1986).
342. K. Mohan, S. Ananda, and D. S. Mahadevappa, *Ind. J. Chem., Sect. A*, **25**, 666 (1986).
343. B. T. Gowda and R. V. Rao, *Ind. J. Chem., Sect. A*, **25**, 578 (1986).
344. K. C. Gupta and K. Gupta, *Int. J. Chem. Kinet.*, **17**, 769 (1985).
345. M. S. Ramachandran, T. S. Vivekanandam, and R. Nithyanandhan, *J. Chem. Soc., Perkin Trans. 2*, 1507 (1985).
346. B. T. Gowda, B. S. Sherigara, D. S. Mahadevappa, and K. S. Rangappa, *Ind. J. Chem., Sect. A*, **24**, 932 (1985).
347. M. K. Reddy and E. V. Sundaram, *Ind. J. Chem., Sect., A*, **25**, 471 (1986).
348. V. Sharma and K. K. Banerji, *J. Chem. Res (S)*, 340 (1985).

349. K. K. Banerji, *J. Org. Chem.*, **51**, 4764 (1986).
350. D. S. Mahadevappa, K. Mohan, and S. Ananda, *Tetrahedron*, **42**, 4857 (1986).
351. S. Kothari and K. K. Banerji, *Can. J. Chem.*, **63**, 2726 (1985).
352. S. Kothari, V. Sharma, and K. K. Banerji, *J. Chem. Res (S)*, 234 (1985).
353. B. Singh and R. Shrivastav, *Tetrahedron*, **42**, 2749 (1986).
354. M. Jambulingam, P. Nanjappan, K. Natarajan, J. Nagalingam, M. Palaniswamy, N. Sivakumar, V. Prekumar, and K. Ramarajan, *J. Chem. Soc., Perkin Trans. 2*, 1699 (1985).
355. S. Mittal, V. Sharma, and K. K. Banerji, *Int. J. Chem. Kinet.*, **18**, 689 (1986).
356. B. Singh, A. K. Singh, and N. B. Singh, *Kinet. Catal. (Engl. Transl.)*, **27**, 83 (1986).
357. P. A. Delaney and R. A. W. Johnstone, *Tetrahedron*, **41**, 3845 (1985).
358. S. Perumal, S. Alagumalai, S. Selvaraj, and N. Arumugam, *Tetrahedron*, **42**, 4867 (1986).
359. K. Ganapathy and S. Kabilan, *Ind. J. Chem., Sect. A*, **25**, 681 (1986).
360. K. Ganapathy, R. Gurumurthy, N. Mohan, and G. Sivagnanam, *Ind. J. Chem. Sect. A*, **25**, 478 (1986).
361. B. Singh and A. K. Singh, *J. Ind. Chem. Soc.*, **62**, 523 (1985).
362. S. Kothari, A. Agarwal, and K. K. Banerji, *Ind. J. Chem., Sect. A*, **25**, 722 (1986).
363. R. A. Bednar and W. P. Jencks, *J. Am. Chem. Soc.*, **107**, 7135 (1986).
364. R. A. Bednar and W. P. Jencks, *J. Am. Chem. Soc.*, **107**, 7117 (1986).
365. R. A. Bednar and W. P. Jencks, *J. Am. Chem. Soc.*, **107**, 7126 (1986).
366. N. E. Briffert, F. Hibbert, and R. J. Sellens, *J. Am. Chem. Soc.*, **107**, 6712 (1985).
367. P. Pruszynski and A. Jarczewski, *J. Chem. Soc., Perkin Trans. 2*, 1117 (1986).
368. J. Chrisment, J. J. Delpuech, W. Rajerson, and C. Selve, *Tetrahedron*, **42**, 4743 (1986).
369. J. L. Kurz, M. W. Daniels, K. S. Cook, and M. M. Nasr, *J. Phys. Chem.*, **90**, 5357 (1986).
370. Y. Chiang, A. J. Kresge, S. Van Do, and D. P. Weeks, *J. Org. Chem.*, **51**, 4035 (1986).
371. M. Tsuhako, C. Sueyoshi, T. Miyajima, S. Ohashi, H. Nariai, and I. Motooka, *Bull. Chem. Soc. Jpn.*, **59**, 3091 (1986).
372. M. Tsuhako, A. Nakajima, T. Miyajima, S. Ohashi, H. Nariai, and I. Motooka, *Bull. Chem. Soc. Jpn.*, **58**, 3092 (1985).
373. R. N. Bose, R. D. Cornelius, and R. E. Viola, *Inorg. Chem.*, **24**, 3989 (1985).
374. S. H. Gellman, R. Petter, and R. Breslow, *J. Am. Chem. Soc.*, **108**, 2388 (1986).
375. P. Hendry and A. M. Sargeson, *Aust. J. Chem.*, **39**, 1177 (1986).
376. R. S. Brown and M. Zamkanei, *Inorg. Chim. Acta*, **108**, 201 (1985).
377. R. W. Hay, A. K. Basak, M. P. Pujari, and A. Perotti, *J. Chem. Soc., Dalton Trans.*, 2029 (1986).
378. P. Hendry and A. M. Sargeson, *Inorg. Chem.*, **25**, 865 (1986).
379. D. B. Collum, J. A. Klang, and R. T. Depue, *J. Am. Chem. Soc.*, **108**, 2333 (1986).
380. F. Tafesse, S. S. Massoud, and R. M. Milburn, *Inorg. Chem.*, **24**, 2591 (1985).
381. E. Z. Utyanskaya and A. E. Shilov, *Kinet. Catal. (Engl. Transl.)*, **27**, 113 (1986).
382. P. G. Yohannes, M. P. Mertes, and K. B. Mertes, *J. Am. Chem. Soc.*, **107**, 8288 (1985).
383. G. C. Mei and C. D. Gutsche, *J. Am. Chem. Soc.*, **107**, 7959 (1985).
384. C. D. Gutsche and G. C. Mei, *J. Am. Chem. Soc.*, **107**, 7964 (1985).
385. D. Herschlag and W. P. Jencks, *J. Am. Chem. Soc.*, **108**, 7938 (1986).
386. M. W. Hosseini and J.-M. Lehn, *J. Chem. Soc., Chem. Commun.*, 1155 (1985).
387. B. V. L. Potter, *J. Chem. Soc., Chem. Commun.*, 21 (1986).
388. F. R. Spitz, J. Cabral, and P. Haake, *J. Am. Chem. Soc.*, **108**, 2802 (1986).
389. K. M. Inani, P. D. Sharma, and Y. K. Gupta, *J. Chem. Soc., Dalton Trans.*, 2571 (1985).
390. X. Creary and M. E. Mehrsheikh-Mohammadi, *J. Org. Chem.*, **51**, 7 (1986).
391. X. Creary and P. A. Inocencio, *J. Am. Chem. Soc.*, **108**, 5979 (1986).
392. R. Wolfenden and R. Williams, *J. Am. Chem. Soc.*, **107**, 4345 (1985).
393. J. Symes and T. A. Modro, *Can. J. Chem.*, **64**, 1702 (1986).
394. R. Ramage, B. Atrash, D. Hopton, and M. J. Parrott, *J. Chem. Soc., Perkin Trans. 1*, 1617 (1985).
395. S. Cocks, K. R. Koch, and T. A. Modro, *J. Org. Chem.*, **51**, 265 (1986).
396. W. B. Knight, P. M. Weiss, and W. W. Cleland, *J. Am. Chem. Soc.*, **108**, 2759 (1986).

397. P. M. Weiss, W. B. Knight, and W. W. Cleland, *J. Am. Chem. Soc.*, **108**, 2761 (1986).
398. A. Williams, *J. Am. Chem. Soc.*, **107**, 6335 (1985).
399. W. P. Jencks, M. T. Haber, D. Herschlag, and K. L. Nazaretian, *J. Am. Chem. Soc.*, **108**, 479 (1986).
400. J. M. Friedman and J. R. Knowles, *J. Am. Chem. Soc.*, **107**, 6126 (1985).
401. P. M. Cullis and A. J. Rous, *J. Am. Chem. Soc.*, **108**, 1298 (1986).
402. G. Lowe and S. P. Tuck, *J. Am. Chem. Soc.*, **108**, 1300 (1986).
403. P. M. Cullis and A. J. Rous, *J. Am. Chem. Soc.*, **107**, 6721 (1985).
404. F. Ramirez, J. Marecek, J. Minore, S. Srivastava, and W. le Noble, *J. Am. Chem. Soc.*, **108**, 348 (1986).
405. P. M. Cullis and A. Iagrossi, *J. Am. Chem. Soc.*, **108**, 7870 (1986).
406. J. H. Cummins and B. V. L. Potter, *J. Chem. Soc., Chem. Commun.*, 851 (1985).
407. J. R. P. Arnold and G. Lowe, *J. Chem. Soc., Chem. Commun.*, 865 (1986).
408. K. W. Y. Abell and A. J. Kirby, *Tetrahedron Lett.*, **27**, 1085 (1986).
409. A. M. Beltran, A. Klaebe, and J. J. Perie, *Tetrahedron Lett.*, **26**, 1715 (1985).
410. T. A. Modro, *Tetrahedron Lett.*, **27**, 3063 (1986).
411. N. Iwamoto, Y. Okamoto, and S. Takamuku, *Bull. Chem. Soc. Jpn.*, **59**, 1505 (1986).
412. Y. Okamoto, N. Iwamoto, and S. Takamuki, *J. Chem. Soc., Chem. Commun.*, 1516 (1986).
413. T. B. Rauchfuss and G. A. Zank, *Tetrahedron Lett.*, **27**, 3445 (1986).
414. M. T. Craw, M. C. Depew, and J. K. S. Wan, *Phosphorus Sulfur*, **25**, 369 (1985).
415. H. W. Roesky, R. Ahlrichs, and S. Brode, *Angew. Chem., Int. Ed. Engl.*, **25**, 82 (1986).
416. S. Freeman and M. J. P. Harger, *J. Chem. Soc., Chem. Commun.*, 1394 (1985).
417. P. A. Frey, W. Reimschüssel, and P. Paneth, *J. Am. Chem. Soc.*, **108**, 1720 (1986).
418. R. D. Cook, S. Farah, L. Ghawi, A. Itani, and J. Rahil, *Can. J. Chem.*, **64**, 1630 (1986).
419. T. Fanni, K. Taira, D. G. Gorenstein, R. Vaidyanathaswamy, and J. G. Verkade, *J. Am. Chem. Soc.*, **108**, 6311 (1986).
420. R. Kluger and G. R. J. Thatcher, *J. Am. Chem. Soc.*, **107**, 6006 (1985).
421. R. Kluger and G. R. J. Thatcher, *J. Org. Chem.*, **51**, 207 (1986).
422. P. M. Cullis, P. B. Kay, and S. Trippett, *J. Chem. Soc., Chem. Commun.*, 1329 (1985).
423. R. A. McClelland, D. A. Cramm, and G. H. McGall, *J. Am. Chem. Soc.*, **108**, 2416 (1986).
424. R. A. McClelland, G. H. McGall, and G. Patel, *J. Am. Chem. Soc.*, **107**, 5204 (1985).
425. N. Lowther and C. D. Hall, *J. Chem. Soc., Chem. Commun.*, 1303 (1985).
426. G. H. McGall and R. A. McClelland, *J. Am. Chem. Soc.*, **107**, 5198 (1985).
427. R. D. Cook and M. Metni, *Can. J. Chem.*, **63**, 3155 (1985).
428. R. J. P. Corriu, G. F. Lanneau, and D. Leclercq, *Tetrahedron*, **42**, 5591 (1986).
429. R. D. Cook, W. A. Daouk, A. N. Hajj, A. Kabbani, A. Kurku, M. Samaha, F. Shayban, and O. V. Tanielian, *Can. J. Chem.*, **64**, 213 (1986).
430. J.-P. Majoral, G. Bertrand, E. Ocando-Mavarez, and A. Baceiredo, *Bull. Soc. Chim. Belg.*, **95**, 945 (1986).
431. K. C. Kumara Swamy and S. S. Krishnamurthy, *Inorg. Chem.*, **25**, 920 (1986).
432. H. Winter and J. C. van de Grampel, *J. Chem. Soc., Dalton Trans.*, 1389 (1986).
433. H. Winter and J. C. van de Grampel, *J. Chem. Soc., Dalton Trans.*, 1269 (1986).
434. C. W. Allen and J. A. MacKay, *Inorg. Chem.*, **25**, 4628 (1986).
435. A.-M. Caminade, E. Ocando, J.-P. Majoral, M. Cristante, and G. Bertrand, *Inorg. Chem.*, **25**, 712 (1986).
436. E. Niecke, D. Gudat, and E. Symalla, *Angew. Chem., Int. Ed. Engl.*, **25**, 834 (1986).
437. A. J. Arduengo III, D. A. Dixon, and D. C. Roe, *J. Am. Chem. Soc.*, **108**, 6821 (1986).
438. H.-P. Abicht, J. T. Spencer, and J. G. Verkade, *Inorg. Chem.*, **24**, 2132 (1985).
439. P. B. Kay and S. Trippett, *J. Chem. Res. (S)*, 292 (1985).
440. A. B. Reitz, S. O. Nortey, A. D. Jordan, Jr., M. S. Mutter, and B. E. Maryanoff, *J. Org. Chem.*, **51**, 3302 (1986).
441. B. E. Maryanoff, A. B. Reitz, M. S. Mutter, R. R. Inners, H. R. Almond, Jr., R. R. Whittle, and R. A. Olofson, *J. Am. Chem. Soc.*, **108**, 7664 (1986).

442. B. E. Maryanoff and A. B. Reitz, *Tetrahedron Lett.*, **26**, 4587 (1985).
443. S. M. Cairns and W. E. McEwen, *Tetrahedron Lett.*, **27**, 1541 (1986).
444. W. E. Mcewen, B. D. Beaver, and J. V. Cooney, *Phosphorus Sulfur*, **25**, 255 (1985).
445. L. Donxia, W. Dexian, L. Yaozhong, and Z. Huaming, *Tetrahedron*, **42**, 4161 (1986).
446. A. Baceiredo, A. Igau, G. Bertand, M. J. Menu, Y. Dartiguenave, and J. J. Bonnet, *J. Am. Chem. Soc.*, **108**, 7868 (1986).
447. A. Baceiredo, G. Bertrand, and G. Sicard, *J. Am. Chem. Soc.*, **107**, 4781 (1985).
448. J. Grobe and J. Szameitat, *Z. Naturforsch.*, **41b**, 974 (1986).
449. J. Grobe, D. Le Van, and J. Nientiedt, *Z. Naturforsch.*, **41b**, 149 (1986).
450. D. V. Griffiths and J. C. Tebby, *J. Chem. Soc., Chem. Commun.*, 871 (1986).
451. S. Alunni, *J. Chem. Res. (S)*, 231 (1986).
452. N. S. Isaacs and O. H. Abed, *Tetrahedron Lett.*, **27**, 1209 (1986).
453. F. Y. Khalil, F. M. Abdel-Halim, M. T. Hanna, and M. El-Batouti, *Bull. Soc. Chim. Belg.*, **94**, 379 (1985).
454. H.-G. Schäfer, W. W. Schoeller, J. Niemann, W. Haug, T. Dabisch, and E. Niecke, *J. Am. Chem. Soc.*, **108**, 7481 (1986).
455. A. H. Cowley, R. A. Kemp, J. G. Lasch, N. C. Norman, C. A. Stewart, B. R. Whittlesey, and T. C. Wright, *Inorg. Chem.*, **25**, 740 (1986).
456. M. Mikolajczyk, P. Kielbasinski, and A. Sut, *Tetrahedron*, **42**, 4591 (1986).
457. J. R. Lloyd, N. Lowther, G. Zsabo, and C. D. Hall, *J. Chem. Soc., Perkin Trans. 2*, 1813 (1985).
458. S. N. Ramos, J. C. Owrutsky, and P. M. Keehn, *Tetrahedron Lett.*, **26**, 5895 (1985).
459. A. Alberti, D. Griller, A. S. Nazran, and G. F. Pedulli, *J. Org. Chem.*, **51**, 3959 (1986).
460. N. J. Winter, J. Fossey, B. Beccard, Y. Berchadsky, F. Vila, L. Werbelow, and P. Tordo, *J. Phys. Chem.*, **90**, 6749 (1986).
461. J.-E. Kessler, C. T. G. Knight, and A. E. Merbach, *Inorg. Chim. Acta*, **115**, 75 (1985).
462. J.-E. Kessler, C. T. G. Knight, and A. E. Merbach, *Inorg. Chim. Acta*, **115**, 85 (1985).
463. A. J. Ashe, III and E. G. Ludwig, Jr., *J. Organomet. Chem.*, **303**, 197 (1986).
464. A. J. Ashe, III and E. G. Ludwig, Jr., *J. Organomet. Chem.*, **308**, 289 (1986).
465. V. K. Gupta, L. K. Krannich, and C. L. Watkins, *Inorg. Chem.*, **25**, 2553 (1986).
466. C. A. Stewart, R. L. Harlow, and A. J. Arduengo III, *J. Am. Chem. Soc.*, **107**, 5543 (1985).
467. C. A. Stewart and A. J. Arduengo III, *Inorg. Chem.*, **25**, 3847 (1986).
468. K. Akiba, A. Shimizu, H. Ohnari, and K. Ohkata, *Tetrahedron Lett.*, **26**, 3211 (1985).
469. M. Kakihana, M. Okamoto, and T. Nagumo, *Z. Naturforsch.*, **40a**, 1085 (1985).
470. K.-D. Merboldt and J. Frahm, *Ber. Bunsenges. Phys. Chem.*, **90**, 614 (1986).
471. R. M. Jarret and M. Saunders, *J. Am. Chem. Soc.*, **108**, 7549 (1986).
472. J. L. Kurz, S. L. Hazen, and L. C. Kurz, *J. Phys. Chem.*, **90**, 543 (1986).
473. P. Scharlin, *Acta Chem. Scand., Ser. A*, **40**, 221 (1986).
474. M. Kakiuchi and S. Matsuo, *J. Phys. Chem.*, **89**, 4627 (1985).
475. Y. Yoshida and S. Nishikawa, *Bull. Chem. Soc. Jpn.*, **59**, 1941 (1986).
476. M. Ueno, K. Ito, N. Tsuchihashi, and K. Shimizu, *Bull. Chem. Soc. Jpn.*, **59**, 1175 (1986).
477. E. Grunwald, *J. Am. Chem. Soc.*, **108**, 5726 (1986).
478. W. L. Jorgensen, J. Gao, and C. Ravimohan, *J. Phys. Chem.*, **89**, 3470 (1985).
479. C.-H. Fischer, E. J. Hart, and A. Henglein, *J. Phys. Chem.*, **90**, 222 (1986).
480. C.-H. Fischer, E. J. Hart, and A. Henglein, *J. Phys. Chem.*, **90**, 3059 (1986).
481. J. Lee, G. W. Robinson, S. P. Webb, L. A. Philips, and J. H. Clark, *J. Am. Chem. Soc.*, **108**, 6538 (1986).
482. E. Pines, D. Huppert, M. Gutman, N. Nachliel, and M. Fishman, *J. Phys. Chem.*, **90**, 6366 (1986).
483. G. W. Robinson, P. J. Thistlethwaite, and J. Lee, *J. Phys. Chem.*, **90**, 4224 (1986).
484. M. A. Muñiz, J. Bertán, J. Andrés, M. Duran, and A. Lledós, *J. Chem. Soc., Faraday Trans. 1*, **81**, 1547 (1985).
485. S. L. Fornili, M. Migliore, and M. A. Palazzo, *Chem. Phys. Lett.*, **125**, 419 (1986).

486. P. M. Hierl, A. F. Ahrens, M. Henchman, A. A. Viggiano, J. F. Paulson, and D. C. Clary, *J. Am. Chem. Soc.*, **108**, 3140 (1986).

487. P. M. Hierl, A. F. Ahrens, M. Henchman, A. A. Viggiano, J. F. Paulson, and D. C. Clary, *J. Am. Chem. Soc.*, **108**, 3142 (1986).

488. K. Ohta and K. Morokuma, *J. Phys. Chem.*, **89**, 5845 (1985).

489. J. D. Madura and W. L. Jorgensen, *J. Am. Chem. Soc.*, **108**, 2517 (1986).

490. S. Scheiner and E. A. Hillenbrand, *J. Phys. Chem.*, **89**, 3053 (1985).

491. E. A. Hillenbrand and S. Scheiner, *J. Am. Chem. Soc.*, **108**, 7178 (1986).

492. E. Grunwald, *J. Am. Chem. Soc.*, **108**, 5719 (1986).

493. Y. Kataoka, *Bull. Chem. Soc. Jpn.*, **59**, 1425 (1986).

494. A. K. Soper and M. G. Phillips, *Chem. Phys.*, **107**, 47 (1986).

495. G. A. Olah, G. K. Surya Prakash, M. Barzaghi, K. Lammertsma, P. von R. Schleyer, and J. A. Pople, *J. Am. Chem. Soc.*, **108**, 1032 (1986).

496. T. Telser and U. Schindewolf, *Ber. Bunsenges. Phys. Chem.*, **89**, 1116 (1985).

497. H. Christensen and K. Sehested, *J. Phys. Chem.*, **90**, 186 (1986).

498. B. Hickel and K. Sehested, *J. Phys. Chem.*, **89**, 5271 (1985).

499. C.-H. Fischer, E. J. Hart, and A. Henglein, *J. Phys. Chem.*, **90**, 1954 (1986).

500. E. J. Hart and A. Henglein, *J. Phys. Chem.*, **89**, 4342 (1985).

501. D. F. Evans and M. W. Upton, *J. Chem. Soc., Dalton Trans.*, 2525 (1985).

502. J. M. Aubry, *J. Am. Chem. Soc.*, **107**, 5844 (1985).

503. S. Anic and L. Kolar-Anic, *Ber. Bunsenges. Phys. Chem.*, **90**, 539 (1986).

504. M. F. Zipplies, W. A. Lee, and T. C. Bruice, *J. Amer. Chem. Soc.*, **108**, 4433 (1986).

505. M. L. Kremer, *Int. J. Chem. Kinet.*, **17**, 1299 (1985).

506. S. Baral, C. Lume-Pereira, E. Janata, and A. Henglein, *J. Phys. Chem.*, **89**, 5779 (1985).

507. W. A. Lee and T. C. Bruice, *Inorg. Chem.*, **25**, 131 (1986).

508. L.-C. Yuan and T. C. Bruice, *J. Am. Chem. Soc.*, **108**, 1643 (1986).

509. M. Inamo, S. Funahashi, and M. Tanaka, *Inorg. Chem.*, **24**, 2475 (1985).

510. M. Inamo, S. Funahashi, Y. Ito, Y. Hamada, and M. Tanaka, *Inorg. Chem.*, **24**, 2468 (1985).

511. A. Kunai, S. Hata, S. Ito, and K. Sasaki, *J. Am. Chem. Soc.*, **108**, 6012 (1986).

512. P.-S. K. Leung and M. R. Hoffmann, *J. Phys. Chem.*, **89**, 5267 (1985).

513. I. B. Afanas'ev and N. S. Kuprianova, *J. Chem. Soc., Perkin Trans. 2*, 1361 (1985).

514. P. Cofré and D. T. Sawyer, *Inorg. Chem.*, **25**, 2089 (1986).

515. P. Cofré and D. T. Sawyer, *Anal. Chem.*, **58**, 1057 (1986).

516. K. Yamaguchi, T. S. Calderwood, and D. T. Sawyer, *Inorg. Chem.*, **25**, 1289 (1986).

517. J. L. Roberts, Jr., H. Sugimoto, W. C. Barrette, Jr., and D. T. Sawyer, *J. Am. Chem. Soc.*, **107**, 4556 (1985).

518. B. H. J. Bielski, D. E. Cabelli, R. L. Arudi, and A. B. Ross, *J. Phys. Chem. Ref. Data*, **14**, 1041 (1985).

519. H. A. Schwarz and B. H. J. Bielski, *J. Phys. Chem.*, **90**, 1445 (1986).

520. J. D. Rush and B. H. J. Bielski, *J. Phys. Chem.*, **89**, 5062 (1985).

521. J. R. Kanofsky, *J. Org. Chem.*, **51**, 3386 (1986).

522. J. R. Kanofsky, *J. Am. Chem. Soc.*, **108**, 2977 (1986).

523. A. M. Caminade, F. El Khatib, M. Koenig, and J. M. Aubry, *Can. J. Chem.*, **63**, 3203 (1985).

524. E. J. Corey, M. M. Mehrotra, and A. U. Khan, *J. Am. Chem. Soc.*, **108**, 2472 (1986).

525. W. C. Eisenberg, K. Taylor, and R. W. Murray, *J. Am. Chem. Soc.*, **107**, 8299 (1985).

526. E. L. Clennan and R. P. L'Esperance, *J. Am. Chem. Soc.*, **107**, 5178 (1985).

527. E. L. Clennan and R. P. L'Esperance, *J. Org. Chem.*, **50**, 5424 (1985).

528. E. L. Clennan, R. P. L'Esperance, and K. K. Lewis, *J. Org. Chem.*, **51**, 1440 (1986).

529. E. L. Clennan and K. K. Lewis, *J. Org. Chem.*, **51**, 3721 (1986).

530. Y.-Y. Chan, C. Zhu, and H.-K. Leung, *J. Am. Chem. Soc.*, **107**, 5274 (1985).

531. T. Akasaka, M. Nakagawa, and W. Ando, *J. Org. Chem.*, **51**, 4477 (1986).

532. N. Raja, P. K. Arora, J. P. S. Chatha, and K. G. Vohra, *Int. J. Chem. Kinet.*, **17**, 1315 (1985).

533. R. Sato, H. Sonobe, T. Akasaka, and W. Ando, *Tetrahedron*, **42**, 5273 (1986).

534. T. Akasaka, M. Nakagawa, Y. Nomura, R. Sato, K. Someno, and W. Ando, *Tetrahedron*, **42**, 3807 (1986).

535. K. Gollnick and A. Schnatterer, *Tetrahedron Lett.*, **26**, 5029 (1985).

536. T. Akasaka and W. Ando, *Tetrahedron Lett.*, **26**, 5049 (1985).

537. W. Ando, H. Sonobe, and T. Akasaka, *Tetrahedron Lett.*, **27**, 4473 (1985).

538. H. H. Wasserman and T.-J. Lu, *Rec. Trav. Chim. Pays-Bas*, **105**, 345 (1986).

539. T. Akasaka, R. Sato, and W. Ando, *J. Am. Chem. Soc.*, **108**, 5539 (1986).

540. S. L. Wilson and G. B. Schuster, *J. Org. Chem.*, **51**, 2056 (1986).

541. J. C. Mitchell, *Chem. Soc. Rev.*, **14**, 399 (1985).

542. S. F. Nelsen, M. F. Teasley, and D. L. Kapp, *J. Am. Chem. Soc.*, **108**, 5503 (1986).

543. J. A. Hutchinson, J. DiBenedetto, and P. M. Rentzepis, *J. Am. Chem. Soc.*, **108**, 6517 (1986).

544. A. A. Frimer, T. Farkash-Solomon, and G. Aljadeff, *J. Org. Chem.*, **51**, 2093 (1986).

545. R. N. McDonald and A. K. Chowdhury, *J. Am. Chem. Soc.*, **107**, 4123 (1985).

546. J. H. Bowie, C. H. DePuy, S. A. Sullivan, and V. Bierbaum, *Can. J. Chem.*, **64**, 1046 (1986).

547. S. Yamauchi and N. Hirota, *J. Am. Chem. Soc.*, **107**, 5021 (1985).

548. D. E. Falvey and G. B. Schuster, *J. Am. Chem. Soc.*, **108**, 7419 (1986).

549. R. E. Huie and P. Neta, *Int. J. Chem. Kinet.*, **18**, 1185 (1986).

550. Y. Tanimoto, M. Nishino, and M. Itoh, *Bull. Chem. Soc. Jpn.*, **58**, 3365 (1985).

551. T. Komai and K. Matsuyama, *Bull. Chem. Soc. Jpn.*, **58**, 2207 (1985).

552. T. Komai and S. Suyama, *Bull. Chem. Soc. Jpn.*, **58**, 3045 (1985).

553. T. Komai, H. Ishigaki, and K. Matsuyama, *Bull. Chem. Soc. Jpn.*, **58**, 2431 (1985).

554. H. Sawada, H. Hagii, K. Aoshima, M. Yoshida, and M. Kobayashi, *Bull. Chem. Soc. Jpn.*, **58**, 3448 (1985).

555. M. J. Bourgeois, B. Maillard, and E. Montaudon, *Tetrahedron*, **42**, 5309 (1986).

556. H. Tomiyasu, H. Fukutomi, and G. Gordon, *Inorg. Chem.*, **24**, 2962 (1985).

557. T. C. Yang and W. C. Neely, *Anal. Chem.*, **58**, 1551 (1986).

558. E. J. Hart and A. Henglein, *J. Phys. Chem.*, **90**, 3061 (1986).

559. K. Griesbaum, G. Zwick, S. Agarwal, H. Keul, B. Pfeffer, and R. W. Murray, *J. Org. Chem.*, **50**, 4194 (1985).

560. H. Keul, H.-S. Choi, and R. L. Kuczkowski, *J. Org. Chem.*, **50**, 3365 (1985).

561. K. Griesbaum and W. Volpp, *Angew. Chem., Int. Ed. Engl.*, **25**, 81 (1986).

562. M. K. Painter, H.-S. Choi, K. W. Hillig, II, and R. L. Kuczkowski, *J. Chem. Soc., Perkin Trans. 2*, 1025 (1986).

563. M. Matsui, A. Konda, and K. Shibata, *Bull. Chem. Soc. Jpn.*, **58**, 2829 (1985).

564. M. G. Matturro, R. P. Reynolds, R. V. Kastrup, and C. F. Pictroski, *J. Am. Chem. Soc.*, **108**, 2775 (1986).

565. W. Sander, *Angew. Chem., Int. Ed. Engl.*, **24**, 988 (1985).

566. W. Sander, *Angew. Chem., Int. Ed. Engl.*, **25**, 255 (1986).

567. W. Adam, H. Dürr, W. Haas, and B. Lohray, *Angew. Chem., Int. Ed. Engl.*, **25**, 101 (1986).

568. W. Adam and B. B. Lohray, *Angew. Chem., Int. Ed. Engl.*, **25**, 188 (1986).

569. K. Ishiguro, K. Tomizawa, Y. Sawaki, and H. Iwamura, *Tetrahedron Lett.*, **26**, 3723 (1985).

570. D. H. Giamalva, D. F. Church, and W. A. Pryor, *J. Am. Chem. Soc.*, **108**, 6646 (1986).

571. D. H. Giamalva, D. F. Church, and W. A. Pryor, *J. Am. Chem. Soc.*, **108**, 7678 (1986).

572. B. Cohen and S. Weiss, *J. Phys. Chem.*, **90**, 6275 (1986).

573. T. S. Chen and P. L. Moore Plummer, *J. Phys. Chem.*, **89**, 3689 (1985).

574. R. I. Gelb, L. M. Schwarz, and L. J. Zompa, *Inorg. Chem.*, **25**, 1527 (1986).

575. N. Bourne, A. Hopkins, and A. Williams, *J. Am. Chem. Soc.*, **107**, 4327 (1985).

576. G. Lowe and M. J. Parratt, *J. Chem. Soc., Chem. Commun.*, 1075 (1985).

577. J. N. Drummond and A. J. Kirby, *J. Chem. Soc., Perkin Trans. 2*, 579 (1986).

578. M. Nakagaki and S. Yokoyama, *Bull. Chem. Soc. Jpn.*, **59**, 935 (1986).

579. S. Fronaeus, *Acta Chem. Scand., Ser. A*, **40**, 572 (1986).

580. C. Srinivasan and K. Pitchumani, *Can. J. Chem.*, **63**, 2285 (1985).

581. C. Srinivasan, S. Perumal, and N. Arumugam, *J. Chem. Soc., Perkin Trans. 2*, 1855 (1985).

582. S. Nishida and M. Kimura, *J. Chem. Res. (S)*, 336 (1986).
583. M. T. Abilov and V. A. Golodov, *Kinet. Catal. (Engl. Transl.)*, **26**, 1276 (1985).
584. C. R. Dennis, J. G. Leipoldt, S. S. Basson, and A. J. Van Wyk, *Inorg. Chem.*, **25**, 1268 (1986).
585. R. Renganathan and P. Maruthamuthu, *J. Chem. Soc., Perkin Trans. 2*, 285 (1986).
586. R. Renganathan and P. Maruthamuthu, *Int. J. Chem. Kinet.*, **18**, 49 (1986).
587. M. S. Ramachandran, T. S. Vivekanandam, and V. Arunachalam, *Bull. Chem. Soc. Jpn.*, **59**, 1549 (1986).
588. G. Manivannan and P. Maruthamuthu, *J. Chem. Soc., Perkin Trans. 2*, 565 (1986).
589. M. J. Davies, B. C. Gilbert, C. B. Thomas, and J. Young, *J. Chem. Soc., Perkin Trans. 2*, 1199 (1985).
590. D. A. Horner and R. E. Connick, *Inorg. Chem.*, **25**, 2414 (1986).
591. W. L. Reynolds and Y. Yuan, *Polyhedron*, **5**, 1467 (1986).
592. U. Deister, R. Neeb, G. Helas, and P. Warneck, *J. Phys. Chem.*, **90**, 3213 (1986).
593. T. M. Olson, S. D. Boyce, and M. R. Hoffmann, *J. Phys. Chem.*, **90**, 2482 (1986).
594. M. P. Youngblood, *J. Org. Chem.*, 51, 1981 (1986).
595. M. R. Crampton, C. Greenhalgh, G. Machell, and D. P. E. Williams, *J. Chem. Res. (S)*, 240 (1986).
596. R. E. Huie and P. Neta, *J. Phys. Chem.*, **89**, 3918 (1985).
597. P. Neta and R. E. Huie, *J. Phys. Chem.*, **89**, 1783 (1985).
598. T. Ozawa and T. Kwan, *Polyhedron*, **5**, 1531 (1986).
599. T. Ozawa and T. Kwan, *Polyhedron*, **4**, 1425 (1985).
600. L. I. Simándi, M. Jáky, C. R. Savage, and Z. A. Schelly, *J. Am. Chem. Soc.*, **107**, 4220 (1985).
601. R. C. Thompson, *Inorg. Chem.*, **25**, 184 (1986).
602. A. Rodriguez, S. López, M. C. Carmona-Guzmán, and F. Sánchez, *J. Chem. Soc., Dalton Trans.*, 1265 (1986).
603. R. N. Bose, N. Rajasekar, D. M. Thompson, and E. S. Gould, *Inorg. Chem.*, **25**, 3349 (1986).
604. A. A. El-Awady and G. M. Harris, *Inorg. Chem.*, **25**, 1323 (1986).
605. V. K. Joshi, R. van Eldik, and G. M. Harris, *Inorg. Chem.*, **25**, 2229 (1986).
606. D. W. Neumann and S. Lynn, *Ind. Eng. Chem., Process Des. Dev.*, **25**, 248 (1986).
607. C. Paradisi and J. F. Bunnett, *J. Am. Chem. Soc.*, **107**, 8223 (1985).
608. X. Creary, *J. Org. Chem.*, **50**, 5080 (1985).
609. S. Thea, G. Guanti, A. R. Hopkins, and A. Williams, *J. Org. Chem.*, **50**, 3336 (1985).
610. S. Thea, G. Guanti, A. R. Hopkins, and A. Williams, *J. Org. Chem.*, **50**, 5592 (1985).
611. S. Thea, G. Cevasco, G. Guanti, and A. Williams, *J. Chem. Soc., Chem. Commun.*, 1582 (1986).
612. J. M. Allen and K. S. Venkatasubban, *J. Org. Chem.*, **50**, 5108 (1985).
613. J. D. Halliday and P. E. Bindner, *Can. J. Chem.*, **63**, 2821 (1985).
614. S. Licht, G. Hodes, and J. Manassen, *Inorg. Chem.*, **25**, 2486 (1986).
615. A. Wawer, *J. Chem. Soc., Perkin Trans. 2*, 1295 (1986).
616. R. Steudel, G. Holdt, and R. Nagorka, *Z. Naturforsch.*, **41b**, 1519 (1986).
617. T. Takata and W. Ando, *Bull. Chem. Soc. Jpn.*, **59**, 1275 (1986).
618. M. M. Ito, T. Ogikubo, H. Ueda, K. Hagino, and T. Endo, *Bull. Chem. Soc. Jpn.*, **59**, 1263 (1986).
619. G. C. Pillai and E. S. Gould, *Inorg. Chem.*, **25**, 3353 (1986).
620. M. Bonifacic, J. Weiss, S. A. Chaudri, and K.-D. Asmus, *J. Phys. Chem.*, **89**, 3910 (1985).
621. J. Mönig, R. Goslich, and K.-D. Asmus, *Ber. Bunsenges. Phys. Chem.*, **90**, 115 (1986).
622. M. Bonifacic and K.-D. Asmus, *J. Org. Chem.*, **51**, 1216 (1986).
623. H. J. Shine, D. H. Bae, A. K. M. Mansurul Hoque, A. Kajstura, W. K. Lee, R. W. Shaw, and M. Soroka, *Phosphorus Sulfur*, **23**, 111 (1985).
624. D. N. Kevill and S. W. Anderson, *J. Org. Chem.*, **51**, 5029 (1986).
625. F. D. Saeva, *Tetrahedron*, **42**, 6123 (1986).
626. K. Kikuchi, N. Furukawa, and S. Oae, *Phosphorus Sulfur*, **24**, 291 (1985).
627. C. W. Perkins and J. C. Martin, *J. Am. Chem. Soc.*, **108**, 3211 (1986).
628. C. W. Perkins, R. B. Clarkson, and J. C. Martin, *J. Am. Chem. Soc.*, **108**, 3206 (1986).

629. T. J. Burkey, J. A. Hawari, F. P. Lossing, J. Lusztyk, R. Sutcliffe, and D. Griller, *J. Org. Chem.*, **50**, 4966 (1985).
630. A. Alberti, D. Griller, A. S. Nazran, and G. F. Pedulli, *J. Am. Chem. Soc.*, **108**, 3024 (1986).
631. C. Chatgilialoglu, *J. Org. Chem.*, **51**, 2871 (1986).
632. A. L. J. Beckwith and D. R. Boate, *J. Chem. Soc., Chem. Commun.*, 189 (1986).
633. J. Drabowicz and M. Mikolajczyk, *Tetrahedron Lett.*, **26**, 5703 (1985).
634. M. Mikolajczyk, J. Drabowicz, and B. Bujnicki, *Tetrhedron Lett.*, **26**, 5699 (1985).
635. M. Julia, H. Lauron, J.-P. Stacino, and J.-N. Verpeaux, *Tetrahedron*, **42**, 2475 (1986).
636. P. R. Young and P. E. McMahon, *J. Am. Chem. Soc.*, **107**, 7572 (1985).
637. P. R. Young and P. E. McMahon, *J. Org. Chem.*, **51**, 4078 (1986).
638. P. R. Young, A. P. Zygas, and I.-W. E. Lee, *J. Am. Chem. Soc.*, **107**, 7578 (1985).
639. E. Besenyei, G. K. Eigendorf, and D. C. Frost, *Inorg. Chem.*, **25**, 4404 (1986).
640. H. Bock, B. Solouki, and H. W. Roesky, *Inorg. Chem.*, **24**, 4425 (1985).
641. T. Chivers, J. F. Richardson, and N. R. M. Smith, *Inorg. Chem.*, **25**, 47 (1986).
642. P. Fritz, R. Bruchhaus, R. Mews, and H.-U. Höfs, *Z. Anorg. Allg. Chem.*, **525**, 214 (1985).
643. T. Chivers, *Chem. Rev.*, **85**, 341 (1985).
644. U. K. Kläning and K. Sehested, *J. Phys. Chem.*, **90**, 5460 (1986).
645. K. Rashid and H. R. Krouse, *Can. J. Chem.*, **63**, 3195 (1985).
646. T.-Y. Luh, W.-H. So, K. S. Cheung, and S. W. Tam, *J. Org. Chem.*, **50**, 3051 (1985).
647. G. H. Schmid and D. G. Garratt, *Tetrahedron*, **41**, 4787 (1985).
648. E. S. Lewis, M. L. McLaughlin, and T. A. Douglas, *J. Am. Chem. Soc.*, **107**, 6668 (1985).
649. J. L. Kice, S. Chiou, and L. Weclas, *J. Org. Chem.*, **50**, 2508 (1985).
650. J. L. Kice and S. Chiou, *J. Org. Chem.*, **51**, 290 (1986).
651. S.-I. Kang and J. L. Kice, *J. Org. Chem.*, **51**, 287 (1986).
652. S.-I. Kang and J. L. Kice, *J. Org. Chem.*, **50**, 2968 (1985).
653. S.-I. Kang and J. L. Kice, *J. Org. Chem.*, **51**, 295 (1986).
654. A. S. Secco, K. Alam, B. J. Blackburn, and A. F. Janzen, *Inorg. Chem.*, **25**, 2125 (1986).
655. M. R. Detty and H. R. Luss, *Organometallics*, **5**, 2250 (1986).
656. M. R. Detty, H. R. Luss, J. M. McKelvey, and S. M. Geer, *J. Org. Chem.*, **51**, 1692 (1986).
657. M. A. K. Ahmed, W. R. McWhinnie, and P. Granger, *Polyhedron*, **5**, 859 (1986).
658. B. Gautheron, G. Tainturier, and C. Degrand, *J. Am. Chem., Soc.*, **107**, 5579 (1985).
659. M. M. Carnell, F. Grein, M. Murchie, J. Passmore, and C.-M. Wong, *J. Chem. Soc., Chem. Commun.*, 225 (1986).
660. M. J. Collins, R. J. Gillespie, J. F. Sawyer, and G. J. Schrobilgen, *Inorg. Chem.*, **25**, 2053 (1986).
661. R. C. Burns, M. J. Collins, R. J. Gillespie, and G. J. Schrobilgen, *Inorg. Chem.*, **25**, 4465 (1986).
662. P. Boldrini, I. D. Brown, M. J. Collins, R. J. Gillespie, E. Maharajh, D. R. Slim, and J. F. Sawyer, *Inorg. Chem.*, **24**, 4302 (1985).
663. R. C. Thompson, *Adv. Inorg. Bioinorg. Mechanisms*, **4**, 65 (1985).
664. K. O. Christe, *Inorg. Chem.*, **25**, 3721 (1986).
665. M. Diksic, S. Farrokhzad, and L. D. Colebrook, *Can. J. Chem.*, **64**, 424 (1986).
666. *Chem. Eng. News*, **64**, 22 (1986).
667. L. Drougge and L. I. Elding, *Inorg. Chem.*, **24**, 2292 (1985).
668. E. Choshen, R. Elits, and C. Rav-Acha, *Tetrahedron Lett.*, **27**, 5989 (1986).
669. I. Nagypál, I. R. Epstein, and K. Kustin, *Int. J. Chem. Kinet.*, **18**, 345 (1986).
670. O. Valdes-Aguilera, D. W. Boyd, I. R. Epstein, and K. Kustin, *J. Phys. Chem.*, **90**, 6696 (1986).
671. O. Valdes-Aguilera, D. W. Boyd, I. R. Epstein, and K. Kustin, *J. Phys. Chem.*, **90**, 6702 (1986).
672. R. H. Simoyi, *J. Phys. Chem.*, **89**, 3570 (1985).
673. D. J. McLennan, A. R. Stein, and B. Dobson, *Can. J. Chem.*, **64**, 1201 (1986).
674. A. G. Turner, *Inorg. Chem. Acta*, **111**, 157 (1986).
675. P. J. MacDougall, *Inorg. Chem.*, **25**, 4400 (1986).
676. C. L. Brooks, III, *J. Phys. Chem.*, **90**, 6680 (1986).
677. T. P. Lybrand, I. Ghosh, and J. A. McCammon, *J. Am. Chem. Soc.*, **107**, 7793 (1985).
678. Y. Katsuragi, O. Neda, K. Yamauchi, and T. Masuda, *Bull. Chem. Soc. Jpn.*, **59**, 3971 (1986).

679. G. Schmitz and H. Rooze, *Can. J. Chem.*, **64**, 1747 (1986).
680. F. Ariese and Z. Nagy, *J. Phys. Chem.*, **90**, 1496 (1986).
681. F. Ariese and Z. Ungvárai-Nagy, *J. Phys. Chem.*, **90**, 1 (1986).
682. R. J. Field and H. Dieter-Fösterling, *J. Phys. Chem.*, **90**, 5400 (1986).
683. R. H. Simoyi, P. Masvikeni, and A. Sikosana, *J. Phys. Chem.*, **90**, 4126 (1986).
684. M.-F. Ruasse, J. Aubard, B. Galland, and A. Adenier, *J. Phys. Chem.*, **90**, 4382 (1986).
685. M.-F. Ruasse, J. Aubard, and P. Monjoint, *J. Chim. Phys.*, **82**, 539 (1985).
686. R. W. Ramette and D. A. Palmer, *J. Solution Chem.*, **15**, 387 (1986).
687. G. Bellucci, R. Bianchini, R. Ambrosetti, and G. Ingrosso, *J. Org. Chem.*, **50**, 3313 (1985).
688. H. Slebocka-Tilk, R. G. Ball, and R. S. Brown, *J. Am. Chem. Soc.*, **107**, 4504 (1985).
689. D. W. Margerum, P. N. Dickson, J. C. Nagy, K. Kumar, C. P. Bowers, and K. D. Fogelman, *Inorg. Chem.*, **25**, 4900 (1986).
690. D. A. Palmer and R. van Eldik, *Inorg. Chem.*, **25**, 928 (1986).
691. J. C. Wren, J. Paquette, S. Sunder, and B. L. Ford, *Can. J. Chem.*, **64**, 2284 (1986).
692. J. Paquette and B. L. Ford, *Can. J. Chem.*, **63**, 2444 (1985).
693. M. T. Beck and G. Rábai, *J. Phys. Chem.*, **90**, 2204 (1986).
694. R. Pascual and M. A. Herraez, *Can. J. Chem.*, **63**, 2349 (1985).
695. B. M. Derakhshan, A. Finch, and P. N. Gates, *Polyhedron*, **5**, 1543 (1986).
696. I. Sanemasa, Y. Nishimoto, A. Tanaka, and T. Deguchi, *Bull. Chem. Soc. Jpn.*, **59**, 2269 (1986).
697. W. B. Farnham and J. C. Calabrese, *J. Am. Chem. Soc.*, **108**, 2449 (1986).
698. P. H. De Ryck, L. Verdonck, and G. P. van der Kelen, *Bull. Soc. Chim. Belg.*, **94**, 621 (1985).
699. P. H. De Ryck, S. Hoste, and G. P. van der Kelen, *Bull. Soc. Chim. Belg.*, **95**, 217 (1986).
700. M. Varga and E. Körös, *J. Phys. Chem.*, **90**, 4373 (1986).
701. B. Schwitters and P. Ruoff, *J. Phys. Chem.*, **90**, 2497 (1986).
702. M. Varga, L. Györgyi, and E. Körös, *J. Am. Chem. Soc.*, **107**, 4780 (1985).
703. R. J. Field and P. M. Boyd, *J. Phys. Chem.*, **89**, 3707 (1985).
704. Z. Noszticzius, P. Stirling, and M. Wittmann, *J. Phys. Chem.*, **89**, 4914 (1985).
705. P. Ševčík and Ľ. Adamčíková, *J. Phys. Chem.*, **89**, 5178 (1985).
706. A. B. Rovinsky, *J. Phys. Chem.*, **90**, 217 (1986).
707. V. Gáspár, G. Bazsa, and M. T. Beck, *J. Phys. Chem.*, **89**, 5495 (1985).
708. R. M. Noyes, *J. Phys. Chem.*, **90**, 5407 (1986).
709. P. Ruoff, *J. Phys. Chem.*, **90**, 6744 (1986).
710. M. F. Crowley and R. J. Field, *J. Phys. Chem.*, **90**, 1907 (1986).
711. M. Menzinger and P. Jankowski, *J. Phys. Chem.*, **90**, 1217 (1986).
712. P. Ruoff and R. M. Noyes, *J. Phys. Chem.*, **90**, 4700 (1986).
713. C. K. McKinnon and R. J. Field, *J. Phys. Chem.*, **90**, 166 (1986).
714. W. P. Huskey and I. R. Epstein, *J. Phys. Chem.*, **90**, 4699 (1986).
715. R. J. Field, *J. Phys. Chem.*, **90**, 4700 (1986).
716. P. Ševčík and J. Dubovská, *Coll. Czech. Chem. Commun.*, **50**, 1450 (1985).
717. Ľ. Adamčíková and I. Hainárová, *Coll. Czech. Chem. Commun.*, **50**, 1588 (1985).
718. V. Gáspár and P. Galambosi, *J. Phys. Chem.*, **90**, 2222 (1986).
719. J. L. Hudson, P. Lamba, and J. C. Mankin, *J. Phys. Chem.*, **90**, 3430 (1986).
720. S. Vajda and T. Turányi, *J. Phys. Chem.*, **90**, 1664 (1986).
721. R. H. Simoyi, *J. Phys. Chem.*, **90**, 2802 (1986).
722. G. Bazsa and I. Fábián, *J. Chem. Soc., Dalton Trans.*, 2675 (1986).
723. O. Citri and I. R. Epstein, *J. Am. Chem. Soc.*, **108**, 357 (1986).
724. M. T. Beck and G. Rábai, *J. Phys. Chem.*, **89**, 3907 (1985).
725. G. Rábai and M. T. Beck, *J. Chem. Soc., Dalton Trans.*, 1669 (1985).
726. J. Harrison and K. Showalter, *J. Phys. Chem.*, **90**, 225 (1986).
727. V. Gáspár and M. T. Beck, *J. Phys. Chem.*, **90**, 6306 (1986).
728. E. C. Edblom, M. Orbán, and I. R. Epstein, *J. Am. Chem. Soc.*, **108**, 2826 (1986).
729. M. Alamgir and I. R. Epstein, *J. Phys. Chem.*, **89**, 3611 (1985).
730. J. Szamosi and S. J. Lasky, *J. Phys. Chem.*, **90**, 1995 (1986).

731. I. Nagypál, G. Bazsa, and I. R. Epstein, *J. Am. Chem. Soc.*, **108**, 3635 (1986).
732. G. Bazsa and I. R. Epstein, *J. Phys. Chem.*, **89**, 3050 (1985).
733. P. Gray and S. K. Scott, *Ber. Bunsenges. Phys. Chem.*, **90**, 985 (1986).
734. S. M. Kaushik, R. L. Rich, and R. M. Noyes, *J. Phys. Chem.*, **89**, 5722 (1985).
735. Z. Yuan, P. Ruoff, and R. M. Noyes, *J. Phys. Chem.*, **89**, 5726 (1985).

References for Chapter 5

1. R. J. Cross, *Chem. Soc. Rev.*, **14**, 197 (1985).
2. L. H. Skibsted, *Adv. Inorg. Bioinorg. Mech.*, **4**, 137 (1986).
3. R. van Eldik, *Angew. Chem., Int. Ed. Engl.*, **25**, 673 (1986).
4. H. M. Colquhoun, J. F. Stoddart, and D. J. Williams, *Angew. Chem., Int. Ed. Engl.*, **25**, 487 (1986).
5. W. E. Hill, J. G. Taylor, C. P. Falshaw, T. J. King, B. Beagley, D. M. Tonge, R. G. Pritchard, and C. A. MacAuliffe, *J. Chem. Soc., Dalton Trans.*, 2289 (1986).
6. A. J. Blake, R. O. Gould, A. J. Lavery, and M. Schröder, *Angew. Chem., Int. Ed. Engl.*, **25**, 274 (1986).
7. S. Kawaguchi, *Coord. Chem. Rev.*, **70**, 51 (1986).
8. C. T. Mortimer, *Rev. Inorg. Chem.*, **6**, 233 (1984).
9. a. Y.-T. Fanchiang, *Coord. Chem. Rev.*, **68**, 131 (1985). b. Y.-T. Fanchiang, *Inorg. Chim. Acta*, **105**, 209 (1985).
10. E. K. Byrne, D. S. Richeson, and K. H. Theopold, *J. Chem. Soc., Chem. Commun.*, 1491 (1986).
11. A. Turco, U. Vettori, and C. Giancaspro, *Gaz. Chim. Ital.*, **116**, 193 (1986).
12. R. Romeo, D. Minniti, G. Alibrandi, L. de Cola, and M. L. Tobe, *Inorg. Chem.*, **25**, 1944 (1986).
13. M. Becker and H. Elias, *Inorg. Chim. Acta*, **116**, 47 (1986).
14. L. Canovese, M. L. Tobe, G. Annibale, and L. Cattalini, *J. Chem. Soc., Dalton Trans.*, 1107 (1986).
15. P. Cavoli, R. Graziani, U. Casellato, and P. Uguagliati, *Inorg. Chim. Acta*, **111**, L35 (1986).
16. D. D. Gummin, E. M. A. Ratilla, and N. M. Kostić, *Inorg. Chem.*, **25**, 2429 (1986).
17. M. Bonivento, L. Cattalini, G. Chessa, G. Michelon, and M. L. Tobe, *J. Coord. Chem.*, **15A**, 153 (1986).
18. a. R. T. Boeré and C. J. Willis, *Can. J. Chem.*, **63**, 3530 (1985). b. R. T. Boeré and C. J. Willis, *Can. J. Chem.*, **64**, 492 (1986). c. R. T. Boeré, N. C. Paine, and C. J. Willis, *Can. J. Chem.*, **64**, 1474 (1986).
19. E. Kimura, Y. Lin, R. Machida, and H. Zenda, *J. Chem. Soc., Chem. Commun.*, 1020 (1986).
20. L. I. Elding and L.-F. Olsson, *Inorg. Chim. Acta*, **117**, 9 (1986).
21. M. Kotowski and R. van Eldik, *Inorg. Chem.*, **25**, 3896 (1986).
22. G. Mahal and R. van Eldik, *Inorg. Chem.*, **24**, 4165 (1985).
23. J. K. K. Sarhan, S.-W. F. Murray, H. M. Asfour, M. Green, R. M. Wings, and M. Parra-Haake, *Inorg. Chem.*, **25**, 243 (1986).
24. A. D. Ryabov, I. K. Sakodinskaya, and A. K. Yatsimirsky, *J. Chem. Soc., Dalton Trans.*, 2629 (1985).
25. V. A. Polyakov and A. D. Ryabov, *J. Chem. Soc., Dalton Trans.*, 589 (1986).
26. J. Ventegodt, B. Øby, and L. H. Skibsted, *Acta Chem. Scand., Ser. A*, **39**, 453 (1985).
27. L. Cattalini, G. Chessa, G. Michelon, B. Pitteri, M. L. Tobe, and A. Zanardo, *Inorg. Chem.*, **24**, 3409 (1985).
28. W. M. Rees, M. R. Churchill, J. C. Fettinger, and J. D. Atwood, *Organometallics*, **4**, 2179 (1985).
29. B. Corrain, B. Longato, R. Angeletti, and G. Valle, *Inorg. Chim. Acta*, **104**, 15 (1985).
30. M. Fuchs, W. Kuchen, and W. Peters, *Chem. Ber.*, **119**, 1569 (1986).
31. R. T. Boeré, C. D. Montgomery, N. C. Payne, and C. J. Willis, *Inorg. Chem.*, **24**, 3680 (1985).

32. L. E. Nivorozhkin, A. L. Nivorozhkin, M. S. Korobov, L. E. Konstantinovsky, and V. I. Minkin, *Polyhedron*, **4**, 1701 (1985).
33. M. Schumann and H. Elias, *Inorg. Chem.*, **24**, 3187 (1985).
34. W. R. Kennedy and D. W. Margerum, *Inorg. Chem.*, **24**, 2490 (1985).
35. Q. Zhang, Y. Li, J. Chen, R. Chen, Z. Zhang, and X. Wang, *Wuli Huaxeu Xuebao*, **1**, 430 (1985); *Chem. Abstr.*, **104**, 40565k.
36. I. S. Crick, R. W. Gable, B. F. Hoskins, and P. A. Tregloan, *Inorg. Chim. Acta*, **111**, 35 (1986).
37. E. K. Barefield, A. Bianchi, E. J. Billo, P. J. Connolly, P. Paoletti, S. J. Summers, and D. G. van Derveer, *Inorg. Chem.*, **25**, 4197 (1986).
38. R. W. Hay and M. A. Ali, *Inorg. Chim. Acta*, **103**, 23 (1985).
39. E. Kimura, T. Koike, M. Yamaka, and M. Kodama, *J. Chem. Soc., Chem. Commun.*, 1341 (1985).
40. C. Kratky, R. Waditschatka, C. Angst, J. E. Johansen, J. C. Plaquevent, J. Schreiber, and A. Eschenmoser, *Helv. Chim. Acta*, **68**, 1312 (1985).
41. J. Browning, K. A. Beverage, G. W. Bushnell, and K. R. Dixon, *Inorg. Chem.*, **25**, 1987 (1986).
42. F. D. Rochon and P. C. King, *Can. J. Chem.*, **64**, 1894 (1986).
43. J. A. Davies and A. Sood, *Inorg. Chem.*, **24**, 4213 (1985).
44. H. Kurosawa, A. Urabe, and M. Emoto, *J. Chem. Soc., Dalton Trans.*, 891 (1986).
45. R. H. Crabtree and A. Habib, *Inorg. Chem.*, **25**, 3698 (1986).
46. A. Albinati, H. Moriyama, H. Rüegger, P. S. Pregosin, and A. Togni, *Inorg. Chem.*, **24**, 4430 (1985).
47. L. I. Elding, B. Norén, and A. Oskarsson, *Inorg. Chim. Acta*, **114**, 71 (1986).
48. T. G. Appleton, J. R. Hall, and S. F. Ralph, *Inorg. Chem.*, **24**, 4685 (1985).
49. S. S. M. Ling and R. J. Puddephatt, *Polyhedron*, **5**, 1423 (1986).
50. B. Lippert, H. Schöllhorn, and U. Thewalt, *J. Am. Chem. Soc.*, **108**, 525 (1986).
51. G. Annibale, L. Cattalini, L. Canovese, B. Pitteri, A. Tiripicchio, M. Tiripicchio Camellini, and M. L. Tobe, *J. Chem. Soc., Dalton Trans.*, 1101 (1986).
52. A. Scrivanti, A. Benton, L. Toniolo, and C. Botteghi, *J. Organomet. Chem.*, **314**, 369 (1986).
53. S. P. Roe, J. O. Hill, and R. J. Magee, *Inorg. Chim. Acta*, **115**, L15 (1986).
54. G. Farraglia, L. Sindellari, and V. Cherchi, *Transition Met. Chem.*, **11**, 98 (1986).
55. L. V. Rybin, E. A. Petrovskaya, M. I. Rubinskaya, L. G. Kuz'mina, Yu. T. Struchkov, V. V. Kaverin, and N. Yu. Koneva, *J. Organomet. Chem.*, **288**, 119 (1985).
56. A. N. Nyrkova, G. A. Kukina, O. N. Temkin, and M. A. Porai-Koshits, *Koord. Khim.*, **11**, 954 (1985).
57. H. C. Clark and M. J. Hampden Smith, *J. Am. Chem. Soc.*, **108**, 3829 (1986).
58. a. R. H. Crabtree and M. Lavin, *J. Chem. Soc., Chem. Commun.*, 1661 (1985). b. R. H. Crabtree, M. Lavin, and L. Bonneviot, *J. Am. Chem. Soc.*, **108**, 4032 (1986).
59. H. Azizian, K. R. Dixon, C. Eaborn, A. Pidcock, N. M. Shuaib, and J. Vinaixa, *J. Chem. Soc., Chem. Commun.*, 1020 (1982).
60. D. R. Eaton, M. J. McGlinchey, K. A. Moffat, and R. J. Buist, *J. Am. Chem. Soc.*, **106**, 8110 (1984).
61. a. D. L. Packett, C. M. Jensen, R. L. Cowan, C. E. Strouse, and W. C. Trogler, *Inorg. Chem.*, **24**, 3578 (1985). b. D. L. Packett and W. C. Trogler, *J. Am. Chem. Soc.*, **108**, 5036 (1986).
62. a. S. Obara, K. Kitamura, and K. Morokuma, *J. Am. Chem. Soc.*, **106**, 7482 (1984). b. J. J. Low and W. A. Goddard, *Organometallics*, **5**, 609 (1986).
63. E. Grimley and D. W. Meek, *Inorg. Chem.*, **25**, 2049 (1986).
64. M. S. Ram, A. Bakac, and J. H. Espenson, *Inorg. Chem.*, **25**, 3267 (1986).
65. M. S. Ram, J. H. Espenson, and A. Bakac, *Inorg. Chem.*, **25**, 4115 (1986).
66. D. J. Brauer, F. Gol, S. Hietkamp, H. Peters, H. Sommer, O. Stelzer, and W. S. Sheldrick, *Chem. Ber.*, **119**, 349 (1986).
67. J. Granell, D. Sainz, J. Sales, X. Solans, and M. Font-Altaba, *J. Chem. Soc., Dalton Trans.*, 1785 (1986).
68. Z. Taira and S. Yamazaki, *Bull. Chem. Soc. Jpn.*, **59**, 649 (1986).

69. S. Yamazaki, *Polyhedron*, **4**, 1915 (1985).
70. V. G. Albano, F. Demartin, A. de Renzi, G. Morelli, and A. Saporito, *Inorg. Chem.*, **24**, 2032 (1985).
71. O. Wernberg, *J. Chem. Soc., Dalton Trans.*, 725 (1986).
72. M. S. Holt and J. H. Nelson, *Inorg. Chem.*, **25**, 1316 (1986).
73. M. S. Holt, J. H. Nelson, and N. W. Alcock, *Inorg. Chem.*, **25**, 2283 (1986).
74. M. A. Murphy, B. L. Smith, G. P. Torrence, and A. Aguilo, *Inorg. Chim. Acta*, **101**, L47 (1985).
75. F. R. Hartley, S. G. Murray, D. M. Potter, and J. R. Chipperfield, *J. Organomet. Chem.*, **306**, 133 (1986).
76. G. Alibrandi, D. Minniti, R. Romeo, P. Uguagliati, L. Calligaro, and U. Belluco, *Inorg. Chim. Acta*, **112**, L15 (1986).
77. A. Höhn and H. Werner, *Angew. Chem., Int. Ed. Engl.*, **25**, 737 (1986).
78. E. Rotondo and F. Priolo-Cusmano, *J. Organomet. Chem.*, **292**, 429 (1985).
79. J. A. M. van Beek, G. van Koten, W. J. J. Smeets, and A. L. Spek, *J. Am. Chem. Soc.*, **108**, 5010 (1986).
80. J. Terheijden, G. van Koten, W. P. Mul, D. J. Stufkens, F. Muller, and C. H. Stam, *Organometallics*, **5**, 519 (1986).
81. P. Courtot, R. Pichon, R. Rumin, and J. Y. Salaün, *Inorg. Chim. Acta*, **103**, L7 (1985).
82. P. Courtot, R. Pichon, R. Rumin, and J. Y. Salaün, *Nouv. J. Chem.*, **10**, 447 (1986).
83. R. L. Batstone-Cunningham, J. P. Hunt, and D. M. Roundhill, *J. Organomet. Chem.*, **289**, 431 (1985).
84. R. J. Blau, J. H. Espenson, S. Kim, and R. A. Jacobson, *Inorg. Chem.*, **25**, 757 (1986).
85. S. Muralidharan, J. H. Espenson, and S. A. Ross, *Inorg. Chem.*, **25**, 2557 (1986).

References for Chapter 6

1. D. A. House, in: *Mechanism of Inorganic and Organometallic Reactions* (M. V. Twigg, ed.), Vol. 4, Plenum Press, New York (1986).
2. R. Colton, *Coord. Chem. Rev.*, **62**, 85 (1985).
3. J. E. Newbury, *Annu. Rep. R. Soc. Chem.*, **81A**, 185 (1984).
4. F. A. Cotton and R. A. Walton, *Struct. Bonding (Berlin)*, **62**, 1 (1985).
5. H. Ogino, *Kagaku (Kyoto)*, **40**, 278 (1985); *Chem. Abstr.*, **103**, 93491 (1985).
6. R. L. Carlin, *Science*, **227**, 1291 (1985).
7. E. P. Kundig, *Kagaku Zokan (Kyoto)*, 175 (1985); *Chem. Abstr.*, **103**, 142041 (1985).
8. A. Solladie-Cavallo, *Polyhedron*, **4**, 901 (1985).
9. R. N. Bose and E. S. Gould, *Inorg. Chem.*, **24**, 2645 (1985).
10. R. N. Bose and E. S. Gould, *Inorg. Chem.*, **24**, 2832 (1985).
11. R. N. Bose and E. S. Gould, *Inorg. Chem.*, **25**, 94 (1986).
12. R. N. Bose, V. D. Neff, and E. S. Gould, *Inorg. Chem.*, **25**, 165 (1986).
13. R. N. Bose, N. Rajasekar, D. M. Thompson, and E. S. Gould, *Inorg. Chem.*, **25**, 3349 (1986).
14. Y. T. Fanchiang, R. N. Bose, E. Gelerinter, and E. S. Gould, *Inorg. Chem.*, **24**, 4679 (1985).
15. S. K. Gosh and E. S. Gould, *Inorg. Chem.*, **25**, 3357 (1986).
16. A. Rajavelu, B. S. Srinivasan, and S. Vangalur, *Indian J. Chem., Sect. A.*, **24**, 936 (1985).
17. E. S. Gould, *Acc. Chem. Res.*, **19**, 66 (1986).
18. K. Nag and S. N. Bose, *Struct. Bonding (Berlin)*, **62**, 153 (1985).
19. J. T. Groves and W. J. Kruger, *Isr. J. Chem.*, **25**, 148 (1985); *Chem. Abstr.*, **103**, 104289 (1985).
20. M. Ardon and A. Bino, *Inorg. Chem.*, **24**, 1343 (1985).
21. P. Andersen, K. M. Nielsen, and A. Petersen, *Acta Chem. Scand., Ser. A*, **38**, 593 (1984).
22. F. Rotzinger, H. Stunzi, and W. Marty, *Inorg. Chem.*, **25**, 489 (1986).
23. D. A. House, E. V. McKee, and P. J. Steel, *Inorg. Chem.*, **25**, 4884 (1986).
24. S. Kaizaki and M. Mizu-uchi, *Inorg. Chem.*, **25**, 2732 (1986).

25. K. Kanamori and K. Kawai, *Inorg. Chem.*, **25**, 3711 (1986).
26. R. J. Bianchini and J. I. Legg, *Inorg. Chem.*, **25**, 3263 (1986).
27. W. E. Broderick and J. I. Legg, *Inorg. Chem.*, **24**, 3724 (1985).
28. N. Koine, R. J. Bianchini, and J. I. Legg, *Inorg. Chem.*, **25**, 2835 (1986).
29. W. D. Wheeler and J. I. Legg, *Inorg. Chem.*, **24**, 1292 (1985).
30. C. A. Green, H. Place, R. D. Willett, and J. I. Legg, *Inorg. Chem.*, **25**, 4672 (1986).
31. S. Kaizaki, *Kagaku (Kyoto)*, **40**, 550 (1985); *Chem. Abstr.*, **104**, 27541 (1986).
32. S. O. Oh, S. H. Lee, and J. W. Lim, *Taehan Hwahakhoe Chi*, **30**, 307 (1986); *Chem. Abstr.*, **104**, 231435 (1986).
33. J. W. Vaughn and E. L. King, *Inorg. Chem.*, **24**, 4221 (1985).
34. A. D. Kirk, C. Namasivayam, and T. Ward, *Inorg. Chem.*, **25**, 2225 (1986).
35. J. Casabo, X. Solans, C. Diaz, J. Ribas, A. Segui, and M. Corbella, *Transition Met. Chem.*, **10**, 128 (1985).
36. D. T. Fagan, J. S. Frigerio, and J. W. Vaughn, *Inorg. Synth.*, **24**, 185 (1986).
37. J. Ribas, M. Monfort, and J. Casabo, *Transition Met. Chem.*, **9**, 407 (1984).
38. J. W. Vaughn and J. S. Frigerio, *Inorg. Chem.*, **24**, 2110 (1985).
39. M. Corbella and J. Ribas, *Inorg. Chem.*, **25**, 4390 (1986).
40. N. A. P. Kane-Maguire, K. C. Wallace, and D. G. Speece, *Inorg. Chem.*, **25**, 4650 (1986).
41. F. H. Herbstein, M. Kapon, and G. M. Reisner, *Z. Krist.*, **171**, 209 (1985).
42. C. Diaz, A. Segui, J. Ribas, X. Solans, and M. Font-Altaba, *Transition Met. Chem.*, **9**, 469 (1984).
43. R. J. Bianchini, U. Geiser, H. Place, S. Kaizaki, Y. Morita, and J. I. Legg, *Inorg. Chem.*, **25**, 2129 (1986).
44. J. V. Brencic and I. Leban, *Vestn. Slov. Kem. Drus.*, **32**, 209 (1985); *Chem. Abstr.*, **104**, 13305 (1985).
45. J. W. Vaughn and R. D. Rogers, *J. Cryst. Spect. Res.*, **15**, 281 (1985).
46. K. A. Beveridge, G. W. Bushnell, and A. D. Kirk, *Acta Crystallogr., Sect. C*, **41**, 899 (1985).
47. W. C. Kupferschmidt and R. B. Jordan, *Inorg. Chem.*, **24**, 3357 (1985).
48. A. Bakar bin Baba, V. Gold, and F. Hibbert, *J. Chem. Soc., Perkin Trans. 2*, 1039 (1985).
49. A. Rotman, H. Cohen, and D. Meyerstein, *Inorg. Chem.*, **24**, 4158 (1985).
50. P. Kita and R. B. Jordan, *Inorg. Chem.*, **25**, 4791 (1986).
51. M. Shoji, M. Shimura, H. Ogino, and T. Ito, *Chem. Lett.*, 995 (1986).
52. H. Cohen and D. Meyerstein, *Angew. Chem.*, **97**, 785 (1985); *Chem. Abstr.*, **103**, 129825 (1985).
53. H. Cohen, D. Meyerstein, and A. J. Shusteman, *J. Chem. Soc., Chem. Commun.*, 424 (1985).
54. K. Crouse and L. Y. Goh, *Inorg. Chim. Acta*, **99**, 199 (1985).
55. K. Crouse and L. Y. Goh, *Inorg. Chem.*, **25**, 478 (1986).
56. J. D. Melton, A. Bakac, and J. H. Espenson, *Inorg. Chem.*, **25**, 3360 (1986).
57. M. Shimura, H. Ogino, and N. Tanaka, *Nippon Kagaku Kaishi*, 370 (1985); *Chem. Abstr.*, **104**, 50940 (1986).
58. P. Kita and R. B. Jordan, *Inorg. Chem.*, **24**, 2701 (1985).
59. M. J. Sisley and R. B. Jordan, *Inorg. Chem.*, **25**, 3547 (1986).
60. S. Bruhn, A. Bakac, and J. H. Espensen, *Inorg. Chem.*, **25**, 535 (1986).
61. P. W. Moore, in: *Mechanism of Inorganic and Organometallic Reactions* (M. V. Twigg, ed.), Vol. 3, Plenum Press, New York (1985).
62. R. F. Johnston and R. A. Holwerda, *Inorg. Chem.*, **24**, 3176, 3181 (1985).
63. L. Mønsted and O. Mønsted, *Acta Chem. Scand., Ser. A*, **39**, 615 (1985).
64. M. F. Amira, S. A. El-Shazly, M. M. Khalil, and F. M. Abdel-Haim, *Transition Met. Chem.*, **11**, 72 (1986).
65. L. Mønsted, T. Ramasami, and A. G. Sykes, *Acta Chem. Scand., Ser. A*, **39**, 437 (1985).
66. A. D. Kirk, S. Arunachalam, and D. Kneeland, *Inorg. Chem.*, **25**, 3551 (1986).
67. W. L. Reynolds, L. Reichley-Yinger, and Y. Yuan, *Inorg. Chem.*, **24**, 4273 (1985).
68. G. A. Lawrance and A. M. Sargeson, *Inorg. Synth.*, **24**, 250 (1986).
69. N. J. Curtis, G. A. Lawrance, P. A. Lay, and A. M. Sargeson, *Inorg. Chem.*, **25**, 484 (1986).

70. D. A. House and W. T. Robinson, *Inorg. Chim. Acta*, **141**, 211 (1988).

71. D. A. House, *Inorg. Chim. Acta*, **121**, 223 (1986).

72. R. G. Swisher and E. L. Blinn, *Inorg. Chem.*, **25**, 1320 (1986).

73. C. Narayanaswamy, T. Ramasami, and D. Ramaswamy, *Inorg. Chem.*, **25**, 4052 (1986).

74. H. Ogino, A. Masuka, S. Ito, N. Miura, and M. Shimura, *Inorg. Chem.*, **25**, 709 (1986).

75. J. C. Chang, *Inorg. Chem.*, **25**, 1725 (1986).

76. D. A. House, *Inorg. Chem.*, **25**, 1671 (1986).

77. D. A. House, *Inorg. Chim. Acta*, **121**, 167 (1986).

78. D. J. Hodgson, E. Pedersen, H. Toftlund, and C. Weiss, *Inorg. Chim. Acta*, **120**, 177 (1986).

79. N. A. P. Kane-Maguire, K. C. Wallace, D. P. Cobranchi, J. M. Derrick, and D. G. Speece, *Inorg. Chem.*, **25**, 2101 (1986).

80. R. Banerjee, *Coord. Chem. Rev.*, **68**, 145 (1985).

81. R. Banerjee, *Transition Met. Chem.*, **10**, 147 (1985).

82. R. Tamilarasan and J. F. Endicott, *J. Phys. Chem.*, **90**, 1027 (1986).

83. G. H. Searle and D. A. House, *Aust. J. Chem.*, **40**, 361 (1987).

84. I. K. Adzamli and E. Deutsch, *Inorg. Chem.*, **24**, 4086 (1985).

85. A. C. Ukwueze, *Inorg. Chim. Acta*, **120**, 21 (1986).

86. L. Spiccia and W. Marty, *Inorg. Chem.*, **25**, 266 (1986).

87. L. Mønsted, O. Mønsted, and J. Springborg, *Inorg. Chem.*, **24**, 3496 (1985).

88. F. C. Xu, H. R. Krouse, and T. W. Swaddle, *Inorg. Chem.*, **24**, 267 (1985).

89. B. K. Niogy and G. S. De, *Indian J. Chem., Sect. A*, **24**, 208 (1985).

90. V. I. Korner, L. Trubacheva, and I. A. Prozorava, *Zh. Neorg. Khim.*, **31**, 1418 (1986); *Chem. Abstr.*, **105**, 49818 (1986).

91. J. Maslowska and L. Churscinski, *Acta Univ. Lodz. Folia Chim.*, **3**, 23 (1984); *Chem. Abstr.*, **102**, 192223 (1985).

92. S. I. Ali and Z. Murtaza, *Polyhedron*, **4**, 463 (1985).

93. Kabir-Ud-Din and J. G. Khan, *Ann. Chim. (Rome)*, **75**, 279 (1985).

94. M. J. Hynes, C. O'Mara, and D. F. Kelly, *Inorg. Chim. Acta*, **120**, 131 (1986).

95. G. J. Khan and Kabir-Ud-Din, *Indian J. Chem., Sect. A*, **24**, 739 (1985).

96. H. Stieger, G. M. Harris, and H. Kelm, *Ber. Bunsenges. Phys. Chem.*, **74**, 262 (1970).

97. S. Mazumdar and K. De, *Curr. Sci.*, **55**, 437 (1986).

98. M. Satyanarayana and V. A. Raman, *Bull. Pure Appl. Sci., Sect. C*, **2**, 52 (1983).

99. M. E. Sosa and M. L. Tobe, *J. Chem. Soc., Dalton Trans.*, 427 (1986).

100. D. A. House and D. Yang, *Inorg. Chim. Acta*, **74**, 179 (1983).

101. K. Wieghardt, M. Guttmann, D. Ventur, and W. Gebert, *Z. Anorg. Allg. Chem.*, **527**, 33 (1985).

102. P. Kita, *Pol. J. Chem.*, **59**, 685 (1985).

103. P. J. Marriott and Y. H. Lai, *Inorg. Chem.*, **25**, 3680 (1986).

104. P. Biscarini, *Inorg. Chim. Acta*, **99**, 183, 189 (1985).

105. A. D. Kirk, G. B. Porter, and D. K. Sharma, *Chem. Phys. Lett.*, **123**, 584 (1986).

106. G. E. Rojas, C. Dupuy, D. A. Sexton, and D. Magde, *J. Phys. Chem.*, **90**, 87 (1986).

107. N. Serpone and M. Z. Hoffman, *Chem. Phys. Lett.*, **123**, 551 (1986).

108. F. Wasgestian and E. Gowin, *Inorg. Chim. Acta*, **120**, L17 (1986).

109. J. Lilie, W. L. Waltz, S. H. Lee, and L. L. Gregor, *Inorg. Chem.*, **25**, 4487 (1986).

110. E. Gowin and F. Wasgestian, *Inorg. Chem.*, **24**, 3106 (1985).

111. A. D. Kirk, G. B. Porter, and M. A. Rampi Scandola, *Inorg. Chim. Acta*, **90**, 161 (1984).

112. S. H. Lee, W. L. Waltz, D. R. Demmer, and R. T. Walters, *Inorg. Chem.*, **24**, 1531 (1985).

113. J. F. Endicott, R. Tamilarasan, and G. R. Brubaker, *J. Am. Chem. Soc.*, **108**, 5193 (1986).

114. J. F. Endicott, R. Tamilarasan, and R. B. Lessard, *Chem. Phys. Lett.*, **112**, 381 (1984).

115. B. S. Brunschwig, P. J. De Laive, A. M. English, M. Goldberg, H. B. Gray, S. L. Mayo, and N. Sutin, *Inorg. Chem.*, **24**, 3743 (1985).

116. R. E. Gamache, R. A. Rader, and D. R. McMillin, *J. Am. Chem. Soc.*, **107**, 1141 (1985).

117. P. Comba, I. I. Creaser, L. R. Gahan, J. MacB. Harrowfield, G. A. Lawrance, L. L. Martin, A. W. H. Mau, A. M. Sargeson, W. H. F. Sasse, and M. R. Snow, *Inorg. Chem.*, **25**, 384 (1986).

118. L. S. Forster and O. Mønsted, *J. Phys. Chem.*, **90**, 5131 (1968).
119. N. A. P. Kane-Maguire, N. Helwic, and J. M. Derrick, *Inorg. Chim. Acta*, **102**, L21 (1985).
120. A. Ghaith, L. S. Forster, and J. V. Rand, *Inorg. Chim. Acta*, **116**, 11 (1986).
121. A. Marchaj and F. Wasgestian, *Inorg. Chim. Acta*, **102**, L13 (1985).
122. W. Strek and I. Trabjerg, *Physica C*, **141**, 323 (1986).
123. A. F. Fucaloro, L. S. Forster, S. G. Glover, and A. D. Kirk, *Inorg. Chem.*, **24**, 4242 (1985).
124. M. Mvele and F. Wasgestian, *Spectrochim. Acta, Part A*, **42**, 775 (1986).
125. H. Riesen and H. U. Guedel, *Mol. Phys.*, **58**, 509 (1986).
126. A. Ditze and F. Wasgestian, *Ber. Bunsenges. Phys. Chem.*, **90**, 111 (1986).
127. S. Howard and K. I. Hardcastle, *J. Cryst. Spect. Res.*, **15**, 643 (1985).
128. J. V. Brencic, J. Leban, and J. Zule, *Z. Anorg. Allg. Chem.*, **521**, 199 (1985).
129. E. Forsellini, T. Parasassi, G. Bombieri, M. L. Tobe, and M. E. Sosa, *Acta Crystallogr., Sect. C*, **42**, 563 (1986).
130. W. Clegg, P. Leupin, D. T. Richens, A. G. Sykes, and E. S. Raper, *Chem. Abstr.*, **102**, 212998 (1984).
131. T. W. Hambley and P. A. Lay, *Inorg. Chem.*, **25**, 4553 (1986).
132. P. Comba, A. M. Sargeson, L. M. Engelhardt, J. MacB. Harrowfield, A. H. White, E. Horn, and M. B. Snow, *Inorg. Chem.*, **24**, 2325 (1985).
133. U. Casellato, R. Graziani, G. Maccarrone, and A. J. Di Bilio, *J. Cryst. Spect. Res.*, **16**, 695 (1986).
134. S. Larsen, K. Michelsen, and E. Pedersen, *Acta Chem. Scand., Ser. A*, **40**, 63 (1986).
135. I. I. Yashina and Y. N. Shevchenko, *J. Therm. Anal.*, **30**, 319 (1985).
136. M. Serra, A. Eseuer, M. Monfort, and J. Ribas, *Thermochim. Acta*, **92**, 493 (1985).
137. Y. H. Shevchenko, V. A. Logvinenko, N. I. Yaschina, E. A. Pisarev, and G. V. Gavrilova, *J. Therm. Anal.*, **30**, 365 (1985).
138. J. M. Lopez-Alcala, M. C. Puerta, and F. Gonzalez-Vilchez, *Thermochim. Acta*, **96**, 1 (1985).
139. F. Hanic, I. Horvath, G. Plesch, and L. Galikova, *J. Solid State Chem.*, **59**, 190 (1985); *Chem. Abstr.*, **103**, 201501 (1985).
140. E. A. Pisarev, Y. K. Shevchenko, and V. A. Logvineko, *Teor. Eksp. Khim.*, **21**, 637 (1985); *Chem. Abstr.*, **104**, 11115 (1986).
141. J. Ribas, A. Escuer, and M. Monfort, *Inorg. Chem.*, **24**, 1874 (1985).
142. J. Ribas, M. Monfort, and X. Solans, *An. Quim., Ser. B*, **80**, 288 (1984); *Chem. Abstr.*, **102**, 124467 (1985).
143. D. Banerjea and J. C. Bailar Jr, *Transition Met. Chem.*, **10**, 331 (1985).
144. A. D. Kirk and P. A. Warren, *Inorg. Chem.*, **24**, 720 (1985).
145. P. Connolly and J. H. Espenson, *Inorg. Chem.*, **25**, 2684 (1986).
146. R. J. Balahura and A. Johnston, *Can. J. Chem.*, **64**, 841 (1986)
147. K. Tsukahara, *Inorg. Chim. Acta*, **120**, 11 (1986).
148. E. F. Hills, M. Moszner, and A. G. Sykes, *Inorg. Chem.*, **25**, 339 (1986).
149. P. Acott, G. Ali, and N. A. Lewis, *Inorg. Chim. Acta*, **99**, 169 (1985).
150. A. G. Lappin, C. A. Lewis, and W. J. Ingledew, *Inorg. Chem.*, **24**, 1446 (1985).
151. G. D. Armstrong, T. Ramasami, and A. G. Sykes, *Inorg. Chem.*, **24**, 3230 (1985).
152. F. P. Rotzinger, *Inorg. Chem.*, **25**, 4570 (1986).
153. L. M. Wilson and R. D. Cannon, *Inorg. Chem.*, **24**, 4366 (1985).
154. S. Jagannathan and R. C. Patel, *Inorg. Chem.*, **24**, 3634 (1985).
155. S. D. Nadler, J. G. Dick, and C. H. Langford, *Can. J. Chem.*, **63**, 2732 (1985).
156. A. Bakac, V. Butkovic, J. H. Espenson, R. Marcec, and M. Orhanovic, *Inorg. Chem.*, **25**, 2562 (1986).
157. R. Curci, S. Giannattasio, O. Sciacovelli, and L. Troisi, *Tetrahedron*, **40**, 2763 (1984).
158. C. K. Ranganathan, T. Ramasami, D. Ramaswamy, and M. Santappa, *Inorg. Chem.*, **25**, 915 (1986).
159. G. P. Haight, G. M. Jursich, M. T. Kelso, and P. J. Merrill, *Inorg. Chem.*, **24**, 2740 (1985).
160. L. C. Yuan, T. S. Calderwood, and T. C. Bruice, *J. Am. Chem. Soc.*, **107**, 8273 (1985).

161. B. De Poorter and B. Meurier, *Nouv. J. Chim.*, **9**, 393 (1985).
162. E. G. Samsel, K. Srinivasan, and J. K. Kochi, *J. Am. Chem. Soc.*, **107**, 7606 (1985).
163. K. Srinivasan and J. K. Kochi, *Inorg. Chem.*, **24**, 4671 (1985).
164. V. P. Singh, I. M. Pandey, and S. B. Sharma, *J. Indian Chem. Soc.*, **62**, 64 (1985).
165. M. P. Alvarez Macho, *An. Quim.*, **80**, 687 (1984); *Chem. Abstr.*, **103**, 27988 (1985).
166. T. J. Mason, J. A. Turrell, and C. W. Went, *J. Chem. Educ.*, **62**, 344 (1985).
167. H. B. Davis, R. M. Sheets, W. W. Paudler, and G. L. Gard, *Heterocycles*, **22**, 2029 (1984); *Chem. Abstr.*, **102**, 6415 (1985).
168. M. P. Alvarez Macho, *An. Quim.*, **80**, 672 (1984); *Chem. Abstr.*, **103**, 27986 (1985).
169. K. K. Sen Gupta, S. Sen Gupta, S. K. Mandel, and S. N. Basu, *Carbohydr. Res.*, **139**, 185 (1985); *Chem. Abstr.*, **103**, 178516 (1985).
170. P. H. Connett and K. E. Wetterhahn, *J. Am. Chem. Soc.*, **107**, 4282 (1985).
171. P. H. Connett and K. E. Wetterhahn, *J. Am. Chem. Soc.*, **108**, 1842 (1986).
172. V. Balaiah and P. V. V. Satyanarayana, *Curr. Sci.*, **54**, 384 (1985); *Chem. Abstr.* **103**, 87330 (1985).
173. B. Bhattacharjea, M. N. Bhattacharjea, M. Bhattacharjea, and A. K. Bhattacharjea, *Int. J. Chem. Kinet.*, **17**, 629 (1985).
174. A. A. Woolf, *J. Fluorine Chem.*, **27**, 285 (1985); *Chem. Abstr.*, **102**, 192282 (1985).
175. M. Dasgupta and M. K. Mahanti, *Afinidad*, **43**, 353 (1986); *Chem. Abstr.*, **105**, 103413 (1986).
176. E. Baumgartner, M. A. Blesa, R. Larotonda, and A. J. G. Maroto, *J. Chem. Soc., Faraday Trans. 1*, **81**, 1113 (1985).
177. E. Baumgartner, J. H. Garcia, and R. Larotonda, *J. Chem. Educ.*, **63**, 907 (1986).
178. M. Wawrzenczyk, *Mater. Sci.*, **9**, 263 (1985); *Chem. Abstr.*, **104**, 136787 (1986).
179. P. Uma, P. K. Rao, and M. N. Sastri, *Indian J. Chem.*, *Sect. A*, **24**, 539 (1985).
180. G. Michel and R. Cahay, *J. Raman Spectrosc.*, **17**, 79 (1986); *Chem. Abstr.*, **104**, 156745 (1986).
181. S-Y. Tong and L. Li, *Talanta*, **33**, 775 (1986).
182. E. N. Rizkalla, S. S. Amis, and M. S. Antonious, *Inorg. Chim. Acta*, **97**, 165 (1985).
183. C. G. Roffey, *J. Oil Colour Chem. Assoc.*, **68**, 116 (1985); *Chem. Abstr.*, **103**, 45668 (1985).
184. K. J. Martin and T. J. Pinnavaia, *J. Am. Chem. Soc.*, **108**, 541 (1986).
185. T. J. Cardwell, *J. Chem. Educ.*, **63**, 90 (1986).
186. N. D. Yardamov and Y. Karadzhor, *Transition Met. Chem.*, **10**, 15 (1985).
187. A. Okumura, N. Takeuchi, and N. Okazaki, *Inorg. Chim. Acta*, **102**, 127 (1985).
188. D. A. Brown, W. K. Glass, M. Rasul Jan, and R. M. W. Mulders, *Environ. Technol. Lett.*, **7**, 283, 289 (1986).

References for Chapter 7

1. R. Banerjee, *Coord. Chem. Rev.*, **68**, 145 (1985).
2. G. A. Lawrance, *Chem. Rev.*, **86**, 17 (1986).
3. N. E. Dixon, G. A. Lawrance, P. A. Lay, A. M. Sargeson, and H. Taube, *Inorg. Synth.*, **24**, 243 (1985).
4. S. Fallab and P. R. Mitchell, *Adv. Inorg. Bioinorg. Mech.*, **3**, 311 (1984).
5. R. van Eldik, *Adv. Inorg. Bioinorg. Mech.*, **3**, xx (1984).
6. J. D. Atwood, *Inorganic and Organometallic Reaction Mechanisms*, Brooks/Cole Publishing Co., Monterey, CA (1985).
7. N. J. Curtis and G. A. Lawrance, *Inorg. Chem.*, **25**, 1033 (1986).
8. P. A. Lay, *J. Chem. Soc., Chem. Commun.*, 1422 (1986).
9. T. W. Swaddle, *Adv. Inorg. Bioinorg. Mech.*, **2**, 1 (1983), and references cited therein.
10. G. A. Lawrance, K. Schneider, and R. van Eldik, *Inorg. Chem.*, **23**, 3922 (1984).
11. N. J. Curtis, G. A. Lawrance, P. A. Lay, and A. M. Sargeson, *Inorg. Chem.*, **25**, 484 (1986).
12. Y. Kitamura, T. Itoh, and M. Takeuchi, *Inorg. Chem.*, **25**, 3887 (1986).

13. G. A. Lawrance, *Polyhedron*, **5**, 2113 (1986).
14. G. Schiavon and F. Marchetti, *Polyhedron*, **4**, 1143 (1985).
15. A. Okumura, N. Takeuchi, and N. Okazaki, *Inorg. Chim. Acta*, **102**, 127 (1985).
16. D. A. House, *Inorg. Chim. Acta*, **121**, 167 (1986).
17. M. Harnett, D. A. House, and W. T. Robinson, *Inorg. Chim. Acta*, **102**, 87 (1985).
18. D. A. House, *Inorg. Chem.*, **25**, 1671 (1986).
19. V. McKee, M. Harnett, and D. A. House, *Inorg. Chim. Acta*, **102**, 83 (1985).
20. D. A. House, *Helv. Chim. Acta*, **68**, 1872 (1985).
21. A. Reisen, M. Zehnder, and D. A. House, *Inorg. Chim. Acta*, **113**, 163 (1986).
22. F. P. Rotzinger and W. Marty, *Helv. Chim. Acta*, **68**, 1914 (1985).
23. D. A. Buckingham, W. Marty, and A. M. Sargeson, *Inorg. Chem.*, **13**, 2165 (1974).
24. F. P. Rotzinger and W. Marty, *Inorg. Chem.*, **24**, 1617 (1985).
25. M. Iida and H. Yamatera, *J. Chem. Soc., Dalton Trans.*, 2049 (1986).
26. H. Millar, *Polyhedron*, **5**, 1965 (1986).
27. M. Iida and H. Yamatera, *Polyhedron*, **4**, 623 (1985).
28. S. I. Amer and J. A. McLean Jr., *Inorg. Chim. Acta*, **101**, 1 (1985).
29. D. A. House, *Helv. Chim. Acta*, **68**, 1872 (1985).
30. V. McKee, M. Harnett, and D. A. House, *Inorg. Chim. Acta*, **102**, 83 (1985).
31. M. E. Sosa and M. L. Tobe, *J. Chem. Soc., Dalton Trans.*, 427 (1986).
32. J. Lichtig, M. Sosa, and M. L. Tobe, *J. Chem. Soc., Dalton Trans.*, 581 (1984).
33. S. Balt, H. J. Gamelkoorn, and S. Oosterink, *Inorg. Chem.*, **24**, 2384 (1985).
34. S. Balt and H. J. Gamelkoorn, *Inorg. Chem.*, **24**, 3347 (1985).
35. N. J. Curtis, G. A. Lawrance, P. A. Lay, and A. M. Sargeson, *Inorg. Chem.*, **25**, 484 (1986).
36. G. A. Lawrance, *Polyhedron*, **4**, 599 (1985).
37. M. C. Gomez-Vaamonde and K. B. Nolan, *Inorg. Chim. Acta*, **101**, 67 (1985).
38. A. A. Watson, M. R. Prinsep, and D. A. House, *Inorg. Chim. Acta*, **115**, 95 (1986).
39. P. C. Tripathy, P. Mohanty, and R. K. Nanda, *Transition Met. Chem.*, **10**, 338 (1985).
40. E. Baraniak, D. A. Buckingham, C. R. Clark, and A. M. Sargeson, *Inorg. Chem.*, **25**, 1956 (1986).
41. See, for example, M. D. Alexander and D. H. Busch, *Inorg. Chem.*, **5**, 602 (1966); D. A. Buckingham, D. M. Foster, and A. M. Sargeson, *J. Am. Chem. Soc.*, **90**, 6032 (1968); C. J. Boreham, D. A. Buckingham, and C. R. Clark, *Inorg. Chem.*, **18**, 1990 (1979); D. A. Buckingham, D. M. Foster, and A. M. Sargeson, *J. Am. Chem. Soc.*, **91**, 4102 (1969).
42. E. Baraniak, D. A. Buckingham, C. R. Clark, and A. M. Sargeson, *Inorg. Chem.*, **25**, 1952 (1986).
43. R. W. Hay, V. M. C. Reid, and D. P. Piplani, *Transition Met. Chem.*, **11**, 302 (1986).
44. R. A. Henderson and M. L. Tobe, *Inorg. Chem.*, **16**, 2576 (1977).
45. M. L. Tobe, *Adv. Inorg. Bioinorg. Mech.*, **2**, 1 (1983).
46. M. J. Gaudin, C. R. Clark, and D. A. Buckingham, *Inorg. Chem.*, **25**, 2569 (1986).
47. A. C. Dash and R. C. Nayak, *Inorg. Chem.*, **25**, 2237 (1986).
48. G. Calvaruso, F. P. Cavasino, and E. Di Dio, *Inorg. Chim. Acta*, **119**, 29 (1986).
49. R. van Eldik, Y. Kitamura, and C. P. P. Mac-Coll, *Inorg. Chem.*, **25**, 4252 (1986).
50. R. W. Hay, R. Bembi, and B. Jeragh, *Transition Met. Chem.*, **11**, 385 (1986).
51. G. Bombieri, E. Forsellini, A. Del Pra, C. J. Cooksey, M. Humanes, and M. L. Tobe, *Inorg. Chim. Acta*, **61**, 43 (1982).
52. R. J. Geue, M. G. McCarthy, A. M. Sargeson, P. Jorgensen, R. G. Hazell, and F. K. Larsen, *Inorg. Chem.*, **24**, 2559 (1985).
53. G. M. Miskelly, C. R. Clark, and D. A. Buckingham, *J. Am. Chem. Soc.*, **108**, 5202 (1986).
54. S. K. Saha and P. Banerjee, *Transition Met. Chem.*, **10**, 252 (1985).
55. P. Bhattacharya, M. C. Ghosh, and P. Banerjee, *Transition Met. Chem.*, **11**, 76 (1986).
56. P. Bhattacharya and P. Banerjee, *Bull. Chem. Soc. Jpn.*, **58**, 3593 (1985).
57. H. F. Rexroat and N. S. Rowan, *Polyhedron*, **4**, 1357 (1985).
58. J. Casabo, T. Flor, F. Teixidor, and J. Ribas, *Inorg. Chem.*, **25**, 3166 (1986).

59. S. K. Saha, M. C. Ghosh, and P. Banerjee, *J. Chem. Res.*, 186 (1986).
60. I. M. Sidahmed and A. M. Ismail, *Transition Met. Chem.*, **11**, 288 (1986).
61. G. S. Groves and C. F. Wells, *Transition Met. Chem.*, **11**, 20 (1986).
62. R. van Eldik and G. M. Harris, *Inorg. Chem.*, **18**, 1997 (1979).
63. R. van Eldik, *Inorg. Chim. Acta*, **44**, L197 (1980).
64. M. B. Davies and J. W. Lethbridge, *J. Inorg. Nucl. Chem.*, **43**, 1579 (1981).
65. S. Aygen, H. Hanssum, and R. van Eldik, *Inorg. Chem.*, **24**, 2853 (1985).
66. S. C. Chan and M. L. Tobe, *J. Chem. Soc.*, 966 (1963).
67. G. Schiavon and F. Marchetti, *Polyhedron*, **5**, 761 (1986).
68. G. Schiavon, F. Marchetti, and M. Patrikaki, *Polyhedron*, **5**, 765 (1986).
69. K. Koshy and T. P. Dasgupta, *Inorg. Chim. Acta*, **117**, 111 (1986).
70. M. Ishikawa, K-I. Okamoto, J. Hidaka, and H. Einaga, *Polyhedron*, **5**, 1345 (1986).
71. D. E. Murray, E. O. Schlemper, S. Siripaisarnpipat, and R. K. Murman, *J. Chem. Soc., Dalton Trans.*, 1759 (1986).
72. M. H. M. Abou-El-Wafa, M. G. Burnett, and J. F. McCullagh, *J. Chem. Soc., Dalton Trans.*, 2083 (1986).
73. K. D. Onan, G. Davies, M. A. El-Sayed, and A. El-Toukhy, *Inorg. Chim. Acta*, **119**, 121 (1986).
74. Y. Kitamura, S. Nariyuki, and K. Yoshitani, *Inorg. Chem.*, **24**, 3021 (1985).
75. K. Nakajima, K. Tozaki, M. Kojima, and J. Fujita, *Bull. Chem. Soc. Jpn.*, **58**, 1130 (1985).
76. D. Banerjee and J. C. Bailar, Jr., *Transition Met. Chem.*, **10**, 331 (1985).
77. Y. J. Wong, J. D. Petersen, and J. F. Geldard, *Inorg. Chem.*, **24**, 3352 (1985).
78. H. Kawaguchi, M. Matsuki, T. Ama, and T. Yasui, *Bull. Chem. Soc. Jpn.*, **59**, 31 (1986).
79. M. A. Cox, R. H. Griffiths, P. A. Williams, and R. S. Vagg, *Inorg. Chim. Acta*, **103**, 155 (1985).
80. X. Ling and K. E. Hyde, *Polyhedron*, **5**, 1559 (1986).
81. G. Sadler and T. P. Dasgupta, *Inorg. Chem.*, **25**, 3593 (1986).
82. M. Ebihara, Y. Sasaki, and K. Saito, *Inorg. Chem.*, **24**, 3831 (1985).
83. T. Ohishi, K. Kashiwabara, J. Fujita, S. Ohba, T. Ishii, and Y. Saito, *Bull. Chem. Soc. Jpn.*, **59**, 385 (1986).
84. J. Kraft and R. van Eldik, *Inorg. Chem.*, **24**, 3391 (1985).
85. See, for example, K. C. Koshy and G. M. Harris, *Inorg. Chem.*, **22**, 2947 (1983) and references cited therein.
86. A. A. El-Awady and G. M. Harris, *Inorg. Chem.*, **25**, 1323 (1986).
87. V. K. Joshi, R. van Eldik, and G. M. Harris, *Inorg. Chem.*, **25**, 2229 (1986).
88. M. Ishikawa, K-I. Okamoto, J. Hidaka, and H. Einaga, *Helv. Chim. Acta*, **68**, 2015 (1985).
89. M. Kimura, M. Yamashita, and S. Nishida, *Inorg. Chem.*, **24**, 1527 (1985).
90. P. Ramamurthy and P. Natarajan, *J. Chem. Soc., Chem. Commun.*, 1554 (1985).
91. M. Kikkawa, Y. Sasaki, S. Kawata, Y. Hatakeyama, F. B. Ueno, and K. Saito, *Inorg. Chem.*, **24**, 4096 (1985).
92. H. Maecke, V. Houlding, and A. W. Adamson, *J. Am. Chem. Soc.*, **102**, 6888 (1980).
93. V. Houlding, H. Maecke, and A. W. Adamson, *Inorg. Chem.*, **20**, 4279 (1981).
94. W. Weber, H. Maecke, and R. van Eldik, *Inorg. Chem.*, **25**, 3093 (1986).
95. K. L. Brown and Z. Szeverenyi, *Inorg. Chim. Acta*, **119**, 149 (1986).
96. L. G. Marzilli, M. F. Summers, E. Zangrando, N. Bresciani-Pahor, and L. Randaccio, *J. Am. Chem. Soc.*, **108**, 4830 (1986).
97. S. Balt and A. M. Van Herk, *Inorg. Chim. Acta*, **125**, 27 (1986).
98. G. C. Pillai and E. S. Gould, *Inorg. Chem.*, **25**, 3353 (1986).
99. See, for example, C. R. Clark, R. F. Tasker, D. A. Buckingham, D. R. Knighton and D. R. K. Hancock, *J. Am. Chem. Soc.*, **103**, 7023 (1981); H. Wautier, D. Marchal, and J. Fastrez, *J. Chem. Soc., Dalton Trans.*, 2484 (1981); H. Wautier, V. Daffe, M. Smets, and J. Fastrez, *J. Chem. Soc., Dalton Trans.*, 2479 (1981).
100. N. Mensi and S. S. Isied, *Inorg. Chem.*, **25**, 147 (1986).
101. S. Isied, J. Lyon and A. Vassilian, *Int. J. Pept. Protein Res.*, **19**, 354 (1982).
102. S. Isied and C. Kuehn, *J. Am. Chem. Soc.*, **100**, 6752 (1978).

103. D. A. Buckingham and C. R. Clark, *Inorg. Chem.*, **25**, 3478 (1986).
104. D. A. Buckingham, D. M. Foster, and A. M. Sargeson, *J. Am. Chem. Soc.*, **90**, 6032 (1968).
105. D. A. Buckingham, D. M. Foster, and A. M. Sargeson, *J. Am. Chem. Soc.*, **92**, 5701 (1970).
106. E. Baraniak, D. A. Buckingham, C. R. Clark, B. H. Moynihan, and A. M. Sargeson, *Inorg. Chem.*, **25**, 3466 (1986).
107. L. R. Gahan, J. M. Harrowfield, A. J. Herlt, L. F. Lindoy, P. O. Whimp, and A. M. Sargeson, *J. Am. Chem. Soc.*, **107**, 6231 (1985).
108. See, for example, P. W. A. Hubner and R. M. Milburn, *Inorg. Chem.*, **19**, 1267 (1980); J. M. Harrowfield, D. R. Jones, L. F. Lindoy, and A. M. Sargeson, *J. Am. Chem. Soc.*, **102**, 7733 (1980); P. R. Norman and R. D. Cornelius, *J. Am. Chem. Soc.*, **104**, 2356 (1982); S. H. McClaugherty and C. M. Grisham, *Inorg. Chem.*, **21**, 4133 (1982); R. A. Kenley, R. H. Fleming, R. M. Laine, D. S. Tse, and J. S. Winterle, *Inorg. Chem.*, **23**, 1870 (1984).
109. R. M. Milburn, M. Gautam-Basak, R. Tribolet, and H. Sigel, *J. Am. Chem. Soc.*, **107**, 3315 (1985).
110. F. Tafesse, S. S. Massoud, and R. M. Milburn, *Inorg. Chem.*, **24**, 2591 (1985).
111. O. Vollárová, J. Benko, and E. Skalná, *Transition Met. Chem.*, **10**, 401 (1985).
112. G. A. Lawrance, *Transition Met. Chem.*, **11**, 396 (1986).
113. Y. Kitamura, *Bull. Chem. Soc. Jpn.*, **58**, 2699 (1985).
114. K. Nakajima, M. Kojima, and J. Fujita, *Bull. Chem. Soc. Jpn.*, **59**, 3505 (1986).

References for Chapter 8

1. M. F. Summers, J. van Rijn, J. Reedijk, and L. G. Marzilli, *J. Am. Chem. Soc.*, **108**, 4254 (1986).
2. G. A. Lawrance, *Chem. Rev.*, **86**, 17 (1986).
3. T. W. Hambley and P. A. Lay, *Inorg. Chem.*, **25**, 4553 (1986).
4. Y. Harel and S. W. Adamson, *J. Phys. Chem.*, **90**, 6690 (1986).
5. J.-S. Kim, W.-S. Jung, H. Tomiyasu, and H. Fukutomi, *Bull. Chem. Soc. Jpn.*, **59**, 613 (1986).
6. S. Christie, I. F. Fraser, A. McVitie, and R. D. Peacock, *Polyhedron*, **5**, 35 (1986).
7. P. A. Agaskar and F. A. Cotton, *Inorg. Chem.*, **25**, 15 (1986).
8. U. Abram, S. Abram, J. Stach, and R. Kirmse, *J. Radioanal. Nucl. Chem.*, **100**, 325 (1986).
9. J. Baldas, J. Bonnyman, and G. A. Williams, *Inorg. Chem.*, **25**, 150 (1986).
10. R. Kirmse, J. Stach, and U. Abram, *Inorg. Chim. Acta*, **117**, 117 (1986).
11. U. Mazzi, F. Refosco, F. Tisato, G. Bandoli, and M. Nicolini, *J. Chem. Soc., Dalton Trans.*, 1623 (1986).
12. M. V. Mikelsons and T. C. Pinkerton, *Anal. Chem.*, **58**, 1007 (1986).
13. F. Grases, J. Palou, and E. Amat, *Transition Met. Chem.*, **11**, 253 (1986).
14. Z. Proso and P. Lerch, *J. Radioanal. Nucl. Chem.*, **92**, 237 (1985).
15. P. M. Treichel, *J. Organomet. Chem.*, **318**, 83, 121 (1987).
16. F. F. Kashani and R. K. Murmann, *Int. J. Chem. Kinet.*, **17**, 1007 (1985).
17. N. E. Stacy, K. A. Connor, D. R. McMillin, and R. A. Walton, *Inorg. Chem.*, **25**, 3649 (1986).
18. J. Benko, O. Vollarova, O. Grancicova, and V. Holba, *J. Coord. Chem.*, **14**, 175 (1985).
19. P. A. M. Williams and P. J. Aymonino, *Transition Met. Chem.*, **11**, 213 (1986).
20. P. A. M. Williams and P. J. Aymonino, *Inorg. Chim. Acta*, **113**, 37 (1986).
21. H. E. Toma and A. B. P. Lever, *Inorg. Chem.*, **25**, 176 (1986).
22. M. H. M. Abou-El-Wafa, M. G. Burnett, and J. F. McCullagh, *J. Chem. Soc., Dalton Trans.*, 2083 (1986).
23. H. E. Toma, *Quim. Nova*, **7**, 305 (1984); *Chem. Abstr.*, **104**, 140854x (1986).
24. D. R. Eaton, J. M. Watkins, and R. J. Buist, *J. Am. Chem. Soc.*, **107**, 5604 (1985).
25. S. Ašperger, I. Murati, D. Pavlović, and A. Šustra, *J. Chem. Soc., Chem. Commun.*, 814 (1986).
26. A. B. Nikol'skii and Y. Yu. Kotov, *Vestn. Leningr. Univ., Fiz., Khim.*, 110 (1986); *Chem. Abstr.*, **104**, 177569j (1986).

27. G. Stochel, *Zest. Nauk. Uniw. Jagiellon., Pr. Chem.*, **29**, 97 (1985); *Chem. Abstr.*, **103**, 113221m (1985); G. Stochel and Z. Stasicka, *Polyhedron*, **4**, 1887 (1985).

28. G. Stochel, R. van Eldik, and Z. Stasicka, *Inorg. Chem.*, **25**, 3663 (1986).

29. E. Hankiewicz, C. Stradowski, and M. Wolszczak, *J. Radioanal. Nucl. Chem.*, **105**, 291 (1986); *Chem. Abstr.*, **105**, 162061j (1986).

30. A. R. Butler, C. Glidewell, A. R. Hyde, and J. McGinnis, *Inorg. Chem.*, **24**, 2931 (1985).

31. J. Casado, M. Mosquera, M. F. Rodriguez Prieto, and J. Vazquez Tato, *Ber. Bunsenges. Phys. Chem.*, **89**, 735 (1985).

32. A. R. Butler, C. Glidewell, V. Chaipanich, and J. McGinnis, *J. Chem. Soc., Perkin Trans. 2*, 7 (1986).

33. S. Tachiyashiki and H. Yamatera, *Inorg. Chem.*, **25**, 3209 (1986).

34. V. Gutmann and G. Resch, *Monatsh.*, **116**, 1107 (1985).

35. A. Yamagishi, *Inorg. Chem.*, **25**, 55 (1986).

36. A. Kobayashi, M. Kobayashi, T. Ohno, and S. Mizusawa, *Nippon Kagaku Kaishi*, 1055 (1986); *Chem. Abstr.*, **105**, 140679q (1986).

37. C. C. Deb, D. K. Hazra, and S. C. Lahiri, *Z. Phys. Chem. (Frankfurt am Main)*, **267**, 769 (1986).

38. P. Krumholz, *Nature*, **163**, 724 (1949); *J. Phys. Chem.*, **60**, 87 (1956).

39. J. H. Baxendale and P. George, *Trans. Faraday Soc.*, **46**, 736 (1950).

40. F. Basolo, J. C. Hayes, and H. M. Neumann, *J. Am. Chem. Soc.*, **76**, 3807 (1954); J. E. Dickens, F. Basolo, and H. M. Neumann, *J. Am. Chem. Soc.*, **79**, 1286 (1957); R. Davies, M. Green, and A. G. Sykes, *J. Chem. Soc., Dalton Trans.*, 1171 (1972).

41. E. Rodenas and F. Ortega, *J. Chem. Educ.*, **63**, 448 (1986).

42. F. Ortega and E. Rodenas, *Transition Met. Chem.*, **9**, 331 (1984).

43. G. Nord, *Comments Inorg. Chem.*, **4**, 193 (1985).

44. E. C. Constable, *Inorg. Chim. Acta*, **117**, L33 (1986).

45. O. S. Tee, J. Pika, M. J. Cornblatt, and M. Trani, *Can. J. Chem.*, **64**, 1267 (1986), and reference cited therein.

46. J. Grodkowski, P. Neta, C. J. Schlesener, and J. K. Kochi, *J. Phys. Chem.*, **89**, 4373 (1985).

47. P. Neta, J. Silverman, V. Markovic, and J. Rabani, *J. Phys. Chem.*, **90**, 703 (1986), and references cited therein.

48. D. K. Geiger and G. Ferraudi, *Inorg. Chim. Acta*, **117**, 139 (1986).

49. R. D. Gillard and H.-U. Hummel, *J. Coord. Chem.*, **14**, 315 (1986).

50. D. W. Clack and R. D. Gillard, *Transition Met. Chem.*, **10**, 419 (1985).

51. J. Burgess, J. Fawcett, S. Radulović, and D. R. Russell, unpublished observations.

52. A. G. Maddock, *J. Chem. Soc., Dalton Trans.*, 2349 (1986).

53. G. A. Lawrance, *Austr. J. Chem.*, **38**, 1117 (1985).

54. J.-M. Lucie, D. R. Stranks, and J. Burgess, *J. Chem. Soc., Dalton Trans.*, 245 (1975).

55. S. Grammenudi and F. Vögtle, *Angew. Chem., Int. Ed. Engl.*, **25**, 1122 (1986).

56. Ya. Z. Voloshin and A. Yu. Nazarenko, *Dopov. Akad. Nauk Ukr. R.S.R., Ser. B: Geol., Khim. Biol. Nauki*, 33 (1985); *Chem. Abstr.*, **103**, 221755x (1985).

57. M. J. Blandamer, J. Burgess, B. Clark, P. P. Duce, A. W. Hakin, S. Radulovic, P. Guardado, F. Sanchez, C. D. Hubbard, and E. A. Abu-Gharib, *J. Chem. Soc., Faraday Trans. 1*, **82**, 1471 (1986).

58. *J. Burgess*, this series, Vol. 1, pp. 149–151.

59. M. Cazanga, J. G. Santos, and F. Ibanez, *J. Chem. Soc., Dalton Trans.*, 465 (1986).

60. S. Tachiyashiki and H. Yamatera, *Inorg. Chem.*, **25**, 3043 (1986).

61. J. Ige and O. Soriyan, *J. Chem. Soc., Faraday Trans. 1*, **82**, 2011 (1986).

62. F. Ortega and E. Rodenas, *J. Phys. Chem.*, **90**, 2408 (1986).

63. F. Ortega and E. Rodenas, *Transition Met. Chem.*, **11**, 351 (1986).

64. M. J. Blandamer, J. Burgess, B. Clark, A. W. Hakin, M. W. Hyett, S. Spencer, and N. Taylor, *J. Chem. Soc., Faraday Trans. 1*, **81**, 2357 (1985).

65. S. Funahashi, N. Uchiyama, and M. Tanaka, *Bull. Chem. Soc. Jpn.*, **59**, 161 (1986).

66. N. Siddiqui and D. V. Stynes, *Inorg. Chem.*, **25**, 1982 (1986).

67. Xuening Chen and D. V. Stynes, *Inorg. Chem.*, **25**, 1173 (1986).
68. H. E. Toma and A. C. C. Sylva, *Can. J. Chem.*, **64**, 1280 (1986).
69. D. V. Stynes, D. Fletcher, and Xuening Chen, *Inorg. Chem.*, **25**, 3483 (1986).
70. D. Lexa, M. Momenteau, J. M. Saveant, and Feng Xu, *J. Am. Chem. Soc.*, **108**, 6937 (1986).
71. C. Lecomte, R. H. Blessing, P. Coppens, and A. Tabard, *J. Am. Chem. Soc.*, **108**, 6942 (1986).
72. C. Gueutin, D. Lexa, M. Momenteau, J.-M. Saveant, and Feng Xu, *Inorg. Chem.*, **25**, 4294 (1986).
73. D. Lexa, M. Momenteau, J. M. Saveant, and Feng Xu, *Inorg. Chem.*, **25**, 4857 (1986).
74. E. Tsuchida, Shen-guo Wang, M. Yuasa, and H. Nishide, *J. Chem. Soc., Chem. Commun.*, 179 (1986).
75. K. Caldwell, L. J. Noe, J. D. Ciccone, and T. G. Traylor, *J. Am. Chem. Soc.*, **108**, 6150 (1986).
76. M. Hoshino, *Inorg. Chem.*, **25**, 2476 (1986).
77. U. Keppeler and M. Hanack, *Chem. Ber.*, **119**, 3363 (1986).
78. L. L. Costanzo, S. Giuffrida, G. de Guidi, and G. Condorelli, *Inorg. Chim. Acta*, **101**, 71 (1985).
79. L. L. Costanzo, S. Giuffrida, G. de Guidi, V. Cucinotta, and G. Condorelli, *J. Organomet. Chem.*, **289**, 81 (1985).
80. L. Christiansen, D. N. Hendrickson, H. Toftlund, S. R. Wilson, and Chuan-Liang Xie, *Inorg. Chem.*, **25**, 2813 (1986).
81. E. G. Jäger, *Z. Chem.*, **25**, 446 (1985).
82. M. W. Fuller, K.-M. F. Le Brocq, E. Leslie, and I. R. Wilson, *Aust. J. Chem.*, **39**, 1411 (1986).
83. R. M. Naik and P. C. Nigam, *Transition Met. Chem.*, **10**, 227 (1985).
84. R. M. Naik and P. C. Nigam, *Inorg. Chim. Acta*, **114**, 55 (1986).
85. R. M. Naik and P. C. Nigam, *Transition Met. Chem.*, **11**, 337 (1986).
86. F. Arifuku, K. Ujimoto, and H. Kurihara, *Bull. Chem. Soc. Jpn.*, **59**, 149 (1986).
87. Meng Qing-jin, G. A. Tondreau, J. O. Edwards, and D. A. Sweigart, *J. Chem. Soc., Dalton Trans.*, 2269 (1985).
88. C. E. Castro, M. Jamin, W. Yokoyama, and R. Wade, *J. Am. Chem. Soc.*, **108**, 4179 (1986).
89. T. C. Woon, C. M. Dicken, and T. C. Bruice, *J. Am. Chem. Soc.*, **108**, 7990 (1986).
90. S. A. Kretchmar and K. N. Raymond, *J. Am. Chem. Soc.*, **108**, 6212 (1986).
91. D. Baldwin, *Biochim. Biophys. Acta*, **623**, 183 (1980).
92. Y. Maeda, H. Oshio, Y. Takashima, M. Mikuriya, and M. Hidaka, *Inorg. Chem.*, **25**, 2958 (1986).
93. C. Glidewell and A. R. Hyde, *Polyhedron*, **4**, 1155 (1985).
94. A. R. Butler, C. Glidewell, A. R. Hyde, and J. C. Walton, *Polyhedron*, **4**, 797 (1985).
95. L. A. Gentil, H. O. Zerga, and J. A. Olabe, *J. Chem. Soc., Dalton Trans.*, 2731 (1986).
96. J. M. A. Hoddenbagh and D. H. Macartney, *Inorg. Chem.*, **25**, 380 (1986).
97. J. M. A. Hoddenbagh and D. H. Macartney, *Inorg. Chem.*, **25**, 2099 (1986).
98. O. H. Bailey and A. Ludi, *Inorg. Chem.*, **24**, 2582 (1985); M. Stebler-Röthlisberger and A. Ludi, *Polyhedron*, **5**, 1217 (1986).
99. W. Weber and P. C. Ford, *Inorg. Chem.*, **25**, 1088 (1986).
100. J. C. do Nascimento Filho and D. W. Franco, *Inorg. Chim. Acta*, **113**, 55 (1986).
101. E. Trabuco and D. W. Franco, *Polyhedron*, **5**, 1503 (1986).
102. C. M. Lieber and N. S. Lewis, *J. Am. Chem. Soc.*, **107**, 7190 (1985).
103. C. M. Lieber, M. H. Schmidt, and N. S. Lewis, *J. Am. Chem. Soc.*, **108**, 6103 (1986).
104. K. Kalyanasundaram, *J. Phys. Chem.*, **90**, 2285 (1986).
105. L. J. Henderson, M. Ollino, V. K. Gupta, G. R. Newkome, and W. R. Cherry, *J. Photochem.*, **31**, 199 (1985).
106. W. F. Wacholtz, R. A. Auerbach, and R. H. Schmehl, *Inorg. Chem.*, **25**, 227 (1986); A. Juris, P. Belser, F. Barigelletti, A. von Zelewsky, and V. Balzani, *Inorg. Chem.*, **25**, 256 (1986).
107. V. Balzani, N. Sabbatini, and F. Scandola, *Chem. Rev.*, **86**, 319 (1986).
108. E. C. Constable, *J. Chem. Soc., Dalton Trans.*, 2687 (1985).
109. J. M. Kelly, C. M. O'Connell, and J. G. Vos, *J. Chem. Soc., Dalton Trans.*, 253 (1986).
110. B. P. Sullivan, D. Conrad, and T. J. Meyer, *Inorg. Chem.*, **24**, 3640 (1985).

111. R. K. Coll, J. E. Fergusson, and Teow Sian Keong, *Austr. J. Chem.*, **39**, 1161 (1986).
112. K. Aoyagi, Y. Yukawa, K. Shimizu, M. Mukaida, T. Takeuchi, and H. Kakihana, *Bull. Chem. Soc. Jpn.*, **59**, 1493 (1986).
113. K. Aoyagi, H. Nagao, Y. Yukawa, M. Ogura, A. Kuwayama, F. S. Howell, M. Mukaida, and H. Kakihana, *Chem. Lett.*, 2135 (1986).
114. L. L. Whinnery, Hong Jun Yue, and J. A. Marsella, *Inorg. Chem.*, **25**, 4136 (1986).
115. M. E. Sosa and M. L. Tobe, *J. Chem. Soc., Dalton Trans.*, 427 (1986).
116. H. E. Toma, E. Giesbrecht, and R. L. Espinoza Rojas, *J. Chem. Soc., Dalton Trans.*, 2469 (1985).
117. D. P. Fairlie and H. Taube, *Inorg. Chem.*, **24**, 3199 (1985).
118. W. D. Harman, D. P. Fairlie, and H. Taube, *J. Am. Chem. Soc.*, **108**, 8223 (1986).
119. J. A. Baumann and T. J. Meyer, *Inorg. Chem.*, **19**, 345 (1980).
120. J. C. Dobson, K. J. Takeuchi, D. W. Pipes, D. A. Geselowitz, and T. J. Meyer, *Inorg. Chem.*, **25**, 2357 (1986).
121. J. A. Gilbert, D. Geselowitz, and T. J. Meyer, *J. Am. Chem. Soc.*, **108**, 1493 (1986).
122. E. C. Constable and K. R. Seddon, *J. Chem. Soc., Chem. Commun.*, 34 (1982).
123. O. Wernberg, *J. Chem. Soc., Dalton Trans.*, 1993 (1986).
124. A. R. Siedle, R. A. Newmark, and L. H. Pignolet, *Inorg. Chem.*, **25**, 3412 (1986).
125. S. Joss, P. Bigler, and A. Ludi, *Inorg. Chem.*, **24**, 3487 (1985).
126. M. D. Fryzuk, P. A. MacNeil, and R. G. Ball, *J. Am. Chem. Soc.*, **108**, 6414 (1986).
127. S. Wieland, J. DiBenedetto, R. van Eldik, and P. C. Ford, *Inorg. Chem.*, **25**, 4893 (1986).
128. L. H. Skibsted, *Inorg. Chem.*, **24**, 3791 (1985).
129. H. Miller, *Polyhedron*, **5**, 1965 (1986).
130. M. E. Frink and P. C. Ford, *Inorg. Chem.*, **24**, 3494 (1985).
131. M. Hoshino, K. Yasufuku, H. Seki, and H. Yamazaki, *J. Phys. Chem.*, **89**, 3080 (1985).
132. M. Hoshino, H. Seki, K. Yasufuku, and H. Shizuka, *J. Phys. Chem.*, **90**, 5149 (1986).
133. D. C. Thackray, S. Ariel, T. W. Leung, K. Menon, B. R. James, and J. Trotter, *Can. J. Chem.*, **64**, 2440 (1986).
134. J. G. Leipoldt and H. Meyer, *Polyhedron*, **4**, 1527 (1985).
135. C. Chatterjee and A. S. Bali, *Indian J. Chem.*, *Sect. A*, **25**, 439 (1986).
136. C. Chatterjee and A. S. Bali, *Bull. Chem. Soc. Jpn.*, **59**, 3233 (1986).
137. J. Ribas, A. Escuer, and M. Serra, *Thermochim. Acta*, **102**, 137 (1986).
138. R. Dreos Garlatti, G. Tauzher, M. Blaschich, and G. Costa, *Inorg. Chim. Acta*, **105**, 129 (1985).
139. R. D. Garlatti, G. Tauzher, and G. Costa, *Inorg. Chim. Acta*, **121**, 27 (1986).
140. N. M. Samus', G. I. Shpakov, and S. P. Vlasenko, *Russ. J. Inorg. Chem.*, **30**, 1788 (1985).
141. N. M. Samus', G. I. Shpakov, and S. P. Vlasenko, *Russ. J. Inorg. Chem.*, **31**, 543 (1986).
142. P. Hendry and A. M. Sargeson, *J. Chem. Soc., Chem. Commun.*, 164 (1984); *Aust. J. Chem.*, **39**, 1177 (1986).
143. A. J. Deeming, G. P. Proud, H. M. Dawes, and M. B. Hursthouse, *J. Chem. Soc., Dalton Trans.*, 2545 (1986).
144. L. Mønsted and L. H. Skibsted, *Acta Chem. Scand.*, *Ser. A*, **40**, 590 (1986).
145. Liangshiu Lee, S. F. Clark, and J. D. Petersen, *Inorg. Chem.*, **24**, 3558 (1985).
146. R. McGrindle, G. Ferguson, A. J. McAlees, M. Parvez, B. L. Ruhl, D. K. Stephenson, and T. Wieckowski, *J. Chem. Soc., Dalton Trans.*, 235 (1986).
147. N. A. P. Kane-Maguire, K. C. Wallace, D. P. Cobranchi, J. M. Derrick, and D. G. Speece, *Inorg. Chem.*, **25**, 2101 (1985).
148. L. Volponi, G. Favero, and A. Peloso, *Gazz. Chim. Ital.*, **115**, 237 (1985).
149. F. Christensson and J. Springborg, *Inorg. Chem.*, **24**, 2129 (1985).
150. J. Telser and R. S. Drago, *Inorg. Chem.*, **25**, 2989 (1986).
151. B. R. James, R. H. Morris, and P. Kvintovics, *Can. J. Chem.*, **64**, 897 (1986).
152. M. Serra, A. Escuer, M. Monfort, and J. Ribas, *Thermochim. Acta*, **92**, 493 (1985).
153. J. Ribas and M. Monfort, *Thermochim. Acta*, **91**, 115 (1985).
154. E. F. Hills, D. T. Richens, and A. G. Sykes, *Inorg. Chem.*, **25**, 3144 (1986).

155. H. Elias, H.-T. Macholdt, K. J. Wannowius, M. J. Blandamer, J. Burgess, and B. Clark, *Inorg. Chem.*, **25**, 3048 (1986).
156. W. Preetz and H. J. Steinebach, *Z. Naturforsch.*, **40B**, 745 (1985).
157. M. A. Bennett and G. T. Crisp, *Organometallics*, **5**, 1792, 1800 (1986).
158. M. F. Finlayson, P. C. Ford, and R. J. Watts, *J. Phys. Chem.*, **90**, 3916 (1986).
159. R. M. Naik and P. C. Nigam, *Transition Met. Chem.*, **11**, 11 (1986).
160. N. Jubran, D. Meyerstein, and H. Cohen, *Inorg. Chim. Acta*, **117**, 129 (1986).
161. R. W. Hay, R. Bembi, and B. Jeragh, *Transition Met. Chem.*, **11**, 385 (1986).
162. R. W. Hay, M. P. Pujari, and R. Bembi, *Transition Met. Chem.*, **11**, 261 (1986).
163. M. G. Fairbank and A. McAuley, *Inorg. Chem.*, **25**, 1233 (1986).
164. V. K. Polovnyak, T. E. Busygina, and N. S. Akhmetov, *Russ. J. Inorg. Chem.*, **30**, 222 (1985).
165. S. A. Vinogradov, K. P. Balashev, and G. A. Shagisultanova, *Koord. Khim.*, **11**, 959 (1985); *Chem. Abstr.*, **103**, 113224q (1985).
166. I. I. Blinov, K. P. Balashev, and G. A. Shagisultanova, *Koord. Khim.*, **11**, 1121 (1985); *Chem. Abstr.*, **103**, 169726x (1985).
167. L. Drougge and L. I. Elding, *Inorg. Chem.*, **24**, 2292 (1985); L. Drougge and L. I. Elding, *Inorg. Chim. Acta*, **121**, 175 (1986); Pongchan Chandayot and Yueh-Tai Fanchiang, *Inorg. Chem.*, **24**, 3535 (1985).
168. R. El-Mehdawi, S. A. Bryan, and D. M. Roundhill, *J. Am. Chem. Soc.*, **107**, 6282 (1985).
169. E. W. Abel, S. K. Bhargava, and K. G. Orrell, *Prog. Inorg. Chem.*, **32**, 1 (1984).
170. E. W. Abel, K. Kite, and P. S. Perkins, *Polyhedron*, **5**, 1459 (1986).
171. E. W. Abel, P. K. Mittal, K. G. Orrell, and V. Šik, *J. Chem. Soc., Dalton Trans.*, 961 (1986).
172. E. W. Abel, T. E. MacKenzie, K. G. Orrell, and V. Šik, *J. Chem. Soc., Dalton Trans.*, 2173 (1986).
173. E. W. Abel, T. E. MacKenzie, K. G. Orrell, and V. Šik, *J. Chem. Soc., Dalton Trans.*, 205 (1986).
174. D. D. Gummin, E. M. A. Ratilla, and N. M. Kostić, *Inorg. Chem.*, **25**, 2429 (1986).
175. E. W. Abel, S. K. Bhargava, P. K. Mittal, K. G. Orrell, and V. Šik, *J. Chem. Soc., Dalton Trans.*, 1561 (1985); E. W. Abel, P. K. Mittall, K. G. Orrell, and V. Šik, *J. Chem. Soc., Dalton Trans.*, 1569 (1985).
176. K. Nakajima, M. Kojima, and J. Fujita, *Bull. Chem. Soc. Jpn.*, **59**, 3505 (1986).

References for Chapter 9

1. M. Shamsipur and A. I. Popov, *J. Phys. Chem.*, **90**, 5997 (1986).
2. W. Jung, H. Tomiyasu, and H. Fukutomi, *Inorg. Chem.*, **25**, 2582 (1986).
3. Y. Ikeda and H. Fukutomi, *Inorg. Chim. Acta*, **115**, 223 (1986).
4. P. Fux, J. Lagrange, and P. Lagrange, *J. Am. Chem. Soc.*, **107**, 5927 (1985).
5. S. Gangopadhyay, R. N. Banerjee, and D. Banerjea, *Transition Met. Chem.*, **10**, 325 (1985).
6. A. Hioki, S. Funahashi, M. Ishii, and M. Tanaka, *Inorg. Chem.*, **25**, 1360 (1986).
7. A. Hioki, S. Funihashi, and M. Tanaka, *J. Phys. Chem.*, **89**, 5057 (1985).
8. S. F. Lincoln, A. M. Hounslow, and A. N. Boffa, *Inorg. Chem.*, **25**, 1038 (1986).
9. R. Mohr and R. van Eldik, *Inorg. Chem.*, **24**, 3396 (1985).
10. K. Kojima, T. Inoue, M. Izaki, and R. Shimozawa, *Bull. Chem. Soc. Jpn.*, **59**, 139 (1986).
11. T. Inoue, K. Kojima, and R. Shimozawa, *Bull. Chem. Soc. Jpn.*, **59**, 1683 (1986).
12. P. Dasgupta and R. B. Jordan, *Inorg. Chem.*, **24**, 2717 (1985).
13. R. L. Reeves, *Inorg. Chem.*, **25**, 1473 (1986).
14. R. L. Reeves and J. A. Reczek, *Inorg. Chem.*, **25**, 4452 (1986).
15. P. Dasgupta and R. B. Jordan, *Inorg. Chem.*, **24**, 2721 (1985).
16. S.-H. Liu and C.-S. Chung, *Inorg. Chem.*, **24**, 2368 (1985).
17. M. J. Hynes and J. Walsh, *J. Chem. Soc., Dalton Trans.*, 2565 (1985).
18. S. Gangopadhyay, R. N. Banerjee, and D. Banerjea, *Transition Met. Chem.*, **10**, 310 (1985).

19. J. Walsh and M. J. Hynes, *J. Chem. Soc., Dalton Trans.*, 2243 (1986).
20. J. K. Beattie, M. T. Kelso, W. E. Moody, and P. A. Tregloan, *Inorg. Chem.*, **24**, 415 (1985).
21. S. F. Lincoln, J. H. Coates, B. G. Doddridge, and A. M. Hounslow, *Aust. J. Chem.*, **39**, 367 (1986).
22. M. Schumann and H. Elias, *Inorg. Chem.*, **24**, 3187 (1985).
23. K. E. Gilmore and G. K. Pagenkopf, *Inorg. Chem.*, **24**, 2436 (1985).
24. G. C. Mei and C. D. Gutsche, *J. Am. Chem. Soc.*, **107**, 7959 (1985).
25. K. J. Butenhof, D. Cochenour, J. L. Banyasz, and J. E. Stuehr, *Inorg. Chem.*, **25**, 691 (1986).
26. A. Das, S. Gangopadhyay, and D. Banerjea, *Transition Met. Chem.*, **11**, 259 (1986).
27. S. F. Lincoln, I. M. Brereton, and T. M. Spotswood, *J. Chem. Soc., Faraday Trans. 1*, **81**, 1623 (1985); **82**, 1999 (1986).
28. S. F. Lincoln, I. M. Brereton, and T. M. Spotswood, *J. Am. Chem. Soc.*, **108**, 8134 (1986).
29. B. O. Strasser and A. I. Popov, *J. Am. Chem. Soc.*, **107**, 7921 (1985).
30. A. Delville, H. D. H. Stover, and C. Detellier, *J. Am. Chem. Soc.*, **107**, 4172 (1985).
31. B. O. Strasser, M. Shamsipur, and A. I. Popov, *J. Phys. Chem.*, **89**, 4822 (1985).
32. R. J. Adamic, B. A. Lloyd, E. M. Eyring, S. Petrucci, R. A. Bartsch, M. J. Pugia, B. E. Knudsen, Y. Liu, and D. H. Desai, *J. Phys. Chem.*, **90**, 6571 (1986).
33. B. G. Cox, P. Firman, and H. Schneider, *J. Am. Chem. Soc.*, **107**, 4297 (1985).
34. R. M. Izatt, J. S. Bradshaw, S. A. Nielsen, J. D. Lamb, and J. J. Christensen, *Chem. Rev.*, **85**, 271 (1985).
35. S. Petrucci, R. J. Adamic, and E. M. Eyring, *J. Phys. Chem.*, **90**, 1677 (1986).
36. G. Busse and H. Strehlow, *Ber. Bunsenges. Phys. Chem.*, **89**, 977 (1985).
37. A. Walsleben and H. Strehlow, *J. Soln. Chem.*, **14**, 881 (1985).
38. T. E. Eriksen, I. Grenthe, and I. Puigdomenech, *Inorg. Chim. Acta*, **121**, 63 (1986).
39. I. Ando and G. A. Webb, *Mag. Reson. Chem.*, **24**, 557 (1986).
40. L. Helm, S. F. Lincoln, A. E. Merbach, and D. Zbinden, *Inorg. Chem.*, **25**, 2550 (1986).
41. P. Martinez, R. Mohr, and R. van Eldik, *Ber. Bunsenges. Phys. Chem.*, **90**, 609 (1986).
42. K. Ishihara, S. Funahashi, and M. Tanaka, *Inorg. Chem.*, **25**, 2898 (1986).
43. K. Ishihara, S. Funahashi, and M. Tanaka, *Inorg. Chem.*, **22**, 194, 3589 (1983).
44. J.-E. Kessler, C. T. G. Knight, and A. E. Merbach, *Inorg. Chim. Acta*, **115**, 85 (1986).
45. A. Hioki, S. Funahashi, and M. Tanaka, *Inorg. Chem.*, **25**, 2904 (1986).
46. R. J. Deeth and M. A. Hitchman, *Inorg. Chem.*, **25**, 1225, 3720 (1986).
47. L. I. Elding and L. F. Olsson, *Inorg. Chim. Acta*, **117**, 9 (1986).
48. T. W. Swaddle and A. E. Merbach, *Inorg. Chem.*, **20**, 4212 (1981).
49. M. Grant and R. B. Jordan, *Inorg. Chem.*, **20**, 55 (1981).
50. R. B. Wilhelmy, R. C. Patel, and E. Matijevic, *Inorg. Chem.*, **24**, 3290 (1985).
51. P. Chaudhri and H. Diebler, *J. Chem. Soc., Dalton Trans.*, 1693 (1986).
52. A. Hugi. Doctoral thesis, Universite'de Lausanne, 1984.
53. S. F. Lincoln and A. White, *Polyhedron*, **5**, 1351 (1986).
54. R. Geue, S. H. Jacobson, and R. Pizer, *J. Am. Chem. Soc.*, **108**, 1150 (1986).
55. B. Metz, D. Moras, and R. Weiss, *J. Chem. Soc; Perkin Trans. 2*, 423 (1976).
56. D. Moras and R. Weiss, *Acta Crystallogr.; Sect. B*, **29**, 396 (1973). B. Metz, D. Moras and R. Weiss, *Acta Crystallogr.; Sect. B*, **29**, 1377, 1382 (1973).
57. H. Schneider, S. Rauh, and S. Petrucci, *J. Phys. Chem.*, **85**, 2287 (1981).
58. B. G. Cox, D. Knop, and H. Schneider, *J. Amer. Chem. Soc.*, **100**, 6002 (1978). B. G. Cox and H. Schneider, *J. Chem. Soc; Perkin Trans. 2*, 1293 (1979).
59. G. Wipff and P. Kollman, *Nouv. J. Chim.*, **9**, 457 (1985).
60. P. A. Kollman, G. Wipff, and U. C. Singh, *J. Am. Chem. Soc.*, **107**, 2212 (1985).
61. S. F. Lincoln, E. Horn, M. R. Snow, T. W. Hambley, I. M. Brereton, and T. M. Spotswood, *J. Chem. Soc.; Dalton Trans.*, 1075 (1986).
62. B. G. Cox, J. Garcia-Rosas, H. Schneider, and Ng. van Truong, *Inorg. Chem.*, **25**, 1165 (1986).
63. J. Rebek, S. V. Luis, and L. R. Marshall, *J. Am. Chem. Soc.*, **108**, 5011 (1986).
64. G. Ranghino, S. Romano, J. M. Lehn, and G. Wipff, *J. Am. Chem. Soc.*, **107**, 7873 (1985).

65. A. E. Merbach, _Pure Appl. Chem._, **54**, 1479 (1982).
66. S.-H. Liu and C.-S. Chung, _Inorg. Chem._, **25**, 3890 (1986).
67. J. A. Drumhiller, F. Montavon, J.-M. Lehn, and R. W. Taylor, _Inorg. Chem._, **25**, 3751 (1986).
68. C.-T. Lin, D. B. Rorabacher, G. R. Cayley, and D. W. Margerum, _Inorg. Chem._, **14**, 919 (1975).
69. R. W. Hay, R. Bembi, and B. Jeragh, _Transition Met. Chem._, **11**, 385 (1986).
70. R. W. Hay, M. P. Pujari, and R. Bembi, _Transition Met. Chem._, **11**, 261 (1986).
71. Y. Miyake, M. Shigeto, and M. Teramoto, _J. Chem. Soc., Faraday Trans. 1_, **82**, 1515 (1986).
72. M. Fischer, W. Knoche, B. H. Robinson, and J. H. M. Wedderburn, _J. Chem. Soc., Faraday Trans. 1_, **75**, 119 (1979).
73. V. C. Reinsborough and B. H. Robinson, _J. Chem. Soc., Faraday Trans. 1_, **75**, 2395 (1979).
74. P. D. I. Fletcher and B. H. Robinson, _J. Chem. Soc., Faraday Trans. 1_, **79**, 1959 (1983).
75. A. K. Yatsimirsky, O. I. Kavetskaya, and I. V. Berezin, _J. Chem. Soc., Faraday Trans. 1_, **82**, 319 (1986).
76. E. Mentasti, C. Baiocchi, and L. J. Kirschenbaum, _J. Chem. Soc., Dalton Trans._, 2615 (1985).
77. B. Perlmutter-Hayman, F. Secco, and M. Venturini, _Inorg. Chem._, **24**, 3828 (1985).
78. J. Konstantatos, G. Kalatzis, E. Vrachnou-Astra, and D. Katakis, _J. Chem. Soc., Dalton Trans._, 2461 (1985).
79. R. J. Cross, _Chem. Soc. Rev._, **14**, 197 (1985).
80. J. G. Leipoldt and E. C. Grobler, _Transition Met. Chem._, **11**, 110 (1986).
81. R. van Eldik, S. Aygen, H. Kelm, A. M. Trzeciak, and J. J. Ziolkowski, _Transition Met. Chem._, **10**, 167 (1985).
82. K. Wieghardt, M. Kleine-Boymann, W. Swiridoff, B. Nuber, and J. Weiss, _J. Chem. Soc., Dalton Trans._, 2493 (1985).
83. H. Elias, C. Hasserodt-Taliaferro, L. Hellriegel, W. Schonherr, and K. J. Wannowius, _Inorg. Chem._, **24**, 3192 (1985).
84. Y. Wu and T. A. Kaden, _Helv. Chim. Acta_, **68**, 1611 (1985).
85. J. M. A. Hoddenbagh and D. H. Macartney, _Inorg. Chem._, **25**, 380 (1986).
86. F. L. Dickert and M. F. Waidhas, _Angew. Chem., Int. Ed. Engl._, **24**, 575 (1985).
87. D. L. Pisaniello and S. F. Lincoln, _Aust. J. Chem._, **32**, 715 (1979).
88. J. J. Led, _J. Am. Chem. Soc._, **107**, 6755 (1985).
89. P. J. Breen, W. DeW. Horrocks, and K. A. Johnson, _Inorg. Chem._, **25**, 1968 (1986).
90. P. J. Breen, K. A. Johnson, and W. DeW. Horrocks, _Biochemistry_, **24**, 4997 (1985).
91. S. R. Martin, A. Andersson Teleman, P. Bayley, T. Drakenberg, and S. Forsen, _Eur. J. Biochem._, **151**, 543 (1985).
92. L. H. Skibsted in: _Advances in Inorganic and Bioinorganic Mechanisms_ (A. G. Sykes, ed.), Academic Press, London (1986) Vol. 4, p. 137.
93. L. I. Elding and L. H. Skibsted, _Inorg. Chem._, **25**, 4084 (1986).
94. S. Yamada and M. Tanaka, _Bull. Chem. Soc. Jpn._, **58**, 2234 (1985).
95. M. Birus, Z. Bradic, N. Kujundzic, M. Pribanic, P. C. Wilkins, and R. G. Wilkins, _Inorg. Chem._, **24**, 3980 (1985).
96. L. L. Fish and A. L. Crumbliss, _Inorg. Chem._, **24**, 2198 (1985).
97. R. Bechtold, M. B. Gardineer, A. Kazmi, B. van Hemelryck, and S. S. Isied, _J. Phys. Chem._, **90**, 3800 (1986).
98. L. G. Marzilli, M. F. Summers, N. Bresciani-Pahor, E. Zangrando, J.-P. Charland, and L. Randaccio, _J. Am. Chem. Soc._, **107**, 6880 (1985).
99. W. O. Parker, N. Bresciani-Pahor, E. Zangrando, L. Randaccio, and L. Marzilli, _Inorg. Chem._, **25**, 1303 (1986).
100. W. O. Parker, E. Zangrando, N. Bresciani-Pahor, L. Randaccio, and L. Marzilli, _Inorg. Chem._, **25**, 3489 (1986).
101. N. K. Kildahl and G. Antonopoulos, _J. Coord. Chem._, **14**, 293 (1986).
102. A. A. el-Awady and G. M. Harris, _Inorg. Chem._, **25**, 1323 (1986).
103. M. G. Fairbank and A. McAuley, _Inorg. Chem._, **25**, 1233 (1986).
104. V. C. Sekhar and C. A. Chang, _Inorg. Chem._, **25**, 2061 (1986).

105. C. A. Chang, V. O. Ochaya, and V. C. Sekhar, *J. Chem. Soc., Chem. Commun.*, 1724 (1985).
106. S. F. Lincoln, in: *Advances in Inorganic and Bioinorganic Mechanisms* (A. G. Sykes, ed.) Academic Press, London (1986) Vol. 4, p. 217.
107. H. Ogino and M. Shimura, in: *Advances in Inorganic and Bioinorganic Mechanisms* (A. G. Sykes, ed.) Academic Press, London (1986) Vol. 4, p. 107.
108. H. Ogino, A. Masuko, S. Ito, N. Miura, and M. Shimura, *Inorg. Chem.*, **25**, 708 (1986).
109. N. Fujiwara, H. Tomiyasu, and H. Fukutomi, *Bull. Chem. Soc. Jpn.*, **58**, 1386 (1985).
110. N. Fujiwara, H. Tomiyasu, and H. Fukutomi, *Bull. Chem. Soc. Jpn.*, **57**, 1576 (1984).
111. H. Fukotomi, H. Ohno, and H. Tomiyasu, *Bull. Chem. Soc. Jpn.*, **59**, 2303 (1986).
112. G. D. Armstrong and A. G. Sykes, *Inorg. Chem.*, **25**, 3135 (1986).
113. M. Inamo, S. Funahashi, and M. Tanaka, *Inorg. Chem.*, **24**, 2475 (1985).
114. M. Inamo, S. Funanashi, Y. Ito, Y. Hamada, and M. Tanaka, *Inorg. Chem.*, **24**, 2468 (1985).
115. B. Wang, Y. Sasaki, K. Okazaki, K. Kanesato, and K. Saito, *Inorg. Chem.*, **25**, 3745 (1986).
116. J.-S. Kim, W.-S. Jung, H. Tomiyasu, and H. Fukutomi, *Bull. Chem. Soc. Jpn.*, **59**, 613 (1986).
117. J. G. Leipoldt, R. van Eldik, S. S. Basson, and A. Roodt, *Inorg. Chem.*, **25**, 4639 (1986).
118. I. Tabushi, A. Yoshizawa, and H. Mizuno, *J. Am. Chem. Soc.*, **107**, 4585 (1985).
119. I. Tabushi and A. Yoshizawa, *Inorg. Chem.*, **25**, 1541 (1986).

References for Chapter 10

1. R. B. Hitam, K. A. Mahmoud, and A. J. Rest, *J. Organomet. Chem.*, **291**, 321 (1985).
2. R. M. Kowaleski, D. O. Kipp, K. J. Stauffer, P. N. Swepston, and F. Basolo, *Inorg. Chem.*, **24**, 3750 (1985).
3. M. E. Rerek and F. Basolo, *Organometallics*, **2**, 372 (1983).
4. L.-N. Ji, D. L. Kershner, M. E. Rerek, and F. Basolo, *J. Organomet. Chem.*, **296**, 83 (1985).
5. R. J. Angelici and W. Lowen, *Inorg. Chem.*, **6**, 682 (1967).
6. J. A. Belmont and M. S. Wrighton, *Organometallics*, **5**, 1421 (1986).
7. G. T. Palmer, F. Basolo, L. B. Kool, and M. D. Rausch, *J. Am. Chem. Soc.*, **108**, 4417 (1986).
8. P. Mura, B. G. Olby, and S. D. Robinson, *Inorg. Chim. Acta*, **98**, L21 (1985).
9. F. R. Estevan, P. Lahuerta, and J. Latorre, *Inorg. Chim. Acta*, **116**, L33 (1986).
10. T. R. Herrinton and T. L. Brown, *J. Am. Chem. Soc.*, **107**, 5700 (1985).
11. N. N. Turaki and J. M. Huggins, *Organometallics*, **5**, 1703 (1986).
12. R. M. Kowaleski, F. Basolo, W. C. Trogler, and R. D. Ernst, *J. Am. Chem. Soc.*, **108**, 6046 (1986).
13. R. M. Kowaleski, W. C. Trogler, and R. Basolo, *Gazz. Chim. Ital.*, **116**, 105 (1986).
14. D. J. Darensbourg, M. Y. Darensbourg, R. L. Gray, D. Simmons, and L. W. Arndt, *Inorg. Chem.*, **25**, 880 (1986).
15. D. J. Sikara, D. W. Macomber, and M. D. Rausch, *Adv. Organomet. Chem.*, **25**, 317 (1986).
16. D. A. Edwards, *Organomet. Chem.*, **14**, 196 (1986).
17. I. S. Butler and A. A. Ismail, *Inorg. Chem.*, **25**, 3910 (1986).
18. J. A. S. Howell, D. T. Dixon, J. C. Kola, and N. F. Ashford, *J. Organomet. Chem.*, **294**, C1 (1985).
19. T. I. Odiaka, *Inorg. Chem. Acta.*, **103**, 9 (1985).
20. R. M. Kowaleski, A. L. Rheingold, W. C. Trogler, and F. Basolo, *J. Am. Chem. Soc.*, **108**, 2460 (1986).
21. A. M. McNair and K. R. Mann, *Inorg. Chem.*, **25**, 2519 (1986).
22. T. G. Traylor and K. J. Stewart, *J. Am. Chem. Soc.*, **108**, 6977 (1986).
23. J. A. S. Howell, *Organomet. Chem.*, **14**, 294 (1986).
24. S. P. Nolan, R. L. de la Vega, and C. D. Hoff, *Organometallics*, **5**, 2529 (1986).
25. C. D. Hoff, *J. Organomet. Chem.*, **246**, C53 (1983).
26. K. H. Whitmire, T. R. Lee, and E. S. Lewis, *Organometallics*, **5**, 987 (1986).
27. R. H. Crabtree and A. Habib, *Inorg. Chem.*, **25**, 3698 (1986).
28. C. D. Falk and J. Halpern, *J. Am. Chem. Soc.*, **87**, 3003 (1965).

29. J. M. Buchanan, J. M. Stryder, and R. C. Bergman, *J. Am. Chem. Soc.*, **108**, 1537 (1986).
30. H.-T. Macholdt, R. van Eldik, and G. R. Dobson, *Inorg. Chem.*, **25**, 1914 (1986).
31. G. R. Dobson, C. S. Binzet, and J. E. Cortes, *J. Coord. Chem.*, **14**, 215 (1986).
32. G. R. Dobson, C. B. Dobson, and S. E. Mansour, *Inorg. Chem.*, **24**, 2179 (1985).
33. G. R. Dobson, S. S. Basson, and C. B. Dobson, *Inorg. Chim. Acta*, **105**, L17 (1985).
34. G. R. Dobson, H. H. Awad, and S. S. Basson, *Inorg. Chim. Acta*, **118**, L5 (1986).
35. X. Chen and D. V. Stynes, *Inorg. Chem.*, **25**, 1173 (1986).
36. M. Poliakoff and E. Weitz, *Adv. Organomet. Chem.*, **25**, 277 (1986).
37. S. P. Church, H. Hermann, F.-W. Grevels, and K. Schaffner, *Inorg. Chem.*, **24**, 418 (1985).
38. S. P. Church, F.-W. Grevels, H. Hermann, and K. Schaffner, *J. Chem. Soc., Chem. Commun.*, 30 (1985).
39. S. P. Church, F.-W. Grevels, H. Hermann, and K. Schaffner, *Inorg. Chem.*, **23**, 3830 (1984).
40. R. K. Upmacis, M. P. Poliakoff, and J. J. Turner, *J. Am. Chem. Soc.*, **108**, 3645 (1986).
41. I. R. Dunkin, P. Harter, and C. J. Shields, *J. Am. Chem. Soc.*, **106**, 7248 (1984).
42. S. P. Church, H. Hermann, F.-W. Grevels, and K. Schaffner, *J. Chem. Soc., Chem. Commun.*, 785 (1984).
43. S. Firth, P. M. Hodges, M. Poliakoff, and J. J. Turner, *Inorg. Chem.*, **25**, 4608 (1986).
44. S. C. Fletcher, M. Poliakoff, and J. J. Turner, *Inorg. Chem.*, **25**, 3597 (1986).
45. A. J. Dixon, M. A. Healy, M. Poliakoff, and J. J. Turner, *J. Chem. Soc., Chem. Commun.*, 994 (1986).
46. B. D. Moore, M. Poliakoff, and J. J. Turner, *J. Am. Chem. Soc.*, **108**, 1819 (1986).
47. M. Basato, *J. Chem. Soc., Dalton Trans.*, 217 (1986).
48. N. N. Turaki and J. M. Huggins, *Organometallics*, **4**, 1766 (1985).
49. C. E. Philbin, C. A. Granatir, and D. R. Tyler, *Inorg. Chem.*, **25**, 4806 (1986).
50. A. E. Stiegman, A. S. Goldman, C. E. Philbin, and D. R. Tyler, *Inorg. Chem.*, **25**, 2976 (1986).
51. S. P. Schmidt, R. Basolo, C. M. Jensen, and W. C. Trogler, *J. Am. Chem. Soc.*, **108**, 1894 (1986).
52. M. W. Kokkes, D. J. Stufkens, and A. Oskam, *Inorg. Chem.*, **24**, 2934 (1985).
53. R. J. Blau and J. H. Espenson, *J. Am. Chem. Soc.*, **108**, 1962 (1986).
54. A. J. McLennan and R. J. Puddephatt, *Organometallics*, **5**, 811 (1986).
55. B. D. Martin, S. A. Matchett, J. R. Norton, and O. P. Anderson, *J. Am. Chem. Soc.*, **107**, 7952 (1985).
56. K. Dahlinger, A. J. Poë, P. K. Sayal, and V. C. Sekhar, *J. Chem. Soc., Dalton Trans.*, 2145 (1986).
57. N. Brodie, A. Poë, and V. Sekhar, *J. Chem. Soc., Chem. Commun.*, 1090 (1985).
58. A. Poë and V. C. Sekhar, *Inorg. Chem.*, **24**, 4376 (1985).
59. A. J. Poë and C. V. Sekhar, *J. Am. Chem. Soc.*, **108**, 3673 (1986).
60. M. F. Desrosiers, D. A. Wink, R. Trautman, A. E. Friedman, and P. C. Ford, *J. Am. Chem. Soc.*, **108**, 1917 (1986).
61. D. J. Taube and P. C. Ford, *Organometallics*, **5**, 99 (1986).
62. D. M. Dalton, D. J. Barnett, T. P. Duggan, J. B. Keister, P. T. Malik, S. P. Modi, M. R. Shaffer, and S. A. Smesko, *Organometallics*, **4**, 1854 (1985).
63. C. S. Browning, D. H. Farrar, R. R. Gukathasan, and S. A. Morris, *Organometallics*, **4**, 1750 (1985).
64. K. Knoll, G. Huttner, L. Zsolnai, I. Jibril, and M. Wasincionek, *J. Organomet. Chem.*, **294**, 91 (1985).
65. C. E. Housecroft and T. P. Fehlner, *Inorg. Chem.*, **25**, 404 (1986).
66. C. E. Housecroft and T. P. Fehlner, *J. Am. Chem. Soc.*, **108**, 4867 (1986).
67. D. J. Darensbourg and D. J Zalewski, *Organometallics*, **4**, 92 (1985).
68. D. J. Darensbourg, D. J. Zalewski, A. L. Rheingold, and R. L. Durney, *Inorg. Chem.*, **25**, 3281 (1986).
69. K. Dahlinger, F. Falcone, and A. J. Poë, *Inorg. Chem.*, **25**, 2654 (1986).
70. D. C. Sonnenberger and J. D. Atwood, *Inorg. Chem.*, **20**, 3243 (1981).
71. H. H. Ohst and J. K. Kochi, *Inorg. Chem.*, **25**, 2066 (1986).

72. J. K. Kouba, E. L. Muetterties, M. R. Thompson, and V. W. Day, *Organometallics*, **2**, 1065 (1983).
73. H. H. Ohst and J. K. Kochi, *J. Am. Chem. Soc.*, **108**, 2897 (1986).
74. M. G. Richmond and J. K. Kochi, *Inorg. Chem.*, **25**, 1334 (1986).
75. M. G. Richmond and J. K. Kochi, *Inorg. Chem.*, **25**, 656 (1986).
76. W. M. Rees, M. R. Churchill, Y.-J. Li, and J. D. Atwood, *Organometallics*, **4**, 1162 (1985).
77. G. K. Anderson and G. J. Lumetta, *Organometallics*, **4**, 1542 (1985).
78. N. Koga and K. Morokuma, *J. Am. Chem. Soc.*, **107**, 7230 (1985).
79. N. Koga and K. Morokuma, *J. Am. Chem. Soc.*, **108**, 6136 (1986).
80. H. E. Bryndza, *Organometallics*, **4**, 1686 (1985).
81. H. Werner, A. Hohn, and M. Dziallas, *Angew. Chem., Int. Ed. Engl.*, **25**, 1090 (1986).
82. W. D. McGhee and R. G. Bergman, *J. Am. Chem. Soc.*, **108**, 5621 (1986). 361 (1986).
83. J. D. Cotton and H. A. Kimlin, *J. Organomet. Chem.*, **294**, 213 (1985).
84. S. L. Webb, C. M. Giandomenico, and J. Halpern, *J. Am. Chem. Soc.*, **108**, 345 (1986).
85. C. R. Jablonski and Y.-P. Wang. *J. Organomet. Chem.*, **301**, C49 (1986).
86. C. R. Jablonski, *Inorg. Chem. Acta*, **112**, L19 (1986).
87. R. J. Ruszczyk, B.-L. Huang, and J. D. Atwood, *J. Organomet. Chem.*, **299**, 205 (1986).
88. G. D. Vaughn, K. A. Krein, and J. A. Gladysz, *Organometallics*, **5**, 936 (1986).
89. S. P. Nolan, R. L. de la Vega, S. L. Mukerjee, and C. D. Hoff, *Inorg. Chem.*, **25**, 1160 (1986).
90. R. S. Paonessa, N. C. Thomas, and J. Halpern, *J. Am. Chem. Soc.*, **107**, 4333 (1985).
91. H. Fujimoto, T. Yamasaki, H. Mizutani, and N. Koga, *J. Am. Chem. Soc.*, **107**, 6157 (1985).
92. D. J. Darensbourg, R. Kudaroski Hanckel, C. G. Bauch, M. Pala, D. Simmons, and J. N. White, *J. Am. Chem. Soc.*, **197**, 7463 (1985).
93. D. J. Darensbourg and G. Grotsch, *J. Am. Chem. Soc.*, **107**, 7473 (1985).
94. D. J. Darensbourg and M. Pala, *J. Am. Chem. Soc.*, **107**, 5687 (1985).
95. B. P. Sullivan and T. J. Meyer, *Organometallics*, **5**, 1500 (1986).
96. L. J. Newman and R. G. Bergman, *J. Am. Chem. Soc.*, **107**, 5314 (1985).

References for Chapter 11

1. *The Nature and Cleavage of Metal–Carbon Bonds*, Volume 2 of *The Chemistry of the Metal–Carbon Bond* (F. R. Hartley and S. Patai, eds.), John Wiley and Sons, New York, 1985.
2. R. H. Crabtree, *Chem. Rev.*, **85**, 245 (1985).
3. R. D. Adams and I. T. Horvath, *Prog. Inorg. Chem.*, **33**, 127 (1985).
4. Λ. Yamamoto, *J. Organomet. Chem.*, **300**, 347 (1986).
5. P. J. Brothers and J. P. Collman, *Acc. Chem. Res.*, **19**, 209 (1986).
6. Y.-T. Fanchiang, *Coord. Chem. Rev.*, **68**, 131 (1985).
7. M. I. Bruce, *Pure Appl. Chem.*, **58**, 553 (1986).
8. M. D. Vargas and J. N. Nicholls, *Adv. Inorg. Chem. Radiochem.*, **30**, 123 (1986).
9. R. J. Cross, *Chem. Soc. Rev.*, **14**, 197 (1985).
10. J. Halpern, *Pure Appl. Chem.*, **58**, 575 (1986).
11. G. J. Kubas, R. R. Ryan, B. I. Swanson, P. J. Vergamini, and H. J. Wasserman, *J. Am. Chem. Soc.*, **106**, 451 (1984).
12. K. W. Zilm, R. A. Merrill, M. W. Kummer, and G. J. Kubas, *J. Am. Chem. Soc.*, **108**, 7837 (1986).
13. G. J. Kubas, C. J. Unkefer, B. I. Swanson, and E. Fukushima, *J. Am. Chem. Soc.*, **108**, 7000 (1986).
14. G. J. Kubas, R. R. Ryan, and D. A. Wrobleski, *J. Am. Chem. Soc.*, **108**, 1339 (1986).
15. H. J. Wasserman, G. J. Kubas, and R. R. Ryan, *J. Am. Chem. Soc.*, **108**, 2294 (1986).
16. R. K. Upmacis, M. Poliakoff, and J. J. Turner, *J. Am. Chem. Soc.*, **108**, 3645 (1986).

17. G. E. Gadd, R. K. Upmacis, M. Poliakoff, and J. J. Turner, *J. Am. Chem. Soc.*, **108**, 2547 (1986).
18. R. L. Sweany, *J. Am. Chem. Soc.*, **108**, 6986 (1986).
19. R. L. Sweany, *Organometallics*, **5**, 387 (1986).
20. R. H. Morris, J. F. Sawyer, M. Shiralian, and J. D. Zubkowski, *J. Am. Chem. Soc.*, **107**, 5581 (1985).
21. F. M. Conroy-Lewis and S. J. Simpson, *J. Chem. Soc., Chem. Commun.*, 506 (1986).
22. R. H. Crabtree and M. Lavin, *J. Chem. Soc., Chem. Commun.*, 1661 (1985).
23. R. H. Crabtree, M. Lavin, and L. Bonneviot, *J. Am. Chem. Soc.*, **108**, 4032 (1986).
24. Y. Jean, O. Eisenstein, F. Volatron, B. Maouche, and F. Sefta, *J. Am. Chem. Soc.*, **108**, 6587 (1986).
25. E. J. Moore, J. M. Sullivan, and J. R. Norton, *J. Am. Chem. Soc.*, **108**, 2257 (1986).
26. R. G. Pearson, *Chem. Rev.*, **85**, 41 (1985).
27. B. D. Martin, K. E. Warner, and J. R. Norton, *J. Am. Chem. Soc.*, **108**, 33 (1986).
28. M. Y. Darensbourg and M. M. Ludvig, *Inorg. Chem.*, **25**, 2894 (1986).
29. J. B. Schilling, W. A. Goddard, and J. L. Beauchamp, *J. Am. Chem. Soc.*, **108**, 582 (1986).
30. J. L. Elkind and P. B. Armentrout, *Inorg. Chem.*, **25**, 1078 (1986).
31. D. L. Packett and W. C. Trogler, *J. Am. Chem. Soc.*, **108**, 5036 (1986).
32. D. A. Wink and P. C. Ford, *J. Am. Chem. Soc.*, **107**, 5566 (1985).
33. D. A. Wink and P. C. Ford, *J. Am. Chem. Soc.*, **108**, 4838 (1986).
34. M. Hillman, S. Michaile, S. W. Feldberg, and J. Eisch, *Organometallics*, **4**, 1258 (1985).
35. C. S. Bajgur, S. B. Jones, and J. L. Petersen, *Organometallics*, **4**, 1929 (1985).
36. P. Zhou, A. A. Vitale, J. San Filippo, and W. H. Saunders, *J. Am. Chem. Soc.*, **107**, 8049 (1985).
37. A. J. Kunin, R. Farid, C. E. Johnson, and R. Einsenberg, *J. Am. Chem. Soc.*, **107**, 5315 (1985).
38. C. E. Johnson and R. Eisenberg, *J. Am. Chem. Soc.*, **107**, 6531 (1985).
39. W. A. G. Graham, *J. Organomet. Chem.*, **300**, 81 (1986).
40. M. L. Deem, *Coord. Chem. Rev.*, **74**, 101 (1986).
41. M. L. H. Green and D. O'Hare, *Pure Appl. Chem.*, **57**, 1897 (1985).
42. M. Ephritikhine, *Nouv. J. Chim.*, **10**, 9 (1986).
43. J. Schwartz, *Acc. Chem. Res.*, **18**, 302 (1985).
44. H. Rabba, J.-Y. Saillard, and R. Hoffmann, *J. Am. Chem. Soc.*, **108**, 4327 (1986).
45. P. J. Watson and G. W. Parshall, *Acc. Chem. Res.*, **18**, 51 (1985).
46. M. A. Tolbert, M. L. Mandich, L. F. Halle, and J. L. Beauchamp, *J. Am. Chem. Soc.*, **108**, 5675 (1986).
47. D. B. Jacobson and B. S. Freiser, *J. Am. Chem. Soc.*, **107**, 4373 (1985).
48. M. F. Asaro, S. R. Cooper, and N. J. Cooper, *J. Am. Chem. Soc.*, **108**, 5187 (1986).
49. F. Timmers and M. Brookhart, *Organometallics*, **4**, 1365 (1985).
50. D. L. Lichtenberger and G. E. Kellogg, *J. Am. Chem. Soc.*, **108**, 2560 (1986).
51. K. Kanamori, W. E. Broderick, R. F. Jordan, R. D. Willett, and J. I. Legg, *J. Am. Chem. Soc.*, **108**, 7122 (1986).
52. W. J. Evans, D. K. Drummond, S. G. Bott, and J. L. Atwood, *Organometallics*, **5**, 2389 (1986).
53. J. W. Park, P. B. Mackenzie, W. P. Schaefer, and R. H. Grubbs, *J. Am. Chem. Soc.*, **108**, 6402 (1986).
54. N. J. Fitzpatrick and M. A. McGinn, *J. Chem. Soc., Dalton Trans.*, 1637 (1985).
55. A. Demolliens, Y. Jean, and O. Eisenstein, *Organometallics*, **5**, 1457 (1986).
56. S. L. Latesky, A. K. McMullen, I. P. Rothwell, and J. C. Huffman, *J. Am. Chem. Soc.*, **107**, 5981 (1985).
57. L. R. Chamberlain, I. P. Rothwell, and J. C. Huffman, *J. Am. Chem. Soc.*, **108**, 1502 (1986).
58. R. N. McDonald and M. T. Jones, *J. Am. Chem. Soc.*, **108**, 8097 (1986).
59. W. D. Jones and J. A. Maguire, *Organometallics*, **5**, 590 (1986).
60. W. D. Jones and M. Fan, *Organometallics*, **5**, 1057 (1986).
61. R. G. Bergman, P. F. Seidler, and T. T. Wenzel, *J. Am. Chem. Soc.*, **107**, 4358 (1985).
62. T. T. Wenzel and R. G. Bergman, *J. Am. Chem. Soc.*, **108**, 4856 (1986).
63. A. H. Klahn-Oliva, F. D. Singer, and D. Sutton, *J. Am. Chem. Soc.*, **108**, 3107 (1986).

64. M. L. H. Green and D. O'Hare, *J. Chem. Soc., Dalton Trans.*, 2469 (1986).
65. C. G. Kreiter, K. H. Franzreb, and W. Michels, *Z. Naturforsch.*, **40b**, 1188 (1985).
66. M. V. Baker and L. D. Field, *J. Am. Chem. Soc.*, **108**, 7433 (1986).
67. M. V. Baker and L. D. Field, *J. Am. Chem. Soc.*, **108**, 7436 (1986).
68. M. Antberg and L. Dahlenburg, *Angew. Chem., Int. Ed. Engl.*, **25**, 260 (1986).
69. P. J. Desrosiers, R. S. Shinomoto, and T. C. Flood, *J. Am. Chem. Soc.*, **108**, 1346 (1986).
70. P. J. Desrosiers, R. S. Shinomoto, and T. C. Flood, *J. Am. Chem. Soc.*, **108**, 7964 (1986).
71. H. Werner and K. Zenkert, *J. Chem. Soc., Chem. Commun.*, 1607 (1985).
72. J. P. Blaha, J. C. Dewan, and M. S. Wrighton, *Organometallics*, **5**, 899 (1986).
73. K. J. Del Rossi and B. B. Wayland, *J. Am. Chem. Soc.*, **107**, 7941 (1985).
74. K. J. Del Rossi and B. B. Wayland, *J. Chem. Soc. Chem. Commun.*, 1653 (1986).
75. P. O. Stoutland and R. G. Bergman, *J. Am. Chem. Soc.*, **107**, 4581 (1985).
76. J. M. Buchanan, J. M. Stryker, and R. G. Bergman, *J. Am. Chem. Soc.*, **108**, 1537 (1986).
77. R. A. Periana and R. G. Bergman, *J. Am. Chem. Soc.*, **108**, 7332 (1986).
78. W. D. Jones and F. J. Feher, *J. Am. Chem. Soc.*, **108**, 4814 (1986).
79. D. M. Haddleton, *J. Organomet. Chem.*, **311**, C21 (1986).
80. D. M. Haddleton and R. N. Perutz, *J. Chem. Soc., Chem. Commun.*, 1734 (1986).
81. H. Werner, A. Hohn, and M. Dziallas, *Angew. Chem., Int. Ed. Engl.*, **25**, 1090 (1986).
82. W. D. McGhee and R. G. Bergman, *J. Am. Chem. Soc.*, **108**, 5621 (1986).
83. M. J. Burk, R. H. Crabtree, and D. V. McGrath, *J. Chem. Soc., Chem. Commun.*, 1829 (1985).
84. C. J. Cameron, H. Felkin, T. Fillebeen-Khan, N. J. Forrow, and E. Guittet, *J. Chem. Soc., Chem. Commun.*, 801 (1986).
85. C. Bianchini, D. Masi, A. Meli, M. Peruzzini, M. Sabat, and F. Zanobini, *Organometallics*, **5**, 2557 (1986).
86. R. S. Dickson, G. D. Fallon, S. M. Jenkins, B. W. Skelton, and A. H. White, *J. Organomet. Chem.*, **314**, 333 (1986).
87. D. H. Berry and R. Eisenberg, *J. Am. Chem. Soc.*, **107**, 7181 (1985).
88. J. W. Suggs, M. J. Wovkulich, P. G. Williard, and K. S. Lee, *M. Organomet. Chem.*, **307**, 71 (1986).
89. J. M. Cogen and W. F. Maier, *J. Am. Chem. Soc.*, **108**, 7752 (1986).
90. M. Hackett, J. A. Ibers, P. Jernakoff, and G. M. Whitesides, *J. Am. Chem. Soc.*, **108**, 8094 (1986).
91. J. A. Davies and C. T. Eagle, *Organometallics*, **5**, 2149 (1986).
92. C. B. Lebrilla and W. F. Maier, *J. Am. Chem. Soc.*, **108**, 1606 (1986).
93. A. D. Ryabov, I. K. Sakodinskaya, and A. K. Yatsimirsky, *J. Chem. Soc., Dalton Trans.*, 2629 (1985).
94. J. W. Bruno, G. M. Smith, T. J. Marks, C. K. Fair, A. J. Schultz, and J. M. Williams, *J. Am. Chem. Soc.*, **108**, 40 (1986).
95. C. M. Fendrick and T. J. Marks, *J. Am. Chem. Soc.*, **108**, 425 (1986).
96. G. M. Smith, J. D. Carpenter, and T. J. Marks, *J. Am. Chem. Soc.*, **108**, 6805 (1986).
97. M. H. Chisholm, *Polyhedron*, **5**, 25 (1986).
98. M. A. Biddulph, R. Davis, C. H. J. Wells, and F. I. C. Wilson, *J. Chem. Soc., Chem. Commun.*, 1287 (1985).
99. I. Kovacs, C. D. Hoff, F. Ungvary, and L. Marko, *Organometallics*, **4**, 1347 (1985).
100. K. E. Warner and J. R. Norton, *Organometallics*, **4**, 2150 (1985).
101. R. T. Edidin and J. R. Norton, *J. Am. Chem. Soc.*, **108**, 948 (1986).
102. W. J. Carter, S. J. Okrasinski, and J. R. Norton, *Organometallics*, **4**, 1376 (1985).
103. T. G. Schenck, C. R. C. Milne, J. F. Sawyer, and B. Bosnich, *Inorg. Chem.*, **24**, 2338 (1985).
104. R. S. Paonessa, N. C. Thomas, and J. Halpern, *J. Am. Chem. Soc.*, **107**, 4333 (1985).
105. R. J. Puddephatt and J. D. Scott, *Organometallics*, **4**, 1221 (1985).
106. B. Kellenberger, S. J. Young, and J. K. Stille, *J. Am. Chem. Soc.*, **107**, 6105 (1985).
107. H. Schmidbaur, C. Hartmann, J. Riede, B. Huber, and G. Muller, *Organometallics*, **5**, 1652 (1986).

108. J. D. Basil, H. H. Murray, J. P. Fackler, J. Tocher, A. M. Mazany, B. Trzcinska-Bancroft, H. Knachel, D. Dudis, T. J. Delord, and D. O. Marler, *J. Am. Chem. Soc.*, **107**, 6908 (1985).
109. M. L. Steigerwald and W. A. Goddard, *J. Am. Chem.*, **107**, 5027 (1985).
110. M. J. Calhorda, A. R. Dias, A. M. Galvao, and J. A. Martinho Simoes, *J. Organomet. Chem.*, **307**, 167 (1986).
111. P. Kita and R. B. Jordan, *Inorg. Chem.*, **25**, 4791 (1986).
112. K. Crouse and L.-Y. Goh, *Inorg. Chem.*, **25**, 478 (1986).
113. A. Rotman, H. Cohen, and D. Meyerstein, *Inorg. Chem.*, **24**, 4158 (1985).
114. R. J. Bernhardt, M. A. Wilmoth, J. J. Weers, D. M. LaBrush, D. P. Eyman, and J. C. Huffmann, *Organometallics*, **5**, 883 (1986).
115. P. K. Rush, S. K. Noh, and M. Brookhart, *Organometallics*, **5**, 1745 (1986).
116. G. D. Vaughn, C. E. Strouse, and J. A. Gladysz, *J. Am. Chem. Soc.*, **108**, 1462 (1986).
117. J. C. Selover, G. D. Vaughn, C. E. Strouse, and J. A. Gladysz, *J. Am. Chem. Soc.*, **108**, 1455 (1986).
118. G. D. Vaughn and J. A. Gladysz, *J. Am. Chem. Soc.*, **108**, 1473 (1986).
119. D. Lexa, J.-M. Saveant, and D. L. Wang, *Organometallics*, **5**, 1428 (1986).
120. S. I. Hommeltoft and M. C. Baird, *Organometallics*, **5**, 190 (1986).
121. A. M. Stolzenberg and E. L. Meutterties, *Organometallics*, **4**, 1739 (1985).
122. L. J. Sanderson and M. C. Baird, *J. Organomet. Chem.* **307**, C1 (1986).
123. I. Kovacs, F. Ungvary, and L. Marko, *Organometallics*, **5**, 209 (1986).
124. E. A. Betterton, S. M. Chemaly, and J. M. Pratt, *J. Chem. Soc., Dalton Trans.*, 1619 (1985).
125. Y.-T. Fanchiang, *J. Chem. Soc., Dalton Trans.*, 1375 (1985).
126. Y.-T. Fanchiang, *Organometallics*, **4**, 1515 (1985).
127. B. P. Hay and R. G. Finke, *J. Am. Chem. Soc.*, **108**, 4820 (1986).
128. G. Costa, A. Puxeddu, and C. Tavagnacco, *J. Organomet. Chem.* **296**, 161 (1985).
129. E. G. Samsel and J. K. Kochi, *Inorg. Chem.* **25**, 2450 (1986).
130. E. G. Samsel and J. K. Kochi, *J. Am. Chem. Soc.*, **108**, 4790 (1986).
131. R. C. McHatton, J. H. Espenson, and A. Bakac, *J. Am. Chem. Soc.*, **108**, 5885 (1986).
132. N. Bresciani-Pahor, L. Randaccio, E. Zangrando, M. F. Summers, J. H. Ramsden, P. A. Marzilli, and L. G. Marzilli, *Organometallics*, **4**, 2086 (1985).
133. E. Zangrando, N. Bresciani-Pahor, L. Randaccio, J.-P. Charland, and L. G. Marzilli, *Organometallics*, **5**, 1938 (1986).
134. L. G. Marzilli, M. F. Summers, N. Bresciani-Pahor, E. Zangrando, J.-P. Charland, and L. Randaccio, *J. Am. Chem. Soc.*, **107**, 6880 (1985).
135. W. L. Mock and C. Bieniarz, *Organometallics*, **4**, 1917 (1985).
136. M. A. Murphy, B. L. Smith, G. P. Torrence, and A. Aguilo, *Inorg. Chim. Acta.* **101**, L47 (1985).
137. S. S. Basson, J. G. Leipoldt, A. Roodt, J. A. Venter, and T. J. Van Der Walt, *Inorg. Chim. Acta*, **119**, 35 (1986).
138. E. Lindner, R. Fawzi, and H. A. Mayer, *Z. Naturforsch.*, **40b**, 1333 (1985).
139. D. Milstein, *J. Am. Chem. Soc.*, **108**, 3525 (1986).
140. R. A. Periana and R. G. Bergman, *J. Am. Chem. Soc.*, **108**, 7346 (1986).
141. S. L. Van Voorhees and B. B. Wayland, *Organometallics*, **4**, 1887 (1985).
142. B. B. Wayland, S. L. Van Voorhees, and C. Wilker, *Inorg. Chem.*, **25**, 4039 (1986).
143. J. E. Anderson, C.-L. Yao, and K. M. Kadish, *Inorg. Chem.*, **25**, 718 (1986).
144. J. P. Collman, J. I. Brauman, and A. M. Madonik, *Organometallics*, **5**, 215 (1986).
145. J. P. Collman, J. I. Brauman, and A. M. Madonik, *Organometallics*, **5**, 218 (1986).
146. J. P. Collman, J. I. Brauman, and A. M. Madonik, *Organometallics*, **5**, 310 (1986).
147. D. Milstein, J. C. Calabrese, and I. D. Williams, *J. Am. Chem. Soc.*, **108**, 6387 (1986).
148. L. J. Newman and R. G. Bergman, *J. Am. Chem. Soc.*, **107**, 5314 (1985).
149. W. M. Rees, M. R. Churchill, Y.-J. Li, and J. D. Atwood, *Organometallics*, **4**, 1162 (1985).
150. M. A. Bennett and G. T. Crisp, *Aust. J. Chem.*, **39**, 1363 (1986).
151. M. A. Bennett and G. T. Crisp, *Organometallics*, **5**, 1792 (1986).
152. M. A. Bennett and G. T. Crisp, *Organometallics*, **5**, 1800 (1986).

153. K. Osakada, M. Maeda, Y. Nakamura, T. Yamamoto, and A. Yamamoto, *J. Chem. Soc., Chem. Commun.*, 442 (1986).
154. A. Bakac and J. H. Espenson, *J. Am. Chem. Soc.*, **108**, 713 (1986).
155. A. Bakac and J. H. Espenson, *J. Am. Chem. Soc.*, **108**, 719 (1986).
156. M. S. Ram, J. H. Espenson, and A. Bakac, *Inorg. Chem.*, **25**, 4115 (1986).
157. R. J. Mckinney and D. C. Roe, *J. Am. Chem. Soc.*, **108**, 5167 (1986).
158. J. J. Low and W. A. Goddard, *Organometallics*, **5**, 609 (1986).
159. J. J. Low and W. A. Goddard, *J. Am. Chem. Soc.*, **108**, 6115 (1986).
160. L. Chassot, A. von Zelewsky, D. Sandrini, M. Maestri, and V. Balzani, *J. Am. Chem. Soc.*, **108**, 6084 (1986).
161. K. Osakada, T. Chiba, Y. Nakamura, T. Yamamoto, and A. Yamamoto, *J. Chem. Soc., Chem. Commun.*, 1589 (1986).
162. J. A. M. van Beek, G. van Koten, W. J. J. Smeets, and A. L. Spek, *J. Am Chem. Soc.*, **108**, 5010 (1986).
163. G. Ferguson, P. K. Monaghan, M. Parvez, and R. J. Puddephatt, *Organometallics*, **4**, 1669 (1985).
164. P. K. Monaghan and R. J. Puddephatt, *Organometallics*, **5**, 439 (1986).
165. J. D. Scott and R. J. Puddephatt, *Organometallics*, **5**, 2522 (1986).
166. H. Kurosawa and M. Emoto, *Chem. Lett.*, 1161 (1985).
167. H. Kurosawa, M. Emoto, A. Urabe, K. Miki, and N. Kasai, *J. Am. Chem. Soc.*, **107**, 8253 (1985).
168. F. Ozawa, M. Fujimori, T. Yamamoto, and A. Yamamoto, *Organometallics*, **5**, 2144 (1986).
169. E. Gretz and A. Sen, *J. Am. Chem. Soc.*, **108**, 6038 (1986).
170. G. Alibrandi, D. Minniti, R. Romeo, P. Uguagliati, L. Calligaro, and U. Belluco, *Inorg. Chim. Acta.*, **112**, L15 (1986).
171. R. L. Brainard, T. M. Miller, and G. M. Whitesides, *Organometallics*, **5**, 1481 (1986).

References for Chapter 12

1. G. Henrici-Olivè and S. Olivè, in: *The Chemistry of the Metal-Carbon Bond*, Vol. 3, Ch. 9 (F. R. Hartley and S. Patai, eds.), John Wiley, New York (1985).
2. G. Blyholder, K-M. Zhao, and M. Lawless, *Organometallics*, **4**, 1371 (1985).
3. R. F. Fenske and M. C. Milletti, *Organometallics*, **5**, 1243 (1986).
4. R. F. Fenske, M. C. Milletti, and M. Arndt, *Organometallics*, **5**, 2316 (1986).
5. B. A. Narayanan, C. Amatore, and J. K. Kochi, *Organometallics*, **5**, 926 (1986).
6. D. S. Barratt and D. J. Cole-Hamilton, *J. Organomet. Chem.*, **306**, C41 (1986).
7. A. P. Ayscough and S. G. Davies, *J. Chem. Soc., Chem. Commun.*, 1648 (1986).
8. J. I. Seeman and S. G. Davies, *J. Am. Chem. Soc.*, **107**, 6522 (1985).
9. G. Consiglio, F. Morandini, G. F. Ciani, and A. Sironi, *Organometallics*, **5**, 1976 (1986).
10. A. Yamashita and A. Toy, *Tetrahedron Lett.*, **27**, 3471 (1986).
11. M. P. Doyle, *Acc. Chem. Res.*, **19**, 348 (1986).
12. L. Bencze and L. Prókai, *J. Organomet. Chem.*, **294**, C5 (1985).
13. V. Guerchais and D. Astruc, *J. Chem. Soc., Chem. Commun.*, 835 (1985), and references cited therein.
14. S. G. Davies and T. R. Maberly, *J. Organomet. Chem.*, **296**, C37 (1985).
15. S. G. Davies, I. M. Dordor-Hedgecock, K. H. Sutton, and J. C. Walker, *Tetrahedron*, **42**, 5123 (1986).
16. S. G. Davies, I. M. Dordor-Hedgecock, K. H. Sutton, and J. C. Walker, *Tetrahedron Lett.*, **27**, 3787 (1986).
17. S. G. Davies, J. I. Seeman, and I. H. Williams, *Tetrahedron Lett.*, **27**, 619 (1986).

18. S. L. Brown, S. G. Davies, D. F. Foster, J. I. Seeman, and P. Warner, *Tetrahedron Lett.*, **27**, 623 (1986).

19. S. G. Davies, I. M. Dordor-Hedgecock, K. H. Sutton, J. C. Walker, C. Bourne, R. H. Jones, and K. Prout, *J. Chem. Soc., Chem. Commun.*, 607 (1986).

20. S. G. Davies and J. C. Walker, *J. Chem. Soc., Chem. Commun.*, 495 (1986).

21. S. G. Davies and J. C. Walker, *J. Chem. Soc., Chem. Commun.*, 609 (1986).

22. L. R. Beanan and J. B. Keister, *Organometallics*, **4**, 1713 (1985).

23. R. C. Bush and R. J. Angelici, *J. Am. Chem. Soc.*, **108**, 2735 (1986).

24. K. Zaw, M. Lautens, and P. M. Henry, *Organometallics*, **4**, 1286 (1985).

25. J. E. Bäckvall and A. Heumann, *J. Am. Chem. Soc.*, **108**, 7107 (1986).

26. F. Mares, S. E. Diamond, F. J. Regina, and J. P. Solar, *J. Am. Chem. Soc.*, **107**, 3545 (1985).

27. H. Fujimoto and T. Yamasaki, *J. Am. Chem. Soc.*, **108**, 578 (1986).

28. H. T. Kalinoski, U. Hacksell, D. F. Barofsky, E. Barofsky, and G. D. Daves, *J. Am. Chem. Soc.*, **107**, 6476 (1985).

29. G. S. Silverman, S. Strickland, and K. M. Nicholas, *Organometallics*, **5**, 2117 (1986).

30. a. M. Rosenblum, M. M. Turnbull, and B. M. Foxman, *Organometallics*, **5**, 1062 (1986).
    b. M. Rosenblum, *J. Organomet. Chem.*, **300**, 191 (1986).

31. D. Wilhelm, J.-E. Bäckvall, R. E. Nordberg, and T. Norin, *Organometallics*, **4**, 1296 (1985).

32. D. L. Reger, S. A. Klaeren, and L. Lebioda, *Organometallics*, **5**, 1072 (1986).

33. S. G. Davies, M. L. H. Green, and D. M. P. Mingos, *Tetrahedron*, **34**, 3047 (1978).

34. L. A. P. Kane-Maguire, in: *Mechanisms of Inorganic and Organometallic Reactions* (M. V. Twigg, ed.), Vols. 3 and 4, Ch. 12, Plenum Press, New York (1985, 1986).

35. J.-E. Bäckvall, R. E. Nordberg, and D. Wilhelm, *J. Am. Chem. Soc.*, **107**, 6892 (1985).

36. B. Åkermark and A. Vitagliano, *Organometallics*, **4**, 1275 (1985).

37. W. E. van Arsdale, R. E. K. Winter, and J. K. Kochi, *Organometallics*, **5**, 645 (1986).

38. W. E. van Arsdale, R. E. K. Winter, and J. K. Kochi, *J. Organomet. Chem.*, **296**, 31 (1985).

39. J. W. Faller and B. C. Whitmore, *Organometallics*, **5**, 752 (1986).

40. H. M. Asfour and M. Green, *J. Organomet. Chem.*, **292**, C25 (1985).

41. G. R. Wiger, S. S. Tomita, M. F. Rettig, and R. M. Wing, *Organometallics*, **4**, 1157 (1985).

42. O. Eisenstein and R. Hoffmann, *J. Am. Chem. Soc.*, **102**, 6148 (1980).

43. D. J. Evans and L. A. P. Kane-Maguire, *J. Organomet. Chem.*, **312**, C24 (1986).

44. L. S. Barinelli, K. Tao, and K. M. Nicholas, *Organometallics*, **5**, 588 (1986).

45. T. I. Odiaka, *J. Chem. Soc., Dalton Trans.*, 2707 (1986).

46. T. I. Odiaka and L. A. P. Kane-Maguire, *J. Chem. Soc., Dalton Trans.*, 1162 (1981).

47. G. R. John, L. A. P. Kane-Maguire, and R. Kanitz, *J. Organomet. Chem.* **312**, C21 (1986).

48. D. A. Brown, N. J. Fitzpatrick, and M. A. McGinn, *J. Organomet. Chem.*, **293**, 235 (1986)

49. D. A. Brown, N. J. Fitzpatrick, and M. A. McGinn, *J. Chem. Soc., Dalton Trans.*, 701 (1986).

50. A. J. Pearson and Y. Yoon, *J. Chem. Soc., Chem. Commun.*, 1467 (1986).

51. Y. K. Chung and D. A. Sweigart, *J. Organomet. Chem.*, **308**, 223 (1986).

52. E. D. Honig and D. A. Sweigart, *J. Organomet. Chem.*, **308**, 229 (1986).

53. E. D. Honig and D. A. Sweigart, *J. Chem. Soc., Chem. Commun.*, 691 (1986).

54. H. K. Bae, I. N. Jung, and Y. K. Chung, *J. Organomet. Chem.*, **317**, C1 (1986).

55. V. V. Litvak, P. P. Kun, I. I. Oleynik, O. V. Volkov, and V. D. Shteingarts, *J. Organomet. Chem.*, **310**, 189 (1986).

56. R. L. Harris, *J. Organomet. Chem.*, **299**, 105 (1986).

57. N. A. Vol'Kenau, P. V. Petrovskii, L. S. Shilovtseva, and D. N. Kravtsov, *J. Organomet. Chem.*, **303**, 121 (1986).

58. P. K. Rush, S. K. Noh, and M. Brookhart, *Organometallics*, **5**, 1745 (1986).

59. K. R. Lane and R. R. Squires, *J. Am. Chem. Soc.*, **107**, 6403 (1985).

60. W. R. Jackson, I. D. Rae, and M. G. Wong, *Aust. J. Chem.*, **39**, 303 (1986).

61. R. G. Sutherland, R. L. Chowdhury, A. Piórko, and C. C. Lee, *J. Chem. Soc., Chem. Commun.*, 1296 (1985).

62. F. Rose-Munch, E. Rose, and A. Semra, *J. Chem. Soc., Chem. Commun.*, 1108 (1986); 1551 (1986).
63. R. P. Alexander and G. R. Stephenson, *J. Organomet. Chem.*, **314**, C73 (1986).
64. D. A. Brown, N. J. Fitzpatrick, M. A. McGinn, and T. H. Taylor, *Organometallics*, **5**, 152 (1986).
65. D. A. Brown, N. J. Fitzpatrick, W. K. Glass, and T. H. Taylor, *Organometallics*, **5**, 158 (1986).
66. P. Powell, M. Stephens, and K. H. Yassin, *J. Organomet. Chem.*, **301**, 313 (1986).
67. E. Rotondo, F. P. Cusmano, G. Neri, A. Donato, and R. Pietropaolo, *J. Organomet. Chem.*, **292**, 429 (1985).
68. G. L. Crocco and J. A. Gladysz, *J. Chem. Soc., Chem. Commun.*, 1154 (1986).
69. D. E. Smith and J. A. Gladysz, *Organometallics*, **4**, 1480 (1985).
70. S. Georgiou and J. A. Gladysz, *Tetrahedron*, **42**, 1109 (1986).
71. D. Mandon, L. Toupet, and D. Astruc, *J. Am. Chem. Soc.*, **108**, 1320 (1986).
72. P. C. Heah, A. T. Patton, and J. A. Gladysz, *J. Am. Chem. Soc.*, **108**, 1185 (1986).
73. R. M. G. Roberts and A. S. Wells, *J. Organomet. Chem.*, **317**, 233 (1986).

References for Chapter 13

1. W. A. Herrmann and J. Okuda, *Angew. Chem., Int. Ed. Engl.*, **25**, 1092 (1986).
2. R. H. Crabtree, M. Lavin, and L. Bonneviot, *J. Am. Chem. Soc.*, **108**, 4032 (1986).
3. D. L. Packett, C, M. Jensen, R. L. Cowan, C. E. Strouse, and W. C. Trogler, *Inorg. Chem.*, **24**, 3578 (1985).
4. J. Boyer, R. J. P. Corriu, A. Kpoton, M. Mazhur, M. Poirier, and G. Royo, *J. Organomet. Chem.*, **301**, 131 (1986).
5. R. J. P. Corriu, M. Mazhar, M. Poirier, and G. Royo, *J. Organomet. Chem.*, **306**, C5 (1986).
6. S. I. Bailey, D. Colgan, L. M. Englehardt, W. P. Leung, R. I. Papasergio, C. L. Raston, and A. H. White, *J. Chem. Soc., Dalton Trans.*, 603 (1986).
7. J. Arnold, D. N. Shina, T. D. Tilley, and A. M. Arif, *Organometallics*, **5**, 2037 (1986).
8. S. I. Vdovenko, I. I. Guerus, and Yu. L. Yagupolskii, *J. Organomet. Chem.*, **301**, 195 (1986).
9. P. Janser, L. M. Venanzi, and F. Bachechi, *J. Organomet. Chem.*, **296**, 229 (1985).
10. R. Damrauer and S. E. Danahey, *Organometallics*, **5**, 1490 (1986).
11. R. Colton and D. Dakternieks, *Inorg. Chim. Acta*, **102**, L17 (1985).
12. R. Willem, M. Gielen, H. Pepermans, J. Brocas, D. Fastenakel, and P. Finocchiaro, *J. Am. Chem. Soc.*, **107**, 1146 (1985).
13. R. Willem, M. Gielen, H. Pepermans, K. Hallenga, A. Recca, and P. Finocchiaro, *J. Am. Chem. Soc.*, **107**, 1153 (1985).
14. A. A. Ismail, F. Sauriol, J. Sedman, and I. S. Butler, *Organometallics*, **4**, 1914 (1985).
15. J. E. Salt, G. Wilkinson, M. Motevalli, and M. B. Hursthouse, *J. Chem. Soc., Dalton Trans.*, 1141 (1986).
16. R. S. Herrick and J. L. Templeton, *Inorg. Chem.*, **25**, 1270 (1986).
17. C. G. Kreiter, M. Kotzian, U. Schubert, R. Bau, and M. A. Bruck, *Z. Naturforsch, Teil B*, **39**, 1553 (1985).
18. T. Ito, H. Tosaka, S.-j. Yoshida, K. Mita, and A. Yamamoto, *Organometallics*, **5**, 735 (1986).
19. M. L. Luetkens, Jr., D. J. Santure, J. C. Huffman, and A. P. Sattelberger, *J. Chem. Soc., Chem. Commun.*, 552 (1985).
20. D. J. Darensbourg, M. Y. Darensbourg, R. L. Gray, D. Simmons, and L. W. Arndt, *Inorg. Chem.*, **25**, 880 (1986).
21. B. Longato, B. D. Martin, J. R. Norton, and O. P. Anderson, *Inorg. Chem.*, **24**, 1389 (1985).
22. L. R. Martin, F. W. B. Einstein, and R. K. Pomeroy, *Inorg. Chem.*, **24**, 2777 (1985).
23. J. L. Davidson and G. Vasapollo, *J. Chem. Soc., Dalton Trans.*, 2231 (1985).
24. N. N. Greenwood, J. D. Kennedy, I. Macpherson, and M. Thornton-Pett, *Z. Anorg. Allg. Chem.*, **540/541**, 45 (1986).

25. F. A. Boltino and P. Finocchiaro, *Polyhedron*, **4**, 1507 (1985).

26. L. Christiansen, D. N. Hendrickson, H. Toftlund, S. R. Wilson, and C.-L. Xie, *Inorg. Chem.*, **25**, 2813 (1986).

27. M. Fuchs, W. Kuchen, and W. Peters, *Chem. Ber.*, **119**, 1569 (1986).

28. C. Arlen, M. Pfeffer, O. Bars, and G. Le Borgne, *J. Chem. Soc., Dalton Trans.*, 359 (1986).

29. J. G. Kraaijkamp, G. Van Koten, T. A. Van der Knaap, F. Bickelhaupt, and C. H. Stam, *Organometallics*, **5**, 2014 (1986).

30. L. E. Nivorozhkin, M. S. Korobov, N. I. Borisenko, G. S. Borodkin, Yu. E. Chernysh, and V. I. Minkin, *Zh. Obshch. Khim.*, **55**, 2149 (1985); *Chem. Abstr.*, **104**, 27827 (1986).

31. S. I. Klein and J. F. Nixon, *J. Organomet. Chem.*, **302**, 87 (1986).

32. A. J. Deeming, G. P. Proud, H. M. Dawes, and M. B. Hursthouse, *J. Chem. Soc., Dalton Trans.*, 2545 (1986).

33. S. J. B. Price, C. Brevard, A. Pagelot, and P. J. Sadler, *Inorg. Chem.*, **25**, 596 (1986).

34. J. R. Weir and R. C. Fay, *Inorg. Chem.*, **25**, 2969 (1986).

35. D. C. Brower, T. L. Tonker, J. R. Morrow, D. S. Rivers, and J. L. Templeton, *Organometallics*, **5**, 1093 (1986).

36. R. W. Moore, jun, A. Streitwieser, jun., and H. K. Wang, *Organometallics*, **5**, 1418 (1986).

37. I. I. Schuster, W. Weissensteiner, and K. Mislow, *J. Am. Chem. Soc.*, **108**, 6661 (1986).

38. M. L. H. Green and D. O'Hare, *J. Chem. Soc., Dalton Trans.*, 2469 (1986).

39. K. M. Larsson, J. Kowalewski, and U. Henriksson, *J. Magn. Reson.*, **62**, 260 (1985).

40. A. V. Khristoforov, F. I. Bashirov, and L. K. Yuldasheva, *Zh. Strukt. Khim.*, **26**, 180 (1985); *Chem. Abstr.*, **104**, 148987 (1986).

41. M. J. McGeary, T. L. Tonker, and J. L. Templeton, *Organometallics*, **4**, 2102 (1985).

42. R. H. Cragg, T. J. Miller, and D. O. Smith, *J. Organomet. Chem.*, **291**, 273 (1985).

43. S. Kerschl and B. Wrackmeyer, *J. Chem. Soc., Chem. Commun.*, 403 (1986).

44. R. H. Cragg, Y. J. Miller, and D. O'N. Smith, *J. Organomet. Chem.*, **302**, 19 (1986).

45. P. Kölle and H. Nöth, *Chem. Ber.*, **119**, 313 (1986).

46. P. R. Sharp and M. T. Rankin, *Inorg. Chem.*, **25**, 1508 (1986).

47. H. J. R. de Boer, O. S. Akkermann, F. Bickelhaupt, G. Erker, P. Czisch, R. Mynott, J. M. Wallis, and C. Krüger, *Angew. Chem., Int. Ed. Engl.*, **25**, 639 (1986).

48. M. Herberhold, H. Kniesel, L. Haumaier, and U. Thewalt, *J. Organomet. Chem.*, **301**, 355 (1986).

49. S. Murahashi, Y. Kitani, T. Uno, T. Hosokawa, K. Miki, T. Yonezawa, and N. Kasai, *Organometallics*, **5**, 356 (1986).

50. A. R. Siedle, W. B. Gleason, R. A. Newmark, and L. H. Pignolet, *Organometallics*, **5**, 1969 (1986).

51. D. J. Cardin and A. C. Sullivan, *J. Chem. Soc., Dalton Trans.*, 2321 (1986).

52. A. M. Crespi and D. F. Shriver, *Organometallics*, **4**, 1830 (1985).

53. W. B. Studabaker and M. Brookhart, *J. Organomet. Chem.*, **310**, C39 (1986).

54. J. R. Lisko and W. M. Jones, *Organometallics*, **5**, 1890 (1986).

55. G. J. Kubas, C. J. Unkefer, B. I. Swanson, and E. Fukushima, *J. Am. Chem. Soc.*, **108**, 7000 (1986).

56. G. J. Kubas, R. R. Ryan, and D. A. Wrobleski, *J. Am. Chem. Soc.*, **108**, 1339 (1986).

57. R. H. Morris, J. F. Sawyer, M. Shiralian, and J. Zubkowski, *J. Am. Chem. Soc.*, **107**, 5581 (1985).

58. R. H. Crabtree and M. Lavin, *J. Chem. Soc., Chem. Commun.*, 1661 (1985).

59. H. C. Clark and M. J. H. Smith, *J. Am. Chem. Soc.*, **108**, 3829 (1986).

60. L. G. Marzilli, M. F. Summers, E. Zangrando, N. Bresciani-Pahor, and L. Randaccio, *J. Am. Chem. Soc.*, **108**, 4830 (1986).

61. C. H. Bushweller, C. D. Rithner, and D. J. Butcher, *Inorg. Chem.*, **25**, 1610 (1986).

62. C. D. Rithner and C. H. Bushweller, *J. Phys. Chem.*, **90**, 5023 (1986).

63. C. G. Kreiter, W. Michels, and M. Wenz, *Chem. Ber.*, **119**, 1994 (1986).

64. H. Berke, G. Huttner, C. Sontag, and L. Zsolnai, *Z. Naturforsch., B,* **40**, 799 (1985); *Chem. Abstr.,* **104**, 88744 (1986).
65. P. B. Winston, S. J. N. Burgmayer, T. L. Tonker, and J. L. Templeton, *Organometallics,* **5**, 1707 (1986).
66. A. Mayr, K. S. Lee, M. A. Kjelsberg, and D. Van Engen, *J. Am. Chem. Soc.,* **108**, 6079 (1986).
67. J. L. Davidson and G. Vasapollo, *J. Chem. Soc., Dalton Trans.,* 2239 (1985).
68. J. L. Davidson, *J. Chem. Soc., Dalton Trans.,* 2434 (1986).
60. M. Scotti, M. Valderrama, S. Rojas, and W. Kläui, *J. Organomet. Chem.,* **301**, 369 (1986).
70. M. Mlekuz, P. Bougeard, B. G. Sayer, M. J. McGlinchey, C. A. Rodger, M. R. Churchill, J. W. Ziller, S. K. Kang, and T. A. Albright, *Organometallics,* **5**, 1656 (1986).
71. D. G. Allen, S. B. Wild, and D. L. Wood, *Organometallics,* **5**, 1009 (1986).
72. F. Alvarez, E. Carmona, J. M. Marín, M. L. Poveda, E. Gutiérrez-Puebla, and A. Monge, *J. Am. Chem. Soc.,* **108**, 2286 (1986).
73. M. Herberhold and B. Sdhmidkonz, *J. Organomet. Chem.,* **308**, 35 (1986).
74. R. P. Hughes, J. W. Reisch, and A. L. Rheingold, *Organometallics,* **4**, 1754 (1985).
75. G.-H. Lee, S.-M. Peng, T.-W. Lee, and R.-S. Liu, *Organometallics,* **5**, 2378 (1986).
76. P. K. Baker, S. Clamp, N. G. Connelly, M. Murray, and J. B. Sheridan, *J. Chem. Soc., Dalton Trans.,* 459 (1986).
77. J. R. Bleeke and A. J. Donaldson, *Organometallics,* **5**, 2401 (1986).
78. M. K. Andari, A. V. Kryov, and A. P. Belov, *Khim. Svyaz Str. Mol.,* 208 (1984); *Chem. Abstr.,* **104**, 149111 (1986).
79. K. Nakasuji, M. Yamaguchi, I. Murata, and H. Nakanishi, *J. Am. Chem. Soc.,* **108**, 325 (1986).
80. G. M. Smith, H. Suzuki, D. C. Sonnenberger, V. W. Day, and T. J. Marks, *Organometallics,* **5**, 549 (1986).
81. G. Erker, T. Mühlenbernd, R. Benn, and A. Rufińska, *Organometallics,* **5**, 402 (1986).
82. G. Erker, T. Mühlenbernd, R. Benn, A. Rufińska, G. Tainturier and B. Gautheron, *Organometallics,* **5**, 1023 (1986).
83. S. Ozkar and C. G. Kreiter, *J. Organomet. Chem.,* **303**, 367 (1986).
84. C. G. Kreiter, J. Kögler, and K. Nist, *J. Organomet. Chem.,* **310**, 35 (1986).
85. E. J. Probitts and R. J. Mawby, *J. Organomet. Chem.,* **310**, 121 (1986).
86. J. R. Bleeke, G. G. Stanley, and J. J. Kotyk, *Organometallics,* **5**, 1642 (1986).
87. M. C. L. Trimarchi, M. A. Green, J. C. Huffman, and K. G. Caulton, *Organometallics,* **4**, 514 (1985).
88. J. R. Bleeke and D. A. Moore, *Inorg. Chem.,* **25**, 3522 (1986).
89. T. Dave, S. Berger, D. Bilger, H. Kaletsch, J. Pebler, J. Knecht, and K. Dimroth, *Organometallics,* **4**, 1565 (1985).
90. G. M. Williams, R. A. Fisher, and R. H. Heyn, *Organometallics,* **5**, 818 (1986).
91. R. T. Baker and T. H. Tulip, *Organometallics,* **5**, 839 (1986).
92. G. Schmid and F. Schmidt, *Chem. Ber.,* **119**, 1766 (1986); *Chem. Abstr.,* **105**, 24310 (1986).
93. V. A. Polyakov and A. D. Ryabov, *J. Chem. Soc., Dalton Trans.,* 589 (1986).
94. M. Farra-Hake, M. F. Rettig, J. L. Williams, and R. M. Wing, *Organometallics,* **5**, 1032 (1986).
95. T. Chivers, C. Lensink, and J. F. Richardson, *Organometallics,* **5**, 819 (1986).
96. P. J. Hammond, P. D. Beer, C. Dudman, I. P. Danks, C. D. Hall, J. Knychala, and M. C. Grossel, *J. Organomet. Chem.,* **306**, 367 (1986).
97. P. D. Beer and A. D. Keefe, *J. Organomet. Chem.,* **306**, C10 (1986).
98. T. M. Gilbert and R. G. Bergman, *J. Am. Chem. Soc.,* **107**, 3502 (1985).
99. R. D. Thomas, M. T. Clarke, R. M. Jensen, and T. C Young, *Organometallics,* **5**, 1851 (1986).
100. H. Günther, D. Moskau, R. Dujardin, and A. Maercker, *Tetrahedron Lett.,* **27**, 2251 (1986).
101. E. L. Eliel, M. Manoharan, S. G. Levine, and A. Ng, *J. Org. Chem.,* **50**, 4978 (1985).
102. K. W. Nugent and J. K. Beattie, *J. Chem. Soc., Chem. Commun.,* 186 (1986).
103. R. Benn, H. Lehmkuhl, K. Mehler, and A. Rufińska, *J. Organomet. Chem.,* **293**, 1 (1985).
104. C. P. Casey, J. M. O'Connor, and K. J. Haller, *J. Am. Chem. Soc.,* **107**, 1241 (1985).

105. A. G. Davies, J. P. Goddard, M. B. Hursthouse, and N. P. C. Walker, *J. Chem. Soc., Dalton Trans.*, 471 (1985).
106. P. Jutzi, *Chem. Rev.*, **86**, 983 (1986).
107. B. E. Mann, *Chem. Soc. Rev.*, **15**, 125 (1986).
108. F. J. Manganiello, S. M. Oon, M. D. Radcliffe, and W. M. Jones, *Organometallics*, **4**, 1069 (1985).
109. R. F. Childs and A. Varadarajan, *Can. J. Chem.*, **63**, 418 (1985).
110. J. A. Bandy, M. L. H. Green, and D. O'Hare, *J. Chem. Soc., Dalton Trans.*, 2477 (1986).
111. U. Kölle and B. Fuss, *Chem. Ber.*, **119**, 116 (1986).
112. W. E. Geiger, T. Gennett, M. Grzeszczuk, G. A. Lane, J. Moraczewski, A. Salzer, and D. E. Smith, *J. Am. Chem. Soc.*, **108**, 7454 (1986).
113. M. Grassi, B. E. Mann, B. T. Pickup, and C. M. Spencer, *J. Magn. Reson.*, **69**, 92 (1986).
114. K. Schlögl, A. Werner, and M. Widhalm, *Monatsh. Chem.*, **117**, 1423 (1986).
115. I. D. Kalikhman, O. B. Bannikova, B. A. Gostevskii, O. A. Vyazankina, N. S. Vyazankin, and V. A. Pestunovich, *Izv. Akad. Nauk SSSR, Ser. Khim.*, 1688 (1985); *Chem. Abstr.*, **104**, 148977 (1986).
116. H. H. Karsch, A. Appelt, and G. Hanika, *J. Organomet. Chem.*, **312**, C1 (1986).
117. H. H. Karsch, A. Appelt, and G. Müller, *Organometallics*, **5**, 1664 (1986).
118. J. Browning, K.A. Beveridge, G. W. Bushnell, and K. R. Dixon, *Inorg. Chem.*, **25**, 1987 (1986).
119. R. K. Chadha, J. E. Drake, and A. B. Sarkar, *Inorg. Chem.*, **25**, 2201 (1986).
120. G. Michael, J. Kaub, and C. G. Kreiter, *Chem. Ber.*, **118**, 3944 (1985).
121. R. Benn, S. Holle, P. W. Jolly, R. Mynott, and C. C. Romão, *Angew. Chem., Int. Ed. Engl.*, **25**, 555 (1986).
122. F. Timmers and M. Brookhart, *Organometallics*, **4**, 1365 (1985).
123. C. G. Kreiter, M. Leyendecker, and W. S. Sheldrick, *J. Organomet. Chem.*, **302**, 217 (1986).
124. D. Baudry, P. Boydell, and M. Ephritikhine, *J. Chem. Soc., Dalton Trans.*, 525 (1986).
125. G. F. Schmidt and M. Brookhart, *J. Am. Chem. Soc.*, **107**, 1443 (1985).
126. R. A. Bell, S. A. Cohen, N. M. Doherty, R. S. Threlkel, and J. E. Bercaw, *Organometallics*, **5**, 972 (1986).
127. C. A. Ghilardi, P. Innocenti, S. Midollini, and A. Orlandini, *J. Chem. Soc., Dalton Trans.*, 605 (1985).
128. M. Lusser and P. Peringer, *Z. Naturforsch., B*, **40**, 1417 (1985); *Chem. Abstr.*, **105**, 79081 (1986).
129. D. A. Wrobleski, D. T. Cromer, J. V. Ortiz, T. B. Rauchfuss, R. R. Ryan, and A. P. Sattelberger, *J. Am. Chem. Soc.*, **108**, 174 (1986).
130. H. Köpf and T. Klapötke, *Chem. Ber.*, **119**, 1986 (1986).
131. H. Köpf and T. Klapötke, *J. Organomet. Chem.*, **310**, 303 (1986).
132. H. Köpf and T. Klapötke, *J. Chem. Soc., Chem. Commun.*, 1192 (1986).
133. E. W. Abel, S. K. Bhargava, P. K. Mittal, K. G. Orrell, and V. Šik, *J. Chem. Soc., Dalton Trans.*, 1561 (1985).
134. E. W. Abel, P. K. Mittal, K. G. Orrell, and V. Šik, *J. Chem. Soc., Dalton Trans.*, 1569 (1985).
135. E. W. Abel, K. M. Higgins, K. G. Orrell, V. Šik, E. H. Curzon, and O. W. Howarth, *J. Chem. Soc., Dalton Trans.*, 2195 (1985).
136. A. L. Crumbliss, R. J. Topping, J. Szewczyk, A. T. McPhail, and L. D. Quin, *J. Chem. Soc., Dalton Trans.*, 1895 (1986).
137. R. T. Boeré and C. J. Willis, *Can. J. Chem.*, **63**, 6530 (1985).
138. R. T. Boeré and C. J. Willis, *Can. J. Chem.*, **64**, 492 (1986).
139. E. W. Abel, T. E. MacKenzie, K. G. Orrell, and V. Šik, *J. Chem. Soc. Dalton Trans.*, 205 (1986).
140. E. W. Abel, T. P. J. Coston, K. G. Orrell, V. Šik, and D. Stephenson, *J. Magn. Reson.*, **70**, 34 (1986).
141. N. Wiberg and G. Wagner, *Chem. Ber.*, **119**, 1467 (1986).
142. M. H. Chisholm and R. J. Tatz, *Organometallics*, **5**, 1590 (1986).
143. T. W. Coffindaffer, W. M. Westler, and I. P. Rothwell, *Inorg. Chem.*, **24**, 4565 (1985).
144. K.-W. Lee, J. M. Hanckel, and T. L. Brown, *J. Am. Chem. Soc.*, **108**, 2266 (1986).

145. W. Arbiel and J. Heck, *J. Organomet. Chem.*, **302**, 363 (1986).
146. R. Boese, W. B. Tolman, and K. P. C. Volhardt, *Organometallics*, **5**, 582 (1986).
147. R. G. Ball, F. Edlemann, G.Yu. Kiel, J. Takats, and R. Drews, *Organometallics*, **5**, 829 (1986).
148. B. B. Wayland, B. A. Woods, and V. L Coffin, *Organometallics*, **5**, 1059 (1986).
149. S. Aime, R. Gobetto, G. Nicola, D. Osella, L. Milone, and E. Rosenberg, *Organometallics*, **5**, 1829 (1986).
150. C. G. Kreiter, K. H. Franzreb, W. Michels, U. Schubert, and K. Ackermann, *Z. Naturforsch.*, *B*, **40**, 1188 (1985); *Chem. Abstr.*, **105**, 97647 (1986).
151. H. Patin, B. Misterkiewicz, J.-Y. Le Marouille, and A. Mousser, *J. Organomet. Chem.*, **314**, 173 (1986).
152. R. H. Voegeli, H. C. Kang, R. G. Finke, and V. Boekelheide, *J. Am. Chem. Soc.*, **108**, 7010 (1986).
153. M. I. Altbach, C. A. Muedas, R. P. Korswagen, and M. L. Ziegler, *J. Organomet. Chem.*, **306**, 375 (1986).
154. O. Bars, P. Braunstein, G. L. Geoffroy, and B. Metz, *Organometallics*, **5**, 2021 (1986).
155. H. H. Karsch, B. Milewski-Mahrla, J. O. Besenhard, P. Hofman, P. Stauffert, and T. A. Albright, *Inorg. Chem.*, **25**, 3811 (1986).
156. M. A. Freeman and D. A. Young, *Inorg. Chem.*, **25**, 1556 (1986).
157. A. L. Balch, M. M. Olmstead, and D. E. Oram, *Inorg. Chem.*, **25**, 298 (1986).
158. A. Lagadec, B. Misterkiewicz, H. Patin, A. Mousser, and J.-Y. Le Marouille, *J. Organomet. Chem.*, **315**, 201 (1986).
159. R. J. Blau, J. H. Espenson, S. Kim, and R. A. Jacobson, *Inorg. Chem.*, **25**, 757 (1986).
160. C. G. Pitt, A. P. Purdy, K. T. Higa, and R. L. Wells, *Organometallics*, **5**, 1266 (1986).
161. E. W. Abel, P. K. Mittal, K. G. Orrell, and V. Šik, *J. Chem. Soc., Dalton Trans.*, 961 (1986).
162. E. W. Abel, T. E. MacKenzie, K. G. Orrell, and V. Šik, *J. Chem. Soc., Dalton Trans.*, 2173 (1986).
163. C. Wynants, G. Van Binst, C. Mügge, K. Jurkschat, A. Tzschach, H. Pepermans, M. Gielen, and R. Willem, *Organometallics*, **4**, 1906 (1985).
164. J. H. Wengrovius, M. F. Garbauskas, E. A. Williams, R. C. Goint, P. E. Donahue, and J. F. Smith, *J. Am. Chem. Soc.*, **108**, 982 (1986).
165. M. R. Shaffer and J. B. Keister, *Organometallics*, **5**, 561 (1986).
166. P. A. Bates, S. S. D. Brown, A. J. Dent, M. B. Hursthouse, G. F. M. Kitchen, A. G. Orpen, I. D. Salter, and V. Šik, *J. Chem. Soc., Chem. Commun.*, 600 (1986).
167. L. R. Martin, F. W. B. Einstein, and R. K. Pomeroy, *J. Am. Chem. Soc.*, **108**, 338 (1986).
168. R. D. Adams and S. Wang, *Organometallics*, **5**, 1272 (1986).
169. R. D. Adams, I. T. Horvath, and S. Wang, *Inorg. Chem.*, **25**, 1617 (1986).
170. J. H. Osborne, R. C. P. Hill, and D. M. Ritter, *Inorg. Chem.*, **25**, 372 (1986).
171. T. Beringhelli, G. D'Alfonso, H. Molinari, B. E. Mann, B. T. Pickup, and C. M. Spencer, *J. Chem. Soc., Chem. Commun.*, 796 (1986).
172. W. L. Blohm, D. E. Fjare, and W. L. Gladfelter, *J. Am. Chem. Soc.*, **108**, 2301 (1986).
173. S. Aime, D. Osella, A. J. Deeming, A. J. Arce, M. B. Hursthouse, and H. M. Dawes, *J. Chem. Soc., Dalton Trans.*, 1459 (1986).
174. M. Cree-Uchiyama, J. R. Shapley, and G. M. St. George, *J. Am. Chem. Soc.*, **108**, 1316 (1986).
175. J. Evans and B. P. Gracey, *J. Chem. Res. (S)*, 42 (1986).
176. L. J. Farrugia, *J. Organomet. Chem.*, **310**, 67 (1986).
177. F. W. B. Einstein, L. R. Martin, R. K. Pomeroy, and P. Rushman, *J. Chem. Soc., Chem. Commun.*, 345 (1985).
178. S. Aime, G. Gobetto, D. Osella, L. Milone, G. E. Hawkes, and E. W. Randall, *J. Magn. Reson.*, **65**, 308 (1985).
179. E. C. Lisic and B. E. Hanson, *Inorg. Chem.*, **25**, 812 (1986).
180. R. A. Gancarz, M. W. Baum, G. Hunter, and K. Mislow, *Organometallics*, **5**, 2327 (1986).
181. I. T. Horváth, *Organometallics*, **5**, 2333 (1986).

446 References

182. A. L. Balch, L. A. Fossett, R. R. Guimerans, M. M. Olmstead, P. E. Reedy, jun., and F. E. Wood, *Inorg. Chem.*, **25**, 1248 (1986).
183. C. Allevi, B. T. Heaton, C. Seregni, L. Strona, R. J. Goodfellow, P. Chini, and S. Martinengo, *J. Chem. Soc., Dalton Trans.*, 1375 (1986).
184. X. L. R. Fontaine, H. Fowkes, N. N. Greenwood, J. D. Kennedy, and M. Thornton-Pett, *J. Chem. Soc., Chem. Commun.*, 1165 (1985).
185. S. F. T. Froom, M. Green, R. J. Mercer, K. R. Nagle, A. G. Orpen, and S. Schwiegk, *J. Chem. Soc., Chem. Commun.*, 1666 (1986).
186. M. D. Curtis, L. Messerle, J. J. D'Errico, W. M. Butler, and M. S. Hay, *Organometallics*, **5**, 2283 (1986).
187. L. Y. Hsu, W. L. Hsu, D. Y. Jan, and S. G. Shore, *Organometallics*, **5**, 1041 (1986).
188. T. Venäläinen and T. A. Pakkanen, *J. Organomet. Chem.*, **316**, 183 (1986).
189. F. W. B. Einstein, K. G. Tyers, A. S. Tracey, and D. Sutton, *Inorg. Chem.*, **25**, 1631 (1986).
190. A. A. Aitchison and L. J. Farrugia, *Organometallics*, **5**, 1103 (1986).
191. G. D. Williams, M.-C. Lieszkovszky, C. A. Mirkin, G. L. Geoffroy, and A. L. Rheingold, *Organometallics*, **5**, 2228 (1986).
192. S. Cartwright, J. A. Clucas, R. H. Dawson, D. F. Foster, M. M. Harding, and A. K. Smith, *J. Organomet. Chem.*, **302**, 403 (1986).
193. S. Aime, R. Bertoncello, V. Busetti, R. Gobetto, G. Granozzi, and D. Osella, *Inorg. Chem.*, **25**, 4004 (1986).
194. E. Roland, W. Bernhardt, and H. Vahrenkamp, *Chem. Ber.*, **118**, 2858 (1985).
195. R. Ros, A. Scrivanti, and R. Roulet, *J. Organomet. Chem.*, **303**, 273 (1986).
196. G. Ferguson, B. R. Lloyd, L. Manojlović-Muir, K. W. Muir, and R. J. Puddephatt, *Inorg. Chem.*, **25**, 4190 (1986).
197. S. S. D. Brown, P. J. McCarthy, and I. D. Salter, *J. Organomet. Chem.*, **306**, C27 (1986).
198. W. Bos, R. P. F. Kanters, C. J. van Halen, W. P. Boswan, H. Behm, J. M. M. Smits, P. T. Beurskens, J. J. Bour, and L. H. Pignolet, *J. Organomet. Chem.*, **307**, 385 (1986).
199. P. D. Boyle, B. J. Johnson, A. Buchler, and L. H. Pignolet, *Inorg. Chem.*, **25**, 5 (1986).

References for Chapter 14

1. L. Marko, *J. Organomet. Chem.*, **305**, 333 (1986).
2. T. G. Southern, *J. Mol. Catal.*, **30**, 267 (1985).
3. P. Escaffre, A. Thorez, P. Kalck, B. Besson, R. Perron, and Y. Colleuille, *J. Organomet. Chem.*, **302**, C17 (1986).
4. P. Kalck, A. Thorez, M. Pinillos, and L. Oro, *J. Mol. Catal.*, **31**, 311 (1985).
5. A. M. Trzeciak and J. J. Ziolkowski, *J. Mol. Catal.*, **34**, 213 (1986).
6. A. M. Trzeciak, J. Ziolkowsi, S. Aygen, and R. Van Eldik, *J. Mol. Catal.*, **34**, 337 (1986).
7. R. Choukroun, A. Iraqui, and D. Gervais, *J. Organomet. Chem.*, **311**, C60 (1986).
8. P. Van Leeuwen and C. F. Roobeek, *J. Mol. Catal.*, **31**, 345 (1985).
9. K. Prokai-Tatrai, S. Toros, and B. Heil, *J. Organomet. Chem.*, **315**, 231 (1986).
10. J. Brown, S. Cook, and R. Khan, *Tetrahedron*, **42**, 18, 5015 (1986).
11. I. Kovacs, F. Ungvary, and L. Marko, *Organometallics*, **5**, 209 (1986).
12. F. Ungvary and L. Marko, *Organometallics*, **5**, 2341 (1986).
13. M. Hidai, A. Fukoka, Y. Koyasu, and Y. Uchida, *J. Mol. Cat.*, **35**, 29 (1986).
14. R. Dubois, P. Garrou, K. Lavin, H. Allock, *Organometallics*, **5**, 460 (1986).
15. R. Dubois and P. Garrou, *Organometallics*, **5**, 466 (1986).
16. A. Scrivanti, G. Cavinato, L. Toniolo, and C. Botteghi, *J. Organomet. Chem.*, **286**, 115 (1985).
17. A. Scrivanti, A. Berton, L. Toniolo, and C. Botteghi, *J. Organomet. Chem.*, **314**, 369 (1986).
18. P. Haelg, G. Consiglio, and P. Pino, *J. Organomet. Chem.*, **296**, 281 (1985).

19. P. W. N. M. van Leeuwen, C. Roobeek, R. Wife, and J. Frijns, *J. Chem. Soc., Chem. Commun.*, 31 (1986).
20. T. D. Dekleva and D. Forster, *Adv. Catal.*, **34**, 81 (1986).
21. T. D. Dekleva and D. Forster, *J. Am. Chem. Soc.*, **107**, 3565 (1985).
22. T. D. Dekleva and D. Forster, *J. Mol. Catal.*, **33**, 269 (1985).
23. S. B. Dake and R. V. Chaudhari, *J. Mol. Catal.*, **35**, 119 (1986).
24. J. Hjortkjaer and J. Jorgenson, *J. Chem. Soc., Dalton Trans.*, **2**, 763 (1978).
25. T. Dekleva and D. Forster, *J. Am. Chem. Soc.*, **107**, 3568 (1987). T. Dekleva and D. Forster, *J. Chem. Educ.*, **63**, 204 (1986).
26. S. W. Pollchnowski, *J. Chem. Educ.*, **63**, 206 (1986).
27. M. Murphy, B. Smith, G. P. Torrence, and A. Aguilo, *Inorg. Chim. Acta.*, **101**, L47 (1985).
28. T. Dekleva and D. Forster, *Adv. Catal.*, **34**, 81 (1986).
29. G. Braca, A. Raspolli Galletti, G. Sbrana, and R. Lazzaroni, *J. Organomet. Chem.*, **289**, 107 (1985).
30. M. Lutgendorf, E. Elvevoll, and M. Roper, *J. Organomet. Chem.*, **289**, 97 (1985).
31. J. Zoeller, *J. Mol. Catal.*, **37**, 377 (1986).
32. G. Braca, A. Galletti, G. Sbrana, and F. Zanni, *J. Mol. Catal.*, **34**, 183 (1986).
33. J. Zoeller, *J. Mol. Catal.*, **37**, 117 (1986).
34. G. Jenner and P. Andrianary, *J. Organomet. Chem.*, **307**, 263 (1986). G. Jenner, G. Bitsi, and P. Andrianary, *Appl. Catal.*, **24**, 319 (1986).
35. Y. Sugi, K. Takeuchi, H. Arakawa, T. Matsuzaki, and K. Bando, C_1 *Mol. Chem.*, **1**, 423 (1986).
36. R. W. Wegman, *Organometallics*, **5**, 707 (1986).
37. D. Milstein, *J. Chem. Soc., Chem. Commun.*, **11**, 817 (1986).
38. N. Chatani, S. Fujii, Y. Yamasaki, S. Murai, and N. Sonoda, *J. Am. Chem. Soc.*, **108**, 7361 (1986).
39. J. A. Kampmeier, S. Mahalingam, and T. Liu, *Organometallics*, **5**, 823 (1986).
40. M. Gomez, J. Kisenyl, G. Sunley, and P. Maitlis, *J. Organomet. Chem.*, **296**, 197 (1985).
41. S. Murashashi, T. Naota, and N. Nakajma, *J. Org. Chem.*, **51**, 898 (1986).
42. K. Hori, M. Ando, N. Takaishi, and Y. Inamoto, *Tetrahedron Lett.*, **27**, 4615 (1968).
43. H. Ishida, K. Tanaka, M. Morimoto, and T. Tanaka, *Organometallics*, **5**, 724 (1986).
44. T. Venalainen, E. Iiskola, J. Pursianinen, T. Pakanen, and T. Pakkanen, *J. Mol. Catal.*, **34**, 293 (1986).
45. R. Sanchez-Delgado and B. Oramas, *J. Mol. Catal.*, **36**, 283 (1986).
46. H. J. Wasserman, G. J. Kubas, and R. R. Ryan, *J. Am. Chem. Soc.*, **108**, 2294 (1986).
47. G. J. Kubas, C. J. Unkefer, B. I. Swanson, and E. Fukushima, *J. Am. Chem. Soc.*, **108**, 7000 (1986).
48. G. J. Kubas and R. R. Ryan, *Polyhedron*, **5**, 473 (1986).
49. G. J. Kubas, R. R. Ryan, and P. Wrobleski, *J. Am. Chem. Soc.*, **108**, 1339 (1986).
50. G. E. Badd, R. K. Upmacis, M. Poliakoff, and J. J. Turner, *J. Am. Chem. Soc.*, **108**, 2547 (1986).
51. R. K. Upmacis, M. Poliakoff, and J. J. Turner, *J. Am. Chem. Soc.*, **108**, 3645 (1986).
52. R. H. Morris, J. F. Sawyer, M. Shiralian, and J. D. Zubkowski, *J. Am. Chem. Soc.*, **107**, 5581 (1985).
53. R. H. Crabtree and M. Lavin, *J. Chem. Soc., Chem. Commun.*, 1661 (1985).
54. R. H. Crabtree, M. Lavin, and L. Bonneviot, *J. Am. Chem. Soc.*, **108**, 4032 (1986).
55. R. H. Crabtree and D. G. Hamilton, *J. Am. Chem. Soc.*, **108**, 3124 (1986).
56. Y. Jean, O. Eisenstein, F. Volatron, B. Maouche, and F. Sefta, *J. Am. Chem. Soc.*, **108**, 6587 (1986).
57. R. H. Crabtree and M. W. Davis, *J. Org. Chem.*, **51**, 2655 (1986).
58. a. J. M. Brown, A. E. Derome, and S. A. Hall, *Tetrahedron* **41** (20), 4647 (1985). b. J. M. Brown and S. A. Hall, *Tetrahedron*, **41** (20), 4639 (1985). c. J. M. Brown, I. Cutting, P. L. Evans, and P. J. Maddox, *Tetrahedron Lett.*, **27** (28), 3307 (1986).
59. M. Castiglioni, R. Giordano, E. Sappa, A. Turipicchio, and M. T. Camellini, *J. Chem. Soc., Dalton Trans.*, 23 (1986).

60. a. J. L. Zuffa, M. L. Blohm, and W. L. Gladfelter, *J. Am. Chem. Soc.*, **108**, 552 (1986).
    b. J. L. Zuffa and W. L. Gladfelter, *J. Am. Chem. Soc.*, **108**, 4669 (1986).
61. L. N. Lewis, *J. Am. Chem. Soc.*, **108**, 743 (1986).
62. T. Suarez, B. Fontal, and D. Garcia, *J. Mol. Catal.*, **34**, 163 (1986).
63. W. A. Fordyce, R. Wilcgyski, and J. Halpern, *J. Organomet. Chem.* **296**, 115 (1985).
64. R. H. Fish, J. L. Tan, and A. D. Thormodsen, *Organometallics*, **4**, 1743 (1985).
65. A. M. Stolzenberg and F. L. Muetterties, *Organometallics*, **4**, 1739 (1985).
66. M. O. Albers, E. Singleton, and M. M. Viney, *J. Mol. Catal.*, **33**, 77 (1985).
67. S. I. Hommeltoft, D. H. Berry, and R. Eisenberg, *J. Am. Chem. Soc.*, **108**, 5345 (1986).
68. B. R. Sutherland and M. Cowie, *Organometallics*, **4**, 1801 (1985).
69. a. U. Matteoli, G. Menchi, M. Bianchi, and F. Piacenti, *J. Organomet. Chem.*, **299**, 233 (1986).
    b. U. Matteoli, G. Menchi, M. Bianchi, P. E. Frediana, and F. Piacenti, *Gazz. Chim. Ital.*, **115**, 603 (1985).
70. C. J. Casewit, D. E. Coons, L. L. Wright, W. K. Miller, and M. K. DuBois, *Organometallics*, **5**, 951 (1986).
71. G. J. Kubas and R. R. Ryan, *J. Am. Chem. Soc.*, **107**, 6138 (1985).
72. A. Baranyi, F. Ungvary, and L. Marko, *J. Mol. Catal.*, **32**, 343 (1985).
73. C. J. Longley, T. J. Goodwin, and G. Wilkinson, *Polyhedron*, **5** (10), 1625 (1986).
74. J. E. Bercaw, D. L. Davies, and P. T. Wolczanoki, *Organometallics*, **5**, 443 (1986).
75. R. Sariego, I. Carkovic, M. Martinez, and M. Valderrama, *J. Mol. Catal.*, **35**, 161 (1986).
76. S. Shinoda, H. Itagoki, and Y. Saito, *J. Chem. Soc., Chem. Commun.*, 860 (1985).
77. T. A. Smith and P. M. Maitlais, *J. Organomet. Chem.*, **289**, 385 (1985).
78. D. Milstein, *J. Am. Chem. Soc.*, **108**, 3526 (1986).
79. D. Barrat and D. Cole-Hamilton, *J. Organomet. Chem.*, **306**, C41 (1986).
80. D. B. Dombek, *Organometallics*, **4**, 1707 (1985).
81. a. M. Tamura, M. Ishino, T. Deguchi, and S. Nakamura, *J. Organomet. Chem.*, **312**, C75 (1986). b. H. Tanaka, Y. Hara, E. Watanable, K. Wada, and T. O'Nada, *J. Organomet. Chem.*, **312**, C71 (1986).
82. T. Takano, T. Deguchi, M. Ishino, and S. Nakamura, *J. Organomet. Chem.*, **309**, 209 (1986).
83. M. Tanaka, T. Sakakura, T. Hayashi, and T. Kobayashi, *Chem. Lett.*, 39 (1986).
84. J. T. Groves ane R. Quinn, *J. Am. Chem. Soc.*, **107**, 5790 (1985).
85. D. P. Riley and P. E. Correa, *J. Chem. Soc., Chem. Commun.*, 1097 (1986).
86. F. Mares, S. E. Diamond, F. J. Regina, and J. P. Solar, *J. Am. Chem. Soc.*, **107**, 3545 (1985).
87. J. E. Bäckvall and A. Heumann, *J. Am. Chem. Soc.*, **108**, 7107 (1986).
88. P. K. Wong, M. K. Dickson, and L. L. Sterna, *J. Chem. Soc., Chem. Commun.*, 1565 (1985).
89. B. L. Feringa, *J. Chem. Soc., Chem. Commun.*, 909 (1986).
90. G. Read and M. Urgellis, *J. Chem. Soc., Dalton Trans.*, 1591 (1985).
91. M. Faraj, J. Martin, C. Martin, J.-M. Bregeault, and J. Mercier, *J. Mol. Catal.*, **31**, 57 (1985).
92. C. Martin, M. Faraj, J. Martin, J.-M. Bregeault, J. Mercier, J. Fillaux, and P. Dizaba, *J. Mol. Catal.*, **37**, 201 (1986).
93. a. C. E. Summer and G. R. Steinmetz, *J. Am. Chem. Soc.*, **107**, 6124 (1985). b. G. R. Steinmetz and C. E. Sumner, *J. Catal.*, **100**, 549 (1986).
94. Z. Cvengrosova, M. Hronec, J. Kizlink, T. Pasternakova, S. Holotik, and J. Ilavsky, *J. Mol. Catal.*, **37**, 349 (1986).
95. A. Kunai, S. Hata, S. Ito, and K. Sasaki, *J. Org. Chem.*, **51**, 3471 (1986).
96. A. Kunai, S. Hata, S. Ito, and K. Sosaki, *J. Am. Chem. Soc.*, **108**, 6012 (1986).
97. T. Funabiki, A. Mizoguchi, T. Sugimoto, S. Tada, M. Tsuji, H. Sakamoto, and S. Yoshida, *J. Am. Chem. Soc.*, **108**, 292 (1986).
98. G. Speier, *J. Mol. Catal.*, **37**, 259 (1986).
99. S. Tsuruya, S. Yanai, and M. Masai, *Inorg. Chem.*, **25**, 141 (1986).
100. L. S. White and L. Qua, Jr., *J. Mol. Catal.*, **33**, 139 (1985).
101. J. Vcelak, M. Klimova, and V. Chvalovsky, *Coll. Czech. Chem. Commun.*, **51**, 847 (1986).
102. H. Minoun, M. Mignard, P. Brechot, and L. Saussine, *J. Am. Chem. Soc.*, **108**, 3711 (1986).

103. E. G. Samsel, K. Srinivasan, and J. K. Kocki, *J. Am. Chem. Soc.*, **107**, 7606 (1985).
104. K. Srinivasan, S. Perrier, and J. K. Kocki, *J. Mol. Catal.*, **36**, 297 (1986).
105. J. C. Dobson, W. K. Seok, and T. J. Meyer, *Inorg. Chem.*, **25**, 1514 (1986).
106. J. A. S. J. Razenberg, A. W. Van Der Made, J. W. H. Smeets, and R. J. M. Nolte, *J. Mol. Catal.*, **31**, 271 (1985).
107. G. Strukul and R. A. Michelin, *J. Am. Chem. Soc.*, **107**, 7563 (1985).
108. H. Sugimoto and D. T. Sawyer, *J. Am. Chem. Soc.*, **107**, 5712 (1985).
109. B. Meunier, *Bull. Soc. Chim. Fr.*, (4), 578 (1986).
110. J. P. Collman, T. Kodadek, S. A. Raybuck, J. I. Brauman, and L. M. Papazian, *J. Am. Chem. Soc.*, **107**, 4343 (1985).
111. J. T. Groves and Y. Watanabe, *J. Am. Chem. Soc.*, **108**, 507 (1986).
112. T. G. Traylor, T. Nakano, and B. E. Dunlap, *J. Am. Chem. Soc.*, **198**, 2782 (1986).
113. J. P. Collman, T. Kodadek, and J. I. Brauman, *J. Am. Chem. Soc.*, **108**, 2588 (1986).
114. Y. Tatsuno, A. Sekiya, K. Tani, and T. Saito, *Chem. Lett.*, 889 (1986).
115. T. G. Traylor, Y. Samamoto, and T. Nakano, *J. Am. Chem. Soc.*, **108**, 3529 (1986).
116. J. T. Groves and Y. Watanabe, *J. Am. Chem. Soc.*, **108**, 7834 (1986).
117. *J. Mol. Catal.*, **36**(1, 2) (1986).
118. J. Kress, A. Aguero, and J. A. Osborn, *J. Mol. Catal.*, **36**(1, 2), 1 (1986).
119. E. F. Lutz, *J Chem. Educ.*, **63**(3), 202 (1986).
120. V. Dragutan, A. T. Belatan, and M. Dimonie, *Olefin Metathesis and Ring-Opening Polymerization of Cyclo-Olefins*, John Wiley and Sons, New York (1985).
121. C. J. Schaverien, J. C. Dewan, and R. R. Schrock, *J. Am. Chem. Soc.*, **108**, 2771 (1986).
122. L. Bencze, A. Kraut-Vass, and L. Prokai, *J. Chem. Soc., Chem. Commun.*, 911 (1985).
123. J. H. Freudenberger and R. R. Schrock, *Organometallics*, **4**, 1937 (1985).
124. C.-C. Han and T. J. Katz, *Organometallics*, **4**, 2186 (1985).
125. L. R. Gilliom and R. H. Grubbs, *J. Am. Chem. Soc.*, **108**, 733 (1986).
126. J. D. Morrison (ed.), *Asymmetric Synthesis*, 171 (1986).
127. H. Kanai, S. Choe, and K. Klabunde, *J. Am. Chem. Soc.*, **108**, 2019 (1986).
128. Q. Yanlong, Luand Jiaqui, and X. Weihua, *J. Mol. Catal.*, **34**, 31 (1986).
129. J. Podlaha, M. Prochazka, *Coll. Czech. Chem. Commun.*, 1274 (1985).
130. I. Matsuda, T. Kato, S. Sato, and Y. Izumi, *Tetrahedron Lett.*, **27**, 5747 (1986).
131. S. Pillai, M. Ravindranathan, and S. Sivaram, *Chem. Rev.*, **86**, 353 (1986).
132. H. Lekmkuhl, *Pure Appl. Chem.*, **58**, 495 (1986).
133. R. McKinney and M. Colton, *Organometallics*, **5**, 1080 (1986).
134. R. McKinney, *Organometallics*, **5**, 1752 (1986).
135. A. Shitani, H. Itatani, and T. Inagaki, *J. Mol. Catal.*, **34**, 57 (1986).
136. G. Oehme, I. Grassert, H. Mennenga, and H. Baudisch, *J. Mol. Catal.*, **37**, 53 (1985).
137. R. F. Jordan, C. S. Bajgur, R. Willet, and B. Scott, *J. Am. Chem. Soc.*, **108**, 7410 (1986).
138. J. Hamilton, K. Ivin, G. McCann, and J. Rooney, *Macromol. Chem.*, **186**, 1477 (1985).
139. L. Gillion and R. Grubbs, *J. Am. Chem. Soc.*, **108**, 733 (1986).
140. R. Taube, J.-P. Gehrke, and R. Radeglia, *J. Organomet. Chem.*, **291**, 101 (1985).
141. M. Roper, R. He, and M. Schieren, *J. Mol. Cat.*, **31**, 335 (1985).
142. W. Jolly, R. Mynott, B. Raspel, and K. P. Schick, *Organometallics*, **4**, **5**, 473 (1986).
143. W. J. Richter, *J. Mol. Cat.*, **34**, 145 (1986).
144. L. Carlton, J. Davidson, P. Ewing, L. Manoijlovic-Muir, and K. W. Muir, *J. Chem. Soc., Chem. Commun.*, 1474 (1985).
145. K. Abdulla, B. Booth, and C. Stacey, *J. Organomet. Chem.*, **293**, 103 (1985).
146. B. Bosnich, *Chem. Br.*, **20**(9), 808–11 (1985).
147. L. S. Hegedus, *Chem. Met.-Carbon Bond*, **2**, 401–512 (1985).
148. a. J. Tsuji, *Pure & Appl. Chem.*, **58** (6), 869–878 (1986); and b. J. K. Stille, *Angew. Chem. Int. Ed. Engl.*, **25**, 508–524 (1986).
149. B. Bosnich (ed.), *Asymmetric Catalysis*, Chap. 3, Martinus Nijhoff, The Hague (1986).
150. K. Hiroi, K. Suya, and S. Sato, *J. Chem. Soc., Chem. Commun.*, 469 (1986).

151. T. Hayashi, A. Yamamoto, T. Hagihara, and Y. Ito, *Tetrahedron Lett.*, 191 (1986).
152. Y. Hayasi, M. Riediker, J. S. Temple, and J. Scharwtz, *Tetrahedron Lett.*, **22**, 2629 (1981).
153. H. Matushita and E. I. Negishi, *J. Chem. Soc., Chem. Commun.*, 160 (1982).
154. J. Fiaud and L. Aribi-Zouioueche, *J. Organomet. Chem.*, **295**, 383 (1985).
155. G. Consiglio, O. Piccolo, L. Roncetti, and F. Morandi, *Tetrahedron*, **42**, 2043 (1986).
156. J. Brown and J. MacIntyre, *J. Chem. Soc., Perkin Trans.*, **2**, xxx (1985).
157. G. S. Silverman, S. Strickland, and K. M. Nicholas, *Organometallics*, **5**, 2117 (1986).
158. W. J. Scott and J. K. Stille, *J. Am. Chem. Soc.*, **108**, 3033 (1986).
159. I. Colon and D. R. Kelsey, *J. Org. Chem.*, **51**, 2627 (1986).
160. L. N. Lewis and J. F. Smith, *J. Am. Chem. Soc.*, **108**, 2728 (1986).
161. W. D. Jones and W. P. Kosar, *J. Am. Chem. Soc.*, **108**, 5640 (1986).
162. P. A. Wender and N. C. Ihle, *J. Am. Chem. Soc.*, **108**, 4678 (1986).
163. T. Schenck and B. Bosnich, *J. Am. Chem. Soc.*, **107**, 2058 (1985).
164. P. Auburn, J. Whelan, and B. Bosnich, *Organometallics*, **5**, 1533 (1986).
165. Y. Ito, M. Sawamura, and T. Hayashi, *J. Am. Chem. Soc.*, **108**, 6405 (1986).
166. S. Sato, I. Matsuda, and Y. Izumi, *Chem. Lett.*, 1875 (1985).
167. H. Brunner, R. Becker, and S. Gauder, *Organometallics*, **5**, 739 (1986).
168. E. Keinan and N. Greenspoon, *J. Am. Chem. Soc.*, **108**, 7314 (1986).
169. M. Brockman, H. tom Dieck, and J. Klaus, *J. Organomet. Chem.*, **301**, 209 (1986).
170. C. E. Johnson and R. Eisenberg, *J. Am. Chem. Soc.*, **107**, 6531 (1985).
171. L. N. Lewis and N. Lewis, *J. Am. Chem. Soc.*, **108**, 7228 (1986).
172. R. J. McKinney and D. C. Roe, *J. Am. Chem. Soc.*, **108**, 5167 (1986).
173. J. E. Bäckvall and O. S. Andell, *Organometallics*, **5**, 2350 (1986).
174. E. Puentes, A. F. Noels, R. Warin, A. S. Hubert, P. Tejssie, and D. Y. Waddan, *J. Mol. Catal.*, **31**, 183 (1985).
175. C. A. Tolman, *J. Chem. Educ.*, **63** (3), 199 (1986).
176. a. R. Crabtree, *Chem. Rev.*, **85**, 245 (1985). b. W. Graham, *J. Organomet. Chem.*, **300**, 81 (1986). c. R. L. Augustine, *Catalysis of Organic Reactions*, **22**, 37 (1985). d. N. Aktogu, D. Baudry, D. Cox, and M. Ephritikhine, *Bull. Soc. Chim. Fr.*, **3**, 381 (1985).
177. W. Faller and H. Felkin, *Organometallics*, **4**, 1488 (1985).
178. W. Jones and M. Fan, *Organometallics*, **5**, 1057 (1986).
179. W. Jones and F. Feher, *J. Am. Chem. Soc.*, **108**, 4814 (1986).
180. W. Jones and J. Maguire, *Organometallics*, **5**, 590 (1986).
181. C. Cameron, H. Felkin, T. Fillebeen-Khann, N. Forrow, and E. Guittret, *J. Chem. Soc., Chem. Commun.*, 801 (1986).
182. G. M. Smith, J. D. Carpenter, and T. J. Marks, *J. Am. Chem. Soc.*, **108**, 6805 (1986).

References for Chapter 15

1. R. van Eldik, in: *Mechanisms of Inorganic and Organometallic Reactions* (M. Twigg, ed.), Vol. 4, p. 433, Plenum Press (1986).
2. A. E. Merbach, Plenary Lecture: What high pressure tells us about inorganic substitution mechanisms, 24th ICCC, Athens (1986).
3. R. van Eldik and J. Jonas (eds.), *High Pressure Chemistry and Biochemistry*, D. Reidel Publ. Co., Dordrecht, 469 pp. (1987); Proceedings of the Nato ASI on "Advances in High Pressure Studies of Chemical and Biochemical Systems," Corfu, Greece (1986).
4. A. Rahm and G. Jenner, *Actual. Chim.*, 41 (1985).
5. G. Demazeau, *Actual. Chim.*, 46 (1985).
6. S. S. Batsanov, *Usp. Khim.*, **55**, 579 (1986).
7. R. van Eldik, *Comments Inorg. Chem.*, **5**, 135 (1986).
8. R. van Eldik, *Angew. Chem., Int. Ed. Engl.*, **25**, 673 (1986).

9. R. van Eldik (ed.), *Inorganic High Pressure Chemistry: Kinetics and Mechanisms*, Elsevier, Amsterdam, 448 pp. (1986).
10. J. DiBenedetto and P. C. Ford, *Coord. Chem. Rev.*, **64**, 361 (1985).
11. S. Sueno, I. Nakai, M. Imafuku, H. Morikawa, M. Kimata, K. Ohsumi, M. Nomura, and O. Shimomura, *Chem. Lett.*, 1663 (1986).
12. R. C. Munoz, R. A. Holroyd, and M. Nishikawa, *J. Phys. Chem.*, **89**, 2969 (1985).
13. T. Fukushima, H. Arakawa, and M. Ichikawa, *J. Phys. Chem.*, **89**, 4440 (1985).
14. L. A. Philips, S. P. Webb, S. W. Yeh, and J. H. Clark, *J. Phys. Chem.*, **89**, 17 (1985).
15. W. D. Turley and H. W. Offen, *J. Phys. Chem.*, **89**, 2933 (1985).
16. M. R. Zakin, S. G. Grubb, H. E. King, and D. R. Herschbach, *J. Chem. Phys.*, **84**, 1080 (1986).
17. M. Buback, H. P. Vögele, and H. J. Winkels, *Macromol. Chem., Macromol. Symp.*, **5**, 69 (1986).
18. H.-T. Macholdt, R. van Eldik, H. Kelm, and H. Elias, *Inorg. Chim. Acta.*, **104**, 115 (1985).
19. R. T. Roginski, J. R. Shapley, and H. G. Drickamer, *Chem. Phys. Lett.*, **127**, 185 (1986).
20. T. L. Carroll, J. R. Shapley, and H. G. Drickamer, *J. Am. Chem., Soc.*, **107**, 5802 (1985).
21. T. L. Carroll, J. R. Shapley, and H. G. Drickamer, *Chem. Phys. Lett.*, **119**, 340 (1985).
22. M. T. Fisher, R. E. White, and S. G. Sligar, *J. Am. Chem. Soc.*, **108**, 6835 (1986).
23. R. Benz and F. Conti, *Biophys. J.*, **50**, 99 (1986).
24. P. T. T. Wong, *Physica*, **139/140B**, 847 (1986).
25. K. Heremans and M. Bormans, *Physica*, **139/140B**, 878 (1986).
26. C. Balny, *Physica*, **139/140B**, 878 (1986).
27. M. J. Blandamer, J. Burgess, and A. W. Hakin, *J. Chem. Soc., Faraday Trans. 1*, **82**, 3681 (1986).
28. M. J. Blandamer, J. Burgess, and J. B. N. F. Engberts, *Chem. Soc., Rev.*, **14**, 237 (1985).
29. M. J. Blandamer, J. Burgess, B. Clark, R. E. Robertson, and J. M. W. Scott, *J. Chem. Soc., Faraday Trans. 1*, **81**, 11 (1985).
30. T. W. Swaddle and L. Spiccia, *Physica*, **139/140B**, 684 (1986).
31. V. M. Zhulin, O. B. Radakov, G. A. Stashina, A. V. Ganyuskin, A. V. Yablokov, and N. N. Vainberg, *Izv. Akad. Nauk SSSR, Ser. Khim.*, **9**, 1979 (1985).
32. P. P. S. Saluja, C. Cameron, M. A. Florino, A. Lavergne, G. E. McLaurin, and E. Whalley, *Rev. Sci. Instrum.*, **57**, 2791 (1986).
33. Y. Yoshimura, J. Osugi, and M. Nakahara, *Ber. Bunsenges. Phys. Chem.*, **89**, 25 (1985).
34. Y. Yoshimura and M. Nakahara, *Ber. Bunsenges. Phys. Chem.*, **89**, 426 (1985).
35. Y. Yoshimura and M. Nakahara, *Ber. Bunsenges. Phys. Chem.*, **89**, 1004 (1985).
36. Y. Yoshimura and M. Nakahara, *Ber. Bunsenges. Phys. Chem.*, **90**, 58 (1986).
37. J. T. Hynes, *Ann. Rev. Phys. Chem.*, **36**, 573 (1985).
38. D. L. Hasha, T. Eguchi, and J. Jonas, *J. Am. Chem. Soc.*, **104**, 2290 (1982).
39. J. Jonas, *Acc. Chem. Res.*, **17**, 74 (1984).
40. J. Schroeder and J. Troe, *Chem. Phys. Lett.*, **116**, 453 (1985).
41. J. Troe, *J. Phys. Chem.*, **90**, 357 (1986).
42. G. Maneke, J. Schroeder, J. Troe, and F. Voß, *Ber. Bunsenges. Phys. Chem.*, **89**, 896 (1985).
43. C. J. Cobos and J. Troe, *J. Chem. Phys.*, **83**, 1010 (1985).
44. C. J. Cobos, H. Hippler, and J. Troe, *J. Phys. Chem.*, **89**, 342 (1985).
45. A. Nagasawa, H. Kido, T. M. Hatton, and K. Saito, *Inorg. Chem.*, **25**, 4330 (1986).
46. R. M. Nielson, H. W. Dodgen, J. P. Hunt, and S. E. Wherland, *Inorg. Chem.*, **25**, 582 (1986).
47. S. F. Lincoln, A. M. Hounslow, B. G. Doddridge, J. H. Coates, A. E. Merbach, and D. Zbinden, *Inorg. Chim. Acta.*, **100**, 207 (1985).
48. L. Helm, S. F. Lincoln, A. E. Merbach, and D. Zbinden, *Inorg. Chem.*, **25**, 2550 (1986).
49. J. E. Kessler, C. T. G. Knight, and A. E. Merbach, *Inorg. Chim. Acta.*, **115**, 85 (1986).
50. K. Ishihara, S. Funahashi, and M. Tanaka, *Inorg. Chem.*, **25**, 2898 (1986).
51. M. Inamo, S. Funahashi, and M. Tanaka, *Inorg. Chem.*, **25**, 2475 (1986).
52. H.-T. Macholdt, R. van Eldik, and G. R. Dobson, *Inorg. Chem.*, **25**, 1914 (1986).
53. J. J. Chung, and K. S. Park, *J. Korean Chem. Soc.*, **30**, 341 (1986).
54. R. Mohr and R. van Eldik, *Inorg. Chem.*, **24**, 3396 (1985).
55. K. Yoshitani, *Bull. Chem. Soc., Jpn.*, **58**, 1646 (1985).

56. P. Martinez, R. Mohr, and R. van Eldik, *Ber. Bunsenges. Phys. Chem.*, **90**, 609 (1986).
57. S. F. Funahashi, N. Uchiyama, and M. Tanaka, *Bull. Chem. Soc., Jpn.*, **59**, 161 (1986).
58. R. A. Binstead and J. K. Beattie, *Inorg. Chem.*, **25**, 1481 (1986).
59. K. Yoshitana, *Bull. Chem. Soc. Jpn.*, **58**, 2778 (1985).
60. N. J. Curtis and G. A. Lawrance, *Inorg. Chem.*, **25**, 1033 (1986).
61. Y. C. Park and Y. J. Cho, *J. Korean Chem. Soc.*, **29**, 629 (1985).
62. G. A. Lawrance, *Polyhedron*, **5**, 2113 (1986).
63. Y. Kitamura, T. Itoh, and M. Takeuchi, *Inorg. Chem.*, **25**, 3887 (1986).
64. R. van Eldik, Y. Kitamura, and C. P. Piriz Mac-Coll, *Inorg. Chem.*, **25**, 4252 (1986).
65. T. Inoue, K. Kojima, and R. Shimozawa, *Bull. Chem. Soc. Jpn.*, **59**, 1683 (1986).
66. K. Ishihara, S. Funahashi, and M. Tanaka, 25th High Pressure Conf. Jpn., Tsukuba, 236 (1984).
67. H.-T. Masholdt and R. van Eldik, *Transition Met. Chem.*, **10**, 323 (1985).
68. M. Inamo, S. Funahashi, Y. Ito, Y. Hamada, and M. Tanaka, *Inorg. Chem.*, **24**, 2468 (1985).
69. S. O. Oh, D. Y. Chung, and I. H. Cho, *J. Korean Chem. Soc.*, **30**, 553 (1986).
70. G. Mahal and R. van Eldik, *Inorg. Chem.*, **24**, 4165 (1985).
71. M. Kotowski and R. van Eldik, *Inorg. Chem.*, **25**, 3896 (1986).
72. G. Laurenczy, Y. Ducommun, and A. E. Merbach, Fast Reactions in Solution Discussion Group Meeting, Gargnano, Italy, 101 (1986).
73. J. G. Leipoldt, R. van Eldik, S. S. Basson, and A. Roodt, *Inorg. Chem.*, **25**, 4639 (1986).
74. W. Weber, H. Maecke, and R. van Eldik, *Inorg. Chem.*, **25**, 3093 (1986).
75. R. M. Nielson, J. P. Hunt, H. W. Dodgen, and S. Wherland, *Inorg. Chem.*, **25**, 1964 (1986).
76. I. Krack, P. Braun, and R. van Eldik, *Physica*, **139/140B**, 680 (1986).
77. P. Braun and R. van Eldik, *J. Chem. Soc., Chem. Commun.*, 1349 (1985).
78. Y. Sasaki, K. Endo, A. Nagasawa, and K. Saito, *Inorg. Chem.*, **25**, 4845 (1986).
79. I. Krack and R. van Eldik, *Inorg. Chem.*, **25**, 1743 (1986).
80. K. Ishihara and T. W. Swaddle, *Can. J. Chem.*, **64**, 2168 (1986).
81. G. Stochel, R. van Eldik, and Z. Stasicka, *Inorg. Chem.*, **25**, 3663 (1986).
82. W. Weber and P. C. Ford, *Inorg. Chem.*, **25**, 1088 (1986).
83. S. Wieland, J. DiBenedetto, R. van Eldik, and P. C. Ford, *Inorg. Chem.*, **25**, 4893 (1986).
84. W. Weber and R. van Eldik, *Inorg. Chim. Acta.*, **111**, 129 (1986).
85. M. L. Fetterolf and H. W. Offen, *J. Phys. Chem.*, **90**, 1828 (1986).

Index